Karl Schwarzschild

GESAMMELTE WERKE/COLLECTED WORKS

Karl Schwarzschild

GESAMMELTE WERKE/COLLECTED WORKS

Volume 1

Biography – 1. Celestial Mechanics
2. Sun and Stellar Atmospheres
3. Cometary Tails
4. Structure, Kinematics and Dynamics of Stellar Systems

Volume 2

5. Astronomical Positioning – 6. Photographic Photometry
7. Measuring Techniques, Binary Stars, Variable Stars
and Spectroscopy

Volume 3

8. Optics – 9. Physical Papers
10. Miscellaneous
Bibliography

Karl Schwarzschild

GESAMMELTE WERKE
COLLECTED WORKS

Edited by H. H. Voigt

Volume 1

Springer-Verlag
Berlin Heidelberg New York
London Paris Tokyo
Hong Kong Barcelona
Budapest

Professor Dr. Hans-Heinrich Voigt
Universitäts-Sternwarte, Geismarlandstraße 11
W-3400 Göttingen, Fed. Rep. of Germany

Annotators:

Dr. Rudolf Dvorak
Institut für Astronomie der Universität Wien, Türkenschanzstraße 17
A-1180 Wien, Austria

Professor Dr. Wolfgang Mattig
Kiepenheuer-Institut für Sonnenphysik, Schöneckstraße 6
W-7800 Freiburg, Fed. Rep. of Germany

Dr. Klaus Jockers
Max-Planck-Institut für Aeronomie, Postfach 20
W-3411 Katlenburg-Lindau, Fed. Rep. of Germany

Professor Dr. Roland Wielen
Astronomisches Rechen-Institut, Mönchhofstraße 12–14
W-6900 Heidelberg, Fed. Rep. of Germany

ISBN 3-540-52455-X Springer-Verlag Berlin Heidelberg New York
ISBN 0-387-52455-X Springer-Verlag New York Berlin Heidelberg

Die Deutsche Bibliothek – CIP-Einheitsaufnahme
Schwarzschild, Karl:
Gesammelte Werke = Collected works / Karl Schwarzschild. Ed. by H. H. Voigt. -
Berlin; Heidelberg; New York; London; Paris; Tokyo; Hong Kong; Barcelona; Budapest: Springer
NE: Schwarzschild, Karl: [Sammlung]
Vol.1 (1992)
ISBN 3-540-52455-X (Berlin ....)
ISBN 0-387-52455-X (New York ....)

Reproduction and Printing: Beltz Offsetdruck, W-6944 Hemsbach
Binding: J. Schäffer GmbH & Co. KG, W-6718 Grünstadt

55/3140-5 4 3 2 1 0 – Printed on acid-free paper

Vorwort

Karl Schwarzschild war einer der bedeutendsten Astronomen Deutschlands und zählt zu den Begründern der modernen Astrophysik. Sein Nachlaß wird in der Niedersächsischen Staats- und Universitätsbibliothek in Göttingen aufbewahrt, eine reichhaltige Sonderdrucksammlung befindet sich auch in der Göttinger Universitäts-Sternwarte. Die hier vorgelegte Sammlung beschränkt sich bewußt auf Schwarzschilds veröffentlichte Arbeiten. Manuskripte und Entwürfe wissenschaftlicher Arbeiten aus dem Nachlaß wurden nicht hinzugenommen. Im Nachlaß befinden sich ferner rund 7000 Briefe von und an Karl Schwarzschild mit über 700 verschiedenen Personen. Diese Briefsammlung, die sich als Mikrofilm auch am Center for History of Physics, American Institute of Physics in New York befindet, wurde bisher nicht systematisch bearbeitet sondern nur sporadisch für den Lebenslauf herangezogen. Anläßlich dieser Herausgabe hat die Handschriftenabteilung den Nachlaß neu geordnet und katalogisiert, so daß er auch anderen Interessenten nun übersichtlich zur Verfügung steht. Der Handschriftenabteilung der Niedersächsischen Staats- und Universitätsbibliothek unter Leitung von Herrn Dr. H. Rohlfing möchte ich an dieser Stelle für die ständige Bereitstellung des Nachlasses und andere Unterstützung vielmals danken.

Auf der Tagung der Astronomischen Gesellschaft in Hamburg im September 1986 hielt Professor Dr. Subrahmanyan Chandrasekhar die seit 1959 regelmäßig abgehaltene Karl-Schwarzschild-Vorlesung. Bei dieser Gelegenheit schlug er vor, die in vieler Hinsicht bahnbrechenden Arbeiten Schwarzschilds gesammelt herauszugeben. Der Springer-Verlag hat diese Anregung dankenswerter Weise aufgegriffen. Da die meisten Dokumente in Göttingen aufbewahrt sind, habe ich mich auf Bitten des Verlags gerne bereit erklärt, diese Herausgabe zu übernehmen. An dieser Stelle sei darauf hingewiesen, daß schon Ende der 20er Jahre Arnold Sommerfeld und Schwarzschilds Schwager Robert Emden den Plan faßten, Schwarzschilds Werke gesammelt herauszugeben. Bei der praktischen Arbeit sollte Albrecht Unsöld, damals als Privatdozent am Sommerfeldschen Institut in München tätig, wesentlich mitwirken. Aber die rapide zunehmende Macht der Nationalsozialisten ließ eine Herausgabe der gesammelten Werke von Karl Schwarzschild bald aussichtslos erscheinen.

Die Werke Schwarzschilds wurden in zehn Sachgebiete aufgeteilt, und mehrere Kollegen fanden sich bereit, zu diesen Gebieten einen einleitenden Kommentar zu schreiben, der den damaligen Stand der Wissenschaft, den wesentlichen Inhalt der Schwarzschildschen Arbeiten und deren Wirkung für den weiteren Fortgang der Wissenschaft darstellt. Diesen Autoren, den Herren Rudolf Dvorak, Edward Geyer, Ulrich Haug, Friedrich Hund, Klaus Jockers, Wolfgang Mattig, Günter Schmahl, Günter Seeber und Roland Wielen, möchte ich an dieser Stelle für Ihre Arbeit meinen besonderen Dank sagen.

Es zeigte sich bald, daß die Zahl der Arbeiten sehr viel größer ist als in der Sonderdruck-Sammlung an der Göttinger Sternwarte. Zahlreiche Sternwarten und Bibliotheken haben bereitwillig geholfen, Sonderdrucke oder Kopien aller Arbeiten sowie Photos der Institute, an denen Schwarzschild gearbeitet hat, für diese Sammlung zur Verfügung zu stellen. Auch ihnen sei an dieser Stelle gedankt.

Die Fülle der übernommenen Arbeiten (insgesamt 119) machte es notwendig, das ganze Werk auf drei Bände zu verteilen. Teils aus sachlichen, teils aus buchtechnischen Gründen – die Bände sollten etwa gleich stark werden – wurde folgende Aufteilung vorgenommen: Band I enthält neben den biographischen Abschnitten die astronomischen Veröffentlichungen, die primär nach Objekten zusammengefaßt wurden: Himmelsmechanik, Sonne und Sternatmosphären, Kometen, Sternsysteme. Band II enthält astronomische Arbeiten, die eher nach methodischen Gesichtspunkten zusammengefaßt sind: astronomische Ortsbestimmung, photographische Photometrie, Meßtechnik und Spektroskopie. Band III schließlich enthält die Arbeiten zur Optik, zur Physik und schließlich im letzten Kapitel die populären und die nicht-fachlichen Arbeiten.

Natürlich nahm ich schon sehr bald auch Kontakt auf mit Martin Schwarzschild, Professor für Astronomie in Princeton, dem Sohn Karl Schwarzschilds. Verständlicherweise wollte er sich nicht an der Herausgabe der Werke seines Vaters aktiv beteiligen; aber er verfolgte die Arbeit mit Interesse und gab im mündlichen Gespräch manche zusätzliche Information zu dem Leben seiner Eltern. Für diese sehr persönlichen Gespräche möchte ich auch an dieser Stelle herzlich danken.

Die Übersetzung der zunächst deutsch verfaßten und die Überarbeitung der englisch geschriebenen Kommentare und sonstigen Begleittexte übernahm Herr Dr. Storm Dunlop in Chichester/West Sussex, dem für diese oft nicht leichte Arbeit ein besonderer Dank gilt. Manche Texte gingen mehrfach zwischen Autor und Übersetzer hin und her. Übersetzer und Herausgeber danken in diesem Zusammenhang auch folgenden Personen, die in schwierigen Fällen und beim Finden der richtigen Fachausdrücke dem Übersetzer geholfen haben: Dr. Suzanne Debarbat, Dr. Roger Griffin, Cmdr. Derek Howse, Dr. Constanze la Dous, Professor Sir William McCrea, Dr. Robert Smith.

Schließlich gilt mein Dank dem Verlag, der allen Wünschen gerne entgegenkam, jedwede technische Hilfe leistete und sich in jeder Hinsicht bemüht hat, dieses Werk so zu gestalten, wie es nun vorliegt.

Göttingen, Januar 1992 *Hans-Heinrich Voigt*

Foreword

Karl Schwarzschild was one of the most important German astronomers and was, in addition, one of the founders of modern astrophysics. His personal papers are held at the Niedersächsische Staats- und Universitätsbibliothek in Göttingen, and an extensive collection of reprints can be found in the Observatory in Göttingen. The material presented in these volumes is restricted intentionally to Schwarzschild's published works. Manuscripts and drafts that are among his personal papers are not included. These personal papers contain some 7000 letters, which Karl Schwarzschild wrote to, and received from, about 700 different correspondents. This collection of letters, a microfilm copy of which is also held by the Center for the History of Physics, American Institute of Physics, New York, has not yet been systematically examined but has been consulted only sporadically for biographical details. For this current work, the Manuscript Department of the Niedersachsen State and University Library in Göttingen has collated and catalogued the papers afresh, so that full and clear details are now available to anyone interested. This is an appropriate place for me to express my great appreciation to the Manuscript Department and to the Chief Librarian, Dr H. Rolfing, for constantly supplying material from Schwarzschild's papers and for much other assistance.

At the meeting of the Astronomische Gesellschaft held in Hamburg in September 1986, Professor Subrahmanyan Chandrasekhar gave the Karl Schwarzschild Lecture, which has been held regularly since 1959. On that occasion he suggested that a collected edition of Schwarzschild's frequently pioneering papers should be published. Springer-Verlag very commendably took up this suggestion. As most of the documents are held at Göttingen, when the publishers approached me, I was delighted to be able to undertake the task of preparing this edition. It should be mentioned here that in the late 1920s Arnold Sommerfeld and Schwarzschild's brother-in-law Robert Emden planned to publish Schwarzschild's collected works. Albrecht Unsöld, who was then an external lecturer at the Sommerfeld Institute in Munich, was to have played an important part in seeing this into being. But the rapid rise to power of the National Socialist movement soon made it apparent that it would be futile to attempt to publish such a collective enterprise.

For this current project Schwarzschild's papers were divided into ten categories, which form the chapters of these volumes, and several colleagues agreed to write introductory commentaries for each of these chapters that would describe the then current state of knowledge, the scientific contribution made by Schwarzschild's papers, and their influence on later developments in that particular field. I particularly want to thank these authors, Rudolf Dvorak, Edward Geyer, Ulrich Haug, Friedrich Hund, Klaus Jockers, Wolfgang Mattig, Günter Schmahl, Günter Seeber and Roland Wielen, for their work.

It soon became obvious that the number of papers was far in excess of those held in the Göttingen Observatory's collection of reprints. Numerous observatories and libraries willingly cooperated in providing reprints or copies of various papers, as well as photographs of the various institutes at which Schwarzschild worked, especially for this collection. I should like to express my thanks to them here.

The number of papers included (a total of 119) required the collection to be divided into three volumes. Both for practical and technical reasons the volumes had to be of approximately the same size, so material was divided as follows. Volume I contains, apart from the biographical sections, those astronomical publications that primarily concern individual objects: celestial mechanics, the Sun and stellar atmospheres, comets, and stellar systems. Volume II contains astronomical papers that are more concerned with methods: astronomical position determination, photographic photometry, measuring techniques, and spectroscopy. Finally, Volume III contains papers on optics, physics and, in the last chapter, popular and non-scientific contributions.

Naturally I lost no time in getting in contact with Martin Schwarzschild, Professor of Astronomy at Princeton University and Karl Schwarzschild's son. Quite understandably, he did not want to play an active part in publishing his father's work; but he followed progress with interest and gave a lot of additional, verbal information about his parents' life. I must express my warmest thanks to him for these personal reminiscences. The translation of the commentaries that had been written in German, and the language-editing of those written in English, as well as work on a variety of additional material, was undertaken by Storm Dunlop of Chichester in West Sussex. He deserves particular thanks for what was often no easy task. Many of the sections went back and forth between the authors and translator several times. In this connection, the translator and publishers would like to thank the following persons, who helped to clarify difficult points and to decide upon the appropriate scientific terminology: Dr Suzanne Debarbat, Dr Roger Griffin, Cmdr Derek Howse, Dr Constanze la Dous, Professor Sir William McCrea, and Dr Robert Smith.

Finally, I must thank the publishers, who willingly accommodated all my requests, provided technical assistance, and took considerable pains in ensuring that this work appeared in its present form.

Göttingen, January 1992 *Hans-Heinrich Voigt*

Bemerkung.

Die Groesse meines Vaters Lebenswerk bedarf nicht eines Sohnes Lob. So habe ich es mir entsagt das Unternehmen, meines Vaters gesammelte Arbeiten neu herauszugeben zu beeinflussen. Aber etwas ganz persoenliches darf ich und will ich hier bekennen: Es ist für mich eine ausserordentliche Freude, dass die Errungenschaften meines Vaters durch diese Neuherausgabe so hoch anerkannt werden, und ich bin Professor H. Voigt von ganzem Herzen dankbar, dass er dies Unternehmen mit Energie und feinem Urteil durchgefuehrt hat.

Martin Schwarzschild.

Contents

3. Cometary Tails

4. Structure, Kinematics and Dynamics of Stellar Systems

Biography of Karl Schwarzschild (1873–1916)

Introduction

In the following description of Karl Schwarzschild's life, his scientific work during the various phases of his career will be mentioned, but not discussed in detail. This will be done in the subsequent 10 sections. Schwarzschild's papers are cited as double numbers inside square brackets [...]; the first number gives the section and the second the citation within that particular section. The same double numbers appear in the lists of contents on p. XI, 497 and 501.

Single numbers in square brackets [...] are references for this biographical chapter and can be found on p. 26. The citation [N-...] refers to Karl Schwarzschild's personal papers, which are listed on p. 27.

Childhood and School-years

Karl Schwarzschild was born on 9 October 1873 in Frankfurt am Main, the eldest of seven children [N-24.1]. His father, Moses Martin Schwarzschild, was a successful member of the business community, whose ancestors can be traced back to the 16th century in Frankfurt [20]. His mother was born Henrietta Sabel. Karl had a happy childhood among his brothers and sisters, and grew up in a house where his parents had a keen interest in art and music. He was the first one in the family to become a scientist, and his tendency towards such an interest did not take long to appear. At first his interests tended towards the technical – he built a telescope around an objective lens, using a tube of newspaper blackened inside, and showed his brothers and sister the rings of Saturn. He had piano lessons, but he was just as interested in the mechanism inside the piano and the theory of sound vibrations as he was in music itself.

His brother Alfred, an artist, later wrote his memoirs, covering the period up to about 1907 [N-24], which are an important source of information about Karl Schwarzschild's childhood and student days. Karl obviously inherited his warm-hearted, cheerful nature from his mother, and his ability for persistent, hard work from his father [20].

Karl first attended the Jewish community's school [N-21] and later the municipal gymnasium in Frankfurt [20]. He learnt things easily and with interest in the subjects.

Apart from pure mathematics, which he grasped with the greatest ease, astronomy soon became a major interest. His father introduced him to his friend J. Epstein, a

Fig. 1. Karl Schwarzschild's family, Frankfurt, about 1906.
left to right: His brother Hermann (agricultural specialist), Karl (astronomer), his brother Alfred (portrait painter), his sister-in-law Else (wife of Robert), his father Moses Martin (stock broker), his mother Henrietta Francisca, his sister Klara (wife of Robert Emden), his brother Otto (financier in New York), his brother Robert (industrial manufacturer)

mathematician with a private observatory. With Epstein's son Paul, later Professor of Mathematics at Strasburg, Karl learnt to use the telescope, and also learnt mathematics far beyond the level taught at his school. Alfred describes an excursion that they made into the Harz mountains, when Karl carried a notebook containing an essay on elliptical series. In later years he retained this habit of carrying a notebook so that he could jot down any sudden inspiration. One of his school exercise books contains thoughts about the equilibrium figures of rotating fluid bodies and the stability of Saturn's rings [5]. Runge describes how, on one occasion at school, he forgot his logarithm tables and calculates the required values himself using a series expansion [7].

Whilst still a 16-year old schoolboy, Karl sent his work on orbital determination [1.1] to the Astronomische Nachrichten. This was accepted, and he gave his father a printed copy on his birthday. The family was naturally very proud of this and read the paper through several times, although – as Albert says – "not one of us understood a word of it". In the same year, 1890, a second paper about orbital determination was also published [1.2].

Fig. 2. Karl Schwarzschild as a boy in Frankfurt

In the last year at school – as Alfred tells us – Karl was fully occupied with studies, preparation for taking the Abitur[1], dance lessons, writing poems and other self-imposed work. On 3 September 1891, he came top in the Abitur [21] and therefore had to give the farewell address in Latin. His report gives his marks as:

Behaviour	1	Greek	1	Mathematics	1	
German	1	French	2	Physics	2	(1 = very good; 2= good; 3 = satisfactory)
Latin	3	History	1	Gymnastics	2	

Karl wanted to study astronomy, and his teachers regretted the fact that he did not want to turn his great abilities to some sphere that would be of general benefit to mankind. A business friend of his father's said: "Gott, was e feiner Beruf" ["God, what sort of a distinguished job is that?"].

Undergraduate Years and Military Service

In the same year, 1891, Karl began studying at the Kaiser Wilhelm University in Strasburg, under Ernst Becker, the Director of the Observatory. Here he learnt the fundamentals of practical astronomy. His work on the observation of variables and of Nova Aurigae [7.9] dates from this period; he also took part in observing the total lunar eclipse of 4 November 1892 [24].

Alfred was studying at the same time in Karlsruhe, and the two brothers frequently took week-end trips to the Black Forest and, in the summer holidays, long excursions in the Alps in Switzerland.

In October 1893, Karl went to Munich to do his military service. He served in the field artillery from 1 October 1893 to 1 October 1894, leaving as "Einjährig freiwilliger Unteroffizier" [one-year, volunteer, non-commissioned officer] [N-22]. His service identity papers record his height as 169 cm. Even as a soldier, Karl always carried his notebook, and during this period published a paper on the influence of secular aberration on the positions of the fixed stars [4.1].

Karl remained in Munich and continued studying at the Ludwig Maximilian University under Hugo von Seeliger. During this period two further papers on cometary orbits appeared [1.3,4]. He also found time for practical work: inspired by Michelson's interferometer, he built a rotatable wire grating, and used it to carry out interferometric measurements of double stars [7.1].

In July 1896, Karl gained his doctorate under Seeliger with the dissertation – judged "summa cum laude" – "The Poincaré Theory of the Equilibrium of a Homogeneous, Rotating, Fluid Body" [9.1]. On his diploma [N-21] the title reads: "Elliptische Gleichgewichtsfigur einer homogenen, rotierenden Flüssigkeit" ["Elliptical Equilibrium Figure of a Homogeneous, Rotating Fluid"].

The holidays again saw him in the mountains, together with his brother Alfred, who had also been living in Munich since 1893.

[1] The school-leaving and university-entrance examination – Trans.

Fig. 3. Karl Schwarzschild as a young man in Munich 1900

Fig. 4. M. von Kuffner's Observatory at Ottakring, Vienna (Photograph by Michael Musil)

Assistant in Vienna

After obtaining his doctorate, in October 1896 Karl Schwarzschild became assistant at the von Kuffner Observatory in Ottakring, Vienna, under the Director, Leo de Ball. The Observatory, a private one, founded by the brewery owner, Moriz von Kuffner (1854–1939), and completed in 1887, was an example of enlightened patronage, and noted at the time for its rapid acquisition of new instruments and by increasing international recognition. It contained the largest heliometer in the world, and the largest meridian circle in the Austro-Hungarian Empire. Of the many scientists who worked here, apart from Karl Schwarzschild, we may mention Johannes Hartmann, Gustav Eberhard, Alexander Wilkens and Carl Wirtz [31, 32].

In Vienna, Schwarzschild published further papers on celestial mechanics, which concerned special aspects of the three-body problem [1.5,6], as well as one paper on measuring techniques [7.2] and one on optics [8.1]. He also took part in de Ball's zonal survey of stars [26]. But this period was primarily notable for the publication of the fundamental papers [6.1–7] concerning photographic photometry (which at that time was still a new science), and in particular the quantitative determination of the departure from the reciprocity law. The telescope and the associated camera with which he carried out his measurements are still in existence and use at the von Kuffner Observatory. For this work he was awarded the Vienna Photographic Society's Medal in 1899. It is preserved among his effects [N-23.2].

Fig. 5. The University Observatory at Munich, about the time of Schwarzschild

University Lecturer [Privatdozent] in Munich

His monograph "Contributions on the Photographic Photometry of Stars" [6.2] was submitted (and accepted) as his post-doctoral thesis [Habilitation] at Munich, and in 1899 Schwarzschild returned to Munich as university lecturer. Included among the five themes that he had to defend for his thesis there were: "The theory that our space is curved should be rejected" and "It is not possible to establish probability theory on a purely logical basis" [N-21].

We learn from Alfred's memoirs that Karl was very excited about his first lecture, and that by the end of the hour, taken at a gallop, he had covered a major part of the material that should have occupied the whole term. He soon learnt the correct technique, however, and his clear lecturing style was later to be widely praised. During his time at Munich he held the following courses:

Summer term 1900:	Nautical Astronomy	(8 students)
Winter term 1900/01:	Astronomical Optics	(7 students)
Summer term 1901:	Mathematical Interpolation	(6 students)

Schwarzschild had large audiences for his public lectures. His lectures were lively and clear, and particularly popular with the ladies; they called him "the lecturer with the sparkling eyes". We learn from the report of the Munich Society for Higher Education [Volkshochschule] [N-21] that he gave a course of 6 lectures on "The Solar System" from January to March 1900. There were 193 participants (made up of 70 workers and craftsmen; 91 teachers, shop-assistants and students; and 32 others). Of these, 128 remained to the end.

As during his studies in Munich and also in Vienna, he continued to write popular articles and book reviews for daily newspapers ([10.1–4] and [10.13–15]), thus augmenting his salary.

In August 1900, the Astronomische Gesellschaft met in Heidelberg. At Seeliger's suggestion, Schwarzschild gave a paper "On the Permissible Scale of Curvature of Space" [9.2], in which he discussed the possibility of non-Euclidean space. The presentation and the subsequent lively discussion were reported in many newspapers.

At the Munich Observatory, Schwarzschild took part in observing the Leonids [25]. He published a paper on the determination of the orbits of binary stars [7.8], an optical paper on the diffraction and polarization of light passing through a slit [8.2]; and a paper about Arrhenius' theory of cometary tails – the last being a work that provoked quite a lot of controversy [3.1].

Göttingen

On 1 July 1901 Wilhelm Schur, the Director of the Göttingen Observatory, died. Alfred's memoirs and, above all, Karl's numerous letters to his parents [N-907/ Suppl.1] provide a lot of information about the events surrounding the appointment of a successor to this post. At that time Göttingen was world-renowned as a mathematical centre, largely through Felix Klein and David Hilbert, and it was felt that the Director of the Observatory should be someone who was experienced in the fields of applied mathematics. Seeliger, in Munich, immediately put forward the name of his pupil

Fig. 6. The University Observatory at Göttingen, about the time of Schwarzschild

Fig. 7. Karl Schwarzschild as a university lecturer in Göttingen

Schwarzschild. Klein had probably asked him for his advice, because Seeliger was regarded as the highest authority among German astronomers in those days, and Klein, who knew Seeliger well, often asked for his advice on astronomical matters. It is possible, however, that Seeliger made the approach to Göttingen. Klein knew Schwarzschild from his lecture in Heidelberg and through subsequent correspondence with him. On 26 July, the faculty sent the Prussian Ministry a triple recommendation containing the names of Hugo von Seeliger (Munich), Max Wolf (Heidelberg) and Karl Schwarzschild (Munich)[2].

As early as the middle of July, Schwarzschild had written to his parents, saying that Seeliger had "dropped in" and had told him about this list, and that naturally

[2] In his memoirs, Alfred writes that Struve from Königsberg was the second name included. This must be a mistake; it was only later that Struve was approached (see below); he was not on the list prepared by the faculty.

everything was still confidential, it being quite uncertain how the very energetic Referent [expert adviser] Althoff in the Ministry in Berlin, and the Minister himself, would react. Schwarzschild's feelings were mixed. He wrote that, above all, he had some qualms of conscience, because he felt that recently he could have done more work than he had actually achieved. He also rather dreaded the envy of his colleagues. He would prefer to be offered something in a couple of years' time. As he said: "I do not know whether to hope that it comes to nothing". He let things take their course.

In August Schwarzschild wrote to his parents, that Wolf at Heidelberg had been offered the post. It was probably already known in Berlin that Seeliger did not want to go to Göttingen. Shortly afterwards Seeliger wrote to Schwarzschild saying that it was unlikely that Wolf would accept; but that he had learnt from a letter from Klein that it was unlikely that he, Schwarzschild, would come next, because it was felt in Berlin that his age was against him. In his letter Schwarzschild added: "I await the outcome calmly, and look forward to a later, more modest, more certain opportunity".

On 21 September, Seeliger wrote to Schwarzschild saying that Wolf wanted to stay in Heidelberg. Apart from this he had heard word suggesting that possibly Schwarzschild would be appointed. Seeliger went on to give some advice on how Schwarzschild should behave. It would be advisable to be discreet in dealing with Althoff, yet at the same time ensuring that the latter would recognize his determination and energy.

At the beginning of October, Schwarzschild was actually invited to Berlin for interview. Althoff told Schwarzschild about the view in the Ministry: Hermann Struve from Königsberg would be made Director at Göttingen, Schwarzschild would become Observer, and assistant professor [Extraordinarius]; Ludwig Ambronn, currently Observer at Göttingen, would be appointed elsewhere[3]. Now they had to wait to see what Struve thought of the plan. Obviously, they wanted Schwarzschild at Göttingen, but felt that he was too young to hold the posts of full professor [Ordinarius] and Director of the Observatory. Schwarzschild himself was happy to agree to this plan. A few days later, however, it became clear that Struve did not want to leave Königsberg. Berlin stuck to its plan, however, and looked elsewhere for a Director, choosing Franz at Breslau. But that also fell through, and on 10 October, Karl sent his parents a telegram: "Extraordinarius and Director. Arrive Monday – Karl". The Ministry had finally been forced to appoint him as Director, but not as a full professor, only as an assistant professor. So, one day after his 28th birthday, he became the youngest university professor in Germany. Under this arrangement, Ambronn remained at Göttingen. Althoff also told Schwarzschild that "he would get the very minimum salary", the combined yearly sum amounting to 3000 Marks. In addition, he was promised around 5000 to 10 000 Marks for building-expansion in 1902, and another 60 000 to 75 000 Marks for 1903. A remarkable point is that Schwarzschild said in a letter that the matter of his religious persuasion – he had not allowed himself to be baptized – had not been explicitly mentioned.

On 19 October, his appointment as assistant professor was announced [27], and on 22 October the investiture took place at Göttingen; his salary was increased to 3600

[3] Ambronn was invited to Berlin for interview at the same time. In a letter to Alfred, Schwarzschild gives a lively account of his first meeting with Ambronn in Althoff's anteroom.

Marks. People in Göttingen were amazed to find that Berlin had insisted that among his duties, he was to hold popular lectures. In Göttingen he began with the obligatory visits (leaving his card with 130 people), with organizing his living quarters in the wing of the observatory - he wrote: "In comparison with Frankfurt everything costs half as much, but I admit that it is not so sturdy" – and with the lectures. He discussed with Felix Klein, at whose house he was a frequent visitor, the planned astronomical meeting in Göttingen. This was the 19th Ordinary Meeting of the Astronomische Gesellschaft, and was held on 4–7 August 1902, with 63 participants, and under the chairmanship of H.v.Seeliger [33].

Exaggerated tidiness was not Schwarzschild's forte, and he wrote: "Schur not only preserved every knick-knack, but also kept everything scrupulously in order. I despair when I look around me". This rather cavalier attitude towards the inventory was to play a part later on in his life. Blumenthal in his obituary [11] wrote that at first glance Schwarzschild's management appeared to have been rather "ingeniously negligent".

Schwarzschild was to spend the happiest years of his life in Göttingen. He was soon fully recognized as a scientist. By 24 May 1902 (before a year had passed), he was appointed to the full professorship [27]. He was in close social and scientific contact with the mathematicians and physicists, and soon played a leading role in the mathematical society and in the physics discussions. In 1904, together with Klein, Hilbert and Minkowski, he organized a seminar on mathematical physics [7]. Under Schur, astronomy in Göttingen had become rather isolated: now, under Schwarzschild, it became fully integrated with general scientific life [11]. Among others, he formed particularly close, friendly relations with Ludwig Prandtl, who was of almost the same age, and whom Klein had persuaded to come to Göttingen at about the same time.

On the personnel side, the Observatory then had, apart from the Director, an Observer, Ambronn (Observer since 1889; qualifying as lecturer in 1892, and promoted to assistant professor in 1902 [28-1897,1902]), an Assistant, Bruno Meyermann (who had been active at the Observatory for quite a few years and who had been an Assistant since 1900 [28-1898, 1900]), and finally, a caretaker, H. Berger, and also a supernumerary computing assistant, E. Jastram [28-1902]. As early as 1898 the department had been divided into astronomy and geophysics, and in the autumn of 1901 and thus just before Schwarzschild took over the post, the geophysics department under Emil Wiechert had moved to a new "Institute for Geophysics" on the Hainberg [34]. Pure astronomy remained at the building on the Geismarlandstraße – then known as the Geismar Chaussee. The Observatory was still outside the town and was fairly isolated, but the Geismar Chaussee had already been surfaced, and was lit by incandescent light, so observations were already beginning to be affected [28-1898]. The Observatory was not particularly well provided with instrumentation. Apart from smaller pieces of equipment, there were two main instruments, the Reichenbach meridian circle, which dated from the time of Gauss, and the large Repsold heliometer with an aperture of 15 cm [28-1900]. Both instruments continued to be used into the 1920s, but were out-of-date even in Schwarzschild's time, and were not capable of meeting modern demands. Schwarzschild left the care of these two instruments to Ambronn and Meyermann. In 1903 a Zeiss Tessar lens, with an aper-

Fig. 8. Schwarzschild with some coworkers in front of the observatory at Göttingen.
Left to right: O. Birck, [name unreadable], [name unreadable], Knopp, W. Dziewulski (*sitting*), Bruns, K., Schwarzschild, [name unreadable], L. Ambronn, B. Meyermann, Lindemann

ture of 45 mm and a focal ratio of f/10 was acquired, which later carried out the Göttinger Aktinometrie. In 1904, Dr Schobloch presented the Observatory with a 17-cm refractor, and a building with a 4.5-m dome was erected in the Observatory garden, specifically for this "Schobloch Refractor".

Schwarzschild had a masterly understanding of the individual characters of his staff, who were generally much older than him, and was able to carry them with him quite irresistibly.

In Göttingen Schwarzschild continued his photographic work. Rather than using out-of-focus images in order to obtain a finite area of blackening on the plates, the Schraffierkasette was developed [6.10], [6.13], which was often referred to as the "Wobbly camera" by members of the Institute [N-907/25]. This led to the first, photographic, precision photometry, the "Göttinger Aktinometrie" [6.17,18], a comprehensive work that continued until Schwarzschild's time at Potsdam. Quite a few other photographic projects ([6.8,9], [6.11,12] and [6.14–16]) arose out of this. Of particular significance is the introduction of the concept of "colour tints" ["Farbentönung"] [6.14], which opened the way to multi-colour photometry. The measurements that were made confirmed, amongst other things, the existence of red giants and the statistical distribution of luminosities found by Hertzsprung, thus providing important material for later studies of stellar statistics.

Another field that had been traditional at Göttingen since the time of Tobias Mayer in the 18th century, was that of geodesy and position determination. Schwarzschild

Fig. 9. Karl Schwarzschild in the library of the observatory at Göttingen, looking at a photographic plate

also became successfully involved with this field, with the development of a suspended zenith camera [5.1–4], which worked automatically for up to 5 hours. Later, in response to a competition on the subject of "Astronomical Position Determination from a Balloon", he developed a bubble-sextant [5.5–7], with which he won first prize in 1909. The balloon flights that he made himself with this instrument appealed to his daredevil spirit.

The development of new instruments frequently led him to optical problems, and in 1905 he took the opportunity to devote himself to producing three fundamental and comprehensive papers on "Geometrical Optics" [8.3–5]. Max Born wrote later [20] that these papers became the backbone of his own book "Optics". In 1905 Born had heard Schwarzschild's lecture on "Astrophysics"; his transcript of it is to be found among Schwarzschild's papers [N-13:1]. A further publication on the ray-tracing of optical systems appeared in 1907 [8.6]. In all these works Schwarzschild put his particular speciality to good use. This was the facility – the almost playful ease – with which he handled mathematics, and also his delight in it, which repeatedly caused him to become involved in non-astronomical problems [10.19]. This also applies to some important papers that appeared in 1903 on electrodynamics [9.3–6], and a paper, with H. Hahn and G. Herglotz, on the flow of water in tubes and channels [9.7].

Schwarzschild's earlier preoccupation with celestial mechanics continued in Göttingen with two papers, one on periodic orbits of the Hecuba type [1.7] and the other on the determination of the orbits of spectroscopic binary stars [7.8]. A lecture entitled "Celestial Mechanics", that Schwarzschild gave in 1903 to the meeting of the Society for German natural scientists and doctors ["Gesellschaft Deutscher Naturforscher und Ärzte"] in Kassel, was widely reproduced [1.8].

In Göttingen Schwarzschild's interest in solar physics was kindled, and with it, general interest in the physics of stellar atmospheres. The first period of papers on solar physics primarily covers the years 1905–1906, opening with the solar eclipse of 30 August 1905. For this, he hastily organized an expedition to Guelma in Algeria. This was highly successful, a considerable amount of valuable material being brought back [2.1,2].

A few minor works on the UV-radiation of the Sun [2.3–5] were also published, in collaboration with his friend W. Villiger, who had helped him earlier (in 1896) in measuring double stars, and who in the meantime had joined the Zeiss company. The measurements were actually carried out in Jena.

In 1906 the important theoretical work on the equilibrium of the Sun's atmosphere [2.6] appeared, and which introduced the important concept of "radiative equilibrium". His first approximation to a solution of the integral equation solved here was very similar to an analogous, but independently discovered solution by Arthur Schuster, and the two later entered the literature as the "Schuster-Schwarzschild Model" for grey atmospheres [20]. A final item related to this particular field is his very thorough review of Robert Emden's book "Gaskugeln" [2.12].

In his last years as Göttingen, Schwarzschild also began to get involved in stellar statistics. In 1907 his important paper on the proper motions of the stars [4.2] appeared, in which the ellipsoidal velocity-distribution was first proposed, rather than the star-stream model of Kapteyn and Eddington. A paper on the determination of the vertex and apex [4.3] should also be mentioned in this context. He also described the ellipsoidal theory in a widely reproduced popular lecture on the system of fixed stars [4.4], given before the Scientific Society ["Wissenschaftlicher Verein"] in Berlin.

Another lecture at a popular level was given to the Royal Society of Sciences in Göttingen – an ordinary member of which he had become in 1905 – and concerned Lambert's cosmological letters, particularly with his concept of a central sun [10.8-II]. Its general theme was therefore the history of science. Other popular lectures dealt with the telescope [10.8-I] and the universe [10.8-IV]. In Berlin he gave a lecture on "Amateur Astronomy and Amateur Astronomers" [N-6:6], and a lecture on "The Physics of the Moon" before the "Society for the Instruction of Women and for Women's Studies" – just to mention a couple [11].

All this shows Schwarzschild's extraordinary breadth of interests. He had tackled practically all of the burning astronomical problems of the day and made fundamental contributions to most of them. This applied to his non-astronomical work as well, particularly in the fields of electrodynamics, optics and, later, of relativity theory and quantum theory. The wide range of his interests is quite apparent when we consider how, even today, we still speak of the Schwarzschild exponent in photography, of the Schuster-Schwarzschild model and the Schwarzschild criterion for convection in stellar atmospheres, of the Schwarzschild equation in stellar statistics, of the Schwarzschild distribution of stellar velocities, and of the Schwarzschild radius of a black hole.

Fig. 10. Karl Schwarzschild in Göttingen

Lectures and publications also show how Schwarzschild became involved with astronomical education in schools [10.9–11]; he signed an appeal for the erection of a tower to Gauss on the Hohenhagen [10.18] and discussed a portrait of Bessel that had previously been wrongly thought to be of Gauss [10.17]. These, not specifically scientific, papers are discussed in more detail in Section 10.

Even whilst at Göttingen Schwarzschild spent most of his holidays in the South, either in skiing or in extended walking tours in the mountains and climbing, for example on the Matterhorn. Apart from his brother Alfred, he was often accompanied by physicists from Munich and elsewhere, such as Robert Emden, who had married Schwarzschild's sister, or Arnold Sommerfeld, Max von Laue, and Wilhelm Wien – to mention just a few. Frequently they were so deep in technical discussions that they saw nothing of the scenery. Alfred tells many anecdotes of these excursions in his memoirs. Schwarzschild and Emden travelled to Sicily in October 1907, to try to find a suitable site for an observatory devoted to solar and geophysical observations, and in 1908 an application for a southern branch was submitted to the ministry [N-907/25].

Schwarzschild liked experimenting, and was often brilliant at it. Whenever he needed some accessory, it was immediately taken from another, temporarily unused, instrument; if a diaphragm was needed immediately, a hole was soon made in a lens cap. The Observatory's inventory list contains numerous entries made by his successor in red link, where such occurrences were deplored in the strongest terms. Even the amusements that went on during one of the parties (of which many were held at the Observatory, and which the daughter of Felix Klein looked back on with enthusiasm fifty years later), made their way into the inventory list in this way [19].

Over the years there were numerous changes of personnel at the Observatory. The most important ones will be mentioned briefly: in 1906, Meyermann left to become Observer at the Naval Observatory at Wilhelmshaven [28-1906], whence he later took over the Directorship of the Observatory at Tsingtau. Arnold Kohlschütter took his place, but left two years later to go to Kiel, being replaced by O. Birck [28-1908]. In 1908, Gustav Herglotz, who had held a special astronomical teaching post, went to Vienna, and Schwarzschild brought Ejnar Hertzsprung to Göttingen to take

Fig. 11. Karl Schwarzschild and Ejnar Hertzsprung as university lecturers in front of the observatory at Göttingen

over [28-1908], having long been interested in his work through his own researches into stellar statistics. Hertzsprung came in April 1909 and was appointed as assistant professor. On 4 February 1909, Schwarzschild had written to his parents: "I now also want to get hold of the Dane, Hertzsprung". On 10 May he wrote that there would be twelve students in the practical astrophysics course that he and Hertzsprung would be taking; Hertzsprung would be both stimulating and an agreeable person: it would have been impossible to find anyone better [N-907:26]. Schwarzschild had introduced this practical astrophysics course in 1907, to run in parallel with the practical astrometry course under Ambronn [28-1907].

The year 1909 brought two great changes in Schwarzschild's life, marriage and his appointment to Potsdam. Soon after his arrival in Göttingen, Schwarzschild had met Else Rosenbach, the daughter of the surgeon, Professor Rosenbach, and his wife, née Merkel, and, together with Prandtl, had been a frequent visitor at their house. But

Fig. 12. Karl Schwarzschild and his wife Else, née Rosenbach

the very strong reservations about the Jewish faith that were held by the Rosenbachs prevented a serious engagement for a long time, causing a lot of distress to Else. But in the early summer of 1909 Schwarzschild wrote: "We are bearing things patiently, so that gradually the people at Göttingen will accept us out of love or curiosity". The engagement was announced in July 1909, and on 22 October 1909, immediately before their move to Potsdam, Karl and Else were married. Else was a very lively person, with a great sense of fun. There were three children: Agathe (20 November 1910), Martin (31 May 1912), and Alfred (24 October 1914).

1907 saw the death of Hermann Vogel, the first director of the Astrophysical Observatory at Potsdam, which was founded in 1879 and which was then the foremost astronomical institution in Germany. It was purely a research institute, not being linked to any university. Some of the members of the Institute would have been pleased to see Johannes Hartmann as successor. He had been at the Institute since 1896, had been appointed assistant professor in 1902, and had been largely responsible for building up the instrumentation. In particular, the double refractor, with objectives of 50 and 80-cm diameter and 12-m focal length, which had been installed in 1899 – and which still remains the fourth largest refractor in the world – was primarily his instrument. But the Prussian ministry wanted to find a theoretician as successor to Vogel, and Hugo von Seeliger was offered the post. But he was already in his 60s, and he declined, suggesting his pupil Schwarzschild instead. In Berlin, however (according to a letter from Schwarzschild to his parents on 4 February 1909 [N-907:26]), they first thought of Runge. Shortly afterwards he wrote to his parents again, saying that as obtaining Runge's services would be too expensive, he was again being considered for Potsdam. Shortly afterwards, at the age of 36, he was officially appointed to head the greatest observatory in Germany.

Although this was the most prominent position that an astronomer could hope to hold in Germany at that time, Schwarzschild was by no means overenthusiastic. His wife's family lived in Göttingen, he had a wide circle of friends and, above all, Göttingen was a mathematical stronghold, and this contact with the mathematicians and physicists was more important to him than the better instrumentation at Potsdam. He had also just gained Hertzsprung as an important collaborator. He was rather intimidated by "cold, severe, Potsdam" [9] and by the much greater load of administrative duties. Nevertheless he finally agreed, with the condition that Hertzsprung should move to Potsdam with him. The only appropriate post for Hertzsprung at Potsdam was the one held by Hartmann. So the ministry had to find something to offer Hartmann that he could hardly refuse, so he was appointed to be Schwarzschild's successor at Göttingen. Hartmann showed no more enthusiasm for the move. He was, above all, a practical, observational astronomer, and Göttingen – whose largest telescope was not even of 20-cm aperture, and which was located on the Leine levels, practically in the town – appeared just a paltry provincial observatory. Göttingen's pre-eminence in mathematics was of no particular attraction to him. He insisted on there being very considerable future investment in instrumentation, and finally agreed on 21 September 1909. The war-years and post-war period that soon followed set most of his plans at naught. Hartmann arrived at Göttingen at the beginning of October, while Schwarzschild, who had been nominated as Director of the Astrophysical Observatory at Potsdam at the end of August [27], was still negotiating

Fig. 13. The Mathematics Club of Göttingen, 1902.
Left to right, front row: Abraham, Schilling, Hilbert, Klein (*at the table*), Schwarzschild, Mrs. Young, Diestel, Zermelo. *Second row*: Fanla, Hansen, C. Müller, Dawney, E. Schmidt, Yoshiye, Epstein, Fleisher, F. Bernstein. *Third row*: Blumenthal, Hamel, H. Müller

about Hertzsprung. So for a few weeks they were both in Göttingen. The day they were married, Else and Karl left for their honeymoon in Switzerland, whence they went to Potsdam at the beginning of November.

Settling in Göttingen proved to be even more difficult for Hartmann than he had expected. He was horrified at the state of the Observatory, and wrote to Schwarzschild on 20 November 1909, seven weeks after he had taken over the post, saying: "Dear Colleague, You are aware that I am not happy with the situation here and, in particular, that I cannot agree as to the state of the inventory. I therefore intend to resign this post. Now you have told me yourself how hard you and your lady wife found it to bid farewell to Göttingen, and I have heard from many people how reluctantly your colleagues here saw you go. I would therefore like to ask you if you would consider being reappointed here... Göttingen, where mathematics is in full flower, is obviously more suited to you than to anyone else; you are in your right place

Fig. 14. The Astrophysical Observatory at Potsdam, about the time of Schwarzschild (Photograph by Mrs. Marie-Louise Strohbusch)

here, I am not...."[4] What human tragedy can be sensed behind this letter! But the exchange envisaged by Hartmann came to nothing. Schwarzschild stayed in Potsdam and Hartmann in Göttingen. In his obituary, however, Runge said that Schwarzschild never put down such deep roots in Potsdam.

Potsdam

Schwarzschild quickly familiarized himself with the various fields of work being carried out at Potsdam, and was soon able to give advice and help. The Institute was naturally far better endowed with personnel, instruments and finances than the observatory at Göttingen. The areas of research being carried out at the Observatory included the observation of the major planets and solar physics, and in particular

[4] "Sehr geehrter Herr Kollege! Es ist Ihnen bekannt, daß es mir hier gar nicht gefällt, und daß ich mich besonders mit dem Zustand des Inventars nicht einverstanden erklären kann. Ich beabsichtige daher, diesen Posten niederzulegen. Nun haben Sie mir selbst gesagt, wie schwer Ihnen und Ihrer Frau Gemahlin der Abschied von Göttingen geworden sei, und ich habe von vielen Seiten gehört, wie ungern die hiesigen Kollegen Sie haben scheiden sehen. Ich möchte Sie daher fragen, ob Sie eine Zurückberufung nach hier annehmen würden... Göttingen, wo die Mathematik in voller Blüte steht, eignet sich für keinen anderen so gut wie gerade für Sie; Sie sind hier am richtigen Platz, ich bin es nicht..." [N-306]

visual photometry, photographic photometry and spectrophotometry. This category also included the photometric survey of the North Polar zone, participation in the photographic Carte du Ciel, and the observation of certain Kapteyn Selected Areas. There was also the AG Catalogue of variables, in connection with which many individual variables were in vestigated. A major project was the measurement of radial velocities.

The numerous staff at the Institute during this period included Hertzsprung, whom Schwarzschild had brought with him from Göttingen; the Chief Observers O. Lohse, who had been in charge of the Institute during the interregnum; Gustav Müller, who became Schwarzschild's successor after his death; Julius Scheiner; and P. Kempf. Observers included Professor Hans Ludendorff; and Professor G. Eberhard, who discovered the photographic adjacency effect named after him (the Eberhard Effect) at about this time. For many years Emanuel von der Pahlen helped at the Institute, as, for a short time, did Kurt Felix Bottlinger, Assistant at the Berlin University Observatory at Babelsberg. Both of the latter were primarily interested in problems of stellar statistics and stellar dynamics. Among the technical staff Bernhard Schmidt must be mentioned, who, twenty years later at Hamburg, was to develop the Schmidt telescope. In Potsdam he corrected the 50-cm objective of the great refractor in 1912, turning it into a first-class objective [29-1914].

Now let us turn to Schwarzschild's own work and projects. In August 1910 he travelled to the USA to attend the Congress of the Astronomical Society of America in Cambridge, Mass. and the Fourth Conference of the International Union for Solar Research at Mount Wilson [2.13]. He used the opportunity to visit many of the large American observatories [10.12]. He returned convinced that Germany definitely needed an out-station south of the equator, like those operated by Harvard College Observatory and Lick Observatory. He proposed Windhoek in German South-West Africa. In 1910 he took part in observing Halley's Comet [3.3] and, together with E. Kron, reduced photographs of the comet that had been obtained by the Potsdam expedition to Teneriffe.

Schwarzschild continued his work on stellar statistics and stellar dynamics in Potsdam, developing his fundamental equations for stellar statistics [4.5–7]. Other statistical papers dealt with the motions and parallaxes of stars, and the frequency distribution of spectral classes [4.8–10]. His last paper on stellar dynamics, concerning the uniform velocity distribution in the Galaxy, was first published posthumously as a fragment in a volume commemorating H.v. Seeliger [4.11]. For many of these studies he required a quick way of determining the radial velocities, so he developed Pickering's method of measuring radial velocities using an objective prism [7.3–5]. In addition to these studies there were minor papers on a new variable in Taurus [7.10,11], on a nova in Gemini [7.12,13] and a short contribution on the spectral classification of stars, which was under much discussion at the time [7.7].

Some of the problems investigated at Göttingen were completed (e.g. the Göttinger Aktinometrie [6.18]), or re-examined, such as a paper on correcting for fogging when using the half-grating method [6.20]; optical papers about astronomical objectives [8.7,8]; and the report on the bubble sextant [5.8]. We may also include here the discussion of the Yerkes Actinometry [6.19]. He also wrote the introduction to a work by Dziewulski on the photographic magnitudes of stars [6.21].

Once again, Schwarzschild became intensely involved with 5 papers on completely different problems concerning the physics of the solar and stellar atmospheres [2.7–11]. He reported the Fifth Conference of the International Union for Solar Research in Bonn in 1913 [2.14], and gave a popular lecture to the Berlin Urania society on "Solar Physics".

Schwarzschild wrote the contributions on "Orbital Determination" [1.9], "Structure of the Universe" [10.6], and "Cosmogony" [10.7] for the fourth and fifth editions of Newcomb-Engelmann's *Popular Astronomy*. He wrote a report on "Astronomy [in 1913]" for the yearbook *The Year 1913: A Survey of Cultural Development* [10.5]. Shortly afterwards a review of books by H. Poincaré and T.J.J. See appeared [10.16].

In 1912, Schwarzschild became an ordinary member of the Prussian Academy of Sciences in Berlin, being proposed by Hermann Struve [30]. On 26 July 1913 he gave his inaugural address, the customary reply being given by the then Secretary Max Planck [10.19]. As a member of the Academy, Schwarzschild had the right to give lectures at the University of Berlin. In the winter term 1913/14 he gave a lecture on "Stellar Astronomy". In February 1916, the University elected him an honorary professor [11]. His wish for renewed contact with students was therefore fulfilled, but unfortunately he was unable to make use of this opportunity.

Fig. 15. Letter of Appointment as a Member of the Royal Prussian Academy of Science, Berlin

Schwarzschild also acted as editor. From 1911 he was one of the collaborators on the leading American journal, the *Astrophysical Journal*. He was also responsible for editing the astronomical volume, (VI, 2), for the *Encyclopedia of Mathematical Sciences*, a work that he was not to complete, however [11].

On 25 June 1914, he gave a lecture to the Deutsche Gesellschaft für Mechanik und Optik [the German Mechanical and Optical Society], on "Precision engineering and scientific research" [7.6]. A few weeks later the First World War broke out, which henceforth had more and more effect on the work of the Institute. Schwarzschild immediately volunteered for service with the army. Because of his age and position, he would not have been drafted, at least not then. As Else Schwarzschild explained later to his children, he felt obliged to take this step, precisely because he was a German Jew. In September 1914 he was sent, as acting officer (later Lieutenant) to Namur in Belgium as head of a field weather station. The whole of 1915 he spent in the field, first with the weather station, and later as a member of the artillery staff, partly in the Argonne region of France and partly in Russia [29-1913, 1914]. During this time he wrote a paper on the influence of wind and air density on the paths of shells [9.13] , which – because of its military implications – was not published until 1920.

In his last years, Schwarzschild again became closely involved with some of the major problems in physics, in particular with the newly developed relativity theory and quantum theory. As early as 1913 he tried to use the cyanogen band at 3883 Å in light from the Sun to detect the redshift of 0.6 km/s predicted by Einstein, but without success [2.11], [29-1913]. In December 1913, Schwarzschild gave two lectures to the Physical Society on the Stark and Zeeman Effects [9.8,9]. In Russia he wrote two papers on the gravitational field of a point mass and of an incompressible fluid sphere, using Einstein's theory [9.11,12]. Einstein submitted these papers to the Academy and they were published in 1916. The first of these papers put forward the first exact solution of Einstein's field equations [20], and the second paper contained the description, later named after him, of the "Schwarzschild Radius". In a letter to Einstein in December 1915, and so only a few months before his death, Schwarzschild gives a rigorous solution to the problem of the rotation of the perihelion of Mercury. This letter, and Einstein's postcard in reply [N-2.2] are reproduced on p. 36. We should like to thank the Jewish National & University Library in Jerusalem for permission to reproduce the card from Einstein here for the first time.

During the Russian campaign, Schwarzschild showed symptoms of pemphigus, a painful and then incurable skin disease. In March 1916 he was invalided home. At the March 1916 meeting of the Academy he presented his last paper "On the Quantum Theory" [9.10], which he had completed in just eight days. He then had to go into hospital, where he died on 11 May 1916. Only a few days before, he read the proofs of his last paper, which appeared in print on the day of his death [11].

The address at the funeral service was given by his successor, G. Müller [9], and the memorial address to the Academy by A. Einstein [3][5]. On 16 May, at his own request, he was interred in the municipal cemetery in Göttingen. The funeral address in Göttingen was given by C. Runge [N-20.2].

[5] An English translation of this memorial address is given on p. 34

Fig. 16. Karl Schwarzschild's grave in the cemetery at Göttingen

Despite the war, the news of Schwarzschild's death spread rapidly among astronomers the world over, where it was greeted with consternation. Many people had heard of his illness, but no one had thought that it would so quickly end fatally. Fifteen obituaries and memorial addresses were published ([1–14] and [N-20.2]). To this may be added about 20 short obituary notices published in various newspapers that are to be found with his personal papers. Some particularly comprehensive evaluations of the man and of his work were made by Hertzsprung [10]; Blumenthal [11], who particularly concentrated on his mathematical skills and who also gives a comprehensive list of Schwarzschild's publications; and also by Sommerfeld [5], who discusses his physical contributions in particular. In addition, a very comprehensive appreciation by Oppenheim [16] appeared in 1923, on the 50th anniversary of Schwarzschild's birth.

All the obituaries stressed the breadth of Schwarzschild's scientific work. Sommerfeld [5] spoke of the stroke of good fortune that combined skill in astronomy, astrophysics and physics in one individual. An obituary in *The Observatory* [1] said: "He was a master of almost all branches of physical and mathematical Science". Runge [7] praised his combination of complete mastery of mathematics, with a deep understanding of various physical processes and his skill in experimentation. His mathematical skills and the role that he played in the group of mathematicians at Göttingen were frequently referred to. Hertzsprung [10] wrote: "Still it was mathematical sense, not mathematical knowledge that he most appreciated in others". Blumenthal [11] emphasized his clear and acute perception, and his objectivity. He was able to work in an incredibly concentrated fashion, without showing any signs of fatigue, and – especially in discussion – he had the facility of rapidly grasping the essentials of an argument and of expressing them clearly. His clear, and readily understandable lectures and presentations were frequently praised. The variety of his scientific contributions was undoubtedly the reason why he did not form part of any particular "school of thought". In later appreciations, the fact that Schwarzschild was one of the great pioneers of astrophysics has been increasingly emphasized. Oort described him in 1962 as being the greatest German astronomer since Kepler [35].

As well as the breadth and depth of his scientific work, his personality was repeatedly mentioned, often in a very effusive manner. He combined youthful exuberance with intellectual brilliance [6]. His life, particularly in the years at Göttingen, was characterised by concentrated work and scientific discussions on the one hand, and by a sensible but cheerful disposition and by boisterous parties on the other. His saying that he could not get on with people who held him in too great a respect, was frequently mentioned [16]. All in all, as Ludendorff said [6], he had a charm that no one could escape.

After his death, Else Schwarzschild and the three children returned to her home town of Göttingen. Schwarzschild had expressed this wish in his will [N-20.2], and doubtless this corresponded with his wife's own feelings. Else Schwarzschild lived in Göttingen for 34 years, and died in 1950, having led a very retired life towards the end. Their daughter Agathe, a classics teacher, married the professor of philosophy Harry Thornton, and lives in New Zealand. Their younger son Alfred was very unlucky and suffered greatly under National Socialism. Martin inherited the talents and much of his father's nature. He studied astronomy at Göttingen, took his doctorate there in 1935, and in 1936, with support from Hans Kienle, the then Director of the Observatory, went first to Oslo, and then emigrated to the USA, where he has since been Professor of Astronomy at Princeton.

Honours

Documents relating to the following awards can be found among Schwarzschild's personal papers [N-21], [N-23.2]:

1899 – Medal of the Photographic Society of Vienna
1905 – Ordinary member of the Kgl. Gesellschaft der Wissenschaften [Royal Society of Sciences] at Göttingen (now known as the Akademie der Wissenschaften [Academy of Sciences]); Mathematical/Physical Section
1909 – Prize for the best instrument (a balloon sextant) in the competition on the theme of "Astronomical position determination from a balloon" organized by the International Airship Exhibition
1909 – Associate of the Royal Astronomical Society, London
1910 – Member of the Kaiserl. Leopoldinisch-Carolinsche Deutsche Akademie der Naturforscher [the Imperial Leopold & Carl German Academy of Natural Scientists] (nowadays known as the Deutsche Akademie der Naturforscher Leopoldina)
1911 – Honorary member of the Phys. Verein [Physical Society], Frankfurt
1912 – Red Order, 4th Class, of his Royal Majesty
1912 – Member of the Kgl. Preuss. Akademie der Wissenschaften in Berlin [Royal Prussian Academy of Sciences, Berlin]
1912 – Member of the Academia Scientarium Borrussia
1914 – Honorary member of the Cambridge Philosophical Society
1914 – Honorary Doctor of the University of Groningen

Biographical References

A. Obituaries and Commemorative Addresses

[1] Anon.: Karl Schwarzschild, *Observatory* **39** (1916) 336–339
Publ. Astron. Soc. Pacific **28** (1916) 269–273

[2] Müller, G.: Karl Schwarzschild [Obituary], *Astron.Nachr.* **202** (1916) 383

[3] Einstein, A.: Memorial lecture on Karl Schwarzschild, *Sitz. Ber. Preuss. Akad. Wiss. Berlin* **1** (1916) 768–770
Reprinted in *Physiker über Physiker*, Vol. II (Akademie-Verlag, Berlin 1979) pp. 243–244

[4] Parkhurst, J.A.: Karl Schwarzschild, *Science*, N.S. **44** (1916) 232–234

[5] Sommerfeld, A.: Karl Schwarzschild, *Die Naturwiss.* **4** (1916) 453–457

[6] Ludendorff, H.: Karl Schwarzschild, *Deutsche Mechaniker-Zeitung (Beibl. zur Zeitschr. f. Instrumentenkunde)* (1916) 107–108

[7] Runge, C.: Karl Schwarzschild, *Phys.Zeitschrift* **17** (1916) 545–547

[8] Kritzinger, H.H.: Karl Schwarzschild, *Sirius* **49** (1916) 129–130

[9] Müller, G.: Funeral address, *Sirius* **49** (1916) 130–132

[10] Hertzsprung, E.: Karl Schwarzschild, *Astrophys. J.* **45** (1917) 285–292

[11] Blumenthal, O.: Karl Schwarzschild, *Jahresber. der dt. Mathematiker-Vereinigung* **26** (1918) 56–75. (With a comprehensive list of publications)

[12] Berndt, G.: Karl Schwarzschild, *Frankfurter Zeitung* No. 135, 16 May 1916, 2nd morning edition, pp. 1–2

[13] Eddington, A.S.: Karl Schwarzschild, *Monthly Not. Royal Astron. Soc.* **77** (1917) 314–319

[14] Geier, L.: Karl Schwarzschild, *Allg. Zeitung Judentum* **81**, No. 44, 2 Nov. 1917, pp. 520–521

B. Additional Biographical References

[15] Sommerfeld, A.: Die Quantentheorie der Spektrallinien und die letzte Arbeit von Karl Schwarzschild, *Umschau* **20** (1916) 941–946

[16] Oppenheim, S.: Karl Schwarzschild. Zur 50. Wiederkehr seines Geburtstages, *Vierteljahrsschrift der Astron. Ges.* **58** (1923) 191–209

[17] ten Bruggencate, P.: Karl Schwarzschild – der Schöpfer der heutigen Astrophysik, in *Forscher und Wissenschaftler im heutigen Europa – Weltall und Erde* (Stelling, Oldenburg); Printed on the 50th anniversary of his death in *Sterne und Weltraum* **5** (1966) 188–190

[18] Wempe, J.: Die Beziehung zwischen Ejnar Hertzsprung und Karl Schwarzschild, in *Geschichte und Popularisierung der Astronomie* (Veröff. Archenhold-Sternwarte, Berlin Treptow, No. 6, 1974) pp. 45–54

[19] Voigt, H.H.: Von Karl Schwarzschild bis Hans Kienle – Der Weg von der klassischen Astronomie zur Astrophysik in Göttingen, *Sterne und Weltraum* **28** (1989) 12–17

[20] Dieke, S.H.: Schwarzschild, Karl, *Dictionary of Scientific Biography* (Charles Scribner's Sons, New York 1975) Vol. 12, pp. 247–253

[21] Fraenkel, A.A.: Jewish mathematics and astronomy, *Scripta Math.* **25** (1960) 45–46

[22] Gerth, E.: Zum 50. Todestag von Karl Schwarzschild, *Bild und Ton* **19** (1966) 339–341

[23] Kienle, H.: Auf den Spuren Karl Schwarzschilds, *Sterne und Weltraum* **3** (1974) 79

C. Miscellaneous Biographical Sources

[24] Beobachtungen der totalen Mondfinsternis 1892 Nov. 4. Zusammengestellt von Herrn Martin, *Astron.Nachr.* **131** (1892) 393–396

[25] Oetel, K.: Beobachtungen der Leoniden- und Bielidenfälle 1899 auf der k. Sternwarte München, *Astron.Nachr.* **151** (1900) 193–196

[26] de Ball, L.: Zonenbeobachtungen der Sterne zwischen $5°50'$ und $10°10'$ südlicher Deklination, *Publ.d. v. Kuffnerschen Sternw.* **6** (1902) Part I

[27] Personalakte Karl Schwarzschild, Archiv der Georg-August-Universität, Göttingen

[28] Jahresberichte der Sternwarten, Göttingen, 1898–1909, *Vierteljahrsschrift der Astron. Ges.* 1899–1910

[29] Jahresberichte der Sternwarten, Potsdam, 1909–1916, *Vierteljahrsschrift der Astron.Ges.* 1909–1916

[30] Struve, H.: Wahlvorschlag für Karl Schwarzschild zum OM, in *Archiv der Akademie der Wiss. der DDR, Berlin*, II-IIIa, Vol. 18, Sheets 215–216 (Typewritten original)
Printed in *Physiker über Physiker, Wahlvorschläge zur Aufnahme von Physikern in die Berliner Akademie 1870–1929* (Akademie-Verlag, Berlin, 1975) pp. 195–196

[31] Weiß, W.W., M. Vyoral-Tschapka: *Sitzber.Österr. Akad. d. Wiss., Group II, Math., Phys., Techn. Wiss.* **191** (1982) 615

[32] Herz, N.: *Publ. v. Kuffner'sche Sternwarte*, Vol. I, 1889

[33] Versammlung der Astronomischen Gesellschaft zu Göttingen 1902 August 4 bis 7, *Vierteljahrsschrift der Astron. Ges.* **37** (1902) 175

[34] Schröder, W.: Emil Wiechert und seine Bedeutung für die Entwicklung der Geophysik zur exakten Wissenschaft, *Archive for History of the Exact Sciences* **27** (1982) 369

[35] Oort, J.H.: On the problem of the origin of spiral structure, Karl Schwarzschild Lecture 1972, *Mitt. Astron. Ges.* **32** (1973) 15

D. Schwarzschild's Personal Papers

[N-...] Nachlaß K. Schwarzschild: Handschr. Abt. der Nieders. Staats- und Univ.-Bibliothek, Göttingen

- -2.2 Eindeutigkeit der Lösungen der Gravitationsgleichungen. Sheets $2, 2^R, 3$: Brief Karl Schwarzschild an Einstein; Sheets $4, 4^R$: Antwortkarte Einsteins
- -6.6 Manuskript zum Vortr. "Amateurastronomie"
- -13.1 Vorlesung "Astrophysik", Nachschrift von Max Born
- -20.2 Persönliches 1898–1916

-21	Zeugnisse und Urkunden
-22	Militärzeit
-23.1	Photos
-23.2	Medaille (Phot. Ges. Wien)
-24.1	Memoiren Alfred Schwarzschild, pp. 1–99
-24.2	... Vol. II, pp. 100–191
-306	Briefe von und an Joh. Hartmann. Sheets 12–13: Brief von J. Hartmann vom 20.11.1909
-907	Briefe von und an die Eltern (/1 bis /34)
/17	Januar bis Juli 1901
/18	Juli bis Dezember 1901
/Beil. 1.:	Transkription der Briefe von 1901 (/17 u. /18) (getippt 1916 für Else)
/24	Briefe 1906
/25	Briefe 1907–1908
/26	Briefe 1909

Karl Schwarzschild Lectures

Inauguration of the Lectures

At the Annual General Meeting of the Astronomische Gesellschaft in Kiel in September 1959, under the chairmanship of Paul ten Bruggencate, the third successor to Schwarzschild's chair, the society inaugurated the "Karl Schwarzschild Lecture" [1]. Since that time twenty astronomers of international repute have given lectures of the very highest class; collectively these have been published in the Society's *Mitteilungen*. The first Karl Schwarzschild Lecture was given by his son, Martin Schwarzschild, on the subject of "The Theory of the Internal Structure of Stars" [2].

We are including in this book the Introduction to the Karl Schwarzschild Lecture given by S. Chandrasekhar, "The Aesthetic Base of the General Theory of Relativity" [3], because it gives a particularly fine insight into Karl Schwarzschild's method of work.

References

[1] ten Bruggencate, P.: Ansprache bei der Eröffnung der Versammlung in Kiel, 1959. Eröffnung der Karl-Schwarzschild-Vorlesung, *Mitt. Astron. Ges.* **12** (1961) 9, 15

[2] Schwarzschild, M.: Die Theorie des inneren Aufbaus der Sterne, Karl-Schwarzschild-Vorlesung 1959, *Mitt. Astron. Ges.* **12** (1961) 15

[3] Chandrasekhar, S.: The Aesthetic Base of the General Theory of Relativity, Karl-Schwarzschild-Vorlesung 1986, *Mitt. Astron. Ges.* **67** (1986) 19

Introduction to the 17th Lecture

The Aesthetic Base of the General Theory of Relativity

By S. Chandrasekhar

I

Karl Schwarzschild is, of course, one of the towering physical scientists of this century. The breadth and range of his contributions are staggering: they cover the entire range of physics, astronomy, and astrophysics of his time.

In physics, they range from electrodynamics and geometrical optics to the then newly developing atomic theory of Bohr and Sommerfeld. In electrodynamics, he derived a variational base for Lorentz's equations of the electron. In geometrical optics, he developed the theory of the aberrations in optical instruments (described by Max Born as "unsurpassed in clarity and rigour by later work") and formulated the principle underlying the optics of the Schmidt telescope. And in the Bohr-Sommerfeld theory, he worked out, in his last published paper, the theory of the Stark effect and of the Deslander term in the rotational-vibrational spectra of diatomic molecules. (In this last paper he introduced, for the first time, the notions of action and angle variables.)

In astronomy and astrophysics, Schwarzschild's contributions are so many and so varied that I shall mention only those of his discoveries to which his name is attached. We have the Schwarzschild exponent in photographic photometry, the Schwarzschild-Milne integral equation in the theory of radiative transfer, the Schwarzschild criterion for the onset of convective instability, the Schwarzschild ellopsoidal distribution of stellar velocities, and, of course, the Schwarzschild solution of Einstein's equations for describing the space-time external to a spherical distribution of mass and of static black-holes.

And all of these in a brief twenty years!

It is possible that the announced title of my lecture has puzzled some of you: it is not addressed to any concrete topic as the earlier lectures in this series have been: and, I am afraid that it will scarcely have any astronomical overtone. My lecture, however, will bear on Schwarzschild's attitude and approach to scientific problems, as I can discern them from his published papers: and it will bear very directly on his solution of the equations of general relativity.

II

I shall consider three examples from Schwarzschild's work which, to my mind, illustrate his approach to scientific problems.

The first relates to his work on star-streaming. The phenomenon was discovered by J. C. Kapteyn: and it was adequately interpreted, almost at once, by Eddington on the basis of his and Kapteyn's hypothesis of two star-streams. Schwarzschild expressed his reaction to this hypothesis as follows [4.2]:

"The magnitudes of the proper motions of the stars in the two streams and along with them, presumably also, their average distances from the sun would appear to be equal. The stars in the two streams, during their motion through each other, must, therefore, share common fluctuations; and it is problematical how this can be brought about.

I have, therefore, believed that the same observational material on which Eddington has based himself should be reworked on a more unified hypothesis concerning stellar motions."

On these grounds, Schwarzschild formulated his ellipsoidal distribution of the peculiar velocities of stars; and this formulation has been the basis of all subsequent discussions bearing on stellar motions and the dynamics of stellar systems. What is remarkable to me, however, is his argument: a description of nature must be natural; it cannot be *ad hoc*.

As a second example, I shall take a still earlier publication of Schwarzschild. At a meeting of this Gesellschaft in 1900, Schwarzschild addressed himself to the question whether the geometry of the three-dimensional space of astronomy might be non-Euclidean. He stated the problem as follows [9.2]:

"As must be known to you, during this century [meaning the 19th century] one has developed non-Euclidean geometry (besides Euclidean geometry), the chief examples of which are the so-called spherical and pseudo-spherical spaces. We can wonder how the world would appear in a spherical or a pseudo-spherical geometry with possibly a finite radius of curvature. ... One would then find oneself, if one will, in a geometrical fairyland; and one does not know whether the beauty of this fairyland may not in fact be realized in nature."

We can only marvel at Schwarzschild's scientific imagination and curiosity in addressing himself to such a question some fifteen years before the founding of general relativity. But to Schwarzschild, it was more than simple imagination. He actually estimated limits to the radius of curvature of the three-dimensional space with the astronomical data available at his time and concluded that if the space is hyperbolic its radius of curvature cannot be less than 64 light years and that if the space is spherical its radius of curvature must at least be 1600 light years.

We need not argue about Schwarzschild's particular estimates. It is far more relevant that Schwarzschild allowed his imagination to contemplate a world that may have features of a fairyland!

My third example bears on Schwarzschild's discovery of the solution of Einstein's vacuum-equations appropriate to the exterior of a central spherical distribution of mass – undoubtedly the most important discovery in relativity after its founding.

Schwarzschild's paper in which he derived his solution was communicated by Einstein to the Berlin Academy in 1916 January 13, just about two months after Einstein himself had published the basic equations of his theory in a short communication – his detailed paper with full derivations was still six months in the future – and had deduced the theoretical rate of the precession of the perihelion of Mercury and of the magnitude of the deflection a light ray will experience as it grazes the limb of the sun. In acknowledging Schwarzschild's paper, Einstein wrote on 1916 January 9,

"I have read your paper with greatest interest. I had not expected that one could obtain the exact solution of the problem so simply. The analytical treatment of the problem appears to me splendid."

The circumstances under which Schwarzschild derived his now famous solution were heroic. During the spring and summer of 1915, Schwarzschild was serving in the German army at the Eastern front. While at the Eastern front with a small technical staff Schwarzschild contracted pemphigus, a fatal disease; and he died on 1916 May 11. It was during this period of illness that Schwarzschild wrote his two papers on general relativity – the second one dealt with the equilibrium of a homogeneous mass and showed that no hydrostatic equilibrium is possible if the radius of the object is less than 9/8 of the Schwarzschild radius, $2GM/c^2$ – and the fundamental one on the Bohr-Sommerfeld theory to which I have already referred.

About Schwarzschild's last illness, Eddington wrote in a moving obituary notice.

"His end is a sad story of long suffering from a terrible illness contracted in the field, borne with great courage and patience."

Parenthetically, I may add a footnote. Richard Courant told me in the late thirties that he had met Karl Schwarzschild proceeding to the Eastern front while he, as a member of the general staff, was with a party retreating from the same front; and Courant said that he was surprised that someone as distinguished as Karl Schwarzschild would be proceeding towards a front that was considered too dangerous for the general staff!

Let me return to Schwarzschild's original paper and his reasons for seeking an exact solution to the problem which Einstein has solved earlier by an approximate procedure. Schwarzschild began with the statement [9.11]:

"It is always satisfying to obtain an exact solution in a simple form. It is even more important, in the present instance, to have the uniqueness of the solution established and remove whatever doubts there may be concerning Mr. Einstein's treatment of the problem, since, as it will appear below, it is difficult, in the nature of this problem, to establish the validity of an approximate procedure."

While Einstein in his letter of acknowledgement to Schwarzschild (quoted earlier) argued that there can be no doubts about the validity of his approximate procedure in solving the equations, it is significant that Schwarzschild, undaunted, wished to solve exactly the problem which he realized was a fundamental one in the newly formulated theory. I said "undaunted" in view of the great much-ado that was to be made soon afterwards (and for some decades, to the detriment of the theory) about the "difficulty" of Einstein's theory in general and of finding exact solutions in particular.

Appendix: Einstein's Memorial Address

On 29 June 1916, Einstein gave a brief memorial address on Karl Schwarzschild at a meeting of the Berliner Akademie. It presents a proper measure of Schwarzschild; and I have thought it worthwhile to append the following translation[1] of Einstein's address.

"On May 11 of this year [1916], Karl Schwarzschild, 42 years old, was by death snatched away. This early demise of so gifted and many-sided a scientist is a grievous loss not only to this body, but also to all his astronomer and physicist friends.

What is specially astonishing about Schwarzschild's theoretical work is his easy command of mathematical methods and the almost casual way in which he could penetrate to the essence of astronomical or physical questions. Rarely has so much mathematical erudition been adapted to reasoning about physical reality. And so it was, that he grappled with many problems from which others shrank away on account of mathematical difficulties. The mainsprings of Schwarzschild's motivations in his restless theoretical quests seem less from a curiosity to learn to deeper inner relationships among the different aspects of Nature than from an artist's delight in discerning delicate mathematical patterns. It is therefore understandable that Schwarzschild's earliest contributions were in celestial mechanics, a branch of science whose foundations are more firmly established than any other. In this area, I may recall his investigations on the periodic solutions of the three-body problem and Poincaré's theory of the equilibrium of rotating fluid masses.

Among the most important of Schwarzschild's astronomical contributions are his investigations on stellar statistics, i.e., a part of science which seeks by statistical methods to relate the observations on the luminosity, the velocity, and the spectral type of stars from a large system of many objects to which the Sun belongs. In this area, astronomers are indebted to him for deepening and widening their understanding of Kapteyn's discovery.

Schwarzschild directed his deep knowledge of theoretical physics to the theory of the Sun. Here, one is grateful for his investigations on the equilibrium of the solar atmosphere and for considerations relating to radiative transfer. To this area also belongs his beautiful investigations on the pressure of light on small spherical particles which provided an exact basis for Arrhenius's theory of comet tails.

[1] Translation by S. Chandrasekhar

These investigations in theoretical physics, while they were motivated by astronomical questions, seem to have led Schwarzschild to be interested in questions purely in physics. We are indebted to him for his interesting contributions to the foundations of electrodynamics. Besides, in his last year he became interested in the new theory of gravitation: he succeeded in obtaining, for the first time, an exact calculation in the new theory of gravitation. And in the very last months of his life, much weakened by a fell disease, he yet succeeded in making some profound contributions to quantum theory.

To Schwarzschild's great theoretical contributions also belong his investigations on geometrical optics in which he refined the theory of aberrations of optical instruments of astronomical importance. These studies will remain an enduring edifice for the perfection of the tools of astronomy.

Schwarzschild's theoretical investigations were carried out simultaneously with his efforts as a practical astronomer. From age 24, he worked at observatories without interruption: 1896–1899 as an Assistant in Wien; 1901–1909 as Director of the Göttingen Observatory; and after 1909 as Director of the Astrophysical Institute at Potsdam. A long series of investigations testify to his efforts as an observer and as a leader of astronomical observations. Moreover, his lively spirit led him to advance his scientific field by charting new methods of observation. He discovered, in experimental physics, what has been named after him, how the blackening of a photographic plate can be used for the purposes of photometry by photographic methods. He also had the brilliant idea of using extra-focal images of stars for measuring their brightness: only through this idea did photographic photometry, besides visual methods, become feasible.

Since 1912, this modest man has been a member of this Academy to whose *Sitzungsberichte* he has, in this short time, communicated many important contributions. Now bitter circumstances have taken him away: but his work will bear fruits and have an enduring influence on Science for which he devoted all his strength."

From the Correspondence of K. Schwarzschild with A. Einstein

Transcript of the draft of the letter of Karl Schwarzschild to Einstein in December 1915

Words that Schwarzschild obviously added afterwards are given in small print.
Illegible words are replaced by ...

Blatt 1

22.XII.15

Verehrter Herr Einstein!

Um mit Ihrer Gravitationstheorie vertraut zu werden, habe
mich weiter mit dem
ich mir die Aufgabe gestellt, das von Ihnen in der Arbeit über
in 1. Näherung beschäftigt
das Merkurperihel gestellte und näherungsweise gelöste Problem
Zunächst machte mich ein Umstand sehr konfus
womöglich vollständig zu lösen. Man kann ja leicht das allgemeinste Linienelement angeben, das die nötigen Symmetrie-
Ich fand hier die erste Näherung der Koeffizienten g_{nv} außer ihrer Lösung noch folgende zweite
eigenschaften hat. Von diesen ausgehend bestimmte ich zunächst die erste Näherung in Ihrer Weise und fand:

$$g_{\varrho\sigma} = -\frac{\beta x_\varrho x_\sigma}{r^5} + \delta_{\varrho\sigma}\frac{\beta}{3r^3}\Big] \qquad g_{44} = 1$$

demnach hätte es außer Ihrem α noch eine zweite gegeben und das Problem wäre
also <u>zwei</u> willkürliche Größen α und β, was sonderbar war.
physikalisch unbestimmt. Daraufhin machte ich einmal auf gut Glück den Versuch einer
Dann ging ich zur vollständigen Lösung über. Eine nicht zu große Rechnerei ergab folgendes Resultat: Es gibt nur ein
nebst Feld- und Determinantengl. erfüllt
Linienelement, das Ihre Bedingungen 1) bis 4) erfüllt und

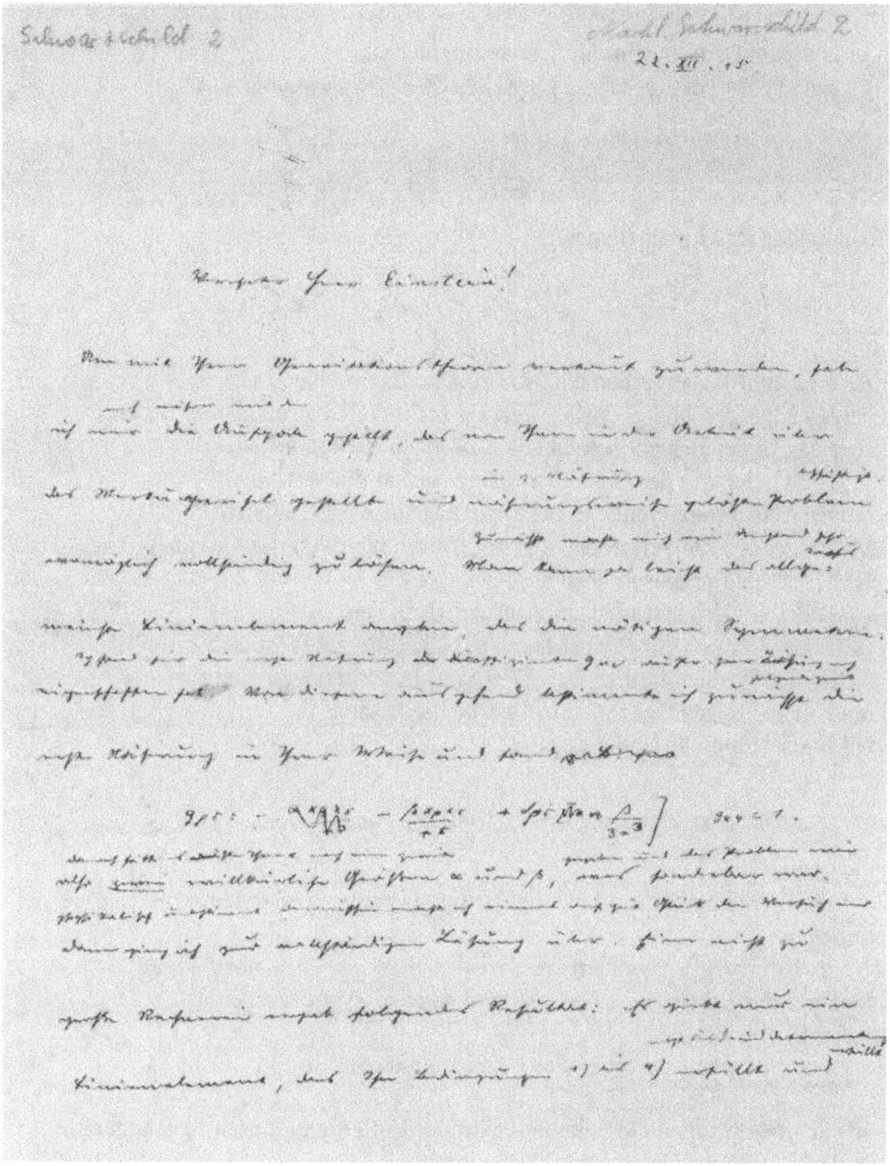
Schwarzschild 2

Karl Schwarzschild 2

22. XII. 15

Verehrter Herr Einstein!

Fig. 1a. First page of a draft of a letter of Karl Schwarzschild to Albert Einstein, December 1915

Blatt 2

im Nullpunkt und nur im Nullpunkt singulär ist.

Sei: $x_1 = r\cos\varphi\cos\vartheta \quad x_2 = r\sin\varphi\cos\vartheta \quad x_3 = r\sin\vartheta$

$$\mathcal{R} = \left(r^3 + \alpha^3\right)^{1/3} = r\left(1 + \frac{1}{3}\frac{\alpha^3}{r^0} + \dots\right)$$

dann lautet das Linienelement:

$$ds^2 = \left(1 - \frac{\gamma}{\mathcal{R}}\right)dt^2 - \frac{d\mathcal{R}^2}{1 - \gamma/\mathcal{R}} - \mathcal{R}^2(d\vartheta^2 + \sin^2\vartheta\, d\varphi^2)$$

$\mathcal{R}$, ϑ, φ sind keine „erlaubten" Koordinaten, mit denen man die Feldgleichungen bilden dürfte, weil sie nicht die Determinante 1 haben, aber das Linienelement schreibt sich in ihnen am schönsten.

Die Gleichung der Bahnkurve bleibt genau die von Ihnen in erster Näherung erhaltene (11), nur muß man unter x nicht $1/r$, sondern $1/\mathcal{R}$ verstehen, was ein Unterschied von der Ordnung 10^{-12} ist, also praktisch absolut gleichgültig.

Die Schwierigkeit mit den zwei willkürlichen Konstanten α und β, welche die erste Näherung gab, löst sich dahin, daß β einen bestimmten Wert von der Ordnung α^4 haben muß, sowie α gegeben ist, sonst würde die Lösung bei Fortsetzung der Näherungen

Blatt 3

divergent.

Es ist also auch die Eindeutigkeit Ihres Problems in schönster Ordnung.

Es ist eine ganz wunderbare Sache, daß von einer so abstrakten Idee aus die Erklärung der Merkuranomalie so zwingend herauskommt.

Wie Sie sehen, meint es der Krieg freundlich mit mir, indem

durchaus...

er mir trotz heftigen Geschützfeuers in der Entfernung diesen Spaziergang in dem … Ihrem Ideenlande (?) erlaubt.

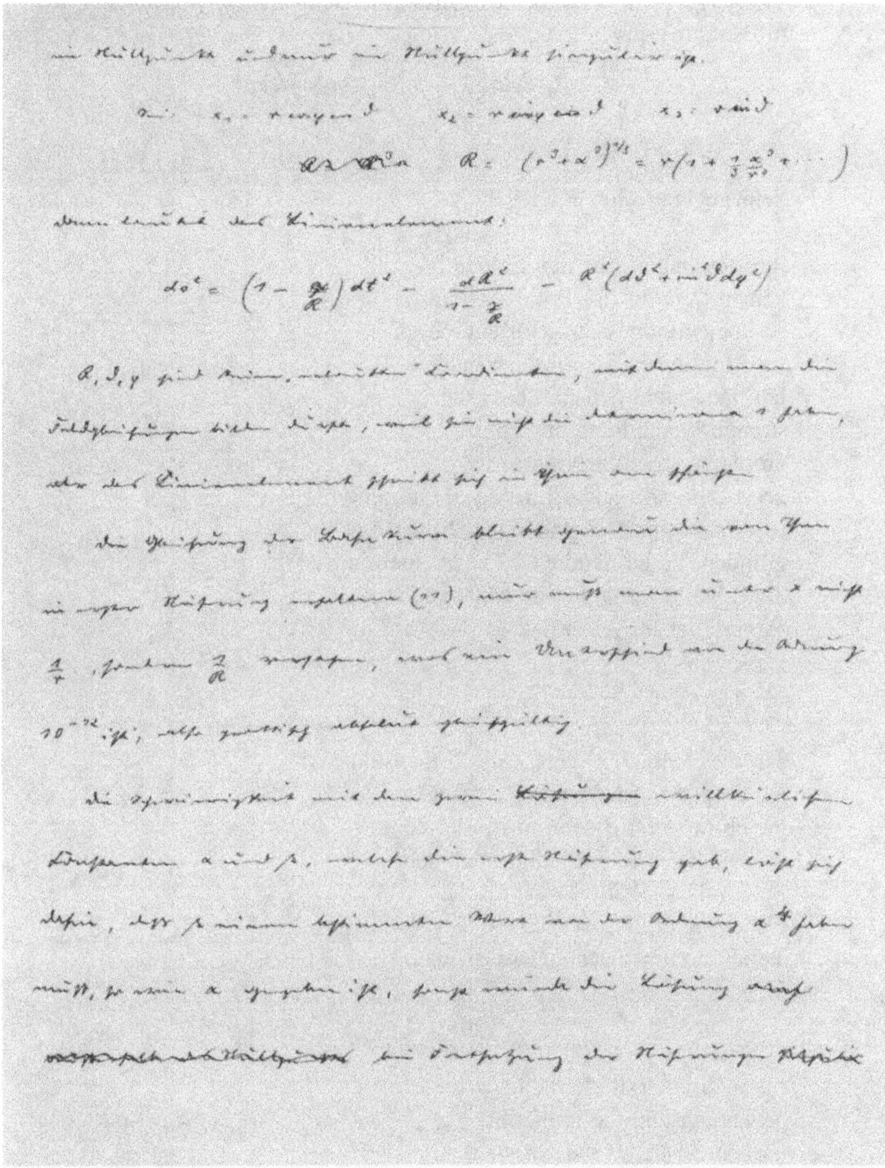

$$R = (r^3+\alpha^3)^{1/3} = r\left(1+\frac{1}{3}\frac{\alpha^3}{r^3}+\cdots\right)$$

$$ds^2 = \left(1-\frac{\alpha}{R}\right)dt^2 - \frac{dR^2}{1-\frac{\alpha}{R}} - R^2(d\vartheta^2+\sin^2\vartheta\, d\varphi^2)$$

Fig. 1b. Second page of a draft of a letter of Karl Schwarzschild to Albert Einstein, December 1915

Transcript of the postcard of Einstein to Schwarzschild, answer to the above given letter

Blatt 1

29.XII.15

Sehr geehrter Herr Kollege!

Ihre Rechnung, die den Eindeutigkeitsbeweis für das Problem liefert, ist höchst interessant. Hoffentlich veröffentlichen Sie dieselbe bald! Ich hätte nicht gedacht, dass die strenge Behandlung des Punktproblems so einfach wäre. Dass Ihre partikuläre Lösung 3. Ordnung ist, erkennt man sogleich aus Dimensionsgründen. Es ist nämlich $k \cdot m/r$ eine dimensionslose Zahl. Da(s) Ihr β von m abhängen muß, so verlangt Ihr Ausdruck

$$g_{\varrho\sigma} = -\beta \frac{x_\varrho x_\sigma}{r^5} + \delta_{\varrho\sigma} \frac{\beta}{3r^3}$$

, in welchem β/r^3 eine Zahl (dimensionslos) ist, dass β bis auf einen Zahlenfaktor gleich $(k \cdot m/r)^3$ sein muss.

Die Theorie befriedigt mich sehr. Schon dass sie die Newton'sche Näherung ergibt, ist nicht selbstverständlich; umso schöner ist es, dass

Blatt 2

sie auch noch die Perihelbewegung und die immerhin noch nicht genügend gesicherte Linienverschiebung liefert. Am wichtigsten ist nun die Frage der Lichtstrahlenkrümmung.

Mit besten Grüssen und Wünschen für das neue Jahr Ihr

Einstein

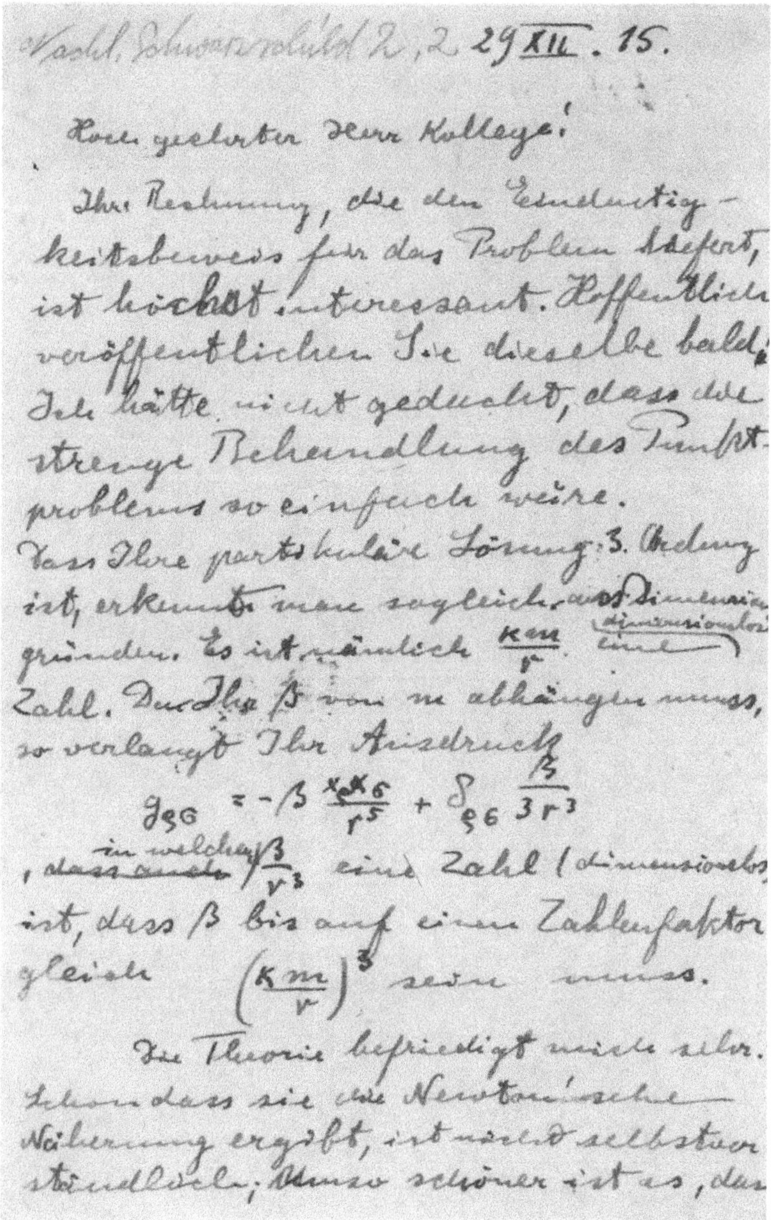

Nachl. Schwarzschild 2, 2 29 XII. 15.

Hoch geehrter Herr Kollege!

Ihre Rechnung, die den Eindeutigkeitsbeweis für das Problem liefert, ist höchst interessant. Hoffentlich veröffentlichen Sie dieselbe bald! Ich hätte nicht gedacht, dass die strenge Behandlung des Punktproblems so einfach wäre.
Dass Ihre partikuläre Lösung 3. Ordnung ist, erkennt man sogleich aus Dimensionsgründen. Es ist nämlich $\frac{km}{r}$ eine dimensionslose Zahl. Da Ihr β von m abhängen muss, so verlangt Ihr Ausdruck

$$g_{\rho\sigma} = -\beta \frac{x_\rho x_\sigma}{r^5} + \delta_{\rho\sigma} \frac{\beta}{3r^3}$$

, in welchem $\frac{\beta}{r^3}$ eine Zahl (dimensionslos) ist, dass β bis auf einen Zahlenfaktor gleich $\left(\frac{km}{r}\right)^3$ sein muss.

Die Theorie befriedigt mich sehr. Schon dass sie die Newton'sche Näherung ergibt, ist nicht selbstverständlich; umso schöner ist es, das

Fig. 2 Postcard of Einstein to Schwarzschild, answer to the above letter

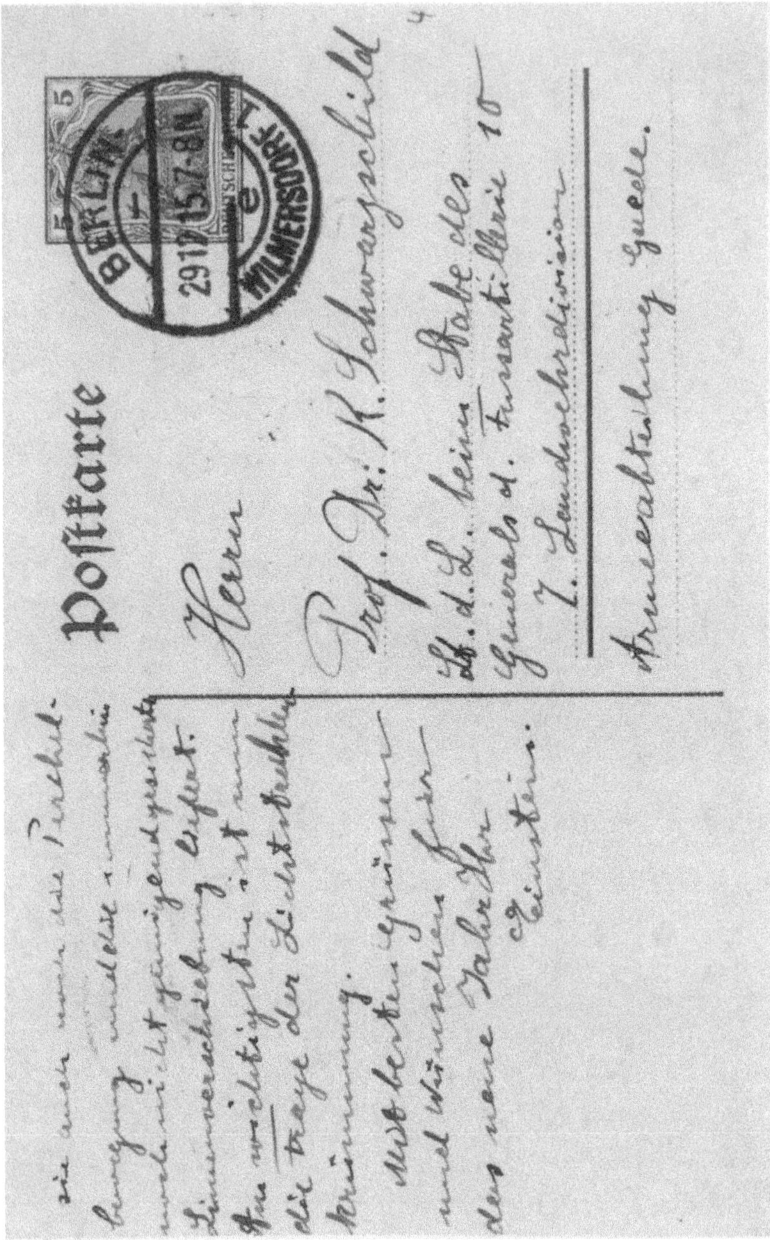
sie auch nun die Perihelbewegung und die immerhin noch nicht genügend gesicherte Linienverschiebung liefert. Am wichtigsten ist nun die Frage der Lichtstrahlenkrümmung.
Mit besten Grüssen und Wünschen für das neue Jahr Ihr
Einstein.

Postkarte

BERLIN-WILMERSDORF 1
29.12.15.7-8N

Herrn
Prof. Dr. K. Schwarzschild
Lt. d. L. beim Stabe des
Generals d. Fussartillerie 10
I. Landwehrdivision
Armeeabteilung Gaede.

Section 1

Celestial Mechanics

Celestial Mechanics

Commentary by R. Dvorak, Vienna

The Karl Schwarzschild's debut in his distinguished publishing career came with two papers on celestial mechanics, which he wrote in 1890 at the age of only 16. This shows that the time when he was working on problems of celestial mechanics was a decisive one, at least for his own scientific development. Fourteen years after these first papers, he wrote an article, the last in a series of eight papers, in which he summarized the state of celestial mechanics then and maked predictions of its future development. (We may for the moment disregard his contribution to Newcomb-Engelmann's "Popular Astronomy", which was written ten years later). In the following years he became busy with one branch of astronomical science then fashionable, namely astrophysics. He wrote his papers on celestial mechanics during the period between 1890 and 1903 when he was in Frankfurt and Munich and later at the von Kuffner Observatory at Vienna.

The time around the turn of the century, which covers this phase of Karl Schwarzschild's work, was perhaps the most productive period for research in celestial mechanics. Beside the monumental achievements of the Frenchman Henri Poincaré (1854–1912), which even now, as the twentieth century comes to an end, have not yet been completely followed up, mathematicians and astronomers such as Moulton, Callendreau, de Sitter, Strømgren, Charlier and others were busy during the same period with celestial mechanics – "Himmlische Mechanik" as it was called in the Astronomischer Jahresbericht until 1905. No wonder that Karl Schwarzschild devoted himself to these questions at the beginning of his scientific career, and just as little wonder that he stopped his work on celestial mechanics soon after 1900, when Planck and Einstein infused new ideas into physics.

Karl Schwarzschild's papers in the area to be reviewed may be grouped into 4 categories:

1. Methods for the determination of orbits and definitive determination of orbits [1.1–3]
2. The paper on the capture of comets by Jupiter [1.4]
3. Periodic orbits [1.5–7]
4. Work on celestial mechanics in general [1.8,9].

These will now be discussed individually, considering their importance for contemporary astronomy as well as for modern research in celestial mechanics.

Determination of Orbits

In paper [1.1], Schwarzschild succeeds very cleverly in putting the finishing touches to a method already developed by Bruns for determining an orbit from three positions: Bruns' elements are regarded as mere first approximations for the computation of further approximations until a satisfactory accuracy is reached. Schwarzschild likewise applies the same method to the determination of the orbital elements of visual binaries [1.2]. Both investigations, carried out while he was still a secondary-school student in Frankfurt, show young Schwarzschild's precocious mathematical talent. The reaction of scientists to these two papers was rather lukewarm, however. Stracke's (1929) book on the determination of orbits (as well as Bauschinger's (1928) book *Die Bahnbestimmung der Himmelskörper* ["Determination of the Orbits of Celestial Objects"]) do not mention Schwarzschild's ideas. Moulton and Fender in their *Einführung in die Himmelsmechanik* ["Introduction to Celestial Mechanics"] (1927) as well Happel (1941) in his book on the three-body problem, ignore Schwarzschild's papers. Aitken (1935) does, however, briefly refer to paper [1.2] in his book on double stars.

Today the methods for the determining a preliminary orbit have already been programmed into fast computers, which can practically instantly utilize every new position measurement; this is especially important for artificial satellites. Even today, however, the determination of definitive orbits for comets and asteroids essentially requires laborious, detailed work if the old observations are to be properly considered according to their importance. Schwarzschild determined the orbit of the comet 1842 II in paper [1.3] and conscientiously re-reduced all available observations of the comet (also partially utilizing more accurate positions of the reference stars). Marsden's (1975) "Catalogue of Cometary Orbits" contains the elements determined by Karl Schwarzschild as definitive. It needs to be pointed out again that in contemporary practice, old observations were sometimes inadequately dealt with in such work. Note that there is NO automatic routine that will correctly prepare old observations for use; this process requires great skill and knowledge on the part of the researcher. Paper [1.3] is therefore a model for a conscientiously and clearly executed determination of a definitive orbit.

Origin of Short-Period Comets

Schwarzschild certainly made a very important contribution toward explaining the origin of short-period comets in the Solar System with his paper [1.4]. This paper is regularly quoted even today in the many numerical and theoretical investigations of this problem. It is also discussed in detail in probably the most comprehensive treaties, which is Hagihara's (1970–1975) five-volume "Celestial Mechanics". Apart from the conclusion that in general a comet caught by Jupiter will escape again, Schwarzschild's method is interesting in itself. Triggered obviously by one of Poincaré's papers, Schwarzschild regards the possible orbits of the individual particles as streamlines in an incompressible liquid and derives the hydrodynamic equation of continuity from the mechanical equations of motion. This approach shows hydrodynamically the result just mentioned, namely that permanent capture is impossible. This is true even for a three-dimensional model in which one takes into account the eccentricity of Jupiter's orbit. Unfortunately, the promised application to the n-body problem is still lacking; Schwarzschild was then obviously already too much involved with other astronomical problems.

Periodic Orbits

This group (publications [1.5–7], all dealing with periodic orbits), is central to Schwarzschild's activity in celestial mechanics. These papers are regularly quoted in the older literature (cf. Charlier 1907) as well as in modern work. The model is the general three-body problem: Sun-planet-asteroid. Schwarzschild [1.5] investigates which particular initial commensurability conditions asteroids must satisfy so that they can travel in periodic orbits that may also have large eccentricities. Furthermore, he shows [1.6] that asteroids can have periodic orbits even when the orbit of the perturbing planet is elliptical and concludes from this that periodic orbits are common and may no longer be regarded as freak occurrences. In addition, he thought that he had come a step closer to the important theorem by Poincaré (1892) – unfortunately never proved – that within any neighbourhood of the initial data that give raise to a bounded orbit one may find initial conditions generating a periodic orbit.

One can see the significance on these results from modern works on celestial mechanics. K. Stumpff (1965) quotes them in the second volume of his "Himmelsmechanik", Hagihara (1975) gives a detailed review in the fourth volume of his "Celestial Mechanics" and V. Szebehely (1970) mentions them, although with some criticism.

Publication [1.7], furthermore, investigates the case of Hecuba in connection with the feasibility of periodic orbits and their stability properties. Dziewulski (1909) used the same method six years later to investigate the perturbations for the Hilda type. At the end of the 1903 paper one finds a stability diagram of periodic orbits. Schwarzschild thus anticipated two important contributions to modern astronomy, namely the exhaustive stability diagrams in the restricted three-body problem by M. Hénon (1965) and those of Contopoulos (1983) in galactic potential fields. I consider it as particularly significant that Schwarzschild pointed out that bifurcations of the periodic orbits cause the corresponding solutions to change their stability character.

Like Poincaré, Karl Schwarzschild therefore recognized, more than 80 years ago, that stability of periodic orbits plays a central role in attempting to give a qualitative analysis of motion in general.

Popular Works

Finally we want to talk about two contributions that were intended for astronomers and for the curious scientific public as well. In the article "Über Himmelsmechanik" [1.8], Karl Schwarzschild discussed what were, in his opinion, the most important questions of celestial mechanics: the problem of secular perturbations, the problem of resonance, and that of the divergence of the trigonometric series used for representing the solutions of the planetary many-body problem. Even today the answers he gave to these three questions, significant for discussions of the stability of the Solar System, can be regarded as substantially correct. Furthermore he hypothesized that the stability of the planetary system was over one million years and "probably also one billion years". This latter statement is made again and again even today, partly on the basis of exhaustive numerical integrations and partly on the basis of qualitative considerations in connection with the so-called KAM theorem, which essentially states that dynamical systems are stable when they are "sufficiently close" to being integrable. Now, as the planetary orbits are "sufficiently near" to the integrable two-body case (Kepler ellipse) and because the masses of the perturbing planets are small compared with that of the Sun, this conclusion, cautiously formulated, is valid. Nevertheless, one of the authors of the theorem, the mathematician Jürgen Moser says: "It the solar system stable? Properly speaking, the answer is still unknown ..." (Moser 1975).

At the end of the article, Schwarzschild considered the close passages of comets to Jupiter and the drastic changes in the shapes of the orbits. In this connection he also conjectured that there could be motions which, as he says with "alternate between the rotating and the oscillating pendulum". With this seminal statement he was far ahead of his time, because, today we recognize the phenomenon of "intermittency", which is precisely just such a spontaneous change between two qualitatively different states of motion.

This is closely connected with the chaotic behavior of non-integrable dynamical systems, which was discovered only very recently and is now the subject of thorough investigations in many disciplines including, of course, celestial mechanics.

For the fourth and fifth editions of Newcomb-Engelmann's "Populäre Astronomie", Karl Schwarzschild wrote (among others) the chapter on the determination of orbits [1.9]; in both editions the text was practically identical. In accordance with the character of this introductory popular book, the author only skims the surface of the problems and does not emphasize any particular points. Still, the principle of how one determines orbits is explained in a well-structured manner and clearly.

What can one say to sum up the effect of his writings on celestial mechanics on the further development of this field? As we have discussed the individual contributions, we have again and again emphasized his forward-looking method of work. However, he undoubtedly left a lot of his own potential unrealized because he turned his attention to other areas of astronomy.

We can thus conclude that his conscientious work, combined with his readiness to accept and adopt new ideas, should serve as a model for later generations of scientists. This, of course, undoubtedly applies not just to his investigations in the field of celestial mechanics but to his entire scientific oeuvre.

References*

Aitken, R.G. (1935) *The Binary Stars* (McGraw-Hill, New York)

Bauschinger, J. (1928) *Die Bahnbestimmung der Himmelskörper*, 2nd edition (Engelmann, Leipzig)

Charlier, C.L. (1907) *Die Mechanik des Himmels*, 2 vols., (Veit, Leipzig) (2nd edition: de Gruyter, Leipzig 1927)

Contopoulos, G. (1983) Ordered and ergodic motions of stars in galaxies, *Astronomy and Astrophysics* **117**: 89–108

Dziewulski, W. (1909) Über die periodischen Bahnen vom Hildatypus, *Astron. Nachr.* **183**: 65–72

Hagihara, Y. (1970–77) *Celestial Mechanics*, Vols. I–V, Vols. I/II (M.I.T. Press, Cambridge, Mass.); Vols. III–V (Japan Soc. for the Promotion of Science, Tokyo)

Happel, H. (1941) *Das Dreikörperproblem* (Köhler, Leipzig)

Hénon, M. (1968) Exploration numérique du problème restreint, *Annales d'Astrophysique* **28**: 499–511

Marsden, B.G. (1975) *Catalogue of Cometary Orbits*, 2nd edition (Smithsonian Astrophys. Obs., Cambridge, Mass.)

Moser, J. (1975) Is the Solar System Stable?, Pauli lecture at the ETH Zürich, January 1975, published in the *Neue Züricher Zeitung*, 14 May 1975

Moulton, R. M., Fender, W. (1927) *Einführung in die Himmelsmechanik* (Teubner, Leipzig)

Poincaré, H. (1892) *Les Méthodes nouvelles de la mécanique céleste*, Vol. 1 (Gauthiers-Villars, Paris)

Stracke, G. (1929) *Bahnbestimmung* (Springer, Berlin)

Stumpff, K. (1965) *Himmelsmechanik* II (VEB, Berlin)

Szebehely, V. (1970) *Theory of Orbits* (Academic, New York)

---

* The reference numbers in the text for Schwarzschild's papers correspond to the contents on page XI and to the bibliography on page 493.

$$v^2 = 2\gamma + \frac{2}{r} + \frac{2\mu}{\Delta} + 2n'\sqrt{p}\cos i - \frac{\mu}{r'^3}(r^2 - \Delta^2) \quad (2)$$

Diese Gleichung ist völlig strenge aus der Voraussetzung einer kreisförmigen Planetenbahn abgeleitet. Führt man noch die augenblickliche Halbaxe a der Cometenbahn ein durch

$$v^2 = \frac{2}{r} - \frac{1}{a}$$

so wird

$$0 = 2\gamma + \frac{1}{a} + \frac{2\mu}{\Delta} + 2n'\sqrt{p}\cos i - \frac{\mu}{r'^3}(r^2 - \Delta^2).$$

Betrachtet man die beiden Momente, wo der Comet in eine Kugel, welche um den Planeten mit dem Radius Δ gelegt wird, ein oder aus ihr austritt und bezeichnet man die diesen Momenten zugehörigen Grössen durch Hinzufügung des Index o bez. 1, so hat man schliesslich

$$\frac{1}{a_0} - \frac{1}{a_1} = 2n'(\sqrt{p_1}\cos i_1 - \sqrt{p_0}\cos i_0) - \frac{\mu}{r'^3}(r_1^2 - r_0^2). \quad (3)$$

Lässt man das in der That stets kleine letzte Glied fort, so erhält man den Tisserand'schen Satz. Für n' kann man den Näherungswerth

$$n' = \frac{\sqrt{a'}}{r'^2}\sqrt{\frac{1+m'}{1+m}}$$

setzen, wo a' die grosse Halbaxe der Planetenbahn ist.

Bogenhausen 1890 Febr. 20. *H. Seeliger.*

Zur Bahnbestimmung nach Bruns.

Von *K. Schwarzschild.*

Die Bahnbestimmungs-Methode, die Herr Prof. Bruns in Nr. 2824 der Astr. Nachr. angegeben hat, löst die Aufgabe nicht vollständig, eine Bahn an drei gegebene Beobachtungen eines Himmelskörpers anzuschliessen. Will man dies aber erreichen, so kann man folgende Methode benutzen, die die Bruns'sche Methode als erste Annäherung giebt und zeigt, wie weitere Annäherungen zu finden sind.

Nach den in der erwähnten Abhandlung angegebenen Bezeichnungen hat man die Gleichungen:

$$\xi'' + \frac{k^2\xi}{r^3} + k^2XP = 0; \quad \eta'' + \frac{k^2\eta}{r^3} + k^2YP = 0; \quad \zeta'' + \frac{k^2\zeta}{r^3} + k^2ZP = 0$$

$$\xi = \varrho . f; \quad \eta = \varrho . g; \quad \zeta = \varrho . h; \quad 1 = f^2 + g^2 + h^2; \quad P = \frac{\mu}{R^3} - \frac{1}{r^3}$$

demnach:

$$\left.\begin{aligned} f\varrho'' + 2f'\varrho' + f''\varrho + k^2\frac{\varrho . f}{r^3} + k^2XP &= 0 \\ g\varrho'' + 2g'\varrho' + g''\varrho + k^2\frac{\varrho . g}{r^3} + k^2YP &= 0 \\ h\varrho'' + 2h'\varrho' + h''\varrho + k^2\frac{\varrho . h}{r^3} + k^2ZP &= 0 \end{aligned}\right\} \quad (1)$$

Aus diesen Gleichungen erhält man durch Elimination von ϱ'' und ϱ':

$$0 = k^2P\left[\left(\frac{X}{f} - \frac{Y}{g}\right)\left(\frac{g'}{g} - \frac{h'}{h}\right) - \left(\frac{Y}{g} - \frac{Z}{h}\right)\left(\frac{f'}{f} - \frac{g'}{g}\right)\right] + \varrho\left[\left(\frac{f''}{f} - \frac{g''}{g}\right)\left(\frac{g'}{g} - \frac{h'}{h}\right) - \left(\frac{g''}{g} - \frac{h''}{h}\right)\left(\frac{f'}{f} - \frac{g'}{g}\right)\right] \quad (2)$$

Diese Gleichung geht in die bekannte Gleichung 8. Grades für r über, wenn man ϱ eliminirt mit Hülfe der Relation: $r^2 = R^2 + \varrho^2 - 2R\varrho\cos\chi$.

Ferner findet man aus den Gleichungen (1) durch Elimination von ϱ'' und P:

$$0 = -2\varrho'\left[\left(\frac{X}{f} - \frac{Y}{g}\right)\left(\frac{g'}{g} - \frac{h'}{h}\right) - \left(\frac{Y}{g} - \frac{Z}{h}\right)\left(\frac{f'}{f} - \frac{g'}{g}\right)\right] + \varrho\left[\left(\frac{f''}{f} - \frac{g''}{g}\right)\left(\frac{Y}{g} - \frac{Z}{h}\right) - \left(\frac{g''}{g} - \frac{h''}{h}\right)\left(\frac{X}{f} - \frac{Y}{g}\right)\right] \quad (3)$$

Durch Differentiation der Gleichungen (1) erhält man:

$$\left.\begin{aligned} f\varrho''' + 3f'\varrho'' + 3f''\varrho' + f'''\varrho + k^2\frac{d}{dt}\left(\frac{\varrho . f}{r^3} + XP\right) &= 0 \\ g\varrho''' + 3g'\varrho'' + 3g''\varrho' + g'''\varrho + k^2\frac{d}{dt}\left(\frac{\varrho . g}{r^3} + YP\right) &= 0 \\ h\varrho''' + 3h'\varrho'' + 3h''\varrho' + h'''\varrho + k^2\frac{d}{dt}\left(\frac{\varrho . h}{r^3} + ZP\right) &= 0 \end{aligned}\right\} \quad (4)$$

dann durch nochmalige Differentiation:

$$
\left.\begin{aligned}
f\varrho'''' + 4f'\varrho''' + 6f''\varrho'' + 4f'''\varrho' + f''''\varrho + k^2\frac{d^2}{dt^2}\left(\frac{\varrho . f}{r^3} + XP\right) &= 0\\
g\varrho'''' + 4g'\varrho''' + 6g''\varrho'' + 4g'''\varrho' + g''''\varrho + k^2\frac{d^2}{dt^2}\left(\frac{\varrho . g}{r^3} + YP\right) &= 0\\
h\varrho'''' + 4h'\varrho''' + 6h''\varrho'' + 4h'''\varrho' + h''''\varrho + k^2\frac{d^2}{dt^2}\left(\frac{\varrho . h}{r^3} + ZP\right) &= 0
\end{aligned}\right\} \qquad (5)
$$

So kann man durch Differenziren beliebig viele Gleichungen aufstellen. Die Grössen X, Y, Z sind stets bekannt. Für die Grössen f, g, h und deren Differentialquotienten werden durch drei Beobachtungen folgende Gleichungen geliefert, wenn man die Grössen, die zu der als Ausgangspunkt genommenen Zeit t gehören, ohne Index lässt, dagegen diejenigen, die zu den Zeiten t_1, t_2, t_3 gehören, durch die Indices 1, 2, 3 kennzeichnet:

$$
\left.\begin{aligned}
f_1 &= f + (t_1 - t)f' + \frac{(t_1 - t)^2}{1.2}f'' + \frac{(t_1 - t)^3}{1.2.3}f''' + \cdots\\
f_2 &= f + (t_2 - t)f' + \frac{(t_2 - t)^2}{1.2}f'' + \frac{(t_2 - t)^3}{1.2.3}f''' + \cdots\\
f_3 &= f + (t_3 - t)f' + \frac{(t_3 - t)^2}{1.2}f'' + \frac{(t_3 - t)^3}{1.2.3}f''' + \cdots
\end{aligned}\right\} \qquad (6)
$$

Entsprechende Gleichungen hat man für g und h.

Die bis hierher aufgestellten Gleichungen genügen zur vollständigen Lösung unserer Aufgabe. Es ist nur noch anzugeben, in welcher Reihenfolge sie zu benutzen sind.

1. Näherung. Man vernachlässigt die dritten und höheren Differentialquotienten von f, g, h und benutzt die Gleichungen (6), (2) und (3).

Aus (6) erhält man f, g, h, f', g', h', f'', g'', h'', die drei letzten Grössen mit einem Fehler der zweiten Ordnung. Mit den hier gefundenen Werthen berechne man die eckigen Klammern in (2) und (3). Dann findet man aus (2) r resp. ϱ und aus (3) ϱ', beide ebenfalls mit einem Fehler der zweiten Ordnung. Mit Hülfe von ϱ und ϱ' lassen sich die heliocentrischen Coordinaten des Himmelskörpers, sowie ihre ersten Differentialquotienten bestimmen und aus diesen kann man bekanntlich Elemente finden, die in unserm Falle natürlich einen Fehler der zweiten Ordnung enthalten.

2. Näherung. Man vernachlässigt die vierten und höheren Differentialquotienten von f, g, h und benutzt die Gleichungen (4) (6) (2) (3).

Man setze zur Abkürzung:

$$
\begin{aligned}
3f'\varrho'' + 3f''\varrho' + k^2\frac{d}{dt}\left(\frac{\varrho . f}{r^3} + XP\right) &= p\\
3g'\varrho'' + 3g''\varrho' + k^2\frac{d}{dt}\left(\frac{\varrho . g}{r^3} + YP\right) &= q\\
3h'\varrho'' + 3h''\varrho' + k^2\frac{d}{dt}\left(\frac{\varrho . h}{r^3} + ZP\right) &= s.
\end{aligned}
$$

Diese drei Grössen lassen sich vermittelst der Elemente der ersten Näherung mit einem Fehler der zweiten Ordnung finden. Die Gleichungen (4) verwandeln sich nun in die folgenden:

$$
\begin{aligned}
f\varrho''' + f'''\varrho + p &= 0\\
g\varrho''' + g'''\varrho + q &= 0\\
h\varrho''' + h'''\varrho + s &= 0.
\end{aligned}
$$

Hieraus findet man vermittelst der Relation:

$$0 = ff''' + gg''' + hh''' + 3(f'f'' + g'g'' + h'h'')$$

durch Elimination von ϱ'':

$$
\left.\begin{aligned}
f''' + 3f(f'f'' + g'g'' + h'h'') - \frac{f}{\varrho}(fp + gq + hs) + \frac{p}{\varrho} &= 0\\
g''' + 3g(f'f'' + g'g'' + h'h'') - \frac{g}{\varrho}(fp + gq + hs) + \frac{q}{\varrho} &= 0\\
h''' + 3h(f'f'' + g'g'' + h'h'') - \frac{h}{\varrho}(fp + gq + hs) + \frac{s}{\varrho} &= 0
\end{aligned}\right\} \qquad (7)
$$

Die aus (7) gefundenen Werthe von f''', g''', h''' setze man in (6) ein und bestimme neue Werthe von f, g, h, f', g', h', f'', g'', h''. Dann findet man wie bei der ersten Näherung ϱ, ϱ' und die Elemente, nun aber mit einem Fehler der dritten Ordnung.

3. Näherung. Man vernachlässigt die fünften und höheren Differentialquotienten und benutzt die Gleichungen (5) (4) (6) (2) (3).

Das Verfahren ist dem der 2. Näherung ganz analog.

Man setze zur Abkürzung:

$$
\begin{aligned}
4f'\varrho''' + 6f''\varrho'' + 4f'''\varrho' + k^2\frac{d^2}{dt^2}\left(\frac{\varrho . f}{r^3} + XP\right) &= p'\\
4g'\varrho''' + 6g''\varrho'' + 4g'''\varrho' + k^2\frac{d^2}{dt^2}\left(\frac{\varrho . g}{r^3} + YP\right) &= q'\\
4h'\varrho''' + 6h''\varrho'' + 4h'''\varrho' + k^2\frac{d^2}{dt^2}\left(\frac{\varrho . h}{r^3} + ZP\right) &= s'.
\end{aligned}
$$

p', q', s' lassen sich aus den Elementen der zweiten Näherung mit einem Fehler der zweiten Ordnung finden. Für f'''', g'''', h'''' erhält man auf dieselbe Weise, wie oben:

$$\left.\begin{aligned} f'''' + f\Big[4(f'f''' + g'g''' + h'h''') + 3([f'']^2 + [g'']^2 + [h'']^2)\Big] - \frac{f}{\varrho}\Big[fp' + gq' + hs'\Big] + \frac{p'}{\varrho} &= 0 \\ g'''' + g\Big[4(f'f''' + g'g''' + h'h''') + 3([f'']^2 + [g'']^2 + [h'']^2)\Big] - \frac{g}{\varrho}\Big[fp' + gq' + hs'\Big] + \frac{q'}{\varrho} &= 0 \\ h'''' + h\Big[4(f'f''' + g'g''' + h'h''') + 3([f'']^2 + [g'']^2 + [h'']^2)\Big] - \frac{h}{\varrho}\Big[fp' + gq' + hs'\Big] + \frac{s'}{\varrho} &= 0 \end{aligned}\right\} \quad (8)$$

Hieraus bestimmt man f'''', g'''', h''''. Ferner hat man die Grössen p, q, s mit den Elementen der zweiten Näherung nochmals zu berechnen, um dann aus (7) f''', g''', h''' mit einem Fehler der dritten Ordnung zu finden. Diese Werthe setze man wieder in die Gleichungen (6) ein und verfahre weiter, wie bei der ersten Näherung. Man erhält dann die Elemente mit einem Fehler der vierten Ordnung.

So kann man beliebig weit in der Näherung fortgehen.

Die Differentialquotienten von $\frac{\varrho . f}{r^3} + XP$ wird man am besten nicht direkt berechnen, sondern sie durch Interpolation aus einer Reihe von Werthen dieser Grösse bestimmen. Eine bedeutende Vereinfachung erhält die Rechnung noch, wenn man in den Gleichungen (6) für t das Mittel aus den drei Beobachtungszeiten t_1, t_2, t_3 nimmt. Man findet dann nämlich gleich in der ersten Näherung f'', g'', h'' nur mit einem Fehler der dritten Ordnung, den nun auch die Elemente erhalten. Daher darf man jetzt die zweite Näherung überspringen und sofort auf die dritte übergehen. Mit den Elementen der ersten Näherung wird man die Verbesserungen der Beobachtungen für Parallaxe etc. berechnen, zumal da die hier nothwendigen Grössen sehr gut zur Bestimmung von p, q, s zu gebrauchen sind. Was die Genauigkeit der Elemente anbetrifft, so wird sie für die dritte Näherung ungefähr $k^3 = 0.00003$, d. h. eine fünfstellige sein.

Frankfurt a. M., 1890 Jan. 8.

K. Schwarzschild.

Methode zur Bahnbestimmung der Doppelsterne.

Von *K. Schwarzschild.*

Die Bruns'sche Bahnbestimmungsmethode für Planeten und Cometen führte mich durch Analogie zu folgender Methode zur Bahnbestimmung bei Doppelsternen.

Das Hauptziel der meisten Methoden zur Bahnbestimmung bei Planeten ist die Aufstellung einer Gleichung für die Entfernung des Planeten von der Erde oder Sonne. Entsprechend kann man sich bei Doppelsternen die Aufgabe stellen, eine Gleichung für die wahre Distanz des Nebensterns vom Hauptstern zu ermitteln.

Man lege ein rechtwinkliges Coordinatensystem durch den Hauptstern, die Axen der x und y in die durch den Hauptstern gehende Tangentialebene an der Himmelskugel und zwar die Axe der x in die Richtung nach Norden, die der y in die Richtung nach Osten. Die Axe der z lege man in die Richtung nach der Erde resp. Sonne. In dem Doppelsternsystem gelten die bekannten Bewegungsgleichungen:

$$x'' + \frac{k^2 x}{r^3} = 0; \quad y'' + \frac{k^2 y}{r^3} = 0; \quad z'' + \frac{k^2 z}{r^3} = 0$$

wo $r^2 = x^2 + y^2 + z^2$ und k^2 die Gauss'sche Constante für das Doppelsternsystem ist. Die Accente bezeichnen die Differentiation nach der Zeit.

Setzt man nun: $\frac{r^3}{k^2} = p^3$, so erhält man an Stelle der Bewegungsgleichungen:

$$\frac{x''}{x} = \frac{y''}{y} = \frac{z''}{z} = -\frac{1}{p^3}. \quad (1)$$

Durch Differentiation der Gleichung $\frac{r^3}{k^2} = p^3$ ergiebt sich:

$$\frac{r'}{r} = \frac{p'}{p}, \qquad \frac{r''}{r} = \frac{p''}{p} \quad (2)$$

durch Differentiation der Gleichung $r^2 = \varrho^2 + z^2$, wo ϱ die scheinbare Distanz der beiden Sterne bezeichnet, findet man:

$$rr' = \varrho\varrho' + zz' \quad (3)$$
$$rr'' + (r')^2 = \varrho\varrho'' + (\varrho')^2 + zz'' + (z')^2.$$

Hier ersetze man r', r'', z, z' und z'' durch die aus den vorhergehenden Gleichungen folgenden Werthe. Dann erhält man:

$$r^2\frac{p''}{p} + r^2\left(\frac{p'}{p}\right)^2 = \varrho\varrho'' + (\varrho')^2 - \frac{r^2 - \varrho^2}{p^3} + \frac{r^4\left(\frac{p'}{p}\right)^2 - 2r^2\varrho\varrho'\frac{p'}{p} + \varrho^2(\varrho')^2}{r^2 - \varrho^2}$$

Durch Multiplication mit $r^2 - \varrho^2$ ergiebt sich schliesslich:

Astron. Nachr. *124* (1890) 215–218

p', q', s' lassen sich aus den Elementen der zweiten Näherung mit einem Fehler der zweiten Ordnung finden. Für f'''', g'''', h'''' erhält man auf dieselbe Weise, wie oben:

$$\left.\begin{aligned} f''''+f\left[4(f'f'''+g'g'''+h'h''')+3([f'']^2+[g'']^2+[h'']^2)\right]-\frac{f}{\varrho}\left[fp'+gq'+hs'\right]+\frac{p'}{\varrho}=0 \\ g''''+g\left[4(f'f'''+g'g'''+h'h''')+3([f'']^2+[g'']^2+[h'']^2)\right]-\frac{g}{\varrho}\left[fp'+gq'+hs'\right]+\frac{q'}{\varrho}=0 \\ h''''+h\left[4(f'f'''+g'g'''+h'h''')+3([f'']^2+[g'']^2+[h'']^2)\right]-\frac{h}{\varrho}\left[fp'+gq'+hs'\right]+\frac{s'}{\varrho}=0 \end{aligned}\right\} \quad (8)$$

Hieraus bestimmt man f'''', g'''', h''''. Ferner hat man die Grössen p, q, s mit den Elementen der zweiten Näherung nochmals zu berechnen, um dann aus (7) f''', g''', h''' mit einem Fehler der dritten Ordnung zu finden. Diese Werthe setze man wieder in die Gleichungen (6) ein und verfahre weiter, wie bei der ersten Näherung. Man erhält dann die Elemente mit einem Fehler der vierten Ordnung.

So kann man beliebig weit in der Näherung fortgehen.

Die Differentialquotienten von $\frac{\varrho \cdot f}{r^3} + XP$ wird man am besten nicht direkt berechnen, sondern sie durch Interpolation aus einer Reihe von Werthen dieser Grösse bestimmen. Eine bedeutende Vereinfachung erhält die Rechnung noch, wenn man in den Gleichungen (6) für t das Mittel aus den drei Beobachtungszeiten t_1, t_2, t_3 nimmt. Man findet dann nämlich gleich in der ersten Näherung f'', g'', h'' nur mit einem Fehler der dritten Ordnung, den nun auch die Elemente erhalten. Daher darf man jetzt die zweite Näherung überspringen und sofort auf die dritte übergehen. Mit den Elementen der ersten Näherung wird man die Verbesserungen der Beobachtungen für Parallaxe etc. berechnen, zumal da die hier nothwendigen Grössen sehr gut zur Bestimmung von p, q, s zu gebrauchen sind. Was die Genauigkeit der Elemente anbetrifft, so wird sie für die dritte Näherung ungefähr $k^3 = 0.00003$, d. h. eine fünfstellige sein.

Frankfurt a. M., 1890 Jan. 8. *K. Schwarzschild.*

Methode zur Bahnbestimmung der Doppelsterne.

Von *K. Schwarzschild.*

Die Bruns'sche Bahnbestimmungsmethode für Planeten und Cometen führte mich durch Analogie zu folgender Methode zur Bahnbestimmung bei Doppelsternen.

Das Hauptziel der meisten Methoden zur Bahnbestimmung bei Planeten ist die Aufstellung einer Gleichung für die Entfernung des Planeten von der Erde oder Sonne. Entsprechend kann man sich bei Doppelsternen die Aufgabe stellen, eine Gleichung für die wahre Distanz des Nebensterns vom Hauptstern zu ermitteln.

Man lege ein rechtwinkliges Coordinatensystem durch den Hauptstern, die Axen der x und y in die durch den Hauptstern gehende Tangentialebene an der Himmelskugel und zwar die Axe der x in die Richtung nach Norden, die der y in die Richtung nach Osten. Die Axe der z lege man in die Richtung nach der Erde resp. Sonne. In dem Doppelsternsystem gelten die bekannten Bewegungsgleichungen:

$$x''+\frac{k^2x}{r^3}=0;\quad y''+\frac{k^2y}{r^3}=0;\quad z''+\frac{k^2z}{r^3}=0$$

wo $r^2 = x^2+y^2+z^2$ und k^2 die Gauss'sche Constante für das Doppelsternsystem ist. Die Accente bezeichnen die Differentiation nach der Zeit.

Setzt man nun: $\frac{r^3}{k^2} = p^3$, so erhält man an Stelle der Bewegungsgleichungen:

$$\frac{x''}{x}=\frac{y''}{y}=\frac{z''}{z}=-\frac{1}{p^3}. \quad (1)$$

Durch Differentiation der Gleichung $\frac{r^3}{k^2}=p^3$ ergiebt sich:

$$\left.\begin{aligned}\frac{r'}{r}&=\frac{p'}{p}\\ \frac{r''}{r}&=\frac{p''}{p}\end{aligned}\right. \quad (2)$$

durch Differentiation der Gleichung $r^2 = \varrho^2+z^2$, wo ϱ die scheinbare Distanz der beiden Sterne bezeichnet, findet man:

$$\begin{aligned} rr' &= \varrho\varrho'+zz' \\ rr''+(r')^2 &= \varrho\varrho''+(\varrho')^2+zz''+(z')^2. \end{aligned} \quad (3)$$

Hier ersetze man r', r'', z, z' und z'' durch die aus den vorhergehenden Gleichungen folgenden Werthe. Dann erhält man:

$$r^2\frac{p''}{p}+r^2\left(\frac{p'}{p}\right)^2=\varrho\varrho''+(\varrho')^2-\frac{r^2-\varrho^2}{p^3}+\frac{r^4\left(\frac{p'}{p}\right)^2-2r^2\varrho\varrho'\frac{p'}{p}+\varrho^2(\varrho')^2}{r^2-\varrho^2}$$

Durch Multiplication mit $r^2-\varrho^2$ ergiebt sich schliesslich:

$$r^4\left[\frac{p''}{p}+\frac{1}{p^3}\right]-r^2\varrho^2\left[\left(\frac{p''}{p}+\frac{1}{p^3}\right)+\left(\frac{\varrho'}{\varrho}-\frac{p'}{p}\right)^2+\left(\frac{\varrho''}{\varrho}+\frac{1}{p^3}\right)\right]+\varrho^4\left[\frac{\varrho''}{\varrho}+\frac{1}{p^3}\right]=0. \tag{4}$$

Dies ist die gesuchte Gleichung für r, da man ϱ', p, p', p'' leicht aus beobachteten Grössen bestimmen kann.

Ich nehme zunächst an, die Beobachtungen seien in rechtwinkligen Coordinaten gegeben oder auf solche reducirt. Dann suche man die beobachteten Werthe von x und y durch Reihen nach Potenzen der Zeit darzustellen, indem man je nach der Zahl der Beobachtungen von einer bestimmten Potenz der Zeit an die höheren Potenzen vernachlässigt. Doch die Beobachtungen müssen hinreichend sein, um die vierte Potenz der Zeit mit berücksichtigen zu können. Aus den Coefficienten der erhaltenen Reihen findet man leicht die Differentialquotienten von x und y.

Nun findet man p aus der Gleichung:

$$-\frac{1}{p^3}=\frac{x''}{x}=\frac{y''}{y}.$$

Ferner die Differentialquotienten von p aus:

$$3\frac{p'}{p}=\frac{x'}{x}-\frac{x'''}{x''}=\frac{y'}{y}-\frac{y'''}{y''}$$

$$3\frac{p''}{p}-3\left(\frac{p'}{p}\right)^2=\frac{x''}{x}-\left(\frac{x'}{x}\right)^2+\left(\frac{x'''}{x''}\right)^2-\frac{x''''}{x''}=\frac{y''}{y}-\left(\frac{y'}{y}\right)^2+\left(\frac{y'''}{y''}\right)^2-\frac{y''''}{y''}$$

Den Werth von ϱ, ϱ', ϱ'' erhält man aus den Gleichungen:

$$\varrho^2=x^2+y^2;\quad \varrho\varrho'=xx'+yy';\quad \varrho\varrho''+(\varrho')^2=xx''+yy''+(x')^2+(y')^2.$$

Nunmehr kann man die Gleichung (4) aufstellen und aus dieser r bestimmen. Dabei bleibt das Vorzeichen von r unbestimmt, was natürlich ist, da man nicht angeben kann, ob der Begleiter näher oder weiter von der Erde ist, als der Hauptstern. Mit Hülfe von r erhält man z und dann aus (2) r' und aus (3) z'. Da man nun für einen Zeitpunkt x, y, z, und x', y', z' kennt, kann man auf bekannte Weise die Elemente berechnen.

Wenn Positionswinkel ohne zugehörige Distanzen gegeben sind, so bestimme man die letzteren mit Hülfe des Gesetzes, dass auch in der scheinbaren Bahn die Flächengeschwindigkeit constant ist. Dieses Gesetz liefert die Gleichung:

$$\varrho^2 v'=c^2,$$

wo v' die scheinbare Winkelgeschwindigkeit ist und c^2 eine Constante bezeichnet, die man mit Hülfe eines einzigen gegebenen Werthes von ϱ und der zugehörigen Winkelgeschwindigkeit finden kann. Da sich v' für jede Zeit durch Interpolation aus einer Anzahl benachbarter Positionswinkel bestimmen lässt, erhält man die zu einem Positionswinkel zugehörige Distanz aus der Gleichung: $\varrho=\frac{c}{\sqrt{v'}}$.

Nunmehr kann man die Reduction auf rechtwinklige Coordinaten ausführen und weiter wie oben verfahren.

Frankfurt a. M., 1890 Febr. 3.

K. Schwarzschild.

Beobachtungen des Mondes und der Mondsterne.

Von Dr. *Roy Matthiessen*, Assistent der Karlsruher Sternwarte.

Nachstehende Beobachtungen sind am 2.5 zölligen, Bamberg'schen Passageninstrument der Karlsruher Sternwarte angestellt; die Durchgänge sind, fast ohne Ausnahme an allen 15 Fäden des Netzes, registrirt worden, und zwar beträgt nach dem Mittel aus einer Anzahl beliebig herausgegriffener Abende:

der wahrscheinliche Fehler eines Fadenantritts eines Sterns	±0ˢ054
» » » des Mittels aus 15 Fäden für einen Stern	±0.014
» » » eines Fadenantritts des Mondrandes	±0.076
» » » des Mittels aus 15 Fäden für den Mondrand	±0.019.

Die dritte Columne der ersten Tafel giebt die Durchgangszeit durch den Meridian für die betreffenden Sterne und Mondränder nach der in den Strom eingeschalteten Uhr Hohwü 41; darauf folgen hinter den Mondsternen die auf den Rectascensionen des Nautical Almanac beruhenden Uhrstände. Die eingeklammerten Werthe von Δt sind nicht mit verwendet; nur in einem Fall, am 19. Sept. 1888 ist der Ort eines Sterns, nämlich der von 27 Piscium nach Loewy »Ephémérides des Etoiles de Culmination Lunaire et de Longitude pour 1888« um $-0^s\!.40$ verbessert worden.

An denjenigen Abenden wo zugleich eine Zeitbestimmung gemacht wurde, geschah die Berechnung des für den Mondort angewendeten Uhrstandes unter Berücksichtigung aller Sterne; an den übrigen Abenden wurden die kleinen und ziemlich constanten Fehler des Instruments in Azimut und Collimation nicht neu berechnet sondern interpolirt.

Eine Vergleichung der aus den Beobachtungen der Ränder sich ergebenden Rectascensionen des Mondcentrums mit der stündlichen Ephemeride des Nautical Almanac lieferte

Astron. Nachr. *137* (1895) 177–190

ASTRONOMISCHE NACHRICHTEN.

Nº 3276.

Band 137. 12.

Definitive Bahnbestimmung des Cometen 1842 II.

Von *Karl Schwarzschild.*

Der von Laugier in Paris entdeckte Comet 1842 II hat bisher zwei Bearbeitungen gefunden, eine vorläufige von Petersen*) und eine eingehende von Kowalczyk†). Gegenwärtige dritte Bearbeitung findet ihre Rechtfertigung in der grösseren Anzahl Beobachtungen und den besseren Vergleichsternpositionen, die zur Verwendung kamen. Es ist nämlich eine ganze Reihe Pariser Beobachtungen aus Bossert's Reduction derselben, sowie eine Göttinger und eine Berliner Beobachtung, bei denen früher Daten für die Vergleichsterne gefehlt hatten, hinzugekommen. Für die Vergleichsterne aber konnten neben den älteren Catalogen fast durchweg AG.-Positionen benutzt werden, wobei die älteren Positionen mit Hülfe der Auwers'schen Tafeln und anderer an verschiedenen Stellen verstreuter Angaben auf das System der AG. übertragen wurden. Um die Betheiligung der einzelnen Vergleichsterne an den Beobachtungen jedes Abends zu erfahren, musste für Göttingen und Altona auf die Manuscripte der Beobachter zurückgegangen werden, während für Cambridge, Greenwich, Berlin, Paris die Veröffentlichungen dieser Sternwarten und für Bonn und Marseille die Mittheilungen in Band 20 der Astr. Nachr. ausreichten. Bei der isolirten Beobachtung zu Kremsmünster fehlt jede Angabe über den Vergleichstern und bei den Hamburger Beobachtungen, wo das Manuscript nicht mehr vorhanden ist, blieb nichts anderes übrig, als das einfache Mittel der Positionsverbesserungen der an jedem Abend angewandten Sterne an die betreffende Beobachtung anzubringen. Die Reduction der Beobachtungen selbst wurde für Göttingen und Berlin von den einzelnen Ring- resp. Fadenantritten aus neu durchgeführt, bei den anderen Sternwarten revidirt, soweit die ursprünglichen Daten vorhanden waren. Für eine Ausschliessung der vier ersten AR.-Vergleichungen in Cambridge Nov. 26, die Kowalczyk vornimmt, war ein Grund nicht ersichtlich. Dagegen wurden an die Bonner Beobachtungen die von Argelander in Bd. 22 der Astr. Nachr. angegebenen Correctionen angebracht, welche Kowalczyk übersehen zu haben scheint. Die Angaben der Beobachter über Minderwerthigkeit der Beobachtung fanden schliesslich dadurch ihre Berücksichtigung, dass die Anzahl der Vergleichungen, für die die betreffende Beobachtung zu zählen hatte, auf die Hälfte oder ein Drittel ihres wirklichen Werthes herabgesetzt wurde.

Es möge zunächst die Tafel der Vergleichsterne folgen. Columne 1 enthält die laufende Nummer, Columne 2 und 3 Rectascension und Declination für das mittlere Aequinoctium 1842.0 und Columne 4 die Betheiligung der verschiedenen Cataloge an dem Sternort.

*	α 1842.0	δ 1842.0	Autorität
1	$16^h 39^m 31^s.58$	$+68° 31' 31''.7$	2 AG. + Par. Mer. + BB. VI
2	17 8 20.53	+65 54 34.5	F. C.
3	17 23 39.08	+60 10 55.6	[1])
4	17 29 4.05	+55 17 39.8	F. C.
5	17 29 9.43	+55 16 58.1	F. C.
6	17 45 13.27	+55 2 28.3	2 AG. Hels. + 2 AG. Camb. + ½ Ll.
7	17 55 50.97	+51 51 9.3	2 AG. + AOe. [2])
8	17 59 40.15	+42 51 8.2	Ll. + 3 Pi. + 3 Gro. + 2 W_2 + 3 Tayl. + 2 Rü. + 3 Rad_1
9	17 59 59.87	+42 56 50.6	Ll. + 3 Pi. + 3 Gro. + 2 W_2 + 3 Tayl. + 2 Rü. + 3 Rad_1
10	18 8 52.05	+42 50 8.2	Ll. + W_2 + 4 AG.
11	18 19 1.64	+39 25 27.7	[3])
12	18 36 9.50	+27 58 29.4	Ll. + W_2 + 4 AG.
13	$18^h 36^m 38^s.35$	$+28° 9 3''.1$	Ll. + W_2 + 4 AG.
14	18 38 6.29	+27 56 39.1	Ll. + W_2 + 4 AG.
15	18 54 45.72	+ 8 9 3.3	½ Ll. + ½ W_1 + Sant. 40 + BB. VI + 2 Gl. + 2 A_2 + 2 AG. + M_1
16	19 6 54.85	− 6 19 9.6	½ Ll. + ½ W_1 + 2 Cord. GC. + M_1
17	18 57 30.20	− 9 51 59.2	½ Ll. + 2 Cord. GC. + M_1
18	19 4 25.41	−12 32 32.5	W_1 + Sant. 60 + 2 Cord. GC. + M_1
19	19 12 30.39	−18 8 20.1	[4])
20	19 9 59.46	−15 48 19.5	[5])
21	19 12 37.84	−18 35 44.0	[6])
22	19 13 19.00	−20 55 57.1	Gr. 1845 + AWe. + Yarn. + ½ M_1
23	17 52 56.38	+51 30 35.6	F. C.
24	18 6 28.16	+45 39 41.5	AOe. + 2 AG.

*) Astr. Nachr. Bd. 20, pag. 273.

†) Astr. Nachr. Bd. 81, pag. 133.

*	α 1842.0	δ 1842.0	Autorität
25	$18^h 12^m 27^s.06$	+42° 44′ 44″.1	Ll.
26	18 13 43.63	+42 30 9.5	½ Ll. + W_2 + 2 AG.
27	18 23 16.78	+35 41 49.8	½ Ll. + 2 AG.
28	17 54 32.22	+52 14 0.8	½ Ll. + AOe. + 3 AG.
29	18 2 14.07	+46 15 23.6	½ Ll. + AOe. + 2 AG.
30	18 9 1.99	+46 10 28.0	AG.
31	18 28 1.32	+32 0 40.2	AG.
32	18 38 40.82	+19 48 32.0	½ Ll. + W_2 + Rü. + BB. VI + 2 AG.
33	18 9 17.69	+43 14 36.0	½ Ll. + W_2 + Rad_1 + 2 AG.
34	18 10 44.23	+42 6 31.3	F. C.
35	18 51 24.06	+ 7 31 14.8	½ W_1 + M_1 + 2 AG.
36	19 3 22.88	− 6 40 25.2	W_1
37	19 4 11.51	− 6 43 4.2	W_1 + M_1 + 2 M_2
38	18 2 33.35	+46 45 31.1	½ Ll. + AOe. + 2 AG.
39	18 17 26.19	+39 10 55.3	½ Ll. + W_2 + 2 AG.
40	18 18 40.00	+38 39 25.7	½ Rü. + 2 AG.
41	18 20 24.19	+39 2 2.8	½ Ll. + Gro. + Rad_1 + 2 AG.[7])
42	18 21 27.93	+39 3 50.0	Gro. + W_2 + Rad_1 + 2 AG.[8])
43	18 24 18.48	+35 17 22.8	½ W_2 + 2 AG.
44	18 25 32.80	+34 56 12.8	½ Ll. + 2 AG. Lund + 2 AG Leid.
45	19 2 28.60	− 6 52 26.9	½ Ll. + W_1 + 2 Cord. GC. + 1½ M_1
46	17 37 58.92	+57 23 20.0	½ Ll. + Rü. + ½ AOe. + 2 AG.
47	17 54 23.89	+54 40 54.8	½ Ll. + 2 Rad_1 + 2 AG.
48	17 49 55.98	+52 25 3.4	½ Ll. + AOe. + 3 AG.
49	18 21 4.37	+35 37 20.3	½ Ll. + 2 AG.
50	$18^h 24^m 39^s.97$	+32° 32′ 8″.4	In α: ½ Ll. + W_2 + 1½ A_2 + 3 AG.; in δ: ½ Ll. + 2 A_2 + 2 AG.
51	18 35 3.87	+27 40 45.0	½ Ll. + W_2 + 2 AG.
52	18 42 22.13	+23 52 55.4	Ll. + 3 AG. + 2 Gr. 1880
53	18 38 51.85	+20 23 58.7	F. C.
54	18 59 5.54	+ 0 24 7.5	½ Ll. + W_1 + Camb. + M_1
55	19 16 33.55	−20 51 13.1	½ Ll. + AWe.
56	19 17 12.59	−20 53 42.7	½ Ll. + AWe. + M_1
57	18 8 42.64	+43 17 26.0	½ Ll. + Gro. + W_2 + Rad_1
58	18 17 50.12	+39 26 18.9	W_2 + 2 AG.
59	18 42 35.61	+15 45 31.7	W_2 + Rü. + 2 AG.
60	18 45 6.49	+15 52 12.9	½ W_2 + Rü. + 2 AG.
61	18 45 15.30	+15 53 46.1	½ W_2 + 2 Rü. + 2 AG.
62	18 47 9.48	+11 51 53.1	½ Ll. + Rü. + 2 AG.
63	18 50 45.74	+11 49 13.8	½ Ll. + 2 AG.
64	18 52 19.92	+11 51 0.5	W_1 + 2 Rü. + 3 AG.
65	18 53 12.86	+12 0 36.0	W_1 + 3 AG.
66	18 55 55.58	+ 8 4 3.0	½ Ll. + ½ W_1 + M_1 + Schj. + 2 Gl. 1870 + + 2 AG.
67	18 56 53.70	+ 8 3 59.7	½ Ll. + ½ W_1 + M_1 + 2 AG.
68	18 52 27.80	+ 7 52 39.5	2 M_1 + 2 AG. + 3 M_2 + Schj.
69	18 52 43.87	+ 7 51 49.5	½ M_1 + 2 AG. + 2 M_2
70	19 3 22.82	− 3 19 15.4	½ Ll. + ½ W_1 + M_1
71	19 6 47.75	− 3 35 44.6	½ Ll. + ½ W_1 + M_1
72	19 6 54.02	− 3 37 15.2	½ Ll. + W_1 + 2 M_1
73	19 7 10.09	− 3 23 0.6	Ll. + W_1 + 2 Rü.
74	19 7 20.77	− 3 18 39.0	½ Ll. + W_1

Bemerkungen.

[1]) Aus den Positionen:

	Epoche	AR.	Gew.	R—B	Epoche	Decl.	Gew.	R—B
Greenw.	1883	$17^h 23^m 39^s.09$	3	+0ˢ.01	1881	+60° 10′ 57″.4	4	+0″.2
AG.	1879	39.16	1	−0.06	1879	57.2	1	+0.3
Greenw.	1875	39.06	2	+0.04	1874	57.1	4	+0.2
»	1863	39.17	1	−0.08	1863	57.9	1	−1.2
»	1859	—	—	—	1859	56.4	1	+0.1
Pulkowa	1855	39.13	2	−0.04	1855	56.9	2	−0.6
Arm_1	1855	38.85	1	+0.24	1855	58.7	1	−2.4
Rad_1	1847	39.21	2	−0.13	1847	55.2	2	+0.7
Kam	1845	39.05	2	+0.03	1845	55.6	2	+0.1
Par. Mer.	1842	38.96	1	+0.12	1842	54.2	1	+1.4

folgt die Eigenbewegung: +0ˢ.0006 +0″.0519.

[7]) Die Declination AOe. +51° 51′ 19″.9 scheint um 10″ zu gross zu sein und wurde weggelassen.

[8]) Die Position ist unter Berücksichtigung der Auwers'schen E. B. −0ˢ.0033 −0″.001 gebildet

für Decl. aus: 2 Auw.-Bradl. + 2 Pi. + ½ d'Agel. + 2 Gro. + Rü. + 2 Rad_1 + 2 Pulk. 1855 + 2 Gr. 1864 + 4 Gr. 1872 + 4 Gr. 1880;

für AR. aus: 3 Auw.-Bradl. + 2 Pi. + 2 Gro. + Rü. + 2 Rad_1 + 2 Pulk. 1855 + 2 Gr. 1864 + 3 Gr. 1872 + 3 Gr. 1880.

[4]) Aus den Positionen:

	Epoche	AR.	Gew.	R—B	Epoche	Decl.	Gew.	R—B
Cap	1879	$19^h 12^m 30^s.27$	2	$-0^s.03$	1875	−18° 8′ 19″.0	2	−0″.1
Greenw.	1886	30.25	2	−0.04	1886	18.7	2	−0.1
Cord. GC.	1877	30.21	2	+0.04	1873	19.5	2	+0.3
Greenw.	1870	30.30	3	−0.02	1870	19.5	3	+0.2
»	1864	30.35	3	−0.05	1864	19.6	3	+0.1
»	1858	30.34	1	−0.01	1858	19.1	1	−0.5
Rad_2	1862	30.24	1	+0.07	1862	17.9	1	−1.7
Yarn.	1857	30.33	1	0.00	1861	19.0	1	−0.6
AWe.	1850	30.29	1/2	+0.07	1850	20.3	1/2	+0.5
Greenw.	1849	30.31	1	+0.05	1849	19.5	1	−0.3
Cap	1853	30.29	1	+0.06	1853	18.6	1	−1.1
Greenw.	1846	30.37	2	+0.01	1846	20.1	2	+0.2
Rad_1	1846	30.43	2	−0.05	1846	18.3	1	−1.6
Greenw.	1841	30.22	1/2	+0.18	—	—	—	—
Rü.	1846	30.34	2	+0.03	1846	20.9	2	+1.0
Taylor	1835	30.41	2	+0.01	1835	19.8	1	−0.4
Pond	1830	30.49	1	−0.05	1830	19.2	1	−1.1
Kam	1829	30.31	2	+0.13	1829	20.2	2	−0.1
Arm_1	1829	30.54	1	−0.09	—	—	—	—
Johnson	1830	30.44	1	0.00	1830	24.0	1	+3.7
Piazzi	1800	30.70	1	−0.14	1800	22.0	1	+0.8

ergiebt sich die Eigenbewegung: $-0^s.00411$ +0″.0278.

[5]) Die Position ist der Abhandlung von Stumpe Astr. Nachr. Nr. 3000 entnommen. E. B. $-0^s.0085$ −0″.266.

[6]) Aus den Positionen:

	Epoche	AR.	Gew.	R—B	Epoche	Decl.	Gew.	R—B
Greenw.	1885	$19^h 12^m 37^s.94$	2	$+0^s.04$	1885	−18° 35′ 47″.0	2	−0″.6
Romberg	1877	38.04	2	−0.09	1877	46.9	2	0.0
Cord. GC.	1878	37.94	2	+0.01	1878	46.9	2	−0.1
Rad_2	1860	37.90	2	0.00	1860	47.6	2	+2.1
Yarn.	1869	37.87	1	+0.06	1873	45.6	1	−0.9
Cap	1853	37.79	1	+0.09	1853	44.8	1	0.0
M_1	1848	37.68	1/2	+0.18	1848	43.1	1/2	−1.3
Greenw.	1843	37.92	2	−0.07	1843	44.5	2	+0.5
»	1840	37.79	1	+0.05	1840	42.8	1	−1.0
Taylor	1835	38.08	1	−0.26	1835	44.7	1	+1.3
Arm_1	1832	37.70	1	+0.11	—	—	—	—
Arg. Abo	1830	37.75	1	+0.06	1830	42.3	1	−0.7
Piazzi	1800	37.68	2	+0.04	1800	39.7	2	−0.7

ergiebt sich die Eigenbewegung: $+0^s.00282$ −0″.0840.

[7]) Die AR. von Gro. $18^h 20^m 24^s.93$ scheint um 1^s zu gross zu sein und wurde weggelassen.

[8]) Die AR. von Gro. $18^h 20^m 28^s.75$ scheint um 1^s zu gross zu sein und wurde weggelassen.

Die für die Vergleichsterne verbesserten Beobachtungen wurden nun auch für Aberrationszeit und Parallaxe corrigirt, welche letztere den Tafeln von v. Rebeur-Paschwitz entnommen wurde, und dann mit einer Ephemeride verglichen, die aus den definitiven Elementen Kowalczyk's unter Anwendung der Leverrier'schen Sonnencoordinaten hergestellt worden war.*) Die ersten Spalten folgender Tafel geben den Beobachtungsort, die für Aberration verbesserte Beobachtungszeit, die corrigirte Beobachtung, die Betheiligung der Vergleichsterne an derselben und die Abweichnng von der Ephemeride im Sinne B — R.

*) Eine Differenz, die zwischen meiner und Kowalczyk's Rechnung auftrat, liess sich nur durch die Annahme erklären, dass Kowalczyk entgegen seiner Bemerkung l. c. pag. 133 nicht direct die Petersen'schen Elemente seiner Ephemeride zu Grunde legte, sondern von vornherein die Petersen'sche Knotenlänge um 15″ vermehrte. Ausserdem ist die AR. des Normalorts IV pag. 136 um 5^s zu klein angegeben, was aber ein blosser Druckfehler zu sein scheint und für die weitere Rechnung Kowalczyk's ohne Bedeutung geblieben ist.

Astron. Nachr. *137* (1895) 177–190

Nr.	Ort	M. Z. Berlin	AR.	Decl.	Vergleichsterne	$\Delta\alpha \cos\delta$	$\Delta\delta$	p_α	p_δ
1	Paris	Oct. 28 13h 48m 53s.4	240° 33′ 25″.4	+68° 33′ 7″.0	(1)	+10″.2	+ 2″.2	1	1
2	»	30 8 6 25.0	255 18 43.5	+65 44 6.7	(2)	− 9.3	+ 6.0	1	1
3	»	Nov. 2 9 11 24.0	262 28 16.6	+59 41 37.3	(3)	+ 9.8	+ 7.7	1	1
4	Marseille	3 9 55 23	264 36 10	+57 16 9	(46)	+ 6.5	−15.3	1/3	1/5
5	»	4 7 59 31	266 23 13	+54 55 39	(47)	−10.8	−59.8	1/3	0
6	Paris	4 7 39 38.4	266 21 47.9	+54 59 25.0	(4) + (5)				
7	»	4 9 27 19.6	—	+54 47 11.2	(4) + (5) + 2 (6)	+18.0	+18.4	1	1
8	»	4 8 31 18.9	266 31 45.0	—	(4) + 2 (6)				
9	Bonn	5 9 36 0.3	268 21 15.6	+52 1 53.5	6 (28)				
10	»	5 11 7 9.8	268 27 53.6	+51 50 57.9	3 (28)	+11.6	− 4.4	3	4
11	»	5 11 24 17.1	268 29 18.1	+51 48 55.0	1 (23)				
12	Paris	5 9 20 4.1	268 20 24.1	+52 4 12.0	(7)	+23.4	+24.4	1	1
13	Berlin	5 7 44 41.5	268 12 38.9	+52 14 53.0	(28)	− 1.3	− 8.8	4	5
14	Marseille	5 7 13 39	268 11 16	+52 18 34	(48)				
15	»	5 10 17 40	268 22 51	+51 56 6	(23)	−10.2	−25.4	1/3	1/5
16	»	5 10 41 40	268 25 21	+51 53 52	(23)				
17	Göttingen	5 10 27 30.9	268 25 1.3	+51 55 53.7	(23)	+10.4	+ 4.8	1	1/4
18	Berlin	7 8 45 42.9	271 37 50.0	+46 4 5.5	2 (29)	− 0.3	− 5.8	3	4
19	»	7 9 6 46.5	271 39 20.1	+46 1 9.3	6 (30)				
20	Altona	7 7 16 54.9	271 31 30.9	+46 16 29.9	(51) + (52)	−24.5	+26.8	1/2	1/2
21	Hamburg	7 6 49 32.8	271 30 45.6	+46 20 9.6	(29)	+18.7	+26.3	1/2	0
22	Göttingen	7 12 22 44.8	—	+45 33 39.9	(24)	—	−65.5	—	0
23	»	7 12 25 16.9	271 52 23.1	—	(24)	+ 6.2	—	1	—
24	Paris	8 8 15 2.0	273 7 19.6	+42 49 31.0	(8) + (9) + (10)	− 6.4	+46.8	1	1
25	Berlin	8 8 49 57.6	273 9 31.8	+42 43 37.9	(25)	− 4.0	− 7.3	4	5
26	Bonn	8 6 53 38.5	273 2 36.7	+43 0 21.7	2 (33)				
27	»	8 7 14 54.2	273 3 49.3	+42 57 21.1	5 (25)	+ 6.9	+ 3.0	3	4
28	»	8 8 54 42.7	273 10 16.4	+42 43 6.8	2 (34)				
29	Altona	8 10 38 11.3	273 16 12.1	+42 28 18.3	[1])	− 2.9	+ 2.2	1	1
30	Hamburg	8 7 26 37.8	273 4 57.2	+42 55 10.2	[2])	+20.3	−27.5	1	1/3
31	Göttingen	8 9 13 55.3	273 10 57.0	+42 40 19.4	[3])				
32	»	8 10 32 49.1	—	+42 29 38.8	5 (26)	+ 4.4	+30.3	1	0*)
33	»	8 10 39 43.7	273 16 33.3	—	5 (26)				
34	Paris	9 8 18 56.9	274 33 40.0	+39 17 30.5	(11)	− 3.5	+ 9.0	1	1
35	Bonn	9 7 33 15.0	274 31 11.7	+39 24 12.4	8 (11)	+ 4.7	− 0.3	3	4
36	Altona	9 10 39 28.4	274 41 39.3	+38 56 19.4	[4])	− 9.0	+ 6.6	1	1
37	Hamburg	9 7 11 44.0	274 30 4.4	+39 27 13.4	(11) + (58)	+ 9.1	−12.9	1	1/3
38	Marseille	9 7 8 6	274 30 24	+39 26 46	(11)	+35.5	−72.9	1/3	0
39	Göttingen	9 8 42 26.4	—	+39 13 21.1	(11)	—	−28.7	—	1/4
40	»	9 8 41 54.5	274 35 0.6	—	(11)	− 2.8	—	1	—
41	Bonn	10 7 45 53.5	275 52 40.9	+35 41 21.6	6 (27)	− 2.1	− 0.7	3	4
42	»	10 8 11 36.4	275 54 13.5	+35 37 23.6	4 (49)				
43	Altona	10 10 5 47.7	276 0 2.4	+35 19 37.0	(61) + (62)	−16.5	+11.1	1/2	1/2
44	Marseille	10 7 8 11	275 50 39	+35 46 48	(49)	− 3.7	−30.3	1/3	1/5
45	Göttingen	10 7 45 27.0	—	+35 41 7.6	(27)	—	−20.5	—	1/4
46	»	10 7 46 37.0	275 52 37.3	—	(27)	− 9.6	—	1	—
47	Berlin	11 7 42 33.8	277 8 36.3	+31 51 53.8	(31)	− 1.8	− 7.8	3	4
48	Marseille	11 7 36 14	277 8 6.9	+31 55 7.6	(50)	−10.2	+124.4	1/3	0
49	Paris	12 6 42 21.3	278 17 1.6	+28 5 54.5	(12) + (13) + (14)	− 5.6	+59.3	1	1
50	Berlin	12 6 56 0.1	278 17 36.4	+28 2 33.3	5 (13)	− 7.0	− 6.6	3	4
51	»	12 7 18 54.8	278 18 53.1	+27 58 39.6	2 (12)				
52	Marseille	12 8 2 17	278 21 39	+27 51 6	(51)	+35.9	−29.7	1/3	1/5
53	Marseille	13 6 58 18	279 25 43	+24 0 50	(52)	+39.7	+45.0	1/3	1/5
54	Berlin	14 7 37 12.9	280 29 15.6	+19 48 8.6	(32)	−12.6	−14.2	3	4

*) Wegen grosser Unterschiede zwischen den Einzelwerthen.

Nr.	Ort	M. Z. Berlin	AR.	Decl.	Vergleichsterne	$\Delta\alpha \cos\delta$	$\Delta\delta$	p_α	p_δ
55	Marseille	Nov. 14 $6^h 28^m 19^s$	280° 26′ 38″	+19° 59′ 43″	(53)	+ 3″.5	−24″.8	1/8	1/5
56	Hamburg	15 6 14 36.0	281 25 8.0	+15 56 46.9	(59) + (60) + (61)	+12.3	−17.2	1	1/3
57	»	16 6 36 52.6	282 20 53.6	+11 49 45.0	5)	+ 7.2	−12.1	1	1/3
58	Paris	17 6 30 10.1	283 11 41.2	+ 7 52 38.8	(15)	− 4.6	+29.2	1	1
59	Bonn	17 6 13 1.8	283 11 11.0	+ 7 54 52.4	4 (15)	} − 0.6	} − 3.6	3	4
60	»	17 7 15 1.7	283 13 16.2	+ 7 44 46.7	4 (35)				
61	Hamburg	17 6 10 7.1	283 11 5.4	+ 7 55 46.7	6)	+ 0.8	+19.6	1	1/8
62	Marseille	19 6 38 4	284 43 59	+ 0 14 44	(54)	+17.9	+24.3	1/8	1/5
63	Hamburg	20 7 20 26.3	285 25 11.5	− 3 25 44.5	7)	−15.1	+43.3	1	1/8
64	Paris	21 6 37 54.9	286 1 35.7	− 6 44 8.5	(16)	+ 3.9	+19.0	1	1
65	Bonn	21 6 15 56.5	286 0 52.7	− 6 41 20.7	4 (36)	} − 7.7	} + 4.3	3	4
66	»	21 6 29 15.0	286 1 10.5	− 6 43 12.6	6 (37)				
67	Altona	21 6 32 46.6	286 1 20.9	− 6 43 37.3	8)	− 3.3	+ 8.3	1	1
68	Hamburg	21 6 4 50.7	286 0 48.8	− 6 40 3.3	(45) + (36) + (37)	+ 5.9	− 9.3	1	1/8
69	Marseille	21 6 21 51	286 1 21	− 6 42 10	(16)	} + 2.3	} + 2.0	1/8	1/5
70	»	21 7 45 51	286 3 4	− 6 53 52	(45)				
71	Paris	22 6 48 16.7	286 35 37.2	− 9 58 43.0	(17)	− 3.1	+19.8	1	1
72	»	23 6 35 23.7	287 5 59.6	−12 59 10.0	(18)	− 0.8	+17.2	1	1
73	Cambridge	24 6 27 23.1	287 33 19.8	—	(20)	+ 6.6	—	1	—
74	»	24 6 44 32.3	—	−15 51 12.6	(20)	—	+15.2	—	1
75	Marseille	24 6 19 26	287 33 2	−15 48 30	(20)	− 2.2	+ 5.6	1/3	1/6
76	Greenwich	24 6 31 3.5	287 33 6.6	—	(20)	− 9.8	—	1	—
77	»	24 6 28 51.9	—	−15 49 51.0	(20)	—	− 9.3	—	1
78	Paris	25 6 21 8.7	287 57 18.9	−18 28 40.0	(19)	+ 4.9	+21.4	1	1
79	Cambridge	25 6 34 53.6	287 57 24.7	—	(21)	− 2.0	—	1	—
80	»	25 6 44 30.1	—	−18 31 28.7	(21)	—	+ 3.4	—	1
81	Greenwich	25 6 21 59.7	287 57 4.5	—	(21)	− 9.6	—	1/2	—
82	»	25 6 23 38.7	—	−18 29 24.0	(21)	—	− 6.6	—	1/2
83	Cambridge	26 6 16 12.0	288 18 16.2	—	(22)	+13.9	—	1	—
84	»	26 6 24 45.3	—	−20 59 15.3	(22)	—	− 5.4	—	1
85	Marseille	26 6 30 6	288 18 45	−20 59 26	(55) + (56)	+30.5	+16.0	1/8	1/5
86	Greenwich	26 6 30 29.0	288 18 8.8	—	(22)	− 3.8	—	1	—
87	»	26 6 30 33.0	—	−20 59 45.2	(22)	—	− 0.4	—	1
88	Kremsmünster	27 5 40 0.9	288 34 57.4	−23 14 49.0	—	−14.2	− 3.1	1	1

Anmerkungen.

Die Zusammenziehung mehrerer Beobachtungen eines Abends hat stets nach Maassgabe der in der Spalte »Vergleichsterne« durch die Ziffer vor der Sternbezeichnung angegebenen Anzahl der Einzelvergleichungen stattgefunden.

1) In AR: (34) + 5 (26); in Decl.: (34) + 10 (26).

2) (57) + (33) + (10) + (25) + (26).

3) In AR.: 2 (25); in Decl.: 1 (25).

4) In AR.: $\frac{1}{9.3}\left(\frac{131}{140}(39) + \frac{153}{140}(40) + \frac{2}{5}(11) + \frac{2129}{420}(41) + \frac{757}{420}(42)\right)$;

in Decl.: $\frac{1}{7.5}\left(\frac{1}{3}(39) + \frac{1}{2}(40) + \frac{19}{3}(41) + \frac{1}{3}(59)\right)$.

5) (62) + (63) + (64) + (65).

6) (15) + (66) + (67) + (68) + (69).

7) (70) + (71) + (72) + (73) + (74).

8) In AR.: $\frac{1}{5}\left(\frac{7}{3}(45) + \frac{1}{3}(36) + \frac{11}{6}(37) + \frac{1}{2}(16)\right)$; in Decl.: $\frac{1}{5}\left(\frac{1}{3}(45) + (36) + \frac{11}{3}(37)\right)$.

Es musste nun eine Wahl der Gewichte der Beobachtungen erfolgen, die deshalb schwierig war, weil die Beobachtungen von sehr verschiedener Güte sind, auf einer sehr wechselnden Anzahl von Einzelvergleichungen beruhen und, wie die Differenzen B — R zeigen, für die verschiedenen Sternwarten beträchtliche Unterschiede in der Auffassung des Cometencentrums vorgekommen sind. So findet man in besonders markanter Weise:

		$\Delta\delta$			$\Delta\delta$
Paris	Nov. 4	+14.″6	Berlin	Nov. 5	−12.″7
	5	+20.5		7	−9.5
	8	+43.3		8	−10.8
	9	+5.7		11	−10.5
	12	+57.0		12	−8.9
	17	+28.5		14	−16.0
Mittel		+28.3	Mittel		−11.4

An Zahlenmaterial dagegen standen nur die Differenzen B − R und für Altona, Cambridge, Paris, Greenwich, Göttingen, Berlin die Unterschiede der Einzelvergleichungen jedes Abends zur Verfügung. Letztere ergaben direct den Fehler einer Einzelvergleichung für die betreffende Sternwarte. Erstere wurden mit Gewichten, die den definitiven hinreichend nahe kamen, zu einigen Mittelwerthen vereinigt, diese graphisch ausgeglichen und die Abweichungen von der Ausgleichungscurve bestimmt. Diese Abweichungen sind dann das Resultat dreier Fehler, des Abendfehlers α, des Fehlers der Einzelvergleichung β und des constanten Auffassungsfehlers γ. Der Fehler γ wurde eliminirt, indem aus den Abweichungen jeder Sternwarte das Mittel genommen und die Unterschiede gegen das Mittel gebildet wurden, die nunmehr die Wirkungssumme allein von α und β darstellten.

Verglich man nun die hier erhaltenen Werthe mit den obigen Werthen des Fehlers der Einzelvergleichung β, so zeigte sich, dass der Fehler β den grösseren Theil des ganzen Betrages ausmachte und zahlenmässig etwa $\beta = 2\alpha$ zu setzen war. Demnach schreibt sich das Gewicht einer Beobachtung, wenn n die Anzahl der Einzelvergleichungen bedeutet und von dem Fehler γ zunächst abgesehen wird:

$$p = \frac{1}{\alpha^2\left(1 + \frac{4}{n}\right)}$$

oder, wenn man den mittl. Fehler α_1 einer auf der durchschnittlichen Anzahl von Einzelvergleichungen n_0 beruhenden Beobachtung einführt:

$$p = \frac{1}{\alpha_1^2} \frac{1 + \frac{4}{n_0}}{1 + \frac{4}{n}}$$

da ferner $\alpha_1 = \frac{\beta}{2}\sqrt{1 + \frac{4}{n_0}}$ ist, so finden sich zwei Werthe für α_1, einer aus den Werthen von β, ein zweiter aus den Differenzen B − R, und zwar lieferte die Rechnung:

Ort	Aus den Einzelvergleichungen				Aus den Differenzen B−R				Def. Gewichte	
	AR.	Decl.	p_α	p_δ	AR.	Decl.	p_α	p_δ	p_α	p_δ
Altona	6″	13″	2.2	0.5	9″	8″	1.0	1.3	1	1
Cambridge	9	7	1.0	1.3	10	14	0.8	0.4	1	1
Greenwich	12	5	0.6	3.2	6	9	2.2	1.0	1	1
Göttingen	14	29	0.4	0.10	8	37	1.3	0.06	1	1/4
Berlin	4	2	5.1	20.2	4	4	5.1	5.1	3	4
Paris	10	6	0.8	2.2	9	17	1.0	0.3	1	1
Bonn	—	—	—	—	5	3	3.2	9.0	3	4
Hamburg	—	—	—	—	9	25	1.0	0.13	1	1/2
Marseille	—	—	—	—	24	32	0.14	0.08	1/2	1/6

Vergleicht man die beiden Reihen der Gewichte p_α für Rectascension und p_δ für Declination, die unter Annahme eines mittl. Fehlers von 9″ für die Gewichtseinheit gebildet sind, so erkennt man zwar einen parallelen Gang, aber im Einzelnen grosse Abweichungen. Da sich auch für die constanten Auffassungsfehler keine so sicheren Werthe ableiten liessen, dass man sie hätte an sämmtliche Beobachtungen einer Sternwarte anbringen dürfen, wurde γ nur durch eine Annäherung aller Gewichte an einander in Rechnung gezogen, was dann im Mittel aus den beiden Gewichtsreihen zu den in der dritten Spalte des Täfelchens angegebenen definitiven Gewichten führte, die freilich mit einer grossen, aber unvermeidlichen Willkür behaftet sind. Da ferner durch die Existenz des Fehlers γ auch der Einfluss der Anzahl der Einzelvergleichungen herabgemindert wird, erfolgte die relative Gewichtsvertheilung für jede Sternwarte nicht nach der obigen Formel, sondern im Verhältniss

$$1 + \frac{2}{n_0} : 1 + \frac{2}{n}.$$

Mit den so entstandenen, in Tafel II Spalte 9 und 10 angegebenen Gewichten wurden die Differenzen B − R in der durch die Horizontalstriche bezeichneten Weise zu vier Normalörtern vereinigt und nach Uebertragung auf die nächstliegenden Ephemeridenzeiten folgende Werthe erhalten:

1842		$\Delta\alpha \cos\delta$	$\Delta\delta$
Nov.	4.0	+6.″2	−0.″7
	9.0	+0.2	−0.2
	18.0	−1.9	+1.1
	25.0	−0.4	+6.0

Die Normalörter selbst werden (scheinb. Aequinoctium):

1842		AR.	Decl.
Nov.	4.0	265° 45′ 28.″3	+55° 48′ 23.″3
	9.0	274 4 32.4	+40 31 37.7
	18.0	283 46 47.7	+ 5 1 59.3
	25.0	287 51 9.2	−17 47 32.8

Die Schönfeld'sche Methode der Elementenverbesserung

führte zu folgenden Bedingungsgleichungen (mit logarithmisch angesetzten Coefficienten), die sich an mehreren willkürlichen Variationen der Elemente als richtig erwiesen:

$$\begin{array}{llllllr}
 & & & & & & \text{Gewicht} \\
d\varkappa\cdot 0.02842 & +dT\cdot 2.91971 & +dq\cdot 5.61661 & +d\lambda\cdot 8.64084 & +d\nu\cdot 9.99518 & = 0.79239 & 15 \\
0.04697 & 3.13137 & 5.67897 & 8.49949_n & 9.89508 & = 9.30103 & 36 \\
9.87303 & 3.40773 & 5.62692 & 8.96572_n & 9.58366 & = 0.27875_n & 18 \\
9.47166 & 3.51691 & 5.45397 & 8.91735_n & 9.23671 & = 9.60206_n & 10 \\
d\varkappa\cdot 9.20098_n & +dT\cdot 3.67559 & +dq\cdot 4.87150_n & +d\lambda\cdot 8.80710_n & +d\nu\cdot 0.16144_n & = 9.84510_n & 15 \\
9.92198_n & 3.89740 & 5.52783_n & 8.73387 & 0.12946_n & = 9.30103_n & 36 \\
0.18903_n & 4.06694 & 5.80493_n & 8.83064 & 9.44858_n & = 0.04139 & 18 \\
0.07558_n & 3.98568 & 5.70082_n & 9.08722_n & 9.40658 & = 0.77815 & 10
\end{array}$$

Für die der Homogeneität wegen eingeführten Unbekannten $x = d\varkappa$, $y = 10^3 dT$, $z = 10^5 dq$, $u = 10^{-2} d\lambda$, $v = d\nu$ ergaben sich hieraus unter Berücksichtigung der beigesetzten Gewichte die Normalgleichungen:

$$\begin{array}{rrrrrl}
15.536\,x & -\ 57.678\,y & +\ 66.357\,z & -\ 39.586\,u & +\ 10.176\,v & = -\ 1.372 \\
-\ 57.678 & +\ 626.968 & -\ 226.451 & +\ 52.692 & -\ 44.644 & = +\ 69.138 \\
+\ 66.357 & -\ 226.451 & +\ 288.393 & -\ 196.128 & +\ 42.968 & = -\ 13.272 \\
-\ 39.586 & +\ 52.692 & -\ 196.128 & +\ 686.008 & -\ 29.113 & = +\ 16.206 \\
+\ 10.176 & -\ 44.644 & +\ 42.968 & -\ 29.113 & +\ 13.878 & = +\ 11.850
\end{array}$$

deren Auflösung folgende Werthe für die Unbekannten und ihre Gewichte, sowie für die Verbesserungen der Elemente lieferte:

$$\begin{array}{lrlll}
x = +11.763 & 1.596 & & dT = +0.000311 & \\
y = +\ 0.3106 & 3296 & & dq = -0.0000282 & \\
z = -\ 2.818 & 30.96 & d\varkappa = +11''.76 & d\omega' = +12''.12 & d\omega = +10''.57 \\
u = -\ 0.04913 & 4325 & d\lambda = -\ 4\ 91 & d\Omega' = -\ 4.60 & d\Omega = -\ 4.20 \\
v = +\ 1.851 & 70.21 & d\nu = +\ 1.85 & di' = +\ 2.56 & di = +\ 3.37
\end{array}$$

Hier beziehen sich ω', Ω', i' auf den Aequator, ω, Ω, i auf die Ekliptik. Die Summe der Fehlerquadrate wird 39.67, woraus für die Gewichtseinheit ein mittlerer Fehler von $11''.5$ folgt, während aus den Differenzen B — R bei Mitnahme des Fehlers γ sich ein Werth von $11''.6$ ergiebt. Es ist also ein genügender Anschluss an die Beobachtungen erreicht.

Die übrig bleibenden Reste im Sinne B — R werden in naher Uebereinstimmung zwischen Rechnung aus den Bedingungsgleichungen und Rechnung nach den unten angegebenen definitiven Elementen:

1842	Aus den Bedingungsgl. $\Delta\alpha\cos\delta$	$\Delta\delta$	Nach den definit. Elem. $\Delta\alpha\cos\delta$	$\Delta\delta$
Nov. 4.0	$+3''.4$	$0''.0$	$+3''.5$	$-0''.1$
9.0	-1.5	$+0.4$	-1.6	$+0.4$
18.0	-0.7	-1.5	-0.8	-1.6
25.0	$+2.4$	$+1.8$	$+2.5$	$+1.6$

Ein Versuch, den Gang in Rectascension durch Einführung der Excentricität als sechster Unbekannter zu beseitigen, scheiterte, da die Determinante der entstehenden Normalgleichungen Null wurde. Es werden somit die folgenden Elemente, die durch Addition der obigen Correctionen aus Kowalczyk's Elementen hervorgehen, als definitive zu betrachten sein:

$$T = 1842 \text{ Dec. } 15.99401 \text{ M. Z. Berlin } \pm 0.000202$$
$$\log q = 9.7027671 \ \pm 0.0000208$$

$$\left.\begin{array}{l}
\omega' = 229^\circ 48'\ 1''.62 \pm 9''.2 \\
\Omega' = 206\ 40\ 22.32 \pm 11.5 \\
i' = 85\ 30\ 46.43 \pm 4.3
\end{array}\right\} \text{Aequator und mittl. Aequ. 1842.0}
\qquad
\left.\begin{array}{l}
\omega = 240^\circ 32'\ 18''.53 \pm 10''.0 \\
\Omega = 207\ 48\ 44.58 \pm 14.4 \\
i = 106\ 25\ 56.73 \pm 4.9
\end{array}\right\} \text{Ekliptik und mittl. Aequ. 1842.0}$$

Die beträchtlichen mittleren Fehler dieser Elemente werden noch vermehrt durch eine eventuelle Abweichung der Excentricität von der Einheit, deren Einfluss nach Maassgabe folgender Zahlen, die aus dem erwähnten System von Normalgleichungen mit verschwindender Determinante abgeleitet wurden:

$$\begin{array}{ll}
dT = -0.000097\, de'' & \\
dq = +0.00000105\, de'' & d\lambda = -0.486\, de'' \\
d\varkappa = -1.034\, de'' & d\nu = +0.111\, de''
\end{array}$$

ein bedeutender sein würde.

Zum Schluss danke ich allen den Herren, die mir durch Uebersendung von Sternpositionen und Beobachtungsmaterial behülflich gewesen sind, vor allem aber Herrn Prof. *Becker* in Strassburg, der mich zu dieser Arbeit veranlasst und mich während ihres Verlaufs bei jeder Gelegenheit mit Rath und That unterstützt hat.

Frankfurt a/M., 1894 Oct. 29. *Karl Schwarzschild.*

Astron. Nachr. 141 (1896) 1–8

ASTRONOMISCHE NACHRICHTEN.

Band 141. № 3361. 1.

Ueber die Stabilität der Bewegung eines durch Jupiter gefangenen Cometen.

Von *Karl Schwarzschild.*

1.

Die Coordinaten und Geschwindigkeitscomponenten eines Cometen in Bezug auf ein durch die Sonne gehendes, der Richtung nach festes Coordinatensystem seien $x_1\ x_2\ x_3$ resp. $x_4\ x_5\ x_6$. Der Comet erleide Störungen durch Jupiter, dessen Bahn als eine ungestörte Ellipse betrachtet werden möge. Dann lassen sich die Bewegungsgleichungen des Cometen in der Form schreiben:

$$\frac{dx_1}{dt} = X_1, \quad \frac{dx_2}{dt} = X_2, \quad \ldots \quad \frac{dx_6}{dt} = X_6 \qquad (1)$$

Hierin sind $X_4\ X_5\ X_6$ die Componenten der von Sonne und Jupiter auf den Cometen ausgeübten Kraft, sie sind abhängig von $x_1\ x_2\ x_3$ und enthalten, wenn man die bekannten Werthe der elliptischen Jupitercoordinaten einführt, die Zeit explicite, sie werden periodische Functionen der Zeit von der Periode des Jupiterumlaufs. Andrerseits sind $X_1\ X_2\ X_3$ resp. gleich den Geschwindigkeitscomponenten $x_4\ x_5\ x_6$. Man sieht also, dass jedes X von dem zugehörigen x unabhängig ist und mithin die Beziehung gilt:

$$\frac{\partial X_1}{\partial x_1} + \frac{\partial X_2}{\partial x_2} + \cdots \frac{\partial X_6}{\partial x_6} = 0 \qquad (2)$$

2.

Wir wollen uns zunächst mit einem einfacheren Problem beschäftigen, wir betrachten nämlich das System mit nur drei Variabeln:

$$\frac{dx_1}{dt} = X_1 \quad \frac{dx_2}{dt} = X_2 \quad \frac{dx_3}{dt} = X_3 \qquad (3)$$

worin $X_1\ X_2\ X_3$ von $x_1\ x_2\ x_3$ abhängen und ausserdem periodische Functionen der Zeit von der Periode T sind (wenn man $x_1\ x_2\ x_3$ constant hält) und wobei die X der Gleichung genügen:

$$\frac{\partial X_1}{\partial x_1} + \frac{\partial X_2}{\partial x_2} + \frac{\partial X_3}{\partial x_3} = 0 \qquad (4)$$

Man fasse nun $x_1\ x_2\ x_3$ als rechtwinklige Coordinaten eines beweglichen Punktes im dreidimensionalen Raum auf. Im Anfangsmoment $t = t_0$ habe der Punkt die Coordinaten $x_1 = a_1$, $x_2 = a_2$, $x_3 = a_3$. Integrirt man die Gleichungen (3) unter Benutzung dieser Anfangswerthe für $x_1\ x_2\ x_3$, so erhält man die x als drei bestimmte Functionen der Zeit und man kann das Resultat der Integration offenbar bildlich darstellen, indem man den beweglichen Punkt von seiner Anfangslage $x_1 = a_1$, $x_2 = a_2$, $x_3 = a_3$ aus längs einer bestimmten Raumcurve mit bestimmter Geschwindigkeit fortschreiten lässt. Verfährt man gerade so für alle möglichen andern Werthe der Integrationsconstanten a_1, a_2, a_3, so erhält man von jedem Punkt des Raumes aus eine bestimmte Curve, längs deren sich ein beweglicher Punkt mit vorgeschriebener Geschwindigkeit bewegt. Man kann natürlich die Gleichungen (3) auch für negative Werthe von $t - t_0$ integriren und erhält dann in unserem Raum eine Fortsetzung der Curven nach rückwärts und eine Bestimmung der Bewegung der beweglichen Punkte für die vor dem Anfangsmoment t_0 liegende Zeit.

Um alle diese gleichzeitigen Bewegungen der Anschauung näher zu bringen, denke man sich den ganzen Raum zur Zeit t_0 von einer überall gleich dichten Flüssigkeit erfüllt.*) Als bewegliche Punkte nehme man jetzt die einzelnen Theilchen der Flüssigkeit und lasse dieselben in der vorher bestimmten Weise fortrücken. Es ist hiermit nichts unmögliches verlangt, weil man den Theilchen einer beliebig compressiblen Flüssigkeit jede beliebige eindeutige Bewegung zuschreiben kann, und wir haben es hier mit einer eindeutigen Bewegung zu thun, weil wir der mechanischen Natur unserer eigentlichen Aufgabe gemäss die Grössen $X_1\ X_2\ X_3$ in (3) natürlich als eindeutige Functionen der x und der Zeit annehmen. Das Resultat der Integration der Gleichung (3) für alle möglichen Integrationsconstanten wird dann bildlich durch die Strömung dieser Flüssigkeit dargestellt. Um nämlich die Werthe von $x_1\ x_2\ x_3$, die für irgend eine Zeit t unter Zugrundelegung der Integrationsconstanten (für $t = t_0$) $x_1 = a_1$, $x_2 = a_2$, $x_3 = a_3$ aus (3) folgen, zu erhalten, braucht man nur nachzusehen, wo sich das Flüssigkeitstheilchen, das zur Zeit t_0 im Punkt $x_1 = a_1$, $x_2 = a_2$, $x_3 = a_3$ war, zur Zeit t befindet, und die Coordinaten dieser Stelle abzulesen.

Es handelt sich darum, eine nähere Einsicht in die Art der Strömung dieser Flüssigkeit zu gewinnen. Die Gleichungen (3) liefern die Veränderung der Coordinaten der Flüssigkeitstheilchen mit der Zeit. Nennt man daher

*) Eine Flüssigkeitsströmung als Darstellung der Integrale des Systems (3) unter der Bedingung (4) findet man benutzt in Herrn Poincaré's Arbeit: Sur le problème des trois corps et les équations de la dynamique (Acta math. Bd. XIII, pag. 54), an die sich überhaupt die vorliegenden Betrachtungen angeschlossen haben.

u_1, u_2, u_3 die Componenten der Strömungsgeschwindigkeit an irgend einer Stelle und zu irgend einer Zeit, so gilt:

$$\frac{dx_1}{dt} = u_1 \quad \frac{dx_2}{dt} = u_2 \quad \frac{dx_3}{dt} = u_3$$
$$\text{und:} \qquad X_1 = u_1 \quad X_2 = u_2 \quad X_3 = u_3 \tag{5}$$

Man sieht also, dass uns zwar nicht die Bewegung der einzelnen Flüssigkeitstheilchen, aber doch die Richtung und Grösse der Strömung in jedem Punkt und zu jeder Zeit unmittelbar gegeben ist. Die Voraussetzungen, die wir über die X gemacht haben, liefern besondere Eigenschaften der Strömung. Setzt man nämlich die aus (5) folgenden Werthe von X_1 X_2 X_3 in (4) ein, so erhält man:

$$\frac{\partial u_1}{\partial x_1} + \frac{\partial u_2}{\partial x_2} + \frac{\partial u_3}{\partial x_3} = 0$$

Dies ist die bekannte Continuitätsgleichung der Hydrodynamik, welche besagt, dass in ein jedes Volumenelement eben so viel Masse ein- wie ausströmt, dass also die in einem Volumenelement enthaltene Masse stets ihren anfänglichen Betrag behält. Da wir aber vorausgesetzt hatten, dass im Anfangsmoment die Flüssigkeit überall gleich dicht war, so folgt hieraus, dass sie während der Bewegung überall gleich dicht bleibt; wir haben es also mit der Strömung einer homogenen incompressiblen Flüssigkeit zu thun. Ferner war vorausgesetzt worden, dass die X Functionen der Zeit von der Periode T sind. Daraus ergiebt sich vermittelst der Gleichungen (5), dass sich im Zeitintervall T stets wieder der nämliche Strömungszustand der ganzen Flüssigkeit herstellt, dass also ein Theilchen, welches sich zur Zeit $t+T$, $t+2T$, $t+3T$, $\cdots$ u. s. w. in einem bestimmten Punkt des Raumes befindet, die nämliche Bahn beschreibt, wie dasjenige Theilchen, welches zur Zeit t von diesem Punkt ausging.

Die beiden eben festgestellten Eigenschaften der Strömung unserer bildlichen Flüssigkeit führen unmittelbar zu folgendem Satz:

»Es ist nicht möglich, dass eine endliche Flüssigkeitsmenge, die aus sehr grosser Entfernung in die Nähe des Coordinatenanfangspunktes gelangt ist, für die Zukunft stets in der Nähe des Coordinatenanfangspunktes bleibt.«

Man fasse ein begrenztes Flüssigkeitsquantum ins Auge, das sich im Verlauf der Strömung einmal dem Coordinatenanfangspunkt nähert, und nehme an, es lasse sich um den Coordinatenanfangspunkt ein endlicher Raum abgrenzen, ausserhalb dessen sich das Flüssigkeitsquantum ursprünglich befand, in den es dann eingeströmt ist und in dem es, was gerade als unmöglich bewiesen werden soll, für alle Zukunft bleibt. Hat die Einströmung des betrachteten Flüssigkeitsquantums in den abgegrenzten Raum zur Zeit t stattgefunden, so wird in Folge der Periodicität der Strömung zur Zeit $t+T$ ein anderes Flüssigkeitsquantum in diesen Raum einströmen, welches genau dieselbe Form und Masse, wie das erste Quantum zur Zeit seines Einströmens, hat, welches aber weder ganz noch zum Theil aus Partikelchen des ersten Quantums bestehen kann, da letztere nach unserer Annahme von der Zeit t an in dem abgegrenzten Raum geblieben sind und daher nicht zur Zeit $t+T$ von neuem in denselben einströmen können. Auch dieses zweite Quantum muss für alle Zukunft in dem abgegrenzten Raum bleiben, da seine Theilchen, wie erwähnt, denen des ersten Quantums nachströmen werden. Derselbe Vorgang wiederholt sich zu den Zeiten $t+2T$, $t+3T$, u. s. w., jedes Mal strömt eine neue Flüssigkeitsmenge von derselben Masse, wie die ursprüngliche, in den abgegrenzten Raum ein, die dann nicht mehr aus diesem Raum heraus kann. Der abgegrenzte Raum würde also schliesslich eine unendliche Flüssigkeitsmasse aufnehmen müssen, was unmöglich ist, da die Flüssigkeit incompressibel ist. Es kann daher schon das erste Quantum nicht für alle Zukunft in dem abgegrenzten Raum bleiben, q. e. d.

Diese Ableitung bleibt bestehen, wie klein man das erste Flüssigkeitsquantum auch wählen möge, wofern es nur nicht direct Null ist.

Der gefundene Satz lässt sich in positiver Form so aussprechen: »Bei einer incompressiblen periodisch strömenden Flüssigkeit wird jede beliebig kleine Flüssigkeitsmenge, die sich ausserhalb irgend eines abgegrenzten endlichen Raumes befunden hat und dann in denselben eingeströmt ist, diesen Raum später wieder verlassen müssen.«

3.

Wir kehren zu dem Problem der Cometenbewegung mit sechs Variabeln zurück und wollen für dasselbe eine ähnliche bildliche Darstellung durchführen. Zu diesem Zweck betrachten wir x_1 x_2 $\cdots$ x_6 als rechtwinklige Coordinaten in einem sechsdimensionalen Raum, führen wieder bewegliche Punkte ein, die den Integralcurven des Systems (1) entlang laufen, und fassen die gleichzeitige Bewegung aller dieser Punkte als Strömen einer Flüssigkeit im sechsdimensionalen Raum auf.

Jetzt bedeutet die Gleichung (2), dass diese Flüssigkeit incompressibel ist. Um dies deutlicher zu machen, sei an einen Satz erinnert, der sich in Herrn Poincaré's Preisschrift: Sur le problème des trois corps et les équations de la dynamique (Acta mathem. Bd. XIII, pag. 57), bewiesen findet und der folgendermaassen lautet:

Man grenze irgend ein sechsdimensionales Gebiet der Variabeln x_1 x_2 $\cdots$ x_6, etwa durch eine Ungleichung der Form:

$$\varphi(x_1 x_2 \cdots x_6) < C$$

ab und berechne das sechsfache Integral

$$\int dx_1\, dx_2\, dx_3 \cdots dx_6$$

über dieses Gebiet ausgedehnt. Man lasse ferner die Punkte dieses Gebietes sich gemäss den Gleichungen (1) bewegen, bis dasselbe in ein neues Gebiet übergegangen ist, das durch eine Ungleichung:

$$\varphi_1(x_1 x_2 \cdots x_6) < C$$

bestimmt sei. Berechnet man dann das Integral

$$\int dx_1\, dx_2 \cdots dx_6$$

über das neue Gebiet ausgedehnt, so erhält dasselbe stets denselben Werth wie bei der Erstreckung über das ursprüngliche Gebiet, wofern die rechten Seiten der Gleichungen (1) der Bedingung (2) genügen.

Das hier auftretende sechsfache Integral ist nichts anderes, als das Volumen des abgegrenzten Gebietes im sechsdimensionalen Raum. Aus der Gleichung (2) folgt mithin, dass das Volumen einer gegebenen Flüssigkeitsmenge bei der Bewegung constant bleibt, dass die Flüssigkeit incompressibel ist.

Da ausserdem die X_1 periodisch von der Jupitersperiode sind und im vorigen Paragraphen nur die Begriffe des Abgrenzens, des Ein- und Ausströmens benutzt wurden, die für eine Mannigfaltigkeit von beliebig hoher Dimension ihren Sinn behalten, so kann man den obigen Satz ohne Weiteres auf den jetzigen allgemeineren Fall übertragen, nur hat man unter Flüssigkeitsmenge den Werth des Integrals

$$\int dx_1\, dx_2 \cdots dx_6$$

über das betreffende Gebiet erstreckt, unter abgegrenztem Raum eine begrenzte sechsdimensionale Mannigfaltigkeit in den $x_1\, x_2 \cdots x_6$ zu verstehen.

Ueberzeugen wir uns nun, was unser Satz bedeutet, wenn man das Bild verlässt und die wirkliche Bewegung eines Cometen betrachtet.

Man grenze ein sechsdimensionales Gebiet durch folgende Ungleichheiten ab:

$$x_1^2 + x_2^2 + x_3^2 \leqq r^2 \tag{6}$$

$$x_4^2 + x_5^2 + x_6^2 \leqq v^2 \tag{7}$$

Für r wähle man eine Grösse, die etwa das Doppelte der Entfernung des Neptun von der Sonne sein mag. Dann fällt jeder Anfangszustand

$$x_1 = a_1\,, \quad x_2 = a_2\,, \quad \cdots \quad x_6 = a_6$$

ausserhalb des abgegrenzten Gebietes, für den die Entfernung des Cometen von der Sonne

$$\sqrt{a_1^2 + a_2^2 + a_3^2}$$

grösser als r ist. Für v wähle man eine so grosse Geschwindigkeit, dass ein Körper, der sie besitzt und der sich innerhalb der Kugel um die Sonne vom Radius r befindet, diese Kugel nothwendig baldigst verlässt, selbst wenn er im Lauf seiner Bewegung die Oberfläche der Sonne und des Jupiter streifen sollte.

Man denke sich ferner einen Cometen, der sich weiter als die doppelte Neptunsdistanz von der Sonne auf dem Wege nach dem Perihel befindet und dessen Geschwindigkeit unter der eben definirten ungeheuren Grösse v liegt. Seine Anfangscoordinaten und Geschwindigkeitscomponenten seien $a_1\, a_2 \cdots a_6$. Man lasse nun die Werthe a auf alle mögliche Art innerhalb gewisser enger Grenzen variiren. Der Comet möge von der ursprünglichen Anfangslage aus nach einiger Zeit mit mässiger Geschwindigkeit in die Kugel r eintreten. Das gleiche würde dann auch gelten, wenn man den Cometen von irgend einem der variirten Anfangszustände ausgehen liesse, falls man nur die Variationen nicht zu gross annimmt. Allen diesen Anfangszuständen entsprechen bestimmte Punkte in unserem sechsdimensionalen bildlichen Raum, die zusammen ein kleines Volumen dieses Raumes erfüllen werden. Um zu erfahren, wie sich der Comet von diesen Anfangszuständen aus bewegen würde, verfolgen wir, wie sich die das entsprechende Volumen des sechsdimensionalen Raumes einnehmende Flüssigkeitsmenge weiter bewegt. Für den Anfang der Bewegung schliessen wir zunächst umgekehrt, dass die Flüssigkeitsmenge in den durch (6) und (7) abgegrenzten Raum einströmt, weil der Comet von allen den bezeichneten Anfangszuständen aus in die Kugel vom Radius r mit einer Geschwindigkeit, die kleiner als v ist, eintritt. Für die Fortsetzung der Bewegung liefert unser Satz, dass die Flüssigkeitsmenge früher oder später wieder den durch (6) und (7) abgegrenzten sechsdimensionalen Raum verlassen wird, oder, auf den Cometen angewandt, dass eine Zeit kommen wird, zu der die Coordinaten und die Geschwindigkeit des Cometen den Ungleichungen (6) und (7) nicht mehr genügen, von welchem der verschiedenen Anfangszustände er auch ausgegangen sein mag.

Wird die Gleichung (6) nicht erfüllt, so bedeutet das, dass der Comet wieder weiter als die doppelte Neptunsdistanz von der Sonne entfernt ist. Wird die Gleichung (7) nicht erfüllt, so besitzt er eine so ungeheure Geschwindigkeit, dass er das Sonnensystem in kürzester Frist verlässt, oder aber — denn auch in diesem Fall kann seine Geschwindigkeit grösser als v sein — er befindet sich näher beim Centrum der Sonne oder des Jupiter, als der Halbmesser dieser Körper beträgt, d. h. in Wirklichkeit: er ist auf einen dieser Körper gestürzt.

Noch einer kleinen Correctur bedürfen diese Betrachtungen. Die Ableitung in § 2 galt für eine beliebig kleine Flüssigkeitsmenge, aber nicht für die Flüssigkeitsmenge Null. Es kann daher einzelne isolirte Punkte des sechsdimensionalen Raumes, die ja die Flüssigkeitsmenge Null enthalten, geben, die beim Fortrücken stets innerhalb eines begrenzten Raumes, in den sie eingetreten sind, bleiben; in anderen Worten, es können gewisse singuläre Anfangs-Lagen und Geschwindigkeiten des Cometen unter den oben bezeichneten Anfangszuständen existiren, von denen aus der Comet eine stabile Bewegung ausführen, etwa stets in einer mässig grossen elliptischen Bahn bleiben würde. Da aber diese Anfangszustände nur singulär sind, da die ihnen entsprechenden Punkte des sechsdimensionalen Raumes niemals ein endliches Volumen desselben erfüllen können, so ist man immer im Stande, durch einen beliebig kleinen Stoss einen solchen Anfangszustand in einen der anderen nicht singulären zu verwandeln. Auch zu jeder um endliches späteren Zeit kann man die Bewegung des Cometen aus einer singulären stabilen vermittelst eines kleinen Stosses in eine zu einem nicht singulären Anfangszustand gehörige instabile überführen, da Bewegungen, die von unendlich benachbarten Anfangszuständen ausgehen, für jede endliche Zeit benachbart verlaufen.

Berücksichtigt man schliesslich, dass man für r in (6) jede Distanz nehmen kann, die kleiner ist als die ursprüngliche Distanz des Cometen von der Sonne, so gewinnt man das Resultat:

»Ist ein Comet, der sich in parabolischer oder hyperbolischer Bahn dem Sonnensystem genähert hat, durch Jupiter in eine elliptische Bahn geworfen worden, so bewirken im Allgemeinen die Jupiterstörungen selbst, dass der Comet auf die Sonne oder auf Jupiter stürzt oder sich wieder in parabolischer oder hyperbolischer Bahn vom Sonnensystem entfernt. Tritt das ausnahmsweise nicht von selbst ein, so kann es doch zu Wege gebracht werden, indem man dem Cometen irgend wann einen beliebig kleinen Stoss ertheilt, das heisst, indem man seine osculirenden Elemente beliebig wenig ändert.« *)

München 1896 April 27.

Karl Schwarzschild.

*) Aehnliche Ueberlegungen, die bei späterer Gelegenheit mitgetheilt werden sollen, lassen sich auch für das allgemeine *n*-Körperproblem und auch in dem Falle, dass man es nicht mit der Bewegung eines aus sehr grosser Entfernung kommenden Körpers zu thun hat, durchführen.

Researches on the orbit of η Coronae Borealis = Σ 1937.

By *T. J. J. See.*

This remarkable star was discovered by Sir William Herschel, Sept. 9, 1781. It proved to be one of the first objects which gave distinct evidence of orbital motion, and the binary character of the system was fully recognized by Herschel in 1803. Since the time of Struve the measures are both numerous and exact. The pair is always rather close, but as the components are nearly equal in magnitude it is generally easy to separate; and while the errors in angle are larger than in case of wider objects the absolute errors of the observed positions are correspondingly reduced by the shortness of the radius vector.

Numerous orbits have been published by previous computers; the following table gives a summary of the elements which is fairly complete. It will be seen that in recent years the elements have not been changed by very large quantities, and yet they are severally by no means identical.

P	T	e	a	Ω	i	λ	Authority	Source
44.242	1806.20	0.26034	0″.8325	220°.6	37°.4	358°.63	Herschel 1833	Mem. R. A. S., VI.156
43.246	1850.23	0.3376	1.0879	24.3	71.13	261.35	Mädler 1842	—
43.310	1815.20	0.3537	1.1912	22.6	71.5	263.17	Mädler 1842	—
42.500	1807.21	0.289	0.9024	20.1	59 47	215.2	Mädler 1842	—
42.501	1805.666	0.4743	1.0125	10.52	65.65	227.17	Villarceau 1842	—
66.257	1780.124	0.4695	1.1108	4.42	58.05	194.62	Villarceau 1852	—
67.309	1779.338	0.4043	1.2015	9.87	59.32	185.0	Villarceau 1852	A. N. 868
43.115	1850.329	0.2865	0.9567	22.3	60.67	215.48	Winnecke	—
41.58	1850.26	0.2625	0.827	26.7	58.0	211.4	Wijkander	—
41.576	1850.26	0.2625	0.827	26.7	58.0	215.6	Dunér 1871	A. N. 1868
40.17	1849.9	0.287	0.985	22.2	60.4	224.1	Flammarion 1874	Cat. ét. doub. 88
41.562	1850.792	0.2667	0.892	25.72	59.68	218.6	Doberck 1880	A. N. 2338
41.6	1892.3	0.33	0.86	26.9	55.0	220.5	Comstock 1893	Proc. Am. Assoc., 1894

Making use of all the measures up to 1895, we find the following elements of η Coronae Borealis:

P = 41.60 years Ω = 27°.10
T = 1892.50 i = 58.50
e = 0.267 λ = 217.57
a = 0″.9165 n = +8°.65385

Apparent orbit:

Length of major axis = 1″.804
Length of minor axis = 0.934
Angle of major axis = 28°.7
Angle of periastron = 229.0
Distance of star from centre = 0″.209.

Comparison of computed with observed places.

t	θ_o	θ_c	ϱ_o	ϱ_c	$\theta_o - \theta_c$	$\varrho_o - \varrho_c$	n	Observers
1781.69	30°.7	27°.4	—	1″.08	+3°.3	—	1	Herschel
1802.69	179.7	174.8	—	0.63	+4.9	—	1	Herschel
1823.27	25.9	27.3	1″.58	1.08	−1.4	+0″.50	2.1	Herschel and South
1826.77	35.3	37.9	1.07	1.09	−2.6	−0.02	4	Struve
1829.55	43.2	47.0	0.96	1.01	−3.8	−0.05	2	Struve
1831.48	51.4	54.5	0.95	0.92	−3.1	+0.03	15.4	Da. 2.0; Herschel 10.1; Σ 3

Astron. Nachr. *147* (1898) 17–24

ASTRONOMISCHE NACHRICHTEN.

Band 147. № 3506. 2.

Ueber eine Classe periodischer Lösungen des Dreikörperproblems.

Von *Karl Schwarzschild.*

Den bereits bekannten periodischen Bewegungen dreier Körper, die ihrer gegenseitigen Anziehung unterliegen, lässt sich eine weitere Classe solcher Bewegungen hinzufügen. Es handelt sich um den Fall, dass ein Planet eine Kreisbahn um die Sonne beschreibt und ein Planetoid von verschwindender Masse, der sich in derselben Ebene bewegt, der Anziehung dieser beiden Körper unterworfen ist. Auf diesen Fall beziehen sich gewisse von den periodischen Lösungen, welche Herr Poincaré in seinem grundlegenden Buche »Les Méthodes Nouvelles de la Mécanique Céleste« als »de la première sorte« bezeichnet, und die zugehörigen, für benachbarte Anfangslagen des Körpersystems geltenden, welche er »du deuxième genre« nennt. Bei allen diesen Lösungen ist die Excentricität der Bahn des Planetoiden klein von der Ordnung der störenden Masse; die hier zu besprechenden hingegen gelten für beliebige Werthe der Excentricität.

Die Masse der Sonne sei $1-\mu$, die Masse des störenden Planeten μ, seine Distanz von der Sonne 1 und seine Umlaufszeit 2π. Hiermit sind die Einheiten gewählt. Ist a_0 die grosse Axe der ungestörten Bahn des Planetoiden in Bezug auf den Schwerpunkt der beiden andern Körper, so wird seine mittlere Bewegung

$$n = \frac{1}{a_0^{3/2}}$$

Die mittlere Bewegung des störenden Körpers ist bei der getroffenen Wahl der Einheiten gleich 1. Die ungestörte Bewegung des ganzen Systems wird also periodisch, wenn n eine rationale Zahl ist:

$$n = \frac{n_1}{n_2}$$

wo n_1 und n_2 zwei ganze Zahlen bedeuten, die relativ prim zu einander sind. Die Periode der Bewegung wird:

$$T_0 = 2\pi n_2$$

Es entfallen auf sie n_1 Umläufe des gestörten und n_2 Umläufe des störenden Körpers.

Wie ändern sich die Verhältnisse, wenn man die Störungen mit in Betracht zieht? Man gehe von einem Zeitmomente aus, in welchem der störende Körper genau die Länge ω_0 des Perihels des Planetoiden habe. Nach Verlauf der Zeit T_0 wäre das System wieder in dieselbe Lage zurückgekehrt, wenn keine Störungen stattfänden, störender Körper und Perihel des Planetoiden hätten dann wieder dieselbe Länge ω_0. In Wirklichkeit aber verschiebt sich das Perihel, es wird also nicht nach der Zeit T_0, sondern nach einer wenig davon verschiedenen Zeit T vom Planeten eingeholt werden und nicht in der Länge ω_0, sondern in einer etwas davon verschiedenen Länge ω. Die mittlere Anomalie l des Planetoiden würde nicht einmal ohne Störungen in der Zeit T genau zum Ausgangswerth zurückkehren, und die Störungen können eine weitere Verschiebung bewirken. Aendert man aber die ursprüngliche grosse Axe a um eine kleine Grösse Δa, so werden dadurch die Störungen des Perihels nur um Grössen von der Ordnung $\mu\Delta a$ geändert, die mittlere Anomalie hingegen um Grössen von der Ordnung Δa. Es lässt sich daher voraussehen, dass man durch geeignete Wahl von Δa bewirken kann, dass nach derselben Zeit T, nach welcher der störende Planet wieder genau die Länge ω des Perihels des Planetoiden hat, auch die mittlere Anomalie wieder zu ihrem Ausgangswerth zurückkehrt. Gäbe es nun Fälle, in welchen auch die Störungen der grossen Axe und der Excentricität e, über die Zeit T summirt, gerade verschwänden, so hätte man Lösungen folgender Art: Nach Verlauf der Zeit T befänden sich alle drei Körper wieder in derselben relativen Lage, nur wäre das ganze System um einen kleinen Winkel, nämlich den Betrag der Perihelstörung, gedreht. Die Principien, welche Herr Poincaré in die Behandlung dieser Fragen eingeführt hat, gestatten in der That, die Existenz solcher periodischen Lösungen zu beweisen.

Die Länge des störenden Planeten wird bei der getroffenen Wahl der Einheiten direct gemessen durch die Zeit t. Setzt man:

$$\left.\begin{aligned} x &= \sqrt{a} & u &= l \\ y &= \sqrt{a(1-e^2)} & v &= \omega - t \end{aligned}\right\} \quad (1)$$

so lauten die Störungsgleichungen (Méthodes Nouv. I § 9):

$$\left.\begin{aligned} \frac{dx}{dt} &= \frac{\partial F}{\partial u} & \frac{du}{dt} &= -\frac{\partial F}{\partial x} \\ \frac{dy}{dt} &= \frac{\partial F}{\partial v} & \frac{dv}{dt} &= -\frac{\partial F}{\partial y} \end{aligned}\right\} \quad (2)$$

Bezeichnet man mit r den Radiusvector, mit u' die Mittelpunktsgleichung des Planetoiden, so ist hierin:

$$F = \frac{1}{2x^2} + y - \frac{1}{r} + \frac{\mu}{\sqrt{(1-\mu)^2 + r^2 - 2(1-\mu)r\cos(u+u'+v)}} + \frac{1-\mu}{\sqrt{\mu^2 + r^2 + 2\mu r\cos(u+u'+v)}} \qquad (3)$$

Setzt man noch:

$$F = F_0 + \mu F' \qquad (4)$$

und:

$$F' = F_1 + \mu F_2 + \mu^2 F_3 + \cdots \qquad (5)$$

so erhält man:

$$F_0 = \frac{1}{2x^2} + y \qquad (6)$$

$$F_1 = \frac{1}{\sqrt{1 + r^2 - 2r\cos(u+u'+v)}} - \frac{r + \cos(u+u'+v)}{r^2}$$

Auf bekannte Weise lässt sich dieser Ausdruck nach den Cosinus der Vielfachen der mittleren Anomalie u entwickeln:

$$F_1 = \Sigma K \cos[i_1 u + i_2 v] \qquad (7)$$

worin i_1 und i_2 ganze Zahlen und die Grössen K Functionen von a und e, resp. x und y bedeuten.

Die periodische Lösung unter Vernachlässigung der Störungen (für $\mu = 0$), von der wir ausgehen, sei:

$$\left.\begin{aligned} x &= x_0 = \sqrt{a_0} & u_0 &= l_0 = nt \\ y &= y_0 = \sqrt{a_0(1-e_0^2)} & v_0 &= \omega_0 - t \\ & n = \frac{1}{x_0^3} = \frac{n_1}{n_2} \end{aligned}\right\} \qquad (8)$$

Diese Ausdrücke stellen in der That eine periodische Bewegung dar. Denn setzt man:

$$t = T_0 = 2\pi n_2$$

so unterscheiden sich sowohl u_0 als v_0 nur um Vielfache von 2π von ihren Werthen zur Zeit $t = 0$. Die Längendifferenz zwischen störendem Körper und Perihel des Planetoiden, wie die mittlere Anomalie des Planetoiden sind wieder zu ihren Ausgangswerthen zurückgekehrt.

Um für $\mu \gtrless 0$ in der Nähe dieser Lösung eine periodische Lösung der beschriebenen Art zu finden, nehme man an, dass für letztere die Ausgangswerthe der Elemente zur Zeit $t = 0$ seien:

$$\begin{aligned} x &= x_0 + \beta_1 & u &= 0 \\ y &= y_0 + \beta_2 & v &= \omega \end{aligned}$$

Hier bedeuten β_1 und β_2 kleine Constanten, ω ist vorerst willkürlich. Man verliert nichts an Allgemeinheit, indem man $u = 0$ setzt, da man als Ausgangspunkt der Zeitrechnung immer den Augenblick eines Periheldurchgangs des Planetoiden wählen kann. Nach Verlauf der Zeit t mögen die Variabeln übergegangen sein in:

$$\begin{aligned} x &= x_0 + \beta_1 + \psi_1 & u &= \frac{n}{1+k}t + \psi_3 \\ y &= y_0 + \beta_2 + \psi_2 & v &= \omega - \frac{t}{1+k} + \psi_4 \end{aligned}$$

Durch Einführung der Constante k wird hier von vorneherein auf den Umstand Rücksicht genommen, dass die Periode der gestörten Bewegung im allgemeinen von der Periode der ungestörten Bewegung verschieden sein wird. Die Grössen ψ verschwinden für $t = 0$ und sind als Functionen der Zeit t, wie der Constanten $k, x_0, y_0, \beta_1, \beta_2, \omega$ aus den Bewegungsgleichungen (2) zu bestimmen. Vernachlässigt man die ψ einen Augenblick, so ergeben die übrig bleibenden Glieder eine periodische Bewegung von der Periode:

$$T = T_0(1+k) = 2\pi n_2(1+k) \qquad (9)$$

Die Werthe, welche die ψ nach dieser Zeit T annehmen, bezeichne man durch Ψ mit dem entsprechenden Index. Die Aufgabe ist dann, womöglich $k, \beta_1, \beta_2, \omega$ so zu wählen, dass

$$\Psi_1 = \Psi_2 = \Psi_3 = \Psi_4 = 0 \qquad (10)$$

wird.

Man führe zunächst an Stelle der Zeit t die neue unabhängige Variable τ, die wir auch kurz als Zeit bezeichnen wollen, ein:

$$t = (1+k)\tau$$

dann wird:

$$\left.\begin{aligned} x &= x_0 + \beta_1 + \psi_1 \\ y &= y_0 + \beta_2 + \psi_2 \\ u &= n\tau + \psi_3 \\ v &= \omega - \tau + \psi_4 \end{aligned}\right\} \qquad (11)$$

und die Ψ sind die Werthe, welche die ψ nach der Zeit $\tau = T_0 = 2\pi n_2$ annehmen. Die Bewegungsgleichungen werden:

$$\begin{aligned} \frac{dx}{d\tau} &= (1+k)\frac{\partial F}{\partial u} \\ \frac{dy}{d\tau} &= (1+k)\frac{\partial F}{\partial v} \\ \frac{du}{d\tau} &= -(1+k)\frac{\partial F}{\partial x} \\ \frac{dv}{d\tau} &= -(1+k)\frac{\partial F}{\partial y} \end{aligned}$$

Führt man hierin die Ausdrücke (4), (6) und (11) ein, so erhält man:

$$\left.\begin{aligned} \frac{d\psi_1}{d\tau} &= \mu(1+k)\frac{\partial F'}{\partial u} \\ \frac{d\psi_2}{d\tau} &= \mu(1+k)\frac{\partial F'}{\partial v} \\ \frac{d\psi_3}{d\tau} &= \frac{1+k}{(x_0+\beta_1+\psi_1)^3} - n - \mu(1+k)\frac{\partial F'}{\partial x} \\ \frac{d\psi_4}{d\tau} &= -k - \mu(1+k)\frac{\partial F'}{\partial y} \end{aligned}\right\} \qquad (12)$$

Man hat daher:

$$\left.\begin{array}{ll} \Psi_1 = \mu(1+k)\int\limits_0^{T_0}\frac{\partial F'}{\partial u}\,d\tau & \Psi_2 = \mu(1+k)\int\limits_0^{T_0}\frac{\partial F'}{\partial v}\,d\tau \\ \Psi_3 = -\,n\,T_0 + (1+k)\int\limits_0^{T_0}\frac{d\tau}{(x_0+\beta_1+\psi_1)^3} & -\,\mu(1+k)\int\limits_0^{T_0}\frac{\partial F'}{\partial x}\,d\tau \\ \Psi_4 = -\,k\,T_0 & -\,\mu(1+k)\int\limits_0^{T_0}\frac{\partial F'}{\partial y}\,d\tau \end{array}\right\} \quad (13)$$

Von dem Systeme (2) ist nun ein Integral bekannt, das Integral der lebendigen Kraft:

$$F = \text{const.}$$

Aus demselben ist zu schliessen, dass, wenn u, v und y zum selben Werthe zurückkehren, auch x zum selben Werthe zurückkehren muss. Sobald daher die drei letzten Gleichungen (10) erfüllt sind, wird die erste $\Psi_1 = 0$ identisch erfüllt sein müssen. Da mithin nur drei Gleichungen zu befriedigen bleiben, setzen wir:

$$\beta_2 = 0$$

und betrachten β_1, k und ω als die zu bestimmenden Unbekannten.

Die Lösungen ψ der Differentialgleichungen (12) und mithin auch die Werthe Ψ, welche sie nach der Zeit $\tau = T_0$ annehmen, lassen sich durch Reihen darstellen, welche nach Potenzen von μ, β_1 und k fortschreiten und convergiren, solange diese Grössen hinreichend klein sind (vgl. Méthodes Nouvelles I, § 27).

Die beiden letzten Gleichungen (10) werden daher in der Form geschrieben werden können:

$$0 = \Psi_3 = [\Psi_3]_0 + \frac{\partial[\Psi_3]_0}{\partial\mu}\mu + \frac{\partial[\Psi_3]_0}{\partial\beta}\beta_1 + \frac{\partial[\Psi_3]_0}{\partial k}k + \Psi_3'$$

$$0 = \Psi_4 = [\Psi_4]_0 + \frac{\partial[\Psi_4]_0}{\partial\mu}\mu + \frac{\partial[\Psi_4]_0}{\partial\beta_1}\beta_1 + \frac{\partial[\Psi_4]_0}{\partial k}k + \Psi_4'$$

wobei Ψ_3' und Ψ_4' Grössen zweiter Ordnung bedeuten, und durch die Klammer mit dem Index 0 angedeutet wird, dass in den betreffenden Grössen $\mu = \beta_1 = k = 0$ zu setzen ist. Nach bekannten Sätzen der Eliminationstheorie haben zwei derartige Gleichungen für kleines μ reelle Lösungen β_1 und k, welche nach Potenzen von μ, mit der ersten beginnend, entwickelt werden können, wenn erstens:

$$[\Psi_3]_0 = [\Psi_4]_0 = 0$$

und zweitens für $\mu = \beta_1 = k = 0$:

$$\begin{vmatrix} \frac{\partial\Psi_3}{\partial\beta_1} & \frac{\partial\Psi_3}{\partial k} \\ \frac{\partial\Psi_4}{\partial\beta_1} & \frac{\partial\Psi_4}{\partial k} \end{vmatrix} \gtrless 0$$

Die Ausdrücke (13) liefern aber sofort für $\mu = k = \beta_1 = 0$ (da dann offenbar auch $\psi_1 = 0$ wird):

$$[\Psi_3]_0 = -\,T_0\left(n - \frac{1}{x_0^3}\right) = 0 \qquad [\Psi_4]_0 = 0$$

Ferner wird für $\mu = \beta_1 = k = 0$:

$$\frac{\partial\Psi_3}{\partial\beta_1} = -\frac{3\,T_0}{x_0^4} \qquad \frac{\partial\Psi_3}{\partial k} = \frac{T_0}{x_0^3}$$

$$\frac{\partial\Psi_4}{\partial\beta_1} = 0 \qquad \frac{\partial\Psi_4}{\partial k} = -\,T_0$$

Die Determinante hat also den Werth $\frac{3\,T_0^2}{x_0^4}$, der stets von Null verschieden ist. Es existiren mithin Entwicklungen:

$$\beta_1 = \mu\,\chi_1(\mu) \qquad k = \mu\,\chi_2(\mu) \qquad (14)$$

welche die Gleichungen:

$$\Psi_3 = \Psi_4 = 0$$

befriedigen. Implicite werden diese Entwicklungen x_0, y_0 und ω enthalten.

Es bleibt die Gleichung, aus der die letzte Unbekannte ω zu bestimmen ist:

$$\Psi_2 = 0$$

Indem wir im Ausdruck (13) für Ψ_2 den stets von Null verschiedenen Factor $\mu(1+k)$ weglassen, aber für den reducirten Werth die Bezeichnung Ψ_2 beibehalten, bleibt:

$$0 = \Psi_2 = \int\limits_0^{T_0}\frac{\partial F'}{\partial v}\,d\tau$$

Ψ_2 ergiebt sich nun ebenso, wie Ψ_3 und Ψ_4, aus den Differentialgleichungen (12) als Potenzreihe nach Potenzen von μ, β_1 und k. Führt man hier die Ausdrücke (14) ein, so entsteht eine Potenzreihe nur nach Potenzen von μ. Da dieselbe convergirt, wenn μ hinreichend klein ist, so kann man setzen:

$$\Psi_2 = [\Psi_2]_0 + \mu\,\Psi_2' \qquad (15)$$

wo Ψ_2' eine endliche Grösse und $[\Psi_2]_0$ der Werth ist, in welchen Ψ_2 für $\mu = k = \beta_1 = 0$ übergeht.

Für $\mu = k = \beta_1 = 0$ wird aber nach (5) und (7):

$$F' = F_1 = \Sigma K\cos[i_1 u + i_2 v]$$

$$\text{mithin:}\quad \frac{\partial F'}{\partial v} = \frac{\partial F_1}{\partial v} = -\,\Sigma i_2 K\sin[i_1 u + i_2 v]$$

und wenn man hier für x, y, u und v die ungestörten Werthe (8) einsetzt:

$$\frac{\partial F'}{\partial v} = -\Sigma i_2 [K]_0 \sin[(i_1 n - i_2) t + i_2 \omega]$$

wobei mit $[K]_0$ der Werth bezeichnet ist, den K für $x = x_0$ und $y = y_0$ annimmt. Um $[\Psi_2]_0$ zu bilden, hat man über diesen Ausdruck von $t = 0$ bis $t = T_0 = 2\pi n_2$ zu integriren. Es ist offenbar, dass dabei alle Glieder fortfallen mit Ausnahme derjenigen, für welche:

$$i_1 n - i_2 = 0$$
$$i_1 = p n_2 \qquad i_2 = p n_1$$
$$p = 1, 2, 3, 4 \cdots$$

ist. Es wird also:

$$[\Psi_2]_0 = \int_0^{T_0} \frac{\partial F_1}{\partial v} dt = -T_0 n_1 \sum_p p [K]_0 \sin p n_1 \omega$$

Betrachten wir diesen Ausdruck als Function von ω. Er wird Null, so oft $n_1 \omega$ ein Vielfaches von π ist, also für die $2 n_1$ Werthe:

$$\left.\begin{array}{c} \omega = \Omega = \dfrac{\pi}{n_1} q \\ q = 0, 1, 2, \cdots, 2n_1 - 1 \end{array}\right\} \qquad (16)$$

Denken wir uns ω als Abscisse, $[\Psi_2]_0$ als Ordinate, so trifft die entstehende Curve die Abscissenaxe in allen Punkten $\omega = \Omega$ und zwar muss sie die Abscissenaxe in diesen Punkten nicht nur berühren, sondern auch schneiden. Setzt man nämlich:

$$\omega = \Omega \pm \alpha$$

so wird:

$$[\Psi_2]_0 = \mp T_0 n_1 \Sigma p (-1)^{pq} [K]_0 \sin p n_1 \alpha$$

Es nimmt also $[\Psi_2]_0$ zu beiden Seiten der Abscisse Ω entgegengesetzt gleiche Werthe an.

Zur Curve Ψ_2 geht man nach (15) über, indem man den Ordinaten $[\Psi_2]_0$ die Grössen $\mu \Psi_2'$ hinzufügt. Je kleiner μ ist, um so weniger wird sich also die Curve Ψ_2 von der Curve $[\Psi_2]_0$ unterscheiden. Sie wird in Folge dessen für kleines μ die Abscissenaxe in der Nähe der Punkte $\omega = \Omega$ schneiden müssen. Das heisst in anderen Worten: Auch für von Null verschiedenes μ, wofern es nur nicht zu gross ist, wird die Gleichung $\Psi_2 = 0$ durch $2 n_1$ Werthe von ω befriedigt, welche in der Nähe der Werthe (16) liegen und bei abnehmendem μ mit diesen Werthen zusammenrücken.

Hiermit ist bewiesen, dass durch geeignete Wahl der Anfangslage alle Ψ zum Verschwinden gebracht werden können, dass periodische Lösungen der beschriebenen Art wirklich existiren. Es soll nun noch einmal zusammengestellt werden, was sich für die Anfangslage ergeben hat.

Willkürlich wählbar ist y_0 und hiermit die Excentricität e_0 der Planetoidenbahn. Die grosse Axe wird um eine Grösse von der Ordnung μ verschieden von dem Ausgangswerth a_0, welcher nur an die Bedingung geknüpft war, dass:

$$n = \frac{1}{a_0{}^3} = \frac{n_1}{n_2}$$

eine rationale Zahl sein musste. Die Länge des Perihels schliesslich unterscheidet sich von einem der Werthe:

$$\omega = \pi \frac{q}{n_1} \qquad q = 0, 1, 2 \cdots 2n_1 - 1$$

nur um eine kleine Grösse, die mit μ zugleich verschwindet. Als Anfangspunkt der Längenzählung ist dabei die Länge des störenden Planeten im Momente eines Periheldurchgangs des Planetoiden gewählt.

Die Lösungen, die zu denselben Werthen a_0 und e_0, aber zu verschiedener Perihellage gehören und daher wesentlich verschiedene Fälle abgeben, wenn man nach den zu einer bestimmten Zeit möglichen Anfangslagen des Systems fragt, sind nicht alle verschieden, wenn man nur die Bahncurven, die von dem System beschrieben werden, in Betracht zieht. Man überzeugt sich leicht, dass die geraden Werthe von q eine erste und die ungeraden eine zweite Bahncurve liefern.

Ist μ so klein, dass man die verschiedenen mit μ zugleich verschwindenden Correctionen der Anfangslage vernachlässigen kann, nicht aber die über lange Zeit sich summirenden Störungen, so kann man die Bedingungen für die Anfangslage so zusammenfassen: Eine periodische Lösung der obigen Art existirt für hinreichend kleines μ immer dann, wenn zu Anfang die grosse Axe a und die Länge des Perihels ω den merkwürdigen Commensurabilitäten:

$$\frac{1}{a^{3/2}} = \frac{p_1}{p_2} \qquad \frac{\omega}{\pi} = \frac{p_3}{p_1} \qquad \begin{array}{l} p_1, p_2, p_3 \text{ ganzzahlig} \\ p_1 \text{ zu } p_2 \text{ relativ prim} \end{array}$$

genüge leisten. Wählt man die p hinreichend gross, so können diese Gleichungen bei gegebenen a und ω stets mit beliebiger Annäherung erfüllt werden.

Wien, v. Kuffner'sche Sternwarte, 1898 Mai 9. *Karl Schwarzschild.*

Beobachtungen von Mira (o) Ceti in den Jahren 1896-98.

Von *W. Stratonoff.*

Unter den zahlreichen Beobachtungen, welche ich im Laufe der letzten drei Jahre über die helleren veränderlichen Sterne nach Argelander's Methode angestellt habe, befindet sich auch eine Anzahl Beobachtungen von Mira Ceti während der beiden letzten Helligkeitsmaxima.

Dieser Stern hat in jüngster Zeit Interesse erregt wegen des Widerspruches des Ganges seiner Lichtcurven mit den Ephemeriden; es ist deshalb vielleicht nicht überflüssig, jetzt schon das Resultat meiner Beobachtungen hier anzuführen, ohne die Veröffentlichung meiner übrigen Beobachtungen der veränderlichen Sterne abzuwarten.

Mit Ausschluss von 1897 Oct. 22 und 25, an welchen

Astron. Nachr. 147 (1898) 289–298

ASTRONOMISCHE NACHRICHTEN.

Band 147. № 3522. 18.

Ueber weitere Classen periodischer Lösungen des Dreikörperproblems.

Von *Karl Schwarzschild.*

1. In meinem vorigen Aufsatze (A. N. 3506) habe ich die Existenz periodischer Lösungen des Dreikörperproblems von folgender Art nachgewiesen: Ein Planetoid von verschwindender Masse und ein Planet von kleiner Masse μ bewegen sich in einer Ebene um einen Centralkörper. Die Bahn des Planeten ist kreisförmig. Die grossen Axen sind so gewählt, dass die Umlaufszeit des Planeten und die des Planetoiden ohne Rücksicht auf Störungen sich verhalten wie zwei ganze Zahlen $n_1 : n_2$. Man geht aus von einem Augenblick, in welchem der Planet die Länge des Perihels des Planetoiden hat, und wartet die Zeit ab, nach welcher der Planet zum n_1^{ten} Male das Perihel des Planetoiden, das sich in Folge der Störungen langsam verschiebt, wieder überholt. Nach dieser Zeit, welche die Periode bestimmt, hat in Folge der Commensurabilität der Umlaufszeiten der Planetoid auch wiederum nahe dieselbe mittlere Anomalie, wie zu Anfang. Wie dort gezeigt worden ist, kann man durch Veränderung der grossen Axe um Grössen von der Ordnung μ bewirken, dass die mittlere Anomalie genau zum Ausgangswerth zurückkehrt, und es verschwinden ausserdem die Störungen der grossen Axe und der Excentricität, über die ganze Periode summirt, falls dieser Ausgangswerth der mittleren Anomalie sich nur um gewisse kleine Grössen von einem der Werthe

$$l = i\frac{\pi}{n_1} \qquad i = 0, 1, 2, \cdots \qquad 2n_1 - 1$$

unterscheidet.

Sobald man annimmt, dass die Bahn des störenden Planeten eine Ellipse von beliebiger Excentricität sei, lässt sich eine periodische Lösung nicht mehr in dieser Weise construiren; denn zu den Zeiten, zu welchen der Planet das sich verschiebende Perihel des Planetoiden überholt, steht er jedesmal in einem andern Theile seiner Bahn, sodass seine Entfernung vom Centralkörper und seine Geschwindigkeit vom Ausgangswerth verschieden ist. Soll in diesem Falle eine periodische Bewegung existiren, so muss ihre Periode ein genaues Vielfaches der Umlaufszeit des Planeten sein, der Planet muss zu Ende der Periode im selben Punkte seiner Bahn stehen wie zu Anfang, und es müssen die vier Elemente des Planetoiden, grosse Axe, Excentricität, mittlere Anomalie und Perihellänge zu ihren Ausgangswerthen zurückkehren. Die Forderung für die Perihellänge zwingt hier von vorneherein zur Annahme einer äusserst langen Periode. Es kann nämlich das Perihel nicht nach einer kurzen Zeit in seine Anfangslage zurückkehren, vielmehr führen es die Störungen entweder mit der Zeit um den ganzen Umkreis herum, oder sie zwingen es zur Oscillation um eine Mittellage, wobei aber die Zeit des Umlaufs resp. der Oscillation für kleine störende Massen jedesmal sehr lang ist. Indem man sich auf die sogenannten Saecularstörungen und auf Glieder zweiter Ordnung in den Excentricitäten beschränkt, bestimmt man nach Laplace und Leverrier, welcher der beiden Fälle eintritt und erkennt, dass die Zeiten, innerhalb deren das Perihel einen vollen Umlauf oder eine vollständige Schwingung ausführt, der störenden Masse μ umgekehrt proportional sind. Um nun hier die Grundlagen für eine periodische Lösung zu gewinnen, wird man zunächst die Periode T_0 des Perihels aus der Theorie der Saecularstörungen genähert zu bestimmen suchen und die Umlaufszeiten T_1 und T_2 des Planetoiden und des Planeten durch geeignete Annahme der grossen Axen so zu wählen haben, dass alle diese drei Perioden in einer noch längeren Periode T enthalten sind. Sei also:

$$T = n_0\, T_0 = n_1\, T_1 = n_2\, T_2 \qquad (1)$$

wo die n ganze Zahlen bedeuten, von denen n_0 verhältnissmässig klein, n_1 und n_2 aber sehr gross sind. Dann würde, wenn man alle Störungen ausser den Saecularstörungen des Perihels vernachlässigt, offenbar das ganze System nach der Periode T in die Anfangslage zurückkehren. Es handelt sich um die Frage, ob und unter welchen Umständen bei Berücksichtigung sämmtlicher Störungen nach der Periode T genaue Rückkehr in die Anfangslage eintreten kann, und es sei mir gestattet, hier kurz den Gang meiner diesbezüglichen Untersuchung anzugeben.

2. Um die gestellte Frage beantworten zu können, bedarf man eines Verfahrens zur Integration des Dreikörperproblems, dessen Convergenzbereich sich über Zeiten von der grossen hier erforderlichen Länge erstreckt. Man erinnere sich an die Methode der Variation der Constanten in der Jacobi'schen Form, wie sie Tisserand in seiner Mécanique céleste darstellt. Es wird dort zunächst die ungestörte Bewegung betrachtet, und es werden die Bewegungsgleichungen eines Planeten in rechtwinkligen Coordinaten auf die canonische Form gebracht. Dann wird die partielle Differentialgleichung für die Hamilton'sche Hauptfunction aufgestellt und ein Integral derselben, welches drei willkürliche Constanten enthält, ermittelt. Die vollständigen Integrale des Bewegungsproblems, in diesem Falle die Gesetze der Kepler'schen Bewegung, lassen sich dann sofort unter Einführung von drei weiteren Constanten anschreiben. Diese sechs Constanten, die »canonischen Elemente«, hängen in einfacher Weise von den üblichen sechs Elementen ab. Um nun die

Störungen zu berücksichtigen, denkt man sich diese Constanten variabel. Ihre Einführung in der Weise, wie sie bei der Jacobi'schen Integrationsmethode erfolgt, hat dabei einen ausserordentlichen Vortheil: Die Gleichungen für die Veränderungen der canonischen Elemente mit der Zeit, die durch die Störungen verursacht werden, lassen sich unmittelbar angeben, sie bilden ein canonisches System, dessen Kräftefunction die Störungsfunction selbst ist.

Man kann — und davon ist schon verschiedentlich Gebrauch gemacht worden — das ganze Verfahren auf einer höheren Stufe wiederholen. Man spalte die Störungsfunction in zwei Theile, von denen der zweite nur Glieder höherer Ordnung enthält. Man bilde das canonische System für die canonischen Elemente, indem man zunächst diesen zweiten Theil der Störungsfunction vernachlässigt, und integrire das System nach der Jacobi'schen Methode. Hierbei erfolgt die Einführung von sechs neuen Constanten, die man als »absolute canonische Elemente« bezeichnen kann. Während nämlich die ursprünglichen Elemente Störungen erster Ordnung unterworfen sind, sind die »absoluten Elemente« in viel höherem Grade constant. Um schliesslich den Rest der Störungsfunction zu berücksichtigen, betrachte man auch die absoluten Elemente als variabel. In Folge ihrer Einführung durch die Jacobi'sche Methode erhält man für ihre Veränderungen mit der Zeit sofort wieder ein canonisches System, dessen Kräftefunction jener Rest der Störungsfunction ist.

Die Spaltung der Störungsfunction, die ich benutzte, war folgende. Die Excentricitäten wurden als Grössen von der Ordnung der vierten Wurzel der störenden Masse betrachtet und es wurden zunächst alle Glieder vernachlässigt, die nach dieser Voraussetzung von der Ordnung der dritten Potenz der störenden Masse wurden, also — da die Störungsfunction an sich den Factor μ enthält — die Glieder von der achten Ordnung in den Excentricitäten. Die Integration des so reducirten Problems erfolgte nach der Jacobi'schen Methode und zwar nach dem Vorbilde, welches Herr Poincaré in Capitel XI seines Werkes: »Les Méthodes nouvelles de la Mécanique céleste« gegeben hat. Dabei war es natürlich erlaubt, die Störungsfunction durch Glieder der Ordnung μ^3 zu ergänzen, die die Integration erleichterten und die dann später von dem Rest der Störungsfunction wieder abgezogen wurden. Um das Auftreten der Zeit ausserhalb der trigonometrischen Zeichen zu vermeiden, musste bei der Integration vorausgesetzt werden, dass innerhalb des ersten Theils der Störungsfunction keine Commensurabilität, keine »kritischen« Glieder, auftreten. Es mussten also die niedersten Commensurabilitäten der Umlaufszeiten von Planet und Planetoid und zwar die Verhältnisse:

$$\frac{T_1}{T_2} = \frac{n_1 \pm \delta}{n_1} \qquad \delta = 0, 1, 2, 3, 4, 5, 6, 7$$

n_1 ganzzahlig

ausgeschlossen werden. Da ferner die Excentricitäten auch im Nenner auftraten, musste vorausgesetzt werden, dass sie weder gar zu gross noch gar zu klein seien. Es genügte, wenn sie, wie das von vorneherein vorausgesetzt war, mit $\mu^{1/4}$ vergleichbar waren. Nach diesen beiden Beschränkungen war die Integration ausführbar und sie führte zur Definition der »absoluten canonischen Elemente«. Der Rest der Störungsfunction, von dem die Aenderungen dieser Elemente abhängig waren, enthielt nunmehr μ^3 als Factor. Daraus war abzusehen, dass die Aenderungen auch für eine Periode, die zu μ umgekehrt proportional war, klein bleiben mussten. Auf das canonische Gleichungssystem, welches für dieselben gilt, kann man sich nun die einfachste Methode, die unmittelbare Entwickelung der Variablen nach Potenzen von μ, angewendet denken, wobei die Zeit bekanntlich aus den trigonometrischen Zeichen heraustritt. Auf die gewöhnlichen Störungsgleichungen angewandt ist dieses Verfahren nur für kurze Zeit convergent, hier aber erscheint die Zeit, wo sie ausserhalb der trigonometrischen Zeichen auftritt, mit einer Potenz von μ multiplicirt, und das bewirkt natürlich eine beträchtliche Vermehrung der Convergenz. Durch Anwendung des »Calcul des limites« liess sich beweisen, dass eine untere Grenze für die Zeit t, innerhalb deren diese Entwickelung nach Potenzen von μ convergirt, gegeben ist durch den Ausdruck:

$$t > \frac{K}{\mu^{5/4}}$$

wo K eine endliche Constante ist. Die Periode eines Perihelumlaufs ist gegeben durch einen Ausdruck der Form:

$$T_0 = \frac{K'}{\mu}$$

wo K' wiederum eine endliche Constante ist, die ganze Periode der gesuchten periodischen Bewegung nach (1) durch:

$$T = \frac{n_0 K'}{\mu}$$

Es folgt:

$$t > T \frac{K}{n_0 K'} \frac{1}{\mu^{1/4}}$$

Das Convergenzbereich des Verfahrens erstreckt sich also über die Periode T hinaus, wenn:

$$\frac{K}{n_0 K'} \frac{1}{\mu^{1/4}} > 1 \qquad \mu < \left(\frac{K}{n_0 K'}\right)^4$$

ist, wenn also nur μ hinreichend klein ist.

3. Auf dieser Grundlage wurde es leicht, Anfangsbedingungen aufzusuchen, von denen aus eine periodische Bewegung von so langer Periode erfolgt.

Um dieselben bezeichnen zu können, muss daran erinnert werden, in welcher Form Laplace die Saecularstörungen der Excentricitäten und Perihele dargestellt hat. Man findet in der Mécanique céleste § 55 die Ausdrücke:

$$e \cos \omega = \sum_i \varrho_i \cos (g_i t + \eta_i)$$

$$e \sin \omega = \sum_i \varrho_i \sin (g_i t + \eta_i)$$

Hierin bedeutet e die Excentricität, ω die Perihellänge des Planetoiden, ϱ_i und η_i sind Constante, deren Werthe aus den anfänglichen Werthen e_0 und ω_0 vermöge der Gleichungen:

$$e_0 \cos \omega_0 = \Sigma \varrho_i \cos \eta_i$$
$$e_0 \sin \omega_0 = \Sigma \varrho_i \sin \eta_i$$

zu bestimmen sind. In unserem speciellen Falle werden diese Formeln, wenn man noch die Länge des Perihels des störenden Planeten zum Nullpunkt der Längenzählung nimmt:

$$\left.\begin{aligned} e_0 \cos \omega_0 &= \varrho \cos \eta + \varrho' \\ e_0 \sin \omega_0 &= \varrho \sin \eta \\ e \cos \omega &= \varrho \cos (gt + \eta) + \varrho' \\ e \sin \omega &= \varrho \sin (gt + \eta) \end{aligned}\right\} \quad (2)$$

Dabei ist ϱ' eine Grösse, die nur von den grossen Axen und der Excentricität e' des störenden Planeten abhängt und von der Grössenordnung der letzteren ist. Man sieht, dass e und ω, wenigstens soweit die Näherung dieser Formeln reicht, nach der Zeit:

$$T_0 = \frac{2\pi}{g}$$

zum Ausgangswerth zurückkehren. Die Grösse g, welche die Periode des Perihels bestimmt, hängt in erster Näherung, wie sie Laplace giebt, vom Verhältniss der grossen Axen ab, auf ihren genaueren Werth sind aber auch ϱ und ϱ' von Einfluss.

Man kann nun solche Werthe der grossen Axe des Planetoiden und der Constante ϱ ausfindig machen, für welche die Periode des Perihels T_0 und die Umlaufszeit T_1 des Planetoiden mit der Umlaufszeit T_2 des Planeten commensurabel werden, sodass also:

$$T = n_0 T_0 = n_1 T_1 = n_2 T_2$$
$$n_0,\ n_1,\ n_2 \text{ ganzzahlig}$$

wird. Dabei soll T_1 aus der sogenannten »wahren« mittleren Bewegung, welche die Saecularstörung der mittleren Länge enthält, berechnet sein und alle merklichen Störungen zweiter Ordnung in den Massen und siebenter Ordnung in den Excentricitäten berücksichtigt sein, so wie es von Leverrier in der Planetentheorie geschehen ist.

Hat man solche Werthe der grossen Axe und der Constante ϱ gefunden, so existiren Lösungen der Periode T für gewisse Anfangswerthe, welche sich von diesen Werthen nur um Grössen der Ordnung μ^2 unterscheiden, und zwar dann, wenn auch die anfängliche mittlere Länge des Planetoiden und die Constante η gewisse besondere Werthe haben. Man wähle als Anfangsmoment die Zeit eines Periheldurchgangs des störenden Planeten. Dann unterscheiden sich die betreffenden Werthe der ursprünglichen mittleren Länge l_0 und der Constanten η nur um kleine, mit abnehmendem μ verschwindende Grössen, von den Werthen eines der Systeme:

$$\left.\begin{aligned} l_0 &= i \cdot \frac{n_1}{n_2} \cdot \pi + \varepsilon_1 \pi \\ \eta &= i \cdot \frac{n_0}{n_2} \cdot \pi + \varepsilon_2 \pi \end{aligned}\right\} \quad (3)$$

$$i \text{ ganzzahlig, } \varepsilon_1 \text{ und } \varepsilon_2 = 0 \text{ oder } 1.$$

Das ist das Resultat der Untersuchung.

Zwei Einschränkungen waren, wie erwähnt, dabei zu machen: es mussten e und e' und hiermit auch ϱ und ϱ' von der Grössenordnung $\mu^{1/4}$ sein und es konnten die niedrigsten Commensurabilitäten, für welche

$$n_1 = n_2 \pm \delta \qquad \delta = 0, 1, 2, 3, 4, 5, 6, 7$$

ist, nicht in die Betrachtung einbezogen werden.

4. Es lässt sich nun sofort die Erweiterung dieser Ueberlegungen auf einen allgemeineren Fall des Dreikörperproblems ersehen. Der Planet bewege sich nach wie vor in einer festen Ellipse, aber die Bahn des Planetoiden liege nicht mehr in derselben Ebene. Das ist ein Fall, wie er im Sonnensystem für die kleinen Planeten und den Erdmond nahezu verwirklicht ist. Es sind dann auch die Veränderungen der Neigung i und des Knotens ☊ zu berücksichtigen. Der Knoten führt in Folge der Saecularstörungen allmählich einen vollen Umlauf aus (der Fall der Oscillation kann hier nicht eintreten). Die Periode dieses Umlaufs sei T_3. Man suche wie oben Werthe des Axenverhältnisses, von ϱ und von i, für welche auch T_3 mit der Periode des Perihels und den Umlaufszeiten beider Körper commensurabel wird. Es sei:

$$T = n_0 T_0 = n_1 T_1 = n_2 T_2 = n_3 T_3$$
$$n_0,\ n_1,\ n_2,\ n_3 \text{ ganzzahlig.}$$

Dann existiren periodische Lösungen für Ausgangswerthe der grossen Axe, von ϱ und von i, welche sich von den erstgefundenen nur um Beträge der Ordnung μ^2 unterscheiden, wiederum unter der Voraussetzung, dass auch die anfängliche mittlere Länge und Knotenlänge und die Constante η gewisse besondere Werthe haben. Letztere Werthe unterscheiden sich nur um kleine, mit abnehmendem μ verschwindende Beträge, von den Werthen:

$$\left.\begin{aligned} l_0 &= i \frac{n_1}{n_2} \pi + \varepsilon_1 \pi \\ \eta &= i \frac{n_0}{n_2} \pi + \varepsilon_2 \pi \\ ☊_0 &= i \frac{n_3}{n_2} \pi + \varepsilon_3 \pi \end{aligned}\right\} \quad (4)$$

$$i \text{ ganzzahlig. } \varepsilon_1,\ \varepsilon_2,\ \varepsilon_3 = 0 \text{ oder } 1.$$

Hier musste zum Beweise angenommen werden, dass ausser e und e' auch i von der Ordnung $\mu^{1/4}$ sei, und es mussten nicht nur, wie oben, die niedrigsten Commensurabilitäten zwischen den Umlaufszeiten beider Körper ausgeschlossen werden, sondern auch die zwischen der Umlaufszeit resp. Oscillation des Perihels und der Umlaufszeit des Knotens, nämlich die Fälle:

$$\frac{n_3}{n_0} = 1/4,\ 1/2,\ 3/4,\ 1,\ 5/4,\ 3/2,\ 2,\ 5/2,\ 3$$

Ueber den Zusammenhang der ganzen Schaar periodischer Lösungen, welche die Formeln (4) liefern, ist eine Bemerkung zu machen, welche in ähnlicher Weise für das System (3) gilt und auch schon bei den in meinem vorigen Aufsatze behandelten periodischen Lösungen angedeutet wurde. Als Anfangspunkt der Zeitrechnung ist der Augenblick eines

Periheldurchgangs des störenden Planeten gewählt worden. Während einer Periode finden aber n_2 solche Periheldurchgänge statt und für jeden Periheldurchgang des störenden Körpers gilt eine andere Lage und Geschwindigkeit des Planetoiden. Indem man also der Reihe nach alle diese n_2 Periheldurchgänge als Anfangsmomente benutzt, erhält man n_2 verschiedene Anfangslagen des Systems, von denen eine periodische Bewegung ausgeht. Natürlich ist es ein und dieselbe Bahncurve, welche von diesen verschiedenen Anfangslagen aus beschrieben wird. Die nähere Discussion der Formeln (4) zeigt, dass die gesammten aus der Willkürlichkeit von i und der ε hervorgehenden Anfangslagen sich auf mindestens vier und höchstens 16 verschiedene Bahncurven vertheilen, deren jede n_2 Anfangslagen enthält. Welches die genaue Anzahl der Bahncurven ist, das hängt von dem Vorkommen des Factors 2 in den Zahlen n_0, n_1, n_2, n_3 ab.

Zur weiteren Verallgemeinerung stände es auch frei, eine beliebige Lage auf einer der Bahncurven des Systems (bei welcher der störende Planet dann nicht mehr im Perihel steht) als Anfangslage zu wählen, und zugleich den Nullpunkt der Längenzählung beliebig anzusetzen.

5. Ein Beispiel wird am besten verdeutlichen, wie gering im Grunde die Specialisirung der Anfangslagen ist, für welche sich streng periodische Lösungen der hier betrachteten Art ergeben. Für den Mond ist die Umlaufszeit

des Perigäums	$T_0 = 3232.57$	Tage
des Mondes	$T_1 = 27.33166$	»
der Sonne	$T_2 = 365.2564$	»
des Knotens	$T_3 = 6793.39$	»

Zwischen diesen Zahlen besteht genähert die Commensurabilität:

$$T = 19\,T_0 = 2246\,T_1 = 168\,T_2 = 9\,T_3$$

Es ist nämlich:

$$19\,T_0 = 61418.8 \text{ Tage}$$
$$2246\,T_1 = 61364.4 \;»$$
$$168\,T_2 = 61363.1 \;»$$
$$9\,T_3 = 61140.5 \;»$$

Durch sehr geringe Aenderung der grossen Axe, Excentricität und Neigung des Mondes (eventuell auch der Sonnenparallaxe) könnte man diese vier Zahlen in genaue Uebereinstimmung mit $168\,T_2$, d. h. mit 168 siderischen Jahren bringen.

Es wird nun für Werthe der grossen Axe, der Excentricität und der Neigung, welche wiederum von diesen Werthen wenig verschieden sind, eine periodische Lösung mit einer Periode von 168 Jahren existiren, wenn die mittlere Anomalie, die Constante η und die Knotenlänge im Augenblick eines Periheldurchgangs der Erde nur um gewisse kleine Grössen verschieden sind von einem der Werthsysteme:

$$\left.\begin{aligned} l_0 &= \frac{2246}{168}\,i\pi + \varepsilon_1\pi = (13\,i + \varepsilon_1)\,\pi + \frac{31}{84}\,i\pi \\ \eta &= \frac{19}{168}\,i\pi + \varepsilon_2\pi \\ \Omega_0 &= -\frac{9}{168}\,i\pi + \varepsilon_3\pi = -\frac{3}{56}\,i\pi + \varepsilon_3\pi \end{aligned}\right\} \quad (5)$$

i ganzzahlig $\varepsilon_1, \varepsilon_2, \varepsilon_3 = 0$ oder 1

Dabei ist die Länge der Sonne im Perigäum als Nullpunkt der Längenzählung gewählt.

Bei dem Knoten erscheint das negative Zeichen, weil seine Bewegung retrograde ist.

Jede Knotenlänge:

$$\Omega_0 = \frac{k}{56}\,\pi\,,$$

wo k ganzzahlig ist, gehört zu einem dieser Systeme. Bedenkt man nämlich, dass man jedem Winkel beliebig oft 2π hinzufügen darf, so erfordert das nur die Erfüllbarkeit der Gleichung:

$$\Omega_0 = \frac{k}{56}\,\pi = \pi\,(2\,i_1 + \varepsilon_3) - \frac{3}{56}\,i\pi$$

durch ganzzahliges i_1 und i. Es folgt aber:

$$k = 56\,(2\,i_1 + \varepsilon_3) - 3\,i$$

oder, indem man: $2\,i_1 + \varepsilon_3 = i_2$

setzt: $k = 56\,i_2 - 3\,i$

und diese diophantische Gleichung hat stets gewisse Lösungen:

$$i = a + 56\,k' \qquad i_2 = b + 3\,k'$$

wo a und b zwei bestimmte ganze Zahlen und k' eine willkürliche ganze Zahl ist. Da:

$$i_1 = \frac{i_2 - \varepsilon_3}{2}$$

ganzzahlig sein muss, so hat man $\varepsilon_3 = 0$ oder 1 zu wählen, je nachdem i_2 grade oder ungrade wird. Man sieht also, dass wirklich alle Werthe:

$$\Omega = \frac{k}{56}\,\pi$$

dem System angehören, und überzeugt sich leicht, dass damit auch alle möglichen Werthe der Knotenlänge erschöpft sind. Hat man einen dieser 112 verschiedenen Werthe gewählt, so ist k festgelegt, aber k', ε_1 und ε_2 sind noch willkürlich. Es folgt:

$$\begin{aligned} l_0 &= (13\,i + \varepsilon_1)\,\pi + \frac{31}{84}\,\pi\,(a + 56\,k') \\ &= (13\,i + 21 + \varepsilon_1)\,\pi + \frac{31}{84}\,a\pi - k'\frac{\pi}{3} \\ \eta &= \varepsilon_2\pi + \frac{19}{168}\,\pi\,(a + 56\,k') \\ &= (6 + \varepsilon_2)\,\pi + \frac{19}{168}\,a\pi + k'\frac{\pi}{3} \end{aligned}$$

Die Willkürlichkeit von k' und ε_2 liefert sechs verschiedene Werthe für η und mithin für die Perihellänge. Wählt man einen von diesen, so ist damit auch k' festgelegt, und die Willkürlichkeit von ε_1 liefert nur mehr zwei verschiedene Werthe für l_0.

Das System (4) enthält also im Ganzen $112 \cdot 6 \cdot 2 = 8 \cdot 168 = 8\,n_2$ verschiedene Anfangslagen. Da, wie oben erläutert, immer n_2 Anfangslagen auf einer Bahncurve zu liegen kommen, so hat man es hier mit acht verschiedenen Bahncurven zu thun.

Unter allen diesen Anfangslagen würden sich solche aussuchen lassen, die der Stellung des Mondes in einem gegebenen Augenblick ziemlich nahe kämen, besonders wenn man sich auch noch eine kleine Verschiebung des Perigäums der Sonne gestatten würde, und man fände so eine streng periodische Bewegung, die sich für geraume Zeit nicht weit von der wirklichen Mondbewegung entfernte.*)

Allerdings ist zu bemerken, dass die Existenz einer solchen Bewegung zwar wahrscheinlich genug, aber doch nicht völlig sicher gestellt ist. Denn der Existenzbeweis erfordert, dass μ »hinreichend« klein ist, ohne dass diese Grenze bis jetzt numerisch angegeben werden könnte, und so ist nicht bewiesen, dass für den Mond, wo übrigens μ gleich der Sonnenmasse dividirt durch den Cubus der Sonnenentfernung zu setzen ist, diese Grösse unter der erforderlichen Grenze liegt. Uebrigens hat man sich bei allen bisherigen praktischen Anwendungen der Hill'schen und Poincaré'schen periodischen Lösungen in derselben Lage befunden.

6. Die begrenzte Convergenz des Beweisverfahrens widersetzt sich auch einem anderen Schluss, den man aus den Gleichungen (4) zu ziehen geneigt sein könnte. Man bemerkt unmittelbar, dass man um so mehr verschiedene Anfangslagen zur Verfügung hat, je höhere Commensurabilitäten man wählt, je grösser die Zahlen n_0, n_1, n_2, n_3 sind, und dass man jeder beliebigen Anfangslage, sowohl was die Werthe der grossen Axen, der Excentricitäten und der Neigung, als was die Winkel l, ω, Ω angeht, beliebig nahe kommen könnte, wenn man nur die n hinreichend vergrösserte. Wäre das nun wirklich erlaubt, so wäre damit ein Satz bewiesen, von dem auch Herr Poincaré sagt, dass er ihn für wahrscheinlich halte, ohne ihn streng begründen zu können: dass nämlich beliebig nahe bei jedem Anfangszustande ein anderer gefunden werden könne, der zu einer periodischen Bewegung von freilich im Allgemeinen sehr langer Periode führte.†) Was aber in aller Strenge behauptet werden kann, ist nur Folgendes: Für ein gegebenes hinreichend kleines μ erstreckt sich das Convergenzbereich des Beweisverfahrens über einen oder mehrere Umläufe des Perihels und des Knotens hinaus. Es kann daher die Existenz einer begrenzten Anzahl periodischer Lösungen behauptet werden, bei denen die Umlaufszeiten des Perihels und des Knotens in einer niederen Commensurabilität stehen, so dass die ganze Periode nur wenige ihrer Umläufe enthält. Nimmt nun μ ab, so wächst das Convergenzbereich wie $\mu^{-5/4}$, die Umlaufszeit von Perihel und Knoten aber nur wie μ^{-1}. Je kleiner also μ wird, um so mehr Umläufe von Perihel und Knoten wird das Convergenzbereich umfassen, um so höhere Commensurabilitäten sind also zulässig, um so zahlreichere periodische Lösungen müssen existiren und um so dichter sind die Anfangslagen gesät, die zu periodischen Bewegungen führen. Dieses Resultat ist wenigstens eine Vorstufe zu dem obigen Satze.

Auf jeden Fall sind auch die periodischen Bewegungen langer Periode, deren Existenz streng bewiesen ist, so zahlreich und die Anfangslagen, die dabei vorkommen, so wenig speciell, dass man die periodischen Lösungen nicht mehr als Ausnahmsfälle, als Curiositäten, betrachten kann.

Wien, v. Kuffner'sche Sternwarte, 1898 Juli 5. *K. Schwarzschild.*

*) Herr Perchot hat in seiner These: Sur les mouvements des Neuds et du Périgée de la Lune. Paris 1892 (Referat im Bulletin Astronomique, Bd. 11, pag. 21) auf Grund der oben erwähnten und von dort entnommenen Commensurabilitäten eine Bewegung von 168jähriger Periode für den Mond construirt, es hat sich dabei aber nicht um eine strenge Lösung gehandelt, weil a priori Unveränderlichkeit der grossen Axe vorausgesetzt wurde.

†) Méthodes nouvelles de la Mécanique Céleste, I, pag. 82.

Württembergische und Badische Coordinaten.

Von Professor *Jordan.*

In diesem 147. Bande der Astronomischen Nachrichten S. 125–142 ist eine längere Abhandlung über Coordinatenberechnung von Herrn Prof. Hammer veröffentlicht worden, in welcher meine Arbeiten auf diesem Gebiete mehrfach citirt sind, weshalb ich mir erlaube, hier ebenfalls meine Ansicht hierüber auszusprechen, zumal es sich um Anwendung auf älteres württembergisches und badisches Material handelt, das wohl kaum einem Anderen ebenso bekannt und geläufig sein dürfte, wie dem Schreiber dieses, vermöge seiner langjährigen Beschäftigung damit in Stuttgart und Karlsruhe bis 1881.

Die Hauptresultate der Landesvermessungen von Württemberg und Baden sind niedergelegt in rechtwinkligen sogenannten Soldner'schen Coordinaten, deren Theorie als bekannt vorauszusetzen ist.

Die württembergischen Coordinaten sind veröffentlicht in dem amtlichen Werke »Die Landesvermessung des Königreichs Württemberg etc. von Kohler, Stuttgart 1858«, aus welchem wir von S. 171–283 folgende 6 Punkte entnehmen:

Württembergische Coordinaten.

	x (württ. Fuss)	y (württ. Fuss)	
1) Mannheim Sternwarte	+375948.62	−149528.71	(1)
2) Katzenbuckel	+369283.70	− 2320.63	
3) Durlach Warte	+185777.58	−144422.26	
4) Hornisgrinde	+ 34789.55	−218515.23	
5) Tübingen Sternwarte	0.00	0.00	
6) Hohentwiel	−293058.26	− 60676.63	

Diese Coordinaten x, y sind mit $+x$ nach Norden und $+y$ nach Osten vom Nullpunkt Tübingen gezählt, und zwar gemessen in württembergischen Fuss, indem 1 Fuss = 126.97 Pariser Linien = 0.2864226161 Meter (log = 9.4570073·071). Die Vermessung ist nicht wie sonst im Meereshorizont genommen, sondern in der Höhe h = 840 Pariser Fuss = 272.865 Meter über dem Meere, wofür $\log\left(1+\frac{h}{r}\right)$ = 0.0000185·744, was zu dem obigen Verwandlungslogarithmus addirt giebt log = 9.4569887·327.

Astron. Nachr. *160* (1903) 385–400

ASTRONOMISCHE NACHRICHTEN.

Band 160. № 3839. 23.

Über die periodischen Bahnen vom Hecubatypus.

Von *K. Schwarzschild.*

1. Wenn man die Bewegung eines kleinen Planeten in der Nähe einer Kommensurabilitätsstelle untersucht, wird man fast mit Notwendigkeit dazu geführt, die Rechnung anzuschließen an die Betrachtung gewisser geschlossener Bahnen (»periodischer Lösungen«), die sich in der Nachbarschaft der wirklichen Bahn des Planeten vorfinden. Herr Simonin [1]) hat, von vornherein ausgehend von einer solchen periodischen Lösung, die Bewegung der Hecuba studiert. Herr Brendel ist in seiner »Theorie der kleinen Planeten« ebenfalls dazu gekommen, die Berechnung einer periodischen Lösung bei der Behandlung »charakteristischer« Planeten an die Spitze zu stellen [2]). Da sich also hier mit dem ursprünglichen theoretischen nun auch ein praktisches Interesse vereinigt, erscheint es wohl angemessen, diese Bewegungsformen, so einfach sie sind, einmal zu isolieren und für sich numerisch zu behandeln. Von diesem Gesichtspunkte ausgehend habe ich — mit wesentlicher Unterstützung des Rechners der Göttinger Sternwarte Herrn C. Jastram — numerische Rechnungen speziell über die periodischen Lösungen in der Nachbarschaft der Hecubabahn, derjenigen Stelle des Asteroidenrings, welche der doppelten mittleren Bewegung des Jupiter entspricht, unternommen und im Anschlusse daran auch die Stabilitätseigenschaften der berechneten Bahnen festgestellt. Die Resultate werden im folgenden mitgeteilt.

2. Die nicht zu umgehende Ableitung der Formeln für die numerische Rechnung möchte ich dadurch instruktiver machen, daß ich einen Weg gehe, der sich tunlichst eng an die alten klassischen Methoden der Himmelsmechanik anschließt.

Denkt man sich einen Planetoiden der Masse Null in der Bahnebene des Jupiter laufend und vernachlässigt die Excentricität der Jupiterbahn, so lassen sich die Bewegungsgleichungen in der Form schreiben [3]):

$$\frac{dL}{dt} = \frac{\partial F}{\partial \lambda} \qquad \frac{d\lambda}{dt} = -\frac{\partial F}{\partial L}$$

$$\frac{dG}{dt} = \frac{\partial F}{\partial \gamma} \qquad \frac{d\gamma}{dt} = -\frac{\partial F}{\partial G}$$

wobei:

$$L = k\sqrt{a} \qquad G = k\sqrt{a}\,(1 - \sqrt{1 - e^2})$$

$$\lambda = \Lambda - \Lambda' \qquad \gamma = \Lambda' - \varpi$$

$$F = \frac{k^4}{2L^2} + n'L - n'G + F'$$

$$F' = k^2 m' \left(\frac{1}{\sqrt{r^2 + a'^2 - 2a'r\cos(v - \Lambda')}} - \frac{r}{a'^2}\cos(v - \Lambda')\right)$$

ist, und die Zeichen folgende Bedeutung haben:

- k Gaußische Konstante
- a Große Achse des Planetoiden
- e Excentricität » »
- ϖ Perihellänge » »
- Λ Mittlere Länge » »
- v Wahre Länge » »
- Λ' Länge des Jupiter
- n' Mittlere Bewegung des Jupiter
- a' Große Achse des Jupiter

Wählt man noch als Ausgangspunkt der Längenzählung die Länge des Jupiter zur Zeit $t = 0$, so gilt:

$$\Lambda' = n' \cdot t\,.$$

Es ist:

$$n' = \frac{k(1 + m')}{a'^{3/2}}$$

und es werde entsprechend definiert:

$$n = \frac{k}{a^{3/2}}\,.$$

Der Ausdruck von F' in eine trigonometrische Reihe entwickelt, laute:

$$F' = k^2 m' \sum_{i,i'} C_{i,i'} \cos(i\lambda + i'\gamma)\,.$$

Ich will nun in der üblichen Weise Störungen erster Ordnung ermitteln, indem ich in F' Werte der Variabeln einsetze, welche ihrerseits bis auf Größen von der Ordnung

[1]) Annales de l'Observatoire de Nice. Tome VI.
[2]) Theorie der kleinen Planeten. I. Teil pag. 128-129.
[3]) Vgl. Tisserand, Méc. céleste. IV. pag. 426.

der störenden Masse richtig sind, und zwar setze ich L und G gleich irgend welchen Konstanten $[L]$ und $[G]$, welche bis auf Größen dieser Ordnung mit den oskulierenden Anfangswerten von L und G übereinstimmen. Was λ und γ angeht, so hätte man ohne Störungen:

$$\lambda = (n - n')\,t + \lambda_0 \qquad \gamma = n'\,t - \varpi\,.$$

Ich will in F' die Werte verwenden:

$$\lambda = N t + \lambda_0 \qquad \gamma = N'\,t + \gamma_0\,,$$

wobei N und N' in der Nähe der Werte von $n - n'$ und n' liegen, welche der ungestörten Bewegung entsprechen und λ_0 und γ_0 in der Nachbarschaft der wirklichen Anfangswerte von λ und γ. Ein Unterschied zwischen N' und n' ist insbesondere deshalb nicht abzuweisen, weil das Perihel ϖ durch die Störungen eine starke Bewegung erfahren kann.

Nachdem ich mir so in dem Ansatze des Integrationsverfahrens eine gewisse Freiheit gewahrt habe, die berechtigt ist, solange man es nur auf Störungen erster Ordnung und auf eine Darstellung der Bewegung für relativ kurze Zeiten abgesehen hat, benutze ich diese Freiheit nicht, um Kommensurabilitäten der mittleren Bewegungen aus dem Wege zu gehen, sondern ich führe umgekehrt Kommensurabilitäten (in dem genäherten Ausdrucke der Störungsfunktion) herbei. Für Planeten vom Hecubatypus gilt nahe:

$$n = 2n'\,.$$

Indem ich mich auf Planeten in der Nachbarschaft dieser Kommensurabilität beschränke, werde ich setzen:

$$N = N'\,.$$

Hiermit nimmt dann die Störungsfunktion die Form an:

$$F' = k^2 m' \,\Sigma\, C_{i,\,i'} \cos\left[N'\,t\,(i + i') + i\,\lambda_0 + i'\,\gamma_0\right].$$

Ich will die »kritischen« Glieder, für welche

$$i + i' = 0$$

ist, von den übrigen Gliedern trennen und die Bezeichnungen einführen:

$$C_{i,-i} = D_i \qquad \lambda_0 - \gamma_0 = \zeta_0\,.$$

Die Störungsfunktion zerfällt damit in einen konstanten Teil:

$$R = k^2 m' \,\Sigma\, D_i \cos i\,\zeta_0$$

und den Rest, den ich durch eine accentuierte Summe kennzeichnen will:

$$T = k^2 m' \,\Sigma'\, C_{i,\,i'} \cos\left[N'\,t\,(i + i') + i\,\lambda_0 + i'\,\gamma_0\right]$$

$$F' = R + T\,.$$

Beachtet man nun, daß $\dfrac{\partial F}{\partial \lambda} = \dfrac{\partial F}{\partial \lambda_0}\,,\ \dfrac{\partial F}{\partial \gamma} = \dfrac{\partial F}{\partial \gamma_0}$ ist, so lauten die Störungsgleichungen:

$$\frac{dL}{dt} = -\,k^2 m' \,\Sigma\, i\,D_i \sin i\,\zeta_0 - k^2 m' \,\Sigma'\, i\,C_{i,\,i'} \sin\left[N'\,t\,(i + i') + i\,\lambda_0 + i'\,\gamma_0\right]$$

$$\frac{dG}{dt} = +\,k^2 m' \,\Sigma\, i\,D_i \sin i\,\zeta_0 - k^2 m' \,\Sigma'\, i'\,C_{i,\,i'} \sin\left[N'\,t\,(i + i') + i\,\lambda_0 + i'\,\gamma_0\right]$$

$$\frac{d\lambda}{dt} = \frac{k^4}{L^3} - n' - k^2 m' \,\Sigma \frac{\partial D_i}{\partial L} \cos i\,\zeta_0 - k^2 m' \,\Sigma' \frac{\partial C_{i,\,i'}}{\partial L} \cos\left[N'\,t\,(i + i') + i\,\lambda_0 + i'\,\gamma_0\right]$$

$$\frac{d\gamma}{di} = n' - k^2 m' \,\Sigma \frac{\partial D_i}{\partial G} \cos i\,\zeta_0 - k^2 m' \,\Sigma \frac{\partial C_{i,\,i'}}{\partial G} \cos\left[N'\,t\,(i + i') + i\,\lambda_0 + i'\,\gamma_0\right]$$

Die erste, zweite und vierte dieser Gleichungen lassen sich unmittelbar integrieren, die dritte, nachdem man das aus der ersten folgende Resultat für L eingesetzt hat. Indem man immer Größen, die mit höheren Potenzen von m' multipliziert sind, vernachlässigt, erhält man das folgende Resultat der Integration:

$$L = L_0 - t\,k^2 m' \,\Sigma\, i\,D_i \sin i\,\zeta_0 + \frac{k^2 m'}{N'} \,\Sigma' \frac{i\,C_{i,\,i'}}{i + i'} \cos\left[N'\,t\,(i + i') + i\,\lambda_0 + i'\,\gamma_0\right]$$

$$G = G_0 + t\,k^2 m' \,\Sigma\, i\,D_i \sin i\,\zeta_0 + \frac{k^2 m'}{N'} \,\Sigma' \frac{i'\,C_{i,\,i'}}{i + i'} \cos\left[N'\,t\,(i + i') + i\,\lambda_0 + i'\,\gamma_0\right]$$

$$\begin{aligned}\lambda = \lambda_0 &+ t\left(\frac{k^4}{L_0^3} - n' - k^2 m' \,\Sigma \frac{\partial D_i}{\partial L} \cos i\,\zeta_0\right) + \frac{3}{2}\,t^2\,\frac{k^6 m'}{L_0^4}\,\Sigma\, i\,D_i \sin i\,\zeta_0 \\ &- \frac{k^2 m'}{N'} \,\Sigma' \left(\frac{3\,k^4}{N'\,L_0^4}\,\frac{C_{i,\,i'}}{(i + i')^2} + \frac{1}{i + i'}\,\frac{\partial C_{i,\,i'}}{\partial L}\right) \sin\left[N'\,t\,(i + i') + i\,\lambda_0 + i'\,\gamma_0\right]\end{aligned} \qquad \text{(A)}$$

$$\gamma = \gamma_0 + t\left(n' - k^2 m' \,\Sigma \frac{\partial D_i}{\partial G} \cos i\,\zeta_0\right) - \frac{k^2 m'}{N'} \,\Sigma' \frac{1}{i + i'}\,\frac{\partial C_{i,\,i'}}{\partial G} \sin\left[N'\,t\,(i + i') + i\,\lambda_0 + i'\,\gamma_0\right].$$

Hierbei sind L_0, G_0, λ_0, γ_0 Integrationskonstante. Soweit λ_0 und γ_0 in den Störungsgliedern auftraten, waren sie bisher noch mit einer gewissen Willkürlichkeit behaftet, aber ich habe diese Größen in den vorstehenden Formeln bereits stillschweigend fixiert, indem ich sie mit den Integrationskonstanten λ_0 und γ_0 identifiziert habe. Ebenso sind aber auch noch die Größen $[L]$, $[G]$ und N', die zur Berechnung der Störungsfunktion verwandt wurden, bis auf kleine Größen willkürlich. Da soll nun $[G]$ wieder mit der Integrationskonstante G_0 identifiziert werden, für $[L]$ will ich den Wert setzen, der einer genauen Kommensurabilität entspricht:

$$[L] = k\sqrt{a'}\,\sqrt[3]{1/2}$$

und schließlich will ich N' mit der mittleren Bewegung von γ genau übereinstimmend machen:

$$N' = n' - k^2 m' \Sigma \frac{\partial D_i}{\partial G} \cos i\zeta_0 .$$

Erinnert man sich, daß $\zeta_0 = \lambda_0 - \gamma_0$ war, so erkennt man hierin ein abgeschlossenes Formelsystem mit den vier willkürlichen Integrationskonstanten L_0, G_0, λ_0, γ_0, welches durchaus geeignet ist, die Bewegung eines beliebigen Planeten in der Nachbarschaft einer Kommensurabilitätsstelle (natürlich unter Vernachlässigung der Neigung und der Excentricität der Jupitersbahn) für einige Jahre hinaus darzustellen.

Übrigens kann man noch eine kleine Vereinfachung erreichen, indem man über den bisher willkürlichen Anfangspunkt der Zeitrechnung so verfügt, daß $\gamma_0 = 0$ wird. Ich werde von jetzt an voraussetzen, daß dies geschehen ist, wonach also gilt:

$$\gamma_0 = 0 \qquad \zeta_0 = \lambda_0 .$$

3. Aus allen den Bewegungsformen, welche die eben aufgestellten Ausdrücke enthalten, sollen nun besonders einfache spezielle Fälle ausgesucht werden. In erster Linie empfehlen sich jedenfalls solche, bei welchen die der Zeit proportionalen »säkularen« Terme in L und G und der quadratische Term in λ verschwinden, für welche also

$$\Sigma\, i\, D_i \sin i\zeta_0 = 0$$

ist. Diese Bedingung ist aber erfüllt, sobald

$$\zeta_0 = \lambda_0 = 0 \text{ oder } \pi$$

ist. Es läßt sich leicht einsehen, was das geometrisch bedeutet. Wenn ζ_0 gleich null oder π ist, gehen nach den Formeln (A) λ_0 und γ_0 direkt in die Anfangswerte von λ und γ über. Unsere Festsetzung, die Zeit von einem solchen Anfangspunkte an zu zählen, daß γ_0 verschwindet, heißt nunmehr, diesen Anfangspunkt so zu wählen, daß $\gamma = \Lambda' - \varpi$ selbst null wird, mit anderen Worten, daß Jupiter in Konjunktion mit dem Perihel des Asteroiden steht. Soll nun dazu noch $\lambda_0 = 0$ oder π sein, so muß in diesem Anfangsmomente der Asteroid selbst in Konjunktion oder Opposition stehen. Es muß also sogenannte »symmetrische« Konjunktion oder Opposition stattfinden, bei welcher die Geschwindigkeit des Asteroiden senkrecht zur Verbindungslinie Sonne–Jupiter steht. Wir können das in den Satz zusammenfassen:

»Die säkularen Terme in L und G verschwinden, sobald man von einer symmetrischen Konjunktion oder Opposition ausgeht.«

Die übrigbleibenden Teile der Änderungen der Elemente sind von folgender Form:

$$\begin{aligned}
L &= L_0 + && \text{rein periodische Funktion der Zeit von der Periode } \frac{2\pi}{N'} \\
G &= G_0 + && \text{» » » » » » » » »} \\
\lambda &= \lambda_0 + Pt + && \text{» » » » » » » » »} \qquad \lambda_0 = 0 \text{ oder } \pi . \\
\gamma &= N't + && \text{» » » » » » » » »}
\end{aligned}$$

Hier ist die Abkürzung

$$P = \frac{k^4}{L_0^3} - n' - k^2 m' \Sigma \frac{\partial D_i}{\partial L} \cos i\zeta_0$$

eingeführt und unter einer »rein« periodischen Funktion eine ohne konstantes Glied verstanden. Man sieht, daß in den solchermaßen spezialisierten Fällen L, G und γ nach der Periode $\frac{2\pi}{N'}$ wieder zu ihren Ausgangswerten zurückkehren, daß, wenn Jupiter wieder mit dem Perihel des Planeten in Konjunktion tritt, auch die große Achse und Excentricität des letzteren wieder dieselben Werte annehmen. Nur λ kehrt nach dieser Periode im allgemeinen nicht zu seinem Ausgangswerte zurück, die Konjunktionen und Oppositionen des Planeten selbst mit Jupiter bleiben späterhin nicht symmetrisch.

Nun wollen wir noch weiter spezialisieren, indem wir die Anfangslage so wählen, daß

$$P = N' = n' - k^2 m' \Sigma \frac{\partial D_i}{\partial G} \cos i\zeta_0 = \frac{k^4}{L_0^3} - n' - k^2 m' \Sigma \frac{\partial D_i}{\partial L} \cos i\zeta_0$$

(dabei $\zeta_0 = 0$ oder π)

wird. Dies bedeutet, daß, nachdem ζ_0 festgelegt ist, nun auch noch eine ganz bestimmte Beziehung zwischen L und G bestehen soll, so daß überhaupt nur noch eine willkürlich wählbare Größe — etwa G — übrig bleibt.

Mit dieser Einschränkung sind wir am Ziele: alle vier Variable L, G, λ, γ werden periodische Funktionen der Zeit von derselben Periode; so oft Jupiter das Perihel des Planeten überholt, steht der Planet selbst im Perihel oder Aphel, es wiederholt sich stets nach der Zeit $\frac{2\pi}{N'}$ dieselbe relative Stellung symmetrischer Opposition oder Konjunktion zwischen Sonne, Jupiter und Planeten, wir haben es mit »periodischen Lösungen« zu tun.

Um den Charakter dieser periodischen Bahnen deutlicher zu erkennen, wollen wir von den kanonischen zu den gewöhnlichen Elementen übergehen. Es folgt dann leicht bis auf Größen von der Ordnung m'^2 genau, indem man

Astron. Nachr. *160* (1903) 385–400

$$L_0 = k\sqrt{a_0} \qquad G_0 = k\sqrt{a_0}\,(1 - \sqrt{1 - e_0^2}) \qquad n_0 = \frac{k^4}{L_0^3} = \frac{k}{a_0^{3/2}}$$

setzt:

$$\begin{aligned} a &= a_0 && + \text{ rein periodische Funktion von } t \text{ der Periode } \frac{2\pi}{N'} \text{ von der Ordnung } m' \\ e &= e_0 && + \text{ » » » » » » » » » » » } \\ \Lambda &= (n_0 + g)\,t && + \text{ » » » » » » » » » » » } \\ \varpi &= f \cdot t && + \text{ » » » » » » » » » » » } \end{aligned} \qquad \text{(B)}$$

wobei:

$$g = -k^2 m' \Sigma \frac{\partial D_i}{\partial L} \cos i\zeta_0 = -\frac{\partial R}{\partial L}, \quad f = k^2 m' \Sigma \frac{\partial D_i}{\partial G} \cos i\zeta_0 = \frac{\partial R}{\partial G}, \quad \zeta_0 = 0 \text{ oder } \pi \qquad \text{(C)}$$

ist. Demnach ist die periodische Bahn bis auf Glieder von der Ordnung der störenden Masse eine Ellipse, bei welcher allerdings die mittlere Bewegung in Länge um den stets kleinen Betrag g von der aus der großen Achse a_0 nach Keplers Gesetz folgenden mittleren Bewegung n_0 verschieden ist, und welche sich außerdem mit der Apsidengeschwindigkeit f in ihrer Ebene dreht. Die Konstanten a_0 und e_0 bedeuten dabei nicht irgend welchen oskulierenden Wert, sondern die zeitlichen Mittelwerte dieser Elemente; zwischen e_0 und a_0 besteht die Relation:

$$0 = n_0 - 2n' + f + g \,. \qquad \text{(D)}$$

4. Die periodischen Bahnen zerfallen nach Herrn Poincaré in die sog. stabilen, welche Librationscentren sind, und die sog. instabilen, an welche sich asymptotische Lösungen anschließen. Setzen wir diese Tatsache als bekannt voraus, so sind wir in der Lage, über die Zugehörigkeit einer jeden unserer periodischen Lösungen zur einen oder anderen Klasse, über ihren Stabilitätscharakter, zu entscheiden. Denn da wir unter (A) Formeln aufgestellt haben, welche die Bewegung von jedem beliebigen Anfangszustande in der Nachbarschaft der Kommensurabilitätsstelle aus für einige Zeit wiedergeben, so können wir natürlich auch aus denselben ableiten, wie sich die Bewegungen speziell in der Nähe der eben gefundenen geschlossenen Bahnen anlassen, und daraus muß sich der Stabilitätscharakter dieser Bahnen erkennen lassen.

Für die Stabilitätsfrage sind offenbar die rein periodischen Teile der Störungen gleichgültig und wir können daher die Gleichungen (A), indem wir nur die säkularen Terme mitnehmen, einfacher schreiben:

$$\begin{aligned} L &= L_0 - t \cdot k^2 m' \Sigma i D_i \sin i\zeta_0 \\ G &= G_0 + t \cdot k^2 m' \Sigma i D_i \sin i\zeta_0 \\ \lambda &= \lambda_0 + t\left(\frac{k^4}{L_0^3} - n' - k^2 m' \Sigma \frac{\partial D_i}{\partial L} \cos i\zeta_0\right) + \frac{3}{2} t^2 \frac{k^6 m'}{L_0^4} \Sigma i D_i \sin i\zeta_0 \\ \gamma &= \gamma_0 + t\left(n' - k^2 m' \Sigma \frac{\partial D_i}{\partial G} \cos i\zeta_0\right) . \end{aligned}$$

Es folgt unter diesen Umständen:

$$L + G = L_0 + G_0 = \text{konst.}$$

$$\lambda - \gamma = \zeta = \zeta_0 + t\left[\frac{k^4}{L_0^3} - 2n' - k^2 m' \Sigma\left(\frac{\partial D_i}{\partial L} - \frac{\partial D_i}{\partial G}\right) \cos i\zeta_0 + \frac{3}{2} \frac{k^4}{L_0^4} (L_0 - L)\right] .$$

Die obigen periodischen Lösungen werden, wenn man wiederum nur säkulare Terme berücksichtigt, gegeben durch:

$$\zeta = \zeta_0 = \text{konst.} = 0 \text{ oder } \pi$$
$$L = L_0 = \text{konst.} \qquad G = G_0 = \text{konst.}$$

Dabei sind L_0 und G_0 durch die Bedingung (D) verknüpft:

$$0 = \frac{k^4}{L_0^3} - 2n' - k^2 m' \Sigma\left(\frac{\partial D_i}{\partial L} - \frac{\partial D_i}{\partial G}\right) \cos i\zeta_0 \,.$$

Die Anfangswerte für irgend eine Bewegung in der Nachbarschaft einer unserer periodischen Lösungen sollen durch

$$L_0 + \partial L_0, \quad G_0 + \partial G_0, \quad \zeta_0 + \partial\zeta_0$$

bezeichnet werden, und spätere Werte für dieselbe Bewegung durch

$$L_0 + \partial L, \quad G_0 + \partial G, \quad \zeta_0 + \partial\zeta \,.$$

Ich will nun die ganzen Bewegungen ordnen nach den Anfangswerten der Größe $L + G$ und immer diejenige periodische Lösung und alle diejenigen ihrer Nachbarbahnen, welche ein und demselben Werte von $L + G$ entsprechen, gemeinsam betrachten. Für diese ganze Gruppe von Bewegungen gilt also dann:

$$\partial L + \partial G = \partial L_0 + \partial G_0 = 0 \,.$$

Damit folgt aus den obigen Gleichungen unter Vernachlässigung höherer Potenzen von $\partial\zeta_0$, $\partial\zeta$, ∂L_0, ∂L:

$$\partial L = \partial L_0 + A\,t\,\partial\zeta_0$$
$$\partial\zeta = \partial\zeta_0 - t\,(C(\partial L_0 + \partial L) + B\,\partial L)$$

wobei zur Abkürzung:

$$A = \frac{\partial^2 R}{\partial \zeta^2} = -k^2 m' \Sigma i^2 D_i \cos i\zeta_0$$

$$B = \frac{\partial^2 R}{\partial G^2} - \frac{\partial^2 R}{\partial L \partial G} = k^2 m' \Sigma \left(\frac{\partial^2 D_i}{\partial G^2} - \frac{\partial^2 D_i}{\partial L \partial G} \right) \cos i\zeta_0 \qquad C = \frac{3}{2} \frac{k^4}{L_0^4}$$

gesetzt ist. Eine einfache Kombination der beiden vorstehenden Gleichungen ergibt:

$$A \partial\zeta^2 + (2C + B) \partial L^2 = A \partial\zeta_0^2 + (2C + B) \partial L_0^2 + A t^2 ([C(\partial L_0 + \partial L) + B \partial L]^2 - AB \partial\zeta_0^2) .$$

Beschränkt man sich auf die Verfolgung der Bewegung für kleine Zeiten t, so darf man das mit t^2 multiplizierte Glied gegen die übrigen vernachlässigen und behält einfach:

$$A \partial\zeta^2 + (2C + B) \partial L^2 = \text{konst.}$$

Hier liegt aber das Stabilitätskriterium offen zu Tage. Man denke sich ∂L und $\partial \zeta$ als rechtwinklige Koordinaten aufgetragen, wobei die periodische Ausgangslösung durch den Nullpunkt $\partial L = \partial L_0 = \partial\zeta = \partial\zeta_0 = 0$ repräsentiert wird. Haben nun die Größen A und $2C + B$ gleiches Vorzeichen, so erscheinen die Nachbarbahnen als kleine Ellipsen, welche die periodische Lösung umschließen. Das bedeutet, daß die periodische Lösung Librationscentrum ist und daß ihr nach Herrn Poincaré die Benennung »stabil« zukommt. Haben andererseits die Größen A und $2C + B$ verschiedenes Vorzeichen, so stellen sich die Nachbarbahnen als zwei durch den Nullpunkt gehende Gerade und die dieselben asymptotisch berührenden Hyperbeln dar. Man hat es in diesem Falle mit einer periodischen Bahn zu tun, an welche sich »asymptotische Lösungen« anschließen und welche als instabil zu bezeichnen ist.

5. Wir gehen zur numerischen Berechnung über. Man kann die, für die periodischen Lösungen charakteristischen Größen f und g nach den Formeln (C) berechnen und den durch die Relation (D) ausgedrückten Zusammenhang zwischen durchschnittlicher großer Achse a_0 und Excentricität e_0 finden, sobald man sich auf irgend eine Weise eine Entwicklung der Störungsfunktion nach den mittleren Anomalien hergestellt hat. Ich benutze zunächst die Leverrier-sche Entwickelung. Aus den Annales de l'Observatoire de Paris, Tome I, entnimmt man die folgende Darstellung des kritischen, von der doppelten Länge des Asteroiden und der einfachen des Jupiter abhängenden Teiles R der Störungsfunktion:

$$R = \frac{k^2 m'}{a'} \cdot s$$

$$s = \frac{1}{2} A^0 + \frac{e^2}{4}(A_1^0 + A_2^0) + 3\frac{e^4}{16}(A_3^0 + A_4^0) + \cos\zeta \left[\frac{e}{2}(-4A^2 - A_1^2) + \frac{e^3}{8}\left(14A^2 + \frac{5}{2}A_1^2 - 6A_2^2 - 3A_3^2\right)\right]$$
$$+ \cos 2\zeta \left[\frac{e^2}{4}(22A^4 + 7A_1^4 + A_2^4) + \frac{e^4}{16}\left(-\frac{596}{3}A^4 - \frac{212}{3}A_1^4 + 8A_2^4 + 16A_3^4 + 4A_4^4\right)\right]$$
$$+ \cos 3\zeta \left[\frac{e^3}{8}\left(-134A^6 - \frac{93}{2}A_1^6 - 10A_2^6 - A_3^6\right)\right] + \cos 4\zeta \left[\frac{e^4}{16}\left(\frac{257}{3}A^8 + \frac{932}{3}A_1^8 + 80A_2^8 + 13A_3^8 + A_4^8\right)\right]$$

(In Leverriers Bezeichnung ist $\zeta = 2l' - \lambda - \varpi$).

Ähnlich erhält man für den Differentialquotienten von R nach a bis auf die zweiten Potenzen der Excentricität genau:

$$a\frac{ds}{da} = \frac{a'}{k^2 m'} \cdot a \frac{\partial R}{\partial a} = \frac{1}{2}A_1^0 + \frac{e^2}{4}(A_1^0 + 4A_2^0 + 3A_3^0) - \cos\zeta \frac{e}{2}(5A_1^2 + 2A_2^2)$$
$$+ \cos 2\zeta \frac{e^2}{4}(29A_1^4 + 16A_2^4 + 3A_3^4) .$$

Zur Berechnung der periodischen Lösung brauchen wir die Differentialquotienten $\frac{\partial R}{\partial L}, \frac{\partial R}{\partial G}$. Man findet leicht:

$$\frac{\partial R}{\partial L} = \frac{2\sqrt{a}}{k}\frac{\partial R}{\partial a} - \sqrt{1-e^2}\,\frac{(1-\sqrt{1-e^2})}{ek\sqrt{a}}\frac{\partial R}{\partial e} \qquad \frac{\partial R}{\partial G} = \frac{\sqrt{1-e^2}}{ek\sqrt{a}}\frac{\partial R}{\partial e}$$

und damit

$$f = \frac{km'}{a'\sqrt{a}}\frac{\sqrt{1-e^2}}{e}\frac{ds}{de} \qquad g = (1-\sqrt{1-e^2})f - \frac{2m'k}{a'}\sqrt{a}\frac{ds}{da} .$$

Zur Untersuchung der Stabilität hat man die Größen zu bilden:

$$A = \frac{\partial^2 S}{\partial\zeta^2} \text{ und } 2C + B = 3\frac{k^4}{L_0^4} + \frac{\partial^2 S}{\partial G^2} - \frac{\partial^2 S}{\partial L \partial G} = 3\frac{1}{a^2} + \frac{\sqrt{1-e^2}}{ek\sqrt{a}}\frac{\partial}{\partial e}(f+g).$$

Die numerische Rechnung selbst verlief folgendermaßen. Für die Kommensurabilitätsstelle hat man:

$$\frac{n_0}{n'} = 2 = \left(\frac{a'}{a_0}\right)^{3/2} \qquad \frac{a_0}{a'} = \alpha = 0.629960 .$$

Mit dem Argumente α lassen sich die Koeffizienten A_p^i der Entwickelung der Störungsfunktion aus den bekannten Tafeln von Runcle entnehmen.

Es ergab sich:

$$s = \tfrac{1}{2}A_0 + 0.38762\,e^2 + 0.41866\,e^4 - \cos\zeta\,(1.19050\,e + 0.19722\,e^3) + \cos 2\zeta\,(1.69583\,e^2 - 1.42090\,e^4)$$
$$- \cos 3\zeta\,(2.97022\,e^3) + \cos 4\zeta\,(5.11476\,e^4) .$$

$$a\frac{\partial s}{\partial a} = 0.348400 + 1.74983\,e^2 - \cos\zeta\, e\cdot 3.29541 + \cos 2\zeta\, e^2\cdot 8.39232 .$$

Daraus folgte mit $n' = 299''.13$ und $m' = \frac{1}{1047.2}$ und indem durchweg nach Potenzen von e entwickelt wurde:

$$f = \mp 0''.42846\frac{1}{e} + 1''.503 \mp 3''.206\,e + 5.172\,e^2$$
$$g = \qquad - 0''.251 \pm 2''.158\,e - 6.550\,e^2 .$$

Hierbei gilt das obere Zeichen für diejenigen periodischen Lösungen, welche aus dem Ansatze $\zeta = 0$ entspringen, das untere für die aus dem Ansatze $\zeta = \pi$ entspringenden.

Ferner fand sich für die Stabilitätskoeffizienten:

$$\frac{a'}{k^2 m'}A = \pm 1.1905\,e - 6.783\,e^2 \pm 26.929\,e^3 - 76.152\,e^4$$

$$(B + 2\,C)\,k\sqrt{a} = 1788''.3 \pm \frac{0''.4285}{e^3} \pm \frac{0''.023}{e} .$$

Wenn man sich auf Werte von $e < 0.2$ beschränkt, da für größere die Entwickelungen nicht weit genug geführt sind, so ergibt sich für die Stabilität unserer periodischen Lösungen innerhalb dieser Grenzen folgendes. Im Falle $\zeta = 0$ (obere Vorzeichen) ist $B + 2\,C > 0$ und $A > 0$. Die Lösungen $\zeta = 0$ sind also stabil. Im Falle $\zeta = \pi$ ist A ständig negativ. $B + 2\,C$ ist ebenfalls negativ, solange e sehr klein ist, es geht jedoch, wie leicht zu errechnen ist, ins Positive über, sobald $e > 0.0621$ wird. Demnach sind die Lösungen für $\zeta = \pi$ stabil, solange $e < 0.0621$ ist, hingegen instabil, sobald $e > 0.0621$ wird.

6. Statt durch Leverriers Entwickelung, kann man den kritischen Teil der Störungsfunktion und ihrer Derivierten auch durch mechanische Quadratur finden. Man hat nach Tisserand Mécanique céleste, Bd. I pag. 433:

$$a\frac{\partial F'}{\partial a} = r\,S$$

$$\frac{\partial F'}{\partial e} = - S\,a\cos w + T\frac{2 + e\cos w}{1 - e^2}\cdot r\sin w ,$$

wo S und T Radial- und Tangentialkomponente der störenden Kraft sind und w die wahre Anomalie des Planetoiden bedeutet. Andererseits ist nach Tisserand IV pag. 192, aber in unserer obigen Bezeichnung:

$$S = k^2 m'\frac{a'\cos(A' - v) - r}{\Delta^3} ,$$

$$T = k^2 m'\frac{a'\sin(A' - v)}{\Delta^3} ,$$

$$\Delta = \sqrt{a'^2 + r^2 - 2\,a' r\cos(A' - v)} .$$

Hierbei habe ich bereits die aus dem indirekten Teile der Störungsfunktion entspringenden Terme, welche keinen Beitrag zu den kritischen Gliedern liefern, weggelassen. Es folgt (wenn E die excentrische Anomalie des Planetoiden bedeutet):

$$a\frac{\partial F'}{\partial a} = k^2 m'\frac{a' r\cos(A' - v) - r^2}{\Delta^3} . \qquad \frac{\partial F}{\partial e} = k^2 m'\frac{a\,a'}{\Delta^3}\left(- \cos(A' - \omega) + \frac{\sin(A' - v)\sin E}{\sqrt{1 - e^2}} + \frac{r}{a'}\cos w\right) .$$

Man erhält nun sehr nahe die kritischen Teile dieser Ausdrücke, also die oben mit $a\frac{\partial R}{\partial a}$ und $\frac{\partial R}{\partial e}$ bezeichneten Größen, indem man $a\frac{\partial F'}{\partial a}$ und $\frac{\partial F'}{\partial e}$ berechnet für eine Anzahl äquidistant über den Jupiterumlauf verteilter Zeitmomente und das Mittel aus allen erhaltenen Werten nimmt. Für die Rechnungen, deren Resultate unten angegeben sind, wurde die Periode des Jupiterumlaufs in 24 Teile geteilt. Hat man erst $a\frac{\partial R}{\partial a}$ und $\frac{\partial R}{\partial e}$, so erhält man nach den Formeln der vorigen Nummer unmittelbar f und g.

7. In folgender Tabelle findet man die Resultate beider Rechnungsarten zusammengestellt. Die erste Spalte gibt die Excentricität, die zweite und dritte die zugehörige Perihelbewegung f, die vierte und fünfte die Störung der mittleren Bewegung in Länge g, die sechste und siebente geben die Größe n_0, welche im Grunde nichts bedeutet als einen Ersatz für Angabe der mittleren großen Achse a_0, mit welcher sie durch die Relation:

$$n_0^2 a_0^3 = k^2$$

zusammenhängt. Die Größe n_0 sowohl, als die wenig davon verschiedene $n_0 + g$ stimmen bis auf wenige Bogensekunden mit der mittleren Bewegung in einer oskulierenden Ellipse überein, sowie auch innerhalb ähnlicher Grenzen mit der mittleren Bewegung, welche man aus einer Bahnbestimmung ohne Berücksichtigung der Störungen erhalten würde.

Periodische Lösungen für $\zeta = 0$.
(Konjunktion mit Jupiter im Perihel des Planetoiden).

e_0	Perihelbewegung f		Störung d. mittl. Bew. g		$n_0 = k a_0^{-3/2}$		$\log \partial_1$	N
	Meth. I	Meth. II	Meth. I	Meth. II	Meth. I	Meth. II	Meth. I	
0.01	−41″.374		−0″.230		639″.862		8.8107	639″.78
0.02	−19.982		−0.210		618.450		8.5093	618.06
0.03	−12.870		−0.192		611.320		8.3233	610.48
0.04	− 9.328	−9″.326	−0.175	−0″.174	607.760	607″.757	8.1861	606.29
0.05	− 7.213		−0.159		605.630		8.0759	603.34
0.06	− 5.812		−0.145		604.214		7.9831	600.92
0.07	− 4.817		−0.132		603.206		7.9023	598.77
0.08	− 4.076	−4.077	−0.120	−0.113	602.453	602.447	7.8303	596.65
0.09	− 3.504		−0.110		601.871		7.7651	594.59
0.10	− 3.050		−0.101		601.408		7.7051	592.46
0.11	− 2.682		−0.093		601.032		7.6496	590.20
0.12	− 2.378	−2.382	−0.086	−0.064	600.721	600.703	7.5975	587.85
0.13	− 2.122		−0.081		600.460		7.5483	585.40
0.14	− 1.905		−0.077		600.239		7.5016	582.80
0.15	− 1.718		−0.075		600.050		7.4568	580.09
0.16	− 1.555	−1.568	−0.073	−0.025	599.886	599.850	7.4137	577.24
0.17	− 1.413		−0.073		599.743		7.3722	574.25
0.18	− 1.287		−0.075		599.619		7.3317	571.12
0.19	− 1.175		−0.077		599.509		7.2922	567.82
0.20	− 1.074	−1.102	−0.081	+0.009	599.402	599.350	7.2533	564.43
0.40		−0.278		+0.065		598.470		470.52
0.60		−0.065		+0.150		598.172		346.16
0.80		+0.016		+0.221		598.020		217.93

Periodische Lösungen für $\zeta = \pi$.
(Konjunktion mit Jupiter im Aphel des Planetoiden).

e_0	Perihelbewegung f		Störung d. mittl. Bew. g		$n_0 = k a_0^{-3/2}$		$\log \partial_1$	N
	Meth. I	Meth. II	Meth. I	Meth. II	Meth. I	Meth. II	Meth. I	
0.01	+44″.382		−0″.273		554″.149		8.9036_n	554″.08
0.02	+22.992		−0.297		575.562		8.6015_u	575.20
0.03	+15.886		−0.322		582.693		8.4346_n	581.89
0.04	+12.351	+12″.346	−0.348	−0″.348	586.254	586″.295	8.3236_n	584.84
0.05	+10.245		−0.375		588.387		8.2409_n	586.15
0.06	+ 8.855		−0.404		589.806		8.1765_n	586.60
0.07	+ 7.874		−0.434		590.818		8.1248_n	586.46
0.08	+ 7.148	+ 7.151	−0.465	−0.477	591.574	591.583	8.0822_n	585.88
0.09	+ 6.594		−0.498		592.161		8.0467_n	585.00
0.10	+ 6.160		−0.532		592.629		8.0168_n	583.81
0.11	+ 5.813		−0.568		593.011		7.9913_n	582.33
0.12	+ 5.533	+ 5.559	−0.604	−0.642	593.329	593.340	7.9697_n	580.65
0.13	+ 5.303		−0.642		593.596		7.9510_n	580.08
0.14	+ 5.114		−0.681		593.825		7.9351_n	576.63
0.15	+ 4.957		−0.722		594.022		7.9214_n	574.33
0.16	+ 4.826	+ 4.891	−0.764	−0.866	594.195	594.232	7.9097_n	571.85
0.17	+ 4.718		−0.807		594.346		7.8997_n	569.16
0.18	+ 4.628		−0.851		594.481		7.8913_n	566.32
0.19	+ 4.554		−0.897		504.601		7.8842_n	563.29
0.20	+ 4.493	+ 4.625	−0.945	−1.172	594.708	594.804	7.8782_n	560.14
0.30		+ 4.873		−2.604		595.988		520.67
0.40		+ 8.713		−5.975		595.519		468.20

Astron. Nachr. *160* (1903) 385–400

Der Vergleich der durch Reihenentwickelung und der durch mechanische Quadratur gefundenen Werte zeigt, daß erstere etwa bis zu $e = 0.16$ bei einer Genauigkeit von 0″1 in der mittleren Bewegung brauchbar bleibt.

Um die obige Tabelle mit der von Herrn Brendel in seiner »Theorie der kleinen Planeten« pag. 128 gegebenen unmittelbar vergleichbar zu machen, ist derselben in einer weiteren Spalte der Logarithmus der Größe

$$\partial_1 = -\frac{f}{n_0}$$

hinzugefügt. Dieses ∂_1 stimmt seiner Bedeutung nach mit der von Herrn Brendel so benannten Größe überein. Ferner ist unser e_0 identisch mit Herrn Brendels β_1. Die numerische Übereinstimmung für ∂_1 und β_1 ist so gut, als bei der Annäherung, mit der sich Herr Brendel bei der Berechnung seiner Tafel begnügt, zu erwarten ist. Die Größe g verschwindet bei dem dort innegehaltenen Genauigkeitsgrade. Außerdem tabuliert Herr Brendel noch den Parameter der Ellipsen $a_0(1 - e^2)$ in doppelter Weise, indem er die beiden Größen n und $\partial = \frac{n - 2n'}{n}$ angibt, wobei n in meiner Bezeichnung gleich

$$\frac{k}{[a_0(1 - e_0^2)]^{3/2}} = \frac{n_0}{(1 - e_0^2)^{3/2}} \text{ ist.}$$

Will man sich die Abhängigkeit der Bestimmungsstücke dieser periodischen Lösungen graphisch verdeutlichen, so empfiehlt es sich, als Abscisse den Wert von

$$L + G = k\sqrt{a}\,(2 - \sqrt{1 - e^2})$$

oder irgend eine Funktion dieses Ausdruckes zu benutzen, z. B. die Größe:

$$N = \frac{k^4}{(L + G)^3} = \frac{n_0}{(2 - \sqrt{1 - e^2})^3},$$

welche oben tabuliert ist. Bei dieser ganz speziellen Wahl der unabhängigen Variabeln gilt nämlich der Satz, daß, wenn man irgend ein anderes Bestimmungsstück der periodischen Lösungen als Ordinate aufträgt, die entstehenden Kurven nur da Verzweigungen haben, wo die entsprechenden periodischen Lösungen ihren Stabilitätscharakter wechseln. Die beistehenden Kurven stellen n_0 und e_0

Darstellung von $n_0 = k a_0^{-3/2}$ (ausgezogene Kurven) und von e_0 (gestrichelte Kurven) als Funktionen von N.

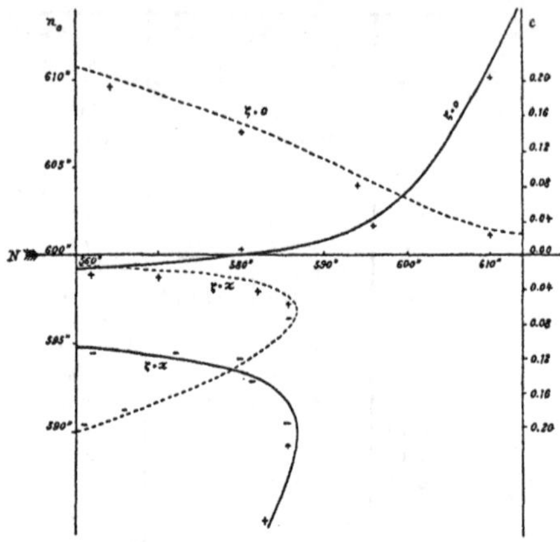

als Funktionen von N dar. Ein Verzweigungspunkt ergibt sich für $N = 586''6$, $e_0 = 0.06$. In der Tat fanden wir oben nahe für diesen Wert von e den Übergang von der Stabilität zur Instabilität. Die stabilen Lösungen sind in der Figur durch +, die instabilen durch — charakterisiert.

Es ist eigentümlich, wie stark man e wachsen lassen muß, um an Werte der großen Achse entsprechend $n_0 = 598''$ einerseits, $n_0 = 596''$ andererseits heranzukommen und daß sich trotz der großen im vorstehenden berücksichtigten Excentricitäten keine periodischen Lösungen für dazwischen liegende Werte von n_0 ergeben haben. Ich bemerke dabei, daß man für $\zeta = \pi$ nach den obigen Methoden nicht gut über $e = 0.4$ hinausgehen kann, da in diesem Falle bereits für $e = 0.59$ der Asteroid mit Jupiter zusammenstoßen müßte.

Auf die viel weitergehende Frage nach einem etwaigen Zusammenhange dieser Verhältnisse mit den Lücken im Asteroidenringe möchte ich bei späterer Gelegenheit zurückkommen.

Göttingen, 1902 Nov. 9.

K. Schwarzschild.

Correzioni al Catalogo AG. Cambridge Mass.

pag. 42 No. 2043 $\delta = +51°\,52'\,6''9$
» » » 2044 $\delta = +51\;\;51\;\;47.8$

pag. 84 No. 4125 $\delta = +51°\,48'\,23''4$
come BB. VI.

E. Millosevich.

Literarische Anzeige.

F. Exner und **E. Haschek.** Wellenlängen-Tabellen für spektralanalytische Untersuchungen auf Grund der ultravioletten Funkenspektren der Elemente. Zwei Teile. Leipzig und Wien 1902.

Die Tabellen werden für alle astrophysikalischen Untersuchungen von großem Werte sein. Sie erstrecken sich über 75 Elemente und gliedern sich in drei Teile: 1) eine Tabelle der Hauptlinien, 2) eine nach Wellenlängen geordnete Tabelle der stärkeren Linien aller Elemente und 3) die vollständigen Spektren der einzelnen Elemente.

Geschlossen 1903 Jan. 17. Herausgeber: H. Kreutz. Druck von C. Schaidt. Expedition: Kiel, Niemannsweg 103.

Ges. dt. Naturforscher und Ärzte, Verhandl. 1903, allg. Teil, 1–14

ÜBER

HIMMELSMECHANIK.

Von

K. SCHWARZSCHILD-GÖTTINGEN.

Wenn die Himmelsmechanik den heutigen Vortragszyklus über mechanische Probleme eröffnet, so entspricht das dem Gange der historischen Entwicklung. Die Bewegung der Himmelskörper ist das Problem, an welchem die Mechanik eigentlich gewachsen ist, an dem sie sich ihre Sporen verdient hat. Aber nicht nur das zeitlich erste grosse Problem der Mechanik ist die Frage nach dem Lauf der Planeten, sondern auch ihr vornehmstes, reinstes Problem. Hier bedarf es keiner künstlichen Idealisierung wegen Ungleichmässigkeit des Materials oder Einflüssen der Reibung, die einfache Formel des NEWTONschen Gesetzes $\frac{m m'}{r^2}$ beherrscht die Bewegungen im Sonnensystem für Jahrtausende vorwärts und rückwärts mit vielstelliger Genauigkeit.

Es wird zwar gegenwärtig vielfach an der alten einfachen Form des NEWTONschen Gesetzes gerüttelt. Der Astronom SEELIGER betont die Schwierigkeiten, die seiner Anwendung auf die im unendlichen Raum verbreiteten Massen der Sternsysteme entgegenstehen, der Physiker LORENTZ konstruiert eine Möglichkeit, die Schwerkraft sich mit Lichtgeschwindigkeit ausbreiten zu lassen. Bei dem Sturmschritt der modernen physikalischen Entwicklung mag der morgige Tag diesen von philosophischen Gesichtspunkten aus so reizvollen Betrachtungen eine praktische Bedeutung geben, heute aber — dies betone ich — befinden wir uns noch in der Periode zunehmender Bestätigung des NEWTONschen Gesetzes. Während vor einigen Jahren noch zur Erklärung der ominösen Bewegung des Merkurperihels angenommen wurde, dass der Exponent 2 im NEWTONschen Gesetz durch 2 + 0.000 000 16 zu ersetzen sei, hat kürzlich E. W. BROWN auf Grund einer genauen Theorie der Mondbewegung gezeigt, dass der Exponent nur um 0.000 000 04 von 2 verschieden sein kann. Er hat die Gültigkeitsgrenze noch um eine Stelle weiter hinausgeschoben. Wo sich noch Abweichungen vom NEWTONschen Gesetz zeigen, wie beim Merkurperihel und in der Acceleration des Mondes, hat man allen Grund, fremdartige störende Einflüsse anzunehmen. Das NEWTONsche Gesetz hat in Praxis gegenwärtig grösseren Anspruch auf absolute Gültigkeit als jemals zuvor.

Wenn man die astronomische Mechanik in ihrer Gesamtheit ins Auge fasst, so enthält sie freilich mehr als die blosse Anwendung des NEWTONschen Gesetzes. Für die Theorie der Polhöhenschwankungen z. B. hat man die Elastizitätstheorie auf die Erde anzuwenden. Bei der Theorie der Sonne kommt die Thermodynamik ins Spiel. Ganz neue Probleme erwachsen der astronomischen Mechanik aus der beginnenden Erkenntnis der Formverhältnisse in einzelnen Fixsternsystemen, wo zwei mächtige Sonnen in unmittelbarer Nähe oder selbst im Kontakt mit einander in wenigen Tagen um einander rotieren. Es wäre verlockend genug, zu schildern, was hier die Theorie der Gleichgewichtsfiguren rotierender Flüssigkeiten geleistet hat, und wie viele Probleme noch ausstehen auf diesem Gebiete, wo moderne Mathematik und Physik in engster Verknüpfung arbeiten müssen. Aber wenn man zu einer Expedition in den Himmel eine halbe Stunde Zeit hat, so muss man sich beschränken. Darum will ich bei dem Kardinalproblem der Astronomie bleiben, bei den blossen Folgerungen aus dem NEWTONschen Gesetz, bei dem sogenannten Vielkörperproblem, und will zu zeigen versuchen, dass dieses für die ganze Mechanik ehemals vorbildliche Problem inzwischen nicht abgestorben ist, dass seine Behandlung bis in die letzte Zeit hinein fortgeschritten ist, und dass es noch immer Fragestellungen enthält, die für die gesamte Mechanik fruchtbar werden können.

Das Problem lässt sich so aussprechen: Eine Anzahl von Körpern, deren Massen wir auf ihre Schwerpunkte konzentriert denken, bewegen sich in einem bestimmten Moment an verschiedenen Stellen des Raumes mit bestimmten Anfangsgeschwindigkeiten. Man soll angeben, in welchen Stellungen sie sich zu irgend einer späteren Zeit befinden, wenn sie sich immerfort untereinander genau nach dem NEWTONschen Gesetz anziehen. Das ist offenbar im wesentlichen das Problem der Bewegung der Planeten und Kometen im Sonnensystem.

LAPLACE gibt darauf zu Anfang des vorigen Jahrhunderts in seiner Mécanique céleste, der Bibel der sogenannten „klassischen Himmelsmechanik", folgende Antwort: Der Lauf der Planeten ist unendlich verwickelt, wenn man ihn genau beschreiben soll. Jede Bewegung des Jupiter oder Saturn hat ihren Reflex in einer Bewegung der Erde. Es bedarf seitenlanger Formeln, um diese tausend feinen Ungleichheiten darzustellen. Verzichtet man aber auf teleskopische Genauigkeit und beschränkt sich auf das, was man am Himmel mit blossem Auge sehen

Ges. dt. Naturforscher und Ärzte, Verhandl. 1903, allg. Teil, 1–14

und bei einer Zeichnung des Planetenlaufs auf einem Blatt Papier zur Darstellung bringen kann, so wird der Verlauf einfach. Die Planeten bewegen sich im wesentlichen nach KEPLERS Gesetzen in Ellipsen um die Sonne. Freilich sind diese Ellipsen nicht fest, sondern in langsamer allmählicher Veränderung begriffen. So nimmt die Exzentrizität der Erdbahn in 1000 Jahren um $^1/_{2300}$ ab. In der gleichen Zeit dreht sich die grosse Achse der Erdbahn um 3^0. Ausser diesen langsamen — wie man sagt „säkularen" — Änderungen der Planetenbahnen gibt es nur eine merkliche Änderung von kürzerer Periode, die sogenannte grosse Ungleichung zwischen Jupiter und Saturn, welche diese Planeten im Verlauf von 930 Jahren um $^2/_3$, resp. $1^1/_2$ Vollmondsbreiten gegen ihren mittleren Ort verschiebt.

Ich will diese ursprüngliche Theorie von LAPLACE mit dem Schlagwort „klassische Himmelsmechanik" bezeichnen, obwohl man sonst in diesen Begriff noch allerlei Fortschritte einzuschliessen pflegt, die LAPLACE selbst gemacht oder wenigstens in den theoretischen Teil seiner Mécanique céleste aufgenommen hat. Klassisch heisst diese Theorie mit Recht. Auf ihre vollständigen, wie gesagt, seitenlangen Formeln ist noch fast das ganze System der heute im Gebrauch befindlichen Planetentafeln gegründet, und diese vollständigen Formeln stimmen mit allen aus dem Altertum überlieferten und den seit 150 Jahren vorliegenden, bis auf die Sekunde genauen teleskopischen Beobachtungen innerhalb der Beobachtungsgenauigkeit überein.

Hiermit könnte man einen Bericht über das Vielkörperproblem beschliessen, wenn man sich auf den grob empirischen Standpunkt stellen und sich begnügen wollte, die Bedürfnisse des Kalendermachers, der Chronologie, der Geographie, der Nautik zu befriedigen. Indessen wollen wir neben die reale Frage die ideale stellen, unseren Blick aus der zeitlichen Beschränktheit der menschlichen Erfahrung erheben und das Geschick des Sonnensystems in einer Vergangenheit und Zukunft zu erkennen suchen, die nach Millionen von Jahren zählt. Von diesem höheren, idealen Standpunkt aus betrachtet, sind die Formeln der klassischen Himmelsmechanik falsch und ungenügend. Wenn die Formel für die Exzentrizität der Erdbahn $e = e_0 - \frac{t}{2300}$, welche das der Zeit proportionale Wachsen derselben ausdrückt, der Wirklichkeit entspräche, so folgte, dass in 2 Millionen Jahren sich die Erdbahn in eine Parabel verwandelt haben müsste.

LAGRANGE hat gezeigt, dass diese Konsequenz nicht richtig ist, dass e nur für kurze Zeit nach diesem Gesetze abnimmt und sich in längeren Zeiträumen in ganz anderer Weise verhält. Wäre — um LAGRANGES Resultate an einem einfachen Beispiel zu erläutern — ausser der Erde nur der mächtige Jupiter vorhanden, so würde sich die Veränderung der Ellipse der Erde durch folgenden Mechanismus darstellen. Ein

Rad *r* vom Radius 0.0518 mit einem Zapfen *z* werde gleichförmig in 83 000 Jahren gedreht. Ein Balken *b*, dessen Ende *c* im Abstand 0.0482 von der Radmitte befestigt ist, gleite mit Hilfe einer Nut an dem Zapfen *z*. Dann gibt die Richtung des Balkens jeder Zeit die Richtung der grossen Achse der Erdbahn an, die Länge des Balkens zwischen dem Endpunkt *c* und dem Zapfen *z* ihre Exzentrizität. Es ergibt sich also, dass die Exzentrizität der Erdbahn nicht ins Ungemessene zunimmt, sondern in langen Perioden nur zwischen den Grenzen 0.003 und 0.100 schwankt. Ähnlich halten sich auch die Erstreckungen aller anderen Planetenbahnen in engen Grenzen.

Ist somit die Schwierigkeit der „Säkularstörungen" durch LAGRANGE überwunden, so bleibt eine zweite noch bedenklichere. Wir wollen sie

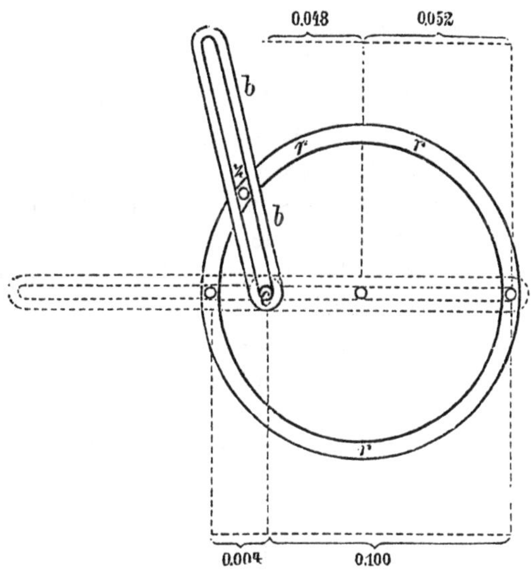

Fig. 1.

gleich mit dem Planetoiden Hekuba verknüpfen, dessen Name mit seinen wehmütigen Associationen seine verhängnisvolle Stellung andeutet. Hekuba umkreist die Sonne in 2101 Tagen, also sehr nahe in der halben Umlaufszeit des Jupiter, die 4323 Tage beträgt. Wir wollen uns Hekuba ein klein wenig weiter hinausgeschoben denken, dann wird sich ihre Umlaufszeit ein wenig verlängern nach dem 3. KEPLERschen Gesetz, dass die Quadrate der Umlaufszeiten den 3. Potenzen der Entfernung proportional sind. Stimmen wir die Verlängerung gerade bis zur halben Umlaufszeit des Jupiter ab und wenden dann die Formeln der klassischen Himmelsmechanik an, so erhalten wir das Resultat: Die grosse Achse der Bahn der Hekuba wächst der Zeit proportional, und zwar in 400 Jahren um $\frac{1}{100}$.

Ges. dt. Naturforscher und Ärzte, Verhandl. 1903, allg. Teil, 1–14

Es lässt sich leicht anschaulich begreifen, wie ein solches Resultat zustande kommen kann. Stehen die Umlaufszeiten zweier Planeten in einem inkommensurablen Verhältnis, so kommen sie im Laufe der Zeit in alle möglichen relativen Stellungen auf ihren Bahnen; die störenden Kräfte, die sie auf einander ausüben, wirken bald in einem, bald im anderen Sinne und heben sich so auf die Dauer zum grossen Teil auf. Anders bei Hekuba. Jupiter und Hekuba kehren nach einem Umlauf des Jupiter, zweien der Hekuba wieder zur selben Stellung zurück. Was sich während dieser Zeit an Störungen ergeben hat, das ergibt sich im selben Betrag und Sinne auch während jedes folgenden Umlaufs des Jupiter. Hier heben sich die Störungen nicht auf, sondern sie summieren sich. Auch die Beobachtungstatsachen sprechen zugunsten der aus der klassischen Himmelsmechanik gezogenen Konsequenz. In dem breit zwischen Mars und Jupiter dahinziehenden Schwarm der Asteroiden bestehen Lücken in allen den Entfernungen von der Sonne, an welchen die nach dem 3. KEPLERschen Gesetz folgende Umlaufszeit in einem einfachen kommensurablen Verhältnis zu der des Jupiter stehen müsste. Und was noch mehr überrascht, dasselbe wiederholt sich beim Saturnring, der bekanntlich ebenfalls als aus einem dichten Schwarm von Felsstücken bestehend gedacht werden muss. Hier befindet sich die CASSINIsche Teilung, welche den inneren Ring vom äusseren trennt, gerade an einer Stelle, entsprechend einer Umlaufszeit, deren 2-, 3-, 4- und 6faches resp. sehr nahe die Perioden der 4 innersten Monde des Saturn liefert. Nichts liegt näher, als zu folgern, dass gemäss den Formeln der klassischen Himmelsmechanik die progressiv wachsenden Störungen alle Körper aus den Kommensurabilitätsstellen herauswerfen.

Trotzdem ist auch diese zweite Konsequenz falsch. Was hier geschieht, und welche neuen Bewegungsformen sich hier einstellen, ist wesentlich unter Führung GYLDÉNS in den letzten 20 Jahren erkannt worden. Es lässt sich das etwa folgendermassen charakterisieren:

Man nenne den Winkel, um den sich ein Planet, von einer bestimmten Anfangslage an gerechnet, um die Sonne gedreht hat, l. Dann ist, wenn wir zur Vereinfachung eine Kreisbahn statt einer Ellipse voraussetzen, l einfach proportional t, also: $l = nt$. Die Grösse n, welche angibt, um welchen Winkel sich der Planet in einem Tage bewegt, wird von den Astronomen „mittlere Bewegung" genannt.

Man hat z. B. für Hekuba:

$$l = 617{,}41'' \cdot t = nt$$

und für Jupiter:

$$l' = 299{,}13'' \cdot t = n't.$$

Nun wollen wir einen Hilfswinkel bilden:

$$\zeta = l - 2l' = (n - 2n')t = 19{,}15'' \cdot t.$$

An diesem Winkel erkennt man am besten, wie nahe man einer genauen

Kommensurabilität der Umlaufszeiten ist. Wäre die Umlaufszeit der Hekuba genau die Hälfte der des Jupiter, so wäre $n = 2n'$, $\zeta = 0$. Nach vorstehender Formel schreitet bei Hekuba ζ noch mit der Zeit fort, aber bereits so langsam, dass es erst in 185 Jahren einmal den Umkreis durchläuft.

Nun gilt diese Formel natürlich nur ohne Rücksicht auf die Störungen. Der Einfluss der Störungen lässt sich aber ebenfalls am besten mit Hilfe des Winkels ζ beschreiben. Das ganze Resultat jener neueren Untersuchungen lässt sich nämlich im wesentlichen dahin aussprechen, dass sich der Winkel ζ verhält wie die Elongation eines Pendels aus seiner Gleichgewichtslage. Ist man weit von der Kommensurabilitätsstelle entfernt, so verhält sich ζ analog einem Pendel mit so viel Schwung, dass es in kurzer Zeit um seine Aufhänge-Axe rotiert. Es lehrt dies zunächst nichts neues, ζ verhält sich im wesentlichen so, wie wenn keine Störungen vorhanden wären. Eine Annäherung an die Kommensurabilitätsstelle entspricht einer Verminderung der Anfangsgeschwindigkeit des Pendels. Man wird an Fälle kommen, wo das Pendel nur noch wenig Kraftüberschuss hat und die höchste Stelle seiner Bahn nur sehr zögernd überwindet. Der Winkel ζ wird dann eine sehr ungleichförmige, einmal rasche, dann langsame Rotation ausführen. Damit wird dann eine entsprechende Schwankung in den Winkeln l und l', ein Tempowechsel in der Umlaufsgeschwindigkeit der Planeten selbst Hand in Hand gehen. Schliesslich gelangt man an die Pendelbewegung von „asymptotischem" Charakter. Das Pendel entfernt sich unendlich langsam von der höchsten Stelle und schwingt unter dem Aufhängepunkt herum, um sich nach unendlich langer Zeit der höchsten Stelle wieder von der andern Seite her anzunähern. Genau so verhält sich ζ bei den asymptotischen Bewegungen im Planetensystem. Mit dieser asymptotischen Bewegung ist ein Grenzfall erreicht. Bei weiterer Annäherung an die Kommensurabilität hat man es mit dem oszillierenden Pendel zu tun. Der Winkel ζ umläuft den Umkreis überhaupt nicht mehr, die störenden Kräfte halten ihn fest und erlauben ihm nur, Schwingungen um einen gewissen Mittelwert auszuführen. Die Astronomen nennen diese Erscheinung „Libration". Sobald Libration eintritt, ändert die Bewegung völlig ihren Charakter insofern, als nun wirklich nicht mehr die betreffenden Körper im Laufe der Zeit unabhängig von einander alle möglichen Stellungen auf ihren Bahnen einnehmen können, vielmehr ist dadurch, dass der Winkel $\zeta = l - 2l'$ innerhalb bestimmter Grenzen zu bleiben gezwungen ist, der eine Körper in gewissem Grade an den anderen gebunden. Es ist eine derartige Libration, eine derartige Bindung, wenn die drei inneren Jupitermonde niemals zugleich auf einer Seite des Jupiter stehen und daher auch niemals zugleich durch den Schatten des Jupiter verfinstert werden können.

Die Amplitude der Libration kann immer kleiner und kleiner werden,

Ges. dt. Naturforscher und Ärzte, Verhandl. 1903, allg. Teil, 1–14

bis man schliesslich zu dem ruhenden Pendel gelangt, ζ wird kostant. die Kommensurabilität ist genau erfüllt, die beiden Körper kehren nach einer bestimmten Zeit genau in dieselbe relative Lage zu einander zurück, man hat es mit einer sogenannten periodischen Lösung des Dreikörperproblems zu tun.

Periodische, asymptotische und librierende Bewegungen sind die neuen Bewegungsformen, die man bei diesem „Problem der Kommensurabilitäten" kennen lernt. Dabei ist das Fundamentale, dass die Bahnen entgegen der klassischen Himmelsmechanik in keinem Falle unbegrenzt progressiv wachsende Störungen erleiden, vielmehr hängen die Veränderungen z. B. der grossen Achse oder der Exzentrizität, abgesehen von teleskopisch kleinen Oszillationen, direkt von dem Winkel ζ ab, und diese Elemente kehren zu ihren Ausgangswerten zurück, wenn der Winkel ζ wieder denselben Wert annimmt, wenn das Pendel eine Rotation oder eine Schwingung vollzogen hat. Allerdings erfahren — soweit behält die klassische Himmelsmechanik Recht — die Elemente an den Kommensurabilitätsstellen ungewöhnlich grosse und rasche Änderungen. Bei Hekuba kann sich die grosse Achse um $^1/_{100}$ ihres Wertes ändern, die Exzentrizität erreicht die Werte 0.066 und 0,15.

Leider wird mit der Erkenntnis dieser Bewegungsverhältnisse die unmittelbare Erklärung der Lücken im Asteroidenschwarm und Saturnring aus der Störungstheorie hinfällig. An den Stellen, wo librierende Bahnen auftreten könnten, fehlen die Asteroiden, doch müssen sie hier von Anfang an gefehlt haben, da die Störungen die Körper nur zeitweilig aus ihrer Anfangslage entfernen, um sie im Laufe von einigen hundert Jahren immer wieder einmal zurückzuführen. Man muss daher mit Herrn Callandreau zur Erklärung der Lücken auf kosmogonische Vorstellungen zurückgehen. Es lässt sich denken, dass das starke Wechseln und Durcheinanderschieben der Bahnen an den Kommensurabilitätsstellen den Kondensationsprozess der Materie behindert hat.

Im übrigen sind uns ausser dem schon erwähnten Fall der Jupitermonde noch drei höchst merkwürdige Fälle von Librationsbewegungen bei den Monden des Saturn bekannt, von denen ich nur die von H. Struve entdeckte Libration zwischen den Saturnmonden Mimas und Thetis anführe. Hier ist der librierende Winkel

$$\zeta = 4l' - 2l - (\Theta + \Theta').$$

Der Zusatz $\Theta + \Theta'$ bezieht sich auf die Lage der Knoten der Bahnebenen und ändert am Wesen der Sache nichts. — Und zwar schwingt das Pendel in äusserst langsamer Bewegung über die Horizontale hinaus. Während die Monde selbst in 22,6, resp. 45,3 Stunden den Saturn umkreisen, schlägt der Winkel ζ in 70,6 Jahren um 97^0 nach beiden Seiten aus.

Mit der Erkenntnis der Bewegungsformen im Fall der Säkularstörungen sowohl, als im Falle der Kommensurabilitäten waren die beiden Barrikaden gefallen. Auf freier Bahn hat sich nun die Störungstheorie rasch zu einem Resultat erhoben, das als der Gipfelpunkt in der fortschreitenden Entwicklung der klassischen Himmelsmechanik zu bezeichnen ist. NEWCOMB, TISSERAND, LINDSTEDT, GYLDÉN, BÖHLIN haben sich darum verdient gemacht, POINCARÉ hat ihm den letzten Schliff gegeben. Es besagt, dass sich die Koordinaten der Planeten in rein trigonometrische Reihen entwickeln lassen. Es treten in der modernen Störungstheorie überhaupt keine der Zeit proportionalen Glieder auf, alle Wirkungen, die so aussehen, als ob sie im Laufe der Zeit ins unendliche wachsen wollten, sind aus unseren Formeln verschwunden, wir sind endlich zur fertigen Ausbildung der Epizykelnmethode der Alten gelangt. Die Alten liessen ja jeden Planeten auf einem kleinen Kreise laufen, dessen Mittelpunkt auf einem grösseren Kreise fortschritt, eventuell musste sich der Mittelpunkt des grösseren Kreises auf einem noch grösseren vorwärts bewegen. Unsere trigonometrische Entwicklung ist nichts anderes als eine Übereinanderlagerung von unendlich vielen kleineren und kleineren Epizykeln. Ich will hinzufügen, dass die Umlaufsgeschwindigkeiten dieser Epizykeln sich aus den Vielfachen von $3n - 1$ Geschwindigkeiten bei n Planeten, also 23 Geschwindigkeiten bei unseren 8 Planeten zusammensetzen.

Wer dieses Resultat unbefangen ansieht, der kann dem bedeutungsvollen Schluss nicht ausweichen, dass das Sonnensystem stabil ist, dass die Planeten ewig dieselben Regionen des Himmels durchkreuzen werden, da sich ihre Bewegung aus der Zusammensetzung von lauter kleinen periodisch zwischen festen Grenzen schwankenden Störungen ergibt. Als man Ende der achtziger Jahre diesen Punkt erreicht hatte, war es in der Tat eine verbreitete Meinung, dass man ausser der schon durch die klassische Himmelsmechanik im wesentlichen erledigten praktischen Frage nun auch die ideale hinreichend beantwortet habe, dass das Vielkörperproblem im Grunde gelöst sei. Aber zum dritten Mal hat die fortschreitende Wissenschaft ihr Verdikt ausgesprochen. In dem Augenblick, wo man das Gebäude der klassischen Himmelsmechanik mit dieser schimmernden Spitze des Stabilitätsbeweises krönen wollte, zeigte sich ein klaffender Riss in den Fundamenten. POINCARÉ wies 1890 nach, dass die Reihen alter und neuer Art, mit denen die Astronomen arbeiten, in gewissem Grade sinnlos, weil divergent, sind, dass der Epizykeln so viele und so grosse sind, dass die Summe ihrer Radien unendlich wird.

Mit diesem revolutionierendem Satze POINCARÉS muss die Bestätigung der Formeln der Astronomen durch den beobachteten Lauf der Gestirne als ein Zufall erscheinen, der Beweis der Stabilität des Planetensystems wird gänzlich hinfällig. Um den Unterschied kräftig zu pointieren — es ist nach den POINCARÉschen Untersuchungen über die Natur des Dreikörperproblems sehr wohl möglich, man darf fast sagen, wahrschein-

Ges. dt. Naturforscher und Ärzte, Verhandl. 1903, allg. Teil, 1–14

lich, dass das Planetensystem auf die Dauer im höchsten Grade instabil ist, dass die Erde einmal mit Jupiter ihren Platz vertauschen, dass unser Mond um den Mars kreisen und wir den Saturnring annektieren würden.

Es ist kaum ein schärferer Gegensatz denkbar, als der zwischen dem Gedanken an eine so extravagante Instabilität des Planetensystems und dem Glauben an jene Reihenentwicklungen, die die Bestätigung an der astronomischen Beobachtung im Rücken haben. Es bildet den Schluss meiner Aufgabe, zu zeigen, wie sich beide Standpunkte vereinigen.

Wenn ich absehe von der nicht gänzlich ausgeschlossenen Möglichkeit, dass jene trigonometrischen Reihen für spezielle, vielleicht „überall dichtliegende" Anfangslagen konvergieren, so hat Poincaré jedenfalls gezeigt, dass jene Reihen den Charakter sogenannter „semikonvergenter" Entwicklungen haben. Zunächst in abstrakt mathematischer Sprache ausgedrückt, lautet sein Satz: Bleibt man bei dem p-ten Gliede der Reihen stehen, so ist der übrig bleibende Fehler kleiner als:

$$\mu^p \,.\, A_p\,(t),$$

wo μ eine Grösse von der Ordnung der störenden Massen und $A_p\,(t)$ eine für jeden endlichen Index p und jede endliche Zeit t von μ unabhängige Grösse ist. Anschaulicher ist gemeint, dass die Reihen der Störungstheorie sich verhalten analog der Reihe

$$1 + \frac{3}{1000} + \frac{3^3}{(1000)^2} + \frac{3^{3^3}}{(1000)^3} + \dots\,.$$
$$= 1 + 0{,}003 + 0{,}000\,027 + 7620 + \dots,$$

deren drei erste Glieder rapide abnehmen, während die folgenden so rasch ansteigen, dass das fünfte Glied bereits eine Zahl mit Billionen von Stellen wäre. Dabei haben Reihen dieser Art das Eigentümliche, was ihre praktische Brauchbarkeit bedingt, dass die Genauigkeit, mit der sie das gewünschte Resultat darstellen, nicht von den unendlich grossen fortgelassenen Gliedern abhängt, sondern durch das letzte mitgenommene Glied wenigstens der Grössenordnung nach gegeben wird. Es würde also an die Astronomen die Vorschrift zu ergehen haben, ihre Reihen nicht unbegrenzt fortzusetzen. Das tun sie aber ohnehin aus praktischen Gründen, und zwar sind sie stets beim dritten Gliede in ihren Reihen stehen geblieben. Was sie dann erhalten, muss nach Poincaré für begrenzte Zeit eine gute Annäherung an die strenge Lösung des Problems geben.

Auf Grund einer einfachen Restabschätzung, die ich für den Fall der Störungen der Erde durch Jupiter ausgeführt habe, kann ich diesen Satz noch konkreter aussprechen. Es lässt sich nachweisen, dass die Formeln, die aus den drei ersten Gliedern der Reihen hervorgehen, für 1000 Jahre in mindestens 5stelliger Genauigkeit gelten, und dass ferner Jupiter in 1 Million Jahre die grosse Achse der Erdbahn sicher nicht mehr als um $^1/_{100}$ ändert. Auch scheint kaum ein Zweifel darüber zu bestehen, dass man durch sorgfältigere Betrachtungen nach vorhandenen Methoden jene Genauigkeit als siebenstellig und diese Zeit als von 100 oder 1000 Millionen Jahren Dauer erweisen könnte.

Damit ist der Zwiespalt behoben. Beide Seiten behalten auf ihre Art Recht. Die Formeln, welche die Astronomen benutzen, müssen für die Zeiten, aus denen Beobachtungen vorliegen, der Beobachtungsgenauigkeit entsprechen, ihre Übereinstimmung mit der Erfahrung ist kein Zufall, solange anders das NEWTONsche Gesetz gültig ist. Die in ihnen enthaltene Zusicherung der Stabilität des Planetensystems ist richtig für 1 Million Jahre, insofern während dieser Zeit nur unbedeutende Änderungen der Bahnen vor sich gehen; sie ist es wahrscheinlich auch noch für 1000 Millionen Jahre. Erst in Billionen oder vielleicht Trillionen Jahren mögen sich die Störungen bis zur Vernichtung der jetzigen Ordnung des Planetensystems angehäuft haben.

Hiermit ist denn auch die ideale Frage beantwortet in einem Umfange, der eine 100000jährige Vergangenheit des Menschengeschlechtes auf der stabilen Erde zulässt und ungemessenen Zukunftsmöglichkeiten seiner Entwicklung für die nächste Million Jahre Raum bietet. Sie ist der Beantwortung wenigstens nahe gerückt in einer Ausdehnung, die selbst den Geologen befriedigt, der für seine Schichtenfolgen 1000 Millionen Jahre in Anspruch nimmt. Nur der Geist des reinen Mathematikers, der alle Knoten lösen will, auch wenn er sie selbst geknüpft hat, ist nicht zufrieden gestellt und fragt logisch weiter: Woran kann es liegen, dass jene vielversprechenden trigonometrischen Reihen nicht konvergieren? Man kann darauf in gewissem Grade eine Antwort geben: Weil sie ihrer ganzen Bauart nach, wenn sie konvergierten, gewiss nicht alle Bewegungsformen darstellten, die im Vielkörperproblem auftreten. Welches diese Bewegungsformen sind, darüber wird die Weiterentwicklung der Himmelsmechanik Aufschluss zu geben haben, man kann Bewegungen vermuten, die zwischen dem rotierenden und dem oszillierenden Pendel abwechseln, man kann sich vergegenwärtigen, dass, wie zwischen den Umlaufszeiten der Planeten selbst, so auch zwischen der Periode der Pendelschwingungen und den Umlaufszeiten Kommensurabilitäten auftreten müssen, welche Librationsbewegungen einer höheren Gattung veranlassen können. Insbesondere wird es erforderlich sein, den Übergang zu finden zwischen den bisher von uns allein besprochenen Bahnformen, welche bei grosser Entfernung und kleinen Massen der störenden Körper auftreten, und

Ges. dt. Naturforscher und Ärzte, Verhandl. 1903, allg. Teil, 1–14

den überraschenden Kurven, die bei starker Annäherung der Körper oder bei im Verhältnis zum Hauptkörper grossen Massen derselben zustande kommen. Ich will zwei Beispiele für die Bahnen letzterer Art anführen. Ein erstes ist der Wirklichkeit entnommen. Der Komet

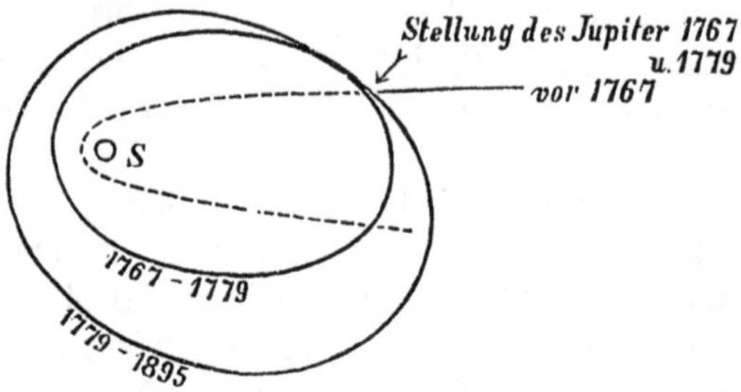

Fig. 2.
Bahn des Kometen Lexell.

Lexell kam im Jahre 1767 in parabolischer Bahn dem Jupiter nahe, wurde von diesem in eine Ellipse geschleudert, die er zweimal in je 5,6 Jahren durchlief, bis er 1779 zum zweiten Male dem Jupiter nahe kam und wiederum eine gänzliche Änderung seiner Bahn erfuhr. Wahr-

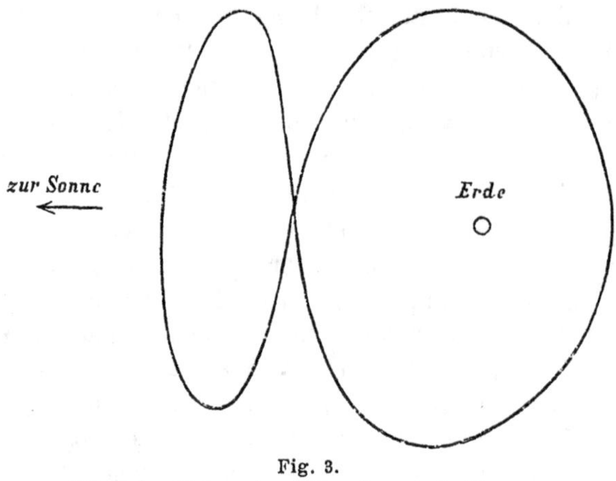

Fig. 3.
Mögliche Bahn eines Mondes nach Darwin.

scheinlich hat er seit jener Zeit in je 7,2 Jahren die zweite in der Figur gezeichnete Ellipse durchlaufen und ist uns im Jahre 1895 als Komet Swift wieder erschienen. Als zweites Beispiel diene die von G. H. Darwin durch mühsame mechanische Quadratur errechnete mögliche Bahn eines Mondes, welche dreimal in einem Umlauf ein Zwischentreten des Mondes zwischen Erde und Sonne, dreimal im Monat Neumond gegen einmal Vollmond aufweist. Der Mathematiker wird ermessen, welche Schwierigkeiten noch einer schmiegsamen Behandlung solcher Bahnformen entgegenstehen.

Senden wir den Blick zum Schluss noch einmal zurück auf die alte Geschichte der Himmelsmechanik in ihrer Verknüpfung mit dem Aufbau der Mechanik überhaupt und denken zugleich an ihre neuen Errungenschaften, die eine lange Erhaltung der jetzigen Erdbewegung verbürgen, so werden wir in doppeltem Sinne sagen dürfen, dass die Himmelsmechanik die Basis geliefert hat, auf der sich die Mechanik des menschlichen Körpers und der menschlichen Werkzeuge, die irdische Mechanik, sicher weiter entwickeln kann.

Newcomb-Engelmann's Populäre Astronomie, 5th ed., ed. by P. Kempf (Leipzig 1914) 89–93

stimmen alle Bewegungen im Sonnensystem mit den Resultaten der Gravitationstheorie überein. Die kleinen Unvollkommenheiten, welche in den astronomischen Tafeln noch bestehen, scheinen hauptsächlich aus Fehlern in den Daten hervorzugehen, von denen der Mathematiker bei der Berechnung der Bewegung eines jeden Planeten ausgehen muß. Weder die Umlaufszeit eines Planeten, noch die Exzentrizität seiner Bahn, die Lage des Perihels oder der Ort in der Bahn zu einer gegebenen Zeit kann aus der Gravitationstheorie berechnet, sondern muß aus den Beobachtungen abgeleitet werden. Wären die Beobachtungen fehlerfrei, so könnten Resultate von jedem gewünschten Genauigkeitsgrade aus ihnen erhalten werden; aber die Unvollkommenheiten aller Instrumente und des menschlichen Auges selbst verhindern, daß die Beobachtungen den Grad der Genauigkeit erreichen, den der Theoretiker verlangt, und sie machen fortdauernd Betrachtungen über die »Beobachtungsfehler« wie über die »Tafelfehler« notwendig.

7. Bahnbestimmung.

So schwierig die Aufgabe ist, der die vorausgehenden Betrachtungen galten, die Bahn eines Planeten mit Rücksicht auf alle Störungen durch andere Planeten zu berechnen und seinen Lauf für Jahrhunderte vorwärts und rückwärts zu verfolgen, so findet sie doch eine gewisse Erleichterung darin, daß die Bahnen der Himmelskörper, die man einem so eingehenden Studium unterwirft, immer bereits nahe bekannt sind, so daß der spätere Berechner jedes Mal nur die Arbeit des früheren auf eine höhere Stufe der Vollkommenheit zu bringen hat. Aber es gibt noch ein anderes Problem der »Bahnbestimmung«, das im Prinzip viel einfacher ist, dessen Schwierigkeit aber darin liegt, daß es plötzlich an den Astronomen herantritt und ohne solche Vorarbeiten gelöst werden muß. Sobald nämlich ein neuer Planet oder Komet entdeckt ist, soll aus den ersten Beobachtungen des Himmelskörpers sofort seine Bahn wenigstens mit solcher Schärfe abgeleitet werden, daß der neue Fund leicht weiter verfolgt und gesichert werden kann. An diese Aufgabe denkt der Astronom zunächst, wenn er von »Bahnbestimmung« spricht, und es soll hier versucht werden, die Natur dieses Problems klar zu machen, ohne daß wir dem rechnenden Astronomen ganz bis in seine von mathematischem Rüstzeug starrende Werkstatt folgen wollen.

Die wesentlichste Vereinfachung, die man sich bei dem Problem der raschen Bahnbestimmung eines neuentdeckten Himmelskörpers erlaubt, ist die, daß man von allen Störungen durch die Planeten absieht und voraussetzt, daß er allein der Anziehung der Sonne folgt. Dann gelten für seine Bewegung die Keplerschen Gesetze in ihrer durch Newton erweiterten Form.

Das erste Keplersche Gesetz sagt aus, daß sich der Himmelskörper in einem Kegelschnitt bewegt, in dessen einem Brennpunkt die Sonne steht. Bei den Planeten ist der Kegelschnitt eine Ellipse, bei den Kometen auch eine Parabel oder eine Hyperbel. Die Kegelschnitte hängen in folgender Weise zusammen.

Wird durch einen geraden Kreiskegel eine Ebene so gelegt, daß sie den Kegelmantel überall schneidet, so entsteht eine Ellipse (*EE* Fig. 27). Die Größe der Ellipse ändert sich mit der Neigung der Schnittebene gegen die Kegelachse. Sie ist für eine gegebene Entfernung von der Spitze des Kegels am kleinsten, und die Ellipse wird zum Kreise (*KK*), wenn die Ebene senkrecht auf der Achse steht; sie wächst, wenn der Neigungswinkel abnimmt. Läßt man nun die Neigung gegen die Achse immer geringer werden, so wird die Schnittebene schließlich einer Seite des Kegels parallel. In diesem Falle haben wir eine Parabel (*PP*). Wird die Neigung noch geringer, so entsteht eine Hyperbel (*HH*), welche, wie man aus der Figur ersieht, aus zwei getrennten Zweigen besteht. In dem in der Figur dargestellten speziellen Fall, wo die Schnittebene der Kegelachse parallel ist, hat man die gleichseitige Hyperbel. Die Ellipse ist eine geschlossene Figur und der Kreis ein spezieller Fall von ihr; Parabel und Hyperbel dagegen sind nicht geschlossene Kurven, sondern haben Arme, die sich in die Unendlichkeit erstrecken (Fig. 28).

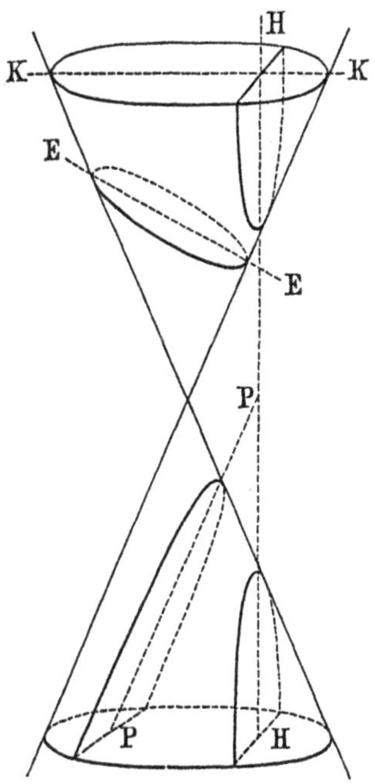

Fig. 27.

Wenn man sich hiernach alle möglichen Kegelschnitte vorstellt, so unterscheiden sich dieselben in zweierlei Hinsicht, erstens durch ihre Form, zweitens durch ihre Größe. Die Form eines Kegelschnitts bezeichnet man am besten, indem man seine Exzentrizität angibt, seine Größe charakterisiert man durch den sogenannten Parameter.

Die *Exzentrizität* (e) der Ellipse ist das Verhältnis des Abstandes FM eines Brennpunktes (F) vom Mittelpunkte (M) zum halben größten Durchmesser oder zur halben großen Achse ($a = ME$, also $e = \frac{FM}{a}$), der *Parameter* ($2p$) die durch den Brennpunkt gehende Senkrechte AB. Der Brennpunkt F entspricht dem Durchschnittspunkte der Kegelachse. Im Kreise fallen Brennpunkt und Mittelpunkt zusammen, die Exzentrizität verschwindet, und der Parameter wird gleich dem Durchmesser des Kreises. In der Parabel, deren Achse unendlich groß ist, und deren Mittelpunkt daher gleichsam auch in unendlicher Entfernung vom Brennpunkte liegt, ist die Exzentrizität gleich eins; in der Hyperbel ist sie größer als eins. Ellipse und Hyperbel haben zwei Brennpunkte, die Parabel hat einen. Die Ellipse hat die Eigenschaft, daß die Summe der zu jedem Punkte gehörigen beiden Radii Vectores gleich der großen Achse ist; dies gibt für sie ein bequemes Konstruktionsmittel.

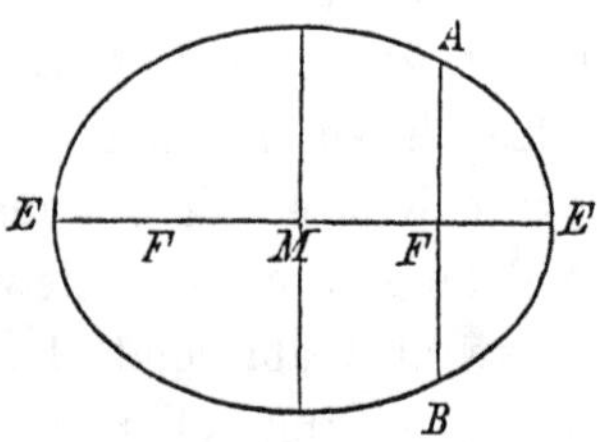

Ellipse.

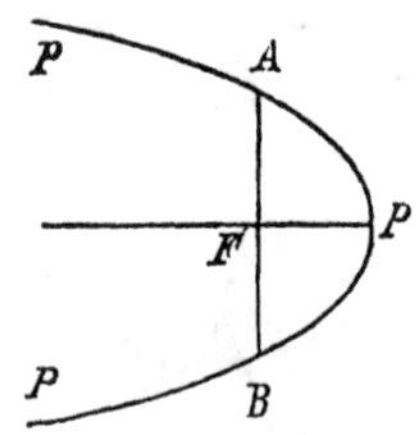

Parabel.

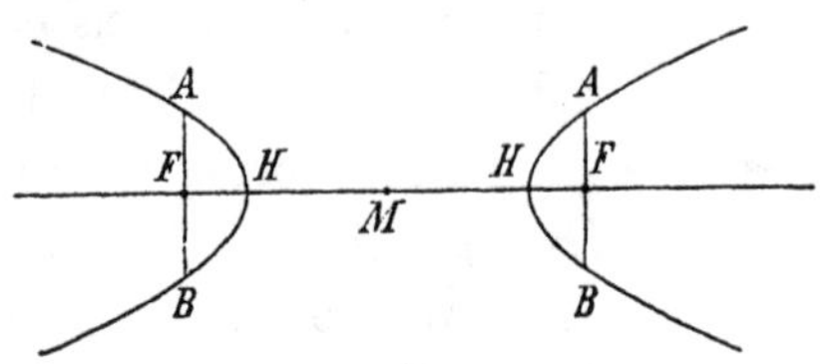

Hyperbel.

Fig. 28.

Wenn man nun die Bahn eines Himmelskörpers bezeichnen will, so hat man zunächst Exzentrizität und Parameter des Kegelschnitts, welchen er beschreibt, anzugeben. Man hat ferner die Lage des Kegelschnitts im Raum zu fixieren. Der eine Brennpunkt liegt freilich fest in der Sonne, aber es kommen Kegelschnitte in allen möglichen durch die Sonne gehenden Ebenen in Betracht. Die Ebenen der Planetenbahnen liegen zwar mehr oder weniger nahe der Ebene der Erdbahn, der Ekliptik, aber die Kometen laufen zum Teil in Ebenen, die fast senkrecht zur Ebene der Erdbahn stehen. Man fixiert jede Bahnebene, indem man ihre *Neigung* gegen die Erdbahn angibt und die *Länge des aufsteigenden Knotens*, d. h. den Winkelabstand desselben vom Frühlingspunkte. Schließlich hat man noch festzustellen, wie der Bahnkegelschnitt innerhalb der Bahnebene orientiert ist. Das geschieht am einfachsten, indem man den Winkel zwischen der großen Achse der Bahn und der Knotenlinie angibt. Man wählt dabei dasjenige

Ende der großen Achse, welches der Sonnennähe, dem Perihel, entspricht, und bezeichnet den Winkel als *Abstand des Perihels vom Knoten.*

Es sind also, um die Bahn eines Himmelskörpers zu fixieren, im ganzen fünf *Elemente* anzugeben: Exzentrizität, Parameter, Knotenlänge, Neigung, Abstand des Perihels vom Knoten.

Es erübrigt nur noch, die Mittel namhaft zu machen, mit denen man nun auch noch für jeden Himmelskörper den Punkt seiner Bahn festlegt, an dem er sich zu einer beliebigen Zeit befindet. Dazu muß man zunächst einmal angeben, an welchem Punkt sich der Himmelskörper zu einer bestimmten Zeit befand. Man pflegt hierfür die Zeit des Perihels zu wählen, gibt also an, zu welcher Zeit der Körper in der Sonnennähe stand. Mit dieser *Perihelzeit* als sechstem Element sind alle notwendigen Angaben beisammen. Denn die späteren Stellungen des Himmelskörpers folgen aus dieser Anfangsstellung mit Hilfe des zweiten und dritten Keplerschen Gesetzes, welche man in die Aussage zusammenfassen kann: Die in der Zeiteinheit vom Radiusvektor überstrichene Fläche ist der Quadratwurzel aus dem Parameter der Bahn proportional.

Auf Grund dieser Regel aus den sechs Elementen den Ort eines Himmelskörpers für eine Reihe von Tagen vorauszusagen, eine *Ephemeride* zu rechnen, ist kein allzuschweres, bereits von Kepler vollständig gelöstes Problem.

Das Problem der »Bahnbestimmung« ist nun die Umkehrung des Problems der Ephemeridenrechnung. Es stellt die Forderung, aus Beobachtungen an einigen Tagen die sechs Elemente des Kegelschnitts abzuleiten. Die geometrisch-mechanische Fragestellung ist dabei folgende. Durch die Beobachtung eines Gestirns wird sein scheinbarer Ort unter den Gestirnen gegeben, mit anderen Worten die Richtung, in welcher es von der Erde aus erscheint, der Sehstrahl, auf welchem es sich befindet. Hat man einen Himmelskörper an drei Tagen beobachtet, so hat man hiermit drei Sehstrahlen, drei gerade Linien im Raume, auf denen sich der Körper zu diesen drei Zeiten befunden haben muß. Sind E_1, E_2, E_3 (Fig. 29) die Stellungen der Erde zu den drei Beobachtungszeiten auf ihrer bekannten Bahn EE, bezeichnen ferner die Geraden E_1A_1, E_2A_2, E_3A_3 die drei Sehstrahlen, so weiß man, daß sich der Himmelskörper zu den Beobachtungszeiten irgendwo auf diesen Sehstrahlen befunden haben muß,

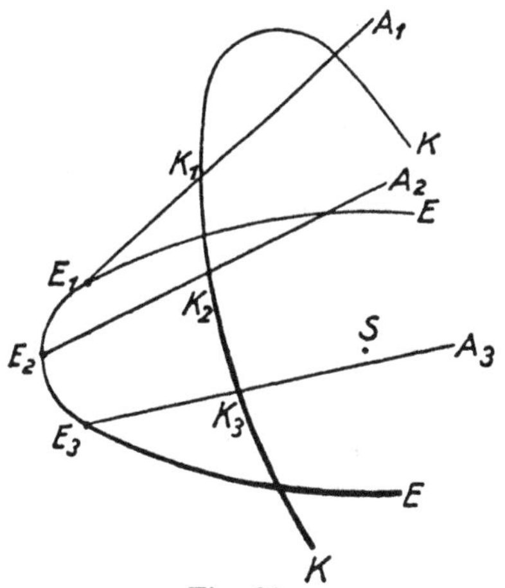

Fig. 29.

Newcomb-Engelmann's Populäre Astronomie, 5th ed., ed. by P. Kempf (Leipzig 1914) 89–93

ohne daß man den genauen Ort auf jedem Sehstrahl, die Entfernung von der Erde, aus der Beobachtung direkt entnehmen könnte. Es bleibt die Aufgabe, einen Kegelschnitt KK mit der Sonne S in einem Brennpunkt so zu legen, daß er erstens durch die drei Sehstrahlen hindurchgeht, und daß zweitens ein in diesem Kegelschnitt nach den Keplerschen Gesetzen sich bewegender Himmelskörper zu den betreffenden Beobachtungszeiten auf den Sehstrahlen in den Punkten K_1, K_2, K_3 ankommt. Drei Beobachtungen ergeben durch die Forderungen, daß die Sehstrahlen von der Bahn geschnitten und zur Beobachtungszeit vom Himmelskörper erreicht werden sollen, gerade sechs Gleichungen für die unbekannten sechs Bahnelemente. Diese Gleichungen sind zu verwickelt, als daß sie direkt aufgelöst werden könnten. Die Auflösung ist vielmehr nur durch eine Reihe aufeinanderfolgender Versuche erreichbar. Olbers ist es zuerst gelungen, diese Versuche für den Fall parabolischer Bahnen, wie man sie bei den meisten Kometen voraussetzen kann, so anzuordnen, daß sie auf kurzem, geradem Wege zum Ziel führen. Die Voraussetzung parabolischer Bahn bedeutet, daß man die Exzentrizität $e = 1$ setzt, so daß statt der sechs nur noch fünf unbekannte Elemente gesucht sind. Auch ist die parabolische Bewegung an sich besonders einfach. Gauß hat in seiner »Theoria motus corporum coelestium« dasselbe für den allgemeinen Fall der Bewegung in einem beliebigen Kegelschnitt geleistet.

So oft bei der astronomischen Zentralstelle in Kiel aus irgendwelchen Teilen der Welt Beobachtungen eines neuentdeckten Himmelskörpers von drei verschiedenen Tagen eingelaufen sind, werden sofort — im Prinzip nach den Methoden von Gauß oder Olbers, denen im Laufe des letzten Jahrhunderts nur unwesentliche Verbesserungen hinzugefügt wurden, — die sechs Bahnelemente gerechnet und nebst der aus ihnen folgenden Ephemeride an die Sternwarten hinausgesandt.

Es war hier natürlich nur möglich, die Methode der Bahnbestimmung von Planeten und Kometen im knappsten Umrisse anzudeuten. Wer sich etwas eingehender mit diesem Gegenstande beschäftigen will, findet in der Schrift von J. Frischauf, »Grundriß der theoretischen Astronomie und der Geschichte der Planetentheorien.« 2. Aufl. Leipzig, Engelmann, 1903, eine ausführlichere, aber immer noch elementare Darstellung. Für ein gründliches Studium können »Die Bahnbestimmung der Himmelskörper« von J. Bauschinger, Leipzig, Engelmann, 1906, oder die Originalwerke von Gauß »Theorie der Bewegung der Himmelskörper usw.« Hannover, 1865, und Olbers »Abhandlung über die leichteste und bequemste Methode, die Bahn eines Kometen zu berechnen.« 3. Aufl., Leipzig, 1864, empfohlen werden.

Section 2

Sun and Stellar Atmospheres

Sun and Stellar Atmospheres

Commentary by W. Mattig, Freiburg

Schwarzschild's scientific contributions to the fields of solar physics and of the physics of stellar atmospheres were concentrated in two periods. The first began with his observation of the total solar eclipse of 1905 August 30 in Algeria, and covered the years 1905–1906. After several years break, he again became active in this field in 1913–1914.

Although his experimental and theoretical contributions set the standard for decades, to some extent it was only years later that they had a real effect. His two theoretical publications "Ueber das Gleichgewicht der Sonnenatmosphäre" ["On the Equilibrium of the Solar Atmosphere"] of 1906 and "Über Diffusion und Absorption in der Sonnenatmosphäre" ["On Diffusion and Absorption in the Solar Atmosphere"] of 1914 remain part of the basis of the physics of stellar atmospheres. The exceptional importance that these papers had, and still have, is easily seen by anyone who flicks through the long-standing, standard work by Albrecht Unsöld "Physik der Sternatmosphären" ["Physics of Stellar Atmospheres"] (1937, 1955). The name of Karl Schwarzschild occurs again and again, and there are 14 references to it in the index to the second edition.

But his large, and primarily experimentally orientated paper "Ueber die totale Sonnenfinsternis vom 30. August 1905" ["On the Total Solar Eclipse of 30 August 1905"] [2.1], contains material that is still instructive and of scientific importance. P. ten Bruggencate (1966) has written of it: "Among Schwarzschild's papers on solar physics, we should not omit to mention his comprehensive research on the total solar eclipse of 30 August 1905. This paper still constitutes a treasure trove for astrophysicists." The situation has not changed today.

The paper is marked by an impressive clarity, each individual operation being comprehensively described. We can see here how photographic photometry was first applied to solar physics. The important observational results are the more remarkable, when we consider the short period of time in which the expedition was prepared. The description begins with the words: "In April 1905, Professor C. Runge and I went to the Ministry with a request for a travel grant in order to observe the total solar eclipse that 30 August. Our original intention was to participate in a foreign expedition in order to become familiar with the phenomena and with observational methods, so that with the experience thus gained, we might be able to plan more extensive projects for a later eclipse. ... As however we were awarded a very generous travel grant, we were unable to refrain from making as many preparations for independent observation of the eclipse as the shortness of the time, the demands of the academic term, and the lack of resources available to us allowed."

During the eclipse the sky was brilliantly clear, all the instruments functioned perfectly, and the expedition, which also included Dr. R. Emden, was a resounding success.

The important results, which are still of significance today, may be summarized briefly:

1) For the first time the changes in brightness of the corona were precisely linked photometrically, by photography, with that of the uneclipsed Sun.
2) Analysis of slitless spectrograms of the chromosphere and corona was carried out. Regarding this, Unsöld wrote in 1955: "As the Moon moves across the Sun at a precisely determined rate during an eclipse, a series of exposures at known intervals provides exceptionally valuable information about the various layers and structure of the chromosphere. Although Schwarzschild made use of this simple concept as long ago as 1905, the paper unfortunately received very little attention, and it was only in 1932 that the procedure was developed further by D.H. Menzel with his "Jumping-Film" spectrograph.

This part of the paper is (a) a textbook example of care in data reduction and (b) outstanding for its exceptionally clear mathematical statement of the problem. The most important result established is that the spectral intensity-distribution in the inner corona in the region between λ 350–500 nm is almost the same as the intensity-distribution at the centre of the Sun's disk. In discussing the interpretation of this result, the hypothesis advanced as an explanation is that the effect is caused by the scattering of sunlight by free electrons. This theory was later fully confirmed.

The brightness at the solar limb is almost constant. At 455 nm it amounts to $I_{\text{(solar limb)}}/I_{\text{(centre of the disk)}} = 0.17$. From the decline in the brightness of the solar crescent with time the variation between the centre and the limb may be determined for the extreme limb. This method is largely free of atmospheric and instrumental effects, but an accurate knowledge of the profile of the lunar limb is required.

Absolute magnitudes of prominences were determined for the first time in the hydrogen $H\beta$–$H\gamma$ lines and the calcium H and K lines. Schwarzschild defined as unit the equivalent width referred to the spectrum of the solar limb. It turned out "that the magnitude of bright prominences in the spectral regions in which they radiate is of the same order of magnitude as that of the Sun itself", to put it in Schwarzschild's own words. From the ratio of intensities it was possible to deduce that in the higher Balmer lines prominences are optically thin, but that this did not apply in the H and K lines.

The observed chromospheric lines, their estimated intensities and range in height were given in a table. This contains 274 measured lines in the region between 334 and 459 nm, 35 of which are unidentified.

As is still normal nowadays, Schwarzschild also described his impressions of the expedition and of the eclipse for a wider audience in a popular journal [2.2]. The account closes with the words that have applied to later generations of astronomers: "As we came back to Germany, it was as if we were having to struggle into clammy clothes that were still damp from the last rain. The week-long grey skies were really depressing after the unchanging, clear blue ones in the south. ... This is why all the astronomers who went on this eclipse trip came to long for an observatory under the favourable southern skies. That is where we should observe and photograph stars; it is easy enough to return to our own cool, less enervating, country for reflective, theoretical work". In recent decades Schwarzschild's wish has been fulfilled, as witnessed by the observatories on Calar Alto in Spain, on the Canary Islands, and in Chile. And the address of the European Southern Observatory in Garching near Munich is – Karl-Schwarzschild-Str. 2.

Almost simultaneously with reducing the eclipse data, Schwarzschild, together with W. Villiger, measured the limb-darkening in the UV at 320 nm. At the time there were few measurements of the overall centre-limb variation, just those of Vogel in the visible, and of Verny in the near infrared at 1.5 μ. The individual measurements, which were published in the *Physikalische Zeitschrift* [2.3], and also, corrected and supplemented, in *Astrophysical Journal* [2.4], are no longer of great significance. Nevertheless, we should note that, once again, Schwarzschild was the first to carry out an important type of observation. The method with which the UV region was selected is interesting. He used an objective of ultraviolet glass, which had been coated with a layer of silver. The transmission of the glass determined the cut-off in the UV and the reflecting properties of the silver layer the cut-off in the red. With careful photographic photometry the authors attained an accuracy of about 1 %, the numerical values being determined to somewhat better than 1 % at the limb.

Together with the limb-darkening measurements at 320 nm, Schwarzschild and Villiger also measured intensities of the umbra and penumbra of large sunspots and published a note of these [2.5]. It is remarkable that they measured intensities as low as about 10 % of that of the undisturbed photosphere; we now know that scattered light considerably degrades such measurements. The intensity of faculae were also measured; their brightness never exceeds 20–30 % of that of the undisturbed photosphere.

At almost the same time as this experimental work was being carried out, Schwarzschild formulated his basic theoretical ideas; he presented his paper "Ueber das Gleichgewicht der Sonnenatmosphäre" ["On the Equilibrium of the Solar Atmosphere"] [2.6] at the meeting of the Göttingen Academy of Sciences on 13 January 1906. The contents of this paper remain one of the pillars of the theory of stellar atmospheres. It formulated the radiative-transfer equation, and introduced the concept of radiative equilibrium, which was consistently applied. In the introduction Schwarzschild wrote: "Until now, most attention has, in general, been given to so-called adiabatic equilibrium, as is found in our atmosphere, when it is thoroughly mixed by ascending and descending currents. Here I should like to call attention to another form of equilibrium, which may be called "radiative equilibrium". Schwarzschild's statement is therefore very cautious, although in the same paper he shows that in the Sun's atmosphere energy transport takes place via radiation. Unsöld (1955) said: "Whereas the older radiation theory was restricted to dealing with the radiation field under thermal equilibrium, Schwarzschild was the first to make a consistent attempt to explain the stable radiative state, and in doing so laid one of the foundations for theoretical astrophysics."

Apart from these general considerations, there are a few details that are of significance:

1) Assuming hydrostatic equilibrium and an isothermal atmosphere, he calculated the "height of the so-called homogeneous atmosphere", or equivalent height, of a pure hydrogen atmosphere as being 212 km. When seen from the Earth, 1 arcsec corresponds to 725 km, so it follows "that the Sun must appear perfectly sharply defined." He also indicates that even with adiabatic equilibrium, the thickness of the atmospheric layer amounts to 300 km.
2) Of at least equal importance to the discussion of radiation transport is the derivation of Schwarzschild's stability criterion. In just a few sentences, he formulates the criterion named after him, and which determines when convection can occur and when the atmosphere is in radiative equilibrium. Convection takes place when the adiabatic temperature gradient is smaller than that for radiative equilibrium. The temperature gradients were calculated, i.e. in the case of radiative equilibrium the relation ship between depth and temperature is given by $T^4 = \frac{1}{2}T_e^4(1 + 2\tau)$, where T_e is the effective temperature and τ the optical depth. From the comparison of the two temperature gradients, i.e. from purely theoretical considerations, Schwarzschild then concluded: "Radiative equilibrium is therefore stable everywhere, as long as the gas that forms the atmosphere is monatomic, diatomic or triatomic. For polyatomic gases, instabilities would arise in the deeper layers (at higher temperatures T)."
3) From the equation of radiative transfer that he had derived, it is possible, by integration, to obtain the amount of radiation emitted at the stellar surface over an angle ϑ, which means that limb-darkening for radiative equilibrium and for adiabatic equilibrium may be compared with observations. This is shown in Fig. 2.1.

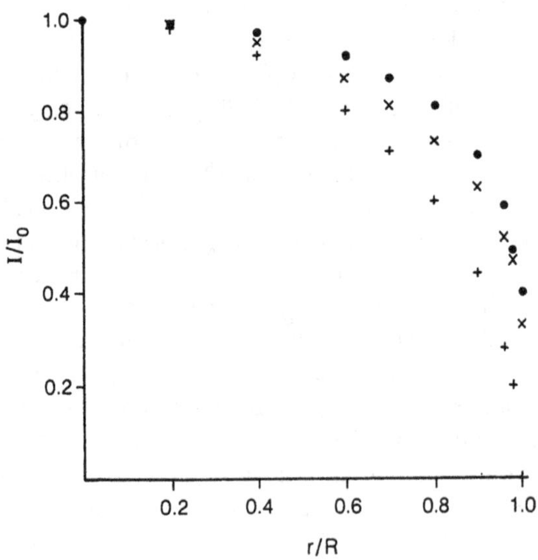

Fig. 2.1. Centre-limb variation in intensity. Observations (• • •); Calculation with radiative equilibrium (xxx); Calculation with adiabatic equilibrium (+++)

This fundamental work closes with the words: "It will be seen that radiative equilibrium describes the brightness distribution across the Sun's disk as well as we can expect with the simplified treatment given here, and that it produces a completely different appearance of the solar disk than that predicted by adiabatic equilibrium. The introduction of radiative equilibrium would appear to be justifiable on empirical grounds".

After this basic treatment of radiative equilibrium, there were no published scientific papers on solar physics for many years. It was 1913 before Schwarzschild discussed radiative equilibrium in the Earth's atmosphere [2.7], where he generalized R. Emden's idealized treatment. Then, in a paper with H. Rubens [2.8], measurements of solar radiation in the far infrared between 100μ and 600μ were discussed, showing that solar radiation in this region would be strongly absorbed in the Earth's atmosphere .

In a short paper with G. Eberhard [2.9], it was pointed out that in some stellar spectra, and in particular in σ Gem, emission cores are found in the H and K absorption lines. The question was raised of whether the strength of the emission lines showed a variation with time, similar to the sunspot cycle. Subsequently a positive answer has been given to this question.

Then in 1914 Schwarzschild presented his second important paper [2.10] concerning problems of the physics of stellar atmospheres to the Prussian Academy of Sciences in Berlin. Whereas the 1906 paper discussed continuum radiation, in this paper "Über Diffusion und Absorption in der Sonnenatmosphäre" ["On Diffusion and Absorption in the Solar Atmosphere"], the problem of the formation of absorption lines is discussed. "In order to make further progress, it would seem to be important for an accurate study of the solar spectrum to be carried out. Apart from the quantity of light available, there is the big advantage that it is possible to study various points of the solar disk, and thus the light emitted at various angles, whereas with stars all that is available is the total light emitted by the whole of their disks". This is what Schwarzschild says in his introduction, again emphasizing the fundamental significance of the intensity variation, this time in the absorption lines, between the centre and the limb. Here Schwarzschild harks back to the important paper by A. Schuster (1905) and generalizes the problem of energy transport, in the sense that this is considered for various angles of incidence. Otherwise the most important of Schuster's propositions are adopted: an absorbing and scattering atmosphere is considered, overlying a radiating black-body at a specific temperature. This shell model is still known as the Schuster-Schwarzschild model. An important point here is the introductory statement that in such a model with a layer that exhibits absorption only, the change in the Fraunhofer lines from centre to limb cannot be explained. The re-emission in the lines therefore arises not just from emission in accordance with Kirchhoff's Law but also from a form of scattering analogous to the Rayleigh scattering of light by small particles. The discussion of this problem led to Schwarzschild's integro-differential equation for the transfer of radiation.

As well as the limiting case of pure absorption, the other limiting case of scattering is particularly discussed. Here E.A. Milne later introduced the concept of monochromatic radiative equilibrium. Schuster's original approximation was improved, but the differences were only small.

Comparison of the theory with observation of the strongest lines in the solar spectrum, the H and K lines, at the centre and at the limb, then offered the possibility of distinguishing whether the radiative transfer in the lines occurs predominantly through absorption or through scattering. In the case of absorption – nowadays we talk of "true absorption", or, following Milne, of "local thermodynamic equilibrium" – the lines vanish at the solar limb. With scattering, the appearance of the spectrum does not change as one moves from the centre to the solar limb. Only the overall brightness of the spectrum changes. This largely corresponds to observation, whence Schwarzschild concluded: "It is therefore probable that scattering of light plays an important part in the phenomena occurring in the solar atmosphere."

Nowadays, this conclusion of Schwarzschild's is no longer regarded as tenable. In discussing radiative-transfer problems a distinction must be made between the influence of radiative transfer and that of the model. Over the years it has been established that, apart from a few exceptions, the model plays a significantly greater role, and that in many cases the shape of the spectral lines can be correctly explained by "pure absorption".

The most important point about this paper is that it provided the basic theoretical tool, which is still of fundamental importance today, in the days of NLTE (non-local thermodynamic equilibrium).

His last paper in the field of solar physics [2.11], an experimental one, was also published in 1914. According to the General Theory of Relativity the Sun has a "relativistic redshift" of 0.635 km/s. Schwarzschild tried to measure this relatively small effect by comparing the wavelengths of the solar oxygen bands at 388.3 nm with a carbon arc. He found notably smaller values for the centre of the Sun than those predicted by theory. This has been dramatically confirmed in later years. He also detected the variation between the centre and the limb, now known as the "limb effect".

Schwarzschild came to the conclusion that "the Einstein Effect, a redshift of 0.63 km/sec, is certainly not clearly detectable in the observations. A particularly telling point against the Einstein Effect is the fact that all the lines investigated in the solar spectrum ... generally show only very weak shifts". We now know that this effect is caused by convection in the solar atmosphere.

Between the two period of particularly fruitful scientific work, we also find a very comprehensive review [2.12] of Emden's book "Gaskugeln". No less than 30 pages are devoted to what amounts to a precis of the exceptionally comprehensive book.

As mentioned earlier, the second phase of Schwarzschild's work on solar physics began in 1913. We cannot exclude the possibility that the observation of the solar eclipse of 16/17 April 1912 gave this a fresh impetus. At all events, H. Struve (1912) recounts how Schwarzschild and Müller attempted to photograph the Moon's shadow from above from a Zeppelin.

Naturally Schwarzschild also recognized the importance of scientific meetings. In short accounts he reported on the fourth meeting of the International Association for Cooperative Solar Research in 1910 [2.13], and the meeting of the Solar Union in 1913 [2.14]. In 1914 Schwarzschild gave a lecture "Zur Physik der Sonne" ["Physics of the Sun"] to the Berlin Urania society, showing that he also took the trouble to impart scientific knowledge to a wider audience.

References*

ten Bruggencate, P. (1966) *Sterne und Weltraum* **5**: 188
Schuster, A. (1905) *Astrophys. J.* **21**: 1
Struve, H. (1912) *Astron. Nachr.* **191**: 187
Unsöld, A. (1937, 1955) *Physik der Sternatmosphären* (Springer, Berlin, Heidelberg) 1st and 2nd editions

* The reference numbers in the text for Schwarzschild's papers correspond to the contents on page XI and to the bibliography on page 493.

Abhandl. Kgl. Ges. d. Wiss. zu Gött. NF 5(2) 1907, 3–73

Ueber die totale Sonnenfinsternis vom 30. August 1905.

Von

K. Schwarzschild.

Vorgelegt in der Sitzung vom 8. Dezember 1906.

Vorbemerkung und Inhaltsverzeichnis.

Der erste und der letzte Abschnitt dieses Berichtes geben eine in sich abgeschlossene Darstellung der Göttinger Sonnenfinsternisexpedition und ihrer Resultate. Die dazwischen liegenden Teile enthalten das Detail der Verarbeitung und sollen dem Leser, der selbst mit einer ähnlichen Unternehmung zu tun hat, zur spezielleren Orientierung dienen.

Der Inhalt gliedert sich folgendermassen:

K. SCHWARZSCHILD,

I. Abschnitt. Allgemeines.

§ 1. Reise.

Im April 1905 wandten sich Professor C. Runge und ich an das Ministerium mit der Bitte um eine Reiseunterstützung zur Beobachtung der totalen Sonnenfinsternis vom 30. August dieses Jahres. Unsere ursprüngliche Absicht war, im Anschluss an irgend eine fremde Expedition die Erscheinungen und Beobachtungsmethoden kennen zu lernen, um dann auf Grund der gewonnenen Erfahrungen für eine spätere Finsternis grössere Unternehmungen planen zu können. Als Beobachtungsstation wählten wir nach einer Korrespondenz zwischen Prof. Runge und Prof. F. Newall in Cambridge Gelma in Algerien, den Ort, den Herr Newall für seine Expedition in Aussicht genommen hatte. Als uns aber jene Reiseunterstützung gütigst gewährt worden war, konnten wir es nicht unterlassen, noch so viel Vorbereitungen für eine selbständige Beobachtung der Finsternis zu treffen, als sich bei der Kürze der Zeit, dem Drange des Semesters und der Ausdehnung der für uns verfügbaren Hülfsmittel ermöglichen liessen. Prof. Runge übernahm für die Göttinger Gesellschaft der Wissenschaften luftelektrische Messungen auf der Fahrt und während der Finsternis. Bei einem gelegentlichen Besuch in Jena stellte die Firma Zeiss uns eine vollständige Prismenkamera aus Ultraviolettdurchlässigem Glas zur Verfügung. Ich beschloss, als Stativ für dieses Instrument die Repsold'sche parallaktische Montierung mitzunehmen, die ursprünglich bei der Venusexpedition vom Jahre 1884 ein Heliometer trug und die seitdem auf der Göttinger Sternwarte als Montierung des Merz-Refraktors gedient hat. Ein kürzlich erworbenes lichtstarkes Petzval-Objektiv von Busch-Rathenau konnte an Stelle des Gegengewichtes noch an diesem Stativ befestigt werden und sollte zu direkten Aufnahmen der Korona dienen. Dr. R. Emden in München wollte in Gelma mit uns zusammentreffen und eine gute Handkamera zu weiteren Koronaaufnahmen mitbringen, bei denen er insbesondere die Wirkung von Gelbscheiben zu studieren beabsichtigte. Nachdem noch in Eile und unter angestrengter Mitarbeit des Assistenten der Sternwarte Dr. Meyermann und des Institutswärters H. Berger ein Beobachtungszelt und die nötigen kleineren Hülfseinrichtungen konstruiert worden waren, erfolgte die Abfahrt am 5. August von Hamburg aus mit dem Dampfer „Byzanz“ der de Freitas-Linie. Unser Dampfer, auf den uns Herr Prof. Schorr in Hamburg freundlichst aufmerksam gemacht,

Abhandl. Kgl. Ges. d. Wiss. zu Gött. NF 5(2) 1907, 3–73

war ein Frachtdampfer und hatte 4 Passagiere an Bord. Die Liebenswürdigkeit des Kapitäns Herrn Röfer ermöglichte uns, auf der ganzen Reise, die zunächst nach Newcastle in England ging und von dort, ohne einen weiteren Hafen anzulegen, nach Algier führte, in die nautische Leitung der Fahrt Einsicht zu nehmen und durch eigene Ortsbestimmungen und Uhrkontrollen uns in astronomischer Nautik umzutun. Wir landeten am 18. August in Algier. Von dort führte uns die Bahn am 21. August nach Gelma.

In Gelma war bereits eine grosse Anzahl von Astronomen versammelt. Die Verteilung der Beobachtungsplätze war von Herrn Trépied, der mit seiner Expedition aus Algier gekommen war, in Gemeinschaft mit dem Sous-maire der Stadt, Herrn Perrette, und dem Stadtbaumeister Herrn Joly in vortrefflicher Weise organisiert. Uns wurde ein geeigneter Platz auf der Bühne des alten römischen Amphitheaters zugewiesen. Das Gewölbe, welches den ehemaligen Eingang des Theaters überdeckte, wurde durch Holztüren verschlossen und diente als Arbeits- und Aufbewahrungsraum. Auch wurden uns als Hülfskräfte ein elsässischer Arbeiter, Mr. Schneider, und ein Neger bestimmt, welch letzterer zur Bewachung auch die Nächte neben unsern Instrumenten schlafen musste.

Am 23. und 24. August stellten Prof. Runge und ich mit diesen Gehülfen das Zelt und die Instrumente auf. Am Abend des 24. traf Dr. Emden ein. An den folgenden Tagen wurde die Aufstellung des Instrumentes justiert. Es wurden Fokusierungen vorgenommen und ein paar kleine Adaptationen ausgeführt, bei denen wir uns der freundlichsten Unterstützung von Seiten des Herrn Newall zu erfreuen hatten. Am 26. riss derselbe Gewittersturm, der auch der Hamburger Expedition im benachbarten Soukh-Ahras Ungelegenheiten bereitete, das Zelt auseinander, ohne aber den Instrumenten Schaden zuzufügen.

Dr. Emden's Handkamera konnte noch auf dem Stativ montiert werden. Da er dieselbe bedienen wollte und Prof. Runge durch luftelektrische Messungen in Anspruch genommen war, so fehlte uns noch ein Gehülfe an den Instrumenten. Wir fanden eine geeignete Persönlichkeit in einem sehr geschickten Schreiner und Photographen aus Gelma namens Henry. Bei der Finsternis selbst half ferner noch ein Schüler des Gymnasiums zu Constantine, Herr R. Vidal, beim Kassettenwechsel und ein junger Beamter der Stadt Gelma, Herr H. Salfati, übernahm das Sekundenzählen. Natürlich wurde am Tage vor der Finsternis das Programm gut eingeübt.

Am 28. und 29. August veranstaltete die Stadt Gelma Festlichkeiten zu Ehren der fremden Gäste. Da die Zahl der täglichen Arbeitsstunden bei einer Temperatur, die bis zu 42° C. im Schatten stieg, eine sehr beschränkte war, so waren wir eben mit allen Vorbereitungen fertig, als der Tag der Finsternis kam. Der 30. August begann völlig klar und blieb auch ohne die Mittagswolken, die sich an den vorhergehenden Tagen regelmässig gebildet hatten. Eine Stunde vor Beginn der Finsternis versagte die Birne des Momentverschlusses, doch konnte sie glücklicher Weise noch ersetzt werden. Die Aufnahmen während der Totalität gingen bis auf kleine Programmwidrigkeiten ordnungsgemäss von statten.

Verlauf und Beobachtung der Finsternis im Einzelnen wird weiter unten ausführlich geschildert. Am Abend des 30. und 31. entwickelten wir unsere Aufnahmen in der Dunkelkammer des Gelmaer Photographen unter Verbrauch von 10 kg Eis. Nach der Finsternis erfolgten lehrreiche gegenseitige Besichtigungen der verschiedenen astronomischen Stationen. Auch hatten wir uns am 31. des Besuches der Hamburger Kollegen aus Soukh-Ahras zu erfreuen.

Nachdem wir am 1. Sept. die Instrumente verpackt und den 2. als Ruhetag gehalten hatten, begann am 3. die Rückreise auf dem Wege über Tunis.

Der vorstehende Bericht deutet an, wie vielfach wir dem Magistrat und den Bürgern von Gelma, sowie den Kollegen, die wir dort trafen, zu Dank verpflichtet sind, ohne dass alle Helfer in der Not genannt wären. Unsere dreiköpfige Truppe selbst hat alle ihre Unternehmungen als gemeinsame betrachtet und da die Hauptarbeit auf die astronomischen Aufgaben fiel, so kann man daraus ermessen, wie sehr diese Auffassung von Seiten Prof. Runge's und Dr. Emden's ein Zeichen freundschaftlicher Uneigennützigkeit war.

§ 2. Instrumentelle Ausrüstung.

Die astronomische Ausrüstung bestand aus folgenden 3 Instrumenten:

1) U. V. Prismenkamera von C. Zeiss (Jena). Das Objektiv war ein sog. Apochromat von 110 mm Oeffnung und 1100 mm Brennweite. Davor befand sich, nahe auf das Minimum der Ablenkung eingestellt, ein Prisma von 200 mm Seitenlänge und 45^0 brechendem Winkel, ebenfalls aus U. V. Glas. Bei der hieraus resultierenden grossen Lichtstärke waren Momentaufnahmen des flash-Spektrums vorzusehen. Es wurde deshalb in der Bildebene ein Steinheilscher Fallverschluss (Altostereoquart) angebracht, der eine Oeffnung von 20 mm auf 75 mm bot und daher für das Sonnenspektrum, welches bei 11 mm Breite nur 50 mm Länge hatte, bequemen Raum liess. Hinter dem Momentverschluss konnte eine Kasette entlang geschoben werden. Die Verschiebung war durch Zapfen, die auf der Rückseite der Kasette angebracht waren, erleichtert und so eingeteilt, dass auf eine $^{13}/_{18}$ cm Platte 4 Spektren aufgenommen werden konnten.

Für die Momentaufnahmen des Flash-Spektrums war indessen eine grössere Zahl von Aufnahmen erwünscht innerhalb einer Zeit, die keinen Plattenwechsel zuliess. Es wurden daher auch Aufnahmen auf einem Rollfilm ermöglicht. In Göttingen war ein Kodak für diesen Zweck adaptiert worden, durch ein Versehen bei der eiligen Herstellung aber so, dass der Film nicht ganz in die Brennebene gebracht werden konnte. Es traf sich daher sehr glücklich, dass Prof. Runge eine Frl. E. du Bois-Reymond gehörige Rollfilmkamera mitgebracht hatte, welche sich den Bedingungen besser fügte. Der Schreiner Henry versah sie mit einer Rückwand, die einen Ausschnitt entsprechend der Oeffnung des Fallverschlusses hatte und die zugleich ein bequemes Ansetzen der ganzen Handkamera an das hintere Ende des grossen Instrumentes ermöglichte. Der Film wurde so eingelegt, dass die empfindliche Schicht nach rückwärts — inbezug auf den gewöhnlichen Gebrauch der Kamera gesprochen — zu liegen kam. Es wurde also

Abhandl. Kgl. Ges. d. Wiss. zu Gött. NF 5(2) 1907, 3–73

von der Handkamera nicht der optische Teil, sondern nur die Rolleinrichtung benutzt. Da sich der Film in 1^s um 3 cm weiterrollen liess und in 1^s Aufziehen und Auslösen des Momentverschlusses erfolgen konnte, so war es hiermit möglich, eine grosse Reihe von Aufnahmen in 1—2^s Intervallen zu erhalten.

2) Petzvalobjektiv von Busch (Rathenau). Oeffnung 156 mm, Brennweite 700 mm. Ueber das Objektiv war als Kamera ein einfaches Blechrohr geschoben, an dessen unterem Ende eine Holzfassung zum Einschieben der Kasette angebracht war.

3) Handkamera von Dr. Emden, ausgerüstet mit Rietschel-Anastigmat von 25 mm Oeffnung, 120 mm Brennweite und mit Gelbscheibe.

Alle 3 Instrumente wurden auf das erwähnte Repsoldsche Stativ aufgesetzt, die Prismenkamera so, dass die Axe ihres Objektivs senkrecht zur Deklinationsaxe stand, an Stelle des Gegengewichts wurde ein Holzklotz auf das andere Ende der Deklinationsaxe geschoben und an diesen wurden die beiden andern Instrumente so angeschraubt, dass sie schräg gegen die Deklinationsaxe standen und dass die Sonne gleichzeitig in allen 3 Instrumenten erschien. Die Nachführung des Ganzen besorgte ein kräftiges Heydesches Uhrwerk. Auf Pointieren wurde verzichtet. Nur war ein kleiner Sucher angebracht, um die Aufstellung justieren und bequemer einstellen zu können.

Zum Sekundenzählen diente ein gewöhnlicher Metronom.

Zwei letzte nicht unwichtige Hülfs-Einrichtungen waren a) ein photometrischer Keil, aus einem braunen Glas und einer photographischen Platte von verlaufender Schwärzung hergestellt, der so in einen Kasettenrahmen eingelassen war, dass man auf den Rand jeder $^{13}/_{18}$ Platte mit Hülfe irgend einer Lichtquelle eine photometrische Vergleichsskala aufdrucken konnte und b) drei Blenden, die das Objektiv der Prismenkamera auf 3 mm, 1 mm und 0.5 mm abzublenden erlaubten.

§ 3. Beobachtungsprogramm.

Das vor der Finsternis eingeübte Programm der Aufnahmen war folgendes. (Als Anfang der Zeitzählung ist der Beginn der Totalität genommen).

Zeiss'sche Prismenkamera (Beobachter Schwarzschild)

Z_a) -20^s bis $+10^s$ Momentaufnahmen des flash-Spektrums auf Rollfilm mit Vergleichsspektren (Orthochromatischer Film von Lumière).

Z_b) 1) $1^m\ 0^s - 1^m\ 1^s$
2) $1\ 30 - 1\ 45$
3) $2\ 0 - 3\ 0$
} Aufnahmen des Koronaspektrums auf einer Platte mit Photometerkeil ($^{13}/_{18}$ cm Schleussner-Platte).

Z_c) $3^m\ 20^s$ – Ende der Totalität } Momentaufnahmen des flash-Spektrums auf einer Platte mit Photometerkeil ($^{13}/_{18}$ cm Schleussner-Platte).

Petzvalobjektiv (Beobachter Henry)

P_1)	$0^m 15^s - 16^s$	ohne Photometerkeil		Aufnahmen der Korona
P_2)	0 45 – 49	mit „		($^9/_{12}$ cm Schleussnerplatte)
P_3)	1 30 – 45	ohne „		
P_4)	2 0 – $3^m 0^s$	mit „		

Rietschel-Anastigmat (Beobachter Emden)

R_1)	$0^m 20^s - 24^s$	mit Gelbscheibe	Aufnahmen der Korona
R_2)	0 45 – 49	ohne „	($^9/_{12}$ cm gelbempfindliche
R_3)	1 0 – 15	mit „	Lumière-Platten)
R_4)	1 30 – 45	ohne „	
R_5)	2 0 – $3^m 0$	mit „	

Die Vergleichsspektren auf dem Film Z_a) sollten durch Momentaufnahme des Sonnenspektrums vor der Totalität bei Abblendung auf die Oeffnung 3 mm, 1 mm und 0,5 mm erhalten werden, die Photometerkeile jeweils durch Exposition von 15^s gegen die freie Sonne.

Die Expositionszeiten für die 4 Instrumente waren natürlich so gewählt, dass während keiner Exposition das Instrument von einem der drei Beobachter erschüttert wurde.

Das Programm, das uns ja im wesentlichen durch die verfügbaren Hülfsmittel vorgeschrieben war, hat vielleicht die Besonderheiten, dass wohl zum 1. Male so kurze Momentaufnahmen (etwa $^1/_{80}{}^s$) des flash erfolgten, dass für die photometrische Vergleichung nicht eine künstliche Lichtquelle, sondern die unverfinsterte Sonne selbst zur Verwendung kam und dass die Photometrie auch auf das Korona- und flash-Spektrum erstreckt wurde.

§ 4. Ausführung des Programms. Entwicklung der Platten.

Vor Eintritt der Totalität wurden die Vergleichsbelichtungen in folgender Weise ausgeführt.

M. Z. Gelma

$0^h 56^m$ Drei Moment-Spektralaufnahmen auf dem Rollfilm Z_a mit den freien Oeffnungen 3 mm, 1 mm, $^1/_2$ mm unmittelbar nach einander.

$1^h 14^m$	Platte	Z_b	Freie Exposition gegen die Sonne 15^s lang durch den photometrischen Keil hindurch.
17	„	Z_c	
20	„	P_2	
23	„	P_4	

Das Zeitsignal für die Aufnahmen sollte in der Weise gegeben werden, dass ich im Sucher die Reduktion der Sichel auf einen Winkel von 90^0 beobachtete und Herr Salfati dann zunächst von 45 auf Null rückwärts und dann vorwärts

Abhandl. Kgl. Ges. d. Wiss. zu Gött. NF 5(2) 1907, 3–73

nach dem Metronom Sekunden zählte. Mir ging nun im Sucher der Massstab für den vollen Sonnendurchmesser verloren, der an der kurzen Sichel nicht mehr recht zu erkennen war, und ich gab das Signal viel zu früh. Ich bemerkte meinen Irrtum aber noch vor Eintritt der Totalität, da bei einem freien Blick auf die Sonne die feine Sichel immer noch blendete, als Herr Salfati sich im Zählen schon der Null näherte, und ich rief daher den andern zu: „Alles um 20^s später“. Die Korrektur war richtig gegriffen. Denn als Herr Salfati im Vorwärtszählen bei 20 angekommen war, rief Prof. Runge, der in einiger Entfernung von uns mit einem Taschenspektroskop beobachtete, „flash“ herüber.

Es wurden nun auch alle Expositionen dementsprechend richtig ausgeführt. Wir können freilich nicht versichern, das wir die Anfangssekunden dem Programm gemäss genommen haben. Ausser jenem Irrtum erwies sich auch das Zählen auf Französisch bei dem angespannten Arbeiten doch als ein Hindernis. Aber die Expositions*dauern* wurden innegehalten und es wurden die Casetten immer nur gewechselt, wenn keiner der Beteiligten exponierte. Die einzige wesentlichere Aenderung des Programms war die, dass wir während der langen Exposition, die um $2^m 20^s$ der Zählung begann, beschlossen, dieselbe um $3^m 0^s$ der Zählung abzubrechen, also die Exposition von 1^m auf 40^s zu reduzieren.

Auf dem Film Z_a wurden um die Zeit des ersten Flash's 16 Expositionen (im folgenden als Nr. 4—Nr. 19 numeriert), auf der Platte Z_e um das Ende der Totalität 4 Expositionen (im folgenden Nr. 1—Nr. 4 numerirt) erhalten.

Die Entwicklung der Platten wurde immer von je zweien von uns gemeinsam ausgeführt. Wie erwähnt, wurde alles Wasser reichlich geeist. Es wurde folgende Entwicklungsart verwandt:

Platte R_1	Glycin	1:30	mit	25	Tropfen	Bromkalium	19^m	lang
R_2	„	„	„	„	„	„	4^m	„
R_3	„	„	„	„	„	„	22^m	„
R_4	„	„	„	„	„	„	4^m	„
R_5	„	„	„	„	„	„	15^m	„
P_1—P_4	„	„	„	30	„	„	je 5^m	„
Platten u. Film Z	Rodinal	1:120	mit	2	ccm	Bromkalium	je 4^m	„

Für den Film hatte Prof. *Runge* eine 1 m lange Schale aus Blech gebogen, sodass der ganze Film in einem entwickelt werden konnte, was für die photometrische Verwertung desselben unerlässlich war.

§ 5. Allgemeines über die erhaltenen Aufnahmen.

Sämtliche Aufnahmen erwiesen sich als im Grossen und Ganzen gelungen, nur hatte ich die Helligkeit aller Phänomene unterschätzt und der Photometerkeil reichte daher auf keiner Platte bis zu den stärksten auf ihr vorkommenden Schwärzungen. Die Vergleichsspektren auf dem Film waren richtig in der Schwärzung, aber z. T. durch einen Streifen falschen Lichts gestört. Es wird sich zeigen, dass trotzdem noch mannigfache — freilich in ihrer Zuverlässigkeit beeinträchtigte — photometrische Resultate aus den Platten gewonnen werden konnten.

Film Z_a. Die Fokusierung ist gut, bei einzelnen der Expositionen sogar vortrefflich. Es scheinen hierbei kleine Durchbiegungen des Films zu Hülfe gekommen zu sein. Die 16 Expositionen zeigen in aller Vollständigkeit den allmählichen Uebergang vom normalen Fraunhofer-Spektrum zum reinen Chromosphärenspektrum. Mit der abnehmenden allgemeinen Erleuchtung des Hintergrundes tritt das Koronaspektrum immer deutlicher hervor. Fünf grosse Protuberanzen erscheinen in der *H*- und *K*-Linie des Calciums als mächtige Wolken, in den Wasserstoff- und Helium-Linien als feinere Sprudel.

Platte Z_b. Alle 3 Aufnahmen geben ausser dem Spektrum der Korona noch das jener grossen Protuberanzen in zahlreichen Linien.

Ein deutlicher und ein schwacher Halbkreis im Ultraviolett entsprechen zwei ultravioletten Strahlungen der Korona. Dagegen ist von der grünen Koronalinie — infolge geringer Grünempfindlichkeit der Platte — nichts zu entdecken.

Bei der 40^s Exposition setzt die Erleuchtung des Hintergrundes bereits der Abbildung zarteren Details ein Grenze.

Die Fokusierung und die Mitbewegung des Instruments sind befriedigend.

Platte Z_c. Das erste der 4 Spektren giebt ausser dem kontinuierlichen Koronaspektrum die Ca, *H* und He-Linien einiger Protuberanzen. Im 2. Spektrum erstrecken sich dieselben Linien über grössere Bögen des Sonnenumfangs und es treten einige andere Linien (*Ti*, *Sr*) hinzu.

Das 3. Spektrum trifft gerade den Zeitpunkt, wo die ganze Chromosphäre an einer Stelle vom Monde freigegeben ist. Es zeigt zahlreiche helle Linien und ist namentlich im Ultraviolett gut im Fokus.

Das 4. Spektrum wurde in dem Momente aufgenommen, wo ich mit blossem Auge den ersten Lichtfunken am Mondrande aufblitzen sah, ein Streifen kontinuierlichen Spektrums durchsetzt die Bögen des Chromosphärenspektrums, welches auch hier besonders im Ultraviolett scharf ist.

Platte P_1. Die Aufnahme zeigt die Korona vom Maximaltypus mit verschiedenen Ausläufern bis zu drei Sonnenradien Erhebung über die Sonnenoberfläche.

Platte P_2. Die Koronaausläufer sind noch etwas deutlicher als auf der vorigen Aufnahme.

Platte P_3 und P_4. Infolge Verschleierung des Hintergrundes zeigen diese Aufnahmen die Korona weniger deutlich, als die kürzeren Expositionen. Man ist offenbar mit der Exposition von 5^s bereits an der für ein so lichtstarkes Objektiv in Rücksicht auf die allgemeine Erleuchtung des Himmels zulässigen Grenze angelangt.

Platten R_1—R_5. Diese Aufnahmen von Dr. Emden zeigten die Wirksamkeit der Gelbscheibe ausserordentlich deutlich. Während die Platten R_3 und R_4

Abhandl. Kgl. Ges. d. Wiss. zu Gött. NF 5(2) 1907, 3–73

(Aufnahmen ohne Gelbscheibe) beim Entwickeln rasch verschleierten, konnten di mit Gelbscheibe exponierten Platten — auch die von der 40ˢ langen Exposition — sehr viel länger entwickelt werden und zeigen überhaupt keinen Schleier. Die liegt nicht etwa an der allgemeinen Herabsetzung der Intensitäten durch di Gelbscheibe, welche durch die Verlängerung der Expositionszeit von 4ˢ auf 40 wieder reichlich ausgeglichen wurde, sondern an der blaueren Färbung des dif fusen Himmelslichtes im Vergleich zur Korona. Man sollte infolge dieser Be seitigung des störenden diffusen Lichtes eine grössere Erstreckung der Korona ausläufer auf den Aufnahmen mit Gelbscheibe erwarten. Aber obwohl die innere Korona infolge der langen Entwicklung bei den letzteren Platten besonders tie geschwärzt ist, zeigen sämtliche Platten doch nur etwa dieselbe Erstreckung der Koronaausläufer, wie die *P*-Platten. Es muss daraus geschlossen werden dass die Helligkeit der Corona ausserhalb der photographierten Teile sehr schnell absinkt oder — in andern Worten — dass die Korona hier praktisch zu Ende ist. Die Frage nach der Farbe der Korona wird unten ausführlich behandelt werden.

§ 6. Erscheinung der Korona für das freie Auge.

Während der 40ˢ Expositionen war es uns möglich, dem Aussehen der Korona mit freiem Auge unsere Aufmerksamkeit zuzuwende.n Beim ersten Anblick erschien mir die Korona, wie ein heller schmaler silberner Ring von wenigen Bogenminuten Breite um die Sonne. Erst bei näherem Zusehen bemerkte ich die außerordentlich viel lichtschwächeren äusseren Ausläufer der Korona, deren nahezu radiale Randlinien in immer mehr abnehmenden Kontrast gegen den hell erleuchteten Himmelshintergrund bis zu etwa einem Sonnendurchmesser Entfernung von der Sonnenoberfläche zu verfolgen waren. Die diffuse Erleuchtung des Himmels war bedeutend heller, als bei Vollmond. Das allgemeine Aussehen der Korona für das freie Auge entspricht durchaus dem der photographischen Aufnahmen — wenn man nur von der überwiegenden Helligkeit der inneren Korona absieht, die auf den Photographien nicht genügend zum Ausdruck kommt. Dr. Emden schätzte die längsten Coronastrahlen auf $1^1/_2$ Sonnendurchmesser, der Schreiner Henry macht die präzise Angabe, der Sonnendurchmesser sei 30 cm und die längsten Koronastrahlen 25 cm lang gewesen.

Die Zeichnungen, welche wir und einige andere zufällig anwesende Beobachter unmittelbar nach der Finsternis anfertigten, bewiesen nur, dass die Auffassung der feineren äusseren Koronaausläufer eine völlig subjektive ist.

II. Abschnitt. Die Helligkeit der äusseren Korona.

Da die Absorption des photometrischen Keils, wie erwähnt, im ganzen zu stark gewählt war, so ist es auf keiner der direkten Koronaaufnahmen mit dem Petzvalobjektiv (Platten P_1—P_4) möglich gewesen, die Helligkeit der inneren Korona zu bestimmen. Hingegen erlaubt Platte P_2 wenigstens die Helligkeit der äusseren Korona und des diffusen Himmelslichtes abzuleiten. Dazu bedarf es der näheren Untersuchung des photometrischen Keiles, der die Beziehung zwischen den Schwärzungen der photographischen Platte und absoluten Intensitäten vermitteln soll.

Zur Messung der Schwärzungen diente stets das Hartmannsche Mikrophotometer.

§ 7. Aichung des photometrischen Keils.

Der photometrische Keil bestand aus einer 5 mm breiten, 50 mm langen photographischen Platte mit kontinuierlich abgestufter Schwärzung, überdeckt von einer braunen Glasscheibe. Eine bestimmte Stelle des Keils soll durch ihre Entfernung von dem stark geschwärzten Ende in mm bezeichnet werden.

Der Keil war so gewählt, dass die hellste Stelle die Sonne etwa auf Vollmondhelligkeit abblendete. Nach Müller (Photometrie der Gestirne S. 315) ist:

$$\frac{\text{Sonne}}{\text{Vollmond}} = 570\,000 = 14{,}4 \text{ Grössenklassen.}$$

Als Vergleichslichtquelle diente stets die Sonne selbst.

Es war zunächst festzustellen, wie stark jede Keilstelle das Sonnenlicht abschwächte. So will ich mich der Kürze wegen ausdrücken. In Wirklichkeit ist dies eine übertragene Redeweise. Denn da der Keil stark gefärbt ist, so werden verschiedene Spektralfarben verschieden stark von ihm absorbiert, es giebt keine einheitliche Abblendungszahl. Doch ist das Resultat dieser verschiedenen Absorptionen eine bestimmte Verminderung der gesamten aktinischen Kraft der Sonne und eine bestimmte entsprechende Schwärzung der photographischen Platte. Man kann nun unabhängig von dem Keil die Sonne ohne Farbenänderung so abblenden, dass sie dieselbe Schwärzung erzeugt, wie eine bestimmte Keilstelle. Den Betrag letzterer Abblendung will ich als Absorption der betreffenden Keilstelle bezeichnen. Diese Absorption ist etwas konstantes, solange die Farbe der Sonne und die relative Empfindlichkeit der Plattensorte für verschiedene Farben sich nicht ändert. Beides soll hier vorausgesetzt werden. Leider war der Keil sehr rot gefärbt, sodass eine Aenderung der Rotempfindlichkeit relativ zur Blauempfindlichkeit von Platte zu Platte einen unkontrollierbaren Einfluss erhält, der bei Benutzung eines Keils von neutraler Farbe weggefallen wäre.

Abhandl. Kgl. Ges. d. Wiss. zu Gött. NF 5(2) 1907, 3–73

a) Aichung mit Hülfe des Mondes. Da der Mond sehr nahe dieselbe Farbe, wie die Sonne hat, so stellt er eine ohne Farbenänderung abgeblendete Sonne dar und kann zur Aichung benutzt werden. Während der Seefahrt nach Algier wurden die folgenden Aufnahmen gemacht. Die Höhen von Sonne und Mond sind mit dem Sextanten gemessen. Für die photographische Extinktion wurde der doppelte Betrag der Potsdamer Extinktion angesetzt. Die Phasenreduktion auf Vollmond wurde der Tabelle in Müllers's Photometrie pag. 342 entnommen unter Verwendung des Mittels der Werte von Bond und Zöllner.

Ich mache zunächst die nötigen allgemeinen Angaben über die Aufnahmen:

Platte	Ort	Ortszeit	Objekt	Höhe	Extinktion	Phasenreduktion
A_1	Im Kanal	10. August 10h	Freier Mond	16°.2	1m.2	1m.0
	Höhe der normannischen Inseln	11. „ 4 0m	Sonne durch Keil	28°.0	0.5	—
A_2	Höhe von Cap Mondego	14. „ 8 0	Freier Mond	14°.1	1.4	0.0
	Höhe von Lissabon	14. „ 21 45	Sonne durch Keil	54°.	0.1	—
A_3	Cap S. Vincent	15. „ 9 20	Freier Mond	23°.7	0.7	0.1
	Strasse von Gibraltar	16. „ 5 0	Sonne durch Keil	25°.0	0.6	—

Die Expositionszeiten und die gemessenen Schwärzungen enthalten die folgenden kleinen Tabellen:

Belichtung durch den freien Mond.

Platte	A_1			A_2		A_3	
Expositionszeit	10s	30s	90s	3s	10s	3s	10s
Schwärzung	65.4	69.0	73.9	41.1	45.0	59.6	64.4

Belichtung mittelst der Sonne durch Keil.

Platte	Exposition	Keilstelle und zugehörige Schwärzung				
		0mm	5	10	15	20
A_1	10s	65.5	65.6	64.0	59.2	54.2
	30	73.9	74.4	72.7	67.2	60.3
A_2	3	39.7	38.7	33.6		
	10	47.3	46.3	44.4	41.2	35.3
A_3	3	54.9	53.9	50.2		
	10	62.1	61.1	58.1	49.0	

Die weitere Bearbeitung dieser Zahlen erfolgt weiter unten.

b) Direkte Aichung. Eine Bestimmung der Keilabsorption ohne Benutzung des Mondes wurde dadurch erhalten, dass die Sonne direkt sehr stark und in messbarer Weise abgeblendet wurde, indem mit einem Objektiv sehr reduzierter Oeffnung ein Sonnenbild entworfen und dies noch mit Hülfe eines Mikroskopobjektivs vergrössert wurde [1]). Es wurden 2 Platten aufgenommen, welche ausser 3 solchen abgeschwächten Sonnenbildern je zwei Expositionen mit dem photometrischen Keil gegen die freie Sonne erhielten. Die Resultate für die Schwärzungen waren folgende:

Direkt abgeblendete Sonnenbilder.

Platte	B_1 (Exposition 60^s)			B_2 (Exposition 15^s)		
Abschwächung	15^m0	16^m8	18^m1	15^m0	16^m8	18^m1
Schwärzung	65.8	59.7	51.6	54.9	47.6	37.0

Aufnahmen der Sonne durch den Keil.

Platte	Exposition	Keilstelle und zugehörige Schwärzung												
		0	1	2.5	5	10	15	20	25	30	35	40	45	50
B_1	15^s	60.0	60.6	59.9	59.3	56.2	52.2	46.7						
	60	64.0	64.7	65.0	64.8	62.5	61.3	57.7	54.4	52.1	51.1	50.9		
B_2	15	53.2	53.2	52.7	52.2	50.8	48.3	42.6	37.6	32.5				
	45	55.8	55.4	56.1	55.3	54.0	52.2	47.9	44.5	41.4	42 0	40.5	40.5	39.5

Es soll zunächst mit Hülfe dieser direkten Vergleichung die Keilaichung durchgeführt werden.

Indem angenommen wurde, dass die Aenderung der Expositionszeit von 15^s auf 60^s resp. 45^s bei der Platte B_1 resp. B_2 wirkt, wie eine proportionale Aenderung der Intensität, wurde nach dem in den Astr. Nachr. Bd. 172. pag. 65 auseinandergesetzten Verfahren die Gradation des Keiles, von der Stelle 2.5 als Nullpunkt an gerechnet, abgeleitet. Es fanden sich so im Mittel aus den beiden Platten B_1 und B_2 die in der ersten Zeile des folgenden Täfelchens angegebenen Werte:

1) Vgl. eine demnächst erscheinende Arbeit von Herrn O. Birck über das Helligkeitsverhältnis der Sonne zu Fixsternen.

Abhandl. Kgl. Ges. d. Wiss. zu Gött. NF 5(2) 1907, 3–73

Tabelle I. (Aichung des photometrischen Keils).

Keilstelle	2.5	5	7.5	10	12.5	15	17.5	20
Relative Absorption	$0\overset{m}{.}0$	0.2	0.5	0.8	1.1	1.4	1.8	2.3
Absolute Absorption	$15\overset{m}{.}3$	15.5	15.8	16.1	16.4	16.7	17.1	17.6

Nunmehr wurden die Schwärzungen der direkt abgeblendeten Sonnenbilder in die Schwärzungen des Keiles interpoliert und für jede Keilstelle durch Benutzung der eben gewonnenen Tabelle auf die Stelle 2.5 reduziert. Es ergaben sich dadurch folgende absolute Absorptionen für die Stelle 2.5.

	Platte B_1			Platte B_2			Mittel
Sonnenbild	1.	2.	3.	1.	2.	3.	
Absorption der Stelle 2.5	$15\overset{m}{.}3$	$15\overset{m}{.}3$	$15\overset{m}{.}1$	$15\overset{m}{.}5$	$15\overset{m}{.}3$	—	$15\overset{m}{.}3$

Die befriedigende Uebereinstimmung der Zahlen ist eine Kontrolle dafür, dass die oben erhaltene Tabelle die innere Gradation des Keiles richtig wiedergiebt. Mit dem Mittelwert $15\overset{m}{.}3$ sind dann die in der zweiten Zeile der Tabelle I angegebenen absoluten Absorptionen erhalten.

Nunmehr wurden zur weiteren Kontrolle die Aufnahmen des Mondes herangezogen. Es wurden die Schwärzungen der Mondbilder auf den Platten A in die Keilschwärzungen jeder Platte interpoliert. Daraus wurde unter Berücksichtigung der verschiedenen Expositionszeiten, der Extinktion und der Phasenreduktion mit Hülfe der Tabelle I der Keilabsorptionen die Vollmondhelligkeit abgeleitet. Es ergaben sich die Werte:

Bild	Platte A_1			Platte A_2		Platte A_3		Mittel
	1.	2.	3.	1.	2.	1.	2.	
Vollmondhelligkeit	$13\overset{m}{.}9$	14.7	13.9	14.1	14.6	14.1	14.6	$14\overset{m}{.}3$

Das Mittel stimmt fast völlig mit dem von Müller angegebenen Werte überein und bestätigt damit die in Tabelle I enthaltenen absoluten Absorptionen, die wir im folgenden verwenden werden.

§ 8. Anwendung auf die Bestimmung der Helligkeit der äusseren Korona nach Platte P_2.

Auf der Koronaaufnahme auf Platte P_2 (Exposition 4^s) habe ich drei annähernd radiale Richtungen unter dem Hartmann'schen Mikrophotometer ver-

Man erkennt, dass sich der Ausläufer nur verhältnismässig wenig über die durchschnittliche Helligkeit der Korona in derselben Entfernung vom Mittelpunkt erhebt.

Mit dem Inhalt von Tabelle III ist alles erschöpft, was sich an quantitativen Angaben aus den Koronaaufnahmen mit dem Petzvalobjektiv ableiten liess. Indessen wird die Bearbeitung der Spektralaufnahmen gestatten, diese Tabelle nach der Seite der inneren, helleren Koronapartieen zu ergänzen.

III. Abschnitt. Spektralphotometrie der Korona.

Die Aufnahmen mit der Prismenkamera gaben zahlreiche Bilder des kontinuierlichen Koronaspektrums. Es sind dies allerdings keine sehr reinen Spektren, da durch die Prismenkamera die Spektren eines ganzen Schnitts durch die Korona übereinandergelagert werden. Doch überragt die innere Korona so sehr an Helligkeit, dass sie doch den Hauptbeitrag liefert und auf den zur Prismenkante senkrechten Tangenten an die Mondscheibe Spektren entstehen, bei denen die Wirkung der Mischung mit genügender Sicherheit abgeschätzt werden kann.

Zur photometrischen Bearbeitung eignete sich besonders die erste Aufnahme des Flash-Spektrums auf Platte Z_c. Dieselbe zeigte ein kräftiges Koronaspektrum zu beiden Seiten des Mondes auf völlig klarem Hintergrund und trug ausserdem einen photometrischen Keil, der durch 15^s freie Exposition gegen die Sonne gewonnen war und dessen Schwärzungen die im Koronaspektrum auftretenden Schwärzungen einschlossen.

§ 9. Photometrisches Prinzip.

Das Prinzip, nach welchem aus dieser Aufnahme die Koronahelligkeit in jeder einzelnen Spektralfarbe abgeleitet werden konnte, war folgendes. Es wurde auf einer Platte, welche dieselbe relative Farbenempfindlichkeit aufwies, wie die damals benutzte Platte, nämlich auf einer Schleussnerplatte derselben Sorte, der Keil wiederum 15_s gegen die freie Sonne exponiert. Dazu wurden auf dieser Platte im bekannten Verhältnis abgeschwächte Sonnenspektren aufgenommen. Schwärzungen auf beiden Platten, welche gleichen Keilstellen korrespondieren und durch gleiche Wellenlängen hervorgebracht sind, bedeuten gleiche Helligkeit für diese Wellenlänge in dem Koronaspektrum einerseits nnd einem der abgeschwächten Sonnenspektren andrerseits.

Abhandl. Kgl. Ges. d. Wiss. zu Gött. NF 5 (2) 1907, 3–73

Die abgeschwächten Sonnenspektren wurden erzeugt, indem ein kleine Kollimator vor die Prismenkamera gesetzt wurde, in dessen Fokalebene sich ein keilförmiger Spalt befand.

§ 10. Theorie.

Es ist zunächst zu überlegen, wie die gesamte photographische Wirkung an irgend einer Stelle des durch die Prismenkamera erzeugten Koronaspektrums zu Stande kommt.

Eine bestimmte Stelle in dem Spektralbilde wollen wir durch zwei rechtwinkliche Koordinaten σ und η bezeichnen, σ parallel, η senkrecht zur Erstreckung des Spektrums gezählt. Als Nullpunkt wählen wir den Punkt des Bildes, in welchen sich der Sonnenmittelpunkt abbilden würde, wenn er nur mit der Wellenlänge 441 $\mu\mu$ (der aktinischsten Stelle des Spektrums) strahlte. Auf der andern Seite möge eine bestimmte Stelle in einem Koronabilde, wie es bei monochromatischer Abbildung erscheinen würde, durch die ebenso gerichteten Koordinaten ξ und η festgelegt werden, welche vom Mittelpunkt eines solchen monochromatischen Bildes aus gezählt werden sollen. Der Sonnenradius sei R, die Entfernung eines Punktes vom Sonnenmittelpunkt $r = \sqrt{\xi^2 + \eta^2}$. Alle diese Grössen sollen in mm auf der photographischen Platte gemessen werden.

Die Korona werde als symmetrisch um den Sonnenmittelpunkt verteilt angenommen und ihre Helligkeit in der Entfernung r für die Wellenlänge λ, bezogen auf die mittlere Flächenhelligkeit der Sonne in der gleichen Wellenlänge, sei:

$$i(r) \cdot \varphi(\lambda).$$

Um Willkürlichkeiten zu vermeiden, werde für $\lambda = 441\,\mu\mu$ der Faktor $\varphi(\lambda) = 1$ gesetzt, sodass für diese Wellenlänge $i(r)$ direkt das Verhältnis der Koronahelligkeit zur Sonnenhelligkeit angiebt. Mit dem Produktansatz ist übrigens ausgedrückt, dass verschiedene Färbung verschiedener Teile der Korona vernachlässigt werden soll.

Statt der Wellenlängen führen wir besser gleich die Abmessungen in dem von der Prismenkamera gelieferten Spektrum ein. Es bezeichne $\mu(\lambda)$ die Länge des Spektrums zwischen der Wellenlänge 441 $\mu\mu$ und der Wellenlänge λ, positiv für die grösseren Wellenlängen gezählt, und es sei:

$$\varphi(\lambda) = \psi(\mu).$$

Schliesslich werde die aktinische Wirkung des Sonnenspektrums für das Wellenlängengebiet λ bis $\lambda + d\lambda$ durch $J(\lambda)\,d\lambda$ gegeben und es sei:

$$J(\lambda)\,d\lambda = K(\mu)\,d\mu,$$

sodass $K(\mu)$ direkt die photographische Intensität der verschiedenen Teile eines mit der Prismenkamera entworfenen reinen Sonnenspektrums darstellt.

Nunmehr lässt sich die gesamte aktinische Wirkung A der Korona an der durch die Koordinaten σ, η bestimmten Plattenstelle leicht ausdrücken. Zu der

Wirkung an dieser Stelle tragen alle Punkte der Korona bei, welche die Koordinate η haben. Die Wellenlänge, mit welcher jeder dieser Punkte wirkt, ist bestimmt durch die Gleichung:

$$\sigma - \zeta = \mu \quad (\mu = \mu(\lambda)).$$

Die Gesamtwirkung giebt das Integral:

$$A(\sigma, \eta) = \int_{-\infty}^{+\infty} i(\sqrt{\eta^2 + \zeta^2})\, \psi(\sigma - \zeta)\, K(\sigma - \zeta)\, d\zeta$$

oder wenn man an Stelle von ζ die Grösse μ als Variable wählt:

$$1) \qquad A(\sigma, \eta) = \int_{-\infty}^{+\infty} i(\sqrt{\eta^2 + (\sigma - \mu)^2})\, \psi(\mu)\, K(\mu)\, d\mu.$$

Unsre zweite Aufgabe ist, eine entsprechende Ueberlegung für die Wirkung der Sonne in dem mit Kollimator und keilförmigem Spalt erzeugten Vergleichsspektrum anzustellen. Auch in diesem wollen wir in der Dispersionsrichtung Koordinaten σ' und senkrecht dazu von der der Keilspitze entsprechenden Grenze des Spektrums an Koordinaten η' rechnen.

Ist ϱ der Sonnendurchmesser in Winkelmass, f die Brennweite des Kollimatorobjektivs, so ist die wirksame Oeffnung des Instrumentes bei der Sonnenaufnahme $f.\varrho$, während sie bei der Koronaaufnahme dem Durchmesser des U. V. Aplanaten D entspricht. Man hat daher für die wirksame Helligkeit im Verhältnis zur vollen Sonnenhelligkeit zu setzen:

$$i.\psi = \left(\frac{f\varrho}{D}\right)^2.$$

Die Grösse der Lichtquelle, welche mit dieser Helligkeit wirkt, ist gegeben durch die Breite des monochromatischen Spaltbildes. Da bei dem verwandten Spalt das Verhältnis von Breite zu Länge 1:30 war, so ist diese Breite für eine Stelle im Vergleichsspektrum, die um die Strecke η' von der Spaltspitze entfernt ist, gleich $\eta'/30$. Die Gesamtwirkung im Vergleichsspektrum wird demnach:

$$B(\sigma', \eta') = \int_0^{\eta'/30} i.\psi.K(\sigma' - \zeta)\, d\zeta = \left(\frac{f\varrho}{D}\right)^2 \int_0^{\eta'/30} K(\sigma' - \zeta)\, d\zeta$$

und wenn wir die Veränderung von K für die kleine Aenderung von ζ vernachlässigen, mit anderen Worten, wenn wir das sehr reine Vergleichsspektrum als völlig rein betrachten, so folgt:

$$2) \qquad B(\sigma', \eta') = \left(\frac{f\varrho}{D}\right)^2 \frac{\eta'}{30} K(\sigma')$$

Jetzt haben wir noch drittens die Helligkeit im Koronaspektrum zu der im Sonnenvergleichsspektrum in Beziehung zu setzen. Bei der Beobachtung habe ich mich im Koronaspektrum auf die Tangentialstreifen am Sonnenrand ($\eta = \pm R$)

Abhandl. Kgl. Ges. d. Wiss. zu Gött. NF 5(2) 1907, 3–73

beschränkt, da diese das hellste und reinste Spektrum geben. Diese Beschränkung soll auch hier gleich eingeführt werden, und es soll zur Abkürzung:

$$A(\sigma, R) = A(\sigma), \quad i(\sqrt{R^2+(\sigma-\mu)^2}) = h(\sigma-\mu)$$

geschrieben werden. Das Beobachtungsverfahren lief darauf hinaus, in dem Koronaspektrum und in dem Vergleichsspektrum Stellen mit gleichem $\sigma (\sigma = \sigma')$ zu betrachten und diejenige Spaltbreite, den Wert von η' heraus zu suchen, für welchen die Helligkeit in beiden Spektren gleich ($B = A$) wurde. Für jede Wellenlänge, jedes σ ergiebt sich ein solches $\eta' = \eta'(\sigma)$ und es gilt dann:

3) $$A(\sigma) = \int_{-\infty}^{+\infty} h(\sigma-\mu)\,\psi(\mu)\,K(\mu)\,d\mu = \left(\frac{f\varrho}{D}\right)^2 \frac{\eta'(\sigma)}{30} K(\sigma).$$

Diese Gleichung stellt das unmittelbare Resultat der Beobachtungen dar. Wir werden im Folgenden zur Abkürzung noch setzen:

4) $$\frac{1}{30}\left(\frac{f\varrho}{D}\right)^2 \eta'(\sigma) = \vartheta(\sigma)$$

und $\vartheta(\sigma)$ als „Helligkeit des unreinen Spektrums" bezeichnen.

Welche Angaben über die Funktionen h und ψ lassen sich auf Grund der Gleichung (3) machen, wenn $\vartheta(\sigma)$ durch die Beobachtung gegeben ist?

Wäre das Koronaspektrum rein, h nur für sehr kleine Werte von $\sigma - \mu$ merklich, die Korona sehr nahe auf den Sonnenrand konzentriert, so dürfte man das Integral ersetzen durch

$$\psi(\sigma)\,K(\sigma)\int h(\sigma-\mu)\,d\mu$$

und es folgte:

5) $$\psi(\sigma)\int_{-\infty}^{+\infty} h(\mu)\,d\mu = \vartheta(\sigma).$$

Speziell für die Wellenlänge 441 $\mu\mu$ ergäbe sich auf Grund unsrer Festsetzung $\psi(0) = 1$:

6) $$\int_{-\infty}^{+\infty} h(\mu)\,d\mu = \vartheta(0)$$

und damit:

7) $$\psi(\sigma) = \frac{\vartheta(\sigma)}{\vartheta(0)}.$$

Damit wäre die Helligkeitsverteilung ψ im Koronaspektrum relativ zum Sonnenspektrum, kurz gesagt die „Farbe" der Korona bestimmt und die Gleichung (6) gäbe noch einen Anhalt für die Gesamthelligkeit der Korona.

Für unser unreines Koronaspektrum kann man diese Lösung als erste Annäherung benutzen und folgendermassen weiter verfahren. Die Gleichung (3) lässt sich schreiben:

$$\psi(\sigma) . \int_{-\infty}^{+\infty} h(\sigma-\mu)\,d\mu = \frac{\vartheta(\sigma)}{g(\sigma)} \tag{8}$$

wobei:

$$g(\sigma) = \frac{\int_{-\infty}^{+\infty} h(\sigma-\mu) \frac{\psi(\mu)}{\psi(\sigma)} \frac{K(\mu)}{K(\sigma)}\,d\mu}{\int_{-\infty}^{+\infty} h(\sigma-\mu)\,d\mu} \tag{9}$$

gesetzt ist. Wäre unser Spektrum völlig rein, so würde sich, unabhängig von der Gestalt der Funktionen h, ψ und K, $g(\sigma) = 1$ ergeben. Bei einem wenig gemischten Spektrum genügt eine ungefähre Kenntnis dieser drei Funktionen, um einen guten Wert von $g(\sigma)$ zu erhalten. Die Aktinität K des Sonnenspektrums lässt sich den Schwärzungen der Vergleichsspektren selbst entnehmen. Was die Abhängigkeit der Koronahelligkeit von der Distanz von der Sonne angeht, so soll gemäss den Untersuchungen von Turner eine Beziehung der Form:

$$i(r) = \frac{i_0}{r^n}$$

vorausgesetzt werden, und zwar wird nach Turner, wie auch nach den unten folgenden Messungen, n etwa gleich 6 anzunehmen sein. Dann ist also:

$$i(\sqrt{R^2+(\sigma-\mu)^2}) = h(\sigma-\mu) = \frac{i_0}{[R^2+(\sigma-\mu)^2]^3}.$$

Die absolute Helligkeit der Korona i_0 fällt aus dem Ausdruck von $g(\sigma)$ offenbar heraus. Für die Funktion ψ schliesslich darf man den aus der ersten Näherung folgenden Wert $\frac{\vartheta(\sigma)}{\vartheta(0)}$ einsetzen.

Hat man mit Hülfe dieser dreierlei Angaben die nahe bei 1 liegende Grösse $g(\sigma)$ berechnet, die man als „Reduktion auf reines Spektrum" bezeichnen kann, so erhält man aus (8), genau so wie vorher aus (5), die Werte der Funktion ψ und eine Angabe über die Gesamthelligkeit der Korona. Die so erhaltenen Grössen können als definitive betrachtet werden.

Die gesamte Rechenvorschrift wird somit diese: man entnehme den Messungen die Verteilung $K(\mu)$ der aktinischen Kraft im Sonnenspektrum und die durch $\eta'(\sigma)$ gegebene Spaltweite, welche das Sonnenspektrum auf die Helligkeit des (unreinen) Koronaspektrums in jeder Wellenlänge reduziert. Damit berechne man die Helligkeit des unreinen Spektrums:

$$\vartheta(\sigma) = \frac{1}{30}\left(\frac{f\varrho}{D}\right)^2 \eta'(\sigma)$$

und daraus die „Reduktion auf reines Spektrum":

Abhandl. Kgl. Ges. d. Wiss. zu Gött. NF 5(2) 1907, 3–73

10) $$g(\sigma) = \frac{\int_{-\infty}^{+\infty} h(\sigma-\mu)\,\vartheta(\mu)\,K(\mu)\,d\mu}{\vartheta(\sigma)\,K(\sigma)\int_{-\infty}^{+\infty} h(\sigma-\mu)\,d\mu}$$

unter Verwendung des Ausdrucks:

$$h(\sigma-\mu) = \frac{i_0}{[R^2+(\sigma-\mu)^2]^3} \quad (i_0 \text{ willkürlich}).$$

Man erhält dann eine Angabe über die absolute Helligkeit der Korona aus der Gleichung:

11) $$\int_{-\infty}^{+\infty} h(\mu)\,d\mu = \frac{\vartheta(0)}{g(0)}$$

und eine Bestimmung der spektralen Intensitätsverteilung der Korona relativ zur Sonne nach der Formel:

12) $$\psi(\sigma) = \frac{\vartheta(\sigma)}{g(\sigma)}\cdot\frac{g(0)}{\vartheta(0)}.$$

§ 11. Aufnahme der Vergleichsspectren und Prüfung des Momentverschlusses.

Da der Keil auf Platte Z_c 15^s frei gegen die Sonne exponiert war, während die Koronaspektren mit dem Momentverschluss erhalten wurden — die verwandte Platte war eine Schleussnerplatte —, habe ich ganz entsprechend die folgenden Vergleichsaufnahmen gemacht:

Platte C_1) Schleussnerplatte. 3. Juli 1906. Göttingen.
$11^h\,33^m$ M. E. Z. Keilexposition frei gegen die Sonne 15^s
40^m M. E. Z. Momentaufnahmen des Sonnenspektrums.

Der Kollimator, mit dem das Sonnenspektrum erzeugt wurde, bestand aus U. V. Glas und hatte eine Brennweite von 320 mm. Da die Oeffnung der U. V. Kamera selbst $D = 110$ mm ist und der Sonnendurchmesser $\varrho = 31'.5$ war, so folgt, wie gleich hier angeführt sei:

$$\vartheta(\sigma) = \frac{1}{30}\left(\frac{f\varrho}{D}\right)^2 \eta'(\sigma) = 2{,}37\,.\,10^{-5}\,\eta'(\sigma)$$

oder wenn man zu Grössenklassen übergeht, indem man den Logarithmus nimmt und mit $-2{,}5$ multipliziert:

13) $$\vartheta^{(m)}(\sigma) = 11^m.56 - 2.5\log\eta'(\sigma).$$

Für die Momentaufnahmen wurde der Steinheilsche Fall-Verschluss genau unter denselben Verhältnissen, wie bei der Sonnenfinsternis, benutzt. Die Verwendung dieses Verschlusses hat eine gewisse Unsicherheit mit sich gebracht. Denn der Verschluss ist eigentlich zum Gebrauch als Objektivverschluss bestimmt und giebt — vor der Platte verwandt — keine gleichmässige Beleuchtung des Gesichtsfeldes, vielmehr ist die Belichtungszeit anfangs am grössten und nimmt allmählich ab.

Um diese Ungleichmässigkeiten zu bestimmen, machte ich Momentaufnahmen des diffusen Himmelslichtes mit freier und mit halb verdeckter Oeffnung der Prismenkamera auf ein und derselben Platte und mass die Schwärzung der entstehenden Bilder aus. Es zeigte sich eine deutliche Abnahme der Schwärzung parallel der Bewegungsrichtung der Verschlussklappen, deren Betrag mit Hülfe der mit der halben Oeffnung aufgenommenen Bilder in Grössenklassen verwandelt werden konnte. Vier Aufnahmen ergaben folgende Helligkeitsüberschüsse gegen den zuletzt belichteten Teil des Gesichtsfeldes. Als Argument ist die Entfernung von dem Rande des Gesichtsfeldes angegeben, an welchem die Bewegung schloss.

Abstand vom Rande des Gesichtsfeldes d	0.7mm	1.2	5.2	9.2	13.2	17.2
Helligkeitsüberschuss 1. Aufnahme	m 0.00	0.05	0.08	0.26	0.54	0.94
2. „	0.00	0.00	0.33	0.36	0.73	1.01
3. „	0.00	0.03	0.03	0.07	0.16	0.28
4. „	0.00	−0.07	−0.03	0.05	0.05	0.16
Mittel	0.00	0.00	0.10	0.18	0.37	0.60
Formel	0.00	0.00	0.06	0.18	0.37	0.62

Die Mittelwerte werden sehr nahe dargestellt durch die Formel

$$\text{Korr.} = 0\overset{\text{m}}{.}0021\, d^2. \tag{14}$$

Man sieht aus den Abweichungen der 4 Reihen von einander, dass der Verschluss für die hier zu stellenden Ansprüche nicht regelmässig genug funktionierte. Indessen ist hervorzuheben, dass die Spektra nahe senkrecht zur Bewegungsrichtung des Verschlusses lagen, dass daher die spektrale Intensitätsverteilung durch Unregelmässigkeit des Verschlusses sehr wenig beeinflusst wird und nur die resultierenden Werte für die absolute Helligkeit der Korona weniger Zutrauen verdienen.

Eine Korrektion nach Formel (14) wurde an die Messungen auf Platte Z_e angebracht.

§ 12. Ausmessung der Spektren.

Das Vergleichsspektrum habe ich für vier Entfernungen von der Kante, nämlich für $\eta' = 0.5$, 1.5, 4.5 und 13.5 mm durchphotometriert. Diesen vier Werten von η' entsprechen nach Formel (13) folgende Werte von $\vartheta(\sigma)$:

$$\vartheta^{\text{m}}(\sigma) = 12\overset{\text{m}}{.}31,\quad 11\overset{\text{m}}{.}12,\quad 9\overset{\text{m}}{.}92,\quad 8\overset{\text{m}}{.}73.$$

Abhandl. Kgl. Ges. d. Wiss. zu Gött. NF 5(2) 1907, 3-73

Bevor indessen diese Zahlen in die obigen Formeln eingeführt werden dürfen, sind folgende Korrektionen an sie anzubringen.

1) Korrektur wegen Ungleichförmigkeit des Verschlusses. Betrag für die vier Stellen: $0^{m}.0$, $0^{m}.0$, $-0^{m}.1$, $-0^{m}.5$.
2) Einfluss der Beugung am Spaltrand. Derselbe wird nur für die schmälste benutzte Spaltstelle ($\eta' = 0.5$) merklich und giebt eine Korrektion von $+0^{m}.2$.
3) Einfluss der Phase der Sonne bei der Keilaufnahme auf Platte Z_e. Die Sonnenhelligkeit betrug nach Bemporad l. c. nur 0.70 der Totalhelligkeit. Daraus folgt eine Korrektion aller 4 Zahlen um $+0^{m}.39$.
4) Reflexion an den 4 Flächen des Kollimatorobjektes. Korrektion aller vier Zahlen um $+0^{m}.25$.
5) Es ist beachtet, dass die Aufnahme des Vergleichsspektrums etwas zerstreutes Licht aufweist. Die Intensität desselben liess sich durch Ausmessung der Schwärzung ausserhalb des Spektrums, wie sie in der Rubrik „falsches Licht“ der folgenden Tabelle angegeben ist, näherungsweise bestimmen. Dieses falsche Licht wirkt so, als ob die Intensität des einfallenden Lichts grösser, die Abblendung geringer wäre. Es ist deswegen an die Abblendungszahlen eine negative Korrektion angebracht, die aber nur an den äussersten Enden des Spektrums merkliche Beträge (bis $0^{m}.4$) erreicht.

Die Einflüsse von selektiven Absorptionen werden weiter unten Berücksichtigung finden.

Die folgende Tabelle IV enthält nun die auf Platte C_1 gemessenen Schwärzungen, die ich mit S bezeichnen will, und dazu die mit den obigen Korrektionen versehenen Werte $\vartheta^{m}(\sigma)$. Die Wellenlängen, welche den Strecken σ im Spektrum entsprechen, sind aus den Fraunhoferschen Linien im Spektrum abgeleitet. Die Zahlen S' werden später ihre Erklärung finden. Das Mikroskop des Mikrophotometers wurde bei dieser Durchmessung unscharf eingestellt, um die durchschnittliche Helligkeit des Sonnenspektrums unter Einschluss der Fraunhoferschen Linien zu erhalten; die breitesten Linien wurden vermieden.

Ganz ähnlich giebt Tabelle V die auf Platte Z_e erhaltenen Schwärzungen, die mit S' bezeichnet werden sollen, einesteils für den Keil auf dieser Platte, andernteils für das Koronaspektrum auf den beiden die Sonne tangierenden Spektralstreifen. Die Orientierung der Wellenlängen ergab sich aus den Chromosphärenlinien, welche die Aufnahme zeigt.

Der Uebergang vom Vergleichsspektrum zur Korona wird durch die beiden Keilbilder vermittelt. Vergleicht man die Schwärzungen S und S' derselben Keilstellen, so erhält man folgende Tabelle zur Reduktion von S auf S':

S	65	60	55	50	45	40	35	30
$S'-S$	10.0	8.7	7.3	6.3	6.2	6.4	6.8	7.2.

Mittelst derselben sind alle Zahlen S der Tabelle IV in Schwärzungen S' verwandelt.

Tabelle IV.

a) Keil.

Keilstellen:	2.5	5	10	15	20	25	30	35
Schwärzung S:	64.8	64.0	60.4	54.9	48.7	43.5	39.0	38.6

b) Vergleichsspektren.

σ in mm		—28.0	—26.0	—24.0	—22.0	—20.0	—18.0	—16.0	—14.0	—12.0	—10.0	—7.8	—6.0	—4.0	—2.0	—0.2	+2.0	+4.0	+6.0	+8.0	+10.0	+12.0	+14.0	+16.0
λ in $\mu\mu$		329	333	337	342	348	355	362	369	376	384	395	405	416	428	440	454	471	489	509	531	556	586	623
$\eta' = 0.5$	S					33.6	39.1	39.5	44.5	44.7	44.5	49.3	51.8	53.1	52.7	53.2	52.7	49.0	43.2	37.9	32.9			
	ϑ					12^{m}8	13.0	13.0	13.1	13.1	13.2	13.2	13.2	13.2	13.2	13.1	13.1	13.1	12.9	12.8	12.7			
	S'					40.5	45.6	46.0	50.7	50.8	50.6	55.6	58.4	60.0	59.5	60.1	59.5	55.2	49.4	44.4	39.9			
$\eta' = 1.5$	S				36.3	39.7	43.8	44.6	49.1	49.4	51.3	55.6	57.7	59.1	58.7	59.6	57.8	54.8	49.4	41.3	36.3			
	ϑ				11^{m}5	11.6	11.7	11.7	11.8	11.8	11.8	11.8	11.8	11.8	11.8	11.8	11.8	11.8	11.7	11.7	11.6			
	S'				43.0	46.1	50.0	50.7	55.3	55.7	57.9	63.1	65.7	67.4	67.0	68.2	65.8	62.0	55.7	47.7	43.0			
$\eta' = 4.5$	S			33.9	40.9	45.2	51.8	53.1	57.8	57.4	59.0	63.0	65.0	66.1	66.3	66.0	64.4	59.8	53.7	46.2	39.6	37.3	35.3	31.9
	ϑ			10^{m}1	10.3	10.4	10.5	10.5	10.5	10.5	10.5	10.5	10.5	10.5	10.5	10.5	10.5	10.5	10.5	10.5	10.4	10.2	10.1	10.0
	S'			40.7	47.3	51.4	58.4	60.0	65.8	65.2	67.3	72.4	75.0	76.5	76.7	76.4	74.2	68.4	60.7	52.4	46.1	43.9	42.1	38.9
$\eta' = 13.5$	S	33.2	36.2	42.0	50.7	55.9	61.3	63.1	67.8	69.0			(zu hell)					69.5	60.8	53.3	47.0	41.7	39.8	37.0
	ϑ	8^{m}5	8.6	8.8	8.9	8.9	8.9	8.9	8.9	8.9								8.7	8.7	8.7	8.7	8.6	8.6	8.5
	S'	40.1	42.9	48.2	56.2	63.4	70.3	72.6	78.7	80.3								81.0	69.6	60.3	53.2	48.0	46.3	43.7
Falsches Licht	S		29.5								29.5		35.2	37.9		37.6	35.6		34.5	30.3				30.3

Tabelle V.

a) Keil.

Keilstellen:	2.5	5	7.5	10	15	20	25	30	35
Schwärzung S':	74.7	73.8	72.5	69.5	62.1	54.6	49.8	45.5	45.5

b) Koronaspektrum.

		σ	— 22.5	— 20.5	— 18.5	— 16.5	— 14.5	— 12.5	— 10.5	— 8.5	— 6.5	— 4.5	— 2.5	— 0.5	+ 1.5	+ 3.5	+ 5.5	+ 7.5
		λ	341	347	353	360	367	374	382	392	402	413	425	437	451	467	485	504
Nördlicher Streifen		S'	39.6	40.4	44.1	47.3	50.6	51.8	54.4	57.7	59.8	62.2	62.3	63.3	63.3	60.3	55.6	49.8
		ϑ	11.8	12.4	12.8	12.4	12.8	12.8	12.4	12.5	13.0	12.6	12.8	12.5	12.3	12.2	11.8	11.7
	Korr. f.	V	0.5	0.5	0.5	0.5	0 4	0.4	0.4	0.4	0.4	0.4	0.4	0.4	0.4	0.4	0.4	0.4
	Endwert	ϑ	12.3	12.9	13.3	12.9	13.2	13.2	12.8	12.9	13.4	13.0	13.2	12.9	12.7	12.6	12.2	12.1
Südlicher Streifen		S'	38.0	40.0	41.2	43.2	45.3	46.7	50.0	52.7	55.0	56.3	57.5	57.0	56.3	54.3	51.0	43.8
		ϑ	12.1	12.6	13.7	13.7	14.0	14.1	13.5	13.7	13.7	13.8	13.7	13.9	13.8	13.6	13.3	13.4

Abhandl. Kgl. Ges. d. Wiss. zu Gött. NF 5(2) 1907, 3–73

Betrachtet man nun die Gesamtheit der zusammengehörigen Zahlen ϑ und S' in Tabelle IV, so hat man eine Tafel mit doppeltem Eingang vor sich. Argumente sind die Wellenlänge, resp. die Stelle σ des Spektrums, und die Schwärzung S', Funktion ist die Helligkeit ϑ. Um das Interpolieren zu erleichtern und kleine Unregelmässigkeiten dieser Werte selbst auszugleichen, habe ich die Kurven ϑ als Funktionen von S' für alle Wellenlängen aufgetragen und weiterhin diese Kurven benutzt.

Mit den Wellenlängen und Schwärzungen S', welche Tafel V b) für die einzelnen Punkte des Koronaspektrums angiebt, bin ich dann in die Kurvenschaar eingegangen und habe damit die Werte ϑ für die Korona interpoliert, wie sie in Tafel V b) eingetragen sind. An diese Werte war dann noch die Korrektion wegen Ungleichförmigkeit des Verschlusses anzubringen, um sie in die „Endwerte“ der Tabelle überzuführen. Für den südlichen Streifen verschwindet die Verschlusskorrektion.

Die hiermit erhaltenen Endwerte stellen genau die Helligkeitswerte des unreinen Spektrums — in Grössenklassen gezählt — dar, welche bei der obigen theoretischen Betrachtung mit ϑ bezeichnet wurden.

§ 13. Aktinität des Sonnenspektrums.

Um zur Helligkeit des reinen Koronaspektrums übergehn zu können, muss bekannt sein, welche photographische Wirkung die verschiedenen Teile des Sonnenspektrums ausüben, welche Gestalt die Funktion $K(\sigma)$ hat. Wir messen $K(\sigma)$, indem wir nachsehn, in welchem Verhältnis das Sonnenspektrum abgeschwächt werden muss, damit die von der wirksamsten Wellenlänge ($\lambda = 441\,\mu\mu$, $\sigma = 0$) erzeugte Schwärzung auf diejenige der betreffenden anderen Wellenlänge reduziert wird. Das Messungsmaterial ist in Tafel IV vorhanden.

Da es sich nur um einen Ueberschlag handelt, können wir die Beziehung zwischen einer in Grössenklassen gemessenen Helligkeit und der Schwärzung S als linear voraussetzen und erhalten aus Tafel IV im Durchschnitt etwa:

$$m = 0.15\,S.$$

Verwandelt man durch Multiplikation mit diesem Faktor 0.15 die Schwärzungen S der Tafel IV in Grössenklassen und subtrahiert die entstehenden Zahlen von dem jeweils für $\lambda = 441\,\mu\mu$ resultierenden Wert, so erhält man folgendes Absinken der photographischen Helligkeit im Sonnenspektrum gegen die wirksamste Wellenlänge:

Tafel VI. Aktinität $K(\sigma)$ des Sonnenspektrums.

σ	−28	−26	−24	−22	−20	−18	−16	−14	−12	−10	−8	−6	−4	−2	0
$\eta' = 0.5$					3.3	2.4	2.3	1.6	1.5	1.5	0.6	0.2	0.0	0.1	0.0
1.5				4.1	3.5	2.8	2.7	1.9	1.9	1.5	0.8	0.4	0.1	0.2	0.0
4.5			5.7	4.6	3.8	2.7	2.5	1.6	1.7	1.4	0.6	0.2	0.0	0.0	0.0
13.5	8.0	7.5	6.1	5.2	4.1	3.0	2.7	1.9	1.5						
m	6.4	6.0	4.8	3.9	3.4	2.6	2.5	1.7	1.7	1.5	0.7	0.3	0.1	0.1	0.0
$K(\sigma)$	0.00	0.00	0.01	0.03	0.04	0.09	0.10	0.21	0.21	0.25	0.52	0.76	0.92	0.92	1.00

σ	+2	4	6	8	10	12	14	16	18
$\eta' = 0.5$	0.1	0.7	1.9	2.7	3.5				
1.5	0.4	0.9	2.0	3.2	4.0				
4.5	0.3	1.2	2.3	3.6	4.6	5.2	5.5	6.1	
13.5		1.5	3.2	4.6	5.6	6.5	6.7	7.2	8.0
m	0.3	0.8	2.0	2.9	3.8	4.3	4.5	5.1	5.7
$K(\sigma)$	0.76	0.44	0.16	0.07	0.03	0.02	0.02	0.01	0.01

Ich habe das Mittel aus den beiden ersten Reihen genommen, welches der Schwärzung der Koronaspektren nahe entspricht, und die letzten Reihen nur zur Fortsetzung an den Enden des Spektrums verwandt. Indem man die so erhaltenen in Grössenklassen ausgedrückten Werte m durch Bildung von 0.4^{m} in Helligkeitsverhältnisse zurückverwandelt, findet man die in der letzten Zeile der vorstehenden Tabelle enthaltene Darstellung der photographischen Wirkung $K(\sigma)$ der Sonne in den verschiedenen Spektralgebieten.

§ 14. Reduktion auf reines Spektrum. Ergebnis.

Nachdem nun durch Tabelle V und VI die Helligkeit des unreinen Koronaspektrums und die Aktinität $K(\sigma)$ des Sonnenspektrums gegeben sind, konnte mit Benutzung der obigen Annahmen über die Helligkeitsverteilung der ganzen Koronastrahlung die „Reduktion auf reines Spektrum" $g(\sigma)$ nach der Formel (10) berechnet werden. Da $\vartheta(\sigma)$ gemäss Tabelle V b) so nahe konstant ist, die Helligkeitsverteilung im unreinen Koronaspektrum der im reinen Sonnenspektrum so nahe liegt, habe ich mir hierbei erlaubt, einfach $\vartheta(\sigma) = \vartheta(0) = \text{const.}$ zu setzen. Ich habe also tatsächlich berechnet, wie sich die Intensität in einem reinen

Abhandl. Kgl. Ges. d. Wiss. zu Gött. NF 5(2) 1907, 3–73

Sonnenspektrum verschieben würde, wenn die Lichtquelle nicht punktförmig, sondern in einem Streifen mit dem der Korona entsprechenden Helligkeitsabfall nach aussen verteilt wäre. Die nötigen Quadraturen wurden für einige Werte von σ mechanisch ausgeführt mit folgendem Ergebnis (es ist $g(\sigma)$ gleich in Grössenklassen ausgedrückt):

σ	−20	−15	−10	−5	0	+5	+10
$g(\sigma)$	$-0^{m}.22$	−0.20	−0.09	+0.10	+0.13	−0.25	−0.9.

Man sieht, dass diese „Reduktion auf reines Spektrum" nur für das rote Ende des Spektrums merkliche Beträge erreicht. Der letzte Wert für $\sigma = 10$ ist unsicher, da er schon stark von einer Aenderung der Kurve $K(\sigma)$ beeinflusst wird.

Bildet man nun entsprechend Formel (12) den Quotienten $\frac{\vartheta(\sigma)}{\vartheta(0)}$, so wie es in folgender Tabelle geschehen ist, und dividiert durch $\frac{g(\sigma)}{g(0)}$ — bei dem Rechnen mit Grössenklassen hat man tatsächlich nur zu subtrahieren — so findet man die unten angegebenen Werte für die Helligkeit des reinen Spektrums $\psi(\sigma)$.

Tabelle VII. Reines Koronaspektrum $\psi(\sigma)$.

σ	−22.5	−20 5	−18.5	−16.5	−14.5	−12.5	−10.5	−8.5	−6.5	−4.5	−2.5	−0.5	+1.5	+3.5	+5.5	+7.5
λ	341	347	353	360	367	374	382	392	402	413	425	437	451	467	485	504
$\frac{\vartheta(\sigma)}{\vartheta(0)}$ Nordstreifen	$-0^{m}.6$	0.0	+0.4	0.0	+0.3	+0.3	−0.1	0.0	+0.5	+0.1	+0.3	0.0	−0.2	−0.3	−0.7	−0.8
$\frac{\vartheta(\sigma)}{\vartheta(0)}$ Südstreifen	−1.8	−1.3	−0.2	−0.2	+0.1	+0.2	−0.4	−0.2	−0.2	−0.1	−0.2	0.0	−0.1	−0.3	−0.6	−0.5
Mittel	−1.2	−0.6	+0.1	−0.1	+0.2	+0.2	−0.2	−0.1	+0.2	0.0	0.0	0.0	−0.2	−0.3	−0.6	−0.6
$g(\sigma)/g(0)$	−0.4	−0.4	−0.4	−0.4	−0.3	−0.3	−0.2	−0.2	−0.1	0.0	0.0	0.0	−0.1	−0.2	−0.4	−0.7
$\psi(\sigma)$	−0.8	−0.2	+0.5	+0.3	+0.5	+0.5	0.0	+0.1	+0.3	0.0	0.0	0.0	−0.1	−0.1	−0.2	+0.1

Aus den ϑ-Werten für den Nord- und Südstreifen habe ich gleich das Mittel genommen und vereinige nun auch je zwei benachbarte ψ-Werte zu folgenden Endwerten:

Tabelle VII a). Endwerte (Mittel aus zweien).

λ	344 $\mu\mu$	356	370	387	408	431	459	494
$\psi(\sigma)$	(−0.50)	$+0^{m}.40$	+0.50	+0.05	+0.15	0.00	−0.10	−0.05

Hier ist die Stelle, um noch auf die selektiven Absorptionen bei den verschiedenen Aufnahmen Rücksicht zu nehmen. Das Sonnenspektrum ist mit eingeschaltetem Kollimatorobjektiv aus U. V. Glas von 8 mm Dicke aufgenommen. Die Absorption desselben beträgt nach freundlichen Angaben von Herrn Prof. Straubel in Jena $0^{m}.2$ für die Wellenlänge 344 $\mu\mu$, $0^{m}.1$ für 356 $\mu\mu$ und verschwindet für grössere Wellenlänge. Ferner ist bei der Koronaaufnahme das ganze U. V. Prisma benutzt, während bei der Vorschaltung des Kollimators, der 30 mm Oeffnung hatte, nur der mittelste Teil desselben zur Verwendung kam. Ein besonderer Versuch ergab folgende Absorptionskoeffizienten des Prismenglases für 4 cm Glasdicke:

λ	335	339	355	373	395	422	477
Absorptionskoeffizient	0.66	0.70	0.83	0.91	0.93	0.91	0.93
Korrektion	$0^m.15$	$0^m.12$	$0^m.07$	$0^m.03$	$0^m.03$	$0^m.03$	$0^m.03$

Mit diesen Absorptionskoeffizienten berechneten sich aus den geometrischen Verhältnissen die darunter gesetzten Beträge, um welche die Vergleichsaufnahmen zu hell waren. Schliesslich ist noch der Einfluss der Erdatmosphäre zu bedenken. Die Zenitdistanz der Sonne bei der Totalität in Gelma betrug $38^0.7$, die bei der Aufnahme des Vergleichspektrums in Göttingen $29^0.5$. Die Extinktion der photographisch wirksamsten Strahlen ($\lambda = 441\,\mu\mu$) beträgt (zum doppelten der Potsdamer optischen Extinktion angesetzt) für diese Zenitdistanzen $0^m.11$ resp. $0^m.05$, Differenz $0^m.06$. Nimmt man die Absorption mit der vierten Potenz der Wellenlänge steigend an, so ergiebt sich damit für $\lambda = 341\,\mu\mu$ eine Absorptionsdifferenz von $0^m.14$. Addiert man alle drei Korrektionen mit dem richtigen Vorzeichen, so heben sie sich zufällig auf. Die Tabelle VII a) stellt somit das Endergebnis der Untersuchung dar. Die nähere Besprechung desselben wird im letzten Abschnitt erfolgen.

IV. Abschnitt. Helligkeit der inneren und der gesamten Korona.

§ 15.

Die Angaben über die absolute Helligkeit der Korona in verschiedener Distanz von der Sonne, welche in den Spektralaufnahmen der Korona enthalten sind, lassen sich nach der Auseinandersetzung im vorigen Abschnitt exakt so angeben. Für einen an die Sonne tangentialen Schnitt durch die Korona (quer zum Spektrum gezählte Koordinate η gleich Sonnenradius R) gilt die Gleichung (11):

$$\int_{-\infty}^{+\infty} h(\mu)\,d\mu = \int_{-\infty}^{+\infty} i(\sqrt{R^2+\mu^2})\,d\mu = \frac{\vartheta(0)}{g(0)},$$

Abhandl. Kgl. Ges. d. Wiss. zu Gött. NF 5(2) 1907, 3–73

wobei $\vartheta(0)$ die Helligkeit des unreinen Spektrums für die Wellenlänge maximaler Intensität ist und $g(0)$ die Reduktion auf reines Spektrum bedeutet. Ganz ebenso gilt für einen Schnitt in der beliebigen Entfernung η vom Sonnenzentrum:

15) $$\int_{-\infty}^{+\infty} i(\sqrt{\eta^2+\mu^2})\, d\mu = \frac{\vartheta(0)}{g(0)}$$

wo $\vartheta(0)$ und $g(0)$ sich wieder auf die Wellenlänge $441\,\mu\mu$, aber auf die betreffende Entfernung η vom Sonnenzentrum beziehn.

Im folgenden beschränke ich mich durchaus auf die Wellenlänge maximaler Intensität, da hier die Reduktion auf reines Spektrum sehr wenig ausmacht und sehr sicher zu bestimmen ist. Ich lasse daher den Index 0 bei ϑ und g weg und führe statt dessen das neue Argument η ein, welches den Abstand des betreffenden Spektralstreifens vom Sonnenmittelpunkt bezeichnet.

Die Aufgabe ist, die Platten in der Wellenlänge maximaler Intensität für verschiedenen Abstand von der zentralen Linie des Spektrums zu photometrieren und zunächst die Helligkeiten des unreinen Spektrums ϑ abzuleiten.

§ 16. Helligkeit am Sonnenrand aus Platte Z_a.

Aus der Untersuchung des § 12 Tabelle V b) können wir noch entnehmen, dass am Sonnenrand im Mittel aus dem nördlichen und südlichen Streifen für die Wellenlänge $441\,\mu\mu$ gilt:

$$\text{Für } \eta = R: \qquad \vartheta = 13^{\text{m}}.4.$$

§ 17. Bearbeitung von Platte Z_b.

Die Platte Z_b gestattet zunächst die Ableitung der **relativen** Helligkeiten des unreinen Spektrums für die innere Korona vom Sonnenrande bis zu etwa 8′ Höhe über demselben. Dle Platte trägt 3 Koronaspektren mit den Expositionen 1^s, 15^s, 40^s. Diesen Aenderungen der Expositionszeit würden Aenderungen der Intensitäten von $2^{\text{m}}.9$ resp. $4^{\text{m}}.0$ entsprechen, wenn das sog. Reziprozitätsgesetz gälte. In Wirklichkeit wirkt die Aenderung der Expositionszeit etwas schwächer, als entsprechende Aenderung der Intensität. Ich multipliziere daher diese Beträge mit 0.9 und setze sie zu $2^{\text{m}}.6$ resp. $3^{\text{m}}.6$ an. Nunmehr stehen 3 Aufnahmen mit bekanntem Abschwächungsverhältnis zur Verfügung und es ist also nach dem Verfahren A. N. Bd. 172 pag. 65 möglich, relative Helligkeiten abzuleiten.

Es ergaben sich folgende Schwärzungen[1]) für die 3 Aufnahmen in ver-

1) Da es sich z. T. um sehr dunkle Stellen der Platte handelte, wurde an dem Mikrophotometer eine besonders starke Durchleuchtung der Platte eingeführt, Die hier folgenden Schwärzungen sind daher mit den sonstigen nicht ohne weiteres vergleichbar. Ich möchte hinzufügen, dass man in der Vergleichung von Schwärzungsangaben, die **nicht** im Texte zueinander in Beziehung gesetzt werden, vorsichtig sein muss, weil die Beleuchtung im Mikrophotometer öfters gewechselt wurde.

schiedenem Abstand η von der Sonnenmitte. (Dabei ist aus den Messungen nördlich und südlich des Sonnenrandes bereits das Mittel genommen).

Abstand η vom Sonnenzentrum und zugehörige Schwärzung S.

Exposition	16′.5	18′.1	19′.6	21′.2	22′.8	24′.4
1^s	63.1	57.3				
15^s	71.0	68.2	65.6	63.7	62.1	59.5
40^s	72.7	71.5	69.3	66.8	65.8	64.6

Die Umwandlung in Grössenklassen konnte nach dem l. c. unter No. 4b) angegebenen Verfahren erfolgen und ergab die Formel:

$$m = 7.7 \log (78 - S).$$

Hiernach verwandelten sich die Schwärzungen in folgende Grössenwerte:

	16′.5	18.1	19.6	21.2	22.8	24.4
1^s	9^m1	10^m2				
15^s	6.6	7.5	8.5	9.0	9.3	9.8
40^s	5.6	6.3	7.3	8.1	8.4	8.7

Der Abfall der Helligkeiten gegen die Werte am Sonnenrande beträgt demnach:

Tabelle VIII. Relative Helligkeiten der inneren Korona.

	16′.5	18.1	19.6	21.2	22.8	24.4
1^s	0^m0	1.1				
15^s	0.0	0.9	1.9	2.4	2.7	3.2
40^s	0.0	0.7	1.7	2.5	2.8	3.1
Mittel	0.0	0.9	1.8	2.4	2.8	3^m2

Es sind weiter diese relativen Werte in absolute zu verwandeln. Das ist möglich mit Hülfe des auf der Platte befindlichen photometrischen Keils, der allerdings infolge seiner zu geringen Schwärzung nur an die mittleren Teile der Korona auf der kurzen Exposition von 1^s heranreicht. Im Uebrigen waren die Aufgabe und das Verfahren ganz dasselbe, wie oben im § 11. Genau wie dort mit demselben Momentverschluss das Sonnenspektrum aufgenommen wurde, der früher zur Koronaaufnahme gedient hatte, so wurde jetzt das 1^s exponierte

Abhandl. Kgl. Ges. d. Wiss. zu Gött. NF 5(2) 1907, 3–73

Koronaspektrum zu einem 1^s exponierten Sonnenspektrum durch Vermittlung der 15^s frei gegen die Sonne exponierten photometrischen Keile in Beziehung gesetzt. Zur Herstellung des Sonnenspektrums wurde derselbe Kollimator verwandt, nur musste, um Ueberlichtung zu vermeiden, noch eigens durch eine geschwärzte vor den Spalt gesetzte Platte abgeblendet werden. Die Extinktion der Platte für die aktinisch wirksamsten Strahlen ergab sich durch einen besonderen Versuch zu $3^m.25$. Eine Aufnahme vom 4. September 1906 $12^h 40^m$ ergab nun, dass die Keilstelle 2.5 bei 15^s Exposition ebenso geschwärzt war, wie das Sonnenspektrum unter Verwendung der Blende, für den Abstand von der Kante $\eta' = 1.0$ mm. Die Helligkeit des unreinen Spektrums ϑ für $\eta' = 1.0$ folgt nach Formel (13) zu $11^m.56$. Diese Zahl ist wegen der Phase der Sonne bei der Aufnahme des Keils auf Platte Z_b um den Intensitätsfaktor 0.75 oder den Betrag in Grössenklassen $0^m.31$, ferner wegen der Reflexion am Kollimatorobjektiv um $0^m.20$ und wegen der vorgesetzten Absorptionsplatte um $3^m.25$ zu vermehren. Das giebt für die Stelle 2.5 des Keils die entsprechende Helligkeit ϑ zu $15^m.3$. Indem man annimmt, dass die aktinischste Wellenlänge dieselbe Gradation giebt, wie das vereinigte Spektrum, darf man die Helligkeiten für die andern Stellen des Keils aus dem durch die obere Zeile von Tab. I gegebenen — ursprünglich für die gesamte photographische Wirkung gültigen — Gradationstäfelchen entnehmen und findet damit nachstehendes Täfelchen für die den einzelnen Keilstellen entsprechenden Helligkeiten des unreinen Spektrums:

Keilstelle:	2.5	5	10	15
Helligkeit ϑ:	$15^m.3$	15.5	16.1	16.7
Schwärzung:	51.2	50.2	45.3	36.2

Dazu gefügt ist gleich die Schwärzung, welche den betreffenden Keilstellen auf der Platte Z_b zu kam.

Für die Korona selbst ergaben sich folgende Schwärzungen (es ist das Mittel aus den sehr nahe übereinstimmenden Werten vom N- und S-Rand gegeben).

Tabelle IX.

Abstand vom Zentrum η:	$21'.2$	22.8	24.4	26.0
Schwärzung:	55.6	51.4	48.4	44.4
Helligkeit ϑ:	$14^m.3$	15.2	15.8	16.2

Der Vergleich mit den obigen Schwärzungen des Keils liefert sofort die darunter gesetzten Helligkeiten ϑ.

Damit ist die absolute Helligkeit des unreinen Koronaspektrums für mittlere Abstände vom Zentrum festgelegt. Man gewinnt die Fortsetzung der Tabelle IX nach der inneren Corona zu, indem man die in Tabelle VIII gegebenen Relativwerte an diese absoluten Helligkeiten anschliesst. Der beste Anschluss ergiebt sich, indem man zu den Relativwerten $12^m.3$ addiert. Man erhält dann folgende Zusammenstellung:

Tabelle X. Helligkeit des reinen Spektrums der inneren Korona.

Abstand vom Zentrum η:	16.5	18.1	19.6	21.2	22.8	24.4	26.0
Aus Tabelle VIII	12ᵐ3	13.2	14.1	14.7	15.1	15.5	
„ „ IX				14ᵐ3	15.2	15.8	16 2
Mittel	12ᵐ3	13.2	14.1	14.5	15.2	15.6	16.2
Helligkeit des reinen Spektrums ζ	13ᵐ0	13.9	14.8	15.2	15.9	16.3	16.9

Aus Platte Z_c hatten wir im § 16 die Helligkeit am Sonnenrande $\vartheta = 13^m\!.4$ abgeleitet, während sich hier $12^m\!.3$ ergiebt. In die erste Zahl geht die Unsicherheit des Verschlusses ein, in letztere ein Fehler in der 1^s Exposition. Indem ich beiden Bestimmungen gleiches Gewicht gebe, vermehre ich die eben erhaltenen Mittelwerte um $0^m\!.6$.

Ausserdem soll hier gleich die Reduktion auf reines Spektrum angebracht werden. Durch die numerische Auswertung der betreffenden zu (10) annalogen Integrale ergab sich für dieselbe in dem ganzen hier in Frage kommenden Gebiet der Wert $0^m\!.1$. Indem auch dieser noch zu obigen Zahlen hinzugefügt wurde, entstanden die definitiven „Helligkeiten des reinen Spektrums ζ“.

§ 19. Die Helligkeitsverteilung in der gesamten Korona.

Es stehen uns nunmehr die folgenden Helligkeitsangaben über die Korona zu gebote: erstens die durch die gesamte aktinische Wirkung bestimmten Helligkeiten der Tabelle III in § 8, zweitens die der Wellenlänge maximaler Wirkung entsprechenden Zahlen der Tabelle X, welche das Integral über einen Schnitt durch die Korona:

$$16)\qquad \int_{-\infty}^{+\infty} i\left(\sqrt{\eta^2+\mu^2}\right) d\mu = \zeta(\eta)$$

geben. Da sich die Farbe der Korona als so nahe übereinstimmend mit der der Sonne erwiesen hat, darf man die Ergebnisse für die Wellenlänge maximaler Wirkung und die Gesamtwirkung miteinander vertauschen, die Helligkeiten aus Tabelle III stellen also direkt die Werte von i für die äusseren Teile der Korona dar.

Um die Helligkeit der inneren Korona aus den Werten von ζ abzuleiten, werde mit Turner der schon erwähnte Ansatz gemacht:

$$17)\qquad i = i_0\left(\frac{R}{r}\right)^n = i_0\left(\frac{R}{\sqrt{\eta^2+\mu^2}}\right)^n$$

sodass i_0 die Koronahelligkeit am Sonnenrand bedeutet. Man hat dann:

Abhandl. Kgl. Ges. d. Wiss. zu Gött. NF 5 (2) 1907, 3–73

18) $$\zeta(\eta) = i_0 \int_{-\infty}^{+\infty} d\mu \cdot \left(\frac{R^2}{\eta^2+\mu^2}\right)^{n/2}$$

was ausgeführt ergiebt:

19) $$\zeta(\eta) = 2\,R\,i_0 \left(\frac{R}{\eta}\right)^{n-1} f(n).$$

Der Faktor $f(n)$ hat den Wert:

$$\frac{n-3}{n-2}\cdot\frac{n-5}{n-4}\cdots\frac{1}{2}\cdot\frac{\pi}{2} \text{ für grades } n$$

$$\frac{n-3}{n-2}\cdot\frac{n-5}{n-4}\cdots\frac{2}{3} \qquad \text{für ungerades } n$$

und es ist z. B. numerisch:

$$f(6) = 0.59 \qquad f(7) = 0.53 \qquad f(8) = 0.49.$$

Schreibt man die Gleichung (19) durch Kombination mit (17) in der Form:

20) $$\frac{1}{2f(n)}\frac{\zeta(\eta)}{\eta} = i(\eta),$$

so wird sie auch für schwach veränderliches n angenähert gelten und eine für unsre Zwecke ausreichende Lösung der Funktionalgleichung (16) liefern.

Indem ich nach Turner $n = 6$ setzte, erhielt ich aus (20) die nachstehenden Helligkeiten (m „beobachtet") der inneren Korona. Die Zahlen aus Tabelle III für die äussere Korona sind hinzugefügt. Alle Helligkeiten sind in Grössenklassen ausgedrückt und es sind die aus Formel (17) hervorgehenden „berechneten" Helligkeiten in Grössenklassen:

$$m = m_0 + 2{,}5\,n \log \frac{r}{R}$$

unter Anwendung der numerischen Werte:

$$m_0 = 15^{\mathrm{m}}.5 \;\; (i_0 = 0{,}64 . 10^{-6}) \qquad n = 6$$

darunter gesetzt.

Tabelle XI. Helligkeit der gesamten Korona.

r	Aus den Spektralaufnahmen							Aus den direkten Aufnahmen				
	16′.5	18.1	19.6	21.2	22.8	24.4	26.0	30′	35′	40′	45′	50′
m beobachtet	$14^{\mathrm{m}}.9$	15.9	16.9	17.3	18.0	18.6	19.2	$19^{\mathrm{m}}5$	20.3	20.9	21.6	22.3
m berechnet	15.5	16.1	16.6	17.1	17.6	18.0	18.5	19.4	20.4	21.3	22.1	22.7
Differenz	+ 0.6	+ 0.2	− 0.3	− 0.2	− 0.4	− 0.6	− 0.7	− 0.1	+ 0.1	+ 0.4	+ 0.5	+ 0.4

Man erkennt in diesen Zahlen eine so gute Bestätigung des Turnerschen Gesetzes, als die geringe Genauigkeit unserer Zahlen und die Unregelmässigkeit der Korona irgend erwarten lassen konnte.

Die Gesamthelligkeit der Korona, verglichen mit der Sonne, folgt aus dem Integral:

$$\frac{2}{R^2}\int_{r=R}^{r=\infty} i(r)\, r dr.$$

Dasselbe giebt unter Benutzung des Ansatzes (17) ausgeführt $\frac{2\, i_0}{n-2}$.

Damit folgt für $n = 6$:

Die Gesamthelligkeit der Korona beträgt

$3{,}2 \cdot 10^{-7}$	oder $16^m_{\cdot}3$	verglichen mit	der Sonne
0,17	„ $1^m_{\cdot}9$	„ „	dem Monde.

Abschnitt V. Die Form der Mondoberfläche beim zweiten Kontakt und die Helligkeit des Sonnenrandes.

§ 20.

Das kontinuierliche Spektrum der schmalen Sonnensichel, welches man bei Aufnahmen mit der Prismenkamera kurz vor Beginn oder kurz nach Ende der Totalität erhält, ist durchsetzt von Längsstreifen wechselnder Helligkeit, welche den Unregelmässigkeiten des Mondrandes entstammen. Jeder Mondberg wird durch eine streifenformige Schwächung oder Unterbrechung, jedes Tal des Mondes durch einen Streifen stärkeren kontinuierlichen Photosphärenlichtes bezeichnet. Die Helligkeitsverteilung in einem solchen Spektrum wird daher bestimmt durch 2 Dinge: erstens durch die Form des Mondrandes, zweitens durch die Helligkeit der hinter dem Mondrand hervorblickenden Randstücke der Sonnenphotosphäre,

Es soll versucht werden, sowohl die Form des Mondrandes, als auch die Randhelligkeit der Sonne aus den Spektren auf Film Z_a, abzuleiten.

§ 21. Photometrische Durchmessung für die Wellenlänge $\lambda = 455\, \mu\mu$.

Die erste Arbeit bestand darin, für eine bestimmte Wellenlänge „Helligkeitsprofile" quer durch die auf Film Z_a befindlichen flash-Spektren hindurch aufzunehmen. Ich wählte speziell die Wellenlänge $\lambda = 455\, \mu\mu$ und folgte der Krümmung der benachbarten Chromosphärenlinien bei der Durchmessung unter dem Mikrophotometer, um genau in einer Wellenlänge zu bleiben. Das Resultat der

Abhandl. Kgl. Ges. d. Wiss. zu Gött. NF 5(2) 1907, 3–73

Messungen der Spektren Nr. 9, 12 und 15 ist in Figur 2 (pag. 46.) dargestellt. Der Messapparat unter dem Mikrophotometer gestattete eine Ablesung auf 0.01 mm = 2″. Man darf an die Genauigkeit der Messungen keine übertriebenen Anforderungen stellen, denn es war oft nötig, feine Spektralstreifen zu photometrieren, die den Photometerfleck keineswegs ausfüllten. Es hätte sich eine etwas abgeänderte Photometerkonstruktion empfohlen, die noch kleinere Bereiche zu messen erlaubte. Es wurde etwa von 0.2 zu 0.2 mm gemessen und das Detail dazwischen schätzungsweise in die Kurven eingezeichnet, wobei ich Maxima und Minima tunlichst messend kontrollierte.

Die Abszissen S der Figur sind in einem gleich zu erläuternden ungleichmässigen Massstab gezeichnete Schwärzungen, die Ordinaten geben Millimeter senkrecht zur Erstreckung des Spektrums und von der Mitte des Mondes an gezählt. Die positive Richtung geht nach Süden. Die Mitte des Mondes liess sich dadurch auf den Aufnahmen festlegen, dass die Aufnahmen beiderseits durch das Koronaspektrum begrenzt waren, welches die Tangenten an den Mondrand bezeichnete.

Für die Beziehungen der Schwärzungen zu den Helligkeiten hat man einen Anhalt in den drei Spektren der abgeblendeten Sonne, welche vor der Finsternis auf demselben Film aufgenommen worden waren. Die Abblendungen berechnen sich aus den Durchmessern der verwendeten Löcher unter Berücksichtigung der Phase der Sonne bei der Aufnahme zu:

Lochgrösse:	0.52	1.06	3.19
Abblendungen:	$11^{m}.7$	10.2	7.8

Die Schwärzungen in den drei Spektren an der Stelle, welcher der Wellenlänge 455 $\mu\mu$ entsprach, und zwar in der Mitte des Streifens, in welchem also der volle Sonnendurchmesser wirkte, waren:

$$S = 62.5 \qquad 66.0 \qquad 71.4.$$

Diese Zahlen geben die lineare Beziehung zwischen Schwärzung S und Grössenklassen

$$S = 89.2 - 2.27m, \quad m = 0.440\,(89.2 - S).$$

Daraus folgt für die Intensitäten $J (= 0.4^{m})$:

$$\log J = 0.176\,(S - 89.2).$$

Ich füge gleich den Ausdruck hinzu:

21) $$\log 2RJ = 0.176\,(S - 70.6),$$

in welchem R den Sonnenradius in Bogensekunden bedeutet. Die Größe $2RJ$ ist es, welche bei der Zeichnung der Helligkeitsprofile als gleichförmige Abscisse verwandt wurde und aus deren Verwendung die in Fig. 2 angegebene ungleichförmige Scale der Schwärzungen hervorging. Die Abscissen in Fig. 2 stellen dadurch bereits Helligkeiten und nicht mehr Schwärzungen dar.

§ 22. Theorie.

Wie hängen die gemessenen Helligkeiten oder Schwärzungen quantitativ mit der Form des Mondrandes und der Helligkeit des Sonnenrandes zusammen?

Man wähle ein rechtwinkliches Koordinatensystem, dessen Nullpunkt der Mittelpunkt des für den Augenblick kreisförmig vorausgesetzten Mondes sei, dessen y-Axe parallel der Dispersionsrichtung des Spektrums liegt, dessen x-Koordinaten also mit den ebenso bezeichneten Werten in Fig. 2 übereinstimmen. Der Mondradius sei r, der Sonnenradius sei R. Der Mittelpunkt der Sonne habe in diesem System die Koordinaten a, b (vgl. Fig. 1). Dann entspricht einer bestimmten x-Koordinate auf dem Mondrand der Punkt:

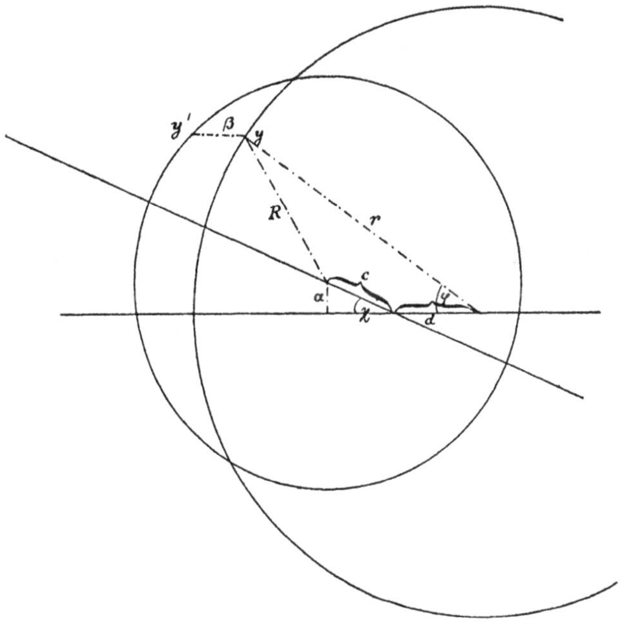

Figur 1.

$$y = \sqrt{r^2 - x^2}$$

und auf dem Sonnenrand der Punkt:

$$y' = b + \sqrt{R^2 - (x-a)^2}.$$

Sei:

$$x = r \sin\varphi, \qquad a = c \sin\chi, \qquad b = c\cos\chi + d, \qquad r - R = \varrho,$$

sodass φ der Positionswinkel am Mondmittelpunkt, χ die Neigung der relativen Bahn von Mond und Sonne gegen die Dispersionsrichtung des Spektrums ist und d die Abweichung der Verfinsterung von der genauen Zentralität misst. Damit ergiebt sich die „Sichelbreite β" an jeder Stelle:

$$\beta = y' - y = d + c\cos\chi - r\cos\varphi + \sqrt{r^2\cos^2\varphi - 2\varrho r + \varrho^2 + 2rc\sin\varphi\sin\chi - c^2\sin^2\chi}$$

oder, solange $r\cos\varphi$ gross gegen ϱ und c ist, durch Taylor-Entwicklung:

Abhandl. Kgl. Ges. d. Wiss. zu Gött. NF 5(2) 1907, 3–73

21) $$\beta = y' - y = d + \frac{c\cos(\chi-\varphi)-\varrho}{\cos\varphi} - \frac{(\varrho\sin\varphi - c\sin\chi)^2}{2r\cos^3\varphi}.$$

Dieser Differenz wird die Helligkeit des Spektrums J proportional sein. Wählt man als Einheit die Helligkeit, die ein Sonnendurchmesser erzeugen würde, so gilt:

$$J = \frac{i(R)}{2R}(y'-y).$$

Dabei ist $i(R)$ die Helligkeit des Sonnenrandes, verglichen mit der durchschnittlichen Helligkeit auf einem Sonnendurchmesser, und es ist angenommen, dass die Helligkeit in der Nachbarschaft des Sonnenrandes konstant ist, eine Voraussetzung, die wir später zu prüfen haben.

Hat nun der Mond an einer Stelle eine Einsenkung h, so ist J zu ersetzen durch:

22) $$J = \frac{i(R)}{2R}\left(y' - y - \frac{h}{\cos\varphi}\right).$$

Für einen Berg ist h positiv zu zählen. Sobald dieser Ausdruck negativ wird, ist er durch Null zu ersetzen, da dies vollständige Verdeckung der Photosphäre bedeutet. Man setze noch $c - \varrho = \xi$. Dann folgt:

$$J = \frac{i(R)}{2R}\left\{d - \varrho\frac{1-\cos(\chi-\varphi)}{\cos\varphi} - \frac{[\varrho(\sin\varphi-\sin\chi)-\xi\sin\chi]^2}{2r\cos^3\varphi} + \frac{\xi\cos(\chi-\varphi)}{\cos\varphi} - \frac{h}{\cos\varphi}\right\}$$

oder, da man in dem vorletzten Glied für die später in Frage kommenden Fälle ξ vernachlässigen darf:

23) $$J = \frac{i(R)}{2R}\left(a\xi + b - \frac{h}{\cos\varphi}\right)$$
$$a = \frac{\cos(\chi-\varphi)}{\cos\varphi} \qquad b = d - \varrho\cdot\frac{1-\cos(\chi-\varphi)}{\cos\varphi} - \frac{\varrho^2}{2r}\frac{(\sin\varphi-\sin\chi)^2}{\cos^3\varphi}.$$

Es sei noch angemerkt, dass der Eintritt der Totalität erfolgt für $\xi = \xi_0$, wo:

$$\xi_0 = -d\cos\chi - \varrho + \sqrt{\varrho^2 - d^2\sin^2\chi} \quad \text{ist.}$$

Für die in den vorstehenden Formeln auftretenden Grössen hat man folgende numerische Angaben. Aus den Daten der Finsternis findet man:

24) $$\xi = \xi_0 - 0''\!.416\, t^s \quad (t \text{ vom Eintritt der Totalität an gerechnet})$$
$$\varrho = 41''\!.5.$$

Ferner ergab die Abmessung am Instrument den Positionswinkel des Spektrums in der Richtung Rot-Violett:

$$p = 106^\circ.$$

Der Positionswinkel der scheinbaren Mondbahn relativ zur Sonne ergab sich rechnerisch zu $115^\circ\!.5$. Daher hatte der Winkel χ den Betrag:

$$\chi = 115^\circ\!.5 - 106^\circ = 9^\circ\!.5.$$

Mit diesen Angaben fand sich folgendes Täfelchen für a und $b-d$:

Tabelle XII.

x	−3.0	−2.5	−2.0	−1.5	−1.0	−0.5	0.0	+0.5	1.0	1.5	2.0	2.5	3.0	3.5	4.0	4.5	5.0
φ	−34°5	−28.2	−22.2	−16.4	−10.9	−5.4	0.0	+5.4	10.9	16.4	22.2	28.2	34.5	41.4	49.0	58.0	70.6
a	0.87	0.90	0.92	0.93	0.95	0.97	0.99	1.00	1.02	1.03	1.05	1.07	1.10	1.13	1.18	1.25	1.55
$b-d$	−16″5	−11.4	−7.8	−5.0	−3.0	−1.6	−0.6	−0.1	0.0	−0.4	−1.2	−2.9	−5.5	−9.9	−17.1	−32.2	−88.4

§ 23. Die Korrektion des Mondortes und die Randhelligkeit der Sonne.

Unbekannt sind noch, abgesehn von den Unebenheiten des Mondrandes h, die Grössen ξ und d, welche den Ort des Mondmittelpunktes gegen den Sonnenmittelpunkt festlegen, und die Randhelligkeit der Sonne i. Sobald man auf die Unregelmässigkeit des Mondrandes Rücksicht nimmt, wird der Mittelpunkt der Mondscheibe in gewissem Grade unbestimmt und man könnte daher ξ und d willkürlich ansetzen und hätte dann unter h eben die Abweichungen des Mondrandes von einem Kreis um diesen willkürlich gewählten Mittelpunkt zu verstehen. Es bliebe dann noch die Grösse i zu bestimmen. Das wäre aber auf keine Weise möglich, wofern nur die Helligkeiten J bekannt sind, denn jede Aenderung von i würde sich einfach auf die ebenfalls unbekannten Grössen h werfen. Das nötige neue Datum hätte man zur Verfügung, wenn die Zeiten der Spektralaufnahmen beobachtet wären. Da diese Beobachtung aber nicht erfolgt ist, so sind wir auf die Benutzung fremder Angaben angewiesen.

Das Profil des Mondrandes in grossen Zügen hat Herr Hayn aus heliometrischen Messungen abgeleitet (Astron. Nachr. Bd. 168, pag. 1). Um von demselben Gebrauch machen zu können, müssen wir den Mondmittelpunkt in Uebereinstimmung mit dem von Hayn gewählten Punkt wählen. Die Genauigkeit der Mondtafeln ist, wie bekannt, nicht gross genug, um aus ihnen etwa den genauen Ort des Mondmittelpunktes zu errechnen. Wir müssen vielmehr den Haynschen Messungen 3 Daten entnehmen, welche zugleich d, ξ und i zu bestimmen gestatten, welche dann aber ausser der Randhelligkeit der Sonne auch noch die Korrektion des Mondortes festlegen.

Herr Hayn giebt die Abweichungen des Mondrandes von der Kreisform als Funktionen der selenographischen Koordinaten P und D, die für die Randstellen bei der Finsternis als Funktionen des Positionswinkels p folgende Werte hatten:

p	P	D	φ
95°	72°	−3°3	−11°
115°	92°	−3.5	+9°
135°	112°	−3.3	+29°
155°	132°	−2.7	+49°.

Abhandl. Kgl. Ges. d. Wiss. zu Gött. NF 5(2) 1907, 3–73

Zwischen dem Positionswinkel p und unserem Winkel φ besteht der Zusammenhang: $p = 106^0 + \varphi$.

Mit diesen Angaben liess sich aus den Haynschen Daten das in der gestrichelten Kurve der Figur 3 gezeichnete Mondprofil ableiten.

Ich entnehme demselben die folgenden drei Höhenwerte:

$$\begin{aligned} \varphi &= 34^\circ\!.5\,(x = +3.0): h = -0''\!.2 \\ \varphi &= 5.4\ \ (x = +0.5): h = +0''\!.5 \\ \varphi &= 10.9\,(x = -1.0): h = +0''\!.1. \end{aligned}$$

Ich will zunächst Spektrum No. 9 an diese Daten anschliessen. Man entnimmt Kurve No. 9 der Figur 2 folgende Werte von $2\,JR$ für die drei Stellen: 0.57, 1.75, 1.15. Damit folgen unter Benutzung des Täfelchens XII nach (23) die 3 Gleichungen:

$$\begin{aligned} 0.57 &= i\,(1.10\,\xi + d - 5.3) \\ 1.75 &= i\,(1.00\,\xi + d - 0.6) \\ 1.15 &= i\,(0.95\,\xi + d - 3.1) \end{aligned}$$

deren Auflösung ergiebt:

$$i = 0.245 \quad \xi = -1''\!.2 \quad d = +8''\!.9.$$

Der hier erhaltene Wert i bedeutet die Helligkeit des Sonnenrandes im Verhältnis zur durchschnittlichen Helligkeit auf einem Sonnenradius. Letztere findet sich aus den Messungen von Vogel für die Wellenlänge 455 $\mu\mu$ zu 0.82 der Zentralhelligkeit der Sonne. Daraus ergiebt sich das Resultat:

$$i = 0.245 \,.\, 0.82 = 0{,}20 \text{ Zentralhelligkeit der Sonne.}$$

Nachdem i gefunden ist, kann man ferner durch Multiplikation mit $1/i$ die Grössen $J.R$ direkt in Bogensekunden verwandeln. Die hieraus sich ergebende Sekundenskale ist in Fig. 2 bereits angegeben. Mit Hülfe dieser Skale wurden nun die mit $\xi = -1''\!.2$ und $d = +8''\!.9$ berechneten Werte $a\xi + b$ für jedes x eingetragen. Es entstand so die „Normallinie" der Figur 2, die angiebt, wie der Abstand zwischen Mond- und Sonnenrand bei genau kreisförmiger Mondgestalt über die Sichel verteilt wäre. Die Abweichungen der beobachteten Kurve No. 9 von der berechneten geben (bis auf den Faktor $\sec\varphi$) die Höhen der Mondberge in Bogensekunden.

Wir fragen schliesslich, welche Verschiebung gegen den Mondort des Nautical Almanac der Wert $d = +8''\!.9$ bedeutet. Es handelt sich hierbei um eine Verschiebung senkrecht zur scheinbaren Mondbahn. Nennt man letztere Verschiebung D (positiv bei Verschiebung des Mondes nach Norden) so ist:

$$D = -d \sin\chi.$$

Nun lag unser Beobachtungsort $2'\!.0$ seitlich der im *N. A.* angegebenen Zentralitätslinie nach Norden zu. Da der Totalitätsstreifen hier im ganzen eine Breite von 104′ hatte und diese Strecke den Mond um den Unterschied Sonnendurchmesser — Monddurchmesser = 2ϱ = 91″ verschiebt, so wird durch die 2′.0 seitliche Entfernung von der Zentralitätslinie der Mond um:

$$D = -\frac{2'.0}{104} \cdot 91'' = -1''.8$$

nach Norden verschoben. Aus unsern Messungen folgt:

$$D = -8''.9 \sin 9^\circ.5 = -1''.5$$

d. h. der Mond befand sich um 0″.3 in der Richtung senkrecht zur scheinbaren Mondbahn nördlicher, als die Ephemeride des Naut. Al. angiebt.

Um eine vollständige Korrektion des Mondortes zu erhalten, nehme man hinzu, dass nach den Beobachtungen der Herren Newall[1]) und Stephan[2]) alle Kontakte um durchschnittlich 20ˢ vor der Vorausberechnung des *N. A.* eingetreten sind. Das giebt eine Verschiebung A des Mondes in der Richtung der scheinbaren Mondbahn vom Betrag 0″.416 . 20 = 8″.3. Da die scheinbare Mondbahn den Positionswinkel 115°.5 hat, so folgte daraus die Korrektion des Mondortes des *N. A.* in *AR.* und δ:

$$\Delta\alpha \cos\delta = 8''.3 \sin 115^\circ.5 - 0.3 \cos 115^\circ.5, \quad \Delta\alpha = +7''.6$$
$$\Delta\delta = 8''.3 \cos 115^\circ.5 + 0.3 \sin 115^\circ.5, \quad \Delta\delta = -3''.3.$$

§ 24. Mängel der Aufnahmen. Versuch einer genaueren Discussion der Randhelligkeit der Sonne.

Die bisberige Discussion ging von der Ausnahme aus, dass die Helligkeit der Randpartien der Sonne, unabhängig von der Entfernung jeder Stelle vom Rande, als konstant betrachtet werden dürfe. Wenn im folgenden der Versuch gemacht wird, das genauere Gesetz der Helligkeitsabnahme der Sonnenscheibe in der Nachbarschaft des Randes abzuleiten, so darf man darin nur eine Erläuterung der Möglichkeit sehen, aus vollkommneren Aufnahmen derselben Art wirklichen Aufschluss über die Randhelligkeit der Sonne zu erhalten. Bei den vorliegenden Aufnahmen scheint leider der Verschluss sehr unregelmässig funktioniert zu haben oder die Empfindlichkeit des Filmbandes ist ungleichmässig gewesen. Denn die grosse Protuberanzengruppe, die nach Ausweis der Aufnahmen in Spanien und Aegypten höchst stabil war, giebt auf den Spektren No. 9—19 eine ständig wachsende Schwärzung. Dazu kommt die ungleichmässige Verschlussbewegung über das Gesichtsfeld hin (vgl. § 11). Diese scheint bei den Filmaufnahmen nicht so ausgeprägt gewesen zu sein, wie bei den späteren Aufnahmen auf Platte Z_c, da die Koronaspektren in Nord und Süd auf dem Film sehr nahe gleich geschwärzt sind. Deswegen habe ich im Vorhergehenden die Korrektion nicht angebracht. Immerhin kann auch hierdurch eine Verfälschung des photometrischen Mondprofils veranlasst worden sein.

1) Proceedings of the Roy. Soc. Vol. 77. 1905. Nov. 17.
2) Comptes rendus. Vol. 141. 1905, pag. 583.

Abhandl. Kgl. Ges. d. Wiss. zu Gött. NF 5(2) 1907, 3–73

Mit allem Vorbehalt, der aus diesen Mängeln entspringt, lässt sich nun aus den Aufnahmen No. 12 und 15 der Schluss ziehn, dass die oben gefundene Randhelligkeit i nicht nur ein Mittelwert für die bei No. 9 in Betracht kommenden 8 letzten Sekunden innerhalb der Sonnenscheibe ist, sondern dass i bis zu 1—2″ vom Sonnenrande nahe diesen Wert beibehält. Die Ueberlegung ist folgende:

Für die kleinen Werte von φ und ξ, die bei No. 12 und 15 nur noch in Betracht kommen, darf man $a = 1$ setzen und hat dann für die Sichelbreiten β:

$$\beta \;= \xi + b - \frac{h}{\cos\varphi}$$

$$\beta' \;= \xi' + b - \frac{h}{\cos\varphi}$$

$$\beta'' = \xi'' + b - \frac{h}{\cos\varphi}$$

wenn sich die Akzente auf No. 12 und 15 beziehen. Daher:

$$\beta'' - \beta' = \xi'' - \xi' \quad \beta' - \beta = \xi' - \xi.$$

In Worten: „Die Sichelbreiten vermindern sich von Aufnahme zu Aufnahme um Beträge, die über die ganze Sichel hin merklich konstant sind.“

Ist nun i konstant, so folgt das, was eben für die Sichelbreiten abgeleitet wurde, auch für J und spricht sich so aus: „die Kurven in Figur 2 müssen Parallelkurven sein.“ In der Tat, schiebt man die Kurve No. 9 um 3.0 resp. 5.0 nach links, so gelangt sie mit den beiden andern Kurven so genau zur Deckung, als die Messschwierigkeiten erwarten lassen.

Würde hingegen i nach dem Sonnenrande hin rasch kleiner, so würde eine Stelle mit an sich geringer Sichelbreite (entsprechend einem Mondberg) bei der Annäherung an die Totalität noch relativ rascher an photogr. Wirksamkeit verlieren, weil nur noch der schwach leuchtende äussere Sonnenrand bei ihr zur Wirkung käme. Betrachtet man z. B. die Stelle $x = 1.3 - 1.7$, so hat man hier bei No. 15 eine Sichelbreite von 2″.0. Wäre die Sonnenhelligkeit i auf diesem Randstück nur halb so gross, wie bei einer Sichelbreite von 4″, so dürfte diese Stelle nur eine Helligkeit J entsprechend 1″.0 unsrer Skale zeigen und nicht mehr denselben ausgeprägten Buckel bilden, wie auf Kurve No. 12. Es folgt also geringe Veränderlichkeit von i dicht am Sonnenrande.

Um dasselbe numerisch zu kontrollieren, habe ich die durchschnittlichen Intensitäten $2JR$ in den verschiedenen Strecken der Sichel abgeschätzt und folgendes erhalten:

x		No. 9	12	15
− 1.0 bis	− 0.5	1.2	0.3	—
− 0.5	0.0	1.8	0.5	—
0.0	+ 0.5	1.7	0.8	—
0.5	1.0	1.8	0.8	0.2
1.0	1.5	1.5	1.0	0.3
1.5	2.0	1.7	0.9	0.3
2.0	2.5	2.6	1.5	0.8.

Es werde nun i nicht konstant, sondern mit dem Abstand σ vom Sonnenrande wachsend vorausgesetzt in der Form:

25) $$i = i_0 + i_1 \sigma.$$

Dann wird die Helligkeit $2RJ$ bei der Sichelbreite β nicht gleich $i\beta$, sondern es gilt:

$$2RJ = \int_0^\beta i d\sigma = \beta\left(i_0 + \frac{i_1}{2}\beta\right)$$

oder bei kleinem i_1 durch Umkehrung

26) $$\beta = \frac{2RJ}{i_0}\left(1 - \frac{i_1}{2i_0^2} 2RJ\right).$$

Diese Gleichung möge für Spektrum No. 9 gelten. Indem wir entsprechende mit ein resp. zwei Akzenten versehene Gleichungen für Spektrum No. 12 und 15 anschreiben und subtrahieren, erhalten wir:

$$\beta' - \beta = \frac{2R(J' - J)}{i_0} - \frac{2R^2}{i_0^3} i_1 (J'^2 - J^2)$$

$$\beta'' - \beta' = \frac{2R(J'' - J')}{i_0} - \frac{2R^2}{i_0^3} i_1 (J''^2 - J'^2).$$

Wir haben oben abgeleitet, dass die Verminderungen $\beta' - \beta$ und $\beta'' - \beta'$ über die ganze Sichel hin, d. h. für alle Werte der Sichelbreite oder der Intensitäten J konstant sein müssen. Daher lässt sich i_1 aus der Bedingung bestimmen, dass die für verschiedene Sichelstellen erhaltenen Werte von $\beta' - \beta$ resp. $\beta'' - \beta'$ möglichst übereinstimmen sollen.

Dies wende man auf die obigen Zahlenwerte an. Man findet zunächst durch Zusammenfassung aus den obigen Zahlen:

x	$2JR$,	$2J'R$,	$2J''R$
2.0 bis 2.5	2.6	1.5	0.8
0.5 „ 2.0	1.7	0.9	0.3
−1.0 „ 0.5	1.6	0.5	

und daraus:

$$\alpha)\quad \beta - \beta' = \frac{1.1}{i_0} - 4.6\frac{i_1}{2i_0^3} \qquad \beta' - \beta'' = \frac{0.7}{i_0} - 1.6\frac{i_1}{2i_0^3}$$

$$\beta)\quad \beta - \beta' = \frac{0.8}{i_0} - 2.1\frac{i_1}{2i_0^3} \qquad \beta' - \beta'' = \frac{0.6}{i_0} - 0.7\frac{i_1}{2i_0^3}$$

$$\gamma)\quad \beta - \beta' = \frac{1.1}{i_0} - 2.3\frac{i_1}{2i_0^3}.$$

Die linken Seiten dieser Gleichungen stimmen schon für $i_1 = 0$, Konstanz der Helligkeit in der Nähe des Sonnenrandes, gut überein. Doch wird die Uebereinstimmung noch ein wenig verbessert, wenn man $\frac{i_1}{2i_0^2} = 0.07$ setzt. Man er-

Abhandl. Kgl. Ges. d. Wiss. zu Gött. NF 5(2) 1907, 3–73

hält dann:

$$\alpha)\quad \beta-\beta' = \frac{0.78}{i_0} \qquad \beta'-\beta'' = \frac{0.59}{i_0}$$

$$\beta)\quad \beta-\beta' = \frac{0.65}{i_0} \qquad \beta'-\beta'' = \frac{0.55}{i_0}$$

$$\gamma)\quad \beta-\beta' = \frac{0.94}{i_0}.$$

Mit dem gefundenen Wert von $\frac{i_1}{2i_0^2}$ lautet Gleichung (26):

$$\beta = \frac{2RJ}{i_0}[1-0.07.2RJ].$$

Um i_0 zu bestimmen, hat man sich daran zu erinnern, dass durch Anschluss an die Hayn'schen Messungen (vgl. Figur 2) für $2RJ = 2.0$ der Wert $\beta = 8''.1$ resultierte. Damit folgt $i_0 = 0.21$ und es ergiebt sich als Schlussresultat für die Helligkeit in der Nähe des Sonnenrandes nach Formel (25):

$$i = 0.21 + 0.006.\sigma$$

oder auf die Zentralhelligkeit der Sonne bezogen:

$$i = 0.17 + 0.005\,\sigma.$$

§ 25. Definitive Ableitung des Mondprofils.

Für das folgende habe ich wieder die kleine Variabilität von i vernachlässigt. Um die Form des Mondprofiles sicherer zu erhalten, habe ich nicht nur die 3 Kurven aus Figur No 1 durch die erwähnten Parallelverschiebungen zur Deckung gebracht, sondern ausserdem noch die entsprechenden Kurven von Spektrum No. 14 und 16 gezeichnet und sie in gleicher Weise behandelt. Es ergaben sich die notwendigen Parallelverschiebungen zu 4''.5 resp. 5.7. Aus den übereinandergeschobenen Kurven habe ich dann eine Art Mittel genommen, indem ich für das Detail an jeder Stelle diejenige Kurve auswählte, die mir die sicherste zu sein schien. Indem ich von den Abscissen der Mittel-Kurve die Werte der Normallinie abzog, erhielt ich in den Differenzen eine Kurve für $\frac{h}{\cos\varphi}$, aus welcher durch Multiplikation mit $\cos\varphi$ die in Figur No. 3 gezeichnete Profillinie des Mondes hervorging.

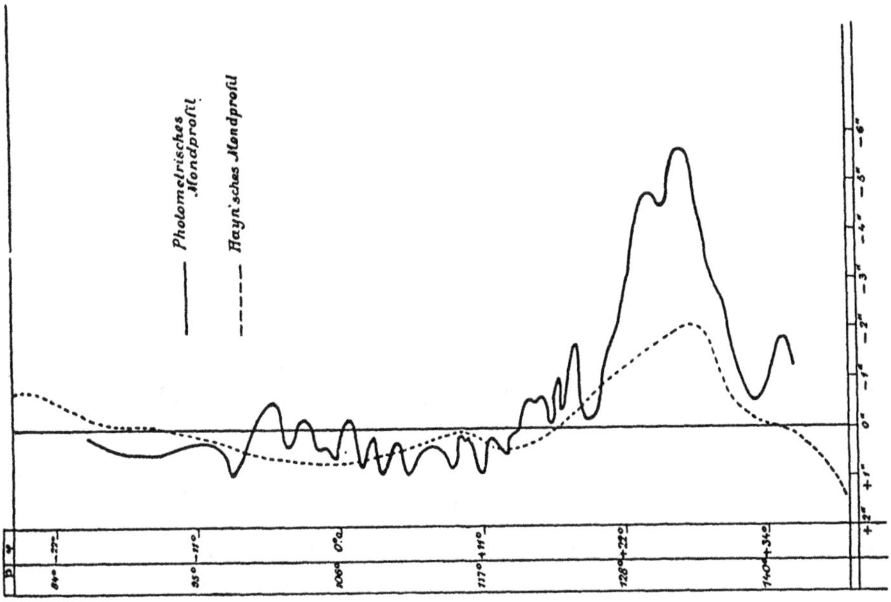
Photometrisches Mondprofil
Hayn'sches Mondprofil

Fig. 3.

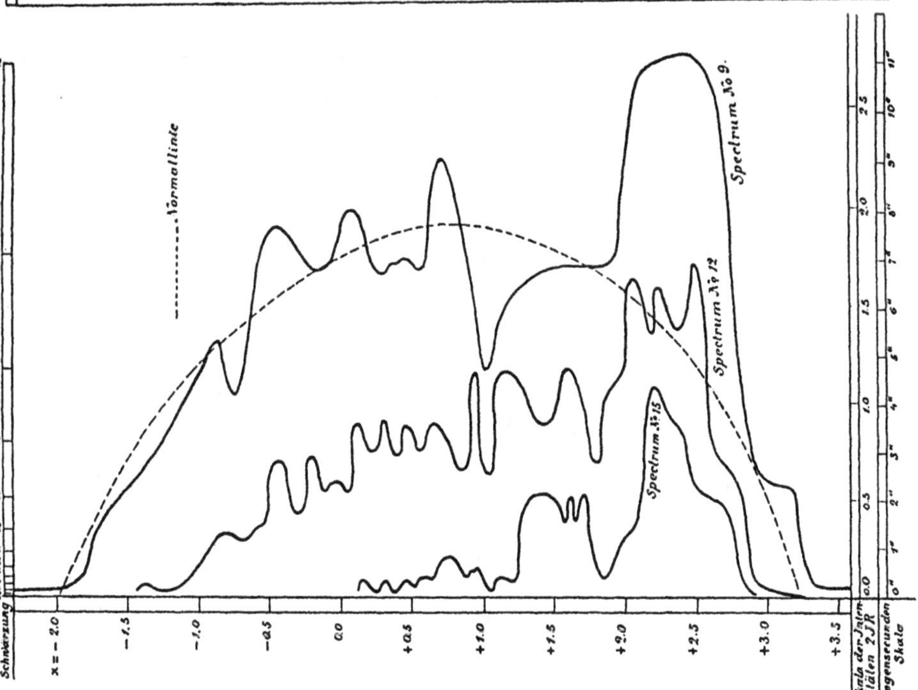
Normallinie
Spectrum No 9.
Spectrum No 12
Spectrum No 15
Skala der Schwärzung
Skala der Intensitäten 2JR
Bogensecunden Skala

Fig. 2.

Abhandl. Kgl. Ges. d. Wiss. zu Gött. NF 5(2) 1907, 3–73

VI. Abschnitt. Spektralphotometrie des Sonnenrandes und der Protuberanzen.

§ 26. Die Helligkeit des Sonnenrandes für verschiedene Wellenlängen.

Die Untersuchung, die bisher für die Wellenlänge $\lambda = 455\,\mu\mu$ vorgenommen wurde, liesse sich für jede andre Wellenlänge wiederholen. Doch habe ich mich begnügt, nur einige Stellen der Sichel herauszugreifen und für diese dafür die Untersuchung auf das ganze zugängliche Spektralgebiet auszudehnen.

Zunächst wurden die Vergleichsspektren No. 1 und 3 (No. 2 war zu sehr von falschem Licht gestört) durch das ganze Spektrum hindurch ausgemessen und zwar nicht nur auf der zentralen Linie, sondern auch auf zwei Sehnen nördlich und südlich vom Mittelpunkt. Ich gebe gleich das Mittel aus diesen 3 Werten für jede Wellenlänge. Zwar war bis zur Wellenlänge $560\,\mu\mu$ auf dem orthochromatischen Film starke Schwärzung vorhanden, doch breche ich bei der Wellenlänge $504\,\mu\mu$ ab, weil die „Reduktion auf reines Spektrum" für die grösseren Wellenlängen wegen des sekundären Maximums der Sensibilitätskurve Schwierigkeiten gehabt hätte. Auch musste der plötzliche Abfall des Vergleichsspektrums im Rot zur Festlegung der Wellenlängen in demselben dienen, da eine andre Orientierung nicht vorhanden war. Ein merklicher Fehler aus dieser Quelle ist für die im folgenden noch berücksichtigten Wellenlängen wegen der starken Zunahme der Dispersion nach Violett zu nicht zu befürchten.

Tabelle XIII. (Schwärzung der Vergleichsspektren No. 1 und 3.)

σ	– 20 0	– 17.5	– 15.0	– 12.5	– 10.0	– 7.5	– 5.0	– 2.5	0	+ 2.5	5.0	7.5
λ	349	357	366	375	386	398	410	424	440	458	480	504
No. 1	63,9	64,9	65,5	65,9	66,2	66,5	67,5	68,0	68,2	68,0	66,8	65,4
No. 3	41	47	50,9	53,0	54,2	56,1	57,4	57,7	57,7	57,8	55,8	51,2

Ich mass nun die Schwärzungen einer Reihe von Stellen des Sonnenrandes, die ich durch ihre x-Koordinaten in Figur 2 bezeichnen kann, auf mehreren Aufnahmen durch. Ich gebe zunächst die Messungen für die Stelle $x = 2{,}40$ auf den Spektren No. 9—12.

Tabelle XIV.

σ	−24.5	−22.0	−19.5	−17.0	−14.5	−12.0	−9.5	−7.0	−4.5	−2.0	+0.5	+3.0	+5.5	+8.0	+10.5	+13.0	β
λ	335	343	351	359	368	377	388	400	413	428	445	464	486	510	539	574	
No. 9	58,3	62,7	65,2	65,6	66,3	67,6	67,7	68,2	68,8	69,0	70,2	69,6	66,6	66,0	68,7	66,5	9″5
10	58,5	62,0	64,3	63,9	65,3	65,4	66,8	67,6	68,7	68,4	69,6	68,7	66,6	64,9	67,0	64,8	7.7
11	58,1	63,2	64,2	64,9	66,7	66,2	67,1	68,6	69,4	70,1	70,0	69,5	67,2	66,0	67,7	66,2	6.4
12	56,7	61,6	64,7	65,1	65,4	66,3	66,4	68,6	68,0	69,9	69,9	69,3	66,7	64,4	66,6	66,4	6.0
Mittel I	57,9	62,4	64,6	64,9	65,9	66,4	67,0	68,3	68,7	69,9	69,9	69,3	66,8	65,3	67,5	66,0	7.4

Unter β finden sich für jede Aufnahme die mit Hülfe des photometrischen Mondprofiles und des betreffenden Wertes von ξ berechneten Sichelbreiten.

Ganz entsprechende Messungen gebe ich ferner noch für einige schmälere Sichelstellen im Wellenlängengebiet 359—464 $\mu\mu$:

Tabelle XIVa).

σ	x	− 17.0	− 14.5	− 12.0	− 9.5	− 7.0	− 4.5	− 2.0	+ 0.5	+ 3.0	β
λ		359	368	377	388	400	413	428	445	464	
No. 14	2,40	63,2	65,6	65,5	66,0	66,4	67,0	67,0	67,3	66,8	4″2
15	2,40	64,6	65,0	65,6	65,7	67,0	67,1	66,0	66,8	66,8	3.6
16	2,40	64,7	66,5	66,4	66,9	68,1	68,2	68,1	67,5	66,3	3.0
15	1,40	62,5	63,8	64,7	64,1	65,0	65,3	65,8	65,5	65,4	2.0
15	0,75	58,5	59,9	60,5	62,2	62,6	63,1	62,8	62,3	62,5	0.8
17	2,40	58,2	58,4	60,3	62,2	61,8	63,2	62,3	62,5	61,2	1.1

Bei der Vergleichung der Tabellen XIII und XIV ist nun zu beachten, dass das Spektrum der schmalen Sichel ein reines ist, während das Vergleichsspektrum nahezu einen vollen Sonnendurchmesser als Spaltweite hat. Der Sonnendurchmesser beträgt 10,5 mm, er überspannt 5 Nachbarwerte der Tabelle XIV. Es liegt also hier die umgekehrte Sachlage vor, wie bei den früheren Koronaspektren, wo das Vergleichsspektrum rein, das zu untersuchende Spektrum unrein war. Im jetzigen Fall ist es am bequemsten, das reine Sichelspektrum so übereinander zu lagern, dass es ebenso unrein wird, wie das Vergleichsspektrum und dann diese beiden unreinen Spektren mit einander zu vergleichen. Die ganze Wirkung der Unreinheit ist sehr klein in dem berücksichtigten Wellenlängengebiet und ein skrupulöses Verfahren nicht am Platze.

Um die Farbenmischung durchzuführen, muss man von Schwärzungen S zu aktinischer Wirksamkeit J übergehn. Aus der Abblendungsdifferenz der Löcher No. 1 und No. 3 von $2^m_.9$ und aus dem durchschnittlichen Unterschied übereinander stehender Schwärzungen in Tabelle XIII ergiebt sich die Formel:

26) $$2{,}5 \log J = \frac{S - 60}{2{,}64}$$

Abhandl. Kgl. Ges. d. Wiss. zu Gött. NF 5(2) 1907, 3–73

oder:

$$\log J = 0{,}151\,(S - 60)$$

(die Zahl 60 ist zur Bequemlichkeit von S abgezogen). Damit verwandeln sich die unter Mittel I gegebenen Schwärzungen in die Aktinitätszahlen J, welche die folgende Tabelle enthält.

Tabelle XV.

σ	−24.5	−22.0	−19.5	−17.0	−14.5	−12.3	−9.5	−7.0	−4.5	−2.0	+0.5	+3.0	+5.5	+8.0
λ	335	343	351	359	368	377	388	400	413	428	445	464	486	510
J	0,48	2,3	4,9	5,5	7,8	9,1	11,5	17,8	20,0	25,1	30,9	25,1	10,7	6,5
J'		2,6	4,2	5,9	7,5	9,7	12,9	16,6	21,6	24,7	25,4	21,3		
S'		62,8	64,2	65,2	65,8	66,6	67,4	68,8	68,8	69,3	69,4	68.8		
m			1,98	1.98	1,94	1.82	1.63	1.48	1.54	1.55	1.52	1.59		
Corr.			−0.08	−0.06	−0.04	−0.03	−0 03	−0.03	−0.03	−0.03	−0.03	−0 03		
m+Corr.			1.90	1.92	1.90	1.79	1.60	1.45	1.51	1.52	1.49	1.56		
i(I)			0.17	0.17	0.17	0.19	0.23	0.26	0.25	0.25	0.25	0.24		

Aus je 5 Nachbarwerten J wurde nunmehr das Mittel genommen, wobei in Rücksicht auf den Abfall der Sonnenhelligkeit nach dem Rande zu die äussersten Werte nur das Gewicht $^1/_4$ erhielten. So entstanden die Zahlen J', die durch Umkehrung der Formel (26) in Schwärzungen S' zurückverwandelt wurden. Die Schwärzungen S' sind nun mit den Zahlen der Tabelle XIII unmittelbar vergleichbar, sie gelten für ein eben so unreines Spektrum, wie es dort vermessen ist.

Die Abschwächung in Grössenklassen für die Vergleichsspektren No. 1 und No. 3 betrüge, bezogen auf den Sonnendurchmesser und mittlere Helligkeit der der Sonne, wie oben angegeben:

$$7^{m}.8 \text{ resp. } 11^{m}.7.$$

Verglichen mit einer Sichelbreite von 7,4 Sekunden ist die Abschwächung im Verhältnis $\frac{1980}{7{,}4}$ d. s. $6^{m}.1$ geringer, beträgt daher: $1^{m}.7$ resp. $5^{m}.6$. Oder wenn wir gleich auf die Zentralhelligkeit der Sonne beziehn, die die mittl. Helligkeit für das hier in Betracht kommende Spektralgebiet im Verhältnis 1,25 übertrifft:

$$2^{m}.0 \text{ resp. } 5^{m}.9.$$

Unter der Annahme, dass die Schwärzungen den Grössenklassen für jede einzelne Farbe proportional sind, ergeben sich die Werte m, wenn man die Zahlen S' zwischen die beiden Zahlenreihen der Tabelle XIII interpoliert und von den beiden eben gefundenen Abschwächungsziffern Gebrauch macht. Verwandelt man schliesslich die Grössenklassen m in Intensitäten, so erhält man für jede Farbe direkt die Randhelligkeit bezogen auf die Zentralhelligkeit der Sonne als Einheit.

Die kleine Korrektion wegen der Absorption des U. V. Prismas ist der Tabelle eingefügt. Die Werte i(I) gelten für eine Sichelbreite von 7″,4 Sekunden, entsprechen also im Mittel einer Entfernung vom Rande von 3″,7.

Von den Messungen an den schmäleren Sichelstellen wollen wir die vier ersten und die zwei letzten zu Mitteln zusammenfassen.

σ	−17.0	−14.5	−12.0	−9.5	−7.0	−4.5	−2.0	+0.5	+3.0	β
λ	359	368	377	388	400	413	428	445	464	
Mittel II	63,7	65,2	65,5	65,7	66,6	66,9	67,0	66,8	66,3	3″,2
Mittel III	58,4	59,2	60,4	62,2	62,2	63,2	62,2	62,4	61,8	1″,0

Diese Schwärzungen vergleicht man am bequemsten nicht mit dem unreinen Spektrum der Tabelle XIII, sondern direkt mit dem reinen Spektrum der mittleren Sichelbreite 7″,5, dessen Helligkeit uns eben durch die Zahlen i(I) bekannt geworden ist.

Zieht man die Mittel II und III von I ab und verwandelt die entstehenden Zahlen mit Hülfe der aus Tabelle XIII hervorgehenden Proportionalitätsfaktoren in Grössenklassen, so erhält man folgende Werte:

λ	359	368	377	388	400	413	428	445	464	β
II.—I.	0^m,30	0,19	0,27	0,43	0,65	0,69	0,88	1,19	1,15	3.2
III.—I.	1^m,6	1,8	1,8	1,6	2,2	2,1	2,6	2,9	2,9	1.0

Reduziert man diese Zahlen auf die Sichelbreite 7″.4, indem man mit $\frac{7.4}{3.2}$ resp. $\frac{7.4}{1.0}$ multipliziert oder in Grössenklassen $0^m.91$ resp. $2^m.3$ subtrahiert, und addiert man dann die für die Sichelbreite 7″.4 erhaltenen Grössen m + Corr. der Tabelle XV, so erhält man die nachstehenden Werte m. In Intensitäten verwandelt geben dieselben die Zahlen i, welche die durchschnittliche Helligkeit für die Sichelbreite 3″.2 resp. 1″.0 verglichen mit der Zentralhelligkeit der Sonne angeben.

Tabelle XVI.

λ	359	368	377	388	400	413	428	445	464	β
m (II)	$1^m.31$	1.18	1.15	1.12	1.19	1.29	1.49	1.77	1.80	3″.2
m (III)	$1^m.3$	1.5	1.4	1.6	1.5	1.4	1.9	2.2	2.3	1″.0
i (II)	0.30	0.34	0.35	0.36	0.33	0.30	0.25	0.20	0.19	3″.2
i (III)	0.30	0.25	0.28	0.23	0.25	0.28	0.17	0.13	0.12	1″.0

Abhandl. Kgl. Ges. d. Wiss. zu Gött. NF 5(2) 1907, 3–73

Die Zahlen *i* (II) und *i* (III) zeigen ein merkwürdiges Verhalten gegenüber der Reihe (I). Während letztere eine Abnahme der Randstrahlung der Sonne mit abnehmender Wellenlänge anzeigt, sowie es den bekannten Ergebnissen für die Helligkeitsverteilung auf der Sonnenscheibe überhaupt entspricht, findet sich für die letzten Sekunden am Sonnenrande in den Reihen *i* (II) und *i* (III) der umgekehrte Gang ausgesprochen. Wir werden weiter unten auf diesen Punkt zurückkommen.

§ 27. Helligkeit der Protuberanzen.

Da noch gar keine Angaben über die absolute Helligkeit von Protuberanzen vorzuliegen scheinen, habe ich einige Feststellungen in dieser Hinsicht versucht. Für die in der Figur 3 (Tafel I) mit *a*, *b*, *b'*, *c* bezeichneten Stellen der grossen Protuberanzengruppe am Ostrand der Sonne habe ich auf Spektrum No. 16 — so gut es ging bei kleinen Gebilden, die den Photometerfleck nicht ganz ausfüllen — in den Wasserstoff- und Kalciumlinien folgende Schwärzungen gemessen

Protub.	H_β	H_γ	H_δ	H_ζ	H	K
b	61.4	62.9	59.0	54.6	60.1	61.0
a	60.1	60.1	56.3		64.2	63.4
c	53.5	57.0	52.9		56.8	57.9
b'	51.8	54.6			58.2	58.8
Mittel I	66.9	69.5	68.7	67.0	67.9	67.5
λ	485	435	412	388	396	393

Unter den gemessenen Zahlen ist die Wellenlänge der betreffenden Linie und die Schwärzung des Photosphärenspektrums für diese Wellenlänge angegeben und zwar sind diese Zahlen dem Spektrum der 7''.4 breiten Sichel entnommen, welches oben unter Mittel I ausgewertet ist. Indem die gemessenen Schwärzungen von Mittel I abgezogen und die Differenzen mit Hülfe der aus Tabelle XIII abgeleiteten Proportionalitätsfaktoren in Grössenklassen verwandelt wurden, ergaben sich folgende Helligkeiten der Protuberanzen in Grössenklassen:

Tabelle XVII.

Protub.	H_β	H_γ	H_δ	H_ζ	H	K	H_β	H_γ	H_δ	H_ζ	H	K
b	$1^m.8$	2.6	3.7	4.1	2.9	2.3	$0^m.8$	0.0	−1.1	−1.5	−0.3	−0.6
a	2.3	3.7	4.8		1.4	1.5	1.4	0.0	−1.1		+2.3	+2.4
c	4.5	4.8	6.1		4.1	3.4	0.3	0.0	−1.3		+0.7	+1.4
b'	5.0	5.8			3.6	3.1	0.8	0.0			+2.2	+2.7

Der zweite Teil der Tabelle giebt die Helligkeiten bezogen auf die $H\gamma$-Linie und zeigt, was auch schon der direkte Anblick der Aufnahmen lehrt, dass die relative Helligkeit der Wasserstofflinien in kräftigen und schwachen Protuberanzen annähernd die gleiche ist, dass dasselbe für die Kalciumlinien untereinander gilt, dass hingegen das Kalcium gänzlich unabhängig von Wasserstoff verteilt ist.

Wir haben noch nach der absoluten Bedeutung dieser Helligkeiten zu fragen. Die Helligkeit einer Protuberanz ist proportional dem Produkt aus der Breite b und der Intensität h ihres Strahlungsbezirkes, wenn wir zur Vereinfachung annehmen, dass sie über einen bestimmten kleinen Wellenbezirk hinweg mit gleichförmiger Intensität strahlt. b werde in Angström-Einheiten, h an der Helligkeit des schwarzen Körpers von 6000° (von Frauenhofer'schen Linien befreites Sonnenspektrum) gemessen. Man überlege, welche Strahlung auf 1″ Längserstreckung unsrer Spektren kommt. Bei der Protuberanz handelt es sich um nahezu homogene Strahlung eines breiten Objekts, bei der Sonne um die inhomogene der schmalen Sichel β. Die Strahlung der Protuberanz wird proportional zu hb. Ist die Anzahl Angström-Einheiten auf 1″ an der betreffenden Stelle des Spektrums α, so wird die Strahlung der Sonne in demselben Maße gemessen $\alpha.\beta''$. Das Helligkeitsverhältnis H, so wie es den Zahlen der obigen Tabelle entspricht, wird daher:

$$H = \frac{h.b}{\alpha.\beta}.$$

Es folgt also aus unsern Messungen für jede Protuberanz die Angabe:

$$h.b = H.\alpha.\beta.$$

Für α hat man auf Grund der im folgenden Paragraphen bestimmten Dispersionskurve die kleine Tabelle:

λ	510	472	441	416	395	377
α	0.580	0.457	0.369	0.304	0.255	0.218

Hiermit ergaben sich nach Verwandlung der Grössenklassen in Intensitäten und unter Benutzung des Wertes 7″.4 für die Breite β der Sonnensichel, folgende Helligkeiten der Protuberanzen:

Tabelle XVIII.

Prot.	H_β	H_γ	H_δ	H_ζ	H	K
b	0.70	0.24	0.073	0.040	0.106	0.22
a	0.44	0.087	0.026		0.54	0.46
c	0.059	0.031	0.008		0.044	0 081
b'	0.037	0.012			0.070	0.11

Abhandl. Kgl. Ges. d. Wiss. zu Gött. NF 5(2) 1907, 3–73

Diese Zahlen werden am anschaulichsten, wenn man sie als „äquivalente Linienbreiten" auffasst, wenn man nämlich $h = 1$ setzt, also annimt, dass die Protuberanz in den Wellenlängen, die sie überhaupt aussendet, strahlt wie die Photosphäre, wie ein schwarzer Körper der Temperatur 6000°. Sie geben dann die Breite der betreffenden Linie in A. E., welche genügt, um die beobachtete Helligkeit zu erzeugen. Kennt man die wirkliche Breite b der Protuberanzenlinie, so giebt $\frac{h'}{b}$ gleich die wirkliche Helligkeit. Da die Breite der Protuberanzenlinien 1—2 A. E. nicht überschreiten dürfte, so sieht man, dass die Helligkeit heller Protuberanzen in dem Gebiete, in welchem sie überhaupt strahlen, von derselben Grössenordnung ist, wie diejenige der Sonne.

Für das Helligkeitsverhältnis der Wasserstofflinien zur H_β-Linie folgt:

Tabelle XIX.

Prot.	H_β	H_γ	H_δ	H_ζ
b	1	0.34	0.10	0.06
a	1	0.20	0.06	
c	1	0.53	0.14	
b'	1	0.32		
Mittel	1	0.35	0.10	0.06

VIII. Abschnitt. Das Chromosphärenspektrum.

§ 28.

Die Wellenlängen der hellen Linien des Chromosphärenspektrums wurden auf Film No. 17 an der Stelle $\varphi = +17^0$ von $\lambda = 4600$ bis $\lambda = 3780$ A. E. vermessen. Ueber dieses Gebiet hin war auf dieser Aufnahme der Focus vortrefflich. Spektrum No. 3 auf Platte Z_6 lieferte die Ergänzung nach dem Ultraviolett zu (von $\lambda = 3780$ A. E. bis $\lambda = 3330$ A. E.), wo es besonders scharfe Bilder gab. Die Messungsergebnisse n (in Millimetern) wurden für beide Spektren zunächst nach den — aus einigen hellen Linien abgeleiteten — Hartmann'schen Formeln

$$\lambda = 1677.28 + \frac{108032.4}{n - 8.893} \text{ resp. } \lambda = 1677.68 + \frac{108032{,}4}{n + 52.600}$$

in Wellenlängen (in A. E.) verwandelt.

Die so erhaltenen vorläufigen Wellenlängen wurden mit dem ausführlichen Verzeichnis der Chromosphärenlinien von Herrn Dyson (Transactions of the London Royal Society 1906) verglichen und auf Grund der sicher identifizierten Linien wurde graphisch folgende Korrektionstabelle (in Hundertstel A. E.) für die vorläufigen Wellenlängen abgeleitet.

λ	460	455	450	445	440	435	430	425	420	415	410	405	400	395	390
Correction:	+3	0	−4	−8	−12	−16	−19	−21	−22	−18	−9	+12	+40	+72	+102

λ	385	380	375	370	365	360	355	350	345	340	335	330
Correction:	+94	+58	+10	−42	−96	−150	−208	−283	−370	−464	−562	−664

Die durch Anbringung dieser Korrektionen erhaltenen definitiven Wellenlängen werden im letzten Abschnitt mitgeteilt.

Besonderer Wert wurde auf die Ableitung der Höhe der einzelnen Chromosphärengase aus der Länge der Sicheln, die sie zeigten, gelegt. Die Rechenvorschrift ist folgende: Behält man ganz die Bezeichnungen der Figur 1 bei, so ergiebt sich für die Endpunkte der Sichel bei einem äusseren Begrenzungsradius $R+\delta R$ die Gleichung:

$$(R+\delta R)^2 = [(r+h)\sin\varphi - c\sin\chi]^2 + [(r+h)\cos\varphi - c\cos\chi - d]^2$$

oder durch Reihenentwicklung für unsere Zwecke genügend genau:

$$\delta R = \varrho[1-\cos(\chi-\varphi)] - d\cos\varphi - \xi\cos(\chi-\varphi) + \frac{[\varrho\sin(\varphi-\chi)+d\sin\varphi]^2}{2r} + h.$$

Sind die Konstanten der Aufnahme r, ϱ, d, χ, ξ und das Mondprofil h bekannt, so ergiebt sich nach dieser Formel durch Beobachtung des Positionswinkels φ des einzelnen Sichelendes der Wert von δR.

Für Spektrum No. 17 fand sich $\xi = -9''.2$. Mit diesem Wert von ξ und den früheren numerischen Daten für die übrigen Grössen wurde δR für jeden Wert von φ bestimmt unter Benutzung des photometrisch erhaltenen Mondprofiles h und — über dessen Grenzen hinaus — des Hayn'schen Mondprofils. Ich trug δR als Funktion von $x = r\sin\varphi$ auf und konnte dann aus der Endkoordinate der Sichel x sofort die Höhe des betreffenden Gases δR der Kurve entnehmen. Benutzt wurden die nördlichen Sichelenden, weil auf der Südseite das Mondprofil zu steil absinkt. Die Resultate werden gleichfalls im letzten Abschnitt mitgeteilt.

Abhandl. Kgl. Ges. d. Wiss. zu Gött. NF 5(2) 1907, 3–73

IX. Abschnitt. Zusammenfassung und Diskussion der Resultate.

§ 29.

Es ist im I. Abschnitt dieses Berichtes erzählt worden, dass die Sonnenfinsternisexpedition ursprünglich nur mit der Absicht der Orientierung unternommen wurde und dass erst im letzten Moment sich die Möglichkeit selbständigerer Beobachtungstätigkeit eröffnete. Von den photometrischen Untersuchungen, die schliesslich die Hauptaufgabe der Bearbeitung bildeten, war vor der Finsternis nur der Gedanke vorhanden, man müsse den Aufnahmen irgend welche durch die Sonne selbst erzeugte photometrische Vergleichsskalen mitgeben. Es ist nicht verwunderlich, wenn sich die Folgen einer so geringen Vorbereitung in den Resultaten zeigen. Die Hauptstörung bildete Unregelmässigkeit des Momentverschlusses, dazu kam ungünstige Anordnung der Vergleichsskalen. Ich möchte daher den Leser bitten, diesen Bericht trotz seines Umfangs mehr als ein Programm für die bei Sonnenfinsternissen möglichen photometrischen Arbeiten und weniger als Ausführung derselben anzusehn.

Wenn ich mich trotzdem entschlossen habe, alle Messungen an den Platten bis zum numerischen Endresultat durchzuführen, auch da, wo sich der mögliche Fehler innerhalb weiter und schwer angebbarer Grenzen befindet, so liegt das an zwei Gründen. Erstens ist zur Zeit noch kein Ueberfluss an — auch nur halbwegs zuverlässigen — numerischen Angaben über die Helligkeiten der Sonnenfinsternisphänomene vorhanden und zweitens wird man durch die genaue Reduktion am ehesten alle künftighin zu beachtenden Fehlerquellen erkennen müssen.

Die Durchführbarkeit der grossen Zahl photometrischer Messungen auf den verhältnismässig kleinen Aufnahmen beruhte auf der Benutzung des mit einem Koordinatenmessapparat verbundenen Hartmann'schen Mikrophotometers. Diese Einrichtung gestattet erst der photographischen Platte zugleich mit den Ortsangaben auch alle die in ihr enthaltenen Helligkeitsangaben zu entlocken und damit das dokumentarische Material, das die Platte aufbewahrt, ganz zu erschöpfen.

§ 30. Die photographische Helligkeit der Korona.

Die direkten Aufnahmen der Korona mit dem Petzvalobjektiv trugen z. T. eine photometrische Vergleichsskala. Dieselbe war, wie hier wiederholt sei, dadurch erzeugt, dass die freie Sonne 15^s durch einen „Keil“ hindurch einen

Streifen der Platte beschienen hatte. Der Keil bestand aus einer photographischen Platte von verlaufender Schwärzung überdeckt mit einem Rauchglas.

Dieses Verfahren scheint mir den Vorzug zu verdienen vor der von den Engländern meist verwandten Methode der Abney-squares, d. i. Erzeugung einer Skala verschieden geschwärzter Quadrate mit einer Normallampe im Laboratorium. Denn man vergleicht unmittelbar mit dem Objekt, auf das man doch bei jeder Auslegung der Resultate beziehen muss, mit der Sonne selbt, und man wird frei von Veränderlichkeit der Atmosphäre und zeitlicher Veränderung der Plattenempfindlichkeit, indem man die Exposition des Keiles unmittelbar vor oder nach der Finsternis vornimmt. Auch der nötige Apparat ist einfacher.

Die Aichung des Keils erfolgt, indem man die Sonne auf irgend eine andre Weise in bekanntem Verhältnis ohne Farbenänderung abblendet und hierdurch dieselbe Schwärzung erzeugt, wie sie eine bestimmte Keilstelle liefert. Färbung des Keils (verschiedene Absorptionskonstante für verschiedene Wellenlänge) ist unschädlich, solange man zur Aichung und zur Aufnahme Platten von der gleichen Sensibilisierungskurve benutzt.

Eine ohne Farbenänderung abgeblendete Sonne stellt der Mond dar. Auch liess sich eine genügend starke Abblendung erreichen, indem mit einem Objektiv sehr reduzierter Oeffnung ein Sonnenbild entworfen und dies noch mit Hülfe eines Mikroskopobjektivs vergrössert wurde. Die Keilaichung wurde sowohl mit Hülfe des Mondes, als mittelst dieser Abblendungsart durchgeführt. Die beiden Aichungen stimmten überein, wenn man die Vollmondhelligkeit zu $14^m.3$ verglichen mit der Sonne ansetzt. Müller (Photometrie pag. 315) giebt für die Vollmondhelligkeit die Zahl $14^m.4$, das bedeutet also eine gute wechselseitige Bestätigung der Keilaichungen.

Da der Keil im ganzen leider zu undurchsichtig gewählt worden war, konnten nur die äussersten Koronapartien und das diffuse Himmelslicht durch Vergleichung mit dem Keil photometrisch festgelegt werden. Unter Berücksichtigung der Oeffnungen und Absorptionen in den verwandten Instrumenten ergaben sich folgende Helligkeiten, bezogen auf die durchschnittliche Flächenhelligkeit der Sonne als Einheit:

a) Diffuses Himmelslicht in der Mitte der Mondscheibe $0.43 \cdot 10^{-8}$ oder $20^m.9$
und in 100′ Entfernung von der Mitte $0.29 \cdot 10^{-8}$ oder $21^m.4$.

Ich habe angenommen, dass das diffuse Licht der Entfernung von der Mitte proportional von dem grösseren auf den kleineren Betrag absinkt. b) Für das Eigenlicht der Korona ergaben sich nach Abzug des diffusen Lichtes folgende Werte (die vermessenen Richtungen sind in Figur 2 der Tafel I bezeichnet). Die Einheit ist 10^{-8} Sonnenhelligkeit.

Abhandl. Kgl. Ges. d. Wiss. zu Gött. NF 5 (2) 1907, 3–73

Entfernung vom Zentrum	30′	35′	40′	45′	50′
Ausläufer *a*	1.66	0.83	0.47	0.26	0.14
Lücke zwischen *a* und *b*	1.54	0.75	0.39	0.20	0 09
Richtung *d*			0.41	0.22	0.12
Mittel	1.60	0.79	0.42	0.23	0.12
In Grössenklassen	19ᵐ5	20ᵐ3	20ᵐ9	21ᵐ6	22ᵐ3

Man erkennt, dass der Ausläufer sich nur verhältnismässig wenig über die durchschnittliche Helligkeit der Korona in derselben Entfernung vom Mittelpunkt erhebt.

§ 31.

Angaben über die Helligkeit der inneren Korona liessen sich aus den Spektralaufnahmen mit der Prismenkamera ableiten. Da die Koroma innerhalb des photographisch wirksamen Spektralgebiets sehr nahe dieselbe Intensitätsverteilung hat wiedie Sonne, so ist es gleichgültig, ob man die photographische Wirkung des gesamten Koronalichtes mit der des gesamten Sonnenlichtes vergleicht oder ob man aus beiden Spektren dasselbe Gebiet und zwar am besten das photographisch wirksamste Gebiet ausblendet. Die dunkelste Stelle der Spektralbilder lag bei der Wellenlänge 441 $\mu\mu$ und in dieser Wellenlänge wurde gemessen.

Während aus der einfachen spektralen Zerlegung des Koronalichtes also keine Schwierigkeit entspringt, ist ein andrer Umstand beschwerlicher. Die von der Prismenkamera gelieferten Spektren sind keine reinen Spektren, sondern eine Uebereinanderlegung von Spektren eines ganzen senkrecht zur Prismenkante gerichteten Schnittes durch die Korona. Wie man diese Superposition wieder entwirrt, soll weiter unten angedeutet werden. (Näheres siehe oben §§ 10. 15. 19) Ich will mir dieselbe für den Augenblick schon ausgeführt denken und nur die weitere Behandlung besprechen.

Die Platte Z_b trug drei Aufnahmen der Korona mit der Prismenkamera von 1^s, 15^s und 40^s Exposition. Indem angenommen wurde, dass die Wirkung mit der 0.9ten Potenz der Zeit steigt, liessen sich aus diesen drei Aufnahmen die relativen Intensitäten der inneren Korona ableiten. Der Uebergang auf absolute Werte gelang mit Hülfe des auf der Platte befindlichen photometrischen Keils, der allerdings infolge seiner zu geringen Schwärzung nur an die mittleren Teile der Korona auf der 1^s Exposition heran reichte. Die Kombination der relativen und absoluten Werte ergab folgendes.

Abstand vom Zentrum:	16′.5	18.1	19.6	21.2	22.8	24.4	26.0
Helligkeit:	14ᵐ3	15.3	16.3	16.7	17.4	18.0	18.6.

Auf der andern Seite ergab sich aus den Spektren auf Platte Z_a durch Vergleichung mit dem auf ihr befindlichen Keil für die Helligkeit im Abstand 16′.5

der Wert $15^{\mathrm{m}}5$ statt des obigen Wertes $14^{\mathrm{m}}3$. In letzteren Wert geht die Unsicherheit des Momentverschlusses ein, in ersteren eine Ungenauigkeit der 1^{s} Exposition, in beide eine Verschiedenheit der Sensibilisierungskurve der zur Keilaichung und zur Aufnahme verwandten Platten. Ich habe das Mittel aus beiden Werten genommen und entsprechend alle Zahlen der obigen Tabelle um $0^{\mathrm{m}}6$ vermehrt.

§ 32.

Im Ganzen ergaben sich damit die folgenden Koronahelligkeiten:

Abstand vom Zentrum r	16'.5	18.1	19.6	21.2	22.8	24.4	26.0	30	35	40	45	50
Beobachtete Helligkeit	$14^{\mathrm{m}}9$	15.9	16.9	17.3	18.0	18.6	19.2	19.5	20.3	20.9	21.6	22.3
Helligkeit nach Turner	$15^{\mathrm{m}}5$	16.1	16.6	17.1	17.6	18.0	18.5	19.4	20.4	21.3	22.1	22.7
Differenz	$+0^{\mathrm{m}}6$	+0.2	−0.3	−0.2	−0.4	−0.6	−0.7	−0.1	+0.1	+0.4	+0.5	+0.4

Herr **Turner** hat aus Aufnahmen vom Jahre 1893 und vom Jahre 1898 gefunden (Monthly Notices Vol. 61), dass die Helligkeit der Korona sich durch das Gesetz:

$$i = i_0 \left(\frac{R}{r}\right)^6 \qquad \left\{ \begin{array}{l} R \text{ Sonnenradius} \\ r \text{ Abstand vom Sonnenmittelpunkt} \\ i_0 \text{ Randhelligkeit der Korona} \end{array} \right.$$

oder in Grössenklassen ausgedrückt:

$$m = m_0 + 15 \log \frac{r}{R}$$

darstellen liesse. Mit $i_0 = 6{,}4 . 10^{-7}$ oder $m_0 = 15^{\mathrm{m}}5$ ergaben sich nach dieser Formel die oben angegebenen „Helligkeiten nach Turner“, welche mit den beobachteten Werten so gut übereinstimmen, als die Ungleichförmigkeit der Korona und die Unsicherheit unsrer Zahlen erwarten lassen konnte.

Die Gesamthelligkeit der Korona, verglichen mit der Sonne, ergiebt sich nach der Turner'schen Formel zu $\frac{i_0}{2}$. Sie beträgt also:

$3{,}2 . 10^{-7}$ oder $16^{\mathrm{m}}3$ verglichen mit der Sonne

0,17 oder $1^{\mathrm{m}}9$ verglichen mit dem Monde.

Es sei für das Folgende gleich noch hinzugefügt, dass das diffuse Licht der oben bestimmten Helligkeit $0{,}3 . 10^{-8}$ über einen Kreis am Himmel von 5^0 Durchmesser gerechnet die Gesamthelligkeit $2{,}5 . 10^{-7}$ verglichen mit der Sonne oder 0,13 verglichen mit dem Monde hat.

Abhandl. Kgl. Ges. d. Wiss. zu Gött. NF 5(2) 1907, 3–73

Ich stelle dieses Resultat mit den mir bekannten neueren Helligkeitsmessungen der Korona zusammen (Einheit die Vollmondhelligkeit).

		Eigentliche Korona	Diffuses Licht eines Kreises von 5° Durchmesser	Summe
Turner (photographisch)	1893	0.6	0.5	1.1
„ „	1898	1.1	1.3	2.4
Schwarzschild („)	1905	0.17	0.13	0.3
Fabry [1]) (optisch)	1905	0.55	0.20	0,75
Knopf [2]) (optisch)	1905	—	—	0,85

Die Zahlen machen den Eindruck, dass die Korona in der Jahren 1893 und 1905 des Sonnenfleckenmaximums schwächer war, als in dem Minimumjahre 1898.

Aus der Stärke der diffusen Beleuchtung des Himmels ist zu schliessen, dass dieselbe hauptsächlich von der Erleuchtung der Luft durch direkte Sonnenstrahlung an den Grenzen des Totalitätskreises und nur zu einem ganz geringen Bruchteil von Diffusion des Koronalichtes herrührt. Denn das diffuse Himmelslicht aus einem Kreis von 5° Durchmesser um die Sonne macht unter gewöhnlichen Umständen bei durchsichtiger Luft weniger als $^1/_{10}$ der direkten Sonnenstrahlung aus [3]). Da die Annahme einer besonderen Trübung der Atmosphäre während der Finsternis durch den Augenschein widerlegt wird, so darf man also auch nur weniger, als ein Zehntel der obigen Zahlen für das diffuse Himmelslicht, auf Rechnung der Korona setzen.

§ 33. Spektralphotometrie der Korona.

Der prinzipiell einfachste Weg, das Spektrum der Korona zu photometrieren, die „Farbe“ der Korona zu bestimmen, bestände darin, mit einem Spaltspektrographen das Koronaspektrum aufzunehmen und mit demselben Apparat auf der-

1) Fabry giebt (Comptes rendus 151, pag. 870) die Helligkeitssumme 0,75, ferner macht er ebenda pag. 941 die Angabe, dass die Helligkeit der Korona 5′ vom Sonnenrand gleich 0,28 der mittleren Flächenhelligkeit des Vollmondes gewesen sei. Daraus ist mittelst des Turner'schen Gesetzes die obige Zahl 0.55 für die eigentliche Korona gerechnet.

2) Vierteljahrsschrift der Astron. Gesellschaft 1906 pag. 158. Nach freundlicher Mitteilung von Herrn Prof. Knopf enthält diese Zahl das diffuse Licht eines Kreises von mehr als 5° Durchmesser.

3) Vgl. K. Schwarzschild und W. Villiger. Astroph. Journal. Vol XXIII, pag. 303.

selben Platte durch eine hinreichend starke Abblendungsvorrichtung hindurch das Sonnenspektrum zu photographieren.

Aufnahmen mit der Prismenkamera geben nun keine reinen Spektren, sondern die Spektren eines ganzen Schnittes durch die Korona übereinandergelagert. Immerhin überwiegt die innere Korona so sehr an Helligkeit, dass sie den Hauptbeitrag liefert und dass namentlich auf den zur Prismenkante senkrechten Tangenten an die Sonnenscheibe ein verhältnismässig reines Spektrum entsteht, bei welchem die Wirkung der Mischung genügend abgeschätzt werden kann.

Bei unsern Aufnahmen ist das Prinzip ausser durch die Verwendung der Prismenkamera noch weiter dadurch kompliziert, dass die Platten nicht ein Vergleichsspektrum der Sonne, sondern nur einen photometrischen Keil tragen. Das Vergleichsspektrum wurde erst nachträglich in Göttingen auf einer andern Platte erzeugt, auf welche derselbe photometrische Keil aufkopiert wurde. Wenn die Finsternisplatte und die Hülfsplatte genau dieselbe Sensibilitätskurve haben, so ist natürlich im Effekt das letztere Verfahren genau so gut, wie die direkte Aufkopierung des Vergleichsspektrums. Es war möglichst für gleiche Farbenempfindlichkeit gesorgt, indem beide Male nicht orthochromatische Schleussner-Platten verwandt wurden. Allerdings wurde auch die Gleichheit der Farbenempfindlichkeit sehr in Anspruch genommen, da zu dem Keil unvorsichtiger Weise ein braunes Glas verwandt worden war, das fast nur die langen Wellenlängen durchliess. Im ganzen war also in unserem Falle folgender Uebergang auszuführen: Abgeschwächtes Sonnenspektrum auf der Hülfsplatte — Keil auf der Hülfsplatte — Keil auf der Finsternisplatte — Koronaspektrum auf letzterer Platte. Es sei noch erwähnt, dass die abgeschwächten Sonnenspektren mit Hülfe eines kleinen Kollimators erzeugt wurden, der in seiner Fokalebene einen keilförmigen Spalt trug und vor die Prismenkamera gesetzt werden konnte.

Das Resultat waren die unten angegebenen Helligkeiten des unreinen Koronaspektrums verglichen mit dem Sonnenspektrum. Da es sich hier nicht mehr um absolute Helligkeiten der Korona handelt, habe ich das Sonnenspektrum als so weit abgeblendet angenommen, dass Sonnen- und Koronaspektrum in der Wellenlänge maximaler photographischer Wirksamkeit (etwa 440 $\mu\mu$) gleich hell sind.

Für die Reduktion auf reines Spektrum kommt folgendes in Betracht. An jeder Stelle der photographischen Platte superponieren sich bei den Aufnahmen mit der Prismenkamera die Wirkungen verschiedener Stellen der Korona und zwar superponieren sie sich mit verschiedenen Wellenlängen. Kurzwelliges Licht der östlichen Koronateile z. B. trifft mit langwelligem der westlichen auf einer Plattenstelle zusammen. Die Gesamtwirkung hängt daher nicht nur von der Helligkeitsverteilung in der Korona, sondern auch von der verschiedenen

Abhandl. Kgl. Ges. d. Wiss. zu Gött. NF 5(2) 1907, 3–73

aktinischen Kraft der sich überlagernden Wellenlängen ab. Für die Helligkeitsverteilung in der Korona habe ich das Turner'sche Gesetz angenommen. Die aktinische Kraft der verschiedenen Wellenlängen konnte aus den Sonnenspektren auf der Hülfsplatte abgeleitet werden. So war die Wirkung der Mischung der Spektren berechenbar, sobald die spektrale Intensitätsverteilung, die Farbe der Korona, bekannt war. Für diese habe ich in erster Näherung die Farbe der Sonne angesetzt, was sich nachträglich als ein völlig ausreichendes Verfahren erwies. Ich habe also tatsächlich berechnet, wie sich die Wirkung des Sonnenspektrums ändert, wenn man von einer punktförmigen Lichtquelle zu einer Lichtquelle übergeht, in welcher die Helligkeitsverteilung eines Schnittes durch die Korona herrscht. Auf diese Weise sind die unten angegebenen Reduktionen auf reines Spektrum gefunden. Im Wellenlängengebiet 350 $\mu\mu$ bis 460 $\mu\mu$ sind dieselben so klein, dass sie von irgend welchen Unsicherheiten der Ableitung nicht merklich beeinflusst werden. Auf ganz entsprechende Art wurden auch die oben erwähnten Angaben über die absolute Helligkeit der Korona aus den spektralen Helligkeiten gewonnen.

Wellenlänge:	344 $\mu\mu$	356	370	387	408	431	459	494
Unreines Spektrum:	(– 0^m.9)	0.0	+ 0.2	– 0.15	+ 0.1	0.0	– 0.25	– 0.6
Reduktion:	(+ 0^m.4)	+ 0.4	+ 0 3	+ 0.2	+ 0.05	0 0	+ 0.15	+ 0.55
Reines Spektrum:	(– 0^m.5)	+ 0.4	+ 0.5	+ 0.05	+ 0.15	0.0	– 0.10	– 0.05.

Die Addition der Reduktionsgrössen zu den Werten für das unreine Spektrum giebt die in der letzten Zeile der vorstehenden Tabelle enthaltene Vergleichung zwischen Korona- und Sonnenspektrum. Das Resultat kann man so aussprechen:

Im Wellenlängengebiet 350—500 $\mu\mu$ stimmt die Intensitätsverteilung im Koronaspektrum mit der im Sonnenspektrum nahe überein. Betrachtet man die Unterschiede näher und sieht man ab von der ersten wegen geringer Schwärzung dieser Wellenlängen unsicheren Zahl, so erscheint die Korona im Ultraviolett schwächer als die Sonne, das Intensitätsmaximum in der Korona gegen das der Sonne noch längeren Wellenlängen zu verlegt.

Das erstere allgemeinere Resultat ist offenbar und stimmt mit der Angabe der Herren Knopf und Fabry, dass bei ihren photometrischen Vergleichungen die Korona sehr genau die Farbe des Mondes d. i. der Sonne gehabt habe. Die Realität des zweiten feineren Ergebnisses möchte ich als in sich noch nicht fest begründet ansehn. Die Hauptunsicherheit entspringt dem schon erwähnten Uebelstand, dass der Keil braun gefärbt war. Die Ergebnisse würden daher von einer nur etwas verschiedenen Empfindlichkeit der beiden verwandten Platten im Rot, verglichen mit ihrer Blauempfindlichkeit, stark betroffen werden.

§ 34. Zur Theorie der Sonnenkorona.

Unter den theoretischen Anschauungen, mit denen man diese Resultate vergleichen kann, steht als eines Extrem die Annahme, dass man es in der innern Korona wesentlich mit selbstleuchtenden Teilchen zu tun hat, die sich im Strahlungsgleichgewicht befinden. Nach den Rechnungen von Arrhenius (Astrophys. J. Bd. XX, pag. 224) erhält man für ein Teilchen 0′.7 über der Photosphäre eine Temperatur von 4620° und die unten angegebenen Abweichungen von der Strahlung der Sonne (des schwarzen Körpers von 6000°), welche sich mit den Messungsergebnissen genau decken.

	344 $\mu\mu$	356	370	387	408	431	459	494
Beobachtung:	(− 0^{m}5)	+ 0.4	+ 0.5	+ 0.05	+ 0.15	0.0	− 0.10	− 0.05
$T = 4620°$:	+ 0.4	+ 0.4	+ 0.3	+ 0.2	+ 0.1	0.0	− 0.2	− 0.3
Rayleigh's Blau:	− 1^{m}0	− 0.9	− 0.7	− 0.5	− 0.3	− 0.1	+ 0.2	+ 0.5.

Das andre Extrem bildet die Annahme, dass die Korona nur im reflektierten Licht leuchtet und aus sehr kleinen Teilchen besteht, welche nach Rayleigh proportional zu λ^{-4} reflektieren und infolge dessen ein mit dem Blau unseres Himmels sehr nahe übereinstimmende Farbe geben. Mit dieser Aufnahme sind unsre Zahlen nicht verträglich. Sie widerspricht dem Augenschein und den erwähnten Beobachtungen der Herren Fabry und Knopf. Sie wird ferner widerlegt durch die Aufnahmen mit Gelbscheibe, wie sie Dr. Emden ausgeführt hat. Dieselben zeigen, dass die Gelbscheibe das diffuse Himmelslicht viel stärker herabsetzt, als die Koronastrahlung, dass die Korona viel gelber ist, als das Himmelslicht. Natürlich — könnte man hier einwerfen — denn das von der Luft diffundierte Licht ist immer blauer als das Licht der Lichtquelle; ist die Korona blau, so muss der Himmel während der Totalität eben noch tiefer blau gefärbt sein. Nach den oben mitgeteilten Messungen stammt aber die allgemeine Erleuchtung des Himmels während der Finsternis nur zum geringsten Teil von der Korona selbst, sie kommt im wesentlichen von der direkten Erleuchtung der Erdatmosphäre ausserhalb des Totalitätskreises und muss daher das gewöhnliche Himmelsblau zeigen, sodass unser Schluss nicht beeinträchtigt wird.

Es stimmt daher alles dahin zusammen, dass die Korona nicht das Rayleigh'sche Blau zeigt.

Wir wollen überlegen, wie man unter diesen Umständen mit dem gesamten optischen Verhalten der Korona zu recht kommt. Man erklärt das eben erhaltene photometrische Resultat und die Tatsache, dass im Koronaspektrum Fraunhofer'sche Linien nicht auftreten, am natürlichsten, indem man annimmt, dass die Korona aus durch die Sonnenstrahlung erhitzten selbstleuchtenden Teilchen besteht, so wie Arrhenius (l. c.) ausgeführt hat. Es herrscht dann, wie erwähnt, völlige Uebereinstimmung mit den photometrischen Ergebnissen, der Uebelstand ist aber, dass die Polarisation des Koronalichtes ganz unerklärt bleibt.

Herr Salet giebt an (Comptes rendus 141, pag. 528), dass die Korona bis zum Mondrand polarisiert gewesen sei und dass das Maximum der Polarisation

Abhandl. Kgl. Ges. d. Wiss. zu Gött. NF 5 (2) 1907, 3–73

in 5′ bis 6′ Abstand von der Sonne gelegen habe. Herr Newall findet (Monthly Notices 1906, pag. 475) ebenfalls starke Polarisation der Korona bis zum Mondrand. Wenn man bedenkt, dass 5′ über der Sonnenfläche die Sonne einen Gesichtswinkel von 140° Oeffnung erfüllt, so sieht man, dass die Polarisation des einzelnen Lichtstrahles am einzelnen Teilchen sehr energisch sein muss, damit bei so diffuser Beleuchtung und bei der Mischung des Lichtes aller auf der Visierlinie liegenden Koronateilchen noch eine auffällige Polarisation übrig bleibt[1]). Es scheint demnach im Gegensatz zu der obigen Auffassung der grössere Teil der Strahlung der Korona aus reflektiertem Licht bestehen zu müssen.

Stellen wir uns daher einmal auf den entgengengesetzten Standpunkt und betrachten die Koronastrahlung nur als reflektiertes Licht, so muss die Korona entweder aus grösseren Teilchen bestehn, oder, wenn sie aus Teilchen besteht, die klein gegen die Wellenlänge des Lichtes sind, so müssen dieselben so beschaffen sein, dass sie trotzdem nicht das Rayleigh'sche Blau geben. Die Annahme grösserer Teilchen scheint mir wenig plausibel. Erstens müssten diese Teilchen sehr stark polarisieren. Zweitens erscheint es unwahrscheinlich, dass sich bei Temperaturen von 4—5000° in einem nahezu absoluten Vakuum, wie es in der Korona herrscht, kleine Teilchen halten können.

Zieht man daher die Annahme molekularen Zerfalls für die Substanz der Korona vor, so hat man nach Molekülen zu fragen, die nicht das Rayleigh'sche Blau geben, aber ebenso stark polarisieren, wie Rayleigh's kleines Teilchen. Wie eine Rechnung lehrt, die an andrer Stelle mitgeteilt werden soll, erfüllen diese Forderung alle Moleküle mit ultraroten Eigenschwingungen. Ebenso wird diese Forderung von einzelnen freien Elektronen, mit andern Worten von Kathodenstrahlen, erfüllt. Denn ein freies Elektron verhält sich, wie eine Substanz von unendlich langer Eigenperiode, es reflektiert resp. diffundiert das Licht aller Wellenlängen gleich stark und polarisiert dabei, wie das Rayleigh'sche Teilchen. Versucht man die in mancher Hinsicht verlockende Annahme durchzuführen, dass die Korona aus negativen Elektronen bestehe, so kommt man auf folgende Schwierigkeit. Man kann berechnen, dass die zur Erzeugung der beobachteten Helligkeit der Korona nötige Zahl von Elektronen gleich 10^{18} über jedem qcm der Sonnenoberfläche ist. Die hieraus resultierende ungeheure negative Ladung der Korona hat man sich durch eine positive Ladung des Sonnenkörpers von ganz ähnlicher Grösse kompensiert zu denken. Das entsprechende elektrische Feld würde aber Elektronen, die fast mit Lichtgeschwindigkeit liefen, schon auf Strecken weit unter dem Millimeter stoppen. Eine grössere Ausdehnung der Korona wäre also undenkbar.

Man kommt daher auf die Annahme von stark polarisierenden Molekülen oder Ionen zurück, welche ultrarote Eigenschwingungen haben oder welche auch in irgend einer andern komplizierteren Weise so gebaut sind, dass sie alle

1) Vgl. den Aufsatz von A. Schuster Monthly Notices Vol. 40. 1880 pag. 35, der hierüber völlige Aufklärung giebt.

Farben ziemlich gleich stark diffundieren. Dass es solche Molekeln giebt, beweisen die ultramikroskopischen Teilchen in colloidalen Lösungen, deren Grösse weit unter den Wellenlänge liegt, die polarisieren, wie Rayleigh's Teilchen, die aber trotzdem ein kontinuierliches Spektrum geben, in welchem häufig Rot und Gelb überwiegt [1]). Das Fehlen der Fraunhofer'schen Linien ist auf grosse Geschwindigkeiten der Teilchen der Korona und entsprechende Dopplersche Verschiebungen zurückführen. Unter den bewegenden Kräften kommt auch bei Molekülen der Lichtdruck in Frage, wofern sie nur Eigenschwingungen haben. Darauf soll an anderer Stelle näher eingegangen werden.

§ 35. Das Mondprofil und die Helligkeit des Sonnenrandes.

Beim Eintritt der Totalität bemerkt man das sog. Phänomen der Baily'schen Tropfen. Der letzte Rest der feinen Sonnensichel, welche hinter dem Monde hervorsieht, löst sich in eine Reihe von kleinen Punkten oder Tropfen auf, die dann rasch nacheinander verschwinden. Die Ursache dieser Erscheinung ist natürlich die Unebenheit der Mondoberfläche. Die Berge unterbrechen zuerst die Sonnensichel, während durch die Thäler der Rand der Photosphäre noch länger hindurchblickt. In dem von einer Prismenkamera entworfenen Bilde wird analog jeder Mondberg durch eine streifenförmige Unterbrechung des Spektrums, jedes Thal durch einen Streifen stärkeren kontinuierlichen Photosphärenlichtes bezeichnet.

Die Helligkeit der Streifen im Spektrum wird also einerseits abhängen von der Form des Mondprofiles, sie wird aber auch abhängen von der Helligkeit der hinter dem Monde hervorsehenden Randpartie der Sonnenphotosphäre. Eine photometrische Bearbeitung solcher Aufnahmen wird daher sowohl eine Bestimmung des Mondprofils, als eine solche der Helligkeit des Sonnenrandes — und zwar letztere für jede einzelne Spektralfarbe — liefern können.

Das gegebene Verfahren, beides zu erreichen, wäre folgendes. Man macht eine Serie von Aufnahmen der verschwindenden Sonnensichel zu chronographisch registrierten Momenten etwa in Sekundenintervallen. Aus der bekannten Geschwindigkeit des Mondes gegen die Sonne kann man dann die Höhen all der Punkte des Mondes bestimmen, welche bei irgend einer der Aufnahmen gerade den Sonnenrand tangieren. Diese Punkte sind durch das Erscheinen der zahlreichen hellen Linien der tiefsten Chromosphäre sehr scharf charakterisiert. Nachdem man so ein Netz von Punkten bekannter Höhe hat, kann man in dasselbe durch Messung der Helligkeiten der Spektralstreifen der überstehenden Photospärenpartien interpolatorisch die Form des dazwischen liegenden Profiles einzeichnen. Ist dann das Mondprofil bekannt, so kann man die Helligkeit des Sonnenrandes mit Hülfe der nunmehr berechenbaren Breiten der Photosphärensichel aus der Schwärzung bestimmen, wofern man auf derselben Platte ein hinreichend abgeblendetes Spektrum der vollen Sonne zum Vergleich hat.

---

1) Prof. W. Biltz (Clausthal) danke ich die Gelegenheit zu dieser Beobachtung.

Abhandl. Kgl. Ges. d. Wiss. zu Gött. NF 5(2) 1907, 3–73

Schliesslich kann man den Kreis bestimmen, welcher sich dem gefundenen Mondprofil am besten anschliesst, und hat in dem Mittelpunkt desselben den korrigierten Mondort.

Indessen ist es auch möglich, ohne Kenntnis der Expositionszeiten mit Hülfe eines einzigen anderen Datums auszukommen — und hierauf war ich angewiesen, da die Expositionszeiten nicht fixiert worden waren. Die Breite der Sichel bei einer Aufnahme an irgend einer Stelle sei β, die Intensität der Photosphäre in der Entfernung β vom Rande sei i. Dann ist die Gesamtintensität der Sichel, wie sie aus der Messung der Schwärzungen und der Benutzung der Vergleichsspektren hervorgeht, an dieser Stelle:

$$J = \int_0^\beta i\,d\beta = \varphi(\beta).$$

Rückt nun der Mond vor, so schneidet er von allen Sichelbreiten das gleiche Stück h ab — das gilt wenigstens für hinreichend schmale Sichel kurz vor Eintritt der Totalität und auf diese wollen wir uns zur Vereinfachung beschränken. Die Intensität derselben Sichelstelle bei einer späteren Aufnahme wird daher:

$$J' = \varphi(\beta - h).$$

Hat man J und J' über die ganzen Sicheln gemessen, so kennt man die Funktionswerte J und J', welche zu um die Konstante h verschiedenen Werten des Argumentes β gehören, und hieraus läßt sich die Funktion φ bestimmen. Es ist dies formal genau dieselbe Aufgabe, wie wenn man eine Sterngruppe erst mit voller Oeffnung und dann mit bekannter Abblendung photographiert und aus der Beziehung der entstehenden Bilder zu einander die Helligkeiten der Sterne ableitet. Für ihre Behandlung verweise ich auf Astron. Nachr. Bd. 172, pag. 65. Hat man mehr als 2 Aufnahmen, so erhält man weitere Kontrollen für die Funktion φ und findet die Stücke h, um die sich der Mond zwischen den Aufnahmen bewegt hat, erhält also eine nachträgliche Zeitbestimmung der Aufnahmen. Voraussetzung ist hier nur, daß man eine Verschiebung h oder irgend ein Datum kennt, welches h nachträglich zu bestimmen gestattet.

Dies eine feststehende Datum ließ sich Herrn Hayn's Angaben über die Form des Mondrandes (Astron. Nachr. Bd. 168, pag. 1) entnehmen. Ich bestimmte aus den Positionswinkeln der Sichelenden auf Spektrum No. 9 zunächst die Lage des Mondmittelpunktes gegen den Sonnenmittelpunkt. (Hierbei habe ich schon die Höhenangaben von Herrn Hayn für die betreffenden Stellen des Mondrandes benutzt, ich habe also genau genommen noch zwei weitere Daten von dort entnommen. Es ist dies deswegen unumgänglich, weil man den an sich nicht scharf definierten Mondmittelpunkt mit dem Hayn'schen Mittelpunkt identifizieren muss, wenn man Herrn Hayn's Daten überhaupt benutzen will.) Nachdem die Lage des Mondmittelpunkts bekannt war, konnte die Sichelbreite für reguläre Begrenzung der Mondscheibe berechnet werden. Für einen Punkt in der Mitte der Sichel habe ich nun auch noch die Höhe der Mondberge von Herrn Hayn übernommen. Damit war eine exakte Sichelbreite bekannt und das war jenes fehlende Datum, mit welchem die übrige Rechnung vor sich gehen konnte.

Die Vergleichsspektren wurden durch direkte Aufnahmen der vollen Sonne bei Abblendung der Prismenkamera auf 3 mm, 1 mm und 0.5 mm Oeffnung erhalten. Sie sind daher keine reinen Spektren, doch macht das für die Wellenlänge 455 $\mu\mu$, in welcher gemessen wurde, wegen ihrer Lage in der Nähe der Wellenlänge maximaler aktinischer Kraft nichts aus.

Das Messverfahren bestand in der Aufnahme von „Helligkeitsprofilen" quer durch die Spektra hindurch, von denen einige in Fig. 2 (pag. 46) wiedergegeben sind. Die Struktur der Längsstreifen in den Spektren ist so fein, dass ein Mikrophotometer für noch kleinere Flächen, als sie das Hartmann'sche Instrument zu messen gestattet, und eine noch schärfere Koordinatenablesung am Platze gewesen wäre.

Das Resultat für das Mondprofil giebt Fig. 3 (pag. 46). Es ist wahrscheinlich, dass das Thal beim Positionswinkel 133° zu sehr vertieft ist, doch muss es nach der ausserordentlichen Helligkeit und langen Erhaltung des entsprechenden Streifens in den Spektren (vgl. Tafel III) tiefer sein, als das Hayn'sche Profil angiebt.

Für die Korrektion des Mondortes des Naut. Almanac ergab sich unter Heranziehung der von den Herren Newall und Stephan beobachteten Kontaktverfrühung von 20^s:

$$\varDelta\alpha = +7''.6 \qquad \varDelta\delta = -3''.3.$$

Ferner geht aus den Aufnahmen hervor, dass ein Absturz der Photosphärenhelligkeit auf die Hälfte des Wertes, den sie 5″ vom Rande besitzt, höchstens innerhalb der letzten Sekunde vom Rande erfolgt. Eine quantitative Untersuchung dieses Punktes wäre natürlich wichtig für die Erkenntnis der Absorption der Sonnenatmosphäre. Ich habe oben auch ein numerisches Resultat zu gewinnen versucht — es ergab sich für die Helligkeit des Sonnenrandes (bezogen auf die Helligkeit der Sonnenmitte) im Abstand β'' vom Rande (Wellenlänge 455 $\mu\mu$):

$$i = 0.17 + 0.005\,\beta$$

Doch kann dasselbe nicht als quantitative Erledigung gelten wegen folgender Missstände: Erstens ist auf die Vergleichsspektren ein Streifen falschen Lichtes gekommen, innerhalb dessen zwar nicht gemessen wurde, der aber unkontrollierbare Wirkungen ausgeübt haben kann. Ferner muss die Empfindlichkeit des Films oder — wahrscheinlicher — die Geschwindigkeit des Momentverschlusses gewechselt haben. Denn die auf den Aufnahmen aus Spanien und aus Aegypten gleichmässig wiederkehrende grosse Protuberanzengruppe erscheint auf unseren Aufnahmen, die innerhalb 20^s erfolgten, an Helligkeit ständig zuzunehmen. Durch Messung habe ich mich überzeugt, dass es sich dabei nicht um Kontrastwirkungen handelt.

Der Wert einer technisch besseren Wiederholung dieser Untersuchung würde meiner Meinung nach weniger in der Festlegung des Mondprofiles zu suchen sein, als in der Beobachtung der Helligkeit des Sonnenrandes. Die Kenntnis des

Abhandl. Kgl. Ges. d. Wiss. zu Gött. NF 5(2) 1907, 3–73

Mondprofils ist für letztere Untersuchung, wie für die genauere Diskussion des Flash-Spektrums, und ausserdem ganz allgemein für eine schärfere Festlegung der Mondörter erforderlich; die Ableitung aus Sonnenfinsternisbeobachtungen ist aber nur ein Notbehelf, solange die Selenographie diese Aufgabe noch nicht im Detail gelöst hat, — zumal in Rücksicht auf eventuelle Fälschung der Resultate durch Sonnenfackeln.

Was hingegen die Helligkeitsbestimmung des äussersten Sonnenrandes angeht, so kann dieselbe ausserhalb der Finsternisse nur mit Schwierigkeiten erfolgen und, da man ohnehin Flash-Spektren aufnehmen wird, so wird es sich immer empfehlen, denselben für diesen Zweck photometrische Vergleichsspektren mitzugeben.

§ 36. Spektralphotometrie des Sonnenrandes.

Die photometrische Durchmessung der kontinuierlichen Spektralstreifen auf den Aufnahmen mit der Prismenkamera, welche den letzten Resten des Photosphärenlichtes entstammten, ergab durch Vergleich mit den erwähnten abgeblendeten Sonnenspektren die spektrale Helligkeitsverteilung für die Randpartieen der Photosphäre. Da jene Vergleichsspektren unrein waren, so musste ganz ähnlich, wie oben, aber in Praxis einfacher, eine Reduktion auf reines Spektrum vorgenommen werden. Ich gebe im Folgenden gleich die Mittelwerte aus den Ergebnissen einer Reihe von Spektralstreifen auf den Filmspektren No. 9—17. Das Spektrum, welches als Norm gewählt ist, ist das Spektrum des Sonnenzentrums.

Intensitäten im Spektrum des Sonnenrandes.

λ	351 $\mu\mu$	359	368	377	388	400	413	428	445	464
$\beta = 7''.4$	0.17	0.17	0.17	0.19	0.23	0.26	0.25	0.25	0.25	0.24
$3''.2$		0.30	0.34	0.35	0.36	0.33	0.30	0.25	0.20	0.19
$1''.0$		0.30	0.25	0.28	0.23	0.25	0.28	0.17	0.13	0.12

Unsicher an diesen Zahlen sind die absoluten Beträge, insofern jede Horizontalreihe vielleicht mit einem von 1 verschiedenen Faktor multipliziert werden muss. Von den Sichelbreiten β, denen die Spektren entsprechen, sind die beiden letzten vermutlich in Wirklickeit kleiner. Gut bestimmt ist hingegen die relative Intensität in jeder Reihe, die Farbe.

Der Streifen der Breite $7''.4$ ergiebt ein wenig verminderte Strahlung des Sonnenrandes im Ultraviolett, verglichen mit Violett und Blau. Dies stimmt mit dem Gang im sichtbaren Spektralgebiet. Hingegen haben Dr. Villiger und ich (Astroph. Journal. Vol. 23, pag. 284) für die Wellenlänge 322 $\mu\mu$ gleiche oder sogar ein wenig grössere Randhelligkeit gefunden, als sie Herr H. C. Vogel für 409 $\mu\mu$ angiebt. Wollte man all diesen Messungen gleichmässig vertrauen, so würde man zu dem Schluss geführt, dass der Abfall der Sonnenbelligkeit nach dem Rande zu etwa für die Wellenlänge 360 $\mu\mu$ ein Maximum hat und für grössere, wie kleinere Wellenlängen geringer ist. Da dieses Resultat sehr auffällig wäre, so wird es wünschenswert sein, das Spektrum des Sonnenrandes im Ultraviolett eventuell mit einem Quarzspektrographen sowohl unter gewöhnlichen Verhältnissen, wie bei totalen Finsternissen, nochmals vollständiger durch zu photometrieren.

Geht man zu der Betrachtung der schmäleren Streifen über, so stösst man auf eine merkwürdige Intensitätsverteilung. Hier überwiegt durchaus das Ultraviolett. Systematische Verfälschungen sind gerade bei diesen Messungen wenig zu befürchten, vielmehr glaube ich, dass die Erscheinung durch Ueberlagerung des Linienspektrums der tiefsten Chromosphäre über das kontinuierliche Spektrum des Photosphärenrestes zu Stande kommt. Denn das Chromosphärenspektrum ist gerade im Ultraviolett durch eine dichte Schaar kräftiger Linien ausgezeichnet. Ist die Erklärung richtig, so könnte die Sonne im Ultraviolett einen um einige Zehntel Sekunden grösseren scheinbaren Durchmesser zeigen, als im sichtbaren Spektralgebiet.

§ 37. Helligkeit der Protuberanzen.

Da noch gar keine Angaben über die absolute Helligkeit von Protuberanzen vorzuliegen scheinen, habe ich einige Feststellungen in dieser Hinsicht versucht. Für die in Fig. 3 (Tafel I) mit a, b, b', c bezeichneten Stellen der grossen Protuberanzengruppe am Ostrand der Sonne habe ich auf Spektrum No. 16 die Schwärzung in einigen Wasserstoff- und Kalciumlinien gemessen — so gut es ging bei so kleinen Gebilden, die den Photometerfleck nicht ganz ausfüllten. Was man von der Strahlung einer Protuberanz aus Aufnahmen unserer Art erfahren kann, ist weder die Breite b der Spektrallinie, die sie erzeugt, noch die Helligkeit h, mit der sie im Durchschnitt leuchtet innerhalb des Spektralstreifens, über welchen sich ihre Strahlung erstreckt, sondern nur das Produkt $H = h \, . \, b$ aus beiden Grössen. Der Wert dieses Produktes liess sich aus den gemessenen Schwärzungen mit Hülfe der erwähnten Vergleichsdaten ableiten. Als Einheit ist für die Linienbreite die Angström-Einheit, für die Helligkeit die Helligkeit der Sonne in der betreffenden Wellenlänge gewählt. Es ergaben sich folgende Zahlen:

Protub.	H_β	H_γ	H_δ	H_ζ	H	K
b	0.70	0.24	0.073	0.040	0.11	0.22
a	0.44	0.087	0.026		0.54	0.46
c	0.059	0.031	0.008		0.044	0.081
b'	0.037	0.012			0.070	0.11

Abhandl. Kgl. Ges. d. Wiss. zu Gött. NF 5(2) 1907, 3–73

Diese Zahlen werden am anschaulichsten, wenn man sie als „äquivalente Linienbreiten" auffasst, wenn man nämlich $h = 1$ setzt, also annimmt, dass die Protuberanz in den Wellenlängen, die sie überhaupt aussendet, strahlt wie die Photosphäre, wie ein schwarzer Körper der Temperatur 6000°. Sie geben dann die Breite der betreffenden Linie in A. E., welche genügt, um die beobachtete Helligkeit zu erzeugen. Kennt man die wirkliche Breite b der Protuberanzenlinie, so giebt $\frac{H}{b}$ gleich die wirkliche Helligkeit. Da die Breite der Protuberanzenlinien 1–2 A. E. nicht überschreiten dürfte, so sieht man, dass die Helligkeit heller Protuberanzen in dem Gebiete, in welchem sie überhaupt strahlen, von derselben Grössenordnung ist, wie diejenige der Sonne.

Für das Helligkeitsverhältnis der übrigen Linien zur H_β-Linie folgt:

Protub.	H_β	H_γ	H_δ	H_ζ	H	K
b	1	0.54	0.10	0.06	0.16	0.31
a	1	0.20	0.06		1.2	1.0
c	1	0.53	0.14		0.74	1.4
b'	1	0.32			1.9	3.0
Mittel	1	0.35	0.10	0.06		

Man sieht — wie das auch aus dem unmittelbaren Anblick der Spektren hervorgeht — dass die Verteilung des Kalciums von der des Wasserstoffs gänzlich verschieden ist. Hingegen ist das Helligkeitsverhältnis der beiden Kalciumlinien unter sich, wie das der Wasserstofflinien unter sich, für Protuberanzen, die sich der absoluten Helligkeit nach um das 10fache unterscheiden, sehr nahe das gleiche.

Um auch für den absoluten Helligkeitsabfall in der Wasserstoffserie einen Anhalt zu gewinnen, habe ich angenommen, dass das Intensitätsmaximum im Sonnenspektrum bei $\lambda = 500\,\mu\mu$ liegt, entsprechend einer schwarzen Temperatur der Sonne von 5770° absolut. Für die Helligkeiten des Sonnenspektrums an der Stelle der vier Wasserstofflinien ergaben sich dann auf Grund der Wien'schen Formel die Zahlen

$$1.00 \quad 0.34 \quad 0.18 \quad 0.089.$$

Durch Multiplikation mit den obigen Mittelwerten erhält man daraus das Verhältnis der Strahlungsenergie der Wasserstofflinien in den Protuberanzen:

$$\begin{array}{cccc} H_\beta & H_\gamma & H_\delta & H_\zeta \\ 1 & 1/10 & 1/60 & 1/190. \end{array}$$

Es scheint kaum zweifelhaft, dass man schon von verhältnismässig rohen)hotometrischen Vergleichungen der Helligkeitsverhältnisse der Wasserstofflinien n Protuberanzen und bei irdischen Experimenten Aufklärungen über die Natur les Leuchtens der Protuberanzen erwarten darf, und ich beabsichtige eine der-ırtige Untersuchung durchzuführen.

§ 38. Das Spektrum der Chromosphäre.

Das Chromosphärenspektrum hat in den letzten Jahren von verschiedenen Seiten eine so eingehende Bearbeitung gefunden, dass eine selbstständige Identifizierung der Linien für unsre mit verhältnismässig kleiner Dispersion erhaltenen Spektren überflüssig erschien. Ich habe mich vielmehr an das umfangreiche Verzeichnis der Chromosphärenlinien gehalten, das Herr Dyson kürzlich publiziert hat (Phil. Trans. of the Lond. Roy. Soc. 1906) — dasselbe giebt zwischen $\lambda = 320$—$590\,\mu\mu$ etwa 1200 Linien mit einer Genauigkeit von einigen Hundertstel A. E. – und habe allein mit diesem Vergleichungen ausgeführt.

Das Chromosphärenspektrum habe ich auf Film Zb) Spektrum No. 17 von $\lambda = 459$—$378\,\mu\mu$ und auf Platte Z c) Spektrum No. 3 von $\lambda = 378 - 333\,\mu\mu$ durchmessen. Diese Strecken waren auf beiden Spektren vortrefflich im Fokus, sodass die Wellenlängen trotz des kleinen Massstabes bis auf wenige Zehntel A. E. sicher waren und die Identifikation mit Herrn Dyson's Linien meist ohne Zweideutigkeit erfolgen konnte. Die Wellenlängen wurden bestimmt, indem zunächst mit Hülfe der Hartmann'schen Formel ein Anschluss der Messungszahlen an markante Linien hergestellt und dann mit Hülfe einer grösseren Zahl sicher identifizierter Linien noch ein graphischer Ausgleich vorgenommen wurde. Einen gewissen selbstständigen Wert habe ich dem kleinen Verzeichnis dadurch zu geben versucht, dass ich die Sichellänge jeder Linie bestimmt und daraus in Rücksicht auf die Unebenheiten des Mondrandes die Erhebung des betreffenden Chromosphärengases über die Photosphäre abgeleitet habe. Für das beim 2. Kontakt aufgenommene Spektrum auf dem Film kamen die nördlichen Enden der Sicheln zur Verwendung, weil auf der Südseite das Mondprofil zu steil abfiel. Bei den Linien zwischen $\lambda = 361 - 333\,\mu\mu$, die nur auf der vom dritten Kontakt stammenden Platte Zc) identifiziert werden konnten, erfolgte eine rohere Einteilung der Erhebung der Gase in 3 Klassen, die etwa $1''$, $2''$ und $3''$ Höhe entsprachen, weil für diese Partie des Mondrandes kein genaueres Mondprofil vorlag.

Die folgende Tabelle enthält die gefundenen Wellenlängen der Linien, dazu die Werte aus Herrn Dyson's Verzeichnis und — in der zweiten Abteilung der Tabelle – die Elemente nach Herrn Dyson's Identifikation. Die beiden letzten Spalten geben eine Schätzung der Schwärzung der Linie und die Höhenerstreckung in Bogensekunden. Ich habe 31 vermessene Linien, denen ich bei der Messung die Intensität 0 gegeben hatte, nicht in die Tabelle aufgenommen, und ich habe in dem Dyson'schen Verzeichnis auch nur solche Linien zur Identifikation herangezogen, die mindestens die Intensität 1 hatten.

Abhandl. Kgl. Ges. d. Wiss. zu Gött. NF 5(2) 1907, 3–73

Wasserstoff 1

4340.3	0.65	30	7″.5
4101.2	1.92	20	7.2
3888.6	9.15	10	5.7
3835.7	5.53	8	5.5
3798.7	8.06	8	5.3
3770.4	0.79	8	4.6
3750.2	0.32	7	4.6
3734.4	4.52	6	4.6
3721.6	2.05	6	4.6
3711.6	2.13	5	4.6
3703.9	4.00	4	4.6
3697.3	7.29	4	4.6
3692.1	1.70	3	4.6
3682.7	2.92	2	1.6
3679.8	9.50	3	1.6
3676.4	6.54	2	1.6
3674.0	3.90	2	1.6
3669.5	9.58	2	1.6
3667.7	7.89	$^1/_2$	1.6

Helium 4

4471 2	1.51	6	5″.5
4144.0	4.04	2	1.4
4025.5	6.34	3	5.5

Kohle 12

38 83.4	3.38	3	1″.3

Magnesium 24

3838.4	8.43	4	5″.5
3832.5	2.46	4	3.2
3829.8	9.51	3	3.2

Aluminium 27

3961.8	1.64	3	1″.3
3943.4	4.14	2	1.6

Kalcium 40

4435.0	5.53	2	1″.0
4226.8	6.85	4	2.0
3968.8	8.60	30	10.0
3932.7	3.84	20	10.0

Scandium 44

4400.7	0.67	2	1″.2
4246.9	7.00	6	3.2
4021.0	0.50	$^1/_2$	$<$0.5
3652.0	1.89	2	1
3645.4	5.48	1	1
3642.9	2.92	1	1
3630.7	0.88	3	2
3613.9	3.98	2	2
3576.5	6.48	2	2

Scandium 44

3572.6	2.68	1	1
3567.4	7.86	1	1
3353.9	3.84	1	1
3590.4	0.64	1	1

Titan 48

4590.3	0.11	1	
4572.0	2.20	4	3″.2
4564.1	3.96	4	3.2
4549.4	9.79	4	1.6
4534.4	4.18	5	1.2
4501.4	1.51	4	3.2
4451.0	0.50	1	1.4
4443.8	3.98	4	1.6
4395.4	5.24	4	1.2
4300.3	0 22	3	1.6
4294.4	4.22	2	1.6
4289.6	0.21	4	1.6
4164.0	3.80	1	0.5
4127.3	7.67	1	0.6
4028.1	8 63	1	0.9
4012.6	2 50	2	1.0
3913.7	3.61	5	1.3
3900.6	0.67	4	1.6
3776.5	6.22	0.5	0.8
3761.1	1.47	6	4.6
3759.0	9.44	6	4.6
3742.2	1.78	3	1.6
3729.5	9.85	1	1
3706.2	6.25	3	1.6
3685.8	5.35	6	4.6
3662.5	2.35	1	1.6
3660.1	9.88	1	1.6
3641.3	1.45	1	2
3624.9	4.97	3	1
3595.9	6.20	1	1
3574.4	3.83	1	1
3535.7	5.58	4	2
3520.3	0.44	1	1
3510.8	1.00	3	2
3505.0	5.02	3	2
3491.4	1.18	3	1
3477.4	7.32	2	3
3462.1	1.68	1	1
3452.7	2.61	2	1
3444.5	4.46	2	1
3402.7	2.53	1	1
3394.5	4.68	2	2
3388.0	7.99	2	2
3383.8	3.91	3	2
3380.3	0.49	2	1
3372.8	2.95	2	2
3361.2	1.34	1	1
3349.3	9.55 9.14	2	3
3342.0	1.98	1	1
3340.4	0.48	2	1
3335.3	5.29	1	1

Vanadin 51

3715.6	5.60	1	1″.2
3545.5	5.27	2	1

Chrom 52

4274.9	4.99	3	1″.6
3678.0	7.92	1	1.6
3593.7	3.60	2	1
3579.0	8.89	1	2
3433.4	3.42	5	2
3423.0	2.86	3	2
3421.3	1.37	3	2
3409.0	8.92	2	2
3403.7	3.45	2	2
3378.5	8.51	1	1
3368.1	8.20	1	2
3358.7	8.65	1	1

Mangan 55

4034.4	4.62	1	1″.0
4030.5	0.87	1	1.0
3497.6	7.67	1	1
3488.8	8.84	2	2
3483.3	2.99	2	2
3474.3	4.30	2	3
3460.4	0.46	2	2
3442.4	2.14	2	2

Eisen 56

4584.0	4.06	3	1″.6
4528.9	8.96	1	0.5
4522.3	2.84	1	1.5
4515.6	5.54	1	0.7
4508.7	8.45	1	0.7
4490.0	1.68 9.36	1	0.8
4482.6	2.14	1	0.7
4405.7	4.89	2	1.5
4384.4	5.63 3.72	4	1.3
4325.9	5.88	3	0.7
4271.6	1.87	2	1.6
4236.2	6.04	1	0.8
4233.2	3.36	3	1.3
4119.1	8.58	1	1.4
4107.8	7.58	1	3.0
4071.7	1.8	2	1.3
4063.7	3.75	4	1.6
4045.5	5.97	5	1.5
4005.3	5.44	2	1.0
3727.6	8.10	2	1.6
3923.2	3.09	2	1.6
3920.1	0.39	2	1.6
3902.9	3.04	1	0.8
3896.0	5.81	2	1.6
3886.4	6.49	1	0.7
3878.8	8.73	4	1.3
3859.9	0.10	4	1.3

Eisen 56

3856.6	6.52	3	1″.3
3851.0	0.91	0.5	1.2
3840.6	0.54	1	1.2
3826.3	6.02	1	1.4
3824.6	4.62	1	1.4
3820.6	0 59	3	1.4
3815.9	6.00	2	1.4
3795.1	5.13	1	1.2
3767.3	7.36	1	1.6
3764.8	3.92	1	1.6
3748.0	8.38	1	1.6
3746.2	5.87	3	1.6
3743.5	3.55	1	1.3
3727.6	7.78	3	1.2
3719.8	0.08	2	1.6
3647.7	7.92	1	1
3618.6	8.88	2	1
3570.2	0.30	1	1
3526.4	6.04	1	1
3441.0	1.20 0.72	2	1

Nickel 58

4067.3	7.23	1	0″.6
3783.6	3.60	1	0.5
3524.6	4.77	1	1

Cobalt 58

3621.6	1.38	1	2″
3465 9	5.83	2	1

Strontium 87

4215.6	5.71	8	5″.0
4077.5	7.86	10	4.8

Yttrium 89

4375.6	5.14	5	1″.6
4309.6	9.86	2	1.6
3950.5	0.45	1	0.6
3788.8	8.86	3	1.3
3774.6	4.45	2	1.3
3710.4	0.43	1	1.0
3664.6	4.78	1	1.6
3629.6	8.89	1/2	1
3611.1	1.15	1	1
3602.2	2.05	1	1
3600.6	0.89	1	1
3585.3	4.67	2	2
3549.1	9.17	2	1

Zirkon 90

4347.8	8.06	1	1″.6
3556.7	6.81	2	1
3552.0	2.11	1	1
3496.2	6.33	3	2
3479.9	9.48	1	1
3438.5	8.38	1	1
3430.6	0.71	1	1

Lanthan 138

4086.9	6.88	2	0″.8
3916.6	6.21	1	1.0

Cer 140

4133.5	3.83	1	1″.4

Neodym 141

4041.2	1.05	1	0″.6
3952.7	2.36	1	0.7

Ytterbium 173

3694.1	4.29	1	0″.7

Linien mehrerer oder unbekannter Elemente.

4554.8	6.05 4.18	Fe Ba	5	1″.6
4461.7	1.44	—	1	1.4
1) 4428.0	9.41 7.30	— Ti	1	1.0
2) 4417.9 4414.5	7.93 6.98 5.65 5.17	Ti — Sc Fe	2 2	1.6 1.6
4368.7	9.86 7.85	Fe Ti	1	0.6
4359.8	9.82	Zr Cr	1	0.6
4352.1	2.00	Mg Cr Fe	1	0.7
4321.3	0.98	Ti Sc	3	0.7
4314.3	5.13 4.31 3.02	Ti Sc Ti	3	1.0
4308.1	8.02	Fe Ca	2	1″.6
4303.0	3.51 2.94	Fe —	3	1.6
4282.7	2.76	—	1	1.6
4261.3	2.14 0.51	Cr Fe	1	1.6
4256.4	8.24 6.31 4.51	Zr — Cr	3	1.6
4205.4	5.21	Eu	1	0.7
4198.3	8.15	—	1	0.7
4187.4	8.18 6.76	— Cl	2	0.8
4178.2	9.03 7.66	Fe Y	3	1.0
4172.8	3.53 2.05	Fe Ti	3	1.0
4156.2	6.33	Zr Nd	2	1 0
4149.7	0.10 9.33	Ce Zr	1	0.8
4146.8	6.36	—	1	0.8
4137.9	7.54	—	1	0.6
4123.1	3.56	Ti La	1	0.6
4110.2	0.64 9.61	Co Nd	1	3.0
4080.5	0.62	—	1	1.3
4053.4	3.85	Ti V	1	0.6
4014.8	4.77	Fe Sc	0.5	0.7
3998.9	9.28	V Zr	1	0.8
3990.9	1.42	Zr Cr	1	0.8
3982.4	2.75 1.87	Y Ti	1	0.8
3958.7	8.29	Zr Ti	1	0.6
3956.4	6.60	Fe Ti	1	0.8
3) 3873.1	3.21 2.63	Co Fe	1	1.0
3866.6	7.60 5.72	He Fe	1	1.0
3813.3	3.57 3.17	Ti Fe	1	0.7
3755.9	5.69	—	2	3
3737.6	7.14	Fe Ca	3	1.6
3700.0	0.41 9.86	V —	1	3
3671.1	1.46	Zr H	2	1.6

1) Wohl 2 schwache Linien.

2) Fliessen zu einem Band zusammen, dessen Kanten eingestellt sind.

3) Offenbar Duplex.

Abhandl. Kgl. Ges. d. Wiss. zu Gött. NF 5 (2) 1907, 3–73

3643.3	3.27	—	1	2″
3609.4	9.79	—	1	1
	9.01	Fe		
3605.8	6.73	—	1	1
	5.48	Cr		
3587.2	7.71	—	1	1
3581.2	1.29	Fe	2	2
	1.08	Sc		
3566.2	6.21	V, Ti	2	1
3532.2	1.85	—	1	1
3530.0	0.14	—	1	1
3513.9	4.33	—	1	1
3458.5	8.99	—	1	1
3456.4	6.14	—	2	1
3407.3	7.46	—	1	1

Nicht identifizierte Linien.

4331.6	1	1.3″
4210.6	1	0.7
4191.2	1	0.7
4018.2	0.5	< 0.5
3906.2	1	0.8
3881.7	0.5	1.0
3791.5	0.5	0.6
3786.4	0.5	0.5
3753.4	1	2
3607.2	1	1
3559.0	1	1
3542.2	1	1
3533.4	1	1
3515.3	1	1
3493.4	1	1″
3436.0	1	1
3425.4	1	1
[1] 3418.6	1	1
[1] 3417.5	1	1
3399.5	1	1
3392.6	1	1
3337.2	1	1

Es ist offenbar, dass das Atomgewicht als solches nicht für das Auftreten eines Elementes in der höheren Chromosphäre massgebend ist. Ebenso ersichtlich ist aber, dass irgend ein andrer innerer Zusammenhang zwischen den vorkommenden Elementen besteht und nicht etwa ein rein zufälliger grösserer Reichtum der Sonne an einem bestimmten Element seine Erstreckung und sein Leuchten bis in grössere Höhen verursacht. Das Bestehen des Zusammenhangs wird deutlicher, wenn man in einem periodischen System der Elemente, das — wie hierunter — ohne Rücksicht auf die sogenannten Nebenperioden geschrieben ist, die in der obigen Tabelle vorkommenden Elemente vermerkt. Dieselben erfüllen dann — abgesehen von Wasserstoff und Helium — ein einziges zusammenhängendes Gebiet, welches hier zurch Umrahmung gekennzeichnet ist.

Elemente der Chromosphäre (Spektralgebiet 460 – 330 $\mu\mu$) im periodischen System.

Li	Be	B	C	N	O	F										
Na	Mg	Al	Si	P	S	Cl										
K	Ca	Sc	Ti	V	Cr	Mn	Fe	Co	Ni	Cu	Zn	Ga	Ge	As	Se	Br
Rb	Sr	Y	Zr	Nb	Mo		Ru	Rh	Pd	Ag	Cd	In	Sn	Sb	Te	J
Cs	Ba	La	Ce	Di									Er			
		Yb		Ta	W		Os	Ir	Pt	Au	Hg	Tl	Pb	Bi		
	Ra		Th		Ur											

Es sei schliesslich noch angeführt, dass sich für die beiden ultravioletten Ringe des Koronaspektrums die Wellenlängen

3451 A. E (schwacher Ring)
3387 A. E. (stärkerer Ring)

ergeben haben.

1) Vielleicht Ränder einer Bande.

Fig. 1. Das Beobachtungszelt. (R. Emden phot.)

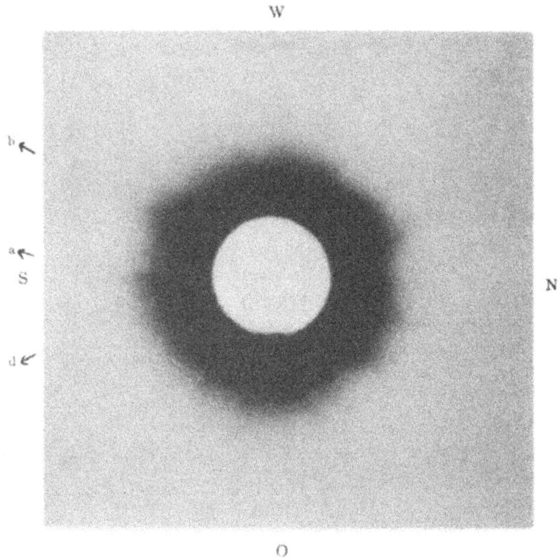

Fig. 2. Die Corona (4 sec. Exposition).

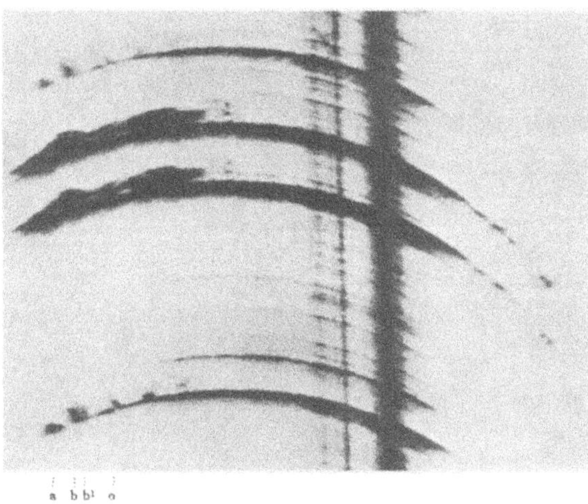

Fig. 3.
Die Protuberanzengruppe am Ostrand in der H δ, H ζ, H und K Linie
nach Film Z b No. 17 vergrößert im Verh. 1 : 15.

9.

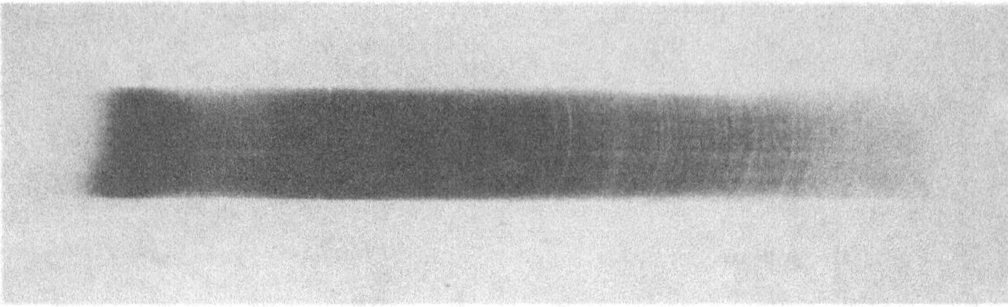

11.

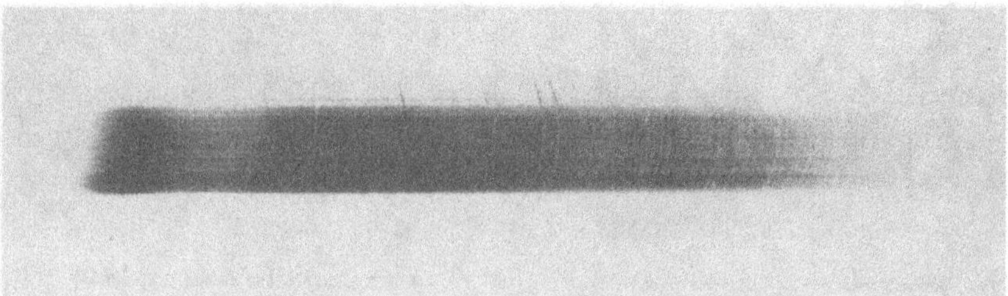

12.

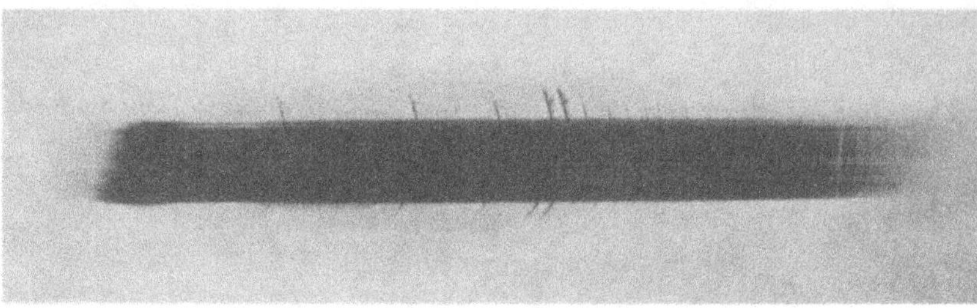

13.

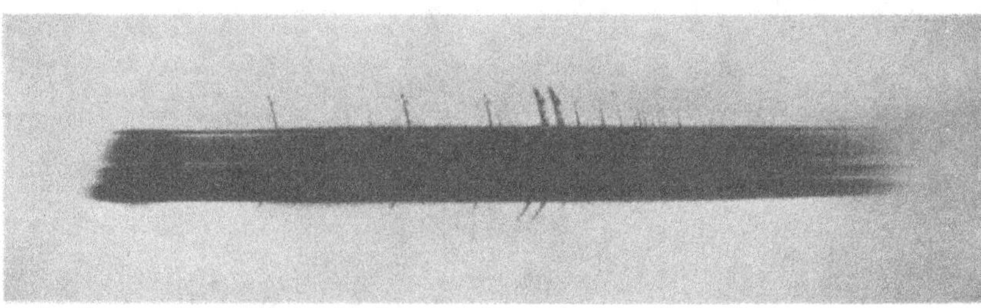

Flashspectren vom II. Kontakt. Nach Film Z b vergrößert im Verh. 1 : 3,5.

14.

15.

17.

19.

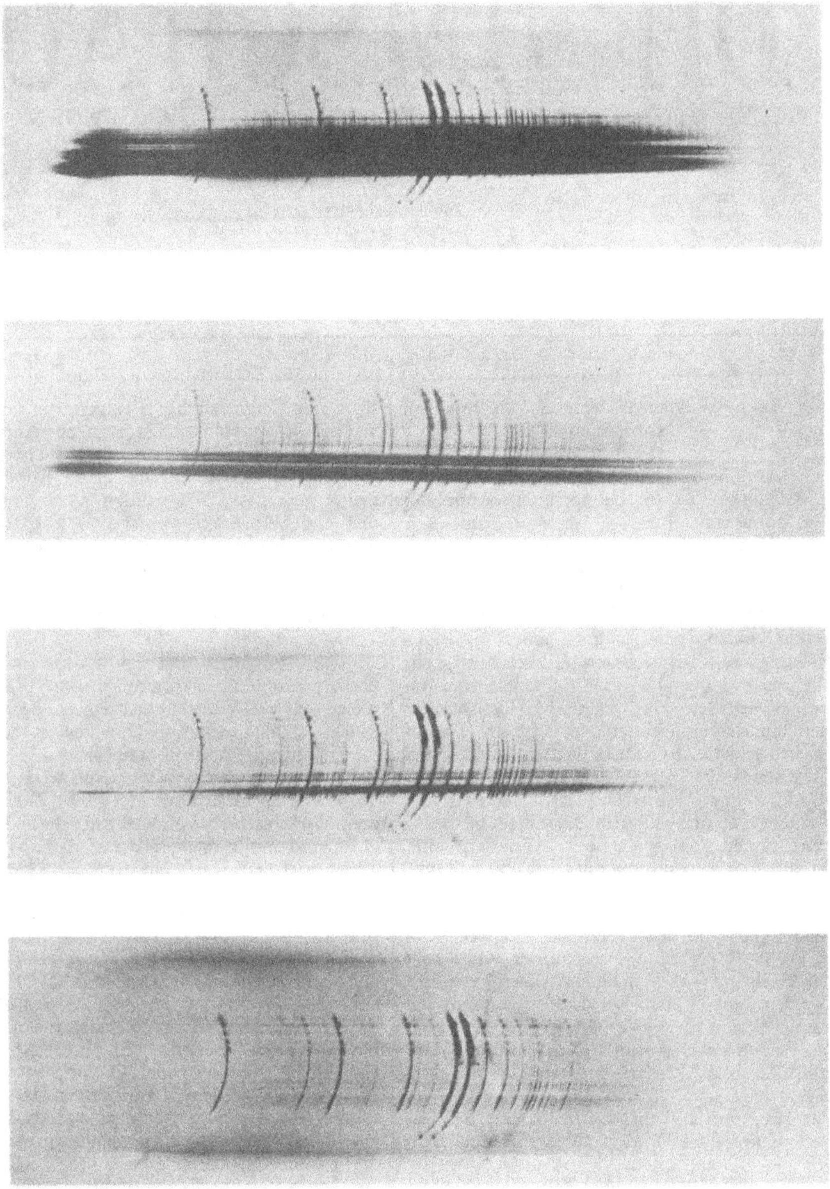

Flashspectren vom II. Kontakt. Nach Film Z b vergrößert im Verh. 1 : 3,5.

DIE UMSCHAU

ÜBERSICHT ÜBER DIE FORTSCHRITTE UND BEWEGUNGEN AUF DEM GESAMTGEBIET DER WISSENSCHAFT, TECHNIK, LITERATUR UND KUNST

Zu beziehen durch alle Buchhandlungen und Postanstalten. Postzeitungspreisliste Nr. 7974.

herausgegeben von

DR. J. H. BECHHOLD.

Erscheint wöchentlich einmal

Geschäftsstelle: H. Bechhold, Verlag Frankfurt a. M.
Redaktionelle Sendungen und Zuschriften zu richten an: Redaktion der »Umschau«, Frankfurt a. M., Neue Kräme 19/21.

№ 46. 11. November 1905. IX. Jahrg.

Zur Sonnenfinsternis vom 29. August in Nordafrika.

Von Prof. Dr. K. SCHWARZSCHILD.

Guelma — unsere Beobachtungsstation bei der Sonnenfinsternis — ist eine Stadt von 7000 Einwohnern halbwegs zwischen Algier und Tunis, 100 km südlich der Küste und schon durch die Nordketten des Atlas vom Meere getrennt. Das kleine von einer alten Mauer umzogene Quadrat heller niedriger Häuser, aus dem sich ein Glockenturm, ein Minaret und der Schornstein des Elektrizitätswerkes erheben, lag jetzt im Hochsommer ziemlich traurig inmitten der ausgetrockneten flachen Hänge. Ein Stück weit hinauf an den Talseiten lagen einzelne Stoppelfelder; auf dem, was im Winter Wiesenland war, kam zwischen den Disteln, der gelbe in der Hitze gesprungene Boden zum Vorschein. Die Bergköpfe im Süden schienen auch mit Gestrüpp bewachsen, im Norden, wo sie an die 1500 m erreichten, begannen die grossen Korkeichenwaldungen, die den Wert des Landes ausmachen. Trotz schöner Konturen wollte sich der Blick auf das Tal und die Bergketten in der trockenen staubigen Luft und bei den kurzen Sonnenuntergängen nie recht verklären.

Viel erfreulicher wurde alles, wenn man ins Einzelne ging. In den feuchteren Bodenfalten lagen hier und da um die Stadt hübsche Farmen, von hohen staubbedeckten Kaktushecken umrahmt. Mitten auf den Distelwiesen vor den Toren hatten sich die Familien der Spahi's in niedrigen grauen Zelten angesiedelt. Die Spahi's sind eingeborene Soldaten, im Dienst französische Bürger, ausser Dienst die alten Nomaden. Zwischen den Zelten weideten Ziegen oder eine Herde von einem halben Dutzend Kameelen frass mit Behagen Disteln und Kaktus. Im Städtchen selbst war immer viel Volk zu sehen, früh und abends in munterer Bewegung, von 11—3 in jedem Schatteneckchen schlafend. Die Vornehmheit der beturbanten Leute, auch derjenigen, die sich nur von Kaktusfeigen und Ziegenmilch nähren, in der vom internationalen Verkehr wenig berührten Landstadt hat uns einen tiefen Eindruck gemacht. Wir wurden nicht angegafft und angebettelt und es ist uns trotz offener Türen nichts gestohlen worden. Das Fremde und Wilde kam erst bei den Feierlichkeiten zum Vorschein, die von den Stadtbehörden den ausländischen Astronomen zu Ehren veranstaltet wurden.

Die *Mozabiten* sind ein Stamm mit viel ureingesessenem afrikanischen Blut, der zugleich als religiöse Sekte einen besonders reinen Islam bewahrt. Sie waren in Guelma die behäbigen Kaufleute und sassen die meiste Zeit in ihren Tuchmagazinen. Nun sahen wir am Tage vor der Finsternis die ganze Bande auf dem Marktplatze, bei ohrenzerreissender Musik, in Pulverdampf gehüllt, abenteuerliche alte Gewehre über dem Kopf schwingend, mitten darin die bleichen schwammigen Gesichter alter Kaufherren, die gewiss die 60 schon überschritten hatten und in aller Würde rasten wie die jüngsten. Abwechselnd von rechts und links stürmte eine Schar an und schoss dicht vor dem Gegner die Gewehre auf den Boden ab, dass der Strassenstaub wider die nackten braunen Beine spritzte. Sie berauschen sich am Pulver, wie wir am Alkohol, sagte ein französischer Offizier. Um 3 fingen sie an und tobten, bis der Sonnenuntergang dem Kampf ein Ende machte. Zum Schluss sprengten die Anführer auf glänzend gezäumten Schimmeln die Strasse hinab. Die schweren scharfkantigen Steigbügel wurden dem Tier gleich als Sporen in den Bauch gesetzt und unten an der Strasse vor dem Hause des Herrn wurde es ebenso plötzlich zusammengerissen und zum Stehen gebracht. Jede Bewegung eine kurze, aber deutliche, nun schon durch Generation vererbte Erinnerung an vergangene Zeiten wilder Freiheit.

Am selben Abend zog wieder der winselnde Ton der arabischen Tuba über die Stadt. Im Hof der Moschee war *Tanz der Aissauahs.* Umringt von allen Schwarzen, Braunen und Weissen, die in Guelma irgend mitzählten, allen Trachten von Frack und Zylinder über Tennisanzug bis zum Turban und Burnus, stand im Fakkelschein eine Kette von etwa 50 meist jüngeren Leuten. Die Arme hatten sie untergefasst. Vor ihnen auf Matten sassen die noch unfertigen Zöglinge und die alten Meister der Sekte, wie uns schien, die rechten mitleidslosen Schächer und Bonzen. Von draussen kam das elektrische Licht über die Hofmauer und glänzte wider die helle Front der Moschee. Die Kette war in unaufhörlicher wiegender Bewegung. Nach dem durchdringenden eintönigen klagenden Gebetsrhythmus der Musik neigten sich die Oberleiber und bei jedem Takt flogen die Köpfe mit einem heftigen Ruck der Halsmuskeln nach vorne. So ging es stundenlang und es war kein Wunder, dass schliesslich jeder in der Kette um einen Teil seiner Besinnung kam. Wer von den Burschen die Hypnose nahe fühlte, zog aufgeregt in der Reihe Weste und Hemd aus und trat in Hose und Gürtel vor. Es begann mit den Nervösen und blöd Aussehenden. Ich sehe noch einen der ersten vor mir, ein kluges dunkles Gesicht mit weit aufgerissenem Munde. Einer der Alten hielt ihm Eisennägel und Glasscherben vor, er fasste sie mit den Zähnen und schluckte sie mit verzweifeltem Ekel hinunter. Andere frassen lebende Skorpione oder stachen sich mit Messern und Schwertern, die ihnen die Alten gaben, in Leib, Schulter, Wange, Augenlider. Blut wurde nicht sichtbar und die Wunden schienen sich gleich zu schliessen. Schien einer Schmerzen zu fühlen, so wurde er von einem Alten untergepackt und durch die rhythmische Halsmassage wieder empfindungslos gemacht. Es muss nach dem, was mir einer der Burschen sagte, ein Zustand sein, wie Angetrunkenheit. Man erkennt noch Bekannte und Umgebung, ist aber unempfindlich und im Nebel. Gegen Mitternacht hatte der Taumel alle ergriffen. Ein ganzer Schwarm kräftiger junger Gestalten fasste brennende Strohwische und spülte mit den Flammen an Brust und Armen entlang. Die Pflichten der morgigen Sonnenfinsternis zogen uns schliesslich von dem unheimlichen Schauspiel hinweg. Die Methode dieser Anästhesie ist eine ausgebildete religiöse Wissenschaft und es ist nicht ausgemacht, ob hier nicht die Anfänge einer medizinischen Kunst liegen. Oder war man selbst hypnotisiert von dem ewig gleichen Schrei der Musik und den roten Flammen, in denen sich die dunklen Leiber bewegten, und meinte Tiefen zu sehen, die nicht vorhanden waren?

↑

Fig. 1. UNSER ZELT IM RÖMISCHEN THEATER

(R. Emden photogr.)

Fig. 2. DAS RÖMISCHE THEATER VON DEM ZELT AUS GESEHEN.

(R. Emden photogr.)

Fig. 3. UNSER BEOBACHTUNGSZELT.
Rechts Runge, in der Mitte unser Diener Ali, links Schwarzschild.
(R. Emden photogr.)

Die Franzosen — insbesondere die Stadtbehörden — begegneten den Astronomen mit ausgezeichneter Liebenswürdigkeit. Der Direktor der Sternwarte zu Algier, Mr. Trépied, und der Stadtsekretär Herr Perrette halfen uns den Beobachtungsplatz aussuchen und stationierten uns schliesslich in dem alten *römischen Amphitheater*. Statt einen freien Platz neben dem Feldlager der Amerikaner vor der Stadt zu wählen, haben wir uns ganz unwissenschaftlich von dem romantischen Gedanken bestechen lassen, auf der einstigen Stätte hoher Kunst den Manen der Vergangenheit mit der Kunst des Beobachtens zu imponieren. Dieses wissenschaftliche Beobachten ist wenigstens eines der Dinge, in denen wir dem Altertume unbedingt überlegen sind. Die tausend Reste römischer Kultur, die uns in Algier und Tunis begegneten, sind im übrigen geeignet, etwaigen Hochmut der Menschen des 20. Jahrhunderts herabzustimmen.

Unsre Instrumente waren rasch aufgebaut. Denn es fanden sich in der kleinen Stadt unerwartet viele geschickte Hilfskräfte. Der Schlosser, der Schreiner, der Maurer, der Photograph, es waren lauter tüchtige Leute, als ob nur bessre Mannschaften auf diesen Aussenposten der europäischen Kultur kommandiert wären. Auf Fig. 1. sieht man die Stellung unsres Zeltes auf der Bühne des Theaters, Fig. 2. zeigt das ehemalige Seitenentree des Theaters, ein gewölbter Gang, der von dem Leiter der Ausgrabungen, Herrn Joly, freundlichst für uns auf beiden Seiten abgeschlossen wurde. Hier bewahrten wir die Hilfsapparate auf und in diese kühle Höhle flüchteten wir in jedem Augenblick, den wir nicht unbedingt draussen in der Sonne sein mussten. Im Zelt (Fig. 3) sieht man uns bei der Prismenkamera beschäftigt, die uns von der Firma Zeiss in Jena für die Finsternis geliehen wurde. Das der Göttinger Sternwarte gehörige eiserne Refraktorstativ, auf dem dieselbe aufsitzt, ist im Jahre 1882 schon bei der Beobachtung des Venusdurchganges auf den Kerguelen gewesen. Prof. Runge zwangen seine luftelektrischen Beobachtungen, sich in der Mitte des Theaters zu postieren. Man muss sich ihn auf Fig. 4. in der weissen Kleidung wie einen einzigen blendenden Sonnenreflex denken, um eine Vorstellung von der Energie zu bekommen, die aus der Zenithsonne niederströmte. Dunkle Brillengläser wurden so heiss, dass ihre Eigenstrahlung das Auge schmerzte.

Der 29. August war ein strahlend blauer Tag und man dachte einen Augenblick: Wäre es doch trüb, dass man die Verantwortung los wäre, wenn irgend etwas an den Instrumenten schief geht! Der Tag begann auch damit, dass die Gummibirne des Momentverschlusses undicht wurde. Zum Glück fand sich noch Ersatz und schliesslich war alles in Ordnung. Ich möchte nun nicht die schwärmerische Schilderung der Finsternis wiederholen, die der Leser schon in allen Variationen in den Zeitungen vor dem 29. August genossen hat, sondern bitte um die Erlaubnis, die ganz persönlichen Empfindungen zu reprodu-

Fig. 4. RUNGE MACHT LUFTELEKTRISCHE MESSUNGEN IM RÖMISCHEN THEATER.
(R. Emden photogr.)

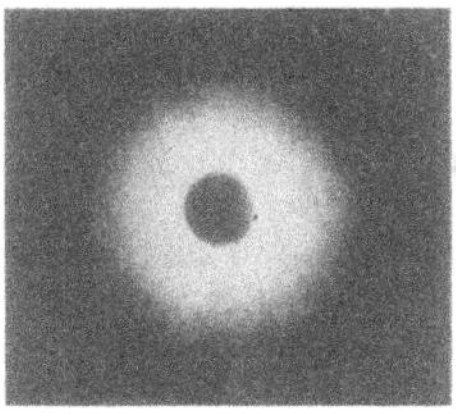

Fig. 5. Die verfinsterte Sonne von der Korona umgeben.

zieren, die dem von der Hitze etwas mitgenommenen und während der $3^1/_2$ Minuten stark beschäftigten Astronomen ins Bewusstsein traten, so, wie ich sie bald nach der Finsternis aufgeschrieben habe. Ich bin überzeugt, dass der glückliche Privatmann, der als blosser Zuschauer einer totalen Sonnenfinsternis beiwohnt, einen viel stärkeren Gefühlseindruck empfangen wird.

Wie ich mit den geladenen photographischen Kasetten im römischen Theater ankomme, hat die partielle Finsternis schon begonnen. Runge sitzt mit weissem Schleier mitten in der Sonne und macht elektrische Messungen. Unsre Hilfsgeister, Mr. Henry, ein Schreiner, ein schwarzer Südfranzose, und die Herren Salfati und Vidal, zwei sechzehnjährige Gymnasiasten, gucken schon durch dunkle Gläser nach der eingekerbten Scheibe. Emden kommt aus der Höhle und wir nehmen unsern Posten am Instrument ein. Wie die Sonne halb verdeckt ist, kann man schon ohne Gefahr Hut und Schleier absetzen. Ich beobachte die abnehmende Sichel im Fernrohr, während die Dunkelheit rasch zunimmt, die Vögel unruhig am Boden flattern und der übliche Wind kommt, und gebe ein Signal, sobald die Sichel nur noch 90° umfasst. Salfati zählt nach dem Metronom 50 Sek. rückwärts. Nach der Rechnung sollte auf Null die Totalität beginnen, ich merke aber sofort, das wir zu früh dran sind, da die Sonne noch blendet, und rufe Emden und dem Schreiner zu: Alles um 20 Sekunden später, was sie trotz meines Französischen auch auffassen. Dann mache ich alle 2 Sekunden auf einem langen Film eine Aufnahme des Spektrums. Runge ruft »flash« (das ist der Anfang der Totalität) — er beobachtet mit einem kleinen Spektroskop — wie Salfati bereits im Vorwärtszählen von Null an bei vingt angekommen ist. Meine Verspätung betrug also wirklich 20 Sekunden. Bei dem Ruf trente höre ich mit den Spektralaufnahmen auf. Emden und der Schreiner machen nun kurze Koronaaufnahmen bis 1 Minute. Ich sehe, dass es unnötig war. Petroleumlampen aufzustellen, weil es viel heller ist, als in einer Vollmondnacht. Von der Stadt kommt Geschrei herüber. Mein geschickter Jüngling Vidal und ich haben inzwischen den Film gegen eine Plattenkasette gewechselt. Zuerst will der Deckel nicht aufgehn — ein böser Moment, weil wir durch zu starkes Ziehen das Instrument erschüttern und den andern alles verderben würden — schliesslich löst er sich. Bei 1 Minute 30 Sekunden machen wir eine kurze Aufnahme des Koronaspektrums, dann wieder bei 1 Minute 50 Sekunden und schliesslich eine lange von 2 Minuten 20 Sekunden bis 3 Minuten. Während der langen Aufnahme hat man endlich Zeit, die Sonne anzusehen — eine kleine schwarze Scheibe, umgeben von einem grünlich-silbernen hellen Ring, der lange Sicheln und Fächer gerade und schräg nach allen Seiten aussendet, die sich dann in dem hellerleuchteten weisslichen Himmelsgrund verlieren. Ich freue mich, etwas sonst immer verdecktes nun enthüllt gesehen zu haben, wundre mich aber nicht, dass die fürchterlich helle Sonne von einer so lebhaften Erscheinung begleitet ist. Es wäre viel verwunderlicher, wenn es um sie herum kahl wäre. Über diesem Gedanken vergesse ich, die Korona durch das bereitstehende Fernröhrchen zu betrachten. Inzwischen erscheint Runge geisterhaft vor dem Zelt, um Emden das kleine Spektroskop zu bringen. Bei 3 Minuten wird wieder die Kasette gewechselt. Bei 3 Minuten 40 Sekunden ruft Emden »flash«. (Ende der Totalität.) Ich reagiere aber nicht gleich darauf, weil die Erscheinung droben unverändert bleibt. Während wir darüber ein paar kurze Worte wechseln, mache ich die letzten Aufnahmen, bis plötzlich rechts unten am Mondrande ein blendender Lichtknoten aufblitzt und alles vorüber ist. —

Die 32 Aufnahmen, die wir in den $3^1/_2$

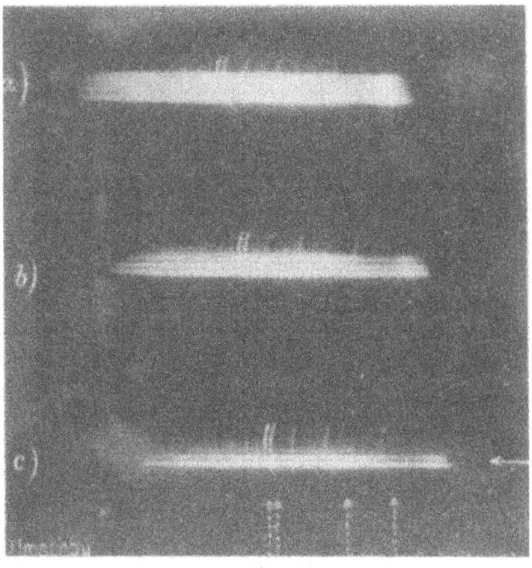

Fig. 6. Unsre Aufnahme des Flashspektrum.

Minuten erhalten haben, sind so gut geraten, als bei unsern kleinen instrumentalen Mitteln erwartet werden konnte. Das hier reproduzierte Bild der Korona zeigt die langen Strahlen, die z. T. sehr stark von der radialen Richtung abweichen. Am besten gelungen sind die Aufnahmen mit der Prismenkamera zu Beginn der Totalität. Man photographiert hier die äusserst schmale Sichel der Sonne, die in den letzten Sekunden übrig bleibt, bevor der Mond die Sonnenscheibe ganz verdeckt, und zwar wird das Licht zu einem Spektrum ausgezogen. Auf Fig. 6 a) sieht man dieses Spektrum der Sichel als breiten Streifen, am einen Ende noch aber weniger heisse und stärker absorbierende Gasatmosphäre hindurchscheint. In diesen Ringen sieht man die von Kirchhoff erschlossene atmosphärische Hülle in ihren einzelnen Bestandteilen leibhaftig vor sich. Hier kommt ihre eigene Leuchtkraft zum Vorschein, da sie nicht mit der übermässigen Helle des eigentlichen Sonnenkerns in Kontrast gesetzt ist. Die Auswüchse der Linien entsprechen Auswüchsen der Atmosphäre, emporgeschleuderten Gasmassen, sog. Protuberanzen. Man sieht bei der Betrachtung unter der Lupe an den Kalciumlinien, dass eine mächtige Kalciumwolke — von drei Erddurchmessern Höhe —

Fig. 7. »Camp Dixie« der Amerikaner. (Nataf photogr.)

von den dunklen der Sichelform entsprechend gekrümmten Fraunhofer'schen Linien durchzogen. Auf dem später aufgenommenen Bild b) ist die Sichel und entsprechend der helle Streifen viel schmäler geworden, auf Bild c) lugt die eigentliche Sonne nur noch an einer Stelle hervor, wo offenbar der Mondrand eine Einsenkung, ein Tal besitzt, und gibt zu dem schmalen Streifen in Richtung des Pfeiles *d* Anlass. Was bedeuten aber die leuchtenden Ringe auf den letzten Bildern? Dieselben entstammen den glühenden Gasmassen, die als eine Art Atmosphäre den eigentlichen Sonnenball umgeben. Jede Sichel entspricht einem bestimmten Gas, so die mit Ca bezeichneten Linien dem Calcium, He dem Helium, H dem Wasserstoff. Kirchhoff hat zuerst die relative Dunkelheit der Fraunhofer'schen Linien durch die Vorstellung erklärt, dass ein äusserst heller Sonnenball durch eine glühende, über einer Stelle der Sonne liegt, dass in derselben Gegend fünf Wasserstoffmassen wie Sprudel in die Höhe getrieben sind. Auch Helium ist in geringerem Masse mit aufwärts gerissen, während die andern — im allgemeinen schwereren — Gase von dem Aufruhr nicht mit betroffen sind. Es sind dies alles schon oft beobachtete Dinge, aber man muss sie selbst gesehen oder photographiert haben, um mit voller Sicherheit an ihrer weiteren Aufklärung arbeiten zu können.

Die andern in Guelma stationierten *Expeditionen*, waren uns grösstenteils an Mitteln weit überlegen. *Die Amerikaner* waren auf dem Kreuzer Dixie herüber gekommen. Sie hausten in einem ganzen Lager (s. Fig. 8) äusserst praktischer Zelte mit doppeltem Dach. Ein grosses Küchenzelt und Messzelt sah nach besserer Verpflegung aus, als uns im Hotel de l'Orient et de l'Univers die dicke französische

Fig. 9. Lange Kamera der Amerikaner für grosse Koronaaufnahmen.
(Nataf photogr.)

Wirtin mit dem Knebelbärtchen gönnte, die aus der Finsternis Kapital schlagen wollte. Die Hauptinstrumente der Amerikaner waren eine lange Kamera für Koronaaufnahmen, ähnlich der von der Hamburger Expedition verwandten (vgl. Umschau 1905 Nr. 35), und ausserdem die in Fig. 9 abgebildete »Polarachse«, die von einem Uhrwerk in einfacher Weise der Rotation des Himmels nachgedreht wurde und die eine ganze Reihe der verschiedensten Spektographen trug. Einer weniger ausgedehnten, aber äusserst fein durchdachten Anordnung hatte Herr Newall aus Cambridge sich bedient. (Bild Nr. 10.) Aus seinen Einrichtungen sprach die Erfahrung von vier Sonnenfinsternissen, denen er in den letzten Jahren bis nach Sumatra und Indien nachgegangen ist. Seine Frau — die einzige, aber umso edlere Vertreterin der Weiblichkeit unter den Astronomen in Guelma — hat ihn überall begleitet. Drei französische Expeditionen unter Herrn Trépied aus Algier, Herrn Bourget aus Toulouse und Herrn Stephan aus Marseille, sowie die schweizerische des Herrn Wolfer

Fig. 9. Polaraxe der Amerikaner mit vielen angesetzten Apparaten.
(Nataf photogr.)

aus Zürich waren ebenfalls mit hübschen Instrumenten ausgestattet.

Was die *Resultate der Beobachtungen* angeht, so kann man schon jetzt sagen, dass eine etwas bestimmtere Fixierung unserer Anschauungen über das äusserst komplexe Phänomen der *Korona* daraus hervorgehen wird. Die Korona entspringt zum Teil dem Sonnenlicht, das an allem möglichen fein verteiltem Staub in der Nachbarschaft der Sonne reflektiert wird, zum andern Teil aber einem auf der Erde noch *unbekannten Gas*, dem sog. *Koronium*, das sich schwach leuchtend noch weit höher über die eigentliche Sonnenatmosphäre erhebt, als die höchsten Protuberanzen aus Kalcium oder Wasserstoff. Die langen Ausläufer der Korona schrieb man den Strömen von diskretem äusserst feinem Staub zu, die durch irgend eine Sonneneruption oder auch einfach durch den Strahlungsdruck der Sonne fortgeschleudert werden. Eine Spektralaufnahme, die uns der Hauptastronom der amerikanischen Expedition, Herr Jewell zeigte, gab aber besonders deutlich zu erkennen, dass auch das Koronium-Gas selbst an dieser strahlenförmigen Struktur, an den Eruptionen teil nimmt, für die man dann auf ganz andere Kräfte angewiesen ist, als wenn es sich nur um das Fortschleudern von Staub handelt. Herr Newall hat eigentümliche Polarisationserscheinungen an der Korona erhalten, und wir alle haben wohl auf unseren Platten noch manche Merkwürdigkeiten mit nach Hause gebracht, die sich erst bei näherem Studium enthüllen.

Fig. 10. DIE INSTALLATION VON MR. NEWALL.
(R. Emden photogr.)

Wenn wir jetzt aus der Heimat an Guelma zurückdenken, so schwebt zwar zu Zeiten immer noch jener kleine Silberring um die dunkle Scheibe über uns am Himmel, aber da das Menschliche auf den Menschen doch immer am stärksten wirkt, so denken wir uns noch lieber unter humorvolle Engländer, höfliche Franzosen und würdige Araber zurückversetzt und starren noch immer auf das rätselhafte bunte Bild dieser orientalischen Kultur mit ihrer äusseren Rückständigkeit und inneren Höhe. Unter den Naturphänomenen bleibt aber auch noch ein nachhaltiger Eindruck. Als man nach Deutschland zurückkam, war es, als ob man in ein vom letzten Regen noch nicht getrocknetes nasskaltes Gewand kröche. Der wochenlang verhangene Himmel wirkte geradezu niederdrückend gegen all die unabänderlich helle südliche Bläue. Und hier soll der Astronom die Sterne beobachten, durch die vereinzelten Löcher der Nebeldecke, die kaum den 5. Teil des Jahres den Durchblick gestatten? Wenn die moderne Technik irgend etwas erreicht hat, so ist es, dass sie uns die Erde klein gemacht hat und dass wir uns für jeden Zweck den geeignetsten Punkt der Kugel aussuchen können. So entspringt für den Astronomen aus dieser Reise zur Sonnenfinsternis vor allem der Wunsch nach *einer Sternwarte unter dem günstigen südlichen Himmel.* Dort sollten wir die Gestirne beobachten und photographieren; zum nachdenklichen theoretischen Arbeiten kann man immer leicht genug in das kühle, Energie spendende Vaterland zurückkehren.

Schumburg: Über Kontaktinfektion.

Der bekannte Hygieniker Oberstabsarzt Prof. Dr. Schumburg, der sich besonders durch seine Forschungen über Wasserdesinfektion für militärische Zwecke verdient gemacht hat veröffentlicht in der »Zeitschr. f. ärztl. Fortbildung« einen Aufsatz, dessen Inhalt weiteste Beachtung verdient und den wir hier auszugsweise wiedergeben.

»Dem jeweiligen Stande der Wissenschaft entsprechend«, sagt Schumburg: »waren es im-

Hiernach zeigen die mitgeteilten Messungen, daß die Ionisierung der Luft in dem geschlossenen Zylinder aus Aluminiumblech sich um etwa 28 Proz. verminderte, wenn er von der Erdoberfläche in einen von Steinsalz in dicken Schichten umgebenen Raum gebracht wurde. Eine einfachere Deutung des Resultats, als durch eine Schirmwirkung des Steinsalzes gegen von außen kommende ionisierende Strahlen erscheint uns, soweit wir die Versuchsbedingungen übersehen können, nicht möglich.

Wenn auch die Gegenwart einer schwachen Becquerelstrahlung wegen der allgemeinen Verbreitung radioaktiver Stoffe überall auf der Erde zu erwarten ist, so ist ihr direkter Nachweis doch so anziehend und von so hohem geophysikalischen Interesse, daß es sich lohnte, Beobachtungen, wie die soeben beschriebenen, auch an anderen günstigen Orten anzustellen. Trotz der Umstände, die mit solchen Arbeiten am Grunde von Bergwerken verbunden sind, möchten wir sie sehr empfehlen, als Ergänzung der Laboratoriumsversuche mittels absorbierender Mäntel von verschiedenem Materiale. Wir müssen gestehen, daß es uns nach letzterer Methode noch nicht gelungen ist, konstante schlagende Unterschiede in der Ionisierung der im Versuchsraume eingeschlossenen Luft zu erhalten, allerdings wurden die Versuche bis jetzt nur mit ziemlich dünnen Hüllen aus Blei- und Zinkblech ausgeführt.

Eine Frage war bei den Untersuchungen im Bergwerke noch von Interesse, ob nämlich in diesen Räumen ebenfalls jene radioaktive Emanation nachweisbar wäre, die sich sonst in der atmosphärischen Luft verbreitet findet. Der oben nach seinen ungefähren Dimensionen angegebene Raum, in dem wir die Beobachtungen anstellten, stieß an einen größeren von gleicher Höhe, 45 m Länge und 12—30 m Breite, letzterer stand nur durch enge, meist mit Türen verschlossene Gänge mit den in Betrieb befindlichen Strecken in Verbindung. Eine Erneuerung der Luft konnte sich, zumal bei dem Fehlen größerer Temperaturdifferenzen, jedenfalls nur langsam vollziehen. Etwa von außen hinein diffundierende Emanation hätte ihre Aktivität voraussichtlich während des Eindringens verlieren müssen, wie ja auch die Luft in nicht ventilierten Räumen der höheren Stockwerke eines Hauses nur sehr geringen Emanationsgehalt zu zeigen pflegt.

Wir waren daher überrascht, als die in bekannter Weise auf einem negativ geladenen Drahte niedergeschlagene Aktivität sich als wohl meßbar ergab, die sogenannte Aktivierungszahl[1]) war 8,5, während für die freie Atmosphäre bei Wolfenbüttel Zahlen gefunden sind, die zwischen 4 und 64 (Mittel 18,6) liegen.

Man ist also auch innerhalb eines Steinsalzlagers nicht völlig frei von der Gegenwart radioaktiver Substanzen und folglich auch nicht von den durch sie erzeugten Strahlen. Diese Erfahrung stimmt übrigens damit überein, daß die von den Solquellen unserer Gegend, die ihren Salzgehalt wohl der Auslaugung von Salzlagern verdanken, abgesetzten Sedimente radioaktive Emanation, und zwar die des Thoriums, entwickeln. Wahrscheinlich wird diese letztere daher auch von den festen Salzmassen im Innern des Bergwerks in äußerst geringen Mengen abgegeben.

Schließlich möchten wir nicht unterlassen, den Direktionen der Vienenburger und Hedwigsburger Kaliwerke, insbesondere den Herren Direktor Schröder und Obersteiger Großmann zu Hedwigsburg, auch hier unsern Dank für freundliches Entgegenkommen und bereitwillige Hilfeleistungen auszusprechen.

Wolfenbüttel, Oktober 1905.

(Eingegangen 14. Oktober 1905.)

1) Vgl. diese Zeitschr. 4, 90, 1902.

Über die Helligkeitsverteilung des ultravioletten Lichtes auf der Sonnenscheibe.

Von K. Schwarzschild und W. Villiger.

Die Abnahme der Helligkeit auf der Sonnenscheibe von der Mitte nach dem Rande ist öfters Gegenstand der Untersuchung gewesen. Aus den letzten Jahrzehnten liegen namentlich drei genauere Messungsreihen vor. Herr H. C. Vogel[1]) hat im Jahre 1871 aus Aufnahmen auf Chlorsilberpapier die Helligkeitsabnahme für die Gesamtheit der photographisch wirksamen Strahlen bestimmt, im Jahre 1877[2]) hat er mit einem Spektralphotometer für eine Reihe von Wellenlängen im sichtbaren Spektrum visuell beobachtet. Herr Very[3]) hat neuerdings die Helligkeitsverteilung durch das sichtbare Spektrum hindurch und darüber hinaus ins Ultrarot bis zu 1,5 μ mit dem Bolometer verfolgt. Die gegenwärtige Untersuchung dehnt das Wellenlängenbereich nach dem Ultraviolett zu bis zur Wellenlänge 0,32 μ aus. Die benutzte Methode war die folgende: In dem Schottschen Glaswerke zu Jena werden seit einiger Zeit Glassorten (U. V. Glas) hergestellt, die in Schichten von mehreren Zentimetern Dicke gute Durchlässigkeit für ultraviolettes Licht bis zu einer Wellen-

1) Sitzungsberichte der Berliner Akademie. 1871.
2) Ebenda. 1877.
3) Astrophysical Journal Vol. 16, 1902.

länge von 0,30 μ herab aufweisen. Auf der anderen Seite ist bekannt, daß dünne Silberschichten, die für längere Wellenlänge völlig undurchsichtig und die besten Spiegel sind, etwa in der Gegend von 0,34 μ ziemlich plötzlich ihre Reflexionsfähigkeit verlieren und durchsichtig werden. Herr Prof. R. Straubel kam auf den Gedanken, einen Lichtfilter für ultraviolettes Licht herzustellen, dadurch, daß er solches ultraviolett-durchlässiges Glas mit einer dünnen Silberschicht überzog.

Gerade für die beabsichtigten Sonnenaufnahmen liefert dieser Gedanke ein außerordentlich einfaches Verfahren, indem nichts weiter nötig war, als das zur Aufnahme dienende Objektiv aus U. V. Glas auf einer oder mehreren — in unserem Falle zwei — Flächen zu versilbern. Um sich zu überzeugen, welche Wellenlängen von dem so entstehenden Lichtfilter durchgelassen wurden, hat Herr Dr. Henker freundlichst einige Aufnahmen des Eisenspektrums mit Hilfe dieses versilberten Objektivs gemacht. Es ergab sich, daß nur ein schmaler Spektralstreif von den Wellenlängen 0,320 bis 0,325 μ durchgelassen wurde, und daß derselbe sich bei starker Überexposition nur auf 0,315 bis 0,327 μ ausdehnte.

Das Objektiv hatte 100 mm freie Öffnung, 3 m Brennweite und wurde mit dem Tragrohr auf den Refraktor der Versuchssternwarte des Zeißwerkes in Jena aufgesetzt. Vor der Platte wurde ein Momentverschluß angebracht. Bei einer Belichtungsdauer von ca. $^1/_{50}$ Sekunden ergaben sich sehr scharfe normal exponierte Sonnenbilder.

Das photometrische Prinzip bestand in der Verwendung von Sektoren-Blenden vor dem Objektiv, die bei der Messung von Flächenhelligkeiten völlig einwandsfrei ist. Es wurden Abblendungen auf $^1/_2$ und $^1/_4$ durch je 3 Sektorenausschnitte von 60° resp. 30° Winkelöffnung verwandt.

Die ersten Aufnahmen wurden bereits vor 2 Jahren gewonnen, doch war eine Reihe von kleinen Hindernissen zu überwinden, bevor brauchbare Ergebnisse erzielt wurden. Anfangs waren 3 und 4 Flächen des Objektivs versilbert, was eine längere Expositionsdauer erforderlich machte, die für die Schärfe der Sonnenbilder ungünstig war. Ferner zeigte sich, daß die Geschwindigkeit des Schlitzverschlusses sich während der Bewegung des Verschlusses über die Platte änderte. Es konnten daher nur Messungen verwandt werden, die sich auf einen Sonnendurchmesser senkrecht zur Bewegung des Verschlusses bezogen. Die Geschwindigkeit des Verschlusses von Aufnahme zu Aufnahme scheint, wie aus den unten folgenden Ergebnissen hervorgeht, eine hinreichend konstante gewesen zu sein. Um den Einflüssen der Absorption der Erdatmosphäre möglichst zu entgehen, erfolgten die Aufnahmen immer dicht beim Meridian, die Bewegung des Verschlusses war eine vertikale, und der Sonnendurchmesser, auf welchem die Verteilung der Helligkeit gemessen wurde, war der horizontale.

Auf jeder einzelnen Platte im Format von 9:12 cm wurden 3 Aufnahmen bei voller Öffnung resp. bei Abblendung auf $^1/_2$ und $^1/_4$ gemacht.

Um ferner den Einflüssen einer verschiedenen Durchlässigkeit der Silberschicht oder des Glases in den verschiedenen Teilen des Objektivs auszuweichen, wurden die Aufnahmen immer in 3 verschiedenen Stellungen der Sektorenblenden wiederholt. Auch wurde die Reihenfolge der Bilder auf der Platte gewechselt, sodaß das bei voller Öffnung exponierte Bild einmal in die Mitte und einmal an den Rand der Platte zu stehen kam, da eine Zeitlang der Verdacht bestand, daß die Randpartien der Platten bei gleicher Belichtung eine etwas stärkere Schwärzung ergäben. Diese Vorsicht wurde beibehalten, obwohl sich jener Verdacht nicht allgemein bestätigt hat.

Die Beobachtungsserie bestand daher schließlich aus 6 Platten mit je 3 Aufnahmen, wobei Blendenstellung und Reihenfolge der Bilder in der unten ersichtlichen Weise abwechselten.

Die Ausmessung der Platten erfolgte auf der Göttinger Sternwarte unter dem Hartmannschen Mikrophotometer, bei welchem die Schwärzung jeder Stelle der Platte mit der Schwärzung eines photographischen Keils verglichen und durch die in mm ausgedrückte Länge des eingeschobenen Keilstücks numerisch bezeichnet wird. Unter das Mikrophotometer war ein Meßapparat von Spindler & Hoyer eingesetzt. Dieser Apparat gestattet mit Sicherheit die Koordinaten der photometrierten Stelle auf $^1/_{100}$ Millimeter abzulesen; der photometrierte Fleck selbst hat in Richtung des zu verfolgenden Sonnendurchmessers einen Durchmesser von 0,15 mm, senkrecht dazu einen Durchmesser von 0,25 mm. Die Schwärzung einer Stelle, die 0.07 mm, das ist bei der Größe der Sonnenbilder 0,005 des Radius, vom Rand entfernt liegt, läßt sich daher noch photometrieren, wenn auch die Sicherheit infolge des raschen Abfalls der Sonnenhelligkeit nach dem Rande zu nur etwa bis zu einer Entfernung von 0,15 mm vom Rande die normale bleibt.

Es wurde zunächst immer auf den Nord- und Südrand der Sonne unter dem Meßapparat eingestellt (die schmale Seite der Platte gab mit genügender Genauigkeit die Nord-Südrichtung an). Das Mittel aus beiden Ablesungen ergab die eine Koordinate des zu verfolgenden Sonnendurchmessers. Dann erfolgte die Einstellung des Ost- und Westrandes der Sonne. Das Mittel aus den beiden Einstellungen ergab die Mitte des Sonnenbildes, und von dieser Mittelstellung aus wurde um die in der unten folgenden Tabelle angegebenen Strecken nach Ost und West weitergegangen. Die Messungen wurden von Herrn Dr. Br. Meyermann und von Herrn E. Jastram ausgeführt.

Die folgende Tabelle gibt die Resultate der Messungen für 2 Serien von Aufnahmen. Unter ϱ ist der Abstand vom Mittelpunkte des Sonnenbildes in Millimetern verstanden; die darauffolgenden Zahlen selbst bezeichnen die Einstellung des Photometerkeiles in Millimetern.

Tabelle I. I. Serie. 28. Mai 1905.

ϱ	Platte 66. Blendenstellung I						Platte 67. Blendenstellung II						Platte 68. Blendenstellung III					
	1		1/2		1/4		1		1/2		1/4		1		1/2		1/4	
	Ost	West	Ost	West	Ost	West	Ost	West	Ost	West	Ost	West	Ost	West	Ost	West	Ost	West
0,0	68,9		64,7		60,2		69,3		65,9		61,4		69,8		65,2		60,7	
4,0	68,7	68,8	63,8	64,7	60,2	60,3	69,3	69,3	66,0	65,8	60,8	60,6	69,9	69,5	64,7	64,4	60,0	60,6
8,0	67,9	67,7	63,2	64,1	59,0	59,3	68,5	69,3	65,1	65,1	60,0	60,1	69,3	68,0	63,8	63,9	59,8	60,0
11,0	67,1	66,7	62,1	63,1	57,3	57,8	66,9	67,5	63,2	63,9	59,0	58,2	67,3	67,6	62,5	62,9	58,2	58,3
13,0	—	65,1	60,3	60,9	55,1	56,6	65,1	65,7	60,8	62,0	56,3	56,6	65,6	65,9	60,8	61,4	56,2	56,4
14,0	—	—	—	—	—	—	63,4	64,1	60,0	60,3	54,3	54,6	64,5	64,0	59,1	59,1	54,3	55,0
14,5	62,3	61,4	56,1	57,2	51,4	52,3	62,0	62,5	57,1	56,4	52,0	52,3	62,6	61,8	56,7	57,2	52,7	53,1
14,7	—	—	—	—	—	—	60,9	61,4	55,3	57,0	50,8	50,5	61,8	60,9	56,1	55,5	51,1	52,1
14,8	60,6	59,4	54,1	55,2	48,9	51,0	60,3	60,3	54,7	56,2	49,5	49,1	61,0	60,0	55,1	55,2	49 9	50,7
14,9	59,7	59,7	52,0	54,2	—	49,2	59,6	59,2	55,8	54,8	—	48,0	60,0	59,4	53,8	54,9	48,4	50,3
Durchmesser:	30,00		29,95		29,81		30,03		29,94		29,78		30,07		29,90		29,90	

ϱ	Platte 69. Blendenstellung III						Platte 70. Blendenstellung II						Platte 71. Blendenstellung I					
	1/2		1		1/4		1/2		1		1/4		1/2		1		1/4	
	Ost	West	Ost	West	Ost	West	Ost	West	Ost	West	Ost	West	Ost	West	Ost	West	Ost	West
0,0	66,5		72,1		63,1		66,1		70,2		62,2		64,4		69,5		59,7	
4,0	66,4	66,5	71,6	71,3	63,0	63,1	66,0	66,4	70,0	70,5	62,4	62,1	64,4	65,0	68,4	68,9	59,1	59,4
8,0	66,0	65,2	70,8	70,2	62,1	62,0	65,3	65,7	68,9	69,0	61,4	60,7	63,1	63,3	68,5	68,5	58,7	58,7
11,0	64,2	64,3	69,3	69,2	60,2	60,8	64,1	64,3	67,6	68,3	59,7	59,9	61,9	62,7	66,0	67,3	57,5	57,2
13,0	63,0	62,5	68,8	68,1	58,1	58,8	62,1	63,0	65,9	67,0	58,0	57,3	60,2	60,4	64,2	65,8	55,6	55,1
14,0	61,1	60,4	67,1	66,1	56,3	56,3	59,8	61,2	64,4	65,1	55,0	55,8	58,1	58,8	62,8	64,0	54,0	54,0
14,5	59,3	58,2	65,2	64,5	54,2	54,8	58,0	59,2	62,7	63,5	54,2	53,9	56,3	57,2	60,3	62,0	52,3	51,4
14,7	58,0	57,5	64,2	63,0	52,9	53,1	57,0	58,2	61,6	62,1	53,2	53,0	55,5	56,1	59,5	61,2	51,0	50,0
14,8	57,1	56,6	63,8	62,2	52,4	51,9	56,0	57,5	61,0	61,5	52,0	51,0	54,9	55,1	59,0	60,4	50,0	49,0
14,9	—	55,4	63,6	62,2	50,9	51,1	55,1	56,6	60,0	60,4	50,3	49,8	53,3	54,2	58,7	59,6	—	—
Durchmesser:	29,95		29,93		29,96		29,98		30,00		29,96		30,00		29,90		29,82	

II. Serie. 27. Juni 1905.

ϱ	Platte 75. Blendenstellung I						Platte 76. Blendenstellung II						Platte 77. Blendenstellung III					
	1/4		1/2		1		1/4		1/2		1		1/4		1/2		1	
	Ost	West	Ost	West	Ost	West	Ost	West	Ost	West	Ost	West	Ost	West	Ost	West	Ost	West
0,0	53,1		58,6		62,6		52,4		57,1		60,3		54,3		58,3		62,6	
4,0	53,4	53,0	57,0	57,6	61,7	62,0	52,0	51,9	56,5	56,7	59,8	60,1	53,7	54,1	57,6	57,9	61,7	61,3
8,0	52,1	52,6	56,2	56,9	61,3	61,1	50,7	51,4	56,0	55,9	58,7	59,5	52,6	53,2	57,2	57,2	61,1	60,6
11,0	51,2	50,6	55,6	55,5	60,9	60,1	50,0	50,0	53,8	54,9	57,4	58,6	51,1	52,2	56,0	55,6	60,1	59,5
13,0	49,3	49,3	53,6	54,3	60,1	58,3	47,8	48,0	52,5	53,2	56,2	56,6	49,6	50,0	54,5	54,1	58,3	57,6
14,0	47,0	47,6	50,9	51,9	57,5	56,8	45,7	46,9	50,8	51,1	54,0	55,4	47,7	48,3	52,2	52,4	56,3	56,1
14,5	45,2	45,4	49,8	50,3	56,2	54,6	45,1	45,1	48,5	49,8	53,0	53,4	46,1	46,9	49,8	50,1	54,6	54,1
14,7	43,6	45,1	48,2	49,2	55,1	53,2	43,3	43,7	47,3	48,2	52,4	52,0	45,0	45,1	49,2	49,3	54,3	53,5
14,8	43,3	43,4	47,6	48,2	54,7	52,5	42,6	43,0	46,9	48,0	51,5	51,8	43,4	43,6	47,6	48,3	53,1	52,2
14,9	41,3	40,5	(45,0)	47,4	54,4	52,3	41,9	41,5	46,4	47,5	51,3	52,5	—	—	—	—	—	—
Durchmesser:	29,76		29,82		29,82		29,70		29,81		29,85		29,80		29,85		29,80	

With W. Villiger in Phys. Zeitschr. *6* (1905) 737–744

ϱ	Platte 78. Blendenstellung III						Platte 79. Blendenstellung II						Platte 80. Blendenstellung I					
	1/2		1		1/4		1/2		1		1/4		1/2		1		1/4	
	Ost	West	Ost	West	Ost	West	Ost	West	Ost	West	Ost	West	Ost	West	Ost	West	Ost	West
0,0	60,0		65,1		55,1		61,3		66,5		56,2		60,1		63,3		52,6	
4,0	59,5	59,1	64,3	64,2	54,3	54,4	60,1	60,8	65,7	65,9	55,9	56,1	59,5	59,6	62,9	63,0	52,4	52,3
8,0	58,3	58,2	62,4	63,0	53,1	53,3	59,6	59,6	65,2	65,1	55,1	55,3	58,2	58,7	61,9	62,1	51,1	51,5
11,0	56,9	56,8	61,0	61,7	51,7	51,8	58,3	58,8	64,1	63,8	53,2	53,2	56,8	57,0	60,8	60,4	49,8	49,9
13,0	54,8	54,7	59,7	60,5	50,1	49,6	56,9	57,1	62,1	62,1	52,5	52,1	55,0	55,7	58,6	58,4	47,6	48,1
14,0	52,9	52,4	57,6	57,9	47,2	48,3	54,3	54,6	60,0	60,1	49,1	49,8	53,1	53,0	56,8	57,0	45,3	46,9
14,5	51,0	50,4	56,3	55,9	45,2	46,0	52,1	52,8	57,6	57,3	47,1	48,3	50,0	51,1	54,4	55,2	43,3	45,4
14,7	49,9	49,3	55,1	54,6	43,4	44,5	50,0	51,1	55,4	56,1	45,8	47,5	49,1	50,1	53,1	54,0	42,6	44,4
14,8	49,3	47,9	54,1	53,7	42,1	44,0	48,5	50,5	54,1	55,3	44,4	46,3	47,8	49,5	51,5	53,9	41,6	43,7
14,9	48,7	—	52,6	51,3	—	43,6	47,0	49,5	52,9	54,5	43,3	46,1	—	49,4	49,1	52,3	—	—
Durchmesser:	29,82		29,88		29,73		29,84		29,78		29,82		29,82		29,80		29,72	

Sie sind das Mittel aus zwei unmittelbar aufeinander folgenden Einstellungen. Es schien nicht erforderlich, die beiden Einzeleinstellungen anzugeben, weil sie stets bis auf wenige Zehntel Millimeter übereinstimmen. Größere Millimeterzahlen bedeuten stärkere Schwärzungen. Unter den Photometereinstellungen jedes Bildes ist der gemessene Durchmesser des Bildes in Millimetern angegeben.

Die weitere Aufgabe besteht darin, die Beziehung zwischen den Photometerablesungen und den Lichtintensitäten, denen die Platte ausgesetzt gewesen ist, zu ermitteln.

Es wurde zunächst das Mittel für gleich weit von der Sonnenmitte nach Ost und West entfernte Punkte gebildet. Die so erhaltenen Werte wurden wieder gemittelt für je drei zusammengehörige Platten, die bei den drei verschiedenen Blendenstellungen exponiert worden waren. Auf diese Weise entstanden die folgenden Zahlenreihen.

Die Abblendung auf 1/2 bedeutet also einen Lichtverlust von 2,5 *log* 1/2 oder 0,753 Größenklassen. Wir wollen nun die Helligkeiten im Sonnenbilde für einen bestimmten Abstand von der Mitte mit m_1, m_2, m_3 bezeichnen, je nachdem es sich um die Abblendung 1, 1/2 oder 1/4 handelt. Die entsprechenden Schwärzungen (Photometerablesungen) seien S_1, S_2, S_3.

Dann ist

$$m_1 = m_2 - 0{,}735, \quad m_3 = m_2 + 0{,}735.$$

Da die Expositionszeit bei allen Aufnahmen dieselbe war, so ist die Schwärzung einfach eine Funktion der Helligkeit: $S = \varphi(m)$. Es gelten daher die Gleichungen:

$$S_1 = \varphi(m_2 - 0{,}735), \quad S_2 = \varphi(m_2),$$
$$S_3 = \varphi(m_2 + 0{,}735)$$

oder durch Umkehrung:

$$m_2 = \psi(S_1) + 0{,}735 = \psi(S_2) = \psi(S_3) - 0{,}735.$$

Wäre die Abhängigkeit der Schwärzung von der Größenklasse eine lineare, so würde folgen:

$$S = a + b\,m, \quad S_1 - S_2 = 0{,}735\,b = S_2 - S_3,$$

Tabelle II.

ϱ	Platte 66—68			Platte 69—71			Platte 75—77			Platte 78—80		
	1	1/2	1/4	1	1/2	1/4	1	1/2	1/4	1	1/2	1/4
0,0	69,3	65,3	60,8	70,6	65,7	61,7	61,8	58,0	53,3	65,0	60,5	54,6
4,0	69,3	65,0	60,4	70,2	65,8	61,6	61,1	57,2	53,0	64,0	59,8	54,3
8,0	68,5	63,8	59,7	69,3	64,8	60,6	60,4	56,6	52,1	63,3	58,8	53,2
11,0	67,2	63,0	58,0	68,0	63,6	59,2	59,4	55,3	50,9	62,0	57,5	51,6
13,0	65,4	61,0	56,2	66,7	61,9	57,2	57,9	53,7	49,0	60,2	55,7	50,0
14,0	64,0	59,3	54,4	64,8	59,9	55,2	56,0	51,6	47,2	58,3	53,4	47,8
14,5	62,1	57,2	52,3	63,1	58,1	53,5	54,3	49,8	45,6	56,1	50,9	45,9
14,7	61,1	55,8	51,0	62,0	57,0	52,2	53,4	48,6	44,3	54,8	49,9	44,7
14,8	60,3	55,1	49,9	61,3	56,2	51,7	52,7	47,8	43,2	53,8	48,9	43,7
14,9	59,6	54,3	—	60,8	55,0	—	52,3	46,8	41,7	52,1	—	—
Durchmesser	30,03	29,93	29,83	29,95	29,97	29,91	29,82	29,83	29,75	29,84	29,80	29,76

Bei der weiteren Bearbeitung wollen wir alle Helligkeitsangaben in astronomischen Größenklassen m machen.

Es besteht zwischen m und der Lichtintensität J die bekannte Definitionsgleichung:

$$m = -2{,}5\,log\,J.$$

der Unterschied der Schwärzungen verschiedener Aufnahmen müßte konstant sein. Das wirkliche Verhalten dieser Differenzen zeigt die folgende Tabelle:

Gleichheit der Differenzen $S_1 - S_2$ und $S_2 - S_3$ wird man nicht allzu strenge erwarten

Tabelle III.

ϱ	Platte 66—68 S_1-S_2	S_2-S_3	Platte 69—71 S_1-S_2	S_2-S_3	Platte 75—77 S_1-S_2	S_2-S_3	Platte 78—80 S_1-S_2	S_2-S_3
0,0	4,0	4,5	4,9	4,0	3,8	4,7	4,5	5,9
4,0	4,3	4,6	4,4	4,2	3,9	4,2	4,6	5,5
8,0	4,5	4,1	4,5	4,2	3,8	4,5	4,5	5,6
11,0	4,2	5,0	4,4	4,4	4,1	4,4	4,5	5,9
13,0	4,4	4,8	4,8	4,7	4,2	4,7	4,5	5,7
14,0	4,7	4,9	4,9	4,7	4,4	4,4	4,9	5,6
14,5	4,9	4,9	5,0	4,6	4,5	4,2	5,2	5,0
14,7	5,3	4,8	5,0	4,8	4,8	4,3	4,9	5,2
14,8	5,2	5,2	5,1	4,5	4,9	—	4,9	5,2
14,9	5,3	—	5,8	—	5,5	—	—	—

dürfen, da ein Wechsel der Verschlußgeschwindigkeit und der Luftdurchsichtigkeit hierauf von Einfluß ist, auch wird man die Messungen unmittelbar am Sonnenrande ($\varrho = 14,8$ und $14,9$) hier besser von der Berücksichtigung ausschließen. Man sieht dann, daß die einzelnen Vertikalreihen eine angenäherte Konstanz aufweisen. Es ist dies eine Folge davon, daß der Photometerkeil mit exponentiell fortschreitender Belichtung hergestellt ist.[1])

Man kann nun versuchen, jeder gemessenen Schwärzung S eine solche Korrektion $\chi(S)$, welche nur eine Funktion der Schwärzung S selbst ist, hinzuzufügen, daß die angenäherte Konstanz der Differenzen in eine vollständige übergeht und die so korrigierten Schwärzungszahlen dann den Größenklassen strenge proportional sind. Mathematisch gefaßt bedeutet dies, daß man die Funktion $\psi(S)$ in der Form

$$\psi(S) = \alpha + \beta[S + \chi(S)]$$

schreibt und $S + \chi(S)$ gleich der korrigierten Schwärzung S' setzt, sodaß:

$$m_2 = \psi(S) = \alpha + \beta S'$$

wird. Die Abweichungen von der linearen Beziehung zwischen Schwärzungszahlen und Größenklassen können so viel verschiedene Ursachen haben (Plattensorte, Entwicklung, Beschaffenheit des Photometerkeils) und so verschieden ausfallen, daß es keinen Zweck hat, dieselben immer durch dasselbe Formelschema ausdrücken zu wollen, vielmehr kann man bei der geringen Größe der hier auftretenden Abweichungen durch unmittelbare Beobachtung des Ganges der Differenzen S_1-S_2 und S_2-S_3 und geringes numerisches Tatonieren die gewünschte Konstanz zuwege bringen. So sieht man, daß für die beiden ersten Plattengruppen ein mit der Schwärzung etwa lineares Abnehmen der Differenzen erfolgt. Einem linearen Gang der Differenzen entspricht ein quadratischer der Schwärzungen selbst. Es wurde daher eine quadratische Korrektion der Schwärzungen zu ermitteln gesucht und durch numerischen Versuch erhalten:

1) Vgl. K. Schwarzschild, Beiträge zur photographischen Photometrie der Gestirne. Publikationen der v. Kuffnerschen Sternwarte, Bd. V, Wien 1900.

$$\begin{aligned} &\text{Platte 66—68: } \chi(S) = +0{,}0085\,(S - 60{,}0)^2 \\ &\text{Platte 69—71: } \chi(S) = +0{,}0055\,(S - 60{,}0)^2. \end{aligned} \quad (1)$$

Ähnlich beobachtet man bei der dritten Plattengruppe 75—77, daß nur die Differenzen S_1-S_2 einen Gang zeigen, während die Differenzen S_2-S_3 konstant bleiben. Für die geringeren bei S_2 und S_3 auftretenden Schwärzungen ist also die lineare Beziehung an sich erfüllt und der Gang in den Differenzen S_1-S_2 muß von dem Verhalten der starken in S_1 auftretenden Schwärzungen herrühren. Daher wurde nur an den starken Schwärzungen eine Korrektion angebracht und gesetzt:

$$\begin{aligned} &\text{Platte 75—77: } \chi(S) = 0 \text{ für } S < 58{,}0, \\ &\chi(S) = 0{,}04\,(S - 58{,}0)^2 \text{ für } S > 58{,}0. \end{aligned} \quad (2)$$

Bei der letzten Gruppe (78—80) schließlich bemerkt man ein Ansteigen in der Reihe S_1-S_2, hingegen ein Absinken in S_2-S_3. Es waren daher sowohl die stärksten als die schwächsten Schwärzungen zu korrigieren und es wurde gesetzt: Platte 78—80:

$$\begin{aligned} &\chi(S) = 0{,}1\,(S - 60{,}0, \text{ für } S > 60{,}0, \\ &\chi(S) = 0 \qquad \text{für } 50{,}0 < S < 60{,}0, \\ &\chi(S) = 0{,}1\,(S - 50{,}0) \text{ für } S < 50{,}0. \end{aligned} \quad (3)$$

Wie weit durch diese Korrektionen wirkliche Konstanz der Differenzen erzielt ist, geht aus der folgenden Tabelle hervor, die die Differenzen der korrigierten Schwärzungen S' enthält:

Tabelle IV.

ϱ	Platte 66—68		Platte 69—71		Platte 75—77		Platte 78—80	
0,0	4,5	4,7	5,3	4,2	4,4	4,7	5,0	5,9
4,0	4,8	4,8	4,8	4,4	4,3	4,2	5,0	5,5
8,0	5,0	4,2	4,9	4,3	4,0	4,5	4,8	5,6
11,0	4,6	5,0	4,7	4,5	4,2	4,4	4,7	5,9
13,0	4,6	4,7	5,0	4,7	4,2	4,7	4,5	5,7
14,0	4,7	4,6	5,0	4,6	4,4	4,4	4,9	5,8
14,5	4,9	4,5	5,0	4,4	4,5	4,2	5,2	5,4
14,7	5,2	4,3	5,0	4,5	4,8	4,3	4,9	5,7
Mittel	4,81	4,60	4,96	4,45	4,35	4,43	4,87	5,69

Unter jeder Spalte ist das Mittel der darüber stehenden Werte (unter Ausschluß der Randwerte für $\varrho = 14,8$ und $14,9$) angegeben. Man sieht, daß die Abweichungen vom Mittel rein zufälliger Natur sind.

Diese Mittelwerte ergeben nun aber auch

den Proportionalitätsfaktor β zwischen den korrigierten Schwärzungen und den Größenklassen, den Skalenwert. Der Unterschied der ersten und dritten Aufnahme beträgt in Größenklassen, stets $2 \cdot 0^m,7525$ oder $1^m,505$. In der korrigierten — den Größenklassen proportionalen — Schwärzungsskala beträgt dieser Unterschied für die Platten 66—68 z. B. 4,81 + 4,60 Skalenteile. Der Skalenwert wird also: $\frac{1^m,505}{9,41} = 0^m,160$. Auf diese Weise ergeben sich für die vier Gruppen die folgenden Skalenwerte:

Platte 66—68: $0^m,160$,
„ 69—71: $0^m,160$,
„ 75—77: $0^m,171$,
„ 78—80: $0^m,143$.

Hiermit sind alle zur Reduktion der Aufnahmen nötigen Daten gegeben. Es wurden nunmehr die sämtlichen Schwärzungen in Tabelle I, nachdem die obigen Korrektionen $\chi(S)$ angebracht waren, von der ersten Zahl in jeder Spalte, welche der Helligkeit im Sonnenzentrum entspricht, subtrahiert und die Differenzen mit dem betreffenden Skalenwert multipliziert. Die entstehenden Zahlen geben unmittelbar die Abnahme der Sonnenhelligkeit gegen die Mitte in Größenklassen an. Indem gleich das Mittel aus den drei Werten für jede Gruppe genommen wurde, ergaben sich folgende vier Zahlenreihen:

Tabelle V.

ϱ	66—68	69—71	75—77	78—80
0,0	$0^m,00$	$0^m,00$	$0^m,00$	$0^m,00$
4,0	0 ,04	0 ,02	0 ,11	0 ,08
8,0	0 ,19	0 ,18	0 ,25	0 ,24
11,0	0 ,40	0 ,40	0 ,46	0 ,44
13,0	0 ,72	0 ,68	0 ,75	0 ,70
14,0	0 ,97	0 ,99	1 ,08	1 ,02
14,5	1 ,29	1 ,27	1 ,37	1 ,34
14,7	1 ,47	1 ,45	1 ,57	1 ,51
14,8	1 ,60	1 ,58	1 ,72	1 ,67
14,9	1 ,72	1 ,69	1 ,88	1 ,90

Um dieselben völlig miteinander vergleichbar zu machen, ist noch auf gleiche Entfernung vom Mittelpunkt in Teilen des Sonnenradius zu reduzieren. Aus den Durchmessern am Fuße von Tab. II sieht man, daß derselbe für die zweite Serie von Aufnahmen etwas kleiner gewesen ist, als für die erste Serie, wohl infolge einer Temperaturdehnung des Fernrohrs. Man bemerkt ferner, daß das schwächste Bild im Durchschnitt um 0,07 mm oder 4″ kleiner gemessen worden ist, als die sehr nahe gleichgroßen helleren Bilder. In der Tat hat man den Eindruck, daß bei der Abblendung des Sonnenbildes auf $^1/_4$ der Sonnenrand noch nicht ganz ausexponiert ist, während das auf $^1/_2$ abgeblendete Bild am Rande gerade die mittlere Schwärzung aufzuweisen scheint, bei der noch keine merkliche Irradiation in der Schicht zu fürchten ist. Unter dem Mikroskop erscheint bei diesen Aufnahmen der Rand auf 0,01 mm scharf definiert und unmittelbar außerhalb der Sonnenscheibe sind die Platten fast völlig glasklar. Es wurde daher der Durchmesser des auf $^1/_2$ abgeblendeten Sonnenbildes als der wahre Durchmesser betrachtet und im Mittel als Wert desselben für die erste Serie 29,95 mm und für die zweite Serie 29,82 angesetzt. Die photometrierten Stellen haben daher folgende Entfernung x vom Sonnenmittelpunkt in Teilen des Sonnenradius:

	ϱ	0,0	4,0	8,0	11,0	13,0
1. Serie .	x	0,000	0,267	0,534	0,735	0,868
2. Serie .	x	0,000	0,269	0,536	0,739	0,872
	ϱ	14,0	14,5	14,7	14,8	14,9
1. Serie .	x	0,935	0,968	0,982	0,988	0,995
2. Serie .	x	0,940	0,973	0,987	0,993	1,000

Wenn man nun die Zahlen in Tabelle V für die beiden letzten Gruppen interpoliert auf die Distanzen x, welche den Messungen für die beiden ersten Gruppen entsprechen, so erhält man die folgenden vier Reihen, deren Mittel als das Endresultat dieser Untersuchung anzusehen ist. Dabei gibt die letzte Spalte die aus dem Größenklassenunterschied folgende Intensität mit der Intensität J_0 im Zentrum der Sonne als Einheit.

Tabelle VI.

Abstand v. Mittelpunkt	66—68	69—71	75—77	78—80	Mittel	$\frac{J}{J_0}$
0,000	$0^m,00$	$0^m,00$	$0^m,00$	$0^m,00$	$0^m,00$	1,000
0,267	0 ,04	0 ,02	0 ,11	0 ,08	0 ,06	0,946
0,534	0 ,19	0 ,18	0 ,25	0 ,24	0 ,21	0,824
0,735	0 ,40	0 ,40	0 ,46	0 ,44	0 ,43	0,673
0,868	0 ,72	0 ,68	0 ,74	0 ,69	0 ,71	0,520
0,935	0 ,97	0 ,99	1 ,05	0 ,99	1 ,00	0,398
0,968	1 ,29	1 ,27	1 ,32	1 ,29	1 ,29	0,305
0,982	1 ,47	1 ,45	1 ,49	1 ,44	1 ,46	0,261
0,988	1 ,60	1 ,58	1 ,60	1 ,54	1 ,58	0,233
0,995	1 ,72	1 ,69	1 ,77	1 ,73	1 ,73	0,203

Fragt man nach der Zuverlässigkeit dieser Werte, so würde man aus der Genauigkeit der Doppeleinstellung unter dem Hartmannschen Mikrophotometer von etwa 2 Proz. in Rücksicht darauf, daß hier schließlich die Messungen an 36 Aufnahmen vereinigt sind, auf einen mittleren Fehler von $0^m,003$ zu schließen haben. Man sieht auf den ersten Blick, daß diese Einstellungsgenauigkeit durch systematische Unterschiede in der Schwärzung der Platten illusorisch gemacht wird. Vergleicht man z. B. die in Tabelle I unmittelbar nebeneinander stehenden Zahlen für die Ost- und Westseite der Sonne, so findet man ein systematisches Überwiegen bald der einen, bald der andern Seite. Da sich im Mittel aus allen Aufnahmen fast völlige Symmetrie der Helligkeitsverteilung der Sonne auf beide Hälften ergibt, so haben diese Differenzen mit der Sonne selbst nichts zu tun, sondern nehmen in den Bedingungen der Aufnahme oder im photographischen Prozeß ihren Ursprung. Die Einflüsse, die für die einzelne Aufnahme systematisch wirken, scheinen aber von Platte zu Platte rein zufälliger Natur zu sein. Wenigstens ist uns nicht gelungen, irgendeine Gesetzmäßigkeit in den Unterschieden der Ost- und Westhälfte der Sonne oder auch der Mittelwerte $S_1 - S_2$ und $S_2 - S_3$ voneinander aufzufinden. Quantitativ sind übrigens die systematischen Einflüsse nicht gar zu bedeutend, indem sie selten $0^m,10$ zu überschreiten scheinen. Bei den starken Unterschieden der Schwärzungen dicht am Ost- und Westrande ist die Schwierigkeit der Messung und die Möglichkeit ungenauer Einstellung der Koordinate des Randes mit in Betracht zu ziehen.

Man wird unter diesen Umständen ein Urteil über die Zuverlässigkeit der obigen Zahlen am besten aus den Unterschieden der vier Reihen in Tabelle VI selbst entnehmen und darnach die Sicherheit der Mittelwerte auf wenige Hundertstel der Größenklasse veranschlagen. Bei den größeren Helligkeitsdifferenzen am Sonnenrande kommt die Unsicherheit des Skalenwertes hinzu, deren Einfluß hier im schlimmsten Falle das Zehntel der Größenklasse erreichen mag. Es sei noch hervorgehoben, daß eine Änderung der Korrektionen $\chi(S)$ innerhalb plausibler Grenzen die Werte der Tabelle VI fast gar nicht berührt, da schon bei jeder einzelnen Platte das Mittel aus drei weit über die Schwärzungsskala verteilten Schwärzungen genommen ist.

Im ganzen ist wohl dieselbe Sicherheit erreicht, wie sie die Vogelschen Messungen im sichtbaren Spektrum besitzen. Hingegen ist die Helligkeitsabnahme weiter nach dem Rande zu verfolgt, als bei den früheren Beobachtern. Während Very bei 0,95, Vogel bei 0,97 des Sonnenradius stehen bleibt, liegt kein Grund vor, warum man die hier für $x = 0{,}982$ und 0,988 gewonnenen Zahlen nicht für einen reellen Ausdruck der Sonnenhelligkeit in diesen Abständen vom Mittelpunkt halten sollte. Das gleiche wird man freilich nicht von dem Werte für $x = 0{,}995$ behaupten dürfen, wo schon Irradiation in der Schicht und Abbildungsfehler des Objektivs vom Betrage einiger Bogensekunden in Betracht kommen.

Vergleichen wir nun die Ergebnisse, die wir für $\lambda = 0{,}320\ \mu$ gewonnen haben, mit den Resultaten für andere Wellenlängen. In der folgenden Figur sind die Werte $\mathfrak{J}/\mathfrak{J}_0$ für $\lambda = 0{,}320\ \mu$ zusammengestellt mit den Kurven, die Herr Vogel für Rot und Violett erhalten hat. Das merkwürdige Ergebnis, das diesen Kurven zu entnehmen ist, lautet: Während beim Übergang von Rot zu Violett der Abfall der Sonnenhelligkeit durchweg ein stärkerer wird, ist im Ultraviolett der Abfall von

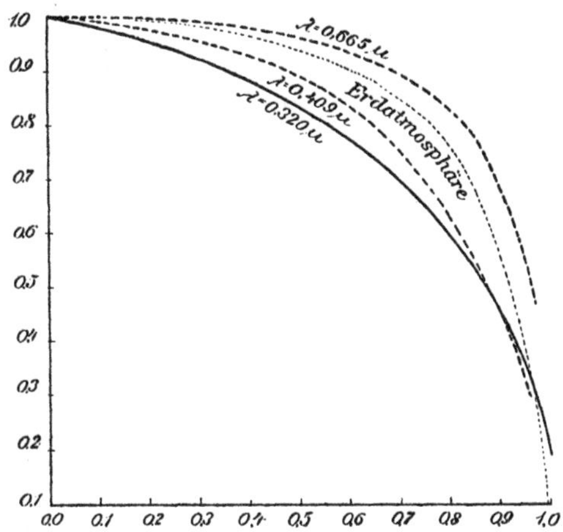

der Sonnenmitte weg zwar zunächst auch noch stärker als im Violett, er verlangsamt sich aber bei der Annäherung an den Rand, sodaß die Abschwächung dicht am Rande schließlich für das Ultraviolett noch geringer wird als für das Violett.

Dies Verhalten widerspricht völlig der Analogie mit der Erdatmosphäre, deren Absorption bekanntlich außerordentlich rasch nach dem Ultraviolett zu zunimmt. Es tritt dies als neues Faktum hinzu zu der bereits von Herrn Vogel bemerkten Tatsache, daß die Form des Abfalls ebenfalls nicht den in der Erdatmosphäre geltenden Gesetzen entspricht. Die punktierte Kurve bezeichnet die Intensitätsverteilung, welche die Erde von außen gesehen für violettes Licht aufweisen würde, wenn ihre Oberfläche ein selbstleuchtender schwarzer Körper wäre (berechnet mit der doppelten Potsdamer Extinktion). Der abweichende Charakter des Abfalls ist offenbar, zumal für das Ultraviolett.

Es liegen zwei Erklärungsversuche für die besondere Form der Kurven vor. Seeliger[1]) schreibt der Sonnenatmosphäre am Boden einen viel stärker von 1 verschiedenen Brechungsexponenten zu, als ihn die Erdatmosphäre besitzt; Schuster[2]) berücksichtigt die Eigenstrahlung der Sonnenatmosphäre. Für beide Fälle ist indessen die Theorie nicht weit genug entwickelt, um eine unmittelbare Vergleichung mit unseren Messungen in der Nähe des Sonnenrandes zu gestatten. Es soll daher die theoretische Bearbeitung der gewonnenen Zahlen einer andern Gelegenheit vorbehalten bleiben.

Wir beabsichtigen, die Messungen fortzusetzen, um über eine etwaige Veränderlichkeit

1) Sitzungsber. der bayr. Akademie der Wissenschaften Bd. 21, München 1891.
2) Astrophysical Journal 16, 1902.

der Absorption der Sonnenatmosphäre Aufschluß zu erhalten.

Göttingen und Jena, September 1905.

(Eingegangen 8. Oktober 1905.)

Vielfach-Telephonie mittels des Telegraphons.

Bemerkung zu einer Arbeit des Herrn J. W. Giltay.

Von Ernst Ruhmer.

In dieser Zeitschrift **6**, 572, 1905 ist eine Betrachtung des Herrn J. W. Giltay über die Möglichkeit einer Vielfachtelephonie mittels des Poulsenschen Telegraphons veröffentlicht, in welcher Herr Giltay zu dem Schluß gelangt, daß eine getreue Wiedergabe mehrerer Gespräche nach der von Pedersen, einem Mitarbeiter Poulsens, angegebenen Anordnung nicht möglich sei.

Zunächst möchte ich richtig stellen, daß Pedersen nicht zwei Hufeisenelektromagneten mit ihren Polverbindungslinien in der Bewegungsrichtung des Stahlbandes gestellt verwendet, wie solches Herr Giltay annimmt, sondern zwei Paar Hufeisenelektromagneten (8 Eisenstäbchen mit Spulen), die mit den Verbindungslinien der zusammengehörigen Elektromagnetpole quer zur Bewegungsrichtung des Stahlbandes stehen, genau so wie ich es in Figur 3 meiner Veröffentlichung im Jahrgang 2 S. 29 (1900) dieser Zeitschrift angegeben habe.

In Fig. 4 derselben Veröffentlichung, welche die Pedersensche Anordnung schematisch in einer Seitenansicht darstellt, sind die hinteren Schenkel der Elektromagneten durch die vorderen verdeckt zu denken. Auch der Übertragungsmagnet und der Löschmagnet ist ein querstehender Doppelmagnet.

Dieser Unterschied zwischen der Pedersenschen Anordnung (mit Quermagnetisierung) und der Giltayschen Annahme (einer Längsmagnetisierung) ist indessen rein technischer Natur und hat selbstverständlich auf die theoretischen Schlüsse des Herrn Giltay keinerlei Einfluß. Auch bei seiner vereinfachten Annahme kommt ja Herr Giltay zu dem Ergebnis, daß das gleichzeitige Aufschreiben zweier Gespräche, sowie deren exakte Übertragung auf das Stahlband der Empfangsstation wohl möglich erscheine, er bezweifelt nur die Möglichkeit der getrennten Wiedergabe in den Telephonen der Empfangsstation.

Da das Stahlband der Empfangsstation aber eine genaue telegraphonische Kopie der magnetischen Schrift des Bandes der Sendestation trägt, so müßte auch an der Sendestation eine getrennte Wiedergabe der beiden magnetischen Fixierungen unmöglich sein.

Herr Giltay legt bei seinen Betrachtungen der Einfachheit wegen nur eine einzige magnetische Fixierung zugrunde und nimmt dann wohl stillschweigend an, daß sich die dabei ergebenden Abweichungen bei jeder einzelnen weiteren Fixierung wiederholen.

Tatsächlich ist der Übelstand bei einer längeren, einem Sprachlaut entsprechenden Gruppe magnetischer Fixierungen nicht so schlimm. Solange sich nämlich die Abhörelektromagneten-Anordnung innerhalb dieser Gruppe befindet, treten keine störenden Induktionsstöße auf, nur beim Eintreten und Verlassen einer Gruppe machen sich dieselben in schwacher Weise geltend und bewirken eine kaum merkliche Verbreiterung des Anfangs und Ende jedes Sprachlautes.

In der Praxis spielen diese Verzerrungen keine bedeutende Rolle, wie aus Versuchen mit einem Telegraphon-Apparat hervorging, der s. Z. zur Demonstration der Pedersenschen Multiplex-Telephonie gebaut wurde und der gestattete zwei auf demselben Stahldraht aufgenommene Gespräche je nach der Schaltung der Elektromagnetenpaare, getrennt gut verständlich wiederzugeben.

Berlin, Physikalisches Laboratorium, 18. September 1905.

(Eingegangen 20. September 1905.)

Zur Technik des Foucaultschen Pendelversuchs.

Von U. Behn.

Im vorigen Jahre diskutierte Herr L. Weinhold[1]) auf Grund eingehender Versuche die Störungen, denen der Foucaultsche Pendelversuch ausgesetzt ist. Er kommt zu dem Schluß, daß die schlimmste in den Luftströmungen besteht, die besonders in hohen Räumen, wie sie für den genannten Versuch benutzt werden, wohl stets vorhanden sind. Dies Resultat kann ich auf Grund wiederholter Versuche vollauf bestätigen. Da es nun aber, wie auch Herr Weinhold betont, wünschenswert ist, den Versuch, wenn man ihn überhaupt anstellt, auch quantitativ einigermaßen korrekt vorzuführen, so entsteht zunächst die Frage, wie man dem genannten Fehler am besten begegnet. Herr Weinhold empfiehlt, die Zugverhältnisse möglichst zu verbessern und die Richtung des stärksten Zuges zur Anfangsrichtung der Pendelschwingungen zu wählen.

In dieser Zeitschrift hat kürzlich Herr Spies[2]) einen andern Weg vorgeschlagen. Er beschränkt

1) L. Weinhold, Zeitschr. f. d. phys. u. chem. Unterr. 17, 198, 1904.

2) P. Spies, diese Zeitschr. **6**, 345, 1905.

ON THE DISTRIBUTION OF BRIGHTNESS OF THE ULTRA-VIOLET LIGHT ON THE SUN'S DISK[1]

By K. SCHWARZSCHILD and W. VILLIGER

I

The decrease of brightness from the center toward the edge on the Sun's disk has frequently been a subject of investigation. From recent decades three precise series of measurements are available. In 1871 H. C. Vogel,[2] by making exposures on silver chloride paper, determined the decrease of brightness for all of the photographically active rays; in 1877[3] he observed visually with the spectral photometer a series of wave-lengths in the visible spectrum. Very[4] later investigated the distribution of brightness throughout the visible spectrum and on into the infra-red to λ 1.5 μ, with the bolometer.

The present investigation extends the range of wave-lengths toward the ultra-violet as far as λ 0.32. The method employed was the following: In Schott's Glass Works at Jena kinds of glass known as "ultra-violet glass" have been manufactured for some time, which show, in layers of several centimeters thickness, good transparency for the ultra-violet light from λ 0.30 downward. On the other side, it is known that thin layers of silver, which are entirely opaque to the longer wave-lengths and which are the best mirrors, lose their power of reflection quite suddenly in the region of about 0.34 μ, and become transparent. Professor R. Straubel conceived the idea of making a light-filter for ultra-violet light by covering such ultra-violet glass with a thin film of silver.

For just such solar photographs as we had in mind, this idea furnished an exceedingly simple procedure, in that nothing further was necessary than to silver the objective of ultra-violet glass, used in the observation, on one or more surfaces, in our case two. In order to be certain which wave-lengths were transmitted by the light-filter

[1] A preliminary communication, based upon a smaller number of plates, appeared in the *Physikalische Zeitschrift*, 6, 737–744, 1905.

[2] *Sitzungsberichte der Berliner Akademie*, 1871.

[3] *Ibid.*, 1877.

[4] *Astrophysical Journal*, 16, 73–91, 1902.

thus produced, Dr. Henker very kindly made several exposures to the iron spectrum with the help of this silvered objective. The result was that only a narrow strip of the spectrum, from wave-length 0.320 to 0.325 μ, was transmitted; and even with a decided over-exposure this only extended from 0.315 to 0.327 μ.

The objective was of 120 mm clear aperture, of 325 cm focus, and was attached with its tube to the refractor of the experimental observatory of the Zeiss Works at Jena. In front of the plate an instantaneous shutter was fastened. With an exposure of about $\frac{1}{50}$ second, very sharp, normally exposed pictures of the Sun were obtained.

The photometric principle consisted in the use of sector diaphragms before the objective, which principle is entirely free from objection in measurements of surface brightness. The light was diaphragmed down to $\frac{1}{2}$ and $\frac{1}{4}$ by three sector openings each of either 60° or 30° angle.

The first photographs were taken two years ago, but there was a series of minor obstacles to be overcome before usable results could be obtained. At first three and four surfaces of the objective were silvered, making a long exposure time necessary which was unfavorable for the sharpness of the image of the Sun. Further, it appeared that the speed of the shutter changed during its passage over the plate. Hence only those measures could be employed which were referred to a diameter of the Sun perpendicular to the motion of the shutter. The speed of the shutter from one exposure to another seems to be sufficiently constant, as is shown in the results below. In order to avoid the effects of absorption of the Earth's atmosphere as much as possible, the observations were taken close to the meridian, the motion of the shutter was vertical, and it was upon the horizontal diameter of the Sun along which the distribution of brightness was measured.

On each plate, of size 9×12 cm, three exposures were made, with full aperture, and with the diaphragms $\frac{1}{2}$ and $\frac{1}{4}$.

In order further to weaken the effect of the different transparency of the silver film, or of the glass in different parts of the objective, the exposures were always repeated in three different positions of the sector diaphragm. The sequence of the images on the plate was also changed, so that the picture exposed with full aperture was sometimes in the center and sometimes on the edge of the plate, since for a little while we had a suspicion that the parts around the edge of the plate were a little more strongly blackened with the same illumination. This precaution was observed later, although the suspicion was not confirmed in general.

A series of observations therefore finally consisted of six plates with three exposures each, in which the position of the diaphragms and the sequence of the pictures were changed in the manner indicated below.

The plates were measured at the Observatory at Göttingen with the Hartmann micro-photometer, with which the blackening of each part of the plate is compared with the blackening of a photographic wedge, and is numerically indicated in millimeters registering the amount of motion of the wedge. A measuring apparatus by Spindler and Hoyer was placed under the micro-photometer. This apparatus made it possible to read with certainty to $\frac{1}{100}$ mm the co-ordinates of the places photometrically measured. The measured spot itself had a diameter of 0.15 mm in the direction of the solar diameter under investigation; perpendicular to that, a diameter of 0.25 mm. It is possible to measure the blackening of a spot which is 0.07 mm distant from the edge, which for the size of the solar image is 0.005 of the radius, although on account of the rapid decrease of light toward the edge only a normal amount of accuracy can be obtained up to a distance of 0.15 mm from the edge.

Settings were first made under the measuring machine on the north and south edges of the Sun (the short side of the plate indicated the north and south directions with sufficient accuracy). The mean of the two readings gave the one co-ordinate of the solar diameter to be followed. Settings were then made on the east and west limbs of the Sun. The mean of these two settings gave the center of the Sun's image, and starting from this central position, settings were carried out at the distances given in the following table from east to west. The measurements were made by Herr E. Jastram.

After a great part of the measurements was completed, a source of error was noticed which occurs with Hartmann's micro-photometer if photometric measurements are made upon a very opaque portion of a plate in the immediate neighborhood of a larger area which is

With W. Villiger in Astrophys. J. *23* (1906) 284–305

very slightly opaque. The very bright light which then passes through the unblackened area is in part reflected at the objective of the observing microscope, and illuminates from in front the dark portion of the plate, so that it appears too bright. In the case of our measurements a very disturbing effect of this error could be noticed near the Sun's limb. It falsified the intensity deduced from the observations by 5 per cent. if the objective was clean, and by from 30 to 50 per cent. if dust had settled upon the objective so that it diffusely reflected still more light. The effect of the error was overcome by always covering the whole plate during the measurement by a black disk laid upon the plate and containing a small circular aperture only slightly larger than the spot to be measured. After the detection of this error, all the measures were repeated with the use of such a disk.

The following table gives the results of the new measurements for the five series of plates obtained last year. The second half of the last series had to be rejected, since the plates had received stray light. Subsequently plate No. 90 was also rejected, since the image obtained with the diaphragm $\frac{1}{2}$ differed too little from the image with the diaphragm $\frac{1}{4}$, and obviously the result was not the mean between the images with diaphragms 1 and $\frac{1}{4}$. In the following table ρ denotes the distance in millimeters from the center of the Sun's disk. The numbers following indicate the settings of the photometer wedge in millimeters. They are the means of two settings, one immediately following the other. It did not seem necessary to give both of the single settings, as they agreed always within a few tenths of a millimeter. Large readings in millimeters signify stronger degrees of blackening. The measured diameter of the image in millimeters is given below the photometer settings of every image.

TABLE I

Series I, May 28, 1905

ρ	Plate 66. Position I of Diaphragm						Plate 67. Position II of Diaphragm						Plate 68. Position III of Diaphragm					
	¼		½		1		¼		½		1		¼		½		1	
	East	West	East	West	East	West	East	West	East	West	East	West	East	West	East	West	East	West
0.0....	52.4		57.4		62.1		53.2		58.8		62.5		52.6		57.3		62.7	
4.0....	52.1	52.1	57.2	56.9	61.8	61.6	52.9	52.1	58.7	58.1	62.1	62.3	52.2	52.1	57.2	56.8	62.6	62.3
8.0....	51.7	50.6	56.1	55.5	61.1	60.8	51.6	51.5	57.7	57.6	61.6	61.2	51.5	51.1	56.9	55.9	61.6	61.6
11.0....	48.8	48.7	54.6	54.3	60.1	60.1	49.4	49.1	55.9	55.1	61.1	60.1	49.2	48.7	54.7	54.6	60.2	60.1
13.0....	46.9	44.6	52.8	51.1	57.7	57.3	47.3	47.3	54.1	52.9	59.2	58.2	46.9	46.8	52.9	52.3	57.9	58.1
14.0....	43.9	41.8	50.5	49.1	56.4	55.2	42.1	42.6	51.5	50.3	57.3	56.1	42.3	42.9	50.1	50.1	56.8	55.4
14.5....	41.4	39.4	47.7	46.5	54.6	54.6	38.9	38.9	49.1	48.6	55.6	54.2	39.9	39.1	48.2	47.5	54.7	54.9
14.7....	37.6	37.3	46.1	44.6	53.3	53.7	35.2	37.6	47.2	46.8	53.6	53.7	38.1	37.5	45.8	46.1	53.2	53.7
14.8....	37.1	31.5	45.9	42.3	52.4	52.2	30.5	35.1	46.3	43.6	53.1	53.1	36.5	33.8	45.6	43.6	52.4	52.8
14.9....	30.1	—	42.8	40.8	50.5	49.6	—	26.1	42.1	39.6	47.5	52.2	33.1	28.1	43.2	38.3	50.1	50.9
15.0....	—	—	29.5	27.6	46.1	45.1	—	—	—	21.1	44.1	47.1	—	—	—	—	42.5	42.3
Diameter..	29.96		29.98		30.01		29.83		29.93		30.00		29.94		29.90		30.06	

ρ	Plate 69. Position III of Diaphragm						Plate 70. Position II of Diaphragm						Plate 71. Position I of Diaphragm					
	½		1		¼		½		1		¼		½		1		¼	
	East	West	East	West	East	West	East	West	East	West	East	West	East	West	East	West	East	West
0.0....	60.1		65.7		56.1		59.1		63.1		53.8		56.8		61.7		51.4	
4.0....	59.3	59.4	65.5	64.9	55.1	55.6	58.2	58.2	62.4	62.8	53.2	53.5	56.1	56.5	61.1	61.3	51.1	51.1
8.0....	58.9	58.9	64.3	64.1	53.8	54.2	57.7	57.6	61.7	62.5	52.6	52.3	55.4	55.7	60.2	61.1	50.3	50.4
11.0....	57.2	57.1	63.1	62.9	51.7	52.3	56.3	56.4	60.8	61.1	51.2	51.1	54.1	54.6	59.1	60.5	48.7	48.1
13.0....	55.6	55.5	61.4	61.5	49.1	49.9	53.8	54.6	58.6	59.6	47.9	48.1	51.2	52.8	57.3	58.3	44.9	45.2
14.0....	54.1	53.1	60.3	59.6	46.4	47.5	51.9	52.1	57.3	57.3	46.7	45.7	49.7	50.3	55.1	56.2	42.3	42.3
14.5....	51.4	50.8	59.1	57.8	44.9	45.1	50.1	50.4	55.2	55.9	43.6	42.6	47.1	48.5	53.6	54.6	40.6	39.1
14.7....	49.1	49.2	57.8	56.9	40.7	42.5	48.5	48.5	54.6	54.3	40.9	39.7	44.6	47.1	52.3	53.1	39.0	37.5
14.8....	48.1	47.6	56.8	54.6	39.4	40.8	47.3	47.3	53.7	53.6	40.6	38.1	44.1	44.3	51.6	51.4	35.1	34.7
14.9....	47.3	45.1	55.8	50.6	29.5	37.1	44.3	42.1	51.9	50.7	37.7	31.6	41.8	41.5	50.1	50.4	30.1	27.1
15.0....	41.1	32.1	47.2	43.4	—	23.5	37.6	23.8	42.9	43.1	—	—	28.9	32.1	42.9	42.3	—	—
Diameter..	29.90		29.90		29.90		29.95		29.95		29.97		29.95		29.96		29.96	

TABLE I—*Continued*

SERIES II, JUNE 27, 1905

ρ	Plate 75. Position I of Diaphragm						Plate 76. Position II of Diaphragm						Plate 77. Position III of Diaphragm					
	¼		½		1		¼		½		1		¼		½		1	
	East	West	East	West	East	West	East	West	East	West	East	West	East	West	East	West	East	West
0.0....	52.6		56.9		61.5		50.8		56.1		60.6		52.6		56.9		60.9	
4.0....	52.4	52.3	56.5	56.3	61.2	61.1	50.1	50.1	55.5	55.6	59.0	59.1	52.5	52.1	56.4	56.2	60.7	60.6
8.0....	52.1	50.9	55.8	55.7	60.5	60.8	49.1	48.7	55.1	54.6	58.6	57.9	51.1	51.4	56.1	55.6	60.1	60.2
11.0....	49.7	50.1	54.6	54.3	58.9	60.1	47.7	47.9	53.2	53.2	58.1	56.8	49.7	49.6	55.6	54.6	58.1	59.1
13.0....	48.2	48.2	52.6	52.6	56.9	58.4	46.1	46.1	52.8	51.7	55.7	55.5	48.2	47.2	52.9	52.9	56.4	56.8
14.0....	46.4	46.1	50.7	50.1	54.7	56.2	43.7	44.1	49.5	49.2	54.1	53.2	45.5	46.3	50.8	50.3	54.7	55.6
14.5....	45.2	43.6	49.2	48.7	53.2	54.8	42.5	41.5	47.6	47.8	52.9	51.6	43.7	43.9	48.4	48.8	53.3	53.3
14.7....	42.3	42.2	47.5	47.9	51.4	54.2	41.3	40.7	46.5	46.4	51.7	50.7	41.9	42.6	47.6	47.3	51.6	52.7
14.8....	41.7	41.3	46.3	47.1	50.1	53.1	40.8	39.9	45.1	45.2	50.9	49.5	41.7	41.7	46.1	46.2	51.1	51.4
14.9....	39.6	39.5	46.1	45.1	47.8	51.1	38.1	37.9	43.1	41.2	47.1	45.8	38.1	40.2	43.1	43.4	47.6	46.5
Diameter..	29.75		29.85		29.85		29.77		29.80		29.82		29.85		29.80		29.85	

ρ	Plate 78. Position III of Diaphragm						Plate 79. Position II of Diaphragm						Plate 80. Position I of Diaphragm					
	½		1		¼		½		1		¼		½		1		¼	
	East	West	East	West	East	West	East	West	East	West	East	West	East	West	East	West	East	West
0.0....	58.3		63.9		53.2		59.7		65.1		54.2		58.1		61.8		51.5	
4.0....	58.1	58.0	63.5	62.8	52.9	53.1	59.8	59.6	64.4	64.9	53.7	53.7	57.9	57.8	61.5	61.9	51.2	50.1
8.0....	57.5	56.5	62.6	62.1	52.2	52.4	58.5	58.5	63.8	63.7	53.2	53.1	57.5	56.6	61.2	60.8	50.1	49.5
11.0....	55.7	55.7	60.7	60.7	51.1	50.8	57.5	57.1	62.5	62.6	51.6	51.8	56.1	55.5	59.9	59.7	48.5	48.4
13.0....	53.2	53.8	59.5	59.1	48.6	48.5	54.8	55.3	60.7	60.8	49.5	49.6	53.2	54.2	57.2	57.2	46.2	46.7
14.0....	51.8	51.5	57.1	56.8	45.8	46.1	52.8	53.2	58.6	59.1	47.3	47.7	51.4	51.8	55.5	56.7	44.1	45.5
14.5....	50.1	49.7	55.7	54.7	44.6	44.6	51.6	50.8	56.8	56.8	45.4	45.4	49.6	49.2	53.2	53.3	42.6	42.8
14.7....	49.6	48.2	54.6	53.5	42.6	42.2	50.1	49.1	54.6	55.5	43.1	43.7	47.7	47.6	52.3	52.6	40.7	41.7
14.8....	47.1	47.2	53.5	52.5	40.9	40.4	48.1	47.7	54.1	54.7	41.9	42.5	47.5	46.4	51.2	51.3	40.2	41.1
14.9....	43.6	43.2	49.3	49.7	38.5	38.2	42.6	43.4	47.6	52.1	39.1	37.5	43.2	42.8	48.9	48.4	37.6	38.1
Diameter..	29.80		29.85		29.79		29.80		29.85		29.85		29.87		29.83		29.82	

TABLE I—*Continued*

SERIES III, JULY 26, 1905

ρ	Plate 87. Position I of Diaphragm						Plate 88. Position II of Diaphragm						Plate 89. Position III of Diaphragm					
	¼		½		1		¼		½		1		¼		½		1	
	East	West	East	West	East	West	East	West	East	West	East	West	East	West	East	West	East	West
0.0....	52.1		59.6		65.3		51.8		58.8		64.4		50.5		56.1		60.7	
4.0....	51.9	51.7	58.9	59.2	64.6	64.5	51.5	51.5	58.2	57.9	64.1	63.8	50.1	49.9	55.8	55.6	60.1	60.1
8.0....	49.6	49.7	58.5	58.3	63.9	63.8	50.8	49.7	57.6	56.7	63.5	63.2	49.1	48.7	54.6	55.2	59.5	59.9
11.0....	48.1	47.8	56.3	56.6	63.2	62.7	48.2	46.7	55.9	55.1	62.6	61.2	46.4	46.2	53.5	53.2	57.8	58.8
13.0....	44.5	45.2	53.2	53.5	60.8	60.2	45.4	43.8	52.8	52.2	60.2	59.6	45.0	42.5	50.9	51.4	55.7	56.8
14.0....	40.6	40.7	51.1	50.9	58.7	58.4	40.9	39.6	50.6	49.1	57.6	58.1	40.7	39.2	48.2	48.2	54.1	55.4
14.5....	35.5	35.5	47.8	48.8	56.1	55.5	34.5	31.5	47.3	44.7	54.7	55.5	36.1	32.5	45.3	45.1	52.3	53.1
14.7....	32.3	33.6	44.7	45.1	55.5	54.1	—	—	44.3	42.3	53.5	53.7	—	22.6	42.2	42.4	50.1	52.4
14.8....	25.7	27.1	44.1	43.1	54.6	52.1	—	—	42.3	39.6	51.6	52.5	—	—	40.9	40.1	49.6	51.1
14.9....	—	—	39.6	38.5	50.5	47.1	—	—	39.1	37.1	48.4	49.7	—	—	36.2	35.5	45.1	48.1
Diameter..	29.85		29.90		29.93		29.85		29.86		29.90		30.00		29.90		29.95	

ρ	Plate 91. Position II of Diaphragm						Plate 92. Position I of Diaphragm					
	½		1		¼		½		1		¼	
	East	West	East	West	East	West	East	West	East	West	East	West
0.0....	60.0		65.3		53.0		59.1		65.8		53.7	
4.0....	59.4	59.6	64.8	65.1	52.7	52.6	58.6	58.2	65.4	65.1	53.6	52.3
8.0....	57.8	58.2	64.3	64.4	51.4	51.1	57.7	56.9	63.9	64.5	51.6	51.8
11.0....	56.2	57.1	63.2	63.4	49.5	48.7	56.3	55.4	63.3	62.8	49.3	49.1
13.0....	53.5	54.1	60.1	61.2	45.3	45.5	53.8	52.6	61.1	60.1	44.8	45.4
14.0....	50.1	51.2	58.1	59.5	40.9	40.1	51.1	49.6	59.3	57.8	40.1	40.8
14.5....	47.6	47.5	55.6	56.3	35.1	33.2	48.1	45.1	56.2	54.7	36.5	34.4
14.7....	45.4	44.1	54.8	55.1	29.8	—	45.9	42.8	54.8	53.7	—	—
14.8....	42.7	41.3	52.9	52.4	—	—	43.9	39.7	53.1	51.9	—	—
14.9....	39.8	34.8	49.1	47.8	—	—	41.1	33.7	49.4	49.2	—	—
Diameter..	29.85		29.85		29.84		30.00		29.95		29.89	

TABLE I—*Continued*

SERIES IV, SEPTEMBER 8, 1905

ρ	PLATE 103. POSITION I OF DIAPHRAGM						PLATE 104. POSITION II OF DIAPHRAGM						PLATE 105. POSITION III OF DIAPHRAGM					
	1/4		1/2		1		1/4		1/2		1		1/4		1/2		1	
	East	West	East	West	East	West	East	West	East	West	East	West	East	West	East	West	East	West
0.0....	49.3		55.3		60.1		48.6		54.3		59.6		49.2		55.6		60.1	
4.0....	49.1	48.5	54.8	54.9	60.1	59.1	48.1	47.9	53.7	53.5	59.3	59.2	48.8	48.5	55.2	54.8	59.8	59.8
8.0....	47.8	47.8	54.1	54.5	59.6	58.5	46.7	47.1	52.6	52.5	57.9	57.8	47.1	47.5	54.6	54.3	58.7	59.4
11.0....	45.6	45.9	52.7	52.6	57.8	57.6	45.2	44.5	51.7	51.8	57.2	56.7	45.6	44.8	54.1	52.1	57.8	57.7
13.0....	42.7	42.6	50.1	51.1	54.9	55.5	42.7	40.8	49.6	49.1	55.1	55.1	41.8	42.7	51.1	51.1	57.0	56.2
14.0....	39.6	40.6	48.8	48.6	54.1	53.7	39.3	38.4	49.1	46.8	53.7	53.7	38.7	39.1	49.7	48.7	55.1	55.1
14.5....	36.2	38.7	46.7	46.2	52.6	52.5	34.2	35.4	45.2	45.1	52.3	52.2	35.5	36.2	48.7	46.5	53.7	54.1
14.7....	32.9	34.5	44.9	44.5	51.5	51.1	31.4	32.1	43.7	42.6	50.5	51.0	32.5	32.1	47.3	44.4	52.6	52.5
14.8....	32.1	33.3	44.3	43.7	51.2	50.7	27.5	27.9	43.1	41.6	50.1	50.5	28.6	30.1	46.1	43.7	51.7	52.0
14.9....	22.5	26.4	42.7	41.8	49.1	49.7	—	19.7	41.8	38.9	49.2	49.1	—	21.4	44.1	41.6	50.8	51.2
15.0....	—	16.3	38.2	38.2	46.7	47.1	—	—	37.5	34.6	46.6	45.6	—	—	40.4	34.1	46.5	46.4
Diameter..	30.10		30.10		30.10		30.03		30.07		30.10		30.00		30.10		30.12	

ρ	PLATE 106. POSITION III OF DIAPHRAGM						PLATE 107. POSITION II OF DIAPHRAGM						PLATE 108. POSITION I OF DIAPHRAGM					
	1/4		1		1/2		1/4		1		1/2		1/4		1		1/2	
	East	West	East	West	East	West	East	West	East	West	East	West	East	West	East	West	East	West
0.0....	52.4		63.2		57.3		52.3		62.5		58.0		53.5		64.1		58.4	
4.0....	52.1	51.7	62.8	62.8	57.1	56.8	52.1	51.8	62.4	62.1	57.6	57.1	53.2	52.3	63.8	63.2	58.2	57.9
8.0....	50.9	50.6	61.3	62.1	56.8	55.9	51.1	50.9	61.3	61.3	56.7	55.6	52.3	51.9	63.1	62.2	57.5	57.3
11.0....	49.3	49.3	60.2	61.1	55.1	54.6	50.1	48.5	59.9	59.8	55.7	55.3	51.1	50.6	61.8	61.1	56.3	55.2
13.0....	46.1	46.1	57.9	59.1	52.3	53.1	46.8	46.9	58.7	57.9	53.6	53.4	48.0	48.2	60.5	59.3	54.9	53.9
14.0....	43.6	42.5	56.1	57.6	50.1	50.1	45.3	44.6	57.0	56.8	51.3	51.3	45.7	46.1	58.7	57.0	53.1	52.1
14.5....	39.1	40.1	54.3	56.1	48.4	48.7	42.4	42.4	55.5	55.5	49.5	49.1	43.8	44.9	57.5	55.7	50.5	50.6
14.7....	37.8	36.8	53.5	55.0	47.2	46.7	41.3	40.2	55.1	54.1	48.3	48.9	41.4	42.1	56.8	54.3	50.1	49.4
14.8....	37.3	35.7	52.8	54.2	45.8	46.2	40.6	38.8	54.1	53.5	47.2	47.5	40.1	41.4	55.7	54.1	48.7	48.3
14.9....	32.6	31.9	51.3	52.8	43.7	43.9	38.9	36.0	53.2	52.1	45.1	46.9	35.8	40.4	54.6	53.2	47.6	46.5
15.0....	29.2	27.3	48.1	50.8	42.5	40.1	32.7	29.1	49.1	49.4	40.1	42.1	31.7	36.2	49.7	50.5	43.1	43.3
Diameter..	30.10		30.10		30.10		30.10		30.10		30.10		30.10		30.10		30.10	

SERIES V. OCTOBER 15, 1905

ρ	PLATE 110. POSITION I OF DIAPHRAGM						PLATE 111. POSITION II OF DIAPHRAGM						PLATE 112. POSITION III OF DIAPHRAGM					
	¼		½		1		¼		½		1		¼		½		1	
	East	West	East	West	East	West	East	West	East	West	East	West	East	West	East	West	East	West
0.0....	50.8		56.7		60.9		50.4		55.8		60.5		50.6		55.7		59.9	
4.0....	50.3	50.1	55.8	56.1	60.9	61.1	50.2	49.8	55.7	55.2	60.1	60.2	50.1	50.1	55.6	54.7	59.2	59.5
8.0....	49.2	49.4	55.5	55.5	60.1	60.1	49.1	49.0	54.6	54.5	60.0	59.5	49.6	49.3	55.1	54.3	58.8	58.7
11.0....	48.1	47.8	53.7	54.4	58.8	59.4	47.6	46.9	53.2	53.1	58.9	59.1	48.1	47.6	54.3	52.6	58.6	57.7
13.0....	45.6	45.3	51.9	52.7	57.7	58.6	45.6	44.6	52.6	51.1	56.6	57.6	46.7	45.1	51.8	50.9	57.1	55.8
14.0....	43.6	42.1	49.5	50.1	56.1	57.2	43.7	42.2	49.5	49.2	55.1	55.9	44.5	42.2	50.6	48.7	55.6	55.1
14.5....	40.8	40.6	47.6	48.3	54.6	55.1	40.5	39.8	48.2	46.6	54.6	55.1	42.7	40.7	48.7	46.9	54.0	53.3
14.8....	40.3	37.6	44.6	47.1	53.2	54.6	38.2	38.3	46.9	44.8	53.2	53.7	40.5	37.5	47.3	45.5	53.1	52.5
15.0....	32.3	33.2	42.8	44.1	52.1	53.5	37.5	35.5	45.0	43.1	51.8	52.8	39.1	35.4	45.2	43.7	51.7	51.5
15.1....	28.7	28.6	42.2	43.2	51.3	53.1	36.9	33.1	43.7	40.6	51.1	51.4	38.2	32.3	44.6	40.5	51.1	50.1
15.2....	—	25.1	36.6	40.1	45.8	49.1	26.2	23.2	36.6	34.1	43.8	45.5	27.9	20.1	37.6	37.2	44.8	46.1
Diameter..	30.45		30.50		30.50		30.50		30.50		30.50		30.50		30.50		30.50	

To compress these numbers, the mean was then taken for the points equally distant from the center east and west. The values thus obtained were again combined into a mean for each set of similar plates which had been exposed with three different settings of the diaphragm. In this way the following series of values were obtained.

TABLE II

ρ	Plates 66–68			Plates 69–71			Plates 75–77			Plates 78–80		
	1	½	¼	1	½	¼	1	½	¼	1	½	¼
0.0.....	62.4	57.8	52.7	63.5	58.7	53.8	61.0	56.6	52.0	63.6	58.7	53.0
4.0.....	62.1	57.5	52.3	63.0	58.0	53.3	60.3	56.1	50.9	63.2	58.6	52.6
8.0.....	61.3	56.6	51.4	62.3	57.4	52.3	59.7	55.5	50.6	62.4	57.5	51.8
11.0.....	60.3	54.9	49.0	61.3	56.0	50.5	58.5	54.3	49.1	61.0	56.3	50.4
13.0.....	58.1	52.7	46.7	59.5	53.9	47.5	56.6	52.6	47.3	59.1	54.1	48.2
14.0.....	56.2	50.4	42.6	57.7	51.9	45.2	54.8	50.1	45.4	57.3	52.1	46.1
14.5.....	54.8	47.9	39.9	56.1	49.7	42.7	53.2	48.4	43.4	55.1	50.0	44.2
14.7.....	53.6	46.1	37.2	54.9	47.9	40.1	52.1	47.2	42.0	53.9	48.7	42.3
14.8.....	52.7	44.7	34.1	53.6	46.5	38.1	51.0	46.0	41.2	52.9	47.4	41.2
14.9.....	50.2	41.1	—	51.6	43.7	32.2	47.7	43.7	38.9	49.4	43.1	38.2
15.0.....	49.5	—	—	43.6	32.6	—	—	—	—	—	—	—
Diameter	30.02	29.94	29.91	29.94	29.93	29.94	29.84	29.82	29.79	29.84	29.82	29.82

TABLE II—*Continued*

ρ	Plates 87–89			Plates 91–92			Plates 103–105		
	1	½	¼	1	½	¼	1	½	¼
0.0............	63.5	58.2	51.5	65.6	59.6	53.4	59.9	55.1	49.0
4.0............	62.9	57.6	51.1	65.2	59.0	52.9	59.6	54.5	48.5
8.0............	62.3	56.8	49.6	64.3	57.7	51.5	58.7	53.8	47.3
11.0............	61.1	55.1	47.3	63.2	56.3	49.2	57.5	52.5	45.3
13.0............	58.9	52.4	44.4	60.7	53.5	45.3	55.6	50.4	42.6
14.0............	56.1	49.7	40.3	58.7	50.5	40.5	54.2	48.6	39.3
14.5............	54.5	46.5	34.3	55.8	47.1	34.9	52.9	46.4	36.1
14.7............	53.2	43.6	—	54.7	44.6	—	51.6	44.6	32.9
14.8............	52.0	41.7	—	52.6	41.9	—	51.1	43.8	29.9
14.9............	48.2	37.7	—	48.9	37.4	—	49.9	41.9	21.9
15.0............	—	—	—	—	—	—	46.5	37.2	—
Diameter........	29.93	29.89	29.90	29.90	29.92	29.87	30.11	30.09	30.04

TABLE II—*Continued*

ρ	Plates 106–108			Plates 110–112		
	1	½	¼	1	½	¼
0.0	63.3	57.9	52.7	60.3	56.1	50.6
4.0	62.9	57.5	52.2	60.2	55.6	50.1
8.0	61.9	56.7	51.3	59.6	54.9	49.3
11.0	60.7	55.4	49.8	58.8	53.6	47.7
13.0	58.9	53.5	47.0	57.3	51.9	45.5
14.0	57.2	51.3	44.7	55.9	49.6	43.1
14.5	55.8	49.5	41.8	54.5	47.7	40.9
14.7	54.8	48.8	40.0	—	—	—
14.8	54.1	47.3	39.0	53.4	46.1	38.7
14.9	52.9	45.6	36.0	—	—	—
15.0	49.6	41.9	31.1	52.2	44.0	35.5
15.1	—	—	—	51.4	42.5	33.0
15.2	—	—	—	45.9	37.1	24.6
Diameter	30.10	30.10	30.10	30.50	30.50	30.48

The further problem consisted in determining the relation between the photometer readings and the light-intensities to which the plate was exposed.

We shall express all data as to brightness in astronomical magnitudes. The relation between m and the light-intensity J is defined by the well-known equation:

$$m = -2.5 \log J .$$

Diaphragming down to ½ signifies therefore a loss of light of 2.5 log ½ or 0.753 magnitudes. We will now designate the brightness in the Sun's image at a certain distance from the center with m_1, m_2, m_3, according as the light is diaphragmed down to 1, ½, or ¼. Let the corresponding blackenings (photometer readings) be S_1, S_2, and S_3.

Then we have:

$$m_1 = m_2 - 0.753, \; m_3 = m_2 + 0.753 . \qquad (1)$$

Since the exposure time for all photographs was the same, the blackening is simply a function of the brightness: $S = \Phi(m)$. The following equations therefore obtain:

$$S_1 = \Phi(m_2 - 0.753), \; S_2 = \Phi(m_2), \; S_3 = \Phi(m_2 + 0.753) ;$$

or, by inversion,

$$m_2 = \Psi(S_1) + 0.753 = \Psi(S_2) = \Psi(S_3) - 0.753 . \qquad (2)$$

The function Ψ is to be determined to satisfy these conditions. This is a problem in interpolation, which we shall treat more fully elsewhere. We now come to our purpose as follows.

Let the three following possibilities be considered:

α) If the relation between magnitudes and opacities were a linear one,

$$m = a - bS \, , \tag{3}$$

it would follow that

$$S_1 - S_2 = S_2 - S_3 = \frac{0.753}{b} \, ,$$

or the difference of the opacities of corresponding points on the Sun would have to be constant for different degrees of diaphragming.

β) Were the relation between magnitudes and opacities logarithmic,

$$m = a \log(\beta + S) \, , \tag{4}$$

we should have

$$\begin{aligned} S_1 &= \beta(\gamma - 1) + \gamma S_2 \, , \\ S_2 &= \beta(\gamma - 1) + \gamma S_3 \, , \end{aligned} \tag{5}$$

where for abbreviation we place

$$\log \gamma = -\frac{0.753}{a} \, . \tag{6}$$

There would thus be a linear relation between the opacities belonging together.

γ) Finally, let us assume that the function Ψ is of such a nature that its third derivatives may be neglected for the interval S_1 to S_2 or S_2 to S_3. Then we may write, in place of (2)

$$\begin{aligned} 0.753 &= \Psi(S_2) - \Psi(S_1) = \Psi^1\left(\frac{S_1 + S_2}{2}\right) \cdot (S_2 - S_1) \\ 0.753 &= \Psi(S_2) - \Psi(S_3) = \Psi^1\left(\frac{S_2 + S_3}{2}\right) \cdot (S_3 - S_1) \, ; \end{aligned} \tag{7}$$

or, if for brevity we make

$$\frac{S_1 + S_2}{2} = T \, , \quad \frac{S_2 + S_3}{2} = T^1 \, , \quad S_2 - S_1 = D \, , \quad S_3 - S_2 = D^1 \, , \tag{8}$$

$$0.753 = \frac{d\Psi}{dT} D = \frac{d\Psi}{dT^1} D^1 \, .$$

By integrating, we obtain

$$\Psi(T) = 0.753 \int^T \frac{dT}{D} = 0.753 \int^T \frac{dT^1}{D^1} \, . \tag{9}$$

We therefore have to treat the reciprocal difference of the related opacities as a function of the mean T of the two opacities, and to integrate in order to obtain the desired function Ψ — the magnitude m belonging to each opacity T. The constant of integration—i. e., the starting-point for computing magnitudes—remains undetermined and a matter of indifference.

The three possibilities just described will in general give the necessary means of determining the function Ψ.

In our case the procedure was as follows:[1] It is easy to see that the related values of the opacities S_1, S_2, S_3, which in Table II are placed next each other, to some extent progress in a linear fashion. If we choose the coefficients β and γ in the relations (5) so that the actual dependence of the opacities is approximately reproduced, and if we then compute for each opacity S a quantity s by the formula

$$s=\alpha \log (\beta+S)\ , \tag{10}$$

then this quantity s will represent a sort of corrected opacity which will correspond very closely to the magnitudes. If we therefore seek further for the precise relation between s and the magnitude m; $m=\psi(s)$. Then we can make the assumption mentioned under γ, that the higher derivatives of ψ may be neglected, and therefore we may find ψ corresponding to (8) and (9) by forming

$$t=\frac{s_1+s_2}{2}\ ,\quad t'=\frac{s_2+s_3}{2}\ ,\quad s_2-s_1=\Delta\ ,\quad s_3-s_2=\Delta', \tag{12}$$

$$\psi(t)=0.753\int^{t}\frac{dt}{\Delta}=0.753\int^{t}\frac{dt'}{\Delta'}\ . \tag{13}$$

In practice the following small modification of this scheme was found necessary. (13) gives two independent determinations of the function ψ, one from the exposures with diaphragms 1 and $\frac{1}{2}$, a second from those with diaphragms $\frac{1}{2}$ and $\frac{1}{4}$. There almost always appeared to be a small systematic difference between the two determinations of the function ψ, due to the fact that the sectors of the diaphragms have not precisely the prescribed dimensions; an exact

[1] The relation between magnitudes and opacities was too far from being linear in the case of several plates to make it advisable to use the simple adjustment employed in our previous communication. We therefore preferred to use the more inconvenient but systematic reduction throughout.

measurement of the sectors used by us yielded the actual values of 0.767 and 0.759 mag.; and that further differences arise from a change in the transparency of the air or in the velocity of the shutter. We shall therefore assume that the true values of the degree of diaphragming from 1 to $\frac{1}{2}$ and from $\frac{1}{2}$ to $\frac{1}{4}$ equal certain values a and b, and that only the total difference between the diaphragming for 1 and $\frac{1}{4}$ equals the prescribed values 0.767 and 0.759. In other words, we assume in place of (1)

$$m_1 = m_2 - a\,, \quad m_3 = m_2 + b\,, \quad a + b = 1.527 \text{ mag.} \tag{14}$$

We therefore have in place of (13)

$$m = \psi(t) = a\int^t \frac{dt}{\Delta} = b\int^t \frac{dt'}{\Delta'}\,. \tag{15}$$

Hence for the differences Δ and Δ' which pertain to equal values of t and t',

$$\frac{\Delta}{a} = \frac{\Delta'}{b}\,. \tag{16}$$

The ratio $\frac{a}{b}$ is therefore to be determined from the average ratio of the differences Δ and Δ'.

In the actual employment of this process the differences Δ and Δ' were entered as the ordinates for the abscissae t and t' in the same figure, which gave two different curves. The ordinates of the second curve were then enlarged in a ratio $\frac{a}{b}$, which brought the two curves into the best possible conformity; and finally a new curve was drawn which followed the mean position between the first curve and the second curve enlarged. From this the differences Δ were then taken off as functions of t, and then by the first of the formulæ (15) m, the function sought, was computed by mechanical quadrature.

We give as an example to render clear the whole process, the group of Plates 66 to 68 in the second and fourth column of Table II. By graphically entering S_1 as a function of S_2, and S_2 as a function of S_3, it was found that with some degree of approximation the following linear relations held good:

$$S_1 = 0.755\,S_2 + 18.14 \text{ and } S_2 = 0.755\,S_3 + 18.14\,.$$

The comparison with (5) and (6) gives

$$\beta(\gamma-1)=18.14\ ;\quad \gamma=0.755\ ;\quad \alpha=-\frac{0.753}{\log\gamma}\ ;$$

whence

$$\beta=-74.0\ ;\quad \alpha=6.18\ .$$

And therefore by (10) $s=6.18\log(S-74.0)$. A "corrected" opacity s was now computed by this formula for every opacity S, with the resulting values s_1, s_2, s_3, given in the following table (for convenience all have been diminished by 6.18):

ρ	s_1	s_2	s_3	t	Δ	t'	Δ'	From Curve t	From Curve Δ
0.0	0.40	1.30	2.03	0.85	0.90	1.66	0.73	0.8	0.900
4.0	0.47	1.35	2.08	0.91	.88	1.72	.74	1.0	.870
8.0	0.64	1.49	2.19	1.07	.84	1.84	.70	1.2	.850
11.0	0.85	1.74	2.46	1.29	.89	2.10	.72	1.4	.830
13.0	1.24	2.03	2.69	1.63	.79	2.36	.66	1.6	.805
14.0	1.54	2.30	3.07	1.92	.76	2.69	.77	1.8	.785
14.5	1.75	2.58	3.29	2.16	.83	2.94	.71	2.0	.790
14.7	1.92	2.76	3.50	2.34	.84	3.13	.74	2.2	.810
14.8	2.03	2.89	3.71	2.46	.86	3.30	.82	2.4	.830
14.9	2.33	3.19		2.76	.86			2.6	.840
15.0	2.40							2.8	.855
								3.0	.855

The mean values t and t', and the appropriate differences Δ and Δ' of the consecutive values of s, were now formed. The graphical process described above gave $\frac{\Delta}{\Delta'}=\frac{a}{b}=1.14$; $a=0.814$ mag.; $b=0.713$ mag.; and the smoothed-out values of Δ given in the table. Then the first integral of (15) was formed simply by the trapezoidal rule, which yielded the function ψ sought for. The following table for ψ resulted:

s	$m=\psi(s)$	s	$m=\psi(s)$
0.4	2.59 mag.	2.0	1.08 mag.
0.6	2.42	2.2	0.87
0.8	2.26	2.4	0.67
1.0	2.07	2.6	0.48
1.2	1.88	2.8	0.29
1.4	1.69	3.0	0.10
1.6	1.49	3.2	−0.09
1.8	1.29	3.4	−0.27

With this we could finally convert the values s_1, s_2, s_3, into magnitudes. The results were as follows:

ρ	m_1	m_2	m_3	$m_1 - m_2$	$m_2 - m_3$
0.0.......	2.59	1.79	1.03	0.80	0.76
4.0.......	2.53	1.74	1.00	.79	.74
8.0.......	2.38	1.60	.88	.78	.72
11.0.......	2.21	1.35	.61	.86	.74
13.0.......	1.84	1.05	.37	.79	.68
14.0.......	1.54	0.76	.03	.78	.73
14.5.......	1.34	0.50	−.17	.84	.67
14.7.......	1.17	0.33	−.37	.84	.70
14.8.......	1.05	0.20	−.56	.85	.76
14.9.......	0.74	−0.09		.83	
15.0.......	0.67				

The differences $m_1 - m_2$ and $m_2 - m_3$ have, in fact, become constant within the limits of accidental errors, which confirms the fact that the opacities corresponding to the conditions (14) are thus transformed into magnitudes.

If desired, the value s can subsequently be rejected, as that only constitutes an intermediate step, and by a combination of (10) and the table for $\psi(s)$, a table may be constructed by which we may pass directly from the opacities S to magnitudes. This procedure was employed in what follows. Table III contains the

TABLE III

S	Plate Numbers								
	66–88	69–71	75–77	78–80	87–89	91–92	103–105	106–108	110–112
	mag.	mag.	mag.	mag.	mag.	mag.	mag.	mag.	mag.
66						2.48			
62	2.51	2.76	3.43	2.64	2.25	1.86		2.64	3.64
58	1.82	2.10	2.76	2.04	1.69	1.37	2.39	2.05	2.80
54	1.23	1.52	2.08	1.46	1.22	0.94	1.80	1.47	2.13
50	0.72	1.03	1.45	0.91	0.80	0.55	1.28	0.98	1.63
46	0.33	0.61	0.82	0.38	0.46	0.24	0.86	0.58	1.19
42	−0.01	0.24	0.17	−0.12	0.17	−0.02	0.52	0.22	0.82
38	−0.31	−0.05			−0.08	−0.28	0.26	−0.06	0.49
34	−0.57	−0.31			−0.27	−0.48	0.06	−0.31	0.21
30							−0.07	−0.54	−0.03
26							−0.18		−0.23
22							−0.27		−0.42
a	0.874	0.825	0.724	0.730	0.763	0.837	0.730	0.784	0.834
b	0.713	0.702	0.803	0.797	0.764	0.690	0.797	0.743	0.693

transformation table of the opacities into magnitudes for all the groups of plates. The values of a and b obtained are given every time at the foot of the table, which gives a certain insight into the trustworthiness of the diaphragming.

Thus all the data needed for the reduction of the plates are given.

III

With the aid of the above tables all of the opacities given in Table II were converted into magnitudes. The numbers in each column were then subtracted from the number at the head of the column, which corresponded to the brightness at the Sun's center, so that the resulting differences gave directly the decrease in the brightness of the Sun from the center. The mean was taken of the three series of numbers resulting for each group. Thus the conversion of the opacities into magnitudes furnished for the group of plates 66–68 the three series given above for m_1, m_2, m_3. The subtraction from the number at the top of each column yielded the values M_1, M_2, M_3, the mean of which is given under M.

ρ	M_1	M_2	M_3	M
0.0	0.00	0.00	0.00	0.00
4.0	0.06	0.05	0.03	0.05
8.0	0.21	0.19	0.15	0.18
11.0	0.38	0.44	0.42	0.41
13.0	0.75	0.74	0.66	0.72
14.0	1.05	1.03	1.00	1.03
14.5	1.25	1.29	1.20	1.25
14.7	1.42	1.46	1.40	1.43
14.8	1.54	1.59	1.59	1.57
14.9	1.85	1.88		1.87
15.0	1.92			1.92

A final correction is to be applied to these measures on account of the varying diameter of the Sun's image. As may be seen from the data at the foot of Table II, the weakest image was smaller by 0.02 mm or 1ʹ.3, the strongest image larger by a similar amount than the average image. The impression is in fact given that, with diaphragm $\frac{1}{4}$, the Sun's limb is not quite sufficiently exposed, that the image with full aperture may suffer a little from irradiation on the film, and that the mean image yields the most trustworthy value of the diameter. On these images the edge is sharply defined to 0.01 mm under the microscope, and just outside the disk there is clear glass. The diameter of the image from the diaphragm $\frac{1}{2}$ is therefore taken as the standard, and its mean value for each group is used, as given at the foot of Table II; for instance, 29.94 for the group 66–68. With this the distances from the center of the disk, ρ, were converted into decimals of the Sun's radius. Finally the brightnesses were interpolated for the arbitrarily chosen standard value x of the distance from the center.

A similar treatment of all the groups of plates yielded the following nine series of values, the mean of which is to be regarded as the final result of this investigation. The differences from the mean are added, and the last column gives the light-intensity corresponding to the difference in magnitude, the intensity of the center of the disk being taken as unity.

TABLE IV

x	66–68	69–71	75–77	78–80	87–89	91–92	103–106	106–108	110–112	Mean
										mag.
0.000..	0.00	0.00	0.00	0.00	0.00	0.00	0.00	0.00	0.00	0.000
0.267..	0.05	0.09	0.13	0.05	0.07	0.06	0.06	0.06	0.05	0.070
0.533..	0.18	0.20	0.21	0.18	0.18	0.21	0.19	0.18	0.16	0.188
0.733..	0.40	0.39	0.42	0.38	0.38	0.38	0.38	0.37	0.36	0.384
0.867..	0.70	0.70	0.69	0.65	0.64	0.68	0.64	0.63	0.63	0.662
0.933..	1.01	0.95	1.01	0.91	0.94	0.95	0.86	0.89	0.91	0.937
0.967..	1.23	1.18	1.29	1.20	1.18	1.25	1.07	1.13	1.15	1.187
0.980..	1.38	1.35	1.45	1.34	1.33	1.36	1.23	1.26	1.34	1.338
0.987..	1.57	1.49	1.54	1.43	1.45	1.51	1.34	1.39	1.45	1.457
0.993..	1.75	1.74	1.72	1.60	1.65	1.76	1.58	1.64	1.64	1.676
0.997..	1.93	1.92	1.92	1.81	1.82	1.94	1.76	1.83	1.96	1.877

Differences from the Mean (Unit 0.01 mag.)									Mean Error	y
66–68	69–71	75–77	78–80	87–89	91–92	103–106	106–108	110–112		
									mag.	
0	0	0	0	0	0	0	0	0	0.000	1.000
−2	+2	+6	−2	0	−1	−1	−1	0	0.008	0.938
−1	+1	+2	−1	−1	+2	0	−1	−3	0.005	0.841
+2	+1	+4	0	0	0	0	−1	−2	0.006	0.702
+4	+4	+3	−1	−2	+2	−2	−3	−3	0.010	0.544
+7	+1	+7	−3	0	+1	−8	−5	−3	0.017	0.422
+4	−1	+10	+1	−1	+6	−12	−6	−4	0.024	0.335
+4	+1	+11	0	−1	+2	−11	−8	0	0.019	0.292
+5	+3	+8	−3	−1	+5	−12	−7	−1	0.021	0.261
+7	+6	+4	−8	−3	+8	−10	−4	−4	0.023	0.214
+5	+4	+4	−7	−6	+6	−12	−5	+8	0.024	0.178

In regard to the question of the reliability of these values, if the precision of the double settings of the Hartmann micro-photometer be taken as 2 per cent., a mean error of 0.002 mag. would result from the measures of the seventy-eight plates. It is evident at a glance that this precision of settings is rendered illusory by systematic differences in the blackening on the plates. For example, a comparison of the corresponding numbers in Table I for the east and west sides of the Sun shows a systematic excess, first on one side, then on the other. Since the mean of all the images shows an almost perfect symmetry in the two halves, the differences do not originate in the Sun, but in the plate or the photographic process. The causes which have a systematic effect on a single plate seem to affect different plates in a purely accidental manner. At least we are unable to trace any law in the differences found by comparing the east and west halves of the Sun or the values a and b.

Under these circumstances an idea of the trustworthiness of the above quantities can be gained from the differences in the nine series. These yield the mean errors given in the next to last column, which reveal the increasing uncertainty with the increase of the observed difference in brightness. The systematic enlargement of the differences in the greater part of the numerical values can be explained from the uncertainty of the scale-value a. From the deviations of the values of a given in Table III from the prescribed value we should derive, with the method of reduction employed, the mean error of the final value of $\pm 0.013\ M$ mag., where M is the difference of magnitude to be measured; and for the effect of other accidental circumstances there would remain a mean error of only ± 0.009 mag. It is probable, all things considered, that an accuracy is attained equal to that of Vogel's measures in the visible spectrum. Furthermore, the decrease in brightness is followed nearer to the limb than by previous observers. While Very's measures stopped at 0.95 and Vogel's at 0.97, there is no reason for doubting that the values obtained for $x=0.980$, 0.987, and even 0.993 are real expressions of the Sun's brightness at those distances from the center of the disk. This will scarcely apply to $x=0.997$, for at this point, in addition to the difficulties of measurement, irradiation in the film and errors in definition of the objective amounting to a second of arc might affect the result.

We must finally meet one more objection, that the decrease in the Sun's brightness from the center of the disk might be rendered inaccurate by the instrument itself. A simple consideration shows that the absorption in the objective and in the silver films does not change appreciably when the inclination to the axis is increased to 15′ (the Sun's radius). At first thought there would seem to be much more danger from the secondary images arising from the double reflection at the silvered surfaces. But, on the one hand, the experiments of E. Hagen and H. Rubens[1] show that the reflecting power of silver in the region solely to be considered here (from λ 315–327 $\mu\mu$) lies below 15 per cent., so that for an image arising from double reflection only 2 per cent. remains. On the other hand, we have also a direct check on this effect. A computation from the known optical constants of the objective showed that the image resulting from the reflection of the two silvered surfaces, which at the same time is the nearest of all the reflected images to the focal plane, lies 85 cm behind that plane. A halo 16 mm wide would thereby be produced around the focal image of the Sun. Its intensity at a distance of 2 mm from the Sun's limb would come out as about 0.2 of the central brightness of the real solar image, if no light was lost by reflection, and if the distribution of brightness on the Sun's disk found above is taken as our basis. Two photographs of the Sun were now made on the same plate, the objective being diaphragmed down to 10 mm and 5 mm respectively; and, moreover, a third exposure with full aperture, for which, however, the image of the Sun itself was covered up by a metal disk placed immediately in front of the plate and extending 2 mm beyond the Sun's limb. The time of exposure was in each case 10 seconds. The comparison of the opacities yielded the result that the brightness at 2 mm distance from the Sun's limb amounted to 0.0008 of the central brightness of the Sun. The brightness of the reflex image must therefore be weakened by the double reflection at least in the ratio 0.0008:0.2 or 0.004:1, and therefore cannot in fact come into consideration.

[1] *Zeitschrift für Instrumentenkunde*, **22**, 42, 1902.

IV

We will now compare the results we have obtained for $\lambda = 0.320$ with those for other wave-lengths. In the diagram the values of J/J_0 for $\lambda = 0.320$ are shown in connection with the curves obtained by Vogel for the red and violet. The noteworthy result to be derived from these curves is this: The decrease of the brightness is not

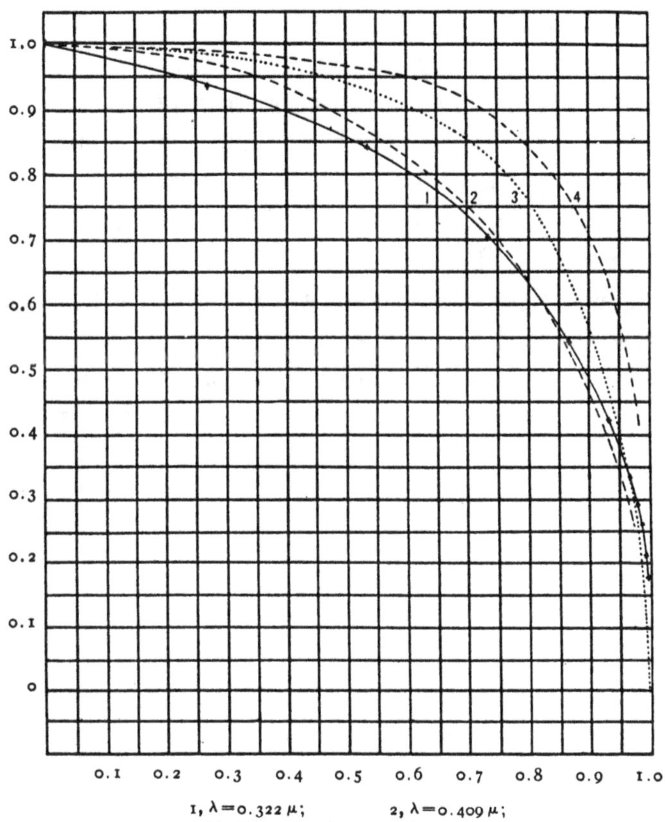

1, $\lambda = 0.322\ \mu$; 2, $\lambda = 0.409\ \mu$;
3, Earth's atmosphere; 4, $\lambda = 0.662\ \mu$.
+observed points.

more marked in the ultra-violet than in the violet; the gradual increase in the rate of falling off which is observed in passing from the red toward the violet, comes to a stop in the ultra-violet. If we were to regard as real the small difference of the curves for $\lambda = 0.320\ \mu$ and $\lambda = 0.409\ \mu$, then we should have to say that in the ultra-violet the decrease, at first stronger than in the violet, is checked on approaching the limb, and finally becomes even less than for the violet.

This behavior directly contradicts the analogy with the Earth's atmosphere, in which the absorption increases very rapidly toward the ultra-violet. This new fact should be added to that noted by Vogel: the shape of the curve does not follow the laws valid for the Earth's atmosphere. The dotted curve shows the distribution in intensity which would be found in the Earth's atmosphere viewed from without, if the surface were a self-luminous black body (calculated by doubling the Potsdam extinction table). The differing character of the decrease is noticeable, especially for the ultra-violet.

Explanation of this difference in curves can be attempted in two different ways. Seeliger[1] ascribes to the Sun's atmosphere a refractive index differing much more from unity than that possessed by the Earth's atmosphere. Schuster[2] takes into account the radiation from the Sun's atmosphere itself. The theoretical treatment of our observations with reference to such views we desire to reserve for another occasion.

We intend to continue the measures in order to obtain data as to a possible variation in the absorption of the solar atmosphere.

GÖTTINGEN AND JENA,
January 1906.

[1] *Sitzungsberichte der bayrischen Akademie der Wissenschaften*, **21**, 1891.

[2] *Astrophysical Journal*, **16**, 320, 1902.

NOTE ON THE ULTRA-VIOLET RADIATION OF SUN-SPOTS AND FACULÆ

In connection with the investigations of Messrs. Hale, Adams, and Abbot on the radiation of sun-spots[1] we beg to communicate a few observations on the ultra-violet radiation of sun-spots and faculæ. These are based on the solar plates taken with light of wave-length 320 $\mu\mu$, upon which we report fully elsewhere in this number. We were struck at the beginning by the fact that the umbræ of the larger spots on these plates exhibit strong contrast to the surroundings. Measurements were made in the following manner. A setting was made under a Hartmann photometer on the opacity of the spot upon a solar image obtained with the full aperture of the instrument, and then upon the image taken with a diaphragm $\frac{1}{4}$, which was situated on the same plate, an undisturbed place was sought out which exhibited the same opacity, and the distance of this point from the limb was determined. Since the falling off in the brightness of the Sun's disk from the center toward the edge was known from the previous investigation, this gave directly the brightness of the spot, which could then be referred to the normal brightness of the Sun at an equal distance from the limb as unity. It was only in case of the exceedingly dark spot of November 16 that no suitable opacity for comparison could be found on the plate, and a small extrapolation had to be made from the opacity-curve.

The measurements were rendered for the most part rather difficult by the fact that the nucleus of the sun-spot did not entirely fill the spot in the photometer.

Date	Distance from Center	Umbra	Penumbra	Remarks
1905				
January 17 . . .	0.72	0.21	0.52	Single spot on west limb
	0.27	0.14	0.64	Great spot group
	0.78	0.40	. . . .	Single spot on east limb
January 18 . . .	0.83	0.35	. . . .	Single spot on west limb
	0.42	0.11	0.47	Great spot group
	0.65	0.47	. . . .	Single spot on east limb
June 28	0.53	0.13	0.49	Great spot group
	0.69	0.11	0.69	Single round spot
October 16 . . .	0.73	0.032	0.52	Great group
November 9 . .	0.83	0.30	. . . .	Single spot
	0.43	0.019	0.60	Single round spot
December 25 . .	0.67	0.12	0.63	Three neighboring spots on west limb
	0.75	0.12	. . . .	
	0.88	0.55	. . .	

[1] *Astrophysical Journal*, **23**, 43, 1906.

The preceding short table gives, in addition to the date, the distance of the spot from the center of the disk in decimals of the radius, the brightness of umbra and penumbra in comparison with the brightness of an undisturbed place at equal distance from the center, and finally a few general remarks to designate the spot.

Faculæ may be followed in many instances on our plates in long curves to the exterior half of the Sun's radius, but in no place does their brightness exceed by more than 20 or 30 per cent. that of the undisturbed neighborhood.

K. SCHWARZSCHILD AND W. VILLIGER.

A THEORY FOR THE DISTRIBUTION OF SPECTRAL LINES IN SERIES

A comparatively recent attempt has been made by Professor F. von Lindemann[1] to account for the distribution of spectral lines in series. His method is to investigate the possible waves which a hypothetical atom of matter will send out into the surrounding luminiferous ether. He assumes that the atom in every case consists of elastic isotropic matter which is the same in kind for all substances, but is different in shape, size, density, and elasticity. The mathematical theory of the various modes of vibration which such a body can assume has been worked out before, but the application of the same to bodies of definite shape has been difficult because it involves the discovery of mathematical functions for each shape. The periodicity of each kind of vibration which the body sets up in the ether always occurs as a root of a certain transcendental equation involving those functions. Such an equation has an infinite number of roots, each real one corresponding to a definite spectral line. Every equation thus gives a series of lines. Each body has a number of such equations and therefore a number of such series.

Lindemann has investigated the various cases which admit of mathematical treatment, namely, where the atom is spherical, ellipsoidal, or ring-shaped. Owing to the extraordinary intricacy of the equations, no detailed calculations have yet been made from them. The solutions, for the most part, have been carried only far enough to indicate what the general type of distribution of the lines is for each case.

[1] *Sitz. Math. phys. Klasse d. Kgl. Bay. Akad.*, 31, 441, 1901; *Ibid.*, 33, 27, 1903; (Translation of Popular Lecture) *Monist*, January 1906.

Nachr. Kgl. Ges. d. Wiss. zu Gött. Math. Phys. Klasse 1906, *No. 1,* 41–53

Ueber das Gleichgewicht der Sonnenatmosphäre

Von

K. Schwarzschild.

Vorgelegt in der Sitzung vom 13. Januar 1906.

I. Inhaltsübersicht.

Die Sonnenoberfläche zeigt uns in Granulationen, Sonnenflecken und Protuberanzen wechselnde Zustände und stürmische Veränderungen. Um die physikalischen Verhältnisse zu begreifen, unter denen diese Erscheinungen stehn, pflegt man in erster Annäherung den räumlichen und zeitlichen Wechsel durch einen mittleren stationären Zustand, ein mechanisches Gleichgewicht der Sonnenatmosphäre zu ersetzen. Im Vordergrunde der Betrachtung stand bisher allgemein das sog. adiabatische Gleichgewicht, wie es in unsrer Atmosphäre herrscht, wenn sie von auf- und absteigenden Strömungen gründlich durchmischt ist. Ich möchte hier auf eine andere Art des Gleichgewichts aufmerksam machen, welche man als „Strahlungsgleichgewicht" bezeichnen kann. Strahlungsgleichgewicht wird sich in einer stark strahlenden und absorbierenden Atmosphäre einstellen, in welcher die durchmischende Wirkung auf- und absteigender Ströme gegenüber dem Wärmeaustausch durch Strahlung zurücktritt. Ob für die Sonne mehr das adiabatische oder das Strahlungsgleichgewicht gilt, wäre aus allgemeinen Gründen schwer zu entscheiden. Indessen giebt es Beobachtungsdaten, welche ein gewisses Urteil ermöglichen. Die Sonnenscheibe ist nicht gleichmässig hell, sondern vom Centrum nach dem Rande zu abgetönt. Aus dieser Helligkeitsverteilung auf der Oberfläche lässt sich unter plausibeln Annahmen zurückschliessen auf die Temperaturverteilung in der Tiefe. Das Resultat ist, daß das

Nachr. Kgl. Ges. d. Wiss. zu Gött. Math. Phys. Klasse 1906, *No. 1*, 41–53

Gleichgewicht der Sonnenatmosphäre im Grossen und Ganzen dem Strahlungsgleichgewicht entspricht.

Die Ueberlegungen, die zu diesem Resultat führen, setzen voraus, daß das Kirchhoff'sche Gesetz gilt oder — in andern Worten — daß die Strahlung der Sonnenatmosphäre reine Temperaturstrahlung ist. Sie setzen ferner voraus, daß man beim Eindringen in den Sonnenkörper einer kontinuierlichen Aenderung des Zustandes begegnet und nicht aus einer ziemlich durchsichtigen Chromosphäre diskontinuierlich in eine aus leuchtenden Wolken gebildete undurchsichtige Photosphäre gerät. Vernachlässigt ist die Wirkung der Streuung des Lichts durch Beugung an den Teilchen der Atmosphäre, auf deren Bedeutung Herr A. Schuster [1]) aufmerksam gemacht hat, ebenso ist die Refraction vernachlässigt, die H. v. Seeliger [2]) zur Erklärung der beobachteten Helligkeitsverteilung heranzieht. Auch wird die verschiedene Absorption verschiedener Wellenlängen, die Abnahme der Schwere mit der Höhe und die kugelförmige Ausbreitung der Strahlung nicht berücksichtigt. Die ganze Betrachtung kann daher keineswegs als abschliessend oder zwingend gelten, doch mag sie weiteren Spekulationen einen Anhalt geben, indem sie einen einfachen Gedanken zunächst in einfachster Form ausführt.

2. Verschiedene Arten des Gleichgewichts.

Man bezeichne den Druck mit p, die absolute Temperatur (in Centesimalgraden) mit t, die Dichte mit ϱ, das Molekulargewicht (im Vergleich zum Wasserstoffatom) mit M, die Schwere mit g, die Tiefe in der Atmosphäre (von einem beliebigen Anfangspunkt nach innen gerechnet) mit h. Die Einheiten entnehme man den Verhältnissen, die an der Erdoberfläche bestehn, wähle also als Einheit von p die Atmosphäre, als Einheit von ϱ die Dichte der Luft bei 273° absoluter Temperatur unter dem Druck einer Atmosphäre, als Einheit von g die Schwere auf der Erdoberfläche und als Einheit von h die Höhe der sog. „homogenen Atmosphäre", welche 8 km beträgt.

Es gilt dann für ein ideales Gas die Beziehung:

$$(1) \qquad \varrho t = \frac{p \,.\, M}{R} \qquad R = 0.106$$

1) Astrophysical Journal. 1905. Bd. 21. pag. 1.

2) Sitzungsberichte der Münchener Akad. d. Wiss., Math.-phys. Classe. 1891. Bd. 21. pag. 264.

und die Bedingung mechanischen Gleichgewichts der Atmosphäre lautet:

(2) $$dp = \varrho g dh.$$

Die Elimination von ϱ aus (1) und (2) ergiebt:

(3) $$\frac{dp}{p} = \frac{M}{R} \cdot \frac{g}{t} dh.$$

a) Isothermes Gleichgewicht. Zur allgemeinen Orientierung betrachte man isothermes Gleichgewicht, setze t als konstant voraus. Es folgt dann:

(4) $$p = p_0 e^{\frac{M}{R}\frac{g}{t}h}, \qquad \varrho = \varrho_0 e^{\frac{M}{R}\frac{g}{t}h}.$$

Die Schwere g ist auf der Sonne 27.7 mal größer, als auf der Erde, die Temperatur (rund 6000°) etwa 20 mal höher. Es folgt also für ein Gas vom Molekulargewicht der Luft etwa dieselbe räumliche Druckverteilung, wie für Luft auf der Erde. Rechnet man genauer, so findet man eine Zunahme von Druck und Dichte auf das 10-fache für ein Gas vom Molekulargewicht der Luft (28.9) auf je 14.7 km, für Wasserstoff auf je 212 km. Da 725 km auf der Sonne einem Winkelwert von einer Bogensekunde von der Erde aus gesehn entsprechen, so ist klar, daß die Sonne völlig scharf begrenzt aussehn muß.

b) Adiabatisches Gleichgewicht. Wenn eine Gasmasse sich adiabatisch ausdehnt, so gelten die Poisson'schen Beziehungen:

(5) $$\frac{p}{p_0} = \left(\frac{\varrho}{\varrho_0}\right)^k = \left(\frac{t}{t_0}\right)^{\frac{k}{k-1}}$$

wobei p_0 und ϱ_0 irgend welche zusammengehörige Ausgangswerte bezeichnen. Die Größe k, das Verhältnis der spezifischen Wärme, ist für ein 1-atomiges Gas gleich 5/3, für ein 2-atomiges gleich 7/5, für ein dreiatomiges 4/3, und nimmt für mehratomige Gase bis auf 1 ab. Das Gleichgewicht einer Atmosphäre heißt adiabatisch, wenn an jeder Stelle die Temperatur herrscht, welche eine von unten aufsteigende, sich adiabatisch abkühlende Gasmasse an dieser Stelle annehmen würde, wenn also innerhalb der ganzen Atmosphäre die Gleichungen (5) erfüllt sind. Es folgt dann aus (3) durch Einsetzen von (5) und Integration:

Nachr. Kgl. Ges. d. Wiss. zu Gött. Math. Phys. Klasse 1906, *No. 1*, 41–53

$$t - t_0 = \frac{k-1}{k} \frac{Mg}{R} (h - h_0). \tag{6}$$

Die Temperatur ändert sich linear mit der Höhe. Der Temperaturgradient berechnet sich für die Erdatmosphäre zu 1° auf je 100 m, für die Sonne wird er 27,7 mal größer als auf der Erde. Die Temperaturzunahme um einen Grad erfolgt dort also für Luft auf je 3,63 m, für Wasserstoff auf je 52 m. Die Atmosphäre hat eine bestimmte äußere Grenze ($t = \varrho = p = 0$). Die Tiefe einer Schicht von 6000° Temperatur unter der äußeren Grenze beträgt auf der Sonne für Luft 22 km, für Wasserstoff 300 km.

c) Strahlungsgleichgewicht. Wenn man sich vorstellt, daß die äußeren Teile der Sonne einen kontinuierlichen Uebergang zu immer heißeren und dichteren Gasmassen bilden, so kann man nicht unterscheiden zwischen strahlenden und absorbierenden Schichten, sondern hat jede Schicht gleichzeitig als absorbierend und strahlend anzusehn. Wir wissen, daß ein mächtiger Energiestrom, aus unbekannten Quellen im Sonneninnern entspringend, die Sonnenatmosphäre durchsetzt und in den Außenraum dringt. Welche Temperatur würden beim Fehlen durchmischender Bewegungen die einzelnen Schichten der Sonnenatmosphäre annehmen müssen, damit sie einen solchen Energiestrom ohne weitere eigene Temperaturänderung stationär befördern?

Man nehme an, jede Höhenschicht dh der Sonnenatmosphäre absorbiere den Bruchteil $a\,dh$ der hindurchgehenden Strahlung. Ist E die Emission eines schwarzen Körpers von der Temperatur dieser Höhenschicht und nimmt man das Kirchhoff'sche Gesetz als gültig an, so folgt, daß diese Höhenschicht die Energie $E \,.\, a\,dh$ nach jeder Seite ausstrahlt.

Man betrachte jetzt die Strahlungsenergie A, die durch die Sonnenatmosphäre an irgend einer Stelle nach außen wandert, und die Strahlungsenergie B, die (infolge der Strahlung der weiter außen liegenden Schichten) nach innen wandert.

Man verfolge zunächst die nach innen wandernde Energie B. Geht man um eine unendlich dünne Schicht dh nach innen, so geht von der von außen kommenden Energie B der Bruchteil $B \,.\, a\,dh$ verloren, auf der anderen Seite kommt durch die nach der einen Seite gehende Eigenstrahlung der Schicht dh der Betrag $a\,E\,dh$ hinzu, es ergiebt sich also im Ganzen:

$$\frac{dB}{dh} = a(E - B). \tag{7}$$

Völlig analog folgt für die nach außen gehende Strahlung:

(8) $$\frac{dA}{dh} = -a(E-A).$$

Indem man sich das Absorptionsvermögen a als Funktion der Tiefe h gegeben denkt, bilde man die über der Tiefe h liegende „optische Masse“ der Atmosphäre:

(9) $$m = \int^h a\,dh.$$

Dann lauten die Differentialgleichungen:

(10) $$\frac{dB}{dm} = E-B, \quad \frac{dA}{dm} = A-E.$$

Wir fragen nach einem stationären Zustand der Temperaturverteilung. Derselbe ist bedingt durch die Forderung, daß jede Schicht ebensoviel Energie empfängt, als ausstrahlt, daß also gilt:

$$aA + a.B = 2aE, \quad A+B = 2E.$$

Führt man dieser Bedingung entsprechend die Hülfsgröße γ ein durch:

$$A = E+\gamma, \quad B = E-\gamma,$$

so gehn die Differentialgleichungen durch Addition und Subtraktion über in:

$$\frac{d\gamma}{dm} = 0, \quad \frac{dE}{dm} = \gamma$$

und integriert:

$$\gamma = \text{const}, \quad E = E_0 + \gamma m,$$
$$A = E_0 + \gamma(1+m), \quad B = E_0 + \gamma(m-1).$$

Die Integrationskonstanten E_0 und γ wurden dadurch bestimmt, daß an der Außengrenze der Atmosphäre ($m = 0$) keine nach innen wandernde Energie B vorhanden ist und die nach außen wandernde Energie den zu beobachtenden Betrag A_0 hat. Es muß also für $m = 0$:

$$B = 0, \quad A = A_0$$

sein. Hiermit ergiebt sich das Resultat:

(11) $$E = \frac{A_0}{2}(1+m), \quad A = \frac{A_0}{2}(2+m), \quad B = \frac{A_0}{2}m.$$

Es läßt sich also nur unter Voraussetzung des Kirchhoff'schen Gesetzes die Abhängigkeit der Strahlung E von der über der betreffenden Stelle liegenden optischen Masse ableiten.

Will man die Verteilung von Druck und Dichte, die bei Strahlungsgleichgewicht herrscht, kennen lernen, so hat man im Grunde eine detailliertere Untersuchung nötig, welche die Strahlung in den einzelnen Wellenlängen betrachtet. Es genüge hier für eine erste Uebersicht die Annahme, daß der Absorptionskoeffizient unabhängig von der Farbe und der Dichte proportional sei:

(12) $$a = \varrho k$$

(k eine Constante). Dann folgt:

(13) $$m = \int a\,dh = k\cdot\int \varrho\,dh = \frac{k}{g}\cdot p.$$

Die Strahlung E des schwarzen Körpers ist nach dem Stefanschen Gesetz:

$$E = c\,.\,t^4 \quad (c \text{ eine Constante}).$$

Setzt man für die nach außen gelangende Energie A_0:

$$A_0 = c\,.\,T^4,$$

so ist T das, was man gewöhnlich als (effektive) Temperatur der Sonne angiebt. Es ist rund $T = 6000^0$. Für Strahlungsgleichgewicht gilt dann nach (11):

(14) $$t^4 = \frac{1}{2}T^4\left[1+\frac{k}{g}p\right]\cdot$$

Unter Einführung der an der äußeren Grenze der Atmosphäre herrschenden Temperatur $\tau\left(= \frac{T}{\sqrt[4]{2}}\right)$ kann man dafür auch schreiben:

(14a) $$t^4 = \tau^4\left[1+\frac{k}{g}p\right]\cdot$$

Geht man hiermit in (3) ein, so erhält man:

(15) $$\frac{M}{R}gh = \int^t \frac{4t^4\,dt}{t^4-\tau^4} = \tau\left[4\frac{t}{\tau} + \log\frac{\frac{t}{\tau}-1}{\frac{t}{\tau}+1} - 2\,\mathrm{arc}\,tg\,\frac{t}{\tau}\right] + \text{const.}$$

Diese Gleichung giebt die Temperatur als Funktion der Tiefe. Die zugehörige Dichte folgt aus:

(16) $$\varrho = \frac{M}{R}\frac{p}{t} = \frac{Mg}{Rk}\,\frac{t^4-\tau^4}{t\tau^4}\cdot$$

Das nachstehende Täfelchen giebt die Werte, die aus (11), (15) und (16) folgen, wenn man die Sonnenatmosphäre aus unserer Luft bestehen läßt. Für den Absorptionskoeffizienten der Luft hat man etwa $k = 0{,}6$ anzusetzen. Aus der Effektivtemperatur $T = 6000^0$ folgt für die Temperatur der äußeren Grenze $\tau = 5050^0$. Die Tiefe h ist von einer Stelle an gerechnet, an welcher die Temperatur das $1^1/_2$fache dieser Grenztemperatur beträgt. Damit sind die Grundlagen der Rechnung gegeben.

Tiefe h	t	m	ϱ
$-\infty$	5050^0	0.000	0.00
-36.9 km	5060^0	0.008	0.02
-19.1	5300^0	0.215	0.51
0.0	7570^0	4.06	6.8
$+12.0$	10100^0	15.0	18.7
$+55.7$	20200^0	255.0	159.4

Um die entsprechende Tabelle für eine Atmosphäre aus Wasserstoff zu erhalten, hätte man die Tiefe h mit 14,4 zu multiplizieren, die Dichte ϱ einerseits durch 14,4 zu dividieren, andrerseits mit einem Faktor zu multiplizieren, der angiebt, um wie viel dieselbe Masse Wasserstoff durchsichtiger ist, als Luft. Die beiden Spalten t und m blieben unverändert.

Man erkennt, daß sich das Strahlungsgleichgewicht mit der Erhebung über die Sonne mehr und mehr dem isothermen Gleichgewicht, welches der Grenztemperatur τ entspricht, nähert und, daß es, wie dieses, theoretisch eine unendliche Erstreckung der Atmosphäre zur Folge hat.

3. Stabilität des Strahlungsgleichgewichts.

Von besonderem Interesse ist ein Vergleich des Temperaturgradienten bei Strahlungs- und bei adiabatischem Gleichgewicht. Ist nämlich der Temperaturgradient kleiner als bei adiabatischem Gleichgewicht, so gerät eine aufsteigende Luftmasse in Schichten, welche wärmer und dünner sind, als sie selbst ankommt, sie erfährt daher einen Druck nach unten. Ebenso erfährt dann eine absteigende Luftmasse einen Druck nach oben. Ein Gleichgewicht mit kleinerem Temperaturgradienten, als das adiabatische, ist daher stabil, umgekehrt eines mit größerem Temperaturgradienten instabil.

Nachr. Kgl. Ges. d. Wiss. zu Gött. Math. Phys. Klasse 1906, *No. 1*, 41–53

Für adiabatisches Gleichgewicht gilt nach (6):

$$\frac{dt}{dh} = \frac{k-1}{k}\frac{Mg}{R},$$

für Strahlungsgleichgewicht nach (15):

$$\frac{dt}{dh} = \frac{1}{4}\left(1-\frac{\tau^4}{t^4}\right)\frac{Mg}{R}.$$

Stabilitätsbedingung ist also:

$$(17) \qquad 1-\frac{\tau^4}{t^4} < 4\frac{k-1}{k}$$

welche für $k > {}^4/_3$ immer erfüllt ist.

Das Strahlungsgleichgewicht ist demnach überall stabil, so lange das Gas, welches die Atmosphäre bildet, ein-, zwei- oder dreiatomig ist. Für mehratomige Gase würde in tieferen Schichten (von höherer Temperatur t) Instabilität eintreten.

Es wird daher hier die Vorstellung nahe gelegt, daß eine äußere Schale der Sonnenatmosphäre sich in stabilem Strahlungsgleichgewicht befindet, während sich vielleicht in der Tiefe eine dem adiabatischen Gleichgewicht angenäherte Zone auf- und absteigender Ströme erstreckt, die dann zugleich die Entnahme der Energie aus ihren eigentlichen Quellen besorgen wird.

4. Helligkeitsverteilung auf der Sonnenscheibe.

Jeder Temperaturverteilung längs der Vertikalen in der Sonnenatmosphäre entspricht bei unseren Annahmen eine bestimmte Verteilung der Helligkeit auf der Sonnenscheibe.

Wir hatten früher die Gesamtenergie A betrachtet, die durch die Sonnenatmosphäre nach außen wandert, ohne die einzelnen Bestandteile zu trennen, welche unter verschiedenen Neigungen gegen die Vertikale laufen, und wir hatten den Absorptionskoeffizienten für die Gesamtenergie mit a bezeichnet. Dieses a giebt einen Durchschnittswert aus den für alle möglichen Neigungen gültigen Absorptionskoeffizienten.

Wir wollen jetzt die in einer bestimmten Richtung wandernde Strahlung isoliert betrachten und mit $F(i)$ grade die Strahlung bezeichnen, die sich unter dem Winkel i gegen die Vertikale bewegt. Es bezeichne α den Absorptionskoeffizienten für Strahlung, welche die Atmosphäre normal durchsetzt. Dann ist offenbar $\frac{\alpha}{\cos i}$

der Absorptionskoeffizient für unter dem Winkel i verlaufende Strahlung. Man erhält daher für F in völliger Analogie zu (8) die Differentialgleichung:

$$\frac{dF}{dh} = -\frac{\alpha}{\cos i}(E-F) \tag{18}$$

oder:

$$\frac{dF}{d\mu} = -\frac{1}{\cos i}(E-F),$$

wenn man die Abkürzung:

$$\mu = \int_{-\infty}^{h} \alpha\, dh \tag{19}$$

einführt. Die Integration ergiebt für die aus der Atmosphäre austretende Strahlung:

$$F(i) = \int_0^\infty E e^{-\mu \sec i}\, d\mu \sec i. \tag{20}$$

Es läßt sich also $F(i)$ berechnen, sobald man die Temperaturverteilung längs der Vertikalen und damit E als Funktion von μ kennt.

Nun hängt aber μ mit der früher eingeführten optischen Masse m zusammen. Man betrachte die Gesamtstrahlung, die auf ein horizontales Flächenelement ds innerhalb der Atmosphäre von unten auffällt. Dieselbe wird gegeben durch das Integral über die aus allen möglichen Richtungen eintreffende Strahlung:

$$J = 2\pi\, ds \int_0^{\pi/2} di \sin i \cos i\, F(i).$$

Die Absorption, welche diese Strahlung innerhalb der Schicht dh erleidet, wird sein:

$$dJ = 2\pi ds \int_0^{\pi/2} di \sin i \cos i\, F(i) \cdot \frac{\alpha\, dh}{\cos i}$$

$$= 2\pi\, ds\, \alpha\, dh \int_0^{\pi/2} di \sin i\, F(i).$$

Der früher verwandte Absorptionskoeffizient für die Gesamtenergie a war definiert durch die Beziehung:

$$\frac{dJ}{J} = a\, dh.$$

Nachr. Kgl. Ges. d. Wiss. zu Gött. Math. Phys. Klasse 1906, *No. 1*, 41–53

Die Vergleichung mit den vorstehenden Formeln ergiebt:

$$a = \alpha \cdot \frac{\int_0^{\pi/2} di \sin i\, F(i)}{\int_0^{\pi/2} di \sin i \cos i\, F(i)}$$

Wenn $F(i)$ für kleine Neigungen i einigermaßen konstant ist und sich erst für i in der Nähe von 90^0 rascher ändert — wie das bei der Sonne nach den unten folgenden empirischen Ergebnissen der Fall ist — so erhält man eine Näherung für a, indem man $F(i)$ überhaupt als unabhängig von i betrachtet, und zwar folgt dann durch Ausführung der Integrale:

(21) $$a = 2\alpha.$$

Von dieser Beziehung wollen wir Gebrauch machen. Es ergiebt sich aus (9) und (19):

$$m = 2\mu$$

und damit statt (20):

(22) $$F(i) = \int_0^\infty E e^{-\frac{m}{2}\sec i} \frac{dm}{2} \sec i.$$

Es ist also nunmehr $F(i)$ bekannt, sobald E als Funktion der optischen Masse m gegeben ist. Die Funktion $F(i)$ liefert aber auch sofort die Helligkeitsverteilung auf der Sonnenscheibe. Denn die Strahlung, die wir auf der Sonnenscheibe im scheinbaren Abstand r vom Centrum der Scheibe beobachten, hat offenbar die Sonnenatmosphäre unter einem Winkel i durchsetzt, der durch die Gleichung bestimmt wird:

(23) $$\sin i = \frac{r}{R}.$$

Hierbei bezeichnet R den scheinbaren Sonnenradius. Die Verbindung von (22) und (23) liefert zu jedem r die zugehörige Helligkeit F.

Sehr übersichtlich wird der Zusammenhang zwischen der Strahlungsverteilung in der Tiefe E und der Helligkeitsverteilung auf der Oberfläche F, wenn sich E in eine Potenzreihe nach m entwickeln läßt:

(24) $$E = \beta_0 + m\beta_1 + m^2\beta_2 + \cdots$$

Dann folgt nämlich sofort aus (22):

$$F = \beta_0 + 2 \cdot 1! \quad \beta_1 \cos i + 4 \cdot 2! \quad \beta_2 \cos^2 i + \cdots \tag{25}$$

Läßt sich E als eine Summe gebrochener Potenzen von m darstellen:

$$E = \sum_n \beta_n \, . \, m^{\gamma_n}, \tag{26}$$

so folgt für F:

$$F = \sum_n \Gamma(\gamma_n) \, \beta_n \, (2 \cos i)^{\gamma_n}, \tag{27}$$

wo Γ die Γ-Funktion bedeutet. Auch hier ist der Uebergang von E zu F also noch einfach zu bewerkstelligen.

Wir wollen insbesondere betrachten, wie sich die Helligkeitsverteilung bei adiabatischem und bei Strahlungsgleichgewicht stellt.

Für Strahlungsgleichgewicht gilt nach (11):

$$E = \frac{A_0}{2}(1 + m).$$

Daraus folgt nach (24) und (25):

$$F(i) = \frac{A_0}{2}(1 + 2 \cos i)$$

oder, wenn man die Helligkeit im Centrum der Sonnenscheibe ($i = 0$) als Einheit nimmt:

$$F(i) = \frac{1 + 2 \cos i}{3}. \tag{28}$$

Für adiabatisches Gleichgewicht galt die Beziehung (5):

$$\frac{t}{t_0} = \left(\frac{p}{p_0}\right)^{\frac{k-1}{k}}$$

Wenn wir ferner, wie oben, die Absorption für alle Farben gleich und der Dichte proportional voraussetzen, so besteht zwischen p und m der Zusammenhang (13):

$$m = \frac{k}{g} p.$$

Damit erhält man für E:

$$E = c \, . \, t^4 = c_1 p^{4 \frac{(k-1)}{k}} = c_2 m^{4 \frac{(k-1)}{k}}$$

Nachr. Kgl. Ges. d. Wiss. zu Gött. Math. Phys. Klasse 1906, *No. 1,* 41–53

wobei c_1 und c_2 neue Konstante bedeuten. Der zugehörige Ausdruck vom F wird nach (26) und (27):

$$F(i) = c_2 \Gamma\left[4\frac{(k-1)}{k}\right](\cos i)^{4\frac{(k-1)}{k}},$$

oder, wenn man wieder die centrale Helligkeit als Einheit wählt:

$$(29) \qquad F(i) = (\cos i)^{4 \cdot \frac{k-1}{k}}.$$

Die Formeln (28) und (29) sind mit der Beobachtung zu vergleichen. Außer den hier nicht in Frage kommenden spektral-photometrischen Untersuchungen für einzelne Farbengebiete giebt es eine Reihe von Messungen, die mit Thermosäulen und Bolometer angestellt sind und die angeben, wie sich die von allen Wellenlängen zugleich gelieferte Gesamtstrahlung über die Sonnenscheibe verteilt. Herr G. Müller hat diese Messungen in seiner „Photometrie der Gestirne" pag. 323 zusammengestellt und zu den Mittelwerten vereinigt, die in der zweiten Spalte des unten folgenden Täfelchens angegeben sind. Die theoretischen Werte für Strahlungsgleichgewicht und adiabatisches Gleichgewicht nach den Formeln (28) und (29) sind daneben gesetzt. Für das adiabatische Gleichgewicht ist $k = 4/3$ gesetzt, was einem 3-atomigen Gase entspricht. Ein- oder zweiatomige Gase würden eine noch schlechtere Uebereinstimmung geben und die physikalische Wahrscheinlichkeit spricht gewiß gegen mehr als 3-atomige Gase in den äußeren Teilen der Sonnenatmosphäre.

$\frac{r}{R}$	Messung.	Strahlungs-gleichgew.	Adiabat. Gleichgew.
0.0	1.00	1.00	1.00
0.2	0.99	0.99	0.98
0.4	0.97	0.95	0.92
0.6	0.92	0.87	0.80
0.7	0.87	0.81	0.71
0.8	0.81	0.73	0.60
0.9	0.70	0.63	0.44
0.96	0.59	0.52	0.28
0.98	0.49	0.47	0.20
1.00	(0.40)	0.33	0.00

Man sieht, daß das Strahlungsgleichgewicht die Helligkeitsverteilung auf der Sonnenscheibe so gut darstellt, als bei den vereinfachten Voraussetzungen, unter denen hier gerechnet worden ist, erwartet werden kann, daß das adiabatische Gleichgewicht hingegen ein ganz anderes Aussehn der Sonnenscheibe zur Folge hätte. Damit hat die Einführung des Strahlungsgleichgewichts eine gewisse empirische Rechtfertigung gefunden.

Meteorolog. Zeitschr. *30* (1913) 454–456

liegenden Schichten ist instabil; die Temperaturen und die Instabilität nehmen nach unten immer rascher zu."

Der Satz ist sehr wichtig: er sagt uns, daß aus dem Strahlungsgleichgewicht heraus konvektiv-, nahe dem Erdboden, auch mechanisch-instabile Verhältnisse entstehen. Die daraus resultierenden Konvektionsströme zerstören wieder den Zustand des Strahlungsgleichgewichtes, sie bringen Temperatursteigerung hervor bis hinauf zur Stratosphäre.

Es werden eine Reihe von Ausblicken gegeben, welche dieser Satz eröffnet; hier sei nur hervorgehoben, daß z. B. die Tatsache, daß Gewitter über den Meeren überhaupt und meist nachts auftreten, durch die tieferen Temperaturen erklärt werden kann, die sich in der Höhe durch Ausstrahlung einstellen können (z. B. in 3000 m extrem — 38°!).

Bei der Berechnung mußte die Annahme gemacht werden, daß die ganze Sonnenstrahlung gleichmäßig über die Erde verteilt werde. Das Resultat der Berechnung gibt dieser Annahme recht, sagt aber damit auch, daß die wirklich beobachteten Temperaturen der höheren Breiten nicht durch Sonnenstrahlung an Ort und Stelle erzeugt sind, sondern daß die allgemeine Zirkulation der Atmosphäre die höheren Breiten wesentlich günstiger stellt, als ihnen auf Grund ihrer Lage zukommen würde.

Als Anhang wird das umgekehrte Problem behandelt: Gegeben seien die Temperaturen der Atmosphäre und die äußeren Strahlungsquellen. Gesucht wird die Strahlung der Atmosphäre; es liegen bereits eine Reihe von experimentellen Ermittelungen vor, mit denen Herr Emden seine Resultate vergleichen konnte. Die fast vollkommene Übereinstimmung zwischen Experiment und Theorie stützt auch die vorausgegangenen Berechnungen des Strahlungsgleichgewichtes und rechtfertigt insbesondere das Vorgehen, die Strahlung nur in kurzwellige und langwellige Strahlung zu zerlegen, innerhalb jeder derselben aber Graustrahlung anzunehmen.

Der letzte Abschnitt weist die ungeheure Bedeutung der Atmosphäre für den Wärmehaushalt der Atmosphäre nach in Überlegungen, die insbesondere wichtig sind für die Frage, welche Veränderungen des Klimas durch eine Veränderung der optischen Konstanten der Atmosphäre eintreten können.

Das Referat konnte nur die wichtigsten Ergebnisse der bedeutungsvollen Arbeit skizzieren. Mit ihr ist die Meteorologie um einen guten Schritt vorwärts gekommen.

A. Schmauß.

Bemerkung zur Berechnung des Strahlungsgleichgewichtes der Atmosphäre. In seiner Arbeit über „Strahlungsgleichgewicht und atmosphärische Strahlung" [1]) hat R. Emden das Strahlungsgleichgewicht der Atmosphäre behandelt, indem er eine Idealisierung des Problems einführte, welche die Wirklichkeit in den Hauptzügen wiedergibt und sich dabei rechnerisch gut behandeln läßt. Der Zweck der folgenden Zeilen ist, eine kleine Verallgemeinerung der Emdenschen Ableitungen mitzuteilen, die sich vielleicht für die weitere Untersuchung der Temperaturverteilung in der Atmosphäre als nützlich erweisen dürfte. Es wird von Emden die ganze die Atmosphäre durchsetzende Strahlung in einen kurzwelligen und einen langwelligen Bestandteil zerlegt. Die Bezeichnungen sind folgende: aufwärtsgehender Energiestrom A, abwärtsgehender B, Absorptionskoeffizienten auf die Einheit der Luftmasse k, Strahlung der Einheit der Luftmasse kE. Für die kurzwellige Strahlung erhalten diese Buchstaben den Index 1, für die langwellige Strahlung den Index 2. Die kurzwellige Strahlung besteht im wesentlichen in der von oben einfallenden Sonnenstrahlung, die an der oberen Grenze der Atmosphäre den Betrag σ haben möge und die nach Durchsetzung der Luftmasse m (von oben gerechnet) auf den Betrag:

$$B_1 = \sigma e^{-\int_0^m k_1 dm} \quad \ldots \ldots \ldots \ldots \ldots \quad (1)$$

abgeschwächt sein wird. Eine aufwärtsgehende kurzwellige Strahlung besteht nicht ($A_1 = 0$). Für die langwellige Strahlung gelten die Differentialgleichungen:

$$\frac{dA_2}{dm} = k_2 A_2 - k_2 E_2, \qquad \frac{dB_2}{dm} = -k_2 B_2 + k_2 E_2 \quad \ldots \ldots \quad (2)$$

[1]) Sitzungsberichte d. Königl. Bayer. Akad. d. Wissensch. München 1913.

welche einfach zum Ausdruck bringen, wieviel auf- oder absteigende Strahlung beim Durchgang durch die Schicht dm absorbiert und wieviel durch Eigenstrahlung der Luft wieder hinzugeliefert wird. An der oberen Grenze der Atmosphäre ($m = 0$) gibt es keine absteigende langwellige Strahlung: Für

$$m = 0 \text{ ist } B_2 = 0 \qquad (3)$$

Das Strahlungsgleichgewicht erfordert, daß die von jeder Schicht (nach beiden Seiten) emittierte Energie $2k_2E_2dm$ gleich der von der Schicht absorbierten Energie $k_2dm(A_2 + B_2) + k_1dmB_1$ ist, daß also die Gleichung besteht:

$$2k_2E_2 = k_2(A_2 + B_2) + k_1B_1 \qquad (4)$$

daß ferner die gesamte in den Raum zurückgestrahlte Energie gleich der Energie der einfallenden Sonnenstrahlung ist: Für

$$m = 0 \text{ ist } A_2 = B_1 \qquad (5)$$

Das durch die Gleichung (1) bis (5) charakterisierte mathematische Problem wird von Emden gelöst, indem für k_1 und k_2 gewisse spezielle Funktionen von m angesetzt werden. Es soll hier gezeigt werden, daß sich auch, wenn k_1 und k_2 beliebige Funktionen von m sind, die strenge Lösung leicht angeben läßt.

Man setze:

$$\int_0^m k_1\,dm = s, \qquad \int_0^m k_2\,dm = r \qquad (\mathrm{I})$$

also: $k_1dm = ds$, $k_2dm = dr$, $\frac{k_1}{k_2} = \frac{ds}{dr}$, $s = r = 0$ für $m = 0$.

Die Gleichungen (1) bis (5) schreiben sich dann:

$$B_1 = \sigma e^{-s} \qquad (1')$$

$$\frac{dA_2}{dr} = A_2 - E_2, \qquad \frac{dB_2}{dr} = E_2 - B_2 \qquad (2')$$

$$B_2 = 0, \text{ für } r = 0 \qquad (3')$$

$$2E_2 = A_2 + B_2 + B_1\frac{ds}{dr} \qquad (4')$$

$$A_2 = B_1, \text{ für } r = 0 \qquad (5')$$

Die Addition und Subtraktion der Gleichungen (2') ergibt:

$$\frac{d(A_2 + B_2)}{dr} = A_2 - B_2, \qquad \frac{d(B_2 - A_2)}{dr} = 2E_2 - A_2 - B_2 = B_1\frac{ds}{dr} \qquad (6)$$

Die Integration der zweiten dieser Gleichungen gibt unter Berücksichtigung von (1') und Einführung der Integrationskonstanten γ:

$$B_2 - A_2 = \gamma - \sigma e^{-s} \quad \text{oder:} \quad B_1 + B_2 - A_2 = \gamma.$$

Bildet man diese Gleichung für $m = 0$, so folgt aus (3') und (5') für die Integrationskonstante: $\gamma = 0$. Es bleibt mithin:

$$A_2 - B_2 = \sigma e^{-s} \qquad (7)$$

Setzt man dies in die erste Gleichung (6) ein und integriert, so folgt unter Einführung der Integrationskonstante δ:

$$A_2 + B_2 = \delta + \sigma\int_0^r e^{-s}dr.$$

Für $r = 0$ folgt hieraus: $A_2 + B_2 = \delta$. Da aber für $r = 0$ wegen $B_2 = 0$ auch $A_2 + B_2 = A_2 - B_2$ und nach (7) $A_2 - B_2 = \sigma$ ist, so folgt $\sigma = \delta$ und somit:

$$A_2 + B_2 = \sigma\left\{1 + \int_0^r e^{-s}dr\right\}.$$

Die Einsetzung in (4') ergibt die Endformel:

$$2E_2 = \sigma\left\{1 + \frac{ds}{dr}e^{-s} + \int_0^r e^{-s}dr\right\} \qquad (\mathrm{II})$$

Aus E_2 ergibt sich die absolute Temperatur T an jeder Stelle der Atmosphäre nach dem Stefanschen Gesetz:

$$E_2 = c\,T^4 \quad \ldots\ldots\ldots\ldots\ldots \quad \text{(III)}$$

wobei numerisch $\frac{2c}{\sigma} = (214^0)^4$ ist.

Die Formeln I, II und III ermöglichen die Berechnung der Temperaturverteilung in der Atmosphäre, wenn die Absorptionskoeffizienten k_1 und k_2 beliebig als Funktionen der Luftmasse m gegeben sind.

An der oberen Grenze der Atmosphäre ($m = r = s = 0$) ergibt sich aus (II):

$$2\,E_2 = \sigma\left\{1 + \frac{ds}{dr}\right\} = \sigma\left\{1 + \frac{k_1}{k_2}\right\} \quad \ldots\ldots\ldots\ldots \quad (8)$$

in Übereinstimmung mit Emdens Formel (86).

Als Beispiel sollen die vorstehenden Formeln für den Ansatz der Absorptionskoeffizienten k_1 und k_2 ausgeführt werden, der den von Emden im Anschluß an Abbot und Fowle entwickelten Anschauungen entspricht. Hiernach hat die Luft an sich äußerst kleine Absorptionskoeffizienten (a_1 und a_2 für die kurzen und langen Wellen), die Absorption wird hauptsächlich bewirkt durch den Wasserdampf, dessen Beitrag zu den Absorptionskoeffizienten nach der empirischen Verteilung des Wasserdampfes durch $b_1 m^3$ bzw. $b_2 m^3$ für kurze bzw. lange Wellen wiedergegeben werden kann. Insgesamt hat man daher für die Absorptionskoeffizienten zu setzen:

$$k_1 = a_1 + b_1 m^3, \qquad k_2 = a_2 + b_2 m^3;$$

damit folgt:

$$s = a_1 m + \frac{b_1}{4} m^4, \qquad r = a_2 m + \frac{b_2}{4} m^4,$$

$$E_2 = \frac{\sigma}{2}\left\{\frac{a_1 + b_1 m^3}{a_1 + b_2 m^3}\, e^{-\left(a_1 m + \frac{b_1}{4} m^4\right)} + 1 + \int_0^m e^{-\left(a_1 m + \frac{b_1}{4} m^4\right)} (a_2 + b_2 m^3)\, dm\right\}.$$

Das Integral ist, wenn man nicht mechanisch quadrieren will, immer in der Weise leicht ausführbar, daß man die Exponentialgröße nach dem Exponenten s entwickelt, der für die Erdatmosphäre klein bleibt.

Man könnte sich die Aufgabe stellen, die Konstanten a_1, a_2, b_1, b_2 so zu bestimmen, daß die Temperaturverteilung in der höheren Atmosphäre möglichst gut wiedergegeben wird. Ich will mich hier mit einem Beispiel begnügen, bei dem die Koeffizienten a_1, a_2 ziemlich willkürlich angesetzt, die Koeffizienten b_1, b_2 aus Emdens Angaben entnommen sind. Es sei:

$$k_1 = 0.005 + 0.4\, m^3, \qquad k_2 = 0.01 + 9.2\, m^3.$$

Dann folgt:

m	Isotherme Höhe	$2\,E_2/\sigma$	T abs.	Cels.
0.0	∞	1.50	236°	—37°
0.2	13 km	1.11	219	—53
0.4	7 „	1.11	219	—53
0.6	4 „	1.36	231	—42
0.8	2 „	1.97	253	—20
1.0	0 „	3.24	287	+14

Man erhält das gewünschte Temperaturminimum in der Stratosphäre. Da für $m = 0$ $\frac{k_1}{k_2} = \frac{a_1}{a_2}$ wird, so hängt die Temperatur an der oberen Grenze der Atmosphäre gemäß der Formel (8) ganz von dem gewählten numerischen Verhältnis $\frac{a_1}{a_2}$ ab.

K. Schwarzschild.

Die Gestalt fallender Regentropfen. Scheint die Sonne auf fallende Regentropfen, so sieht man diese, wie man am besten vor dunklem Hintergrund in den Partien unter der Sonne beobachten kann, nur auf bestimmten Strecken ihrer Bahn aufblitzen. Erklärt wird dieses durch Reflexionsbilder, vorwiegend an der Innenseite der Tropfenrückwand, welche nur in bestimmten Richtungen am hellsten erscheinen. In gerade diesen Richtungen müßte — so sollte man meinen — der Glanz sich bei allen fallenden Tropfen zeigen, man hätte demnach einen hellen Streifen (Kreis) um die Sonne als Mittelpunkt zu erwarten. In

Sind im Sonnenspektrum Wärmestrahlen von großer Wellenlänge vorhanden?

Von H. Rubens und K. Schwarzschild.

Langleys berühmte Messungen im ultraroten Spektrum der Sonne und des Mondes enden bekanntlich bei einer Wellenlänge, welche zwischen 10 und 11 μ liegt[1]. Daß jenseits dieser Grenze keine merkliche Energie spektrobolometrisch nachgewiesen werden konnte, liegt, wie früher gezeigt worden ist, an der Absorption des in der Erdatmosphäre vorhandenen Wasserdampfs. Nach Messungen, welche der eine von uns vor 16 Jahren in Gemeinschaft mit E. Aschkinass ausgeführt hat[2], besitzt der Wasserdampf zwischen 9 und 11 μ ein Gebiet relativ hoher Durchlässigkeit. Aber jenseits 11 μ nimmt die Absorption dieses Gases mit wachsender Wellenlänge wieder rasch zu, und zwar derart, daß eine Wasserdampfschicht von 70 cm Länge und Atmosphärendruck, welche die Strahlung der Wellenlänge 11 μ kaum merklich schwächt, wenig mehr als ein Zehntel derjenigen der Wellenlänge 17.5 μ hindurchläßt. Dabei zeigt die Absorptionskurve eine Reihe stark hervortretender Maxima und Minima, welche von der Existenz einzelner Absorptionsstreifen Zeugnis ablegen. Solche Streifen wurden bei $\lambda = 11.6\mu$, 12.4μ, 13.4μ, 14.3μ und 15.7μ beobachtet. Auch in dem Spektralgebiet zwischen 18 und 20 μ, welches mit Hilfe eines Sylvinprismas noch untersucht werden kann, erwies sich die Wasserdampfabsorption als sehr bedeutend. Ebenso konnte für die Reststrahlen von Flußspat mit der mittleren Wellenlänge 24.4 μ in einer 40 cm dicken Dampfschicht eine Absorption von 69 Prozent festgestellt werden.

Nachdem es durch Untersuchung anderer Reststrahlengruppen gelungen war, im ultraroten Spektrum um einige Oktaven weiter vorzudringen, zeigte es sich, daß auch die langwelligen Reststrahlen von

[1] S. P. Langley, Phil. Mag. I. 26, S. 505, 1888; H. Rubens, Wied. Ann. 53, S. 267, 1894 und 54, S. 476, 1895.
[2] H. Rubens und E. Aschkinass, Wied. Ann. 64, S. 598, 1898.

Steinsalz, Sylvin, Bromkalium, Jodkalium, Chlorsilber, Bromsilber, Bleichlorid und Calomel, welche sich auf das Wellenlängenbereich zwischen 52 und 113 μ verteilen, bereits in einer Wasserdampfschicht von 40 cm Dicke auf einen geringen Bruchteil ihrer Anfangsintensität geschwächt werden[1]. Auch hier erwies sich die Absorption als wesentlich selektiv. Besonders starke Absorption wurde bei den Wellenlängen 50μ, 58μ, 66μ und 79μ beobachtet, relativ hohe Durchlässigkeit dagegen bei den Wellenlängen 47μ, 54μ, 62μ, 75μ und 91μ. Aber auch an diesen Stellen besitzt die Wasserdampfabsorption noch so erhebliche Werte, daß die Strahlen in einer Dampfschicht von einigen Metern Länge vollkommen vernichtet werden. Es ist nach diesem Ergebnis sehr unwahrscheinlich, daß in der auf die Erdoberfläche gelangenden Sonnenstrahlung Wellenlängen des betrachteten Bereichs in merklicher Stärke vorhanden sind, und in der Tat haben Versuche, die Sonne als Lichtquelle zur Erzeugung langwelliger Reststrahlen zu benutzen, zu negativem Ergebnis geführt. Derartige Messungen wurden 1898 für Reststrahlen von Flußspat[2] und 1907 für Reststrahlen von Steinsalz[3] ausgeführt.

Wenn nun auch in einzelnen Spektralgebieten, wie z. B. in dem zwischen 32 und 45 μ gelegenen Teile des Spektrums, bisher keine Messung der Absorption des Wasserdampfs ausgeführt werden konnte, so darf es nach dem Vorausgehenden doch als sehr wahrscheinlich angenommen werden, daß die gesamte Sonnenstrahlung in dem Wellenlängenbereich zwischen 11 und 113 μ durch den Wasserdampf der Atmosphäre zurückgehalten wird. Es bleibt jedoch die Möglichkeit bestehen, daß für Wärmestrahlen von noch erheblich größerer Wellenlänge, welche noch der Messung zugänglich sind, die Atmosphäre wieder durchsichtig ist. Es ist dies schon deshalb nicht unwahrscheinlich, weil der Wasserdampf selbst in Schichten von mehreren Metern Dicke keine merkliche Absorption auf kurze Hertzsche Wellen ($\lambda = 1.75$ cm) ausübt, wie aus einer kürzlich erschienenen Arbeit des Hrn. F. Eckert[4] hervorgeht.

Zur Prüfung der Frage, ob ein merklicher Teil der Sonnenstrahlung des langwelligen Spektralbereichs zur Erdoberfläche gelangt, kann die Quarzlinsenmethode[5] zur Isolierung langwelliger Wärmestrahlen mit Vorteil verwendet werden. Mit Hilfe dieser Methode gelingt es, alle Strahlen

---

[1] H. Rubens und E. Aschkinass, Wied. Ann. 65, S. 241, 1898; H. Rubens und H. Hollnagel, diese Berichte S. 26, 1910 und H. Rubens, diese Berichte S. 513, 1913.

[2] H. Rubens und E. Aschkinass a. a. O.

[3] E. F. Nichols, Astrophysical Journal 26, S. 46, 1907.

[4] F. Eckert, Dissertation, Berlin 1913.

[5] H. Rubens und R. W. Wood, diese Berichte S. 1122, 1910.

einer Lichtquelle, welche jenseits einer gegebenen Grenze liegen, von den kurzwelligen Strahlen abzusondern und zu messen. Beträgt die Dicke der im Strahlengange befindlichen Quarzschicht etwa 2 cm, so werden alle Strahlen, deren Wellenlänge kleiner als 70 μ ist, durch Brechung und Absorption beseitigt. Die langwelligeren Wärmestrahlen dagegen werden, soweit dies bisher untersucht werden konnte, von der Quarzschicht um so besser hindurchgelassen, je größer ihre Wellenlänge ist. Bei einer Wellenlänge von etwa 300 μ beträgt die Absorption der genannten Quarzschicht nur noch wenige Prozente.

Von der Gesamtstrahlung, welche von dem Glühstrumpfe eines Auerbrenners oder von einem auf 500° C erhitzten schwarzen Körper ausgeht, wird durch die Quarzlinsenmethode unter den angegebenen Bedingungen eine Teilstrahlung isoliert, deren Energieverteilungskurve bei 70 μ beginnt, sehr steil ansteigt, bei etwa 100 μ ein Maximum besitzt und dann nach Seite der langen Wellen langsam abfällt. Die Energieverteilung dieser langwelligen Strahlung ist jedoch, wie aus den aufgenommenen Interferometerkurven geschlossen werden muß, eine sehr komplizierte, und zwar konnte gezeigt werden, daß die sich bemerkbar machenden Ausbuchtungen der Energiekurve hauptsächlich durch die selektive Absorption des Wasserdampfs der Zimmerluft hervorgerufen werden. Ferner wurde beobachtet, daß eine Wasserdampfschicht von 40 cm Dicke und Atmosphärendruck mehr als vier Fünftel der isolierten Strahlung absorbiert. Diese Absorption ist offenbar ebenso wie in den kurzwelligeren Spektralgebieten von ausgesprochen selektivem Charakter[1].

Wird statt des Auerstrumpfs bzw. schwarzen Körpers eine Quarzquecksilberlampe als Strahlungsquelle gewählt, so erhält man bei Anwendung der Quarzlinsenmethode bekanntlich eine langwellige Strahlung von ziemlich komplizierter Zusammensetzung[2]. Ein Teil derselben rührt von den Wänden des glühenden Quarzrohrs her. Dieser besitzt angenähert dieselbe Energieverteilung wie die vorher betrachtete langwellige Strahlung des Auerbrenners. Außerdem aber enthält der isolierte Strahlenkomplex eine Teilstrahlung, welche von dem leuchtenden Quecksilberdampf der Lampe ausgesandt wird und eine wesentlich größere mittlere Wellenlänge besitzt. Es wurde gezeigt, daß diese langwellige Quecksilberdampfstrahlung, welche durch Einschalten eines Stücks schwarzer Pappe ziemlich vollständig von der kurzwelligeren Quarzrohrstrahlung befreit werden kann, hauptsächlich aus zwei Strahlungskomplexen besteht, deren Maxima bei den Wellenlängen 218 μ und 343 μ liegen.

[1] Vgl. H. Rubens, diese Berichte a. a. O., S. 540.

[2] H. Rubens und O. von Baeyer, diese Berichte S. 339 und S. 666, 1911.

With H. Rubens in Sitz. Ber. Kgl. Preuss. Akad. d. Wiss. Berlin 1914, 702–708

Auch die langwellige Quecksilberdampfstrahlung wird von dem Wasserdampf beträchtlich absorbiert, und zwar schwächt eine 40 cm dicke Dampfschicht jene Strahlung etwa auf die Hälfte. Es konnte jedoch gezeigt werden, daß diese Schwächung hauptsächlich in einer Verminderung der Energie des kurzwelligeren Streifens besteht, dessen Maximum bei 218 μ gelegen ist, während der langwelligere Streifen, welcher bei 343 μ ein Energiemaximum besitzt, erheblich weniger stark absorbiert wird[1]. Hiernach erschien die Annahme nicht unberechtigt, daß Strahlen von noch größerer Wellenlänge ($\lambda > 400\ \mu$) wieder imstande sein würden, die Erdatmosphäre zu durchdringen, ohne durch den Wasserdampfgehalt derselben bis zur Unmerklichkeit geschwächt zu werden. Zur Entscheidung dieser Frage haben wir in dem Astrophysikalischen Observatorium in Potsdam einige Versuche angestellt, über welche im folgenden kurz berichtet werden soll.

Wir verwendeten einen großen Heliostaten mit einem vorderseitig versilberten Spiegel S_1 von 40 × 50 cm. Die Silberschicht war mit Absicht nicht poliert worden und zeigte eine matte, milchig weiße Oberfläche, welche die kurzwelligen Wärmestrahlen nur sehr unvollkommen reflektierte, während ihr Reflexionsvermögen für die mit Hilfe der Quarzlinsenanordnung isolierte langwellige Strahlung eines Auerbrenners über 95 Prozent betrug. Hierdurch wurde erreicht, daß die in der Sonnenstrahlung etwa vorhandenen langwelligen Wärmestrahlen nahezu ungeschwächt in unseren Apparat gelangten, während die ungemein intensiven kurzwelligen Strahlen demselben nach Möglichkeit ferngehalten wurden. Nach Reflexion an dem Heliostatenspiegel wurden die Strahlen von einem vorderseitig versilberten Hohlspiegel H von 40 cm Durchmesser und 90 cm Brennweite aufgefangen und kurz vor ihrer Vereinigung nochmals von dem Planspiegel S_2 unter etwa 45° reflektiert, so daß sie auf dem Eintrittsdiaphragma a unserer Quarzlinsenanordnung ein Sonnenbild von etwa 8 mm Durchmesser lieferten. Der Strahlengang verlief nun weiter in der üblichen Weise. Die Quarzlinsen L_1 und L_2 sind mit den früher verwendeten identisch[2]; auch im übrigen ist die Versuchsanordnung von a ab mit der früher beschriebenen völlig übereinstimmend, nur betrug der Durchmesser des Diaphragmas a entsprechend der geringen Größe des Sonnenbildes bei diesen Versuchen nur 8 mm. Das Mikroradiometer M befand sich in einer luftdicht schließenden, mit Quarzfenster versehenen Hülle.

[1] H. Rubens und O. von Baeyer, diese Berichte S. 802, 1913.

[2] Die symmetrischen bikonvexen Quarzlinsen hatten 10 cm Durchmesser und 25 cm Brennweite für Lichtstrahlen; ihre Brennweite für lange Wellen berechnet sich hiernach zu etwa 12 cm. Der Abstand beider Linsen von den Diaphragmen a und b bzw. von dem Thermoelement des Mikroradiometers betrug je 24 cm.

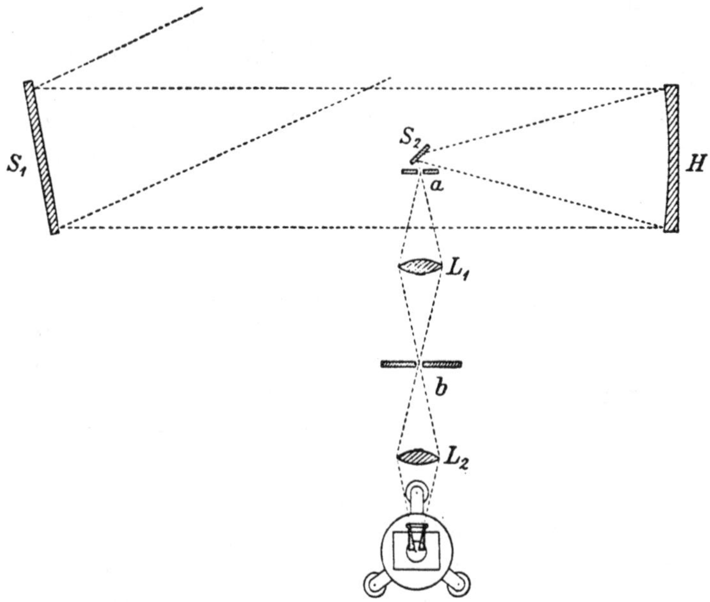

Die temperaturempfindliche Lötstelle des Thermoelements war mit Natronwasserglas und Ruß geschwärzt.

Daß die Empfindlichkeit der uns zur Verfügung stehenden Meßinstrumente zum Nachweis des jenseits 400 μ gelegenen Teiles der Sonnenstrahlung ausreichen würde, wenn man die Sonne als schwarzen Körper betrachtet, läßt sich durch folgende Überschlagsrechnung leicht zeigen. Mit der von uns verwendeten Quarzlinsenanordnung läßt sich aus der Gesamtstrahlung eines schwarzen Körpers von 500° C eine Teilstrahlung der oben beschriebenen Art aussondern, welche einen Ausschlag unseres Mikroradiometers von 15 mm hervorbringt. Mehr als die Hälfte dieser Strahlungsenergie fällt in das Wellenlängenbereich zwischen 100 und 150 μ. — Auf das Spektralgebiet, welches bei viermal größeren Wellenlängen, nämlich zwischen den Grenzen $\lambda = 400\ \mu$ und $\lambda = 600\ \mu$ gelegen ist, würde nach der Planckschen Strahlungsformel[1] ein 64 mal geringerer Energiebetrag entfallen. Berücksichtigt man jedoch, daß die im Strahlengange befindliche Quarzschicht die zwischen 100 und 150 μ vorhandene Strahlung durch Absorption etwa auf ein Drittel reduziert, während sie diejenige des durch 400 und 600 μ begrenzten Wellenlängenbereichs fast ungeschwächt hindurch-

[1] Im Gebiete langer Wellen geht die Plancksche Strahlungsformel in die Rayleighsche über, nach welcher die Intensität für ein gegebenes Wellenlängenintervall $d\lambda$ und für eine gegebene Temperatur T proportional $\frac{T}{\lambda^4}\,d\lambda$ ist.

With H. Rubens in Sitz. Ber. Kgl. Preuss. Akad. d. Wiss. Berlin 1914, 702–708

läßt, so darf man sagen, daß angenähert 2 bis 3 Prozent der beobachteten Strahlungsintensität dem langwelligen Spektralgebiet zwischen 400 und 600 μ angehören. Dieser Energiemenge würde ein Ausschlag unseres Mikroradiometers von etwa 0.3 bis 0.4 mm entsprechen.

Verwenden wir nun die Sonne an Stelle des benutzten schwarzen Körpers von 500° C als Lichtquelle, so müssen wir, um einen quantitativen Vergleich der Strahlung zu ermöglichen, dafür sorgen, daß die Größe des erzeugten Sonnenbildes ausreicht, um das Eintrittsdiaphragma unserer Quarzlinsenanordnung vollständig zu bedecken, und daß ferner die Öffnung der Strahlenkegel genügend groß ist, um die Quarzlinsen vollkommen auszufüllen. Beiden Bedingungen ist in unserem Apparate genügt. Betrachten wir die Sonne als einen schwarzen Körper von 6000° C, d. h. von einer Temperatur, deren Differenz gegen die Zimmertemperatur zwölfmal größer ist als diejenige unseres Versuchskörpers, so haben wir dementsprechend an allen Stellen des langwelligen Spektrums, wenn keine Absorption in der Atmosphäre stattfindet, einen zwölfmal größeren Ausschlag zu erwarten, als wir ihn mit unserem Versuchskörper erhielten, also für das zwischen 400 und 600 liegende Spektralbereich etwa 4 mm. Da nun unser Mikroradiometer einen Ausschlag von 0.2 mm mit Sicherheit erkennen läßt, so würde also noch der zwanzigste Teil der auf jenes langwellige Bereich entfallenden Energiemenge mit unserer Versuchsordnung zweifellos nachzuweisen sein.

Die Versuche wurden an zwei klaren Märztagen zwischen 11 Uhr vormittags und 2 Uhr nachmittags ausgeführt. Das Ergebnis war ein rein negatives. Zwar erhielten wir, sobald das Sonnenbild auf das Diaphragma a unserer Quarzlinsenanordnung fiel, beim Aufziehen eines zwischen L_1 und b eingeschalteten Klappschirms kleine Ausschläge von 1.5 bis 1.7 mm, es ließ sich jedoch leicht zeigen, daß diese Ausschläge von diffuser kurzwelliger Strahlung herrührten; denn ihre Größe verminderte sich um weniger als 10 Prozent, wenn eine 1 cm dicke Steinsalzplatte in den Strahlengang eingeschaltet wurde, dagegen verschwanden die Ausschläge vollkommen, wenn das Diaphragma b durch ein Blatt schwarzen Papiers bedeckt wurde. Während der Versuche wurde die Empfindlichkeit der Anordnung wiederholt dadurch geprüft, daß die Sonnenstrahlung abgeblendet und ein Auerbrenner vor dem Diaphragma a angebracht wurde. Man erhielt dann einen Ausschlag von 20 mm, welcher nur von langwelliger Strahlung herrührte.

Nach dem Vorausgehenden müssen wir entweder annehmen, daß die von der Sonne emittierte Strahlung des zwischen 80 und 600 μ gelegenen Spektralgebiets von der Erdatmosphäre nahezu vollständig absorbiert wird oder daß die Sonne diese langwellige Strahlung nicht in einem mit unseren Instrumenten meßbaren Betrage aussendet. Im letzteren Falle müßte, wie wir gesehen haben, die Energieverteilung der von der Sonne ausgesandten Strahlung von derjenigen eines schwarzen Körpers sehr stark abweichen, was nicht wahrscheinlich ist.

Beschränken wir uns auf die erstgenannte Annahme der atmosphärischen Absorption, so wird man weiterhin mit großer Wahrscheinlichkeit den Wasserdampf als den absorbierenden Bestandteil der Atmosphäre zu betrachten haben, obwohl es nicht völlig ausgeschlossen erscheint, daß in dem äußersten, bisher noch unerforschten Ultrarot auch andere Gase, wie z. B. Kohlensäure, merkliche Absorption besitzen[1].

[1] Frl. Eva von Bahr schließt aus ihren Messungen im kurzwelligen ultraroten Spektrum auf Grund der Bjerrumschen Theorie (Nernst-Festschrift 1912, S. 90), daß die Absorption des Wasserdampfs jenseits 400 μ abnehmen, dagegen diejenige der Kohlensäure in der Nähe von 1100 μ ein Maximum erreichen müsse (Verh. d. Dt. Phys. Ges. S. 710 und 731, 1913).

Ausgegeben am 18. Juni.

With G. Eberhard in Sitz. Ber. Kgl. Preuss. Akad. d. Wiss. Berlin 1913, 308–310

Über Umkehrungen der Calciumlinien H und K in Sternspektren.

Von G. Eberhard und K. Schwarzschild.

Im Jahre 1900 machten die HH. G. Eberhard und H. Ludendorff auf Wunsch von H. C. Vogel Aufnahmen von Sternspektren mit Hilfe des kleinen Spektrographen D, die den Zweck haben sollten, die Sternspektren möglichst weit ins Ultraviolett hinein zu verfolgen. Im sichtbaren Gebiet waren diese Spektren außerordentlich stark überexponiert. G. Eberhard fand auf einer derartigen Aufnahme des Arkturspektrums, daß die Calciumlinie K eine scharfe Umkehr aufwies, daß also in der Mitte der Absorptionslinie eine Calciumemission statthatte, so wie man es bei der Sonne in gestörten Gebieten ihrer Oberfläche beobachtet. Da die Absorption in der breiten Calciumlinie sehr stark und die Emission nicht sehr kräftig war, so kam die Umkehr nur in Spektren zum Vorschein, deren kontinuierlicher Grund bis in die Gegend der H- und K-Linie überexponiert war.

Weitere Beispiele derselben Erscheinung sind neuerdings von K. Schwarzschild mit Hilfe des Objektivprismas aufgefunden worden. Auf den in der vorausgehenden Mitteilung erwähnten Aufnahmen der Hyaden mit dem U. V. Zeißtriplet und U. V. Objektivprisma (150 mm Öffnung, 1500 mm Brennweite, Dispersion Hγ bis K 10 mm) fanden sich stark überexponierte Spektren von Aldebaran, welche gleichfalls die Umkehr der K- — und etwas schwächer — der H-Linie des Calciums aufwiesen.

Bei α Aurigae und α Ursae majoris war auf ähnlichen Aufnahmen keine Umkehr zu erkennen, ebensowenig bei α Orionis und α Cassiopeiae; bei letzteren beiden Sternen war vielleicht allerdings die Exposition noch nicht kräftig genug. Bei einer einstündigen Exposition von β Geminorum mit dem Objektivprisma zeigte auch das Spektrum dieses Sternes keine Umkehr, aber auf derselben Platte fand sich, daß der Nachbarstern σ Geminorum außerordentlich helle, scharfe Emissionslinien inmitten der breiten Absorptionslinien aufwies.

Wenn der bloße Anblick der Spektren auch schon ziemlich überzeugend dartat, daß man es nicht mit einer Aussparung im kontinuierlichen Spektrum, sondern mit wirklicher Emission zu tun habe, so schien es doch gut, durch Bestimmung der Wellenlänge der Umkehrungen ihre Zugehörigkeit zum Calcium zu kontrollieren. Im folgenden sind die Resultate einiger Messungen mitgeteilt, wobei die Lage der Umkehrung durch die Abweichung der Radialgeschwindigkeit gegeben ist, welche aus der Umkehrung, verglichen mit 7 bis 10 benachbarten Sternlinien (Fe und Al) folgt:

α Bootis:	Großer Refraktor. Spektrograph III mit kurzer Kamera (Neuchromat mit 180 mm Brennweite), Spaltbreite 0.10 mm, Breite des Spektrums 0.05 mm. 1913, Febr. 23. Expositionszeit 30^m	H − 3 km/sec.,	K + 6 km/sec.,
α Bootis:	Zeißtriplet mit Objektivprisma. Breite des Spektrums 0.2 mm. 1912, April 23. Zwei Expositionen von 14^m und 7^m. Im Mittel aus beiden	+ 5 »	+ 5 »
α Tauri:	Desgl. 1912, Jan. 14, eine Exposition von 30^m. 1913, Jan. 1, zwei Expositionen von je 30^m. Im Mittel aus allen dreien	− 3 »	+ 3 »
σ Geminorum:	Desgl. 1913, März 9 und 1913, März 12 je 60^m exponiert. Im Mittel aus beiden	− 6 »	− 7 » .

Da in der Gegend von H und K für den Spektographen $1\,\mu = 1.2$ km/sec., für das Objektivprisma $1\,\mu = 2.6$ km/sec. ist, so bedeuten diese Zahlen zunächst nur, daß die Wellenlänge der Umkehrungen innerhalb der Meßgenauigkeit mit der Wellenlänge der H- und K-Linien übereinstimmt. Die Helligkeit der H- und K-Umkehr ist bei σ Geminorum so groß, daß ihre Schwärzung die der hellsten Stellen im kontinuierlichen Spektrum zwischen H und K erreicht. Die Breite der Emissionslinie beträgt auf der Aufnahme vom 12. März etwa 1 A. E., was infolge der Luftunruhe und der Pointierfehler aber kein Maß für die wirkliche — vermutlich viel geringere — Linienbreite ist, sondern nur die Stärke der Erscheinung verdeutlicht. Bei Aldebaran ist, wie aus dem Vergleich mit anderen gleichzeitig auf der Platte erhaltenen Hyadensternen hervorgeht, die Helligkeit der Umkehr bei ähnlicher Breite um etwa 3 Größenklassen relativ zum kontinuierlichen Spektrum schwächer als bei σ Geminorum. Arktur zeigt wenig stärkere Umkehr als Aldebaran. K ist bei allen 3 Sternen kräftiger als H.

Bei der Sonne beobachtet man in Spektren sehr großer Dispersion überall auf ihrer Scheibe eine sehr feine und schwache Umkehr vor allem der K-Linie. In vereinzelten Störungsgebieten wird die Umkehr so stark wie bei σ Geminorum. Um ein Urteil über die durchschnittliche Stärke der Emission auf der Sonne, verglichen mit den Sternen, zu gewinnen, wurde (am 14. März 1913) das Spektrum des diffusen Himmelslichts mit derselben Dispersion, wie die Sternspektren, aufgenommen und ähnlich stark exponiert. Das Spektrum des diffusen Himmelslichts entspricht ja dem durchschnittlichen Spektrum der Sonne. Auf diesen Aufnahmen war von einer Umkehr der H- und K-Linie nichts zu sehen. Daraus ist zu schließen, daß die Emission bei den obigen Sternen sehr viel stärker ist als auf der Sonne.

Es seien noch folgende allgemeine Bemerkungen hinzugefügt. Linienumkehrungen in Sternspektren sind an sich nichts seltenes. Die hier gefundenen Umkehrungen sind aber deshalb interessant, weil sie sich auf Sterne beziehen, die der Sonne in ihrem ganzen Spektraltypus nahe verwandt und uns dadurch im einzelnen verständlicher sind. Dieselbe Art von Eruptionstätigkeit, die die Sonne in Flecken, Flocken und Protuberanzen zeigt, hat man sich offenbar auf Arktur und Aldebaran, vor allem aber auf σ Geminorum in mächtig gesteigerter Intensität vorzustellen. Für Arktur stimmt damit das Ergebnis von Hrn. Adams[1], daß das Spektrum dieses Sterns auch in den Absorptionslinien einen Übergang vom normalen Sonnenspektrum zum Spektrum der Sonnenflecken darstellt.

Für die Folge ergeben sich zwei Probleme. Es werden die Emissionslinien der Sterne auf eine etwaige Veränderlichkeit ihrer Intensität in Analogie zur Sonnenfleckenperiode zu prüfen sein. Ferner ist durch die Untersuchungen der HH. Deslandres[2] und St. John[3] bekannt, daß die Zentren der Calciumemission auf der Sonne eine radial auswärts gerichtete Bewegung von 1 km/sec. haben. Es wird noch durch genauere Wellenlängenbestimmung der Calciumemission der Sterne zu prüfen sein, ob mit der größeren Gesamtintensität der Emission auch eine größere Aufstiegsgeschwindigkeit verbunden ist.

---

[1] Astrophys. Journal Bd. XXIV S. 69.
[2] Comptes rendus, 1905, S. 381.
[3] Astrophys. Journal Bd. 32 S. 36.

Sitz. Ber. Kgl. Preuss. Akad. d. Wiss. Berlin 1914, 1183–1200

Über Diffusion und Absorption in der Sonnenatmosphäre.

Von K. Schwarzschild.

(Vorgelegt von Hrn. Einstein am 5. November 1914 [s. oben S. 979].)

§ 1.

Die Absorptions- und Emissionslinien in den Spektren der Sonne und der Sterne sind außerordentlich verschieden in ihrem Aussehen. Von Linien, die sich nur als geringe Abnahme der Intensität des kontinuierlichen Spektrums über einige hundertstel Angström zu erkennen geben, gibt es alle Übergänge zu Linien von mehreren Ångström Breite, in deren Mitte die Intensität auf 20 Prozent oder weniger der Intensität des benachbarten kontinuierlichen Spektrums absinkt. Auch kommen die verschiedensten Typen der Linienumkehrung vor. Man hat sich in der Astrophysik bisher meist mit recht rohen qualitativen Angaben und Erklärungen des verschiedenen Verhaltens der Spektrallinien begnügt. Nur in einer bedeutsamen Arbeit von Hrn. A. Schuster[1] sind quantitative Überlegungen gegeben. Hr. Schuster setzt voraus, daß die Gase einer Sternatmosphäre erstens strahlen und absorbieren gemäß dem Kirchhoffschen Gesetz, und daß sie zweitens gleichzeitig das Licht streuen (»diffundieren«), so wie es nach Rayleigh die Moleküle unserer Luft tun. Je nach dem Temperaturanstieg bei dem Eindringen in die Sternatmosphäre und je nach dem Verhältnis von Absorptionsvermögen und Streuungsvermögen der Gase ergibt sich sehr verschiedenes Aussehen der Spektrallinien mit mannigfachen Analogien zu den bei Sternen beobachteten Erscheinungen. Einige Messungen des Intensitätsverlaufs in Spektrallinien von Sternen hat Hr. C. F. Bottlinger[2] ausgeführt.

Für den weiteren Fortschritt scheint es wichtig, ein genaueres Studium der Linien im Sonnenspektrum auszuführen, da man hier außer der Lichtfülle noch den großen Vorteil hat, verschiedene Punkte der Sonnenscheibe und damit verschiedene Emissionswinkel des austretenden Lichtes getrennt studieren zu können, während man bei Sternen nur das Gesamtlicht aller Punkte ihrer Scheibe gleichzeitig erhält.

[1] Astrophysical Journal vol. 21 (1905).
[2] Astr. Nachr. Bd. 195.

Vergleicht man das Spektrum der Sonnenmitte mit dem der Nachbarschaft des Sonnenrandes, so ist der erste Eindruck der weitgehendster Übereinstimmung. Man sieht sofort, daß eine ältere Annahme nicht richtig sein kann, welche eine nur absorbierende, nicht strahlende Atmosphäre um die leuchtende Sonnenatmosphäre voraussetzte. Denn bei dieser Annahme müßten die Fraunhoferschen Linien nach dem Rande zu wegen der Zunahme der Schichtdecke immer tiefer und breiter erscheinen.

Was man bei näherer Vergleichung des Spektrums von Sonnenmitte und Sonnenrand unter Verwendung größerer Dispersion konstatiert, ist mit wenig Ausnahmen das Umgekehrte, eine gewisse Abnahme der Tiefe der Fraunhoferschen Linien.

Wir wollen uns nun auf den Standpunkt des Hrn. Schuster stellen und eine gemäß dem Kirchhoffschen Gesetz strahlende und absorbierende, zugleich das Licht streuende Atmosphäre voraussetzen. Wir denken diese Atmosphäre einem schwarzen, mithin unter allen Emissionswinkeln gleich hell strahlenden Körper bestimmter Temperatur, der »Photosphäre«, überlagert. Hr. Schuster hat sich darauf beschränkt, mit einer gewissen Annäherung den gesamten Energietransport durch eine solche Atmosphäre zu behandeln. Wir wollen hier die Energie trennen nach der Richtung, in welcher sie die Atmosphäre durchsetzt, und fragen, wie die austretende Strahlung vom Emissionswinkel abhängt. Das gibt uns an, wie die Intensität auf der Sonnenscheibe verteilt sein wird.

Es soll die Oberfläche der Photosphäre wie die Schichtung der darüberliegenden Gase als eben vorausgesetzt werden. Bei der geringen Höhe der Sonnenatmosphäre im Verhältnis zu ihrem Radius kann das nur in unmittelbarer Nähe des Randes merkliche Fehler geben. Auch soll die Strahlenbrechung in der Sonnenatmosphäre vernachlässigt werden.

Die wirkliche Durchführung der Rechnung wird in dieser Arbeit nur für die beiden extremen Fälle erfolgen, daß entweder nur Absorption und Emission besteht ohne Streuung oder nur Streuung ohne Absorption und Emission.

Eine rohe Messung der Intensitätsverteilung, deren Resultat weiterhin mitgeteilt wird, habe ich an den breitesten Linien im Sonnenspektrum, den Kalziumlinien H und K für die Mitte wie für die Nähe des Randes ausgeführt.

§ 2.

Man betrachte eine homogene, von zwei parallelen Ebenen begrenzte Gasmasse. Die x-Achse möge senkrecht zu den Begrenzungsebenen liegen. Für die Begrenzungsebenen sei $x = 0$ und $x = H$. Auf der »unteren« Seite $x = H$ möge die Strahlung eines schwarzen Körpers eintreten. Die Gasmasse emittiere, absorbiere und diffundiere das Licht. Unser Problem ist, festzustellen, wieviel Strahlung in jeder Richtung aus der anderen »oberen« Begrenzungsebene $x = 0$ austritt. Für Emission und Absorption des Gases werde das Kirchhoffsche Gesetz als gültig vorausgesetzt. Für die Streuung werde angenommen, daß ein bestimmter Bruchteil des auf ein Volumenelement aus irgendeiner Richtung auffallenden Lichtes nach allen Seiten gleichmäßig verteilt wird. Nach Rayleighs Theorie der Streuung des Lichts an kleinen Teilchen schwankt der Streuungskoeffizient für verschiedene Richtungen allerdings im Verhältnis 1 : 2. Diese Schwankung macht aber wenig aus, solange die Strahlung wenigstens aus einer Hälfte der ein diffundierendes Teilchen umgebenden Kugel einigermaßen gleichförmig erfolgt, wie das bei den im folgenden zu behandelnden Problemen in der Hauptsache zutrifft.

Wenn zur Ebene $x = H$ an allen Punkten schwarze Strahlung derselben Temperatur eintritt, wie wir das voraussetzen wollen, so wird der ganze Vorgang von den anderen Koordinaten y und z unabhängig. Die Strahlung in irgendeinem Punkt des Gases in irgendeiner Richtung hängt nur ab von der x-Koordinate des Punktes und dem Winkel i, den diese Richtung mit der x-Achse macht. Die mathematische Formulierung des Problems wird dann folgende.

Wir fassen die Strahlung einer bestimmten Wellenlänge ins Auge.

Wir zählen i nur von 0 bis 90° und unterscheiden absteigende und aufsteigende Strahlung. Es sei die Strahlung des senkrecht zum Strahl stehenden Flächenelements ds im Punkte der Koordinate x für den in Richtung i liegenden Raumwinkel $d\omega$:

$$a(x, i)\, d\omega\, ds \text{ für die absteigende Strahlung,}$$
$$b(x, i)\, d\omega\, ds \text{ für die aufsteigende Strahlung.}$$

Auf die Strecke de möge durch Absorption von der Strahlung a verlorengehen der Betrag: $\varkappa a de$ und durch Streuung analog: $\sigma a de$. $\varkappa$ soll Absorptionskoeffizient, σ Diffusionskoeffizient heißen.

Zwischen den beiden Ebenen x und $x + dx$ geht dann z. B. von der absteigenden Strahlung $a(x, i)$ verloren

$$\text{durch Absorption: } \varkappa a(x, i)\, dx \sec i,$$
$$\text{durch Diffusion: } \sigma a(x, i)\, dx \sec i.$$

Sitz. Ber. Kgl. Preuss. Akad. d. Wiss. Berlin 1914, 1183–1200

Die Emission der Volumeneinheit des Gases habe den Betrag $4\pi\varkappa E$, wo E die Kirchhoffsche Funktion sein wird, wenn die Strahlung des Gases dem Kirchhoffschen Gesetz folgt.

Die vom Volumenelement dv gestreute Strahlung ergiebt sich durch eine Betrachtung analog der Planckschen[1] zu:

$$dv \cdot \sigma \left[\int a(x, i)\, dw + \int b(x, i)\, dw\right]$$

jedes Integral über die betreffende Halbkugel integriert. Das läßt sich in Rücksicht auf die Symmetrie des Problems zur x-Axe umschreiben in:

$$2\pi dv \cdot \sigma \left[\int_0^\pi a(x, i) \sin i\, di + \int_0^\pi b(x, i) \sin i\, di\right].$$

Zur Abkürzung sei:

$$A = \int_0^\pi a(x, i) \sin i\, di + \int_0^\pi b(x, i) \sin i\, di\,, \tag{1}$$

so daß $2\pi\sigma A$ die von der Volumeneinheit gestreute Strahlung bezeichnet. In den Kegel $d\omega$ entsendet das Volumenelement dv bei nach allen Seiten gleichmäßiger Streuung hiernach den Betrag:

$$2\pi dv \sigma A \cdot \frac{d\omega}{4\pi} = dv\, d\omega \frac{\sigma A}{2}.$$

Die gesamte durch Emission und Diffusion von der Volumeneinheit in den Kegel $d\omega$ entsandte Strahlung wird:

$$d\omega\, dv \left[\sigma \frac{A}{2} + \varkappa E\right].$$

Wir werden setzen:

$$\frac{\sigma A}{2} + \varkappa E = J, \tag{2}$$

so daß $4\pi J$ die »Ergiebigkeit« der Volumeneinheit an Strahlung überhaupt (emittierte und diffundierte zusammengerechnet) ist. Betrachtet man nun die Änderung der Strahlung beim Übergang zwischen zwei Nachbarebenen x und $x + dx$, so erhält man leicht:

$$\begin{aligned} \cos i \frac{da}{dx} &= -(\varkappa + \sigma)\, a + J\,, \\ \cos i \frac{db}{dx} &= (\varkappa + \sigma)\, b - J\,. \end{aligned} \tag{3}$$

J ist eine Funktion nur von x, während a und b von x und i abhängen.

[1] Wärmestrahlung, 1. Auflage, § 22.

Das sind die Differentialgleichungen des Problems.

Die Grenzbedingungen sind folgende: Für $x = 0$ findet keine Strahlung von oben statt, da sich oberhalb $x = 0$ keine strahlende Materie befindet, es ist also dort $a = 0$. Für $x = H$ ist die aufwärts gehende Strahlung die des unter dem Gas liegenden schwarzen Körpers: $b = B$, wo B den, von i unabhängigen, der KIRCHHOFFschen Funktion zu entnehmenden Betrag hat. Die Grenzbedingungen sind also:

$$x = 0: \quad a = 0, \tag{4}$$

$$x = H: \quad b = B. \tag{5}$$

Unser Problem ist ein bestimmtes, wenn die Temperatur des einstrahlenden Körpers und damit B, sowie die Temperatur des Gases an jeder Stelle und damit E als Funktion von x gegeben ist.

Es ist bisher immer nur von einer homogenen Atmosphäre die Rede gewesen. Es ist aber unmittelbar ersichtlich, daß die ganze vorstehende Ableitung auch für eine inhomogene Atmosphäre gilt, wenn man unter x nicht die Entfernung von der oberen Grenze, sondern die über der betreffenden Stelle liegende Luftmasse versteht.

Das Problem läßt sich sofort auf eine lineare Integralgleichung zurückführen. Die Integration von (3) in Rücksicht auf die Grenzbedingungen (4) und (5) gibt:

$$a(x, i) = \int_0^x J(\xi)\, e^{(\varkappa+\sigma)(\xi - x)\sec i}\, d\xi \sec i, \tag{6}$$

$$b(x, i) = B e^{(\varkappa+\sigma)(x-H)\sec i} + \int_x^H J(\xi)\, e^{(\varkappa+\sigma)(x-\xi)\sec i}\, d\xi \sec i. \tag{7}$$

Hiermit kann man die Diffusion A des Volumenelements berechnen. Man erhält aus (1) durch Vertauschung der Integrationsordnung:

$$A = \int_0^x d\xi\, J(\xi) \int_0^\pi di \operatorname{tg} i\, e^{(\varkappa+\sigma)(\xi-x)\sec i} + \int_x^H J(\xi)\, d\xi \int_0^\pi di \operatorname{tg} i\, e^{(\varkappa+\sigma)(x-\xi)\sec i}$$

$$+ B \int_0^\pi di \sin i\, e^{(\varkappa+\sigma)(x-H)\sec i}.$$

Setzt man:

$$\sec i = \eta,$$

so kann man hierfür schreiben:

$$A = B \int_1^\infty \frac{d\eta}{\eta^2} e^{(\varkappa+\sigma)(x-H)\eta} + \int_0^H d\xi\, J(\xi) \int_1^\infty \frac{d\eta}{\eta} e^{-(\varkappa+\sigma)|\xi - x|\eta},$$

oder wenn man den Integrallogarithmus einführt:

$$K(z) = \int_1^\infty \frac{d\eta}{\eta} e^{-z\eta} \tag{8}$$

und seine Analoga:

$$K_n(z) = \int_1^\infty \frac{d\eta}{\eta^{n+1}} e^{-z\eta}, \tag{9}$$

so ist:

$$A = B \cdot K_1[(\varkappa+\sigma)(H-x)] + \int_0^H d\xi\, J(\xi)\, K[(\varkappa+\sigma)\,|\xi - x|].$$

Nach (1) folgt daraus die Integralgleichung für die Ergiebigkeit $J(x)$[1]:

$$J(x) - \frac{\sigma}{2}\int_0^H d\xi\, J(\xi)\, K[(\varkappa+\sigma)\,|\xi-x|] = \varkappa E + \frac{\sigma}{2} B K_1[(\varkappa+\sigma)(H-x)].$$

§ 3. Grenzfall der reinen Absorption.

Fällt die Diffusion fort, ist $\sigma = 0$, so gibt die Integralgleichung:

$$J(x) = \varkappa E$$

und nach (7) wird an der äußeren Grenze des Gases, für $x = 0$:

$$b(0, i) = Be^{-\varkappa H \sec i} + \int_0^H \varkappa E e^{-\varkappa\xi \sec i} d\xi \sec i\,.$$

Es verlaufe E als Funktion von x linear gemäß der Formel:

$$E(x) = a + bx, \tag{10}$$

und es sei:

$$B = a + bH,$$

damit an der Photosphäre selbst das Gas die Temperatur der Photosphäre hat. Dann liefert die Integration:

$$b(0, i) = a + \frac{b \cos i}{\varkappa}(1 - e^{-\varkappa H \sec i}) \tag{11}$$

Hiermit ist eine genügende Übersicht möglich.

§ 4. Grenzfall der reinen Diffusion.

Im Fall der reinen Diffusion ($\varkappa = 0$) wird die Integralgleichung:

$$J(x) - \frac{\sigma}{2}\int_0^H d\xi J(\xi) K[\sigma\,|\xi - x|] = \frac{\sigma}{2} BK_1[\sigma(H-x)]\,.$$

[1] Die Integralgleichung und ihre Behandlung für kleine Atmosphärendicken H, wie sie für die Erdatmosphäre in Frage kommen, finden sich bei L. V. King, London Roy. Phil. Soc. Transactions. 1913.

Führt man folgende neue Variable ein:

$$x' = \sigma x, \quad \xi' = \sigma\xi, \quad H' = \sigma H,$$

$$J'(x') = \frac{J(x)}{\sigma B}, \quad a'(x', i) = \frac{a(x, i)}{B}, \quad b'(x', i) = \frac{b(x, i)}{B} \tag{12}$$

und erlaubt sich, da eine Verwechslung nicht zu befürchten ist, die Akzente der neuen Variable gleich wieder wegzulassen, so geht die Integralgleichung über in:

$$J(x) - \frac{1}{2}\int_0^H d\xi J(\xi) K[\xi - x] = \frac{1}{2} K_1[H - x]. \tag{13}$$

Es ist hier $J(x)$ eine Funktion, die außer von x nur von H (dem früheren σH), also von der »effektiven« Dicke der diffundierenden Atmosphäre, abhängt. Für die austretende Strahlung folgt nach (7) in den neuen Variablen:

$$b(0, i) = e^{-H \sec i} + \int_0^H J(\xi) e^{-\xi \sec i} d\xi \sec i. \tag{14}$$

Die Schustersche Näherung. Die Streuung der Strahlung in einer bestimmten Schicht des Gases hängt natürlich sehr stark von der Richtung ab, in der die Strahlung die Schicht durchsetzt, wie sich das in dem Faktor $\cos i$ der Differentialgleichungen (3) auf der linken Seite ausprägt. Hr. Schuster hat in der erwähnten Arbeit noch davon abgesehen, die Strahlung verschiedener Richtung zu trennen und einen durchschnittlichen Absorptions- und Streuungskoeffizienten für die gesamte, jede Schicht durchsetzende Strahlung eingeführt. Dabei wird der Koeffizient für die absteigende und für die aufsteigende Strahlung gleich angesetzt, was natürlich unstreng ist, da die absteigende Strahlung anders über die verschiedenen Richtungen i verteilt ist als die aufsteigende. Namentlich dicht unter der Oberfläche des Gases scheint dieser Ansatz sehr fehlerhaft werden zu müssen, da dort die absteigende Strahlung fast nur aus horizontaler Richtung kommen kann, also einen sehr hohen durchschnittlichen Absorptions- und Diffusionskoeffizienten hat, während die aufsteigende Strahlung gleichmäßiger über alle Richtungen verteilt ist und daher weniger absorbiert und diffundiert wird. Wir wollen alsbald prüfen, wie sich der Fehler numerisch stellt.

Wir führen zunächst den Schusterschen Ansatz speziell für den Fall reiner Diffusion in folgender Weise durch.

Für $\varkappa = 0$ und mit dem unter (12) angegebenen Koordinatenwechsel lauten die Gleichungen (1) (2) (3)

$$\cos i \frac{d}{dx} a(x,i) = -a(x,i) + J \qquad J = \frac{1}{2}\int_0^\pi a(x,i)\sin i di + \frac{1}{2}\int_0^\pi b(x,i)\sin i di$$
$$\cos i \frac{d}{dx} b(x,i) = b(x,i) - J$$

mit den Grenzbedingungen:

$$\text{Für } x = 0: a = 0; \text{ für } x = H: b = 1.$$

Aus den beiden Differenzialgleichungen folgen durch Multiplikation mit $\sin i di$ und Integration von 0 bis π, wenn wir die Abkürzungen einführen:

$$\int_0^\pi a(x,i)\sin i di = \bar{a}(x), \qquad \int_0^\pi b(x,i)\sin i di = \bar{b}(x),$$
$$\int_0^\pi a(x,i)\cos i \sin i di = \bar{\bar{a}}(x), \qquad \int_0^\pi b(x,i)\cos i\sin i di = \bar{\bar{b}}(x),$$

die folgenden Beziehungen:

$$J = \frac{\bar{a}(x) + \bar{b}(x)}{2},$$

$$\frac{d\bar{\bar{a}}(x)}{dx} = \frac{1}{2}\bar{b}(x) - \frac{1}{2}\bar{a}(x), \qquad \frac{d\bar{\bar{b}}(x)}{dx} = \frac{1}{2}\bar{b}(x) - \frac{1}{2}\bar{a}(x).$$

Dies sind noch strenge Gleichungen. Wir ermitteln nun eine genäherte Beziehung zwischen $\bar{a}$ und $\bar{\bar{a}}$ sowie zwischen $\bar{b}$ und $\bar{\bar{b}}$, indem wir diese Integrale auswerten unter der Voraussetzung, daß die Strahlung im Innern des Gases nach allen Richtungen gleichförmig verteilt sei, daß $a(x,i)$ und $b(x,i)$ unabhängig von i gleich $a(x)$ bzw. $b(x)$ seien. Es folgt dann:

$$\bar{a}(x) = a(x), \quad \bar{\bar{a}}(x) = \frac{1}{2}a(x), \quad \bar{b}(x) = b(x), \quad \bar{\bar{b}}(x) = \frac{1}{2}b(x),$$

mithin:
$$\bar{\bar{a}}(x) = \frac{1}{2}\bar{a}(x), \quad \bar{\bar{b}}(x) = \frac{1}{2}\bar{b}(x).$$

Das gibt für $\bar{a}$ und $\bar{b}$ die Differentialgleichungen:

$$\frac{d\bar{a}}{dx} = \bar{b} - \bar{a}, \qquad \frac{d\bar{b}}{dx} = \bar{b} - \bar{a}$$

mit den Grenzbedingungen: Für $x = 0$: $\bar{a} = 0$; für $x = H$: $\bar{b} = 1$.

Die einfache Integration liefert:

$$\bar{a} = \frac{x}{H+1} \qquad \bar{b} = \frac{1+x}{H+1}$$

und damit:

$$J = \frac{x + 0.5}{H + 1} \tag{15}$$

und durch Ausführung des Integrals (14):

$$b(0, i) = \frac{0.5 + \cos i}{H + 1} + e^{-H \sec i} \frac{0.5 - \cos i}{H + 1}. \tag{16}$$

Dies ist die Schustersche Näherung.

Die Korrektion der Schusterschen Näherung. Wir wollen nun die strenge Lösung unserer Integralgleichung in der Form suchen:

$$J(x) = \frac{0.5 + x + L(x)}{H + 1}, \tag{17}$$

so daß also $L(x)/H + 1$ die Korrektion der Schusterschen Näherung ist. Will man diesen Ausdruck in die Integralgleichung einführen, so bedarf man der Kenntnis folgender beider Integralformeln:

$$\int_0^H d\xi\, K|\xi - x| = 2 - K_1(x) - K_1(H - x), \tag{18}$$

$$0 < x < H$$

$$\int_0^H d\xi\, \xi K|\xi - x| = 2x + K_2(x) - K_2(H - x) - H K_1(H - x), \tag{19}$$

die sich leicht durch Einführung des Integrals (8) für den Integrallogarithmus und Vertauschung der Integrationsfolge ergeben.

Mit Benutzung dieser Formeln findet man, daß $L(x)$ folgender Integralgleichung genügt:

$$L(x) - \frac{1}{2}\int_0^H L(\xi) K|\xi - x|\, d\xi = \frac{2K_2(x) - K_1(x)}{4} - \frac{2K_2(H - x) - K_1(H - x)}{4}. \tag{20}$$

Bevor wir den Verlauf von $L(x)$ numerisch bestimmen, wollen wir einige Grenzbetrachtungen anstellen.

Die auf der linken Seite der Integralgleichung stehende Operation wollen wir zur Abkürzung mit F, die rechts stehende gegebene Funktion von x mit $G(x)$ bezeichnen, so daß z. B. die Integralgleichung (13) lautet:

$$F\{J(x)\} = G(x), \quad G(x) = \frac{1}{2} K_1(H - x).$$

Es ergeben sich der Reihe nach folgende Sätze:

1. Aus $G(x) = 0$ folgt $J(x) = 0$. Denn: Man nehme an, $J(x)$ sei irgendwo > 0 und $\bar{x}$ die Stelle des Maximums von $J(x)$, dann wäre

Sitz. Ber. Kgl. Preuss. Akad. d. Wiss. Berlin 1914, 1183–1200

$$F\{J(\bar{x})\} = J(\bar{x}) - \frac{1}{2}\int_0^H d\xi J(\xi) K|\xi - \bar{x}| > J(\bar{x})\left[1 - \frac{1}{2}\int_0^H d\xi K|\xi - \bar{x}|\right].$$

Nun ist nach Formel (18):

$$1 - \frac{1}{2}\int_0^H d\xi K|\xi - x| = \frac{1}{2}K_1(x) + \frac{1}{2}K_1(H-x) > 0\,.$$

Demnach wäre auch $F\{J(\bar{x})\} > 0$. Da das unserer Forderung widerspricht, kann $J(x)$ nirgends positiv sein, ebensowenig negativ, ist also Null. In der Ausdrucksweise der Theorie der Integralgleichungen heißt dies: der Parameterwert $\frac{1}{2}$ unserer Integralgleichung ist kein Eigenwert.

2. Ist für alle Werte von x: $G(x) > 0$, so ist auch für alle Werte von x: $J(x) > 0$. Wäre nämlich $J(x)$ irgendwo negativ, so könnte man den größten negativen Wert nehmen und, wie oben, beweisen, daß an dieser Stelle auch $G(x)$ negativ sein müßte.

Daraus folgt weiter: Hat man die Lösungen zweier Integralgleichungen:

$$F\{J(x)\} = G(x) \quad \text{und} \quad F\{J'(x)\} = G'(x)$$

und es ist für alle x: $G'(x) > G(x)$, so ist auch für alle x: $J'(x) > J(x)$.

3. Wir betrachten speziell $L(x)$. Aus (20) folgt unmittelbar:

$$F\{L(x)\} + F\{L(H-x)\} = 0$$

oder

$$F\{L(x) + L(H-x)\} = 0\,.$$

Daraus folgt nach Satz 1:

$$L(x) + L(H-x) = 0\,. \tag{21}$$

Die Funktion L ist mithin eine ungerade Funktion von $x - \frac{H}{2}$.

4. Wir bestimmen eine obere Grenze für den Absolutwert von $L(x)$.

Es sei zunächst bemerkt, daß $K_1(x)$ und $K_2(x)$ positive, mit wachsendem x ständig abnehmende Größen sind und daß, wie man leicht durch partielle Integrationen der Integralausdrücke nachweist:

$$2\,K_2(x) \geqq K_1(x) \geqq K_2(x)\,. \tag{22}$$

Es ist also auch die Funktion $2\,K_2(x) - K_1(x)$ stets positiv.

Für $x = 0$ wird $K_1(x) = 1$, $K_2(x) = \frac{1}{2}$. Für wachsendes x wird: $K_1(x) = K_2(x) = \frac{e^{-x}}{x}$.

Nun möge $L(x)$ seinen größten positiven Wert an der Stelle $x = \bar{x}$ erhalten. Dann ist:

$$L(\bar{x}) = \frac{2K_2(\bar{x}) - K_1(\bar{x})}{4} - \frac{2K_2(H-\bar{x}) - K_1(H-\bar{x})}{4} + \frac{1}{2}\int_0^H L(\xi)K|\xi - \bar{x}|d\xi .$$

Daraus folgt durch Vergrößerung der rechten Seite:

$$L(\bar{x}) < \frac{2K_2(\bar{x}) - K_1(\bar{x})}{4} + \frac{1}{2}L(\bar{x})\int_0^H K|\xi - \bar{x}|d\xi$$

oder mit Benutzung der Formel (18):

$$L(\bar{x}) \cdot \frac{K_1(\bar{x}) + K_1(H-\bar{x})}{2} < \frac{2K_2(\bar{x}) - K_1(\bar{x})}{4}$$

oder:

$$L(\bar{x}) < \frac{2K_2(\bar{x}) - K_1(\bar{x})}{2K_1(\bar{x}) + 2K_1(H-\bar{x})}$$

oder auch, da $K_1(\bar{x}) > K_2(x)$ ist:

$$L(\bar{x}) < \frac{K_2(\bar{x})}{2K_1(\bar{x})} < \frac{1}{2}.$$

Dieselbe Grenze würde man für den größten negativen Wert von $L(x)$ finden können, es folgt also:

$$|L(x)| < \frac{1}{2}. \qquad (23)$$

Für $J(x)$ selbst folgt daraus nach (17):

$$J(x) = \frac{x + \varepsilon}{H + 1}, \quad 0 < \varepsilon < 1. \qquad (24)$$

5. Hiermit läßt sich der wichtige Satz beweisen, daß mit wachsender effektiver Höhe H der Atmosphäre die Funktion $L(x)$ einer festen Grenze zustrebt.

Man schreibe die Integralgleichung (20) für zwei verschiedene Höhen H und H' ($H' > H$) an:

$$L(x) - \frac{1}{2}\int_0^H L(\xi)K|\xi - x|d\xi = \frac{2K_2(x) - K_1(x)}{4} - \frac{2K_2(H-x) - K_1(H-x)}{4}$$

$$L'(x) - \frac{1}{2}\int_0^{H'} L'(\xi)K|\xi - x|d\xi = \frac{2K_2(x) - K_1(x)}{4} - \frac{2K_2(H'-x) - K_1(H'-x)}{4}.$$

Wir wollen setzen:

$$L'(x) - L(x) = N(x)$$

und zeigen, daß $N(x)$ für wachsendes H verschwindet. Aus den vorstehenden beiden Gleichungen folgt:

$$F\{N(x)\} = \frac{2K_2(H-x) - K_1(H-x)}{4} - \frac{2K_2(H'-x) - K_1(H'-x)}{4} + \frac{1}{2}\int_H^{H'} L'(\xi)\, K|\xi - x|\, d\xi$$

oder, wenn man rechts vergrößert, indem man in dem Integral für $L'(\xi)$ die obere Grenze $\frac{1}{2}$ einführt und die Formel (18) benutzt:

$$F\{N(x)\} < \frac{1}{2}K_2(H-x) - \frac{1}{2}K_2(H'-x)$$

oder erst recht:

$$F\{N(x)\} < \frac{1}{2}K_1(H-x).$$

Die hier auftretende rechte Seite tritt auch auf in der Integralgleichung (13) für $J(x)$:

$$F\{J(x)\} = \frac{1}{2}K_1(H-x).$$

Damit folgt:

$$F\{J(x)\} > F\{N(x)\}$$

und nach Satz (2)

$$J(x) > N(x)$$

oder, wenn wir für $J(x)$ die aus (24) hervorgehende obere Grenze benutzen:

$$N(x) < \frac{x+1}{H+1}.$$

Dieselbe Grenze würde man für $-N(x)$ ableiten können. Es folgt also:

$$|N(x)| < J(x) < \frac{x+1}{H+1}. \tag{25}$$

Das bedeutet, daß für wachsendes H bei festgehaltenem x die Schwankung $N(x)$ der Funktion $L(x)$ der Null zustrebt, daß also $L(x)$ für wachsendes H einer bestimmten Grenzfunktion zustrebt. Nach (17) auf $J(x)$ übertragen heißt dies: Der Wert von $(H+1)J(x)$ strebt mit wachsendem H bei festgehaltenem x einer bestimmten Grenze zu. Und schließlich auf die austretende Strahlung $b(0, i)$ übertragen:

$(H+1)\, b(0,i)$ — oder auch $H\, b(0,i)$ — nähert sich mit wachsendem H einer bestimmten Grenzfunktion von i.

Wenn man es daher nur mit Atmosphären von großer effektiver Schichtdicke zu tun hat, bei welchen die Grenze schon genügend erreicht ist, so ist die austretende Lichtmenge der effektiven Höhe umgekehrt proportional; die Verteilung des austretenden Lichtes auf die verschiedenen Richtungen ist dabei eine ganz bestimmte, von der Höhe der Atmosphäre unabhängige.

Numerische Bestimmung von $L(x)$. Um einen Anhalt über die tatsächlichen Werte von $L(x)$ und $J(x)$ für verschiedene Atmosphärenhöhen zu erhalten, habe ich die Integralgleichung ersetzt durch eine Anzahl gewöhnlicher Gleichungen in folgender Weise. Das Intervall H wurde durch eine Anzahl äquidistanter Punkte $x=0, x=a$, $x=2a, \ldots x=na=H$ in n gleiche Teile zerlegt. Als Unbekannte wurden die Werte von $L(x)$ in diesen Punkten, also die Größen $L(0)$, $L(a)$ usw. angesehen. In den Intervallen zwischen den Punkten wurde linearer Verlauf von $L(x)$ vorausgesetzt. Damit ließ sich der Wert von:

$$\int_0^H d\xi\, L(\xi)\, K|\xi - x|$$

für jeden Wert von x berechnen auf Grund der Formeln (18), (19) und der ähnlich zu erhaltenden:

$$\int_0^\beta d\xi\, K|\xi - x| = K_1(x-\beta) - K_1(x) \qquad x>\beta>0$$

$$\int_0^\beta \xi\, d\xi\, K|\xi - x| = \beta K_1(x-\beta) + K_2(x) - K_2(x-\beta) \qquad x>\beta>0$$

$$= -\beta K_1(\beta - x) + K_2(-x) - K_2(\beta - x) \qquad 0>x.$$

Es war damit die linke Seite der Integralgleichung für jeden Wert von x berechenbar als lineare Funktion der Unbekannten $L(0)$, $L(a)$ usw. Es wurde nun Übereinstimmung der linken und der gegebenen rechten Seite der Integralgleichung gefordert für die Werte $x=0, a, \ldots na$. Das gab ebensoviel Gleichungen als Unbekannte.

Die Werte des Integrallogarithmus und seiner Zugeordneten finden sich tabuliert bei E. GOLD[1].

In der nachstehenden Tabelle sind einige Werte von K_1 und K_2 angeführt und es ist die Differenz $2K_2 - K_1$ hinzugefügt, aus welcher sich die rechte Seite der Integralgleichung für L bildet. Man sieht,

[1] London Roy. Soc. Proceedings Vol. 82, 1909.

Sitz. Ber. Kgl. Preuss. Akad. d. Wiss. Berlin 1914, 1183–1200

daß dieselbe sehr klein ist. Das Intervall a wurde gleich 0.2 gewählt und die Rechnung für $H = 1$, 2, 4 und 8 durchgeführt. Die oben bewiesene Symmetrieeigenschaft $L(x) = -L(H-x)$ reduzierte die Zahl der Unbekannten auf die Hälfte. Blieben so für $H = 8$ noch 20 Unbekannnte, so war die Auflösung der Gleichungen doch nur eine kleine Arbeit, da die meisten Koeffizienten sehr klein waren und zweistellige Rechnung genügend schien. Die Resultate sind in der folgenden Tabelle gegeben.

x	$K_1(x)$	$K_2(x)$	$2K_2(x) - K_1(x)$	$L(x)$ $H=1$	2	4	8	$(H+1)J(x)$ $H=1$	2	4	8
0.0	1.00	0.50	0.00	−0.02	0.00	+0.03	+0.04	0.48	0.50	0.53	0.54
0.2	0.57	0.35	0.13	−0.01	+0.04	0.08	0.09	0.69	0.74	0.78	0.79
0.4	0.39	0.26	0.13	0.00	+0.04	0.08	0.11	0.90	0.94	0.98	1.01
0.6	0.28	0.19	0.11	0.00	+0.03	0.08	0.11	1.10	1.13	1.18	1.21
0.8	0.20	0.14	0.09	+0.01	+0.02	0.07	0.10	1.31	1.32	1.37	1.40
1.0	0.15	0.11	0.07	+0.02	0.00	0.06	0.10	1.52	1.50	1.56	1.60
1.2	0.11	0.08	0.06		−0.02	0.05	0.09		1.68	1.75	1.79
1.4	0.08	0.06	0.05		−0.03	0.04	0.09		1.87	1.94	1.99
1.6	0.06	0.05	0.04		−0.04	0.03	0.08		2.06	2.13	2.18
1.8	0.05	0.04	0.03		−0.04	0.01	0.07		2.26	2.31	2.37
2.0	0.04	0.03	0.02		0.00	0.00	0.06		2.50	2.50	2.56
2.2	0.03	0.02	0.02			−0.01	0.06			2.69	2.76
2.4	0.02	0.02	0.02			0.03	0.05			2.87	2.95
2.6	0.02	0.01	0.01			0.04	0.04			3.06	3.14
2.8	0.01	0.01	0.01			0.05	0.03			3.25	3.33
3.0	0.01	0.01	0.01			0.06	0.03			3.44	3.53
3.2						0.07	0.02			3.63	3.72
3.4						0.08	0.02			3.82	3.92
3.6						0.08	0.01			4.02	4.11
3.8						0.08	0.00			4.22	4.30
4.0						0.03	0.00			4.47	4.50

Die aus der Oberfläche austretende Strahlung wird nach den Formeln (14), (16) und (17):

$$(H+1)\,b(0,i) = 0.5 + \cos i + e^{-H\sec i}(0.5 - \cos i) + \int_0^H L(\xi)\, e^{-\xi \sec i} \sec i\, d\xi$$

Das Integral rechts stellt die Korrektion dar, um welche die Schustersche Näherung zu verbessern ist. Dies Integral wurde durch rohe mechanische Quadratur bestimmt. Das Resultat für einige Werte des Winkels i gibt folgende kleine Tabelle.

$\cos i$ \ H	Austretende Strahlung $(H+1)\,h(0,i)$ Schustersche Näherung 1	2	4	8	Korrektion derselben 1	2	4	8	Resultat 1	2	4	8	$0.5+\cos i$
1.0	1.32	1.43	1.49	1.50	−0.00	+0.01	+0.05	+0.08	1.32	1.44	1.54	1.58	1.50
0.8	1.21	1.28	1.30	1.30	−0.01	+0.02	+0.06	+0.08	1.20	1.30	1.36	1.38	1.30
0.6	1.08	1.10	1.10	1.10	−0.02	+0.02	+0.06	+0.09	1.06	1.12	1.16	1.19	1.10
0.4	0.91	0.90	0.90	0.90	−0.02	+0.02	+0.07	+0.09	0.89	0.92	0.97	0.99	0.90
0.2	0.70	0.70	0.70	0.70	−0.02	+0.02	+0.06	+0.08	0.68	0.72	0.76	0.78	0.70
0.0	0.50	0.50	0.50	0.50	−0.02	0.00	+0.02	+0.04	0.48	0.50	0.52	0.54	0.50

Nach der obigen Grenzbetrachtung würde die Verteilung der austretenden Strahlung für $H=\infty$, wie man leicht aus Formel (25) schließt, sich für jedes i um $\pm\frac{1}{9}$ ihres Wertes noch von der Verteilung für $H = 8$ unterscheiden können. Da die Grenzbetrachtung aber eine zu hohe obere Grenze gibt, dürfte die für $H = 8$ gefundene Verteilung schon eine auf wenige Prozent richtige Annäherung an die Grenzverteilung für $H = \infty$ darstellen.

Die vorstehende Untersuchung lehrt, daß die SCHUSTERsche Annäherung eine sehr gute ist. Für effektive Schichtdicken $H > 2$ gibt auch die einfache Formel:

$$b(0, i) = \frac{0.5 + \cos i}{H + 1}$$

schon eine gute Annäherung, wie die letzte Spalte vorstehender Tabelle zeigt.

Kehren wir wieder zur der ursprünglichen Bezeichnung zurück, indem wir mit H die wirkliche Höhe, mit σ den Streuungskoeffizienten der Atmosphäre bezeichnen, so lautet diese Formel:

$$b(0, i) = \frac{0.5 + \cos i}{1 + \sigma H}. \tag{26}$$

§ 5.

Zur Messung des Intensitätsverlaufs in der H- und K-Linie des Sonnenspektrums wurde das Sonnenspektrum aufgenommen in der zweiten Ordnung eines großen ROWLANDschen Konkavgitters; die Anordnung war durch Einführung einer Zylinderlinse stigmatisch gemacht worden. Die Dispersion war 1 Å = 0.4 mm. Das Sonnenbild wurde durch einen Heliostaten und eine Projektionslinse von 3 m Brennweite auf dem Spalt entworfen. Es wurden auf derselben

Sitz. Ber. Kgl. Preuss. Akad. d. Wiss. Berlin 1914, 1183–1200

Platte die Spektren des Sonnenzentrums und des Sonnenrandes aufgenommen, und zwar sowohl mit der vollen Gitteröffnung, als auch unter Abdeckung der rechten oder der linken Hälfte des Gitters. Man übersieht leicht, daß man damit die nötigen Daten gewinnt, um von den Schwärzungen der Platten zu Intensitätsverhältnissen übergehen zu können. Man muß dabei keineswegs gleiche Intensität des Lichtes von beiden Gitterhälften voraussetzen — was in diesem Falle auch durchaus nicht zutraf; man muß nur wissen, daß die Summe des Lichtes von beiden Hälften gleich dem Licht des vollen Gitters ist.

Die Ausmessung erfolgte für das Zentrum und für das Randspektrum in etwa 12″ Abstand vom Sonnenrand unter dem Hartmannschen Mikrophotometer. Die Einstellung in Richtung des Spektrums konnte auf 0.01 mm genau abgelesen werden. Ich suchte immer die hellsten Stellen in begrenzten Bereichen des Spektrums aus, um von dem Einfluß der zahlreichen anderen Absorptionslinien frei zu werden und möglichst rein den Verlauf der Kalziumabsorption für sich zu messen. Es kam mir wesentlich darauf an, die breiten Flügel der Kalziumlinien zu erhalten; zur Messung der bekannten Umkehrungserscheinungen in der Mitte der Kalziumlinien, die sich auf etwa 0.2 Å beschränken, würde eine größere Dispersion zu verwenden sein.

Man sieht aus den in Fig. 1 wiedergegebenen Resultaten einer allerdings nur einmaligen Messung, daß die Flügel der Kalziumlinie 10 Å weit beiderseitig von der Mitte der Linien aus zu verfolgen sind und halbwegs zwischen beiden Linien fast zusammenkommen.

Fig. 1.

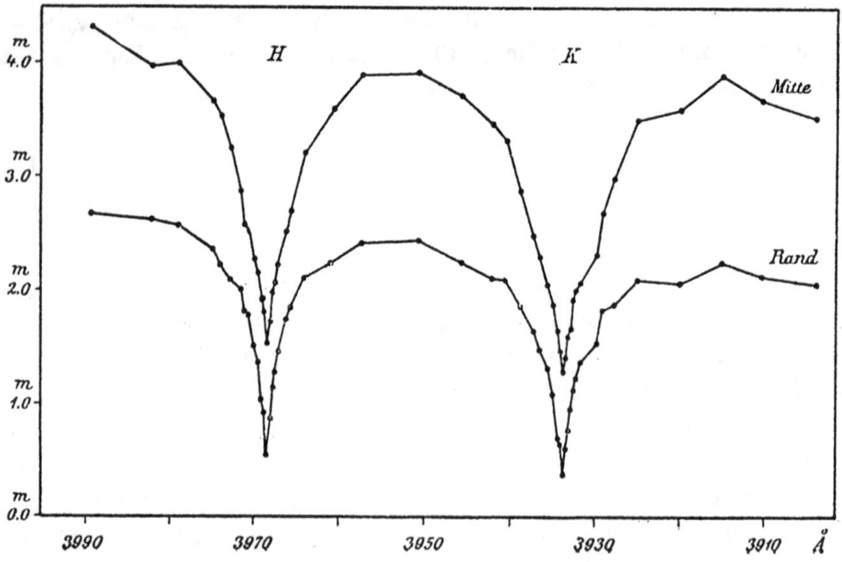

Die K-Linie ist etwas breiter, als die H-Linie. Am Rande ist das kontinuierliche Spektrum außerhalb der Linien um 1.5 Größenklassen schwächer als in der Mitte, die Linien machen diese allgemeine Schwächung im wesentlichen mit, nur sind sie am Rande im Vergleich zum kontinuierlichen Spektrum etwas weniger tief und ihre Flügel etwas weniger mächtig.

§ 6.

Wir wollen nunmehr überlegen, wie das Sonnenspektrum in der Nachbarschaft einer Fraunhoferschen Linie aussehen müßte für den Fall der Absorption einerseits, den der Diffusion anderseits.

Im Falle der Absorption möge die Tiefe H und der Absorptionskoeffizient $\varkappa$ genügend groß sein, um $e^{-\varkappa H}$ vernachlässigen zu können. Im Fall der Streuung möge die Tiefe H und der Streuungskoeffizient σ groß genug sein, um die »effektive« Höhe σH größer als 2 zu machen.

Für die austretende Strahlung haben wir dann im Falle der Absorption nach (11):

$$b(0, i) = a + \frac{b}{\varkappa} \cos i,$$

wobei daran erinnert sei, daß $E(x) = a + bx$ die Kirchhoffsche Funktion für die betreffende Wellenlänge als Funktion der Tiefe bedeutete.

Im Falle der Streuung gilt angenähert nach (26):

$$b(0, i) = \frac{0.5 + \cos i}{1 + \sigma H}.$$

Wir betrachten eine Fraunhofersche Linie und deren Nachbarschaft, im ganzen ein hinreichend schmales Spektralgebiet, um $E(x)$ als identisch für die in Betracht kommenden Wellenlängen ansehen zu können. $\varkappa$ oder σ werden außerhalb der Linie gewisse, der allgemeinen Zusammensetzung der Atmosphäre entsprechende Werte $\varkappa_0$ und σ_0 haben und nach der Mitte der Linie zu erheblich zunehmen bis zu gewissen Maximalwerten $\varkappa_1$ und σ_1.

Im Falle der Absorption ergibt sich damit folgendes: Die Intensität wird:

	in der Mitte ($i = 0°$)	am Rande ($i = 90°$)
außerhalb der Fraunhoferschen Linie	$a + \frac{b}{\varkappa_0}$	a
in der Fraunhoferschen Linie	$a + \frac{b}{\varkappa_1}$	a

Im Falle der Absorption verschwinden die Fraunhoferschen Linien am Sonnenrande, am Sonnenrande herrscht eine

Sitz. Ber. Kgl. Preuss. Akad. d. Wiss. Berlin 1914, 1183–1200

gleichmäßige Intensität a. Die Helligkeit sinkt selbst in der Mitte der Linien nirgends auf der Sonnenscheibe unter diese Randhelligkeit a.

Es erklärt sich dies natürlich daraus, daß bei wachsendem Absorptionskoeffizienten oder bei wachsender Schichtdicke infolge schrägen Durchtritts der Strahlen die Atmosphäre immer mehr wie ein schwarzer Körper von der ihrer höchsten Schicht entsprechenden Temperatur wirkt.

Im Falle der Streuung hat man hingegen für die Helligkeit im Spektrum

	in der Mitte	am Rand der Sonne
außerhalb der Fraunhoferschen Linie	$\frac{1.5}{1+\sigma_0 H}$	$\frac{0.5}{1+\sigma_0 H}$
in der Fraunhoferschen Linie	$\frac{1.5}{1+\sigma_1 H}$	$\frac{0.5}{1+\sigma_1 H}$

Die Helligkeit sinkt am Rande auf ein Drittel der Zentralhelligkeit, sowohl außerhalb wie innerhalb der Linie. Das Aussehen des ganzen Spektrums ändert sich im Falle reiner Streuung von der Mitte bis zum Rand der Sonne nicht, der Kontrast der Fraunhoferschen Linien gegen das umgebende kontinuierliche Spektrum bleibt unverändert. Nur die Gesamthelligkeit des ganzen Spektrums vermindert sich auf ein Drittel.

Offenbar gibt die Annahme der Streuung eine bessere Annäherung an das wirkliche Verhalten des Sonnenspektrums, als die der Absorption und Emission. Der Abfall der Intensität im kontinuierlichen Spektrum folgt für die Gesamtstrahlung sehr nahe dem Gesetz $0.5 + \cos i$; im Rot ist der Abfall etwas geringer, im Violett etwas stärker. Die Fraunhoferschen Linien werden zwar im allgemeinen, und speziell nach den vorstehenden Messungen die H- und die K-Linie, etwas schwächer nach dem Sonnenrande zu; aber die Änderung ist nicht stark und führt keineswegs zu einem Verschwinden der Linien am Sonnenrande, wie es die Absorptionshypothese verlangt.

Es ist daher wahrscheinlich, daß die Streuung des Lichtes eine wesentliche Rolle in den Erscheinungen der Sonnenatmosphäre spielt[1]. Eine genauere Diskussion des Zusammenwirkens von Absorption und Diffusion und der Versuch der Darstellung des wirklichen, scharf gemessenen Intensitätsverlaufs, zunächst in der H- und K-Linie, im Anschluß an die Formeln der Dispersionstheorie ist mir im Augenblick nicht mehr möglich.

[1] Vgl. hierüber auch W. H. Julius, Astrophysical Journal 38, S. 129 (1913).

Ausgegeben am 7. Januar 1915.

Sitz. Ber. Kgl. Preuss. Akad. d. Wiss. Berlin 1914, 1201–1213

Über die Verschiebungen der Bande bei 3883 Å im Sonnenspektrum.

Von K. SCHWARZSCHILD.

(Vorgelegt von Hrn. EINSTEIN am 5. November 1914 [s. oben S. 979].)

1.

Die Wellenlängen der FRAUNHOFERschen Linien im Spektrum der Sonne sind gegen die entsprechenden Linien terrestrischer Lichtquellen infolge der Rotation der Sonne, der Rotation der Erde und der Bewegung der Erde in ihrer exzentrischen Bahn um die Sonne verschoben. Aber auch wenn man den Einfluß dieser drei Bewegungen gemäß dem DOPPLERschen Prinzip berücksichtigt, bleiben noch kleine Verschiebungen übrig, und zwar stimmen die Wellenlängen verschiedener Punkte der Sonnenscheibe weder unter sich noch mit den terrestrischen Wellenlängen überein.

Das Studium all dieser Linienverschiebungen ist nicht nur von Wichtigkeit für die Sonnenphysik, sondern kann nach Hrn. EINSTEIN auch Auskunft über die Relativität der Welt geben. Nach dem von Hrn. EINSTEIN aufgestellten Äquivalenzprinzip[1], welches die Wesensgleichheit eines Gravitationsfeldes mit der Beziehung auf ein beschleunigtes Koordinatensystem behauptet, muß infolge der Gravitation der Sonne eine Verschiebung $\Delta\lambda$ der Wellenlängen λ nach Rot erfolgen vom Betrage:

$$\frac{\Delta\lambda}{\lambda} = \frac{1}{2}\frac{v^2}{V^2},$$

wobei V die Lichtgeschwindigkeit, v die Geschwindigkeit eines Körpers ist, der aus dem Unendlichen in die Sonne fällt. Da $v = 617$ km/sec ist, folgt:

$$\frac{\Delta\lambda}{\lambda} = 2.12 \cdot 10^{-6}.$$

[1] Annalen der Physik 35, S. 905 (1911).

Sitz. Ber. Kgl. Preuss. Akad. d. Wiss. Berlin 1914, 1201–1213

Dies ist übrigens quantitativ dieselbe Verschiebung, wie sie durch eine Fortbewegung der Lichtquelle um 0.635 km/sec hervorgebracht würde.

Auf den ersten Blick scheint es, als ob die Einsteinsche Verschiebung sich an den Linien des Sonnenspektrums deutlich zu erkennen gäbe. Die meisten Linien des Sonnenspektrums sind nach Rot verschoben, verglichen mit den entsprechenden terrestrischen Linien. Man hat diese Verschiebungen bisher meist als Druckverschiebungen gedeutet und daraus die bekannte Angabe von 5 Atmosphären Druck in der umkehrenden Schicht der Sonne abgeleitet. Die Zahlen stimmen aber auch quantitativ merkwürdig gut mit der Einsteinschen Verschiebung, wenn man z. B. das Mittel aus den Eisenlinien nimmt, die Hr. Evershed[1] beobachtet hat[2]. Bei näherem Zusehen erweisen sich aber die Verhältnisse als viel verwickelter. Ich will hier über meine eigenen, durch den Krieg zunächst abgebrochenen, Beobachtungen berichten und die anderweitigen bisher vorhandenen Ergebnisse kurz zusammenstellen.

2.

Im Sonnenspektrum erscheinen recht kräftig die Stickstoffbanden, die man früher meist dem Cyan zuschrieb[3], insbesondere die Bande, die von 3883 Å nach kürzeren Wellenlängen sich erstreckt. Man erhält die Bande sehr bequem terrestrisch im Kohlebogen. Da eine Druckverschiebung dieser Bande in Abrede gestellt wird und die Bandenspektren überhaupt eine besonders große Konstanz haben sollen, schien es mir empfehlenswert, gerade die bei 3883 Å beginnende Bande zu näherem Vergleich herauszugreifen. Vor einem Jahr begann ich Versuche mit Hilfe eines Rowlandschen Konkavgitters. Im letzten Frühjahr kam dann das Observatorium in den Besitz eines vortrefflichen Andersonschen Plangitters. Die definitive Anordnung des Spektralapparates war folgende. Das Licht wurde durch einen Kollimator von 2 m Brennweite und 16 cm Öffnung parallel gemacht und fiel auf das Gitter, von dem eine Fläche von 12×8 cm mit rund 70000 Strichen freigelassen wurde. Das Kameraobjektiv hatte 16 cm Öffnung bei 4 m Brennweite. Die Aufnahmen geschahen stets in der 4. Ordnung. Der Maßstab in der Gegend der Bande ist $1\,Å = 1.15$ mm. Die

[1] Kodoikanal Bulletin Nr. 36.

[2] Vgl. den Aufsatz von Hrn. E. F. Freundlich, Physikalische Zeitschrift 15, S. 369 (1914).

[3] Der Nachweis ihrer Zugehörigkeit zum Stickstoff ist erbracht von C. Runge und W. Grotrian, Physikalische Zeitschrift 15, S. 545 (1914).

Expositionszeiten gingen nicht über drei Minuten. Die stündlichen Temperaturveränderungen im Beobachtungsraum erreichten nur wenige zehntel Grad.

Das Sonnenbild wurde mit Hilfe eines Repsoldschen Heliostaten mit guten Feinbewegungen und eines Objektivs von 20 cm Durchmesser und 3 m Brennweite in der Spaltebene entworfen. Vor dem Spalt befand sich eine Platte mit Millimeterteilung, die eine Einstellung auf den horizontalen Durchmesser des Sonnenbildes auf 0.1 mm genau (was etwa 7" entspricht) gestattete. Die zum Vergleich dienende Bogenlampe wurde mit Hilfe eines Zeissschen Tessars auf den Spalt projiziert. Das Tessar wurde auf dasselbe Öffnungsverhältnis abgeblendet, wie es die zur Projektion der Sonne dienende Linse hatte. Es wurde sorgfältig darauf geachtet, daß bei weitgeöffnetem Spalt der Lichtkegel sowohl für die Sonne wie für das Bogenlicht das Gitter füllte und symmetrisch einschloß. Ein Unterschied der Lichtführung für Sonne und Vergleichslicht besteht dann nur noch insofern, als die verschiedenen Teile des Beleuchtungskegels ein wenig verschiedene Absorptionen erleiden, weil der Absorptionsunterschied zwischen Mitte und Rand bei der Projektionslinse etwas anders als beim Tessar ist. Auch der Heliostatenspiegel kann eine derartige Ungleichförmigkeit geben, wenn seine Versilberung nicht gleichmäßig ist, weshalb auf guten Zustand der Versilberung geachtet wurde. Bei den Aufnahmen wurde der Spalt auf 0.01 mm verengert. Es trat dann eine so starke Streuung des Lichtes durch Beugung ein, daß an genügender Identität der Beleuchtung für Sonne und Vergleichslicht kein Zweifel sein konnte.

Die Spektren bestanden immer aus drei Streifen, wobei die beiden äußeren derselben Lichtquelle entsprangen und gleichzeitig exponiert wurden. Entweder wurden zuerst die beiden äußeren Streifen, dann der mittlere belichtet und dann die Exposition der äußeren Streifen wiederholt, oder es wurde umgekehrt eine Exposition der äußeren Streifen von zwei Expositionen des mittleren zeitlich eingeschlossen. Die Belichtungen wurden mittels einer vom Instrument unabhängigen Spaltblende, ähnlich wie man sie bei Sternspektographen benutzt, ausgeführt.

3.

Zur Erprobung der Aufstellung maß ich zunächst anderweit beobachtete Objekte, nämlich die in Tabelle 1 bezeichneten Eisenlinien. Die Tabelle gibt die Unterschiede Sonne—Bogen in tausendstel Ångström. Der Bogen hatte als unteren negativen Pol eine Eisen-

Sitz. Ber. Kgl. Preuss. Akad. d. Wiss. Berlin 1914, 1201–1213

oxydkugel nach dem Vorschlag von Hrn. Pfund[1]; die obere positive Elektrode bestand aus Kohle. Der Bogen war 8—10 mm lang. Die Stromstärke betrug 3.5 Ampere. Es war dies ein Bogen von sehr geringer Dampfdichte. Das Sonnenspektrum ist das der Sonnenmitte.

Tabelle 1. Sonne—*Fe*.

Platte Nr. λ	96	97	98	99	101	Mittel	Evershed	Royds	Kurzer—langer Bogen	Druckeinfluß	Int. Rowl.	Fabry und Buisson
4187.2		− 8	− 5	+ 3	0	− 3	+ 7	+ 1	23		6	− 3
87.9	− 4	+ 3	+ 2	+ 1	+ 2	+ 1	+ 8	+ 8	24		5	
91.6	+ 5	+ 4	+ 3	+ 3	+ 4	+ 4	− 1	+ 2	20		6	
99.2	0	+ 5	0	+ 4		+ 2	+10		4		5	
4202.2	+ 8	+ 6	+11	+ 5	+ 8	+ 8	+17	+10	2 (*u*)	25	8	−18
16.8	0	+ 3	+ 2	+ 2		+ 2		+ 3	6	15	3 *d*?	
19.5	+17	+13	+14	+26		+17	+15		7		7	
22.3	− 1	0	+ 3	+ 1		+ 1		− 2	27		5	
27.6	+ 2	− 1	+ 6	+ 3		+ 3	− 1	− 5	41	110	4	−29
33.5	− 1	+ 1	− 1	0		0	− 6	− 6	25	90	6	− 8
36.1	+10	0	+ 6	+ 9	+ 7	+ 6	+ 4	+ 3	22	90	8	−19
47.7	− 3			+ 5		+ 1					4	
50.3	+10	+ 5	+11	+10	+10	+ 9	+ 3	+ 7	24	70	8	−14
60.6	+18	+ 9	+10	+11	+15	+13	+ 7		24	51	10	
71.3	− 6	− 9				− 7	+ 4	+ 5	17		6	−18
71.9	+15	+10				+13	+ 9	+17	10 (*u*)	21	15	

Der Einfluß der Erdbewegung und der Erdrotation ist bereits in Rechnung gesetzt. Die Platten (98) und (101) wurden von Hrn. A. von Brunn, die übrigen von mir ausgemessen. Jede Linie wurde in jedem Spektralstreifen zweimal in beiden Lagen der Platte unter einem Toepferschen Meßapparat eingestellt. Das Mittel der Abweichungen jeder Platte vom Gesamtmittel — also der systematische Plattenfehler — beträgt für die fünf Platten 0, —2, 0, +1, +1 tausendstel Ångström. Die Platten stimmen also gut untereinander überein. Bei Platte (96) und (98) liegt das Sonnenspektrum zwischen zwei Eisenspektren, bei den übrigen Platten liegt ein Eisenspektrum zwischen zwei Sonnenspektren. Ein systematischer Unterschied beider Anordnungen, wie sie z. B. durch Spaltkrümmung bewirkt werden könnte, ist nach vorstehenden Zahlen nicht vorhanden. Ferner beträgt das Mittel der Abweichungen der Werte von Evershed[2] und Royds[3] gegen meine +0.7 bzw. +0.5 tausendstel Ångström. Es ist also auch keine systematische Ab-

[1] Astrophysical Journal 27, S. 296 (1908).
[2] Kodaikanal Bulletin Nr. 36.
[3] Kodaikanal Bulletin Nr. 38.

weichung zwischen mir und diesen beiden Beobachtern vorhanden. Evershed und Royds haben mit demselben Instrument und zum Teil gemeinsam gearbeitet. Das erklärt die gute Übereinstimmung zwischen beiden. Meine individuellen Werte weichen weit über ihren mittleren Fehler hinaus von denen beider Beobachter ab.

Um zu prüfen, wie weit die individuellen Abweichungen von der Natur des Eisenbogens abhingen[1], habe ich das Spektrum des beschriebenen Bogens geringer Dampfdichte verglichen mit einem Bogen sehr großer Helligkeit und jedenfalls auch Dampfdichte, der zwischen zwei Eisenstäben von 4 mm Durchmesser in einer Länge von 2 mm bei 7 Ampere brannte, wobei die Elektroden schnell weich wurden. Die auftretenden Linienverschiebungen sind zum Teil ganz auffällig. Allerdings wechseln sie von Platte zu Platte, weil man das Brennen des Bogens nicht ganz in der Gewalt hat. Auf der Platte, auf der die Verschiebungen am stärksten waren, wurden sie gemessen mit den in der Tabelle unter »Kurzer — langer Bogen« angegebenen Resultaten. Diese Verschiebungen gehen offenbar ganz parallel mit den Druckverschiebungen, die nach Adams und Gale[2] ebenfalls in der Tabelle angegeben sind. Man überzeugt sich aber leicht, daß die Unterschiede meiner Werte von denen der englischen Beobachter nicht mit der Druck- oder Dichteverschiebung parallel gehen. Dort ist, nach den Angaben von Hrn. Royds, ebenfalls ein Bogen von geringer Dampfdichte benutzt worden. Ich kann daher eine Erklärung der Abweichungen nur in verschiedener Auffassung der kleinen Unsymmetrien suchen, die fast alle Sonnenlinien durch die Nähe anderer schwacher Linien in diesem linienreichen Spektralgebiet haben.

Die Werte von Fabry und Buisson[3] sind, wie schon Hr. Evershed bemerkt hat, offenbar mit einem kurzen Bogen erhalten. Auch ist die Frage, ob die von Fabry und Buisson benutzte Interferenzmethode bei der Kompliziertheit der Beobachtung im kontinuierlichen Spektrum mit der Verwendung eines guten Gitters konkurrenzfähig ist.

Meine Mittelwerte Sonne — Bogen zeigen eine sehr deutliche Abhängigkeit von der Intensität der Linien:

10 Linien der Intensität 3—6 gaben Werte zwischen -7 und $+4$, im Mittel 0 ± 1 (m. F.) tausendstel Å;

6 Linien der Intensität 7—15 gaben Werte zwischen $+6$ und $+17$, im Mittel $+11 \pm 2$ (m. F.) tausendstel Å.

---

[1] Daß ein solcher Einfluß existiert, hat Hr. Goos, Astrophysical Journal 37, S. 48 (1913), gezeigt.

[2] Astrophysical Journal 35, S. 17.

[3] Astrophysical Journal 31, S. 109.

Es ist dies ein bekanntes Phänomen (s. unten), das sich hier sehr rein ausspricht.

Nach der vorstehenden Probe scheint kein Grund vorzuliegen, an der Freiheit des Instrumentariums von systematischen Fehlern zu zweifeln. Auch ist die Genauigkeit der Messungen befriedigend.

4.

Der Vergleich der Stickstoffbande 3883 Å mit der Sonne wurde zunächst auf zwei Platten Nr. (124) und (125) ausgeführt, wobei erstere das Spektrum des Sonnenzentrums von der Stickstoffbande eingerahmt enthielt, während bei letzterer das Stickstoffspektrum vom Sonnenspektrum eingerahmt war. Ich maß auf beiden Platten ohne die Absicht der Vollständigkeit zunächst 65 Objekte, bei denen die Identifizierung mit Linien der Stickstoffbande sicher zu sein schien. Von den 65 Objekten gibt für 5 die Rowlandsche Tabelle Zusammensetzung aus einer Stickstofflinie und einer Linie eines anderen Elements an. Diese Objekte wurden weggelassen. Ferner wurde das von Rowland gegebene Linienpaar 3861.734, 3861.681 (Intensität 2, 1) weggelassen, weil es unsymmetrisch und die Trennung auf meinen Aufnahmen nicht deutlich war. Viele von den vermessenen Stickstofflinien, namentlich am Kopf der ersten Bande, bilden mit Nachbarlinien zusammen eine Art Band, indem der Zwischenraum zwischen den Linien ebenfalls mit Licht erfüllt ist. Andere Linien haben mehr zufällige Begleiter auf der einen oder anderen Seite. Es zeigte sich, daß jede Unsymmetrie dieser Art den Vergleich mit der Sonne illusorisch macht. Für 11 Linien, welche solche Anhängsel oder Bänder auf der violetten Seite hatten, ergab sich nämlich im Durchschnitt Sonne — $N = -0.002$; für 14 Linien mit Anhängseln nach Rot ergab sich im Durchschnitt $+0.011$. Die Distanz zweier benachbarter Linien ist hiernach von mir im Emissionsspektrum des Stickstoffs größer gemessen worden als in der Sonne. Nach Ausschluß all dieser Linien mit unsymmetrischer Nachbarschaft — nur die beiden eine schwache Unsymmetrie des Hintergrundes aufweisenden, aber sonst sehr scharfen Linien 3868.261 und 3868.539 wurden mitgenommen — blieben 33 Linien, die für die weiteren Messungen beibehalten wurden.

Tabelle 2 gibt die Unterschiede Sonne — N für diese 33 Linien in tausendstel Ångström für 4 Platten. Platte (124) und (125) stammen vom 22. Juli 1914, (137) und (138) vom 12. August 1914. Bei Platte (124) und (138) liegt das Sonnenspektrum zwischen zwei N-Spektren, bei den beiden anderen Platten umgekehrt. Erdbewegung und Erdrotation sind bereits in Rechnung gesetzt.

Tabelle 2.

Rowland \ Platte	(124)	(125)	(137)	(138)	Mittel
3882.828 C 1	+ 2	− 2	+ 3	− 1	+ 1
79.716 C 1	+ 2	+10	+ 6	+ 4	+ 5
77.481 C 1	0	0	− 5	0	− 1
76.448 C 0		0	− 4	− 5	− 3
73.504 C 0	+ 3	+ 5	+ 8	+ 7	+ 6
71.018 C 1	+ 7	+15	+10	+10	+11
68.700 C 0	− 1	+ 6	+ 4	− 5	+ 1
68.539 C 1	+ 2	+ 1	+ 6	+ 1	+ 3
68.261 C 0	+ 8	+ 6	+ 5	− 1	+ 5
67.906 C 1	− 1	+ 0	− 6	− 6	− 3
67.205 C 0	+ 9	+ 6	− 3	+ 3	+ 4
67.118 C 0	+ 4	− 5	− 3	− 1	− 1
66.960 C 2	+ 9	+12	+11	+15	+12
66.122 C? 3Nd?	− 4	0	− 2	− 3	− 2
65.282 C? 3	+ 5	+ 2	0	0	+ 2
65.134 C 0	+ 6	− 3	− 1	− 6	− 1
64.438 C 3	+ 9	+ 6	+ 5	+ 4	+ 6
62.627 C? 2	+ 5	− 2	+ 2	+ 3	+ 2
62.458 C 00	+ 5	− 2	0	− 4	0
61.847 C 2N	+13	+10	+ 4	+ 7	+ 8
58.822 C 2N	+ 7	+ 6	+ 7	0	+ 5
58.722 C 0		+ 3	+ 6	+ 1	+ 3
58.033 C? 1	+ 7	+ 4	− 3	− 4	+ 1
57.063 C? 0	+ 9	+10	+ 7	+ 6	+ 8
54.707 C 2Nd?	+ 3	0	+ 6	+ 7	+ 4
54.191 C 0	+ 5	+ 7	− 2	− 2	+ 2
53.620 C 2d?	+ 2	− 1	− 3	− 8	− 3
52.541 C? 2Nd?	+13	+12	+12	+ 7	+11
51.427 C 2Nd?	+ 7	+ 4	+ 6	0	+ 4
50.300 C 1Nd?	0	0	+ 8	+ 4	+ 3
{46.777 C 1 {46.814 C 1	+ 3	− 6	+ 4	+ 1	+ 1
45.149 C 1	+ 7	+ 3	0	− 2	+ 2
44.378 C 4d	+ 7	+ 4	+ 6	+ 6	+ 6

Ordnet man die Abweichungen Sonne — N nach der Intensität der Linien unter Benutzung des Mittels aus den vier Platten, so findet man:

21 Linien der Intensität 0 und 1

$$+0.0022 \pm 0.0006 \text{ (m. F.) Å} = +0.17 \pm 0.05 \text{ km/sec},$$

13 Linien der Intensität 2 bis 4

$$+0.0045 \pm 0.0014 \text{ (m. F.) Å} = +0.36 \pm 0.11 \text{ km/sec}.$$

Eine stärkere Rotverschiebung der Linien größerer Intensität ist also angedeutet, aber nicht sicher gestellt.

Sitz. Ber. Kgl. Preuss. Akad. d. Wiss. Berlin 1914, 1201–1213

Im Mittel aus allen 33 Linien ergibt sich

Sonne — *N*

Platte	(124)	(125)	(137)	(138)	Gesamtmittel
Verschiebung in Å	+0.0045	+0.0032	+0.0028	+0.0012	+0.0029
Verschiebung in km/sec	+0.35	+0.25	+0.22	+0.09	+0.22

Die Abweichungen der Platten voneinander sind erheblich stärker, als dem aus der Übereinstimmung der Linien folgenden mittleren Fehler von etwa ±0.0005 Å entspricht. Das Gesamtresultat ist:

Die bei 3883 Å beginnende Stickstoffbande im Spektrum des Sonnenzentrums ist, verglichen mit der terrestrischen Bande, nach Rot verschoben um 0.003 Å, entsprechend einer Entfernungsgeschwindigkeit von 0.2 km/sec.

5.

Es wurde weiterhin untersucht, wie die Stickstoffbande sich verschiebt, wenn man vom Sonnenzentrum zu einem anderen Punkt der Sonnenscheibe übergeht. Es wurden bei vertikalem Spalt des Spektrographen stets Punkte des horizontalen Sonnendurchmessers aufgenommen. Der horizontale Durchmesser des Sonnenbildes machte bei den bearbeiteten Aufnahmen einen Winkel bis zu 20° mit dem Sonnenäquator. Zur Elimination der Sonnenrotation erfolgten die Aufnahmen stets paarweise in wenigen Minuten Zeitabstand an zwei symmetrisch zum Sonnenmittelpunkt gelegenen Stellen der Sonnenscheibe. Jede Aufnahme enthielt ein Spektrum des Sonnenzentrums zwischen zwei Spektren eines anderen Punktes der Sonnenscheibe oder umgekehrt ein exzentrisches Spektrum zwischen zwei Spektren der Sonnenmitte. Von den 33 Linien der Tabelle 2 wurde jede zweite vermessen, wobei ich mich mit je einer Einstellung in beiden Lagen der Platte begnügte.

Die Bestimmung der Verschiebungen zwischen Sonnenmitte und Sonnenrand ist deswegen nicht ganz einfach, weil die Linien im Spektrum ihr Aussehen mit der Annäherung an den Sonnenrand erheblich ändern, und zwar die Linien unserer Bande durchweg in der Weise, daß sie flauer werden, daß das Intensitätsverhältnis zwischen Linienmitte und kontinuierlichem Spektrum sich der Einheit nähert. In 0.75 Sonnenradien Abstand von der Mitte ist die Veränderung schon erkennbar, bei 0.85 Radien schon deutlich.

Die Einzelergebnisse für jedes Plattenpaar sind in Tabelle 3 mitgeteilt.

Tabelle 3. Sonnenrand — Sonnenzentrum.

Abstand von der Mitte		0.72	0.86				0.93		0.97	
Plattenpaare / Rowland		(122) (123)	(42) (43)	(58) (59)	(120) (121)	(130) (131)	(116) (117)	(132) (133)	(118) (119)	(134) (135)
3879.716	1	0	0	+1	−6	+1	+2	0	+ 1	− 3
76.448	0	−1	−5	−1	−2	0	+2	+ 1	− 4	− 4
71.018	1	0		+3	+1	+6	+5	− 1	+ 4	
68.539	1	−2	+1	−7	−6	0	−2	+ 6	0	
67.906	1	−4	−5	−5	+1	+2	+2	0	+ 2	+10
67.118	0	−5	+5	−6	−3	−2	−5	0	− 3	+12
66.122	3	−2	−1	−7	0	−2	−5	− 3	− 4	+ 3
65.134	0	0		−5	−7	0	−2	+ 4	+ 2	+ 9
62.627	2	−3	−4	−2	−4	−3	−5	+ 3	− 1	+ 3
61.847	2	−1		−1	−3	−2	+3	+ 4	+ 2	
58.722	0	+1		+5		+2	+3	+ 1	− 1	
57.063	0	−3	+5	+2	−5	+1	+1	+10	+ 3	
54.191	0	+3	0	−2	−3	−2	−1	+ 7	+ 2	+ 1
52.541	2	0	+2	−4	−4	−2	0	− 4	− 5	− 5
50.300	1	0		−6	−4	0	+6	0	+10	+ 8
46.814 } 777 }	1 1	−3	−4	−5	−2	−1	−2	− 2	− 6	− 8
44.378	4	−2	−7	−5	−2	−1	−3	− 1	− 5	+ 2

Im Mittel aus allen Linien sowie im Mittel aus den 6 Linien der Intensität 2 und 3 und den 11 Linien der Intensität 0 und 1 ergibt sich als Verschiebung der Randstellen gegen die Mitte für jedes Plattenpaar:

Tausendstel Ångström

Abstand von der Mitte	0.72	0.86				0.93		0.97	
Alle Linien	−1.3	−1.1	−2.7	−3.1	−0.2	−0.1	+1.5	−0.2	+2.3
Stärkere Linien	−1.8	−2.8	−4.0	−2.5	−1.8	−1.8	−0.5	−3.2	−1.0
Schwache Linien	−1.0	+0.1	−1.9	−3.4	+0.7	+0.9	+2.5	+1.5	+4.7

oder weiter gemittelt für jede Stelle der Sonnenscheibe:

Abstand von der Mitte	Tausendstel Ångström 0.72	0.86	0.93	0.97	km/sec 0.72	0.86	0.93	0.97
Alle Linien	−1.3	−1.8	+0.7	+1.1	−0.10	−0.14	+0.05	+0.09
Stärkere Linien	−1.8	−2.8	−1.1	−2.1	−0.14	−0.22	−0.09	−0.16
Schwache Linien	−1.0	−1.1	+1.7	+3.1	−0.08	−0.09	+0.13	+0.24

Verbunden mit den absoluten Verschiebungen, die in § 4 für die Sonnenmitte gefunden wurden, ergibt sich als absolute Verschiebung für die einzelnen Stellen der Sonnenscheibe:

Sitz. Ber. Kgl. Preuss. Akad. d. Wiss. Berlin 1914, 1201–1213

Verschiebung der *N*-Bande gegen die terrestrischen Wellenlängen für verschiedene Punkte der Sonnenscheibe.

Abstand von der Mitte	Tausendstel Ångström					km/sec				
	0.00	0.72	0.86	0.93	0.97	0.00	0.72	0.86	0.93	0.97
Alle Linien	+2.9	+1.6	+1.1	+3.6	+4.0	0.22	0.12	0.08	0.27	0.31
Stärkere Linien	+4.2	+2.4	+1.4	+3.1	+2.1	0.33	0.19	0.11	0.24	0.17
Schwächere Linien....	+2.2	+1.2	+1.1	+3.9	+5.3	0.17	0.09	0.08	0.30	0.41

Diese Zahlen stellen das Resultat der Untersuchung dar. Sie zeigen zunächst, daß die vorkommenden Verschiebungen durchweg sehr klein sind. Wie man ferner sieht, nimmt die Rotverschiebung von der Mitte weg zunächst ab, es erfolgt relativ zur Lage der Linien in der Sonnenmitte eine Violettverschiebung. Von 0.86 Radien an tritt dann wieder eine Rotverschiebung ein, die als »Randeffekt« bezeichnet werden kann. Für die stärkeren Linien ist der Randeffekt nicht deutlich, es ist wesentlich nur eine Abnahme der Wellenlänge von der Mitte bis in die Nähe des Sonnenrandes zu konstatieren.

Obwohl die Verschiebungen sehr klein und schwer von subjektiver Auffassung frei zu erhalten sind, möchte ich doch glauben, daß die eben geschilderte Abnahme der Wellenlänge von der Mitte nach dem Rande zu, abgelöst von einer Zunahme über die letzten Bogenminuten am Sonnenrande, wenigstens dem Sinne nach reell ist.

Es ist ein solches Verhalten auch bereits bei anderen Linien beobachtet und in folgender Weise gedeutet worden.

Die leuchtenden und absorbierenden — vielleicht ionisierten — Moleküle der Sonnengase befinden sich in ständiger absteigender, zentripetaler Bewegung. Diese Bewegung gibt in der Mitte der Sonne eine Rotverschiebung. Die Verschiebung nimmt nach dem Rande zu ab, weil sich die in die Gesichtslinie fallende Komponente der Bewegung immer mehr verkürzt, sie wird am Rande selbst Null. (Die in die Gesichtslinie fallende Komponente der zentripetalen Geschwindigkeit als Funktion des scheinbaren Abstands von der Sonnenmitte aufgetragen gibt übrigens eine Ellipse.) Über diese Wirkung der absteigenden Geschwindigkeit lagert sich dann noch ein Randeffekt, der eine Rotverschiebung in der Nachbarschaft des Sonnenrandes bewirkt.

Die obigen Resultate für die stärkeren *N*-Linien werden, wie die in Fig. 1 eingezeichnete Ellipse zeigt, genügend dargestellt durch die Annahme einer absteigenden Bewegung des Stickstoffs von 0.3 km/sec ohne Randeffekt. Für die schwächeren Linien (Fig. 2) ergibt sich eine

Fig. 1.

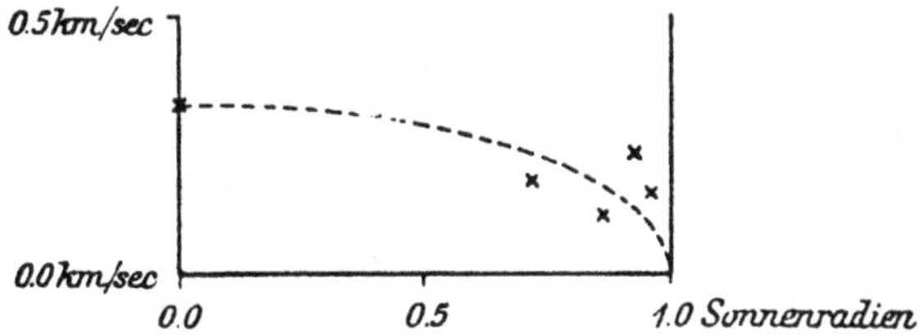

Stärkere *N*-Linien.

Fig. 2.

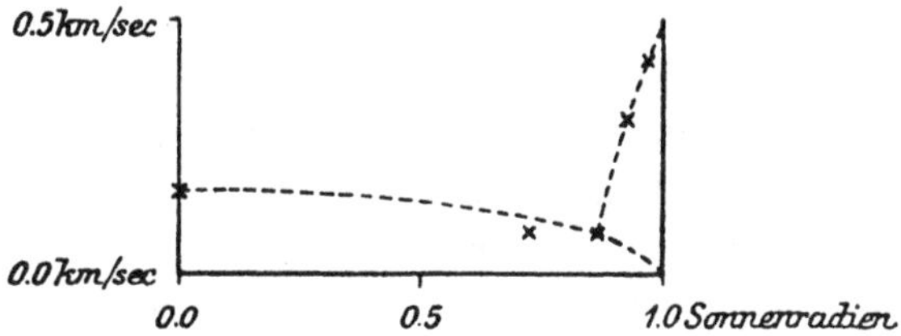

Schwächere *N*-Linien.

Abstiegsgeschwindigkeit von 0.2 km/sec und ein kleiner Randeffekt von 0.86 Radien an.

6.

Es sollen die Ergebnisse für die Stickstoffbande nunmehr mit den sonst vorhandenen Beobachtungen zusammengestellt werden, und zwar zunächst mit Beobachtungen, die sich auf zahlreichere Punkte der Sonnenscheibe beziehen. An Linien ohne Randeffekt liegen folgende Resultate vor:

Autor	Linie	Ursprung	Intensität	Abstiegs-geschwindigkeit	Reduzierte Geschwindigkeit
PEROT u. LINDSTEDT[1] ..	5172.856	Mg	20	1.6 km/sec	+0.3 km/sec
ST. JOHN[2]	3933.825	Ca	1000	1.1	0.0
SCHWARZSCHILD	3883 usw.	N	2—3	0.3	0.0

An Linien mit Randeffekt sind folgende beobachtet. Der Randeffekt beginnt in allen drei Fällen bei 0.85 Sonnenradien. Seine Größe ist für 0.94 Radien angegeben:

Autor	Linie	Ursprung	Intensität	Abstiegs-geschwindigkeit	Randeffekt	Reduzierte Geschwindigkeit
PEROT[3]	4625.227	Fe	5	0.0 km/sec	0.8 km/sec	−0.1 km/sec
PEROT[4]	5896.155	Na	20	1.4	0.9	+0.1
SCHWARZSCHILD	3883 usw.	N	0—1	0.2	0.3	0.0

[1] Comptes rendus 152, S. 1367 (1911).
[2] Astrophysical Journal 32, S. 36 (1910).
[3] Comptes rendus 151, S. 38 (1910).
[4] Comptes rendus 154, S. 326 (1912).

Unter »reduzierter Geschwindigkeit« ist hier folgendes verstanden. Schließt man an Beobachtungen der Punkte der Sonnenscheibe, welche vom Randeffekt noch nicht betroffen sind, eine Ellipse an, so findet man daraus die Abstiegsgeschwindigkeit. Setzt man diese Abstiegsgeschwindigkeit für das Sonnenzentrum in Rechnung, so sollte dadurch Übereinstimmung der solaren mit der terrestrischen Wellenlänge herbeigeführt werden. Der verbleibende Rest, in km/sec umgerechnet, ist als »reduzierte Geschwindigkeit« angegeben. Die Kleinheit dieser Reste würde beweisen, daß der Einstein-Effekt nicht existiert, wenn die Deutung der Verschiebungen in der obigen Weise als einzig mögliche erkannt wäre.

Für eine viel größere Zahl von Linien liegen, wenn auch keine Messungen an vielen Punkten der Sonnenscheibe, so doch wenigstens Vergleichungen der Wellenlängen in der Sonnenmitte mit Punkten in wenigen Sekunden vom Rande vor, in einer Arbeit von Hrn. W. Adams[1]. Die Linien sind danach am Rande fast durchweg ein wenig nach Rot verschoben, doch betragen die Verschiebungen meist nur wenige Zehntel km/sec.

Für die Mg-Linie 5173 und die Ca-Linie 3934 stehen die Messungen von Adams in Widerspruch mit den Ergebnissen von Perot und St. John, da nach deren Angaben wegen Fehlens des Randeffektes eine Violettverschiebung der Linien verglichen mit dem Zentrum bestehen müßte, während Adams in beiden Fällen eine geringe Rotverschiebung mißt.

Schließlich haben die HH. Fabry und Buisson, Evershed und Royds für zahlreiche Eisenlinien eine Vergleichung der Wellenlängen im Sonnenzentrum mit den terrestrischen Linien ausgeführt. Hr. Evershed bemerkte, daß die Rotverschiebung mit der Stärke der Linien geht. Hr. Royds findet aus ausgewählten symmetrischen, von der Dampfdichte wenig abhängigen Linien:

Intensität nach Rowland	2 bis 4	5 bis 7	8 bis 10	über 10
Rotverschiebung	+ 0.2	+ 0.2	+ 0.4	+ 0.8 km/sec.

Man sieht, daß eine dem Einstein-Effekt (0.63 km/sec) entsprechende Verschiebung erst bei Linien von erheblicher Intensität (etwa 10) auftritt.

7.

Überblickt man alle diese Resultate, so muß man zum mindesten sagen, daß der Einsteinsche Effekt, eine Rotverschiebung von 0.63 km/sec, keineswegs klar aus den Beobachtungen hervortritt. Es

[1] Astrophysical Journ. 31, S. 30 (1910).

spricht besonders gegen den Einsteinschen Effekt, daß alle untersuchten schwächeren Linien im Sonnenspektrum, sowohl die des Eisens wie die des Stickstoffs, überhaupt nur sehr geringe Verschiebungen sowohl gegen die terrestrischen Linien wie gegen die Mitte der Sonnenscheibe aufweisen. Daß die von Hrn. Evershed vermessenen Eisenlinien im Durchschnitt gerade die Einsteinsche Verschiebung gaben, beruhte nur auf der zufälligen Mischung starker und schwacher Linien.

Auf der anderen Seite wurden oben Widersprüche zwischen den vorliegenden Angaben erwähnt, und so müssen offenbar die Beobachtungen noch weiter geklärt und vermehrt werden, bevor eine genauere Diskussion der Ursache der Linienverschiebungen möglich ist.

Ausgegeben am 7. Januar 1915.

R. Emden. Gaskugeln. Anwendungen der mechanischen Wärmetheorie auf kosmologische und meteorologische Probleme. Leipzig, B. G. Teubner. 1907. Mit 24 Figuren, 12 Diagrammen und 5 Tafeln im Text.

§ 1. Da der Referent in vielfachem freundschaftlichem Verkehr mit dem Verfasser das Zustandekommen seines Buches miterlebt hat, so steht er demselben vielleicht nicht mit der Objektivität gegenüber, die bei Referaten in dieser Zeitschrift üblich ist; doch kommt es hier wohl auch zunächst nicht auf Kritik an. Ich werde mit der Kritik nicht zurückhalten, wo sie sich mir von selbst darbietet, aber hauptsächlich will dieses Referat eine gedrängte Inhaltsübersicht geben und damit einen Nachteil des Buches etwas kompensieren — nämlich den, daß es dem Verfasser unter der Hand zu lang geworden ist.

Daran ist zum großen Teil die Vielseitigkeit des Inhaltes schuld. Was nur immer mit dem Problem der adiabatischen gravitierenden Gaskugeln oder Gasatmosphären zusammenhing, hat der Verfasser in den Bereich seiner Betrachtungen gezogen. Die umfangreiche Literatur (Helmholtz, Lord Kelvin, Homer Lane, Ritter, Darwin u. a.) ist zusammengefaßt und inhaltlich erweitert. Doch führten auch didaktische Nebenabsichten zu einer gewissen Breite. Das Buch sollte eine Art Einführung in die mathematische Behandlung physikalischer Probleme sein, die den Studierenden lehrt, bis zur rechnerischen Auswertung in den Einheiten des absoluten Maßsystems vorzudringen.

So möchte Ref. für den, dem es sich nur um Gedankengang und Resultate handelt, den Inhalt des Buches wiedergeben. Die Zweiteilung „Theorie" und „Anwendung" ist beibehalten, aber im einzelnen ist im Interesse der Knappheit von der Disposition des Verfassers abgewichen. Das Historische, das beiseite gelassen ist, findet man ausführlich im letzten Abschnitt des Buches dargestellt.

Theorie.

§ 2. Statt der meist betrachteten „adiabatischen", ohne Wärmeaustausch vor sich gehenden Zustandsänderung eines idealen Gases wird die allgemeine sogen. „polytrope" Zustandsänderung zugrunde gelegt, bei welcher durch Kondensation, Wärmestrahlung oder sonst irgend einen Vorgang dem Gase eine der Temperaturänderung proportionale Wärmemenge zugeführt wird. Das Gesetz von der Erhaltung der Energie laute für die Masseneinheit des Gases geschrieben:

$$d\varepsilon = dQ + dW, \tag{1}$$

27

wo ε die (innere) Energie, dQ die zugeführte Wärme, dW die Arbeit äußerer Kräfte ist. Für ein ideales Gas gilt:

$$\begin{aligned} \varepsilon &= c_v\,.\,T, \quad dW = -A p\, dv, \\ p v &= H\,.\,T, \quad c_p - c_v = A H, \end{aligned} \tag{2}$$

wenn man mit T, p, v, c_v, c_p Temperatur, Druck, Volumen, spez. Wärme bei konstantem Volumen resp. konstantem Druck bezeichnet, und H die sogen. Gaskonstante, A das reziproke mechanische Wärmeäquivalent bedeutet. Die Gleichung der Adiabate erhält man, indem man $dQ = 0$ setzt, hingegen entsteht die Polytrope durch den Ansatz:

$$dQ = \gamma\, dT, \tag{3}$$

wo γ eine Konstante ist. Setzt man (2) und (3) in (1) ein, so erhält man durch Integration:

$$T\,.\,v^{k-1} = \text{const.} = \Theta, \quad p\,.\,v^k = H\,.\,\Theta, \quad k = \frac{c_p - \gamma}{c_v - \gamma}; \tag{4}$$

die Größe k wird als die „Klasse" der polytropen Änderung bezeichnet.

Man kommt auf die Adiabate zurück, indem man γ verschwinden läßt. Es wird dann $k = \frac{c_p}{c_v}$, wofür wir $k = \varkappa$ schreiben wollen, indem $\varkappa$ das Verhältnis der spezifischen Wärmen bedeutet. Es sei daran erinnert, daß $\varkappa$ für einatomige Gase den Wert $^5/_3$, für zwei- resp. dreiatomige die Werte $^7/_5$ resp. $^4/_3$ und für größere Moleküle dichter an 1 liegende Werte besitzt.

Es gibt physikalische Vorgänge, wie z. B. die Expansion von Dämpfen und das Aufsteigen Regen abscheidender Luftmassen, wo die Gleichung (3) mit konstantem γ durch größere Strecken hindurch sehr nahe erfüllt ist. Im übrigen sind aber die Polytropen infolge der geschmeidigen Beziehungen (4) ein sehr bequemes Mittel, um beliebige Vorgänge durch sprungweise wechselndes γ angenähert darzustellen. Die bei einer bestimmten polytropen Änderung konstante Größe Θ ist diejenige Temperatur, welche das Gas annimmt, wenn es von seinem augenblicklichen Zustand aus nach der Polytrope von der Klasse k auf die Dichte 1 komprimiert wird. Daher wird Θ von Emden als „polytrope Temperatur (der Klasse k)" bezeichnet.

Für die Verhältnisse von Wärmezufuhr, Arbeitszufuhr und Energieänderung ergibt sich für einen polytropen Vorgang die Beziehung:

$$dQ : dW : d\varepsilon = k - \varkappa : \varkappa - 1 : k - 1. \tag{5}$$

Es sei auch noch der interessante wenig bekannte Satz angeführt, daß ein Kreisprozeß zwischen zwei Isothermen und zwei Polytropen derselben Klasse den gleichen maximalen Nutzeffekt liefert, wie der Carnotsche Kreisprozeß.

§ 3. Soll eine ruhende Gaskugel bei einer bestimmten Verteilung der Gasdichte $\varrho\left(=\frac{1}{v}\right)$ als Funktion der Entfernung r vom Zentrum in mechanischem Gleichgewicht sein, so muß das Druckgefälle überall der Schwere das Gleichgewicht halten, es muß gelten:

$$\frac{dp}{dr} = -\varrho g, \tag{6}$$

und dabei ist die Schwerebeschleunigung g:

$$g = \frac{4\pi G}{r^2}\int_0^r \varrho r^2 dr, \tag{7}$$

wenn man mit G die Gravitationskonstante bezeichnet. Die Gleichungen (6) und (7) ergeben durch Differentiation:

$$\frac{d}{dr}\left(\frac{r^2}{\varrho}\frac{dp}{dr}\right) + 4\pi G r^2 \varrho = 0. \tag{8}$$

Nimmt man nun an, daß ein durch die Gasmasse hindurchgeführtes und überall mit seiner Umgebung in thermisches und mechanisches Gleichgewicht versetztes Gasteilchen polytrope Zustandsänderung der Klasse k erleidet, so hat man gemäß (4): $p = \varrho^k . H . \Theta$ zu setzen. Führt man diesen Wert für p in (8) ein, so erhält man eine Differentialgleichung für die Dichte ϱ, die wir gleich auf die übersichtlichste Form bringen, indem wir statt ϱ die nach (4) der Temperatur proportionale Größe $u = \varrho^{k-1}$ als abhängige Variable einführen und die Abkürzungen benutzen:

$$n = \frac{1}{k-1}, \quad \alpha^2 = \frac{4\pi G}{H\Theta}\frac{k-1}{k}. \tag{9}$$

Die Differentialgleichung lautet dann:

$$\frac{d^2u}{dr^2} + \frac{2}{r}\frac{du}{dr} + \alpha^2 u^n = 0. \tag{10}$$

Es ergibt sich das rein mathematische Problem, Einsicht in die Lösungssysteme dieser nichtlinearen Differentialgleichung zu gewinnen.

4. Setzt man $\mathfrak{r} = \alpha r$, so geht die Differentialgleichung über in:

$$\frac{d^2 u}{d\mathfrak{r}^2} + \frac{2}{\mathfrak{r}}\frac{du}{d\mathfrak{r}} + u^n = 0, \tag{11}$$

so daß in der Differentialgleichung nur noch die Größe n, die Polytropenklasse, außer den Variabeln auftritt; und diese vereinfachte Form werde weiterhin zugrunde gelegt.

Die Differentialgleichung gestattet eine Transformation in sich. Ersetzt man nämlich $\mathfrak{r}$ durch $p\mathfrak{r}$ und u durch $q \cdot u$, so bleibt die Gleichung unverändert, falls:

$$p^2 q^{n-1} = 1 \tag{12}$$

ist. Aus dieser Eigenschaft folgt nach bekannten Sätzen von Lie, daß die Gleichung auf die erste Ordnung reduziert werden kann, und es ist in unserem Falle leicht, bei Kenntnis der Transformation die Reduktion wirklich auszuführen. Das Resultat ist folgendes. Man bestimme r und u mittels der neuen Variabeln y und z gemäß den Gleichungen:

$$\mathfrak{r} = e^{-\int \frac{dz}{y}}, \quad u \cdot \mathfrak{r}^{\frac{2}{n-1}} = z. \tag{13}$$

Dann folgt zwischen y und z die lineare Gleichung:

$$y\frac{dy}{dz} + \frac{5-n}{n-1}y + \frac{2(3-n)}{(n-1)^2}z + z^n = 0, \tag{14}$$

die nun des näheren zu untersuchen ist.

Es seien zunächst die Fälle angeführt, welche ohne weiteres integrabel sind. Die Ausgangsgleichung 2. Ordnung wird linear für $n = 0$ und $n = 1$ und hat dann die einfachen Integrale (die Größen C_1, C_2, C, c in den nächsten Gleichungen bedeuten Integrationskonstanten):

$$n = 0: \quad u = C_1 + \frac{C_2}{\mathfrak{r}} - \frac{\mathfrak{r}^2}{6},$$

$$n = 1: \quad u = C_1 \frac{\sin \mathfrak{r}}{\mathfrak{r}} + C_2 \frac{\cos \mathfrak{r}}{\mathfrak{r}}.$$

Aus (14) sieht man aber, daß auch noch der Fall $n = 5$ sofort integrabel ist, und zwar gibt eine kurze Rechnung:

$$y^2 = C + \frac{z^2}{4} - \frac{z^6}{3}.$$

Die Einführung dieses Wertes y in (13) führt im allgemeinen auf elliptische Integrale. Nur für $C = 0$ ergibt sich einfach:

$$\mathfrak{r} = c\,\frac{1 + \sqrt{1 - {}^4/_3\, z^4}}{\sqrt{{}^4/_3} \cdot z^2}$$

und damit: $$u^2 = \frac{\sqrt{3}\,.\,c}{c^2 + \mathfrak{r}^2}.$$

In allen übrigen Fällen ist man auf die qualitative Diskussion der Lösungskurven von (14) angewiesen, welche als ein spezielles Beispiel zu Poincarés entsprechenden allgemeinen Untersuchungen besonderes Interesse hat. Man beginnt passend mit der Aufsuchung der singulären Punkte.

Die singulären Punkte, in welchen $\frac{dy}{dz}$ unbestimmt wird, sind gegeben durch die Bedingungen:

$$y = 0, \qquad \frac{5-n}{n-1}y + 2\,\frac{(3-n)}{(n-1)^2}z + z^n = 0.$$

Diese Gleichungen haben für $n < 3$ nur die eine reelle Lösung:

$$y = 0, \qquad z = 0;$$

für $n > 3$ außerdem noch die Lösung:

$$y = 0, \quad z = \left[\frac{2(n-3)}{(n-1)^2}\right]^{\frac{1}{n-1}} = Z.$$

Die singulären Punkte — sie sollen mit O und O_1 bezeichnet werden — stellen natürlich selbst singuläre Lösungen der Gleichung (14) dar. Die entsprechenden singulären Lösungen von (11) sind nach (13):

$$u = 0 \text{ und } u = \left[\frac{2(n-3)}{(n-1)^2\mathfrak{r}^2}\right]^{\frac{1}{n-1}}. \qquad (15)$$

Es sind nun die Lösungen von (14) in der Nähe der singulären Punkte zu untersuchen.

Man beginne mit dem Punkte $y = 0$, $z = 0$ und beschränke sich auf Werte $n > 1$. Dann kann man z^n gegen z vernachlässigen und behält die einfachere Differentialgleichung:

$$y\frac{dy}{dz} + \frac{5-n}{n-1}y + \frac{2(3-n)}{(n-1)^2}z = 0,$$

welche nach der Substitution $y = z\,.\,w$ Trennung der Variabeln und Integration erlaubt. Das Resultat der Integration läßt sich in der Form schreiben (C ist Integrationskonstante):

$$\left(y + \frac{a-1}{2}z\right)^{\frac{a-1}{2}} = C\left(y + \frac{a+1}{2}z\right)^{\frac{a+1}{2}}, \quad a = \frac{5-n}{n-1}.$$

Die hierdurch bestimmten Kurven — es sind sogen. W-Kurven — sehen verschieden aus, je nachdem n kleiner oder größer als 3 ist. Im Falle $n < 3$ (Fig. 1) tangieren alle Lösungen im Nullpunkt die Gerade $y + \frac{a-1}{2} z = 0$, welche in den rechten unteren Quadranten verläuft (in der Figur mit ξ bezeichnet). Hiervon ausgenommen ist allein die ausgezeichnete Lösung (die sich für $C = \infty$ ergibt): $y + \frac{a+1}{2} z = 0$ (in der Figur mit η bezeichnet).

Im Falle $n > 3$ (Fig. 2) sind die beiden Geraden

$$y + \frac{a-1}{2} z = 0 \quad \text{und} \quad y + \frac{a+1}{2} z = 0$$

Asymptoten der hyperbelartigen Lösungskurven geworden, und dabei ist die erstere Gerade nach rechts in die obere Halb-

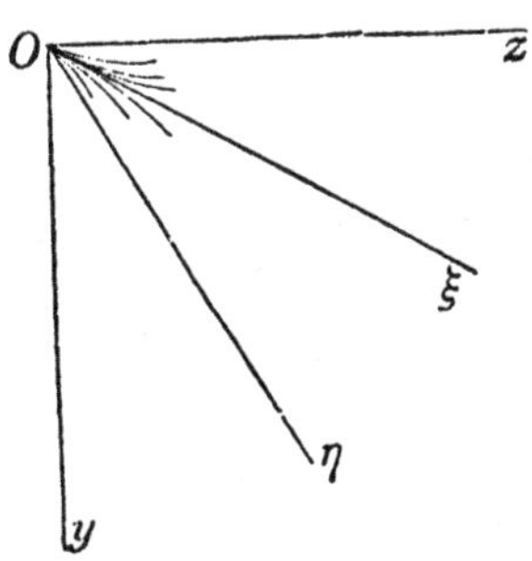

Fig. 1.

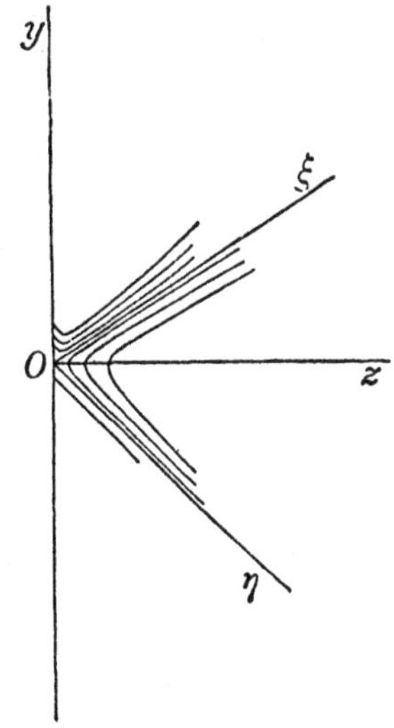

Fig. 2.

ebene übergetreten. Die Asymptoten selbst stellen in der Nähe des singulären Punktes ausgezeichnete Lösungen dar, und zwar wollen wir die Gerade $y + \frac{a+1}{2} z = 0$ als erste ausgezeichnete Lösung, die Gerade $y + \frac{a-1}{2} z = 0$ als zweite ausgezeichnete Lösung bezeichnen.

Man betrachte weiter die Umgebung des zweiten singulären Punktes $y = 0$, $z = Z$. Setzt man $z = Z + z_1$ und vernachlässigt in der Differentialgleichung höhere Potenzen von z_1, so lautet sie:

$$y\frac{dy}{dz_1} + \frac{5-n}{n-1}y - \frac{2(3-n)}{n-1}z_1 = 0.$$

Ähnlich wie oben, erhält man für $3 < n < 3{\cdot}187673$:

$$(y + \alpha z_1)^\alpha = C(y + \beta z_1)^\beta$$

$$\begin{matrix}\alpha \\ \beta\end{matrix} = \left(\frac{5-n}{n-1} \pm \frac{\sqrt{1 + 22n - 7n^2}}{n-1}\right).$$

Ist $n > 3{\cdot}187673$ und damit der Radikand in α und β negativ, so schreibt man das Integral am besten mittels einer Hilfsvariabeln ω in der Form:

$$y = C\left(\sin\omega - \frac{a}{\sqrt{q}}\cos\omega\right)e^{\frac{a\omega}{\sqrt{q}}}, \qquad z_1 = \frac{2C}{\sqrt{q}}\cos\omega\, e^{\frac{a\omega}{\sqrt{q}}},$$

$$a = \frac{5-n}{n-1}, \qquad q = \frac{7n^2 - 22n - 1}{(n-1)^2}.$$

Die Gestalt der Lösungskurven wird im Falle $3 < n < 3{\cdot}18..$ qualitativ durch die Fig. 1 wiedergegeben. Es gehen wieder alle Kurven in Richtung der Geraden ξ $(y + \alpha z_1 = 0)$ vom singulären Punkte ab mit Ausnahme der für $C = \infty$ entstehenden ausgezeichneten Lösung $y + \beta z_1 = 0$.

Für $n > 3{\cdot}18\ldots$ haben die Kurven die Gestalt von Spiralen, die sich in immer engeren Windungen um den singulären Punkt schließen, und zwar nähert sich die Spirale dem singulären Punkt für $n < 5$, wenn man im Sinne des Uhrzeigers, für $n > 5$, wenn man entgegen dem Sinne des Uhrzeigers um den singulären Punkt wandert.

Nachdem man so über den eigentümlichen Verlauf der Lösungskurven in der Nähe der singulären Punkte orientiert ist, wird es verhältnismäßig leicht, über ihren Gesamtverlauf Klarheit zu gewinnen. Man kann nämlich unmittelbar die Kurven gleicher Fortschreitungsrichtung $\frac{dy}{dz} = \text{const.} = h$ zeichnen, da für dieselben aus der Differentialgleichung

$$y = \frac{\frac{2(n-3)}{(n-1)^2}z - z^n}{h + \frac{5-n}{n-1}}$$

folgt.

33

In den dem Emdenschen Buche entnommenen Figuren 3, 4 und 5 sind die Fortschreitungsrichtungen für $n = 2$, 4 und 6 dargestellt. Hält man das, was der Anblick dieser Diagramme unmittelbar lehrt, mit den obigen Betrachtungen über die Nachbarschaft der singulären Punkte zusammen, so ist der Gesamtverlauf der Lösungskurven qualitativ gegeben.

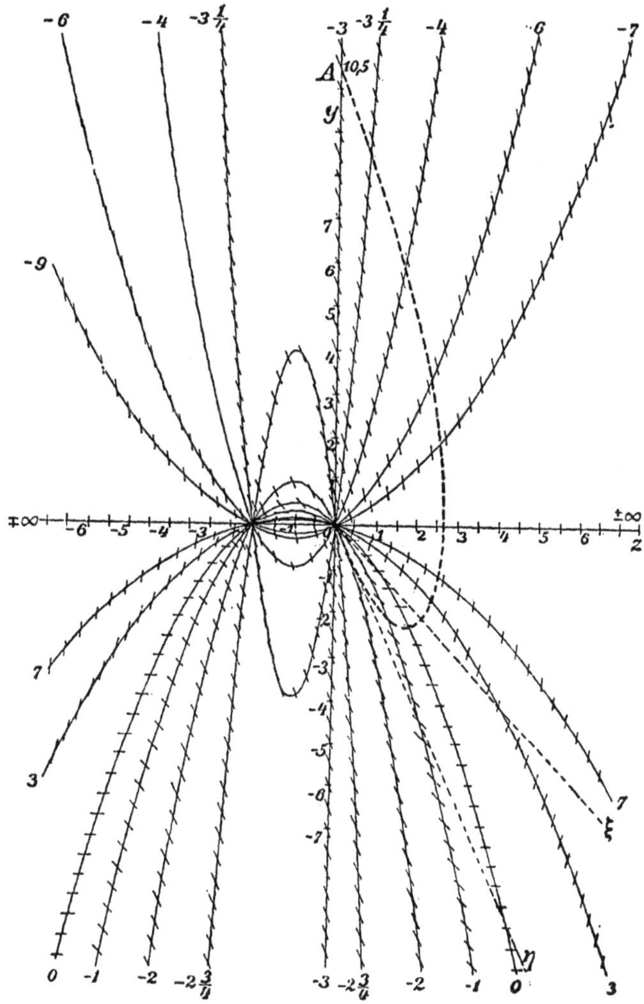

Fig. 3. ($n = 2$)

Es sollen einige charakteristische Momente in dem Verhalten der Lösungskurven, die aus den Figuren 1—5 ersichtlich sind, hervorgehoben werden.

a) $n < 3$. Alle Lösungen gehen vom Nullpunkt oder der $-y$-Achse ab und erreichen, indem sie sich alsbald mehr

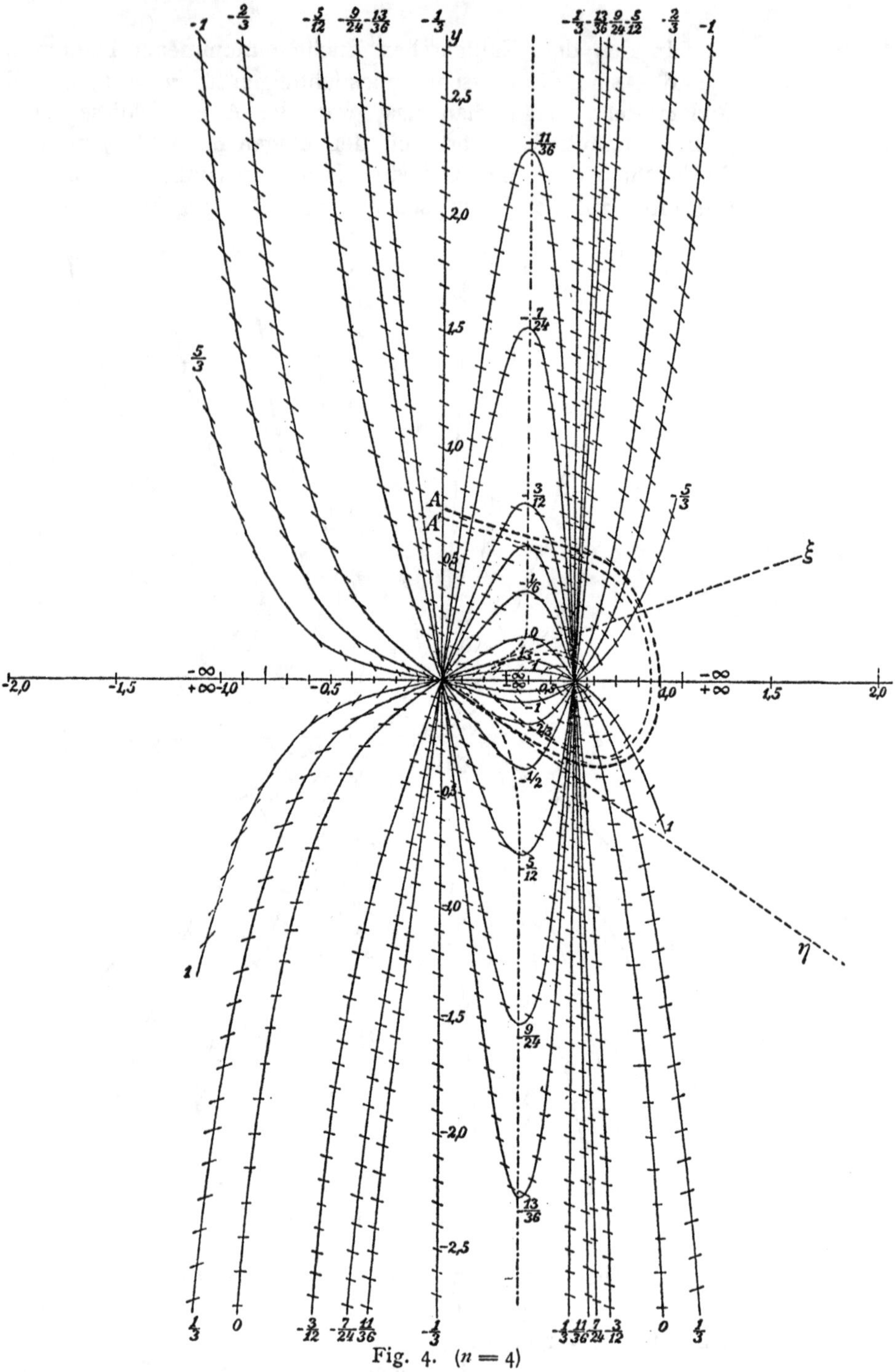

Fig. 4. ($n = 4$)

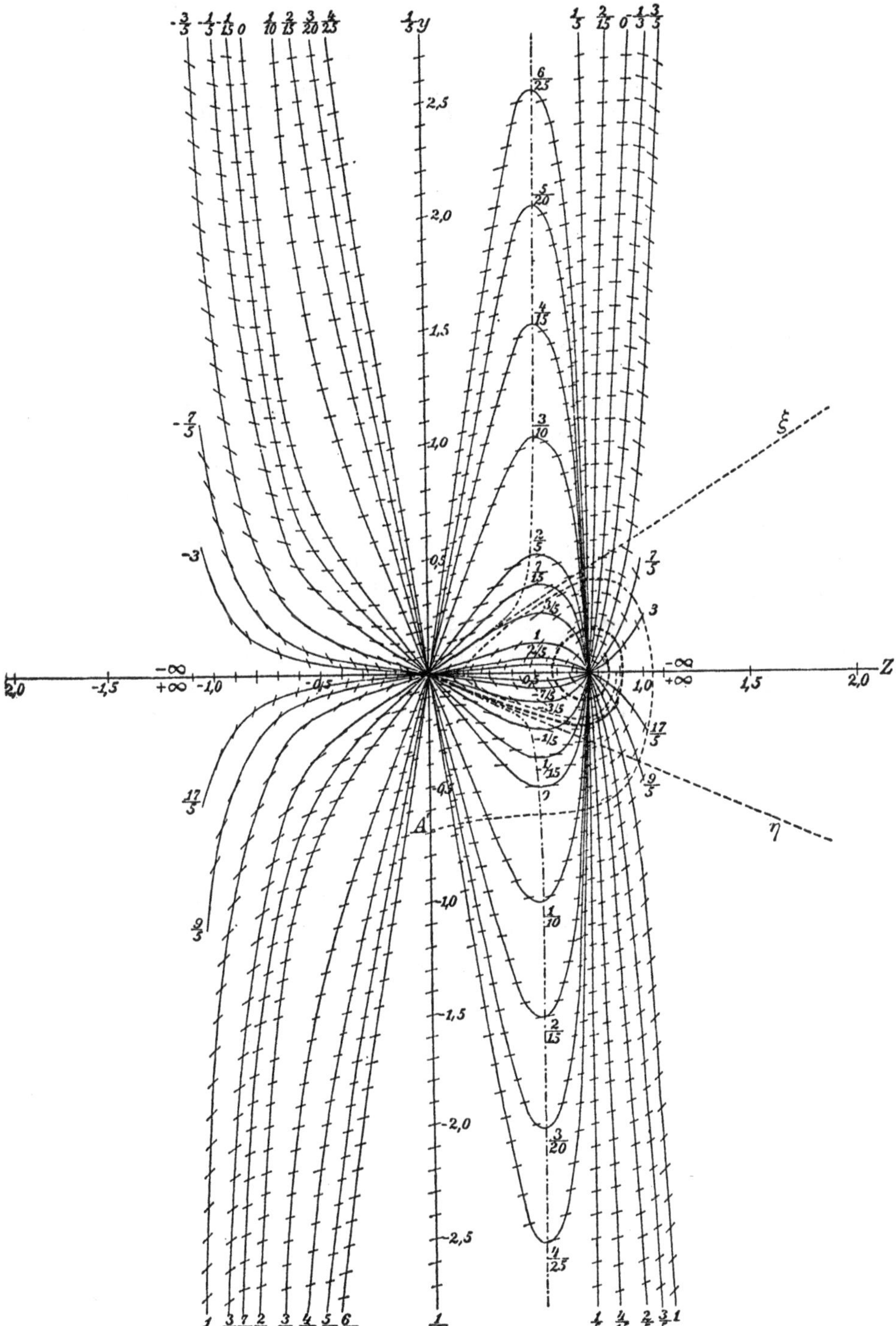

Fig. 5. $(n = 6)$

und mehr links wenden, zunächst die $+z$-Achse, die sie senkrecht durchsetzen, dann die $+y$-Achse. Die im Nullpunkt entspringenden Lösungen gehen alle in Richtung der Geraden $y + \frac{a-1}{2} z = 0$ von ihm ab, die unterhalb des Nullpunkts abgehenden Lösungen beginnen parallel der Geraden $y + \frac{a+1}{2} z = 0$. Den Übergang zwischen beiden Lösungsklassen bildet die schon oben erwähnte „ausgezeichnete Lösung", welche vom Nullpunkt in Richtung der letzteren Geraden η ausgeht.

b) $3 < n < 5$. Die Lösungskurven zerfallen hier in solche, die vom Punkte O_1 ausgehend und ihn entgegengesetzt dem Uhrzeigersinne umlaufend schließlich die $+y$-Achse schneiden, und in solche, die von der $-y$-Achse ausgehend in einfachem Bogen den Punkt O_1 umgehend zur $+y$-Achse gelangen. Der Übergang wird vermittelt durch die beiden „ausgezeichneten Lösungen", die durch den Nullpunkt gehen. Je nachdem $n \gtrless 3{\cdot}18$ ist, erfolgt der Auslauf aus dem Punkt O_1 direkt oder asymptotisch in einer sich erweiternden Spirale.

c) $n > 5$. Die Verhältnisse liegen ähnlich wie unter b), nur daß die obere und die untere Halbebene vertauscht und entsprechend der Umdrehungssinn der Kurven um O_1 der entgegengesetzte geworden ist.

Die Differentialgleichung 1. Ordnung ist hiermit erledigt, und es ist aus ihr zurückzuschließen auf die Lösungen der Differentialgleichung 2. Ordnung, auf die Abhängigkeit zwischen u und $\mathfrak{r}$. Man erhält aus den Gleichungen (13), indem man y als Funktion von z einsetzt, zunächst $\mathfrak{r}$ und u als Funktionen des Parameters z ausgedrückt. Die Elimination von z ergibt dann die Beziehung zwischen $\mathfrak{r}$ und u.

Führt man den Rückschluß zunächst für die Nachbarschaft der singulären Punkte aus, so erhält man in der Nähe des Nullpunktes O im allgemeinen:

$$u = \frac{c_1}{\mathfrak{r}} + c_2 \tag{17}$$

(c_1, c_2 sind Integrationskonstanten), und dies gilt im Falle $n < 3$ für sehr kleines $\mathfrak{r}$, hingegen im Falle $n > 3$ für sehr großes $\mathfrak{r}$ und:

$$u = \text{const.} = c_2$$

für diejenige Lösung der Differentialgleichung 1. Ordnung, die oben als „erste ausgezeichnete Lösung" bezeichnet wurde.

In der Nähe des Punktes O_1 erhält man für $3 < n < 3 \cdot 18$:

$$u = \frac{Z}{\mathfrak{r}^{\frac{2}{n-1}}} + \frac{C_1}{{}^{\frac{1}{2}\left(1+\frac{\alpha-\beta}{2}\right)}} + \frac{C_2}{\mathfrak{r}_1^{\frac{1}{2}\left(1-\frac{\alpha-\beta}{2}\right)}}. \tag{18}$$

Für $3 \cdot 18 < n$ folgt:

$$u = \frac{Z + c_1 \cos\omega\, e^{\frac{a}{\sqrt{q}}\omega}}{\mathfrak{r}^{\frac{2}{n-1}}}, \quad \mathfrak{r} = c_1\, e^{\frac{2}{\sqrt{q}}\omega}$$

$$\frac{a}{\sqrt{q}} = \frac{n-5}{\sqrt{7n^2 - 22n - 1}}, \quad \frac{2}{\sqrt{q}} = \frac{2(n-1)}{\sqrt{7n^2 - 22n - 1}}. \tag{19}$$

Mit wachsendem r und ω schwankt u um die singuläre Lösung $\frac{Z}{\mathfrak{r}^{\frac{2}{n-1}}}$ hin und her, und zwar schmiegt sich u der singulären Lösung für abnehmendes $\mathfrak{r}$ mehr und mehr an, wofern $n < 5$ ist, hingegen für zunehmendes $\mathfrak{r}$, wenn $n > 5$ ist. Im singulären Punkt selbst wird für $n < 5$ $\mathfrak{r} = 0$, $u = \infty$, für $n > 5$ hingegen $\mathfrak{r} = \infty$, $u = 0$.

Aus dem Vorzeichen von y und der ersten Formel (13) ist leicht zu ersehen, daß längs jeder y-Kurve $\mathfrak{r}$ sich ständig im selben Sinne ändert; der Sinn, in welchem $\mathfrak{r}$ wächst, soll als der positive bezeichnet werden.

Ferner sei besonders hervorgehoben, daß ein Schnitt der y,z-Kurven mit der Ordinatenachse $z = 0$ (abgesehen vom Falle $\mathfrak{r} = 0$) nach (13) immer einen Schnitt der $u,\mathfrak{r}$-Kurve mit der $\mathfrak{r}$-Achse: $u = 0$ zur Folge hat. Da u der Temperatur proportional ist, bedeutet dies physikalisch Absinken der Temperatur, wie auch des Druckes, auf Null und damit eine Grenze der betreffenden Gaskugel.

Die Zusammenfassung der angeführten Beziehungen ergibt einen vollständigen Überblick über das Verhalten von u als Funktion von $\mathfrak{r}$. Will man die Abhängigkeit der Größe u von $\mathfrak{r}$ geometrisch zum Ausdruck bringen, so hätte man für jedes u zweifach unendlich viele Kurven zu zeichnen, um alle Lösungen der Differentialgleichung 2. Ordnung darzustellen. Nun gestattet aber, wie unten noch weiter zu besprechen ist, die erwähnte Transformation der Differentialgleichung in sich aus einer Lösung einfach unendlich viele andere abzuleiten. Es genügt daher, zunächst einfach unendlich viele Lösungen zeichnerisch zur Darstellung zu bringen, aus denen dann alle anderen durch die Transformation hervorgehen. Diese einfach unendlich vielen Lösungen wollen wir bestimmen durch die Forderung, daß zu dem größten z, das auf jeder y, z-Kurve erreicht wird, der ein für allemal festzusetzende Wert $\bar{\mathfrak{r}}$ des Radius gehören soll. Die Gleichungen (13) lauten dann:

$$\mathfrak{r} = \bar{\mathfrak{r}}\, e^{-\int_{\bar{z}}^{z} \frac{dz}{y}}, \quad u\,\mathfrak{r}^{\frac{2}{n-1}} = z$$

($\bar{z}$ ist dieser Maximalwert von z) und ordnen jeder y, z-Kurve eindeutig eine $\mathfrak{r}, u$-Kurve zu.

Über das Verhalten der so definierten ∞^1 $\mathfrak{r}, u$-Kurven ($\mathfrak{r}$ sei horizontale Abszisse, u Ordinate) lassen sich nunmehr aus den Darstellungen der y, z-Kurven folgende Aussagen ablesen:

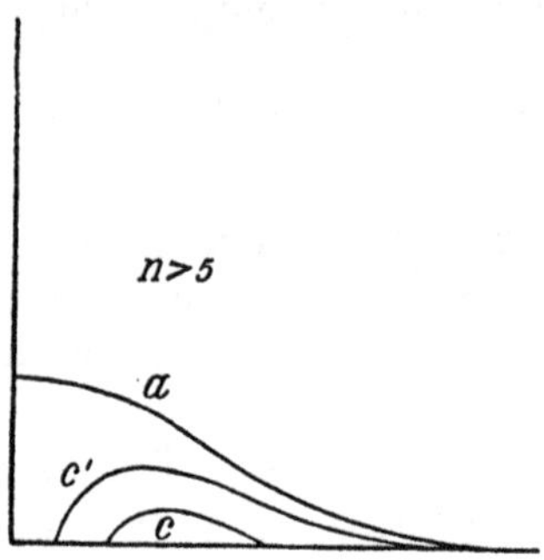

Fig. 6a.

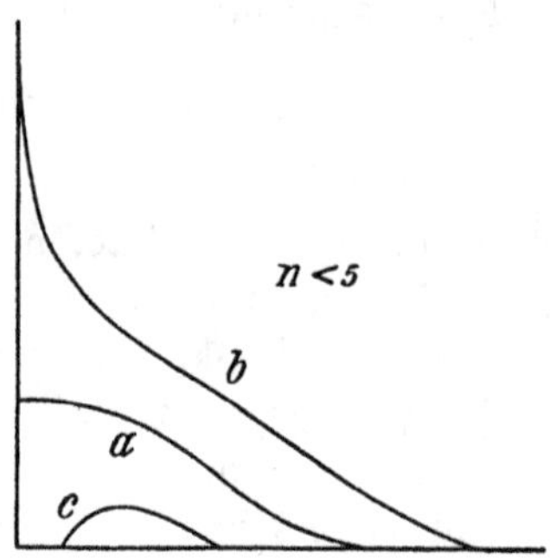

Fig. 6b.

Für jedes n gibt es eine einzige ausgezeichnete $\mathfrak{r}, u$-Kurve, der ersten ausgezeichneten y, z-Kurve entsprechend, die von $\mathfrak{r} = 0$ mit einem endlichen Wert $u = u_0$ horizontal (d. h. mit $\frac{du}{d\mathfrak{r}} = 0$) ausgeht. Die Kurve senkt sich ständig nach der $\mathfrak{r}$-Achse zu, u nimmt ständig ab. Für $n < 5$ schneidet die Kurve die $\mathfrak{r}$-Achse für endliches $\mathfrak{r}$, für $n > 5$ nähert sie sich der $\mathfrak{r}$-Achse asymptotisch in der aus den Gleichungen (19) hervorgehenden Weise in ständiger Oszillation um die singuläre Lösung (15): $u = \text{const.}\ \mathfrak{r}^{-\frac{2}{n-1}}$.

Alle übrigen Lösungen sind im Nullpunkt unstetig oder sie erreichen für positives u die u-Achse überhaupt nicht, verhalten sich also wie die Kurven b und c der Figuren No. 6.

Dabei bestehen je nach dem Werte von n folgende Unterschiede. Für $n < 3$ hat die Unstetigkeit im Nullpunkt den Charakter $\mathfrak{r}^{-1}$, und sämtliche Kurven schneiden die $\mathfrak{r}$-Achse für endliches $\mathfrak{r}$ (abgesehen natürlich von ganz in ∞ liegenden

39

Kurven). Für $3 < n < 5$ hat die Unstetigkeit der Kurven *b* im Nullpunkt den Charakter $\mathfrak{r}^{-\frac{2}{n-1}}$, und es gibt unter ihnen eine (der zweiten ausgezeichneten Lösung entsprechende) Grenzkurve, welche die $\mathfrak{r}$-Achse nicht im Endlichen schneidet, sondern sich asymptotisch der $\mathfrak{r}$-Achse nähert. Im Falle $5 < n$ sind keine Lösungen der Form *b* vorhanden, alle Lösungen, abgesehen von der ausgezeichneten, beginnen bei endlichem $\mathfrak{r}$. Dabei endigen sie zum Teil wieder mit $u = 0$ bei endlichem $\mathfrak{r}$ (Form *c*), zum Teil nähern sie sich asymptotisch der $\mathfrak{r}$-Achse (Form *c'*), indem sie um die ausgezeichnete Lösung $u = \text{const.}\ \mathfrak{r}^{-\frac{2}{n-1}}$ oszillieren.

Was das Quantitative angeht, so ist es bequem, die obige Konstante $\bar{\mathfrak{r}}$ so zu wählen, daß die ausgezeichnete Lösung mit der Ordinate $u = 1$ beginnt. Denkt man das als geschehen, so ist die ausgezeichnete Lösung definiert als diejenige Lösung der Differentialgleichung (11), welche für $\mathfrak{r} = 0$ die Bedingungen: $u = 1$, $\frac{du}{d\mathfrak{r}} = 0$ erfüllt.

Diese wichtigste Lösung hat Emden für die Werte $n = 0$, 0·5, 1, 1·5, 2, 2·5, 3, 4, 4·5, 4·9, 5, 6 nach dem Verfahren von Kutta, das den üblichen Methoden der mechanischen Integration hier weit überlegen ist, mit fünfstelliger Genauigkeit ausgewertet und damit eine numerische Grundlage für die Rechnungen mit der ersten ausgezeichneten Lösung unserer Differentialgleichung geliefert, welche die ungenaueren und unvollständigeren Angaben früherer Autoren überflüssig macht.

Um die Theorie unserer Differentialgleichung zu erschöpfen, haben wir noch hinzuzufügen, wie man aus den ∞^1 oben betrachteten Lösungen die zweifach unendliche Mannigfaltigkeit aller Lösungen erhält. Wie man aus der Transformation (12) oder noch bequemer aus den Gleichungen (13) in Rücksicht auf die Willkürlichkeit der Integrationskonstante in dem dort auftretenden Integral ableitet, geschieht dies so. Ist:

$$u = \psi(\mathfrak{r})$$

eine der oben betrachteten Lösungen, so ist:

$$u = p^{-\frac{2}{n-1}}\, \psi(\mathfrak{r}p) \tag{20}$$

ebenfalls eine Lösung, wobei p eine willkürliche Konstante bedeutet.

§ 5. Aus der Erledigung des abstrakt mathematischen Problems sind nunmehr die anschaulichen physikalischen Konsequenzen zu ziehen.

Man betrachte zunächst eine völlig aus einem Gas aufgebaute Kugel. In derselben kann im Mittelpunkt die Temperatur nicht unstetig sein, ihre Anordnung muß also durch die ausgezeichnete stetige Lösung gegeben werden. Bezeichnet man diese von Emden numerisch tabulierte Lösung durch:

$$u = \psi(\mathfrak{r}), \tag{21}$$

so muß für die Gaskugel gelten:

$$u = p^{-\frac{2}{n-1}}\psi(\alpha p r), \quad \alpha^2 = \frac{4\pi G}{H}\frac{k-1}{k}\cdot\frac{1}{\Theta}, \quad n = \frac{1}{k-1}. \tag{22}$$

Es gibt daher — entsprechend den zwei Größen α und p — im ganzen ∞^2 Gaskugeln aus gegebener Gasart und von vorgeschriebener Polytrope (gegebenem H und k). Eine bestimmte Gaskugel wird demnach durch Angabe von zwei Größen festgelegt. Die eine kann z. B. die polytrope Temperatur Θ sein, dann ist dadurch α festgelegt. Als zweite Bestimmungsgröße kann man die Dichte im Mittelpunkte ϱ_0 wählen. Dann folgt, da für $\mathfrak{r} = 0$ nach Definition $\psi = 1$ ist:

$$u_0 = \varrho_0^{\frac{1}{n}} = p^{-\frac{2}{n-1}}, \qquad p = \varrho_0^{\frac{n-1}{2n}}, \tag{22}$$

womit p und der Aufbau der ganzen Gaskugel festgelegt ist. Ebensowohl kann man aber auch als zweite Bestimmungsgröße den Radius R der ganzen Gaskugel (für endliche Gaskugeln, $n < 5$) wählen. Ist $\mathfrak{R}$ der aus den Emdenschen Tabellen zu entnehmende Wert von $\mathfrak{r}$, für welchen $\psi(\mathfrak{r}) = 0$ wird, so muß in diesem Falle gelten:

$$\alpha p R = \mathfrak{R}, \quad p = \frac{\mathfrak{R}}{R\alpha}, \tag{23}$$

womit wiederum p und die ganze Gaskugel bestimmt ist. Als drittes Beispiel betrachte man die Gesamtmasse M der Gaskugel als gegeben. Für die Masse hat man:

$$M = 4\pi\int_0^R \varrho r^2 dr = 4\pi\int_0^R u^n r^2 dr = 4\pi p^{-\frac{2n}{n-1}}\int_0^R \psi^n(\alpha p r) r^2 dr$$
$$= \frac{1}{\alpha^3 p^3} p^{-\frac{2n}{n-1}} \mathfrak{M}, \tag{24}$$

wofern man zur Abkürzung die wieder aus den Emdenschen Tabellen zu entnehmende Größe:

$$\mathfrak{M} = 4\pi\int_0^{\mathfrak{R}} \psi^n(\mathfrak{r})\mathfrak{r}^2 d\mathfrak{r} \tag{25}$$

einführt. Aus (24) folgt hier bei bekanntem α die Größe p und damit der Aufbau der Kugel.

Viertens wollen wir noch R und M als gegeben betrachten. Wir haben dann α und p aus den beiden Gleichungen (23) und (24) zu bestimmen und erhalten:

$$p = \left[\frac{\mathfrak{M}}{M}\frac{R^3}{\mathfrak{R}^3}\right]^{\frac{n-1}{2n}}, \quad \alpha = \frac{\mathfrak{R}}{Rp},$$

$$\text{sowie noch:} \quad \Theta = \frac{4\pi G}{H}\frac{k-1}{k}\left(\frac{R}{\mathfrak{R}}\right)^{\frac{3-n}{n}}\left(\frac{M}{\mathfrak{M}}\right)^{\frac{n-1}{n}}. \qquad (26)$$

Damit folgt für die Dichteverteilung in der Gaskugel:

$$\varrho = u^n = \frac{M}{\mathfrak{M}}\frac{\mathfrak{R}^3}{R^3}\left[\psi\left(\mathfrak{R}\frac{r}{R}\right)\right]^n, \qquad (27)$$

und für die Temperaturverteilung:

$$T = \Theta \,.\, u = C.\frac{M}{R}\psi\left(\mathfrak{R}.\frac{r}{R}\right), \qquad (28)$$

wobei zur Abkürzung die nur von der Natur des Gases abhängige Größe:

$$C = \frac{4\pi G}{H}\frac{k-1}{k}\frac{\mathfrak{R}}{\mathfrak{M}} \qquad (29)$$

eingeführt ist.

Immer ist die Gaskugel durch zwei unabhängige Daten völlig bestimmt, und es läßt sich daher aus zwei solchen Daten immer jede erwünschte dritte Angabe über die Gaskugel ableiten. Die so entstehenden Beziehungen zwischen je drei Bestimmungsstücken einer Gaskugel werden als „kosmogenetische Flächengleichungen" bezeichnet. Emden verfolgt eine ganze Reihe solcher Beziehungen und fügt numerische Daten in absoluten Maßen hinzu, welche die Anwendung auf einen gegebenen Fall erleichtern.

§ 6. Die Kontraktion einer Gaskugel. Wenn eine Gaskugel von gegebener Masse M ihren Radius R verkleinert, indem sie dabei lauter Gleichgewichtszustände durchläuft, welche ein und derselben Polytropenklasse (demselben k und n) angehören, so leitet man aus den Formeln (26) bis (29) unmittelbar folgendes ab:

Die polytrope Temperatur Θ wächst proportional $R^{\frac{3-n}{n}}$, nimmt also mit der Kontraktion ab für $n<3$, $k>4/3$, für ein- und zweiatomige Gase, nimmt hingegen mit der Kontraktion zu für $n>3$, $k<4/3$, für höheratomige Gase.

Für die Mittelpunktstemperatur T_0 folgt aus (28), da $\psi(0) = 1$ ist:

$$T_0 = C \cdot \frac{M}{R}.$$

Die Mittelpunktstemperatur ist umgekehrt proportional dem Radius der Gaskugel, sie wächst stets bei Kontraktion.

Man kann leicht zeigen, daß bei der Kontraktion die Masse innerhalb solcher Teilkugeln der ganzen Gaskugel konstant bleibt, deren Radius r proportional dem Radius R der Gesamtkugel abnimmt, für welche $\frac{r}{R}$ während des Kontraktionsprozesses konstant bleibt. Damit folgt, daß die Entfernung jedes Teilchens vom Mittelpunkt im selben Verhältnis abnimmt, wie der Radius der Gesamtkugel, daß nach Emdens Bezeichnung die Kontraktion eine gleichförmige ist. Nun ergibt sich aus (27) und (28):

$$T.\varrho^{-1/3} = \frac{4\pi G}{H}\frac{k-1}{k}\left(\frac{M}{\mathfrak{M}}\right)^{2/3}.\ \psi^{\frac{3-n}{3}}\left(\mathfrak{R}\cdot\frac{r}{R}\right).$$

Für das einzelne Gasteilchen bleibt während des Kontraktionsprozesses $\frac{r}{R}$ konstant, daher bleibt auch $T.\varrho^{-1/3} = \text{const.}$ Wie der Vergleich mit der allgemeinen Gleichung der Polytrope

$$T\varrho^{1-k} = \Theta$$

lehrt, kann man dies Resultat auch so aussprechen: „Das einzelne Gasteilchen folgt bei der Kontraktion einer beliebigen Gaskugel stets der Polytrope vom Exponenten $k = 4/3$." Diese Polytrope wird daher auch als „Kosmogenide" bezeichnet.

Auf Grund dieses Satzes läßt sich der Energieaustausch beim Kontraktionsprozeß direkt aus den Formeln (5) ablesen. Für $k = 4/3$ lauten dieselben:

$$dQ : dW : d\varepsilon = 4/3 - \varkappa : \varkappa - 1 : 1/3. \qquad (30)$$

Sie gelten für jedes einzelne Gasteilchen, infolgedessen auch für die Gesamtmasse. Wird der Gasmasse durch Ausstrahlung Wärme entzogen ($dQ < 0$), so folgt aus den Formeln:

$$\frac{dW}{dQ} = \frac{\varkappa - 1}{4/3 - \varkappa}, \qquad \frac{d\varepsilon}{dQ} = \frac{1}{3(4/3 - \varkappa)}$$

folgendes Verhalten: Für $\varkappa > 4/3$, ein- oder zweiatomiges Gas, ist $d\varepsilon$ und dW positiv. Es wird auf Kosten der durch die Kontraktion frei werdenden Gravitationsenergie nicht nur die Ausstrahlung bestritten, sondern dazu die Temperatur der Masse selbst erhöht. Für $\varkappa < {}^4/_3$, hochatomiges Gas, ist hingegen $d\varepsilon$ und dW negativ. Es sinkt nicht nur die Temperatur infolge der Ausstrahlung, sondern es muß auch noch von dem Gase Arbeit gegen die Gravitation geleistet werden.

Wie der Radius der Gaskugel sich mit der Ausstrahlung ändert, entnimmt man der merkwürdig einfachen Formel für das negative Selbstpotential der polytropen Gaskugel:

$$\psi = -\frac{3}{5-n} G . \frac{M^2}{R}. \tag{31}$$

Da die von der Gravitation geleistete Arbeit $dW = -d\psi$ ist, so folgt:

$$dW = \frac{3}{5-n} G \frac{M^2}{R^2} . dR, \quad dQ = \frac{{}^4/_3 - \varkappa}{\varkappa - 1} \frac{3}{5-n} G \frac{M^2}{R^2} dR.$$

Mit der Ausstrahlung ($dQ < 0$) ist also für alle endlichen Gaskugeln ($n < 5$) Kontraktion ($dR < 0$) verbunden, wenn das Gas ein- oder zweiatomig ist, hingegen Dilatation bei höheratomigem Gas. Dies macht die obigen Angaben über das Verhalten der Gravitationsenergie verständlich.

Emden hält die Gaskugeln aus hochatomigem Gas ($\varkappa < {}^4/_3$) für instabil (pag. 121), doch scheint diese Frage noch einer schärferen Diskussion zu bedürfen.

§ 7. Gaskugeln in starrer Hülle. Nimmt man von einer vollständigen Gaskugel die außerhalb einer Kugel vom Radius R liegende Masse hinweg und umgibt den übrig bleibenden Gaskern mit einer starren Hülle, so bleibt der Kern im Gleichgewicht, indem die Hülle den an ihr herrschenden Gasdruck aufnimmt. Man kann auf diese Weise aus jeder vollständigen Gaskugel infolge der Willkürlichkeit des Schnittradius r einfach unendlich viele unvollständige Gaskugeln herstellen. Um eine unvollständige Gaskugel zu bestimmen, sind dem entsprechend nicht zwei, sondern drei Stücke nötig, für welche man am natürlichsten die Masse, den Radius der Hülle und die polytrope Temperatur Θ wählen wird.

Während die unvollständigen Gaskugeln der Polytropenklassen $n < 5$ nichts besonderes bieten, kommen für $n > 5$ die Oszillationseigenschaften in interessanter Weise zur Geltung, die aus der spiraligen Umkreisung des Punktes O_1 durch die Lösung der Differentialgleichung erster Ordnung hervorgehen. Es zeigt sich nämlich, daß durch die drei angeführten Bestimmungsstücke die Gaskugel im allgemeinen nicht eindeutig bestimmt ist, sondern daß man eine gegebene Gasmasse in einer gegebenen Hülle bei bestimmter Temperatur Θ noch auf mehrere — unter Umständen unendlich viele — diskrete Weisen im Gleichgewicht anordnen kann.

Eine genaue Diskussion für den Fall der isothermen Kugel ($n = \infty$) liefert den merkwürdigen Satz: Die Temperatur der Masse M in einer Hülle vom Radius R kann nicht kleiner sein als:

$$0{\cdot}39688\,\frac{G}{H}\cdot\frac{M}{R}.$$

Dabei ist die Oberflächendichte stets größer als 0·26250 der mittleren Dichte, der Energiegehalt ε stets größer als

$$\frac{0{\cdot}39688}{\varkappa - 1}\,G\,\frac{M^2}{R}.$$

§ 8. Gaskugeln mit starrem Kern. Ein Problem, das die vorausgehenden beiden an Schwierigkeiten wesentlich übertrifft, ergibt die Frage nach dem Gleichgewicht der Atmosphären, der einen starren anziehenden Kern umgebenden Gaskugeln. Denn hier fällt die Stetigkeitsbedingung im Mittelpunkt der Kugel weg, und es tritt nicht nur eine ausgezeichnete, sondern jede beliebige Lösung unserer Differentialgleichung zweiter Ordnung auf.

Man fixiere die Verhältnisse (bei gegebener Gasart und Polytropenklasse) durch Angabe der Masse M_0, des Radius R_0 und der Oberflächentemperatur T_0 des Kernes, sowie der polytropen Temperatur Θ der Atmosphäre. T_0 bedeutet dann natürlich auch die Bodentemperatur der Atmosphäre. Man kann dann sofort ableiten, mit welcher Lösung der reduzierten Differentialgleichung erster Ordnung man es zu tun hat. Denn für den Boden der Atmosphäre hat man zunächst (der Index 0 soll sich immer auf den Boden beziehen):

$$z_0 = u_0\,\mathfrak{r}_0^{\frac{2}{n-1}} = u_0 R_0^{\frac{2}{n-1}}\,\alpha^{\frac{2}{n-1}} = \frac{T_0 R_0}{\Theta}^{\frac{2}{n-1}}\,\alpha^{\frac{2}{n-1}}$$

$$= \left[\frac{4\pi G}{H}\,\frac{k-1}{k}\right]^{\frac{1}{n-1}}\frac{T_0 R_0^{\frac{2}{n-1}}}{\Theta^{\frac{n}{n-1}}}. \tag{32}$$

Daraus folgt, da α nach (9) mit Θ gegeben ist, der Wert von z_0. Andererseits hat man für die Schwere am Boden der Atmosphäre g_0, welche durch Masse und Radius des Kerns zu $G\cdot\frac{M}{R^2}$ bestimmt ist:

$$g_0 = -\frac{1}{\varrho}\frac{dp}{dr} = -\frac{H\Theta}{\varrho}\frac{d\varrho^k}{dr},$$

was nach leichter Transformation übergeht in:

$$g_0 = -\alpha H\Theta(n+1)\left(\frac{du}{d\mathfrak{r}}\right)_0.$$

Aus den Gleichungen (13) folgt:

$$\frac{du}{d\mathfrak{r}} = -\mathfrak{r}^{-\frac{n+1}{n-1}}\left(y + \frac{2}{n-1}z\right),$$

und danach gilt:

$$g_0 = G\frac{M_0}{R_0^2} = \alpha^{-\frac{2}{n-1}} H\Theta(n+1) R_0^{-\frac{n+1}{n-1}}\left(y_0 + \frac{2}{n-1}z_0\right) \qquad (33)$$

oder durch Kombination mit (32):

$$\frac{y_0}{z_0} = \frac{k-1}{k}\frac{G}{H}\frac{M_0}{R_0 T_0} - \frac{2}{n-1}. \qquad (34)$$

Da z_0 oben bestimmt wurde, folgt hieraus auch y_0. Man kann also den Punkt y_0, z_0 in das entsprechende Diagramm (Fig. 3—5) eintragen und hat die durch diesen Punkt gehende Lösungskurve zu verfolgen. Die wichtigste Frage — ob die Atmosphäre endlich ist oder sich ins Unendliche erstreckt — wird dann sofort entschieden, indem man nachsieht, ob die betreffende y, z-Kurve (in der positiven Richtung durchlaufen) den Wert $z = 0$ erreicht.

Für $n < 5$ tun dies offenbar alle Kurven, für $n < 5$ ist also jede Atmosphäre bei beliebigem Kern endlich. Für $n > 5$ kommt es darauf an, ob der Punkt y_0, z_0 innerhalb oder außerhalb des von der zweiten ausgezeichneten Lösung umgrenzten Bereiches liegt. Man kann eine einfache hinreichende Bedingung dafür angeben, daß y_0, z_0 außerhalb liegt, daß die Atmosphäre endlich bleibt. Dies gilt nämlich, wenn y_0, z_0 oberhalb der Linie $y + \frac{3-n}{n-1}z = 0$ liegt, welche die ausgezeichnete Lösung im Nullpunkt berührt und dann oberhalb derselben verläuft. Führt man in diese Bedingung die Formel (34) ein, so erhält man den Satz:

Auch für $n > 5$ ist die Atmosphäre endlich, wenn die Bodentemperatur unter der Grenze:

$$T = \frac{M_0}{R_0}\frac{k-1}{k}\frac{G}{H}$$

liegt, gleichgültig, welches der Wert der polytropen Temperatur Θ ist.

Liegt T oberhalb dieser Grenze, so kann man den Punkt y_0, z_0 immer noch dadurch aus dem Innern der ausgezeichneten Lösung entfernen, daß man z_0 hinreichend groß macht. z_0 wächst nach (32) beliebig, wenn man Θ hinreichend klein macht. Es folgt also:

Jede Atmosphäre um einen gegebenen Kern von gegebener Temperatur ist endlich, wenn ihre polytrope Temperatur Θ hinreichend klein ist.

Da die Masse des Kerns in (32) nicht vorkommt, so sieht man, daß man umgekehrt bei zu großer Temperatur Θ durch keine Vermehrung der Gravitation des Kerns die Atmosphäre in eine endliche Begrenzung zwingen kann.

Eine andere Fragestellung ergibt sich, wenn man statt der polytropen Temperatur die Masse der Atmosphäre oder besser die Gesamtmasse M_a von Kern plus Atmosphäre als gegeben betrachtet. Da M_a natürlich endlich sein soll und mit endlicher Masse, wie leicht zu sehen, auch endlicher Radius verbunden ist, so kommen nur die y, z-Kurven außerhalb der zweiten ausgezeichneten Lösung in Frage. Charakterisiert man eine von diesen Kurven durch die Ordinate y_a, mit welcher sie in die $+y$-Achse einläuft, so findet man aus den obigen Formeln den Ausdruck:

$$\frac{M_a}{M} = \frac{y_a}{y_0 + \frac{2}{n-1} z_0} e^{\frac{n-3}{n-1}\int_0^{z_0} \frac{dz}{y}}. \tag{35}$$

Bei gegebener Masse M_a hat man also diese Bedingung mit der Gleichung (34) zu verbinden, um die passende y, z-Kurve zu finden. Um die Natur der Lösung dieser Aufgabe zu überblicken, verfolge man jede y, z-Kurve in negativer Richtung von $z = 0$ an bis zum Punkt y_0, z_0. Im Anfang der Kurve, für $z = 0$, wird $\frac{M_a}{M} = 1$. Man findet ferner durch Differentiation von (35) unter Benutzung der Differentialgleichung (14):

$$\frac{dM_a}{M_a} = \frac{dz_0}{y_0} \frac{z_0{}^n}{y_0 + \frac{2}{n-1} z_0}. \tag{36}$$

Solange man sich also über der Geraden $y_0 + \frac{2}{n-1} z = 0$ befindet — und das ist nach (33) für alle wirklichen Gaskugeln der Fall — wächst M_a immer, wenn man entgegen der Pfeilrichtung die y, z-Kurven durchläuft. Auf der Geraden $y_0 + \frac{2}{n-1} z_0 = 0$ selbst wird M_a unendlich.

Nun ergibt sich folgender interessante Schluß: Ist eine endliche Masse M_a vorgeschrieben, so muß der Punkt y_0, z_0 in endlicher Entfernung oberhalb der Geraden $y_0 + \frac{2}{n-1} z_0 = 0$ liegen, weil beim Anrücken an diese Gerade M_a unendlich würde. Läßt man andererseits die Temperatur T_0 mehr und mehr zunehmen, so muß nach (34) der Punkt y_0, z_0 sich derselben Geraden immer mehr nähern. Es sind also die beiden Forderungen (35) und (34) nicht mehr vereinbar, wenn die Temperatur des Kerns T_0 zu hoch wird, und es folgt der Satz:

Bei gegebenem Kern und gegebener Masse der Atmosphäre existiert eine Temperaturgrenze T_0 („Zerstreuungstemperatur"), oberhalb deren die Atmosphäre sich nicht mehr im Gleichgewicht halten kann.

§ 9. Zusammengesetzte Gaskugeln. Um den physikalischen Verhältnissen wirklicher Gaskugeln gerecht zu werden, wird man annehmen müssen, daß in verschiedenen Tiefen derselben der Bau nach verschiedenen Polytropenklassen ererfolgt. Man hat daher Stücke von Lösungen der Differentialgleichung (10), welche verschiedenem n und α entsprechen, aneinanderzusetzen. Da für jedes neue Lösungsstück nur die Oberflächentemperatur und die Gravitation der innerhalb liegenden, sowie der Druck der außerhalb liegenden Massen in Frage kommt, so reduziert sich die Aufgabe im wesentlichen auf die kombinierte Anwendung der beim starren Kerne und bei der starren Hülle auftretenden Überlegungen. Emden behandelt näher die Atmosphären von der Polytropenklasse $n = 0$, 1 und 1·5 (einatomiges Gas), die einen isothermen Kern umgeben. Für den Kern wird dabei die kleinstmögliche Temperatur (s. oben § 7) vorausgesetzt. Es sei insbesondere der Satz erwähnt: Besitzt eine Gaskugel einen isothermen Kern, so kann die Masse desselben nie mehr betragen als 0·76953 der Masse der ganzen Kugel.

Hiermit sind die wichtigsten Sätze des theoretischen Teiles erschöpft. Zahlreiche hübsche kleinere Bemerkungen und die überall hinzugefügten numerischen Angaben sind weggelassen, um nur das Wesentliche hervortreten zu lassen. Auch ist die Untersuchung des Grenzfalles $n = \infty$ übergangen, welcher die isotherme Gaskugel liefert und von Emden besonders ausführlich behandelt ist.

Anwendungen.

§ 10. Kosmische Staubmassen. G. H. Darwin hat sich zuerst die Frage vorgelegt, ob man nicht den Gasball der Laplaceschen Nebularhypothese durch einen Haufen von Meteoriten ersetzen kann, ohne zu unwahrscheinlichen Vorstellungen zu gelangen. Man darf dabei die Meteoriten nicht als anfänglich ruhend betrachten. Denn ein ursprünglich ruhender Schwarm von Sonnenmasse und Neptunbahnradius würde infolge seiner Gravitation schon in 64 Jahren zusammenstürzen (das Fixsternsystem würde unter der gleichen Voraussetzung die ebenfalls viel zu kurze Zeit von etwa 20 Millionen Jahren zum Zusammensturz brauchen). Vielmehr muß man sich die Meteore genau wie vergrößerte Moleküle mit Geschwindigkeiten nach dem Maxwellschen Verteilungsgesetz begabt denken und ihre Zusammenstöße berücksichtigen. Bei dieser Vorstellung wird die Theorie des Meteorschwarmes identisch mit der der polytropen Gaskugeln. Durch unmittelbare Anwendung der obigen Theorie ließen sich daher die Darwinschen Resultate wiedergewinnen und durch Angabe von noch anders konstituierten Meteorschwärmen erweitern. Die von Emden konstruierten Meteorschwärme haben ebenso vernünftige Anordnung und Geschwindigkeiten, wie Darwins Beispiel, und tragen vielleicht den Verhältnissen, die man in einem Gas als Wärmeleitung und Reibung bezeichnen würde, noch besser Rechnung.

An der Vorstellung der Staubmasse läßt sich auch besonders anschaulich die Frage erledigen, wie man sich die äußere Grenze einer Gaskugel, die in der obigen Theorie mathematisch völlig bestimmt auftritt, nach den Vorstellungen der kinetischen Theorie zu denken hat. Nach der kinetischen Theorie kann es natürlich keine scharfe Begrenzung geben, vielmehr werden in den dünnsten äußeren Partien der Kugel die Zusammenstöße immer seltener werden, die Gasgesetze zu gelten aufhören und die Gasteilchen freie Kegelschnitte beschreiben, die für die schnellsten Moleküle sich bis ins Unendliche erstrecken. Die Dichteverteilung in diesen Grenzgebieten wird mit Darwin unter gänzlicher Vernachlässigung der Zusammenstöße verfolgt mit dem Ergebnis, daß der Dichteabfall außerhalb der theoretischen Grenze ein so schneller ist, daß die theoretische Grenze auch praktisch als solche gelten kann. Auch der Massenverlust durch das endgültige Wegfliegen solcher Moleküle resp. Steine, die zufällig hyperbolische Geschwindigkeit besitzen, wird ein außerordentlich kleiner.

Die Übertragung derselben Überlegungen auf das Fixsternsystem führt zu dem Resultat, daß die freie Weglänge größer ist als die Dimensionen des Systems, d. h. es ist unstatthaft, die Gastheorie auf das Fixsternsystem anzuwenden.

Über das Leuchten von Staubmassen stellt Emden folgende Betrachtung an: Die Zusammenstöße der Meteore mögen zur Verdampfung und Erhitzung kleiner Mengen und zur Ausstrahlung führen. Die Ausstrahlung wird der Zahl der Zusammenstöße und der Energie der Bewegung etwa proportional zu setzen sein. Daraus ergibt sich die Ausstrahlung als mindestens mit der 4. Potenz der „Temperatur" der Staubmasse wachsend, wo unter Temperatur das mittlere Geschwindigkeitsquadrat der ungeordneten Bewegung der Steine verstanden ist. Haben die Steine außer der ungeordneten Bewegung noch eine gemeinsame Geschwindigkeit v und bewegen sie sich in einem Felde konstanten äußeren Potentials, so ist die Summe der Energien der geordneten und der ungeordneten Bewegung konstant. Wo daher die Geschwindigkeit v der geordneten Bewegung klein ist, muß die Energie der ungeordneten Bewegung entsprechend der Änderung von $\frac{v^2}{2}$ zunehmen. Es folgt, daß die Strahlung sehr stark von der geordneten Geschwindigkeit abhängt und daß — entgegen der naiven Vorstellung — die Stellen geringster Geschwindigkeit am hellsten leuchten werden.

§ 11. Nebelflecke. Baut man eine Gaskugel von der Masse der Sonne und dem Radius der Neptunsbahn aus Wasserstoff auf, so ergibt sich im Mittelpunkte:

$$\text{Druck } 25{\cdot}1 \text{ Dynen auf den cm}^2,\ \text{Dichte } 1{\cdot}24 \,.\, 10^{-10},$$
$$\text{Temperatur } 4890^\circ.$$

Dieser Nebelball würde also in den zentralen Partien gewiß heiß genug sein, um durch seine Temperatur leuchten zu können.

Daß er eine beträchtliche Flächenhelligkeit zeigen könnte, läßt sich so überschlagen: Man nehme an, daß die Absorption proportional der Anzahl der Moleküle sei, die das Lichtbündel vom Querschnitt 1 durchsetzt, und wähle als Absorptionseinheit die entsprechende Zahl der Moleküle im Falle der Erdatmosphäre, welche bekanntlich eine Reduktion der Intensität im Verhältnis 1 : 0·8 etwa bewirkt. Es zeigt sich, daß auf den Durchmesser jenes Nebelballs 30 Absorptionseinheiten kommen, so daß er durchgehendes Licht im Verhältnis $0{\cdot}8^{30} = 0{\cdot}0012$ reduzieren würde. Er wäre also fast ganz undurchsichtig und daher auch umgekehrt wohl zu hellem Leuchten befähigt.

In dem Aussehen der Helligkeitsverteilung des Nebels könnte man schließlich nach dem obigen noch einen Ausdruck der Strömungen innerhalb der Nebelmasse sehen.

§ 12. Die Sonne. Setzt man eine Kugel von der Masse und dem Radius der Sonne aus atmosphärischer Luft ($\varkappa = 1{\cdot}4$) in adiabatischem Gleichgewicht ($k = \varkappa = 1{\cdot}4$) zusammen, so erhält man im Mittelpunkt:

Dichte 33·19, Temperatur 45560000°,
Druck $4{\cdot}34 \, . \, 10^{16}$ Dynen / cm²

und kann aus den Tabellen für die Lösung der Differentialgleichung zweiter Ordnung daraus die Verhältnisse in jedem Abstand vom Mittelpunkt durch Multiplikation mit gewissen Proportionalitätsfaktoren entnehmen. Nimmt man statt der Luft dissoziierten (einatomigen) Wasserstoff ($\varkappa = {}^5/_3$), so erhält man im Mittelpunkt:

Dichte 8·27, Temperatur 12070000°,
Druck $8{\cdot}28 \, . \, 10^{15}$ Dynen / cm².

Wie ist die scharfe Begrenzung der Sonnenscheibe zu erklären? Die Vorstellung dieses Gasballs mit seiner stetigen Dichteverteilung fordert heraus zur Anwendung der Schmidtschen Theorie der Strahlenbrechung in Gaskugeln. Emden widmet der Untersuchung der Strahlenbrechung in Gaskugeln ein eigenes Kapitel und gibt Tabellen und Diagramme, aus denen man die optischen Verhältnisse, insbesondere die Radien von Schmidts „kritischen Schichten" durch einfache Rechnung für jede polytrope Gaskugel entnehmen kann. Die Anwendung auf die Sonne liefert in der Tat eine Gaskugel sogen. „zweiter Art", bei welcher unterhalb der „kritischen Schicht" die von Schmidt bemerkten Unstetigkeiten des Strahlengangs auftreten. Halten wir uns an das Beispiel der Kugel aus dissoziiertem Wasserstoff, so liegt die kritische Schicht 7.″1 = 5100 km unter der Oberfläche. Es zeigt sich nun aber, daß ein radial verlaufender Strahl über der kritischen Schicht noch 960 Absorptionseinheiten zu durchsetzen hat, was eine Abschwächung von der Ordnung 10^{-90} bedeutet. Ferner zeigt sich, daß in der Nähe der kritischen Schicht der Strahl 70 Absorptionseinheiten (entsprechend einer Abschwächung der Ordnung 10^{-7}) durchlaufen muß, um sich nur um 1′ zu drehen. Demnach ist klar, daß die Absorption viel zu mächtig ist, als daß die Strahlenbrechung die ihr von Schmidt zugeschriebene Wirkung üben könnte. Dieser Schluß gilt auch für andere Gase und Polytropen. Die Schmidtsche Sonnentheorie muß daher endgültig verworfen werden.

Es ist weiter zu fragen, ob man es in der Photosphäre

mit Kondensationsvorgängen, mit einer Wolkenschicht zu tun hat, was häufig mit einer gewissen Bestimmtheit behauptet wird. Emden kommt zu dem Schluß, daß unsere physikalischen Kenntnisse der Wärmekapazitäten der Dämpfe zu gering sind, um uns eine bestimmte Aussage zu gestatten. Ich möchte aber hierüber hinausgehen und im Gegenteil behaupten, daß die Existenz einer Kondensationsschicht recht unwahrscheinlich ist. Existiert diese Kondensationsschicht, so muß — in Anrechnung der Absorption der über ihr liegenden Gasmassen, die man aus dem Abfall der Helligkeit nach dem Rande der Sonne zu schätzen kann — ihre Temperatur über 7000° betragen. Der Druck der Photosphäre beträgt nach den Messungen der Druckverschiebungen der Spektrallinien von Jewell etwa 5 Atmosphären. Die photosphärischen Wolken müßten also aus einem Stoff bestehen, der bei 7000° unter einem Druck von 5 Atmosphären nicht vergast ist. Nach den Experimenten von Moissan (Comptes rendus, 1906, Bd. 142, pag. 673) sieden alle Metalle im elektrischen Ofen unter der Temperatur des Kohle-Bogens (3500°), und auch von der unschmelzbaren Kohle pflegt man anzunehmen, daß ihr Siedepunkt — in welchem also der Dampfdruck der festen Kohle eine Atmosphäre beträgt — mit der Temperatur des Lichtbogens identisch ist. Es ist nach der bekannten Form sonstiger Dampfdruckkurven höchst unwahrscheinlich, daß die Vermehrung des Druckes auf das Fünffache eine Steigerung der Siedetemperatur bis auf 7000° bewerkstelligen könnte. Somit folgt, daß alle diese Stoffe unter den in der Photosphäre herrschenden Bedingungen gasförmig sind.

Moissan wollte (l. c.) umgekehrt auf eine verhältnismäßig niedrige Sonnentemperatur schließen, indem er von der Voraussetzung der Existenz flüssiger Metalle in der Photosphäre ausging. Hierbei ist wohl auf diese Vorauszusetzung zu sehr vertraut und zu wenig auf die untere Grenze der Photosphärentemperatur geachtet, welche durch die Strahlungsmessungen zweifelsfrei bestimmt ist. Auf der anderen Seite wieder wird von manchen Physikern der Einwand gegen alle Destillationsversuche im elektrischen Ofen erhoben, daß es sich dabei nicht um eigentliche Verdampfung, sondern um elektrische Zerstäubung handle. Dieser Einwand scheint indessen gegenüber dem von Moissan in seinem Traité de Chimie minérale II, pag. 233 beschriebenen Versuch nicht haltbar, in welchem er die Kohle im Innern eines vom Lichtbogen umspülten Rohres zur Verdampfung brachte*).

*) Herr v. Wartenberg hat den Ref. freundlichst hierauf aufmerksam gemacht.

Es bleibt also alle Wahrscheinlichkeit auf seiten der gasförmigen Photosphäre.

Wenn man nach alledem weder die Schmidtsche Theorie noch Wolkenbildung zur Erklärung der scharfen Grenze der Sonnenscheibe heranziehen kann, so bleibt als einfachste Anschauung übrig, daß dieser Umstand überhaupt keiner besonderen Erklärung bedarf. Wie wohl zuerst E. v. Oppolzer (Über die Ursache der Sonnenflecke. Sitzungsber. der Wiener Akad. 1893) hervorgehoben hat, ist der Abfall der Dichte in der Sonnenatmosphäre bei jeder Anordnung in mechanischem Gleichgewicht so schnell, daß für unsere optischen Mittel der Sonnenrand scharf bleiben muß.

§ 13. Um eine größere Annäherung an die Wirklichkeit zu erzielen, als sie die adiabatische Gaskugel gewährt, sucht Emden ein Schema der Sonne aus einer Reihe von Schichten verschiedener Polytropenklassen zusammenzusetzen. Für das Innere behält er die dissoziierte adiabatische Wasserstoffkugel bei, die schon oben erwähnt wurde. Für die Schichten, aus denen die uns zugehende Sonnenstrahlung hauptsächlich entstammt, nimmt er an, daß die Erhitzung durch Strahlung der inneren Partien den adiabatischen Temperaturabfall nach außen mindert, und daß diese Schichten nach der Polytropen $n = 5$ gebaut sind. Auf die Polytrope $n = 5$ kommt Emden, indem er die Strahlung der Sonne und die Helligkeitsverteilung auf der Sonnenscheibe nach denselben Prinzipien überschlägt, die ich in meinem Aufsatz „Über das Gleichgewicht der Sonnenatmosphäre" angewandt habe, und hierbei erweist sich $n = 5$ als der beobachteten Helligkeitsverteilung und dem Strahlungsgleichgewicht am besten entsprechend. Die äußersten Schichten denkt sich Emden fast nach der Isothermen (nach dem sehr großen $n = 3600$) gebaut, um die Grenze der Gaskugel bis zu den Entfernungen hinauszuschieben, bis zu welchen man Coronaausläufer verfolgt hat.

Obwohl der Verfasser dieses Schema nur als Beispiel für eine mechanisch mögliche Anordnung aufgefaßt haben will, so darf ich mir doch gegen die zugrunde liegenden Prinzipien einige Bemerkungen erlauben, die sich zum Teil auch gegen meinen eigenen oben erwähnten Aufsatz richten. Adiabatisches Gleichgewicht, wie Strahlungsgleichgewicht, sind gewiß theoretisch interessante Grenzfälle, aber die Helligkeitsverteilung auf der Sonnenscheibe scheint Eigentümlichkeiten zu haben, zu deren Erklärung Gleichgewichtszustände vielleicht nicht ausreichen. Die die Sonnenoberfläche durchsetzenden Strömungen finden ja in der Granulation der Sonnenoberfläche ihren sichtbaren

Ausdruck. Die Verteilung der Granulation kann sehr wohl auch die durchschnittliche Helligkeit systematisch beeinflussen. Eine Andeutung davon kann man versucht sein in der merkwürdigen Gestalt der Kurve zu finden, welche Dr. Villiger und ich (Astroph. Jour. vol. 23) für die Helligkeitsverteilung auf der Sonnenscheibe für ultraviolettes Licht erhalten haben. Die Kurve, welche sich den Beobachtungen am besten anschließt, hat einen Knick im Sonnenzentrum, genau so wie die Helligkeitskurven von Planeten mit gebirgiger Oberfläche beim Phasenwinkel Null. Der Knick erklärt sich bei den Planeten aus dem Schattenwurf der Gebirge, bei der Sonne aus der Absorption zahlreicher spitzer Flammen, die aus der Photosphäre hervorbrechen und die weniger hell leuchten als die normale Photosphäre selbst. Man hätte dann also in den dunklen Stellen der Granulation die höchsten und kühlsten — vielleicht beim Aufsteigen adiabatisch gekühlten — Massen zu sehen, in den hellen Stellen hingegen den heißeren Untergrund.

Ferner scheint es mir nicht möglich, das Turnersche Gesetz der Helligkeitsverteilung der Corona — Abfall mit der — 6. Potenz der Entfernung vom Sonnenzentrum — durch eine Gasverteilung zu erklären, die nur dem Gasdruck und der Schwere folgt und natürlich nach außen nicht an Temperatur zunimmt. Die Dichteabnahme wird zu rapide. Auch wenn man das Aufhören der Gasgesetze und das Beschreiben freier Kegelschnitte durch die Moleküle in den äußersten Teilen berücksichtigt, wird die Dichteabnahme nicht genügend verringert. Hier müssen andere Kräfte mit im Spiele sein.

§ 14. Chromosphäre und Protuberanzen. Anomale Dispersion. Wie die Schmidtsche Theorie das allgemeine Aussehen der Sonne, so sucht die Auffassung von Julius die speziellen Erscheinungen der Chromosphäre und der Protuberanzen auf die Wirkung der Strahlenbrechung zurückzuführen, indem sie von dem Auftreten anomaler Dispersion in der Nachbarschaft der Eigenschwingungen eines Gases Gebrauch macht. Emden zeigt, daß überall da, wo reguläre Schichtung vorliegt, die anomale Dispersion praktisch keine Rolle spielen kann, weil entweder, genau wie bei der Schmidtschen Theorie, die Absorption zu stark wird auf den Strecken, welche eine merkliche Strahlkrümmung bewirken, oder weil zu große Dichten der Gasmassen erforderlich sind. Ich habe gelegentlich eine ähnliche Überlegung angestellt, indem ich mir ein Natriumprisma auf die Sonne aufgesetzt dachte und nach den Dichtebestimmungen von Jewitt*) und den Messungen der Brechungs-

*) Phil. Mag. 4. 1902.

Vierteljahrsschrift d. Astron. Ges. *43* (1908) 26–55

exponenten von Wood*) die Dichte ausrechnete, welche das in mittleren Protuberanzen auftretende Leuchten der Natriumlinie über 1 A. E. Breite in 1′ Höhe über dem Sonnenrand erklärte. Die Dichte ergibt sich von der Größenordnung unserer Luft, also ein unzulässiger Betrag.

Indessen treffen diese Überlegungen immer nur den Fall regulärer Schichtung. Es kann sich aber sehr wohl auch um Häufung ungeheuer vieler kleiner Brechungen in einer äußerst dünnen schlierigen Gasmasse handeln. Wenn man bedenkt, wie außerordentliche Strahlkrümmungen man mittels einer ganz geringen und dünnen Gasmasse erzeugen kann, indem man die Masse in lauter Prismen von wenigen Wellenlängen Kantenlänge zerlegt, so sieht man, daß man die aus der Annahme regulärer Schichtung berechneten Absorptionen und Dichten um viele Zehnerpotenzen verkleinern kann, indem man zur Voraussetzung zahlloser feiner Schlieren übergeht.

Es wäre daher eine nochmalige quantitative Prüfung der Wirkung der anomalen Dispersion von diesem Gesichtspunkt aus vorzunehmen, von der freilich kaum zu erwarten wäre, daß sie das Endresultat ändern würde.

§ 15. Erhaltung der Sonnenenergie. Nach der bekannten Rechnung von Helmholtz reicht die durch Kontraktion der Sonne gewonnene Gravitationsarbeit aus, um die Strahlung der Sonne für 20 Millionen Jahre zu decken, wenn man annimmt, daß die gesamte gewonnene Energie zur Ausstrahlung verwandt wird. Nun lautet aber nach Formel (30) das Verhältnis zwischen zugeführter Gravitationsarbeit dW und Ausstrahlung $-dQ$ für eine gravitierende Gaskugel:

$$-\frac{dQ}{dW} = \frac{\varkappa - {}^4/_3}{\varkappa - 1}\,.$$

Für ein dissoziiertes Gas ($\varkappa = {}^5/_3$) beträgt dieser Bruch nur ${}^1/_3$, für ein zweiatomiges Gas ($\varkappa = {}^7/_5$) nur ${}^1/_6$. Es wird also in einem Fall die Hälfte, im anderen ${}^5/_6$ der Gravitationsarbeit in der Sonne selbst aufgespeichert, und die Zeit, seit der uns die Sonne im jetzigen Maße Wärme gespendet haben könnte, geht auf 7 Millionen resp. 3 Millionen Jahre herunter. Da die Geologie und Paläontologie für viel längere Zeiten das ungefähre Bestehen der jetzigen Strahlungsverhältnisse fordert, so muß die Sonnenenergie aus anderen Quellen stammen.

§ 16. Zum Schlusse möchte ich noch die beiden Kapitel des Buches nennen, auf deren nähere Besprechung

*) Physik. Zeitschrift. 5. 1904.

hier verzichtet ist. Das eine ist das Kapitel über die „rotierende Sonne", in welchem der Verfasser die früher von ihm entwickelte und in der Literatur schon verbreitete Theorie der Wirbelbildung und ungleichförmigen Rotation der Sonne auseinandersetzt. Das andere ist das Kapitel über die Erdatmosphäre. Für die Meteorologen wird namentlich die Theorie der Zyklone in Betracht kommen. Den Astronomen interessiert darin die Beantwortung der alten Fragen nach der Grenze und der Masse der Erdatmosphäre, die auf Grund der früheren Behandlung der Differentialgleichung (11) in aller Strenge erledigt werden. Es wird gezeigt, daß durch Druck und Temperatur am Boden die Masse der Atmosphäre noch durchaus nicht bestimmt ist, sondern von einer unteren Grenze an beliebig groß sein kann. Die Dichte muß sich aber mit der Entfernung vom Zentrum immer der Grenze Null nähern und kann nicht — wie Zöllner meinte — einem endlichen Grenzwert zustreben.

Göttingen, Juni 1907.

K. Schwarzschild.

Note.

Aprile 20. Sereno impuro a cagione di cirro strati. La cometa vista nel crepuscolo mattutino ebbe apparenza simile a quella della stella di confronto sfocata e ridotta ad un disco ampio una decina di secondi d'arco. Avendo la stella in PD la grandezza 7^m18, e d'altra parte essendo la cometa scomparsa per prima nell'aurora, si può dire che a quest'epoca essa ebbe una grandezza fra la 7^m e la 8^m.

Aprile 23. Sereno ma con orizzonte impuro per nebbia e veli. LP. La cometa si mostrò nel cannocchiale, e nel chiarore del crepuscolo mattutino ampia quanto una stella di prima grandezza, ma d'altra parte non fu più splendente delle due stelle di confronto, che sono di 8^m1 e 7^m0. Al successivo illuminarsi del cielo per l'aurora, e fino a mezz'ora prima del levar del Sole, sparirono quasi insieme tutti tre gli astri. Nei primi istanti di visione, quando il cielo era meno illuminato si potè notare che la macchia cometaria era caudata nella parte australe del cerchio di declinazione.

Aprile 27. Sereno bello. In questa data la cometa si mostrò nel campo dell'Amici, illuminato dalla luce lunare e dall'alba, come una stella di 5^m in 4^m grandezza attorniata da nebulosità di forma parabolica, ampia da 10″ a 20″, diretta lungo il parallelo ed in asc. retta decrescente. Nel cercatore (Ingr. 30 circa) la nebulosità si mostrò estesa qualche primo. Se non fosse stato il chiarore della luna l'estensione della nebulosità si sarebbe riscontrata molto più grande, e probabilmente la cometa sarebbe stata visibile ad occhio nudo.

Maggio 6. Sereno ragnato. Le condizioni atmosferiche sempre avverse dopo il 27 Aprile non permisero fino a questa data alcuna osservazione, ed in questa stessa, soltanto in causa di uno squarcio di nubi all'orizzonte orientale fu concesso di verificare che la cometa era facilmente visibile ad occhio nudo avanti l'alba ed al tempo della levata di Venere. E quanto ampio appariva il disco di questo pianeta, altrettanto si mostrava il nucleo della cometa; e da questo partiva una coda più alta diretta a sud, lunga una quindicina di gradi; ma la stima fu poco sicura essendo che la visione era disturbata dai cirro strati. Lo splendore complessivo dell'astro era bianco latteo, molto inferiore a quello giallo dorato e lucente di Venere; somigliava piuttosto a quello della via lattea.

Maggio 12. Sereno splendido. La cometa fu vista ad occhio nudo alle $2^{3/4^h}$ t. m. EC del 13 Maggio, per via della coda che levava assai tempo prima della testa. Ed era la coda una striscia bianco lattea lunga una trentina di gradi e larga tre o quattro; essa usciva all'incirca dal punto di vero oriente e stava per traverso al cielo diretta verso sud. Il nucleo puntato nel cannocchiale pareva una stella di prima grandezza coronato da una cresta lucente, giallo dorata, opposta alla coda che a sua volta era opposta al Sole.

Maggio 13. Sereno impuro. Osservata fra i cirro strati, e la nebbia dell'orizzonte.

Maggio 14. Sereno bello. La cometa fu vista ad occhio nudo in tutto il suo splendore ancor meglio che il 12. Infatti da questa data, all'altra del 25 le circostanze atmosferiche sempre sfavorevoli non permisero alcuna osservazione malgrado ogni vigilanza.

R. Osservatorio Astronomico di Arcetri-Firenze, 1910 Giugno 6. *A. Abetti.*

Vierte Versammlung der Internationalen Vereinigung für kooperative Sonnenforschung 1910.

Von *K. Schwarzschild.*

Die 85 Teilnehmer der Versammlung verbrachten die Woche vom 27. August bis zum 3. September, zwischen Pasadena und dem Mount Wilson abwechselnd, unter der fürsorglichen Gastfreundschaft von Prof. Hale und seinen Mitarbeitern. Aus den wissenschaftlichen Ergebnissen und Eindrücken sei folgendes hervorgehoben:

Die Instrumente. Auf dem Mount Wilson sind drei Instrumente in voller Tätigkeit. Der täglichen Aufnahme der Sonne gewidmet ist das Snow-Teleskop, ein Cölostat mit horizontal liegendem Fernrohr von 18 m Brennweite. Feinere Untersuchungen an der Sonne erfolgen mittels des wirksameren und dabei in der Konstruktion sehr einfachen sog. »kleinen« Turmteleskops. Hier befindet sich der Cölostat auf einem eisernen Gerüst von 18 m Höhe. Das Gerüst umschließt ein vertikales Fernrohr von 18 m Brennweite. Darunter in einem Schacht befinden sich ein Spektrograph und ein Spektroheliograph von 9 m Brennweite.

Der Photographie des Himmels und der Spektralanalyse der Sterne dient der Reflektor von 150 cm Spiegeldurchmesser. Die Lichtfülle und Bildschärfe dieses Instruments wird durch die Tatsache gekennzeichnet, daß der Stern 1830 Groombridge der Größe 6.5 bei einer Exposition von 15^m mit einem 1 Prismenspektrographen von 40 cm Brennweite des Kollimators und der Kamera ein gutes Spektrum gibt.

In den nächsten Monaten geht das »große« Turmteleskop der Vollendung entgegen. Der Heliostat ist bereits auf dem 50 m hohen Turme aufgestellt und ein Schacht von 25 m Tiefe zur Aufnahme der Spektrographen ausgemauert.

In der Werkstatt in Pasadena ist die Schleifmaschine für den projektierten Spiegel von 250 cm Durchmesser zur Arbeit bereit. Die von den Werken zu St. Gobain gelieferte Glasscheibe von diesem Durchmesser wird zur Zeit probeweise angeschliffen.

Aus den wissenschaftlichen Mitteilungen. Wie Prof. *Hale* in seiner Einführungsrede mitteilte, hat das genauere Studium des magnetischen Feldes der Sonnenflecken ergeben, daß sich zu einem Fleck, der einen positiv magnetischen Pol darstellt, vielfach die negativen Pole in Gestalt eines oder mehrerer benachbarter Flecken nachweisen lassen. Auch die Drehung der Polarisationsebene in den Flecken ist in einzelnen Fällen sehr deutlich.

Herr *Abbot*, dessen bolographisches Observatorium sich dicht bei der Sternwarte am Südabhang des Mount Wilson befindet, gibt als zur Zeit besten Wert der Solarkonstante 1.92 Kalorien an. Die Schwankungen der Tageswerte sind klein, selten die Grenzen 1.90 und 1.95 überschreitend, scheinen aber nach der Übereinstimmung der Stationen Washington, Mount Wilson, Mount Whitney solaren Ursprungs. Andere unabhängige Stationen sind dringend erwünscht, auch empfiehlt sich eine häufigere Kontrolle des Intensitätsabfalls längs des Radius der Sonnenscheibe.

Veranlaßt durch die Mißstimmigkeiten, die sich bei Ableitung der Eigenbewegung der Sonne aus den Radialgeschwindigkeiten der Orionsterne ergaben (Frost u. Kapteyn, Astrophysical Journal Vol. 32 p. 83), hat Prof. *Kapteyn* die Eigenbewegungen der Orionsterne auf Grund des Bossschen Katalogs einer besonderen Untersuchung unterzogen. Er findet einen gemeinsamen Konvergenzpunkt für 90% aller Orionsterne im Gebiet: 12^h bis 18^h RA., 0° bis −60° Dekl. und einen andern gemeinsamen Konvergenzpunkt für 95% der Orionsterne im Gebiet: $2^h 50^m$ bis $4^h 30^m$ RA., +15° bis +55° Dekl. Es sind damit also zwei neue sehr ausgedehnte Sternströme von der Art der Hyaden und der Bärensterne aufgefunden.

Herr *Deslandres* bezeichnet als Hauptresultat seiner Registrierungen der Radialgeschwindigkeit des Calciums auf der Sonne, daß die dunklen Massen aufsteigen, die hellen absteigen.

Aus den Resolutionen: Für die 50 Standards 2. Ordnung im Wellenlängengebiet 4282−6494 Å. E. wird das Mittel aus den drei Bestimmungen von Fabry u. Buisson, Pfund und Eversheim angenommen. Die Zahlen werden demnächst im Astrophysical Journal veröffentlicht.

Als weitere Standards 2. Ordnung werden in der Nähe von $\lambda = 5800$ Å. E., wo die Eisenlinien nicht zahlreich genug sind, Bariumlinien empfohlen.

Die Bestimmung der Standards 3. Ordnung und andrer Spektren im Anschluß an die neuen Standards ist von möglichst vielen Seiten erwünscht. Die — vom Rowlandschen Wert abweichende — Einheit des neuen Systems soll durch I. A. (Internationales Angström) bezeichnet werden.

Die visuellen Beobachtungen des Sonnenspektrums haben bisher zwar nur eine unerwartete Konstanz und Einheitlichkeit des Fleckenspektrums bei den verwandten Dispersionen erkennen lassen, doch sollen dieselben über eine volle Fleckenperiode fortgesetzt werden.

Die Positionswinkel am Sonnenrande werden zukünftig allgemein — auch von den bisher umgekehrt zählenden italienischen Astronomen — im Sinne Nord-Ost-Süd-West gezählt.

Zum Studium der Sonnenrotation, die nach Elementen getrennt durchgeführt werden muß, übernahmen sechs Beobachter je eine bestimmte Strecke des Sonnenspektrums.

Die Vereinigung hat ferner beschlossen, ihr Arbeitsgebiet künftig auf Astrophysik im allgemeinen auszudehnen, und hat bereits eine Kommission ernannt, die sich mit der Frage der Klassifikation der Sternspektren beschäftigen soll. Als Versammlungsort für das Jahr 1913 wurde Bonn gewählt.

Die nebenher tagende Kommission für Sterngrößen der photographischen Himmelskarte einigte sich auf den Vorschlag, daß für die Sterne vom Spektraltypus A_0 der Harvarder Klassifikation und von der Sterngröße 5.5−6.5 photographische und visuelle Sterngröße im Mittel gleich zu setzen sei, wobei die visuellen Größen nach der Harvardskala zu zählen sind. Die Harvardsternwarte übernimmt die Festlegung absoluter photographischer Sterngrößen für 24 Sequenzen in jeder Zone der photographischen Himmelskarte.

K. Schwarzschild.

Effemeride di (472) Roma.

Le perturbazioni di Giove e Saturno su Roma nell' intervallo tra la settima e l'ottava opposizione calcolate di quaranta in quaranta giorni trasformano gli elementi da ultimo pubblicati (A. N. 4325) nei nuovi osculanti:

Epoca ed osculatione 1910 Nov. 8.0 Roma C. R.

$$\begin{aligned} M &= 349^\circ\,15'\,27''.8 \\ \pi &= 61\ \ 55\ \ 10.4 \\ \Omega &= 127\ \ \ 1\ \ 36.1 \\ i &= 15\ \ 51\ \ 35.4 \\ \varphi &= 5\ \ 36\ \ 35.3 \\ \mu &= 875''.41779 \end{aligned} \quad \left.\begin{matrix} \pi \\ \Omega \\ i \end{matrix}\right\} 1910.0$$

Da questi elementi ho tratto una lunga effemeride, per la necessità di osservare il pianeta lungo un grande arco dell'orbita, e ne do ora la prima parte.

Le coordinate sono vere e per 12^h di Roma C. R. di ciascun giorno.

1910		α	δ	log Δ
Sett.	24	$4^h\,2^m\,6^s.80$	−2° 27′ 36″.6	0.214933
	25	2 24.94	2 34 55.9	0.212444
	26	2 41.32	2 42 10.1	0.209969
	27	2 55.93	2 49 45.8	0.207512
	28	3 8.75	2 57 15.8	0.205073
	29	3 19.77	3 4 48.6	0.202653
	30	3 28.96	3 12 24.0	0.200252
Ott.	1	3 36.31	3 20 1.7	0.197874
	2	3 41.81	3 27 41.2	0.195518
	3	3 45.43	3 35 22.3	0.193187
	4	3 47.19	3 43 4.4	0.190881
	5	3 47.09	3 50 47.3	0.188603
	6	3 45.12	3 58 30.3	0.186353
	7	3 41.29	4 6 13.1	0.184135
	8	3 35.60	4 13 55.3	0.181948
	9	3 28.03	4 21 36.4	0.179794
	10	4 3 18.58	−4 29 16.2	0.177673

Gr. $11^m.4 - 11^m.2$.

Roma, Osservatorio al Collegio Romano, 1910 Sett. 14.

G. Zappa.

Naturwiss. 1 (1913) 865–866

Zusammensetzung der umgebenden Körperwelt aber findet der Forscher eine unübersehbare Fülle von Verschiedenheiten. Dazu kommt, daß die chemischen Vorgänge viel subtilerer Art sind und im allgemeinen unmerklicher in der Natur verlaufen. Gerade die reaktionsfähigsten Stoffe, die uns über den Mechanismus eines Vorganges aufklären könnten, sind eben dadurch am schwierigsten zu isolieren und stellen nicht nur an den Verstand des Beobachters, sondern auch an seine praktische Geschicklichkeit im Experimentieren und im Herstellen der Apparatur recht erhebliche Anforderungen. Dies geht soweit, daß die Aufklärung einer Reihe von Erscheinungen erst mit der Fortentwicklung der Technik möglich geworden ist. Denken wir nur an die theoretisch so wichtige Chemie der extrem hohen oder extrem tiefen Temperaturen. Als allmählich auf rein empirischem Wege die Darstellung einer Anzahl von Stoffen und Reagentien gelungen war, boten die scheinbar so wenig zusammenhängenden und verwirrend vielseitigen Phänomene dem Forscher, der sie in ein System zu bringen suchte, schier unüberwindliche Schwierigkeiten. Die Eigenschaften einer chemischen Verbindung sind nicht additiv, und die Stoffe, aus denen sie hervorgeht, haben gar keine Ähnlichkeit mit ihr. Unser Kochsalz läßt sich herstellen aus dem gasförmigen, giftigen, grünen Chlor und dem äußerst reaktionsfähigen leichten Metall Natrium, das auf Wasser verbrennt. Die Beziehung dieser Verbindung zu seinen Elementen besteht lediglich darin, daß sie sich aus jenen zusammensetzen und in sie zerlegen läßt. Eine Ähnlichkeit, die auf einen sonstigen Zusammenhang rein äußerlich schließen läßt, ist nicht vorhanden. Die viel subtileren Erscheinungen des Mikrokosmos waren eben nicht so leicht zu beobachten. Aus vorhandenen Körpern entstanden neue, die ihre Bestandteile wieder der Beobachtung entzogen. Hieraus folgt, daß in der Chemie einerseits das Tatsachen- und Beobachtungsmaterial erst geschaffen werden mußte, andererseits aber bei der Verfolgung und kritischen Durcharbeitung an den Forscher zunächst viel größere Anforderungen gestellt wurden. Erst als der menschliche Geist durch die physikalische Schulung im naturwissenschaftlichen Denken besser erzogen war, konnte er auf dem Gebiete der Chemie Erfolge erringen.

Für die Evolution des menschlichen Geistes in der Entwicklungsgeschichte der Chemie finden wir eine wichtige Stütze auch auf einem anderen Gebiete der Naturwissenschaften. In wie verschiedener Weise der geschulte Verstand, der sich zu wissenschaftlichen Gedanken zu erheben vermag, an eine Aufgabe herantritt, als ein Geist, der sich im Zustand eines naiven Anthropomorphismus befindet, zeigt sich in der Erweiterung unserer Kenntnisse von den elektrischen Erscheinungen. Bis zu den Tagen *Galvanis* und *Voltas* war die Elektrizität das Stiefkind der Physik gewesen. Das kam daher, daß die Beobachtungen auf diesem Gebiet erheblich schwerer waren als diejenigen des Lichts oder der Wärme. Wir besitzen kein Sinnesorgan, das unserem Gehirn die elektrischen Eindrücke direkt übermittelt, und sind bei der Beobachtung auf die Umwandlung der elektrischen Energie in andere unmittelbar wahrzunehmende Energieformen angewiesen. Vor *Volta* besaß man nur einige Kenntnisse von der Reibungselektrizität, und trotz der trefflichen Leistungen von *Otto von Guericke* und *Franklin* machte sie den großen Aufschwung der Physik nicht mit. Erst als man es durch *Volta* lernte, kontinuierliche Ströme zu erzeugen, wurde auch dieses Gebiet durch mathematische und experimentelle Behandlung mit den übrigen Zweigen der Physik verknüpft. Seitdem hat die wissenschaftliche und technische Bearbeitung der Elektrizität im 19. Jahrhundert mit dem gleichen staunenswerten Erfolge eingesetzt wie die der Chemie. Ebenfalls hat die Elektrizität nicht nur ihre Schwesterdisziplin eingeholt, sondern die speziellen theoretischen Vorstellungen, die zur Erklärung der Erscheinungen gebildet wurden, haben gerade in der gegenwärtigen Zeit dazu geführt, daß man die alten mechanischen Bilder aufzugeben und sie durch elektrische zu ersetzen beginnt.

Aus dem Vorhergegangenen ist also ersichtlich, daß sich unsere Erkenntnis in ganz natürlicher Weise entwickelt hat. Vom einfachen Beobachten sind wir allmählich zu den komplizierteren Erscheinungen aufgestiegen, die nur durch das Experiment und die Rechnung erkennbar und verallgemeinerungsfähig waren. Infolge der immer größeren Schulung ist es dem Menschengeist gelungen, schneller in neue Gebiete, die sich ihm darboten, einzudringen und sie auszubauen. Am Ende des 18. Jahrhunderts fügte es sich, daß die Chemie und die Elektrizität gleichzeitig der Bearbeitung zugänglich wurden und in rasendem Siegeslauf die übrigen Zweige der Naturwissenschaften mit sich fortrissen. Dies bewirkte, daß in unserer modernen Existenz die Technik die Führung übernommen hat. Die ganzen Grundlagen unseres Lebens haben sich der Industrie unterwerfen müssen. Die Nationalökonomie, die Finanzpolitik, die Gesetzgebung, die soziale Fürsorge — sie alle sind bestrebt, die Differenzen, die sich zwischen dem rapiden Wachsen unserer Ansprüche und der Behaglichkeit und Ruhe unserer Lebensführung gebildet haben, auszugleichen. Und noch ist es nicht zu übersehen, ob ihnen dies gelingen wird. Die Naturwissenschaften haben eine neue Epoche, die Epoche der Mechanisierung heraufgeführt.

Tagung der Solar Union zu Bonn. 31. Juli — 5. August 1913.

Von Prof. Dr. K. Schwarzschild, Potsdam.

Die Solar Union ist eine von dem amerikanischen Astronomen *G. E. Hale* ins Leben gerufene internationale Vereinigung. Ihr ursprünglicher Zweck war, alle Verabredungen zu treffen, die gemeinsamer Arbeit an den Problemen speziell der Sonnenphysik förderlich sein können; indessen hat sie sich rasch zur maßgebenden internationalen Gesellschaft für das ganze Gebiet der Astrophysik

entwickelt. Vor drei Jahren fand eine Versammlung der Solar Union auf dem Mount Wilson in Kalifornien statt, die zahlreiche europäische Gelehrte mit den Wundern des Landes, wie den wissenschaftlichen Einrichtungen und Erfolgen auf dem dortigen Observatorium bekannt machte. Jetzt sind 60 Ausländer, darunter viele Amerikaner, nebst 40 Deutschen der Einladung der Herren *Kayser* und *Küstner* nach Bonn gefolgt, und es war in Bonn in glücklicher Weise dafür gesorgt, daß diesmal die Vorzüge des alten Europas in Wissenschaft und Kunst, Land und Sitte zur Geltung kamen.

Die Verhandlungen der Union bestehen nicht in Vorträgen, sondern in den Berichten der verschiedenen von der Union eingesetzten Kommissionen und gelegentlich daran anschließenden Mitteilungen einzelner Forscher. Man kann sagen, daß die Verhandlungen dadurch vielleicht weniger interessant, aber um so nützlicher werden. Einen besonderen Erfolg hat die Union bereits auf dem Gebiet der Bestimmung von Wellenlängennormalen zu verzeichnen. Die Bestimmung der Wellenlängen in den Spektren der Elemente geschah bisher fast durchweg in differentiellem Anschluß an *Rowlands* Tafeln des Sonnenspektrums. Es hat sich indessen gezeigt, daß die von *Rowland* benutzte Längeneinheit, statt genau mit dem Meter übereinzustimmen, um etwa 0,03 mm vom Meter abweicht, und daß der Maßstab außerdem für verschiedene Teile des Spektrums etwa entsprechend einem Verhältnis von 0,003 mm auf ein Meter variiert. Das sind bei der Feinheit der spektroskopischen Messungen sehr merkliche Differenzen. Die Solar Union hat nun einen solchen Wetteifer unter den auf diesem Gebiet tätigen Physikern angeregt, daß zurzeit bereits eine das ganze sichtbare Spektrum durchziehende Reihe von Eisenlinien mit einer Genauigkeit, die etwa 0,0003 mm auf einen Meter entspricht, festgelegt ist. Da von Herrn *Goos* geringe Variationen der Wellenlängen im gewöhnlichen Eisenbogen — je nach den Bedingungen, unter denen er brennt — konstatiert worden sind, hat die Union diesmal bestimmte Normen für den Eisenbogen festgesetzt, die für die internationalen Standards der Wellenlängen gelten sollen und die voraussichtlich zu einer Genauigkeit von nahezu $1 : 10^7$ für zukünftige Wellenlängenmessungen führen werden.

Weitere Verabredungen betrafen die Zählweise der Sonnenprotuberanzen für statistische Zwecke, die Beobachtung der Sonnenrotation nach dem Dopplerschen Prinzip, die Klassifikation der Sternspektren, die Beobachtung der nächstjährigen Sonnenfinsternis.

Ausführlichere Mitteilungen gaben: Herr *Abbot* über seine fundamentalen Messungen der Sonnenstrahlung (gleichzeitige Messungen in Kalifornien und Algier scheinen eine Veränderlichkeit der Sonnenstrahlung zu erweisen; die Gesamtstrahlung der Sonne wächst mit dem Fleckenreichtum), ferner Herr *Julius* über die Zerstreuung des Lichts in der Sonnenatmosphäre, Herr *St. John* über die Strömungserscheinungen in und um Sonnenflecken. Eine abendliche „Conversazione" nach englischem Muster brachte reiche Schätze an photographischen Aufnahmen und graphischen Darstellungen aus den Mappen der Teilnehmer zum Vorschein.

Das bifilare Kegelpendel (Instrument für die Aufzeichnung von Erdbeben).

Von Dr. C. Mainka, Straßburg i. E.

Autoreferat.

Diese in den „Mitteilungen der Philomatischen Gesellschaft in Elsaß-Lothringen" Bd. *IV*, Heft 5, Straßburg, 20. Jahrgang pag. 633—667 veröffentlichte Arbeit gibt eine eingehendere Beschreibung des nach den Angaben des Verfassers von der Straßburger Firma für Präzisionsmechanik *J. & A. Bosch* gebauten Seismographen. Außer Schlußwort und Quellennachweis, der u. a. eine Reihe von Veröffentlichungen des Verfassers über dieses Thema nachweist, enthält der Aufsatz sechs Abschnitte, von denen der erste die Geschichte des Instrumentes, wenn auch nicht erschöpfend, behandelt.

Das Prinzip der Aufhängung ist 1832 von *Hengler* zuerst angegeben worden. Ein Arm a bei A und B Fig. 1

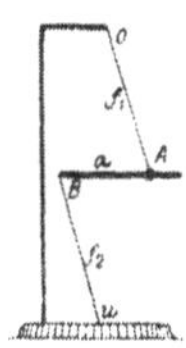

Fig. 1.

durch zwei Drähte f_1 und f_2 bei o und u aufgehängt, zeigt das Urprinzip der Aufhängung, das später Anregung zu mancherlei Umänderungen gegeben hat. Stellt man sich vor: f_2 auf etwa 30 mm verkürzt, um B gedreht, in die horizontale Lage gebracht und auf Zug (eigentlich selbstverständlich) beansprucht, so haben wir, wenn man sich in A noch den Schwerpunkt des Pendelkörpers denkt, eine Anschauung von der hier erfolgten Aufhängung, die also auch „bifilar" ist. Zur Erhöhung der Leistungsfähigkeit des Apparates hat *Mainka* den verkürzten Draht nach der Mitte zu, bis zur Sicherheitsgrenze verjüngen lassen; statt des Drahtes wurde bald ein Stahlband, in gleicher Weise behandelt, angewendet. Diese Herstellung des unteren Drehpunktes ist neu. Wie aus der Figur verständlich, ist die Bewegungsfreiheit des Pendelgewichtes senkrecht zur Ebene, die durch $AoBu$ geht. Um also die horizontalen Bodenbewegungen bei Erdbeben bezüglich ihrer Richtung festzulegen, braucht man zwei solcher, voneinander ganz und gar getrennter Instrumente, deren Ebenen man der Einfachheit wegen senkrecht zueinander, die eine in den Meridian, die andere in den ersten Vertikal stellt.

Der durch ein Erdbeben in einen Schwingungszustand versetzte Boden des Aufstellungsortes und der aufgehängte Pendelkörper sind als zwei miteinander gekoppelte Schwingungssysteme anzusehen. Da die Bodenbewegungen vor allem bei fernen Beben äußerst klein sind, muß also an eine Vorrichtung gedacht werden, die gestattet, die Bewegungen des Pendelgewichtes vergrößert wiederzugeben. Die mathematische Beziehung zwischen zwei solchen Systemen lehrt ferner, daß die Eigenschwingungen des Systems, das die Erregerbewegung richtig wiedergeben soll, eliminiert werden müssen, d. h. aber, der Pendelkörper muß mit einer Dämpfungsvorrichtung ver-

Section 3

Cometary Tails

Cometary Tails

Commentary by K. Jockers, Katlenburg-Lindau

Historical Survey

Before the introduction of the photographic plate as an astronomical observational technique, cometary research in the 19th century had to be confined to the determination of cometary orbits, to visual photometry, and to the study of cometary tails (see Jaegermann 1903). Bessel (1836) showed that the shape of cometary tails could be explained if one assumed that as well as gravity, a force of repulsion that also declined as the square of the distance from the Sun, acted on the material in the tail. This repulsive force, or rather the ratio K between this repulsive force and gravity, depends on the nature of the particles in the tail, and in addition each particle may be described in terms of the time τ at which it leaves the cometary nucleus. Within the framework of this "mechanical theory", curves of equal K were described as syndynes, and curves of equal τ as synchrones. If we know the position of a tail particle on the plane of the comet's orbit, then mechanical theory allows the corresponding parameters K and τ to be determined (although unfortunately not always unambiguously). Bredikhin (see Jaegermann 1903) established that the tail particles may be divided into two types, Type I with $K > 10$ and Type II with $K \leq 3$. He also introduced Type III for very small values of K, but this division is no longer considered to be valid and so plays no part in our discussion. Bessel and Bredikhin assumed that the force of repulsion was electrical in nature. Arrhenius (1900) recognized radiation pressure as being the source of the force of repulsion, and this explanation is still valid for Type-II tails. The first photographic records of comets showed that cometary tails of Type I differ morphologically from those of Type II. In Type I tails, structures can be recognized in exposures made at close intervals of time, and their velocities and accelerations determined directly, without having to go via mechanical theory. This showed that the accelerations were significantly higher than had previously been supposed (frequently $K > 100$), and that the motions of the structures in Type I tails largely contradicted the mechanical theory. It was only in 1951 that L. Biermann recognized interaction with the solar wind as being the cause of the great accelerations found in Type-I tails.

The first spectrographic observations of comets, among them Comet Morehouse 1908 III, and P/Halley 1910 II (Frost, Parkhurst 1909; De la Baume Pluvinel, Baldet 1911; Slipher, Lampland 1911; Bobrovnikoff 1927) showed that tails of Type I and II were different in nature and could be distinguished from their spectra. Type-II tails consist of dust particles and their spectrum is the spectrum of scattered sunlight (Bobrovnikoff erroneously ascribes Type-II tails to the emission of C_2 in the cometary coma). Type-I tails consist of positive ions. Fowler (1909) recognized the connection between the spectrum and CO, but not their ionic nature. The most important ion, H_2O^+ was first identified in 1974 (Wehinger et al. 1974).

Schwarzschild's Papers

We see that at the beginning of the 20th century it was possible, for the first time, to investigate the physical nature of cometary tails both from astronomical observation and from an understanding of atomic and molecular structure. Correspondingly, the two most important papers of Karl Schwarzschild in the field of comets are devoted to this question.

Light Pressure on Reflecting Spheres

The first paper [3.1] is concerned with solar radiation pressure on conductive spheres and is actually negative in nature, in that it shows, contrary to Arrhenius' suggestion (1900) that the high accelerations observed in Type-I tails cannot be explained as the effect of solar radiation pressure. The important physical result is that the interaction between radiation and small particles, which determines the light-pressure and also the scattering efficiency and therefore the visibility of the particles, is greatest for particles whose characteristic size is comparable with the wavelength of light. (See also the contributions of Halm (1902), Cox (1902), and Holst (1905) to the controversy.) The same effect was found a few years later by Mie (1908) in his scattering theory for any dielectric particles, and has since been determined by measurements (see Bohren, Huffman 1983; van de Hulst 1957). Atoms and molecules are also subject to radiation pressure in accordance with the efficiency with which they fluoresce.

Absolute Photographic Photometry of Comet P/Halley

We now see that Schwarzschild and Kron's contribution in this paper [3.2] was in their having recognized, for the first time, that fluorescence induced by sunlight was the excitation mechanism producing the radiation from cometary ions. The mechanism of fluorescence was being thoroughly investigated at the time of their paper, but was still poorly understood. Wood had detected resonant fluorescence in Na-vapour for the first time in 1904. It was only Bohr's atomic model (1913) that enabled a qualitative understanding to be gained of the fact that monochromatic radiation could excite frequencies lower than that of the excitation frequency. A full understanding of fluorescence excitation in molecules only came very much later.

Unfortunately, recognition of fluorescence as being the true cause of the emission from ions in tails was obtained from data that was, in all probability, falsely interpreted. Cometary observations were made at a time when the Earth was very close to the plane of the comet's orbit. As a result the dust tail (Type-II tail) and the gas tail (Type-I tail) lay one behind the other as seen from Earth. Under such conditions, and with a type of comet like Halley, the brightness of the dust would predominate. It can be seen that Schwarzschild and Kron took this possibility into account from the fact that at the end of their paper they give a production rate for dust particles. In a dust tail, conservation of the product of cross-sectional intensity and flow velocity is to be expected – the scattering of sunlight being in an optically thin medium – as it is for fluorescence excitation in the ion tail. This was also known to the authors.

Schwarzschild and Kron's work was the only absolute photometry for Comet Halley's apparition in 1910. So, in principle, it should have been of great significance in determining any variations at the time of the 1986 apparition. Unfortunately, comparison with modern measurements (Lamy et al. 1987a), shows that the absolute intensities given in the paper are at least a factor of 10 too high. Secular changes of this order are improbable, so measurement-errors are probably involved.

The production rates of gas and dust from Comet Halley in 1986 were measured from the ground as well as in-situ during the fly-bys of various spaceprobes. I should like to compare Schwarzschild and Kron's estimates with modern results, considering both hypotheses put forward at the end of their papers. The fluorescence efficiency estimated in their second hypothesis (particles being atoms or molecules) is far too optimistic by several orders of magnitude, so that although the measured intensities are somewhat too large, the gas production rate is several orders of magnitude too low. (The authors also forgot to multiply their flux by 20, the correct value being $E = 3\,\mathrm{kg\,s^{-1}}$.) Using measurements by the Giotto spaceprobe (at a heliocentric distance of 0.90 AU compared with a distance of 0.72 AU in 1910), a value of $3200\,\mathrm{kg\,s^{-1}}$ is obtained for CO production (Eberhardt et al. 1987; Krankowski et al. 1986). If all the CO appears as CO^+ in the tail, then that is the correct value. The authors' first hypothesis (particles represent micron-sized dust) is closer to the truth. It also has a more realistic estimate of the radiative efficiency. In this case the radiation is caused by scattering of the solar continuum. Unfortunately, both neglecting the particle-size distribution – numerous small particles make little contribution to the mass production rate, but significantly affect the intensity of the dust tail – and failing to take the low albedo of the dust particles into account, strongly influence the final result, although these two effects luckily counteract one another. The best modern measurements were obtained by the Vega 1 and 2 spaceprobes (Mazets et al. 1987). Within the particle mass-range 10^{-16} to 10^{-6} g, production rates of $11\,500\,\mathrm{kg\,s^{-1}}$ (Vega 1) and $6000\,\mathrm{kg\,s^{-1}}$ (Vega 2) were found at cometary heliocentric distances of 0.79 AU and 0.83 AU respectively. Schwarzschild and Kron give a value of $1500\,\mathrm{kg\,s^{-1}}$. If we take the average density of the particles as $2.2\,\mathrm{g\,cm^{-3}}$ (Lamy et al. 1987b), then the range over which measurements were made corresponds to particle radii of 0.02 to $47\,\mu\mathrm{m}$. If we only consider particle masses lower than 4×10^{-12} g (Schwarzschild and Kron's particles with a radius of $1\,\mu\mathrm{m}$ and a density of $1\,\mathrm{g\,cm^{-3}}$ have such a mass), the flux amounts to about one thousandth of the value given above (Mazets et al. 1987). From this we see that the largest, invisible particles, which were not recorded on the plates analyzed by Schwarzschild and Kron, contribute the most mass.

Stellar Occultations by Comets; Comet Halley's Transit of the Sun

As far as the question of stellar occultations by comets is concerned (Miethe 1910), which is the subject of the last of the papers reproduced here [3.3], it must be recalled that in Schwarzschild's day the fact that a comet nucleus has a diameter of only a few km and consists of just one solid body was quite unknown. Occultation of a star by the nucleus of a comet would be visible only from a very small area of the Earth, just a few km across. Such an observation would naturally be very valuable, but it is not predictable, and is therefore improbable. Photoelectric observations of the passage of the inner coma of a comet in front of stars have been obtained recently, and there are signs of there being some extinction of starlight by the inner coma of comets (Lecacheux et al. 1984). Schwarzschild's own observations of Halley's Comet (visual observations and obtaining solar spectra during the comet's transit of the solar disk) are discussed by Münch and von Pahlen (1910). A transit of a bright comet across the solar disk is an extremely rare event, so Schwarzschild deserves considerable credit for having obtained solar spectra, during the transit, at the comet's location in front of the solar disk. But from both modern and contemporary points of view, it is not surprising that the comet was not detectable in absorption in front of the Sun. Schwarzschild and Kron [3.2] give the intensity of the comet's tail close to the nucleus as 10^{-11}, where the unit is the average intensity of the solar disk, and as we have seen earlier, this value is somewhat too high. Let us assume that the spectrograph has a resolution of 1/10 Å and the most favourable case – which does not actually occur – that the light from the comet is monochromatic. The intensity of this hypothetical cometary line would then still be about 2×10^{-7} of that of the solar spectrum. Even if the comet close to the nucleus were one thousand times brighter than the value given in [3.2], the intensity of the hypothetical line would still be less than one thousandth of that of the Sun.

References*

Arrhenius, Sv. (1900) *Physikalische Zeitschrift* **2**: 81, 97

De la Baume Pluvinel, A., Baldet, F. (1911) *Astrophys. J.* **34**: 89

Bessel, W. (1836) *Astron. Nachr.* **13**: 185

Biermann, L. (1951) *Z. Astrophys.* **29**: 274

Bobrovnikoff, N.T. (1927) *Astrophys. J.* **66**: 145, 439

Bohr, N. (1913) *Phil. Mag.* **26**: 476, 857

Bohren, C.F., Huffman, D.R. (1983) *Absorption and Scattering of Small Particles* (Wiley, New York)

Cox, J. (1902) *Nature* **66**: 54

Eberhardt, P., Krankowsky, D., Schulte, W., Dolder, U., Lämmerzahl, P., Berthelier, J.J., Woweries, J., Stubbemann, U., Hodges, R.R., Hoffman, J.H., Illiano, J.M. (1987) *Astron. Astrophys.* **187**: 481

Fowler, A. (1909) *Monthly Not. Roy. Astron. Soc.* **70**: 176, 484

Frost, E.B., Parkhurst, J.A. (1909) *Astrophys. J.* **29**: 55

Halm, J. (1902) *Nature* **65**: 415

Holst, H. (1905) *Fys. Tidskr.* **4**: 1

* The reference numbers in the text for Schwarzschild's papers correspond to the contents on page XII and to the bibliography on page 493.

van de Hulst, H.C. (1957) *Light Scattering by Small Particles* (Dover, New York)
Jaegermann, R. (1903) *Prof. D.Th. Bredichin's Mechanische Untersuchungen über Cometenformen. In systematischer Darstellung* (Voss, Leipzig)
Krankowsky, D., Lämmerzahl, P., Herrwerth, I., Woweries, J., Eberhardt, P., Dolder, U., Herrmann, U., Schulte, W., Berthelier, J.J., Illiano, J.M., Hodges, R.R., Hoffman, J.H. (1986) *Nature* **321**: 326
Lamy, P.L., Pedersen, H., Vio, R. (1987a) *Astron. Astrophys.* **187**: 661
Lamy, P.L., Grün, E., Perrin, J.M. (1987b) *Astron. Astrophys.* **187**: 767
Lecacheux, J., Thuillot, W., Encrenaz, T., Laques, P., Rouan, D., Despiau, R. (1984) *Icarus* **60**: 386
Mazets, E.P., Sagdeev, R.Z., Aptekar, R.L., Golenetskii, S.V., Guryan, Yu.A., Dyachkov, A.V., Ilyinskii, V.N., Panov, V.N., Petrov, G.G., Savvin, A.V., Sokolov, I.A., Frederiks, D.D., Khavenson, N.G., Shapiro, V.D., Shevchenko, V.I. (1987) *Astron. Astrophys.* **187**: 699
Mie, G. (1908) *Ann. Phys.* **25**: 377
Miethe, A. (1910) *Astron. Nachr.* **185**: 15
Münch, W., von Pahlen, E. (1910) *Astron. Nachr.* **185**: 136
Slipher, V.M., Lampland, O.C. (1911) *Lowell Obs. Bulletin* **2**, No. 47, 252
Wehinger, P.A., Wyckhoff, S., Herbig, G.H., Herzberg, G., Lew, H. (1974) *Astrophys. J.* **190**: L43
Wood, R.W. (1904) *Proc. Am. Acad.* **15**: 396

Sitz. Ber. math. phys. Classe d. Kgl. bayer. Akad. d. Wiss. 31 (1901) 293–338

Der Druck des Lichts auf kleine Kugeln und die Arrhenius'sche Theorie der Cometenschweife.

Von **Karl Schwarzschild.**

(*Eingelaufen 6. Juli.*)

Die eigentümlichen Formen der Cometenschweife lassen sich erklären durch Annahme einer abstossenden Kraft, die von der Sonne auf die äusserst fein verteilte Schweifmaterie ausgeübt wird. Während früher diese Kraft nur ganz vage als elektrischen Ursprungs bezeichnet wurde, hat vor kurzem Herr Sv. Arrhenius[1]) darauf hingewiesen, dass solche abstossende Wirkungen dem Druck entspringen können, welchen nach der Maxwell'schen Theorie die Sonnenstrahlung auf jeden absorbierenden oder reflektierenden Körper ausübt. Ueber die Existenz des Maxwell'schen Drucks der Lichtstrahlen kann kein Zweifel sein, obwohl er experimentell mit hinreichender quantitativer Sicherheit noch nicht nachgewiesen wurde. Er muss bestehen, wofern die Grundauffassung Maxwell's richtig ist, dass in den Lichtschwingungen eben elektrische Kräfte oscillieren, Kräfte, welche die geladenen Jonenpaare der ponderabeln Massen angreifen. Zudem haben sich die Folgerungen, welche sich aus den Hauptsätzen der mechanischen Wärmetheorie in Verbindung mit dem Maxwell'schen Druck ergaben, an der Erfahrung glänzend bestätigt. Es liegt daher in dem Druck der Lichtstrahlen eine causa realis, eine bei der Bildung der Cometenschweife, wie auch bei anderen Himmels-

[1]) Physikalische Zeitschrift. II. Jahrgang. Heft 6—7.

erscheinungen, zweifellos in irgend einem geringeren oder stärkeren Grade mitwirkende Ursache vor. Die Frage ist nur, ob dieser Druck quantitativ hinreichende Beträge erlangen kann, und das ist ein Punkt, in dem die Darlegungen von Arrhenius nicht ganz befriedigend sind.

Fällt eine ebene Welle normal auf eine vollkommen schwarze Platte, so erleidet letztere einen Druck, der nach Maxwell gleich der in der Volumeneinheit enthaltenen Energie E des Wellenzuges ist. Auf eine vollkommen schwarze Kugel vom Radius a, welche das Licht über eine Fläche von der Grösse πa^2 hin abfängt, wird daher ein Druck $\pi a^2 E$ wirken. Die auf dieselbe Kugel wirkende Schwerkraft hat den Betrag $\frac{4\pi}{3} g \cdot s \cdot a^3$, wo g die Schwerebeschleunigung, s das spezifische Gewicht des Kugelmateriales ist. Je kleiner der Kugelradius a, um so stärker wird daher der Druck im Verhältnis zur Schwerkraft. Unter Annahme der Solarkonstante 2.5 und des spezifischen Gewichtes s gleich 0.8 (Kohlenwasserstoff der Cometen) findet man mit Herrn Arrhenius, dass bei einem Kugeldurchmesser von 1.9 μ der Druck gleich der Schwerkraft wird. Sinkt der Kugeldurchmesser bis auf 0.1 μ herab, so übertrifft die abstossende Kraft der Sonnenstrahlung die Schwerkraft um das 18.5fache, und erreicht damit den Höchstbetrag, welcher zur Erklärung der gestrecktesten Cometenschweife erforderlich ist.[1])

In dieser Berechnung des Drucks liegt nun eine Ungenauigkeit. Der Druckwert $\pi a^2 E$ gilt nur, solange die Kugel das auffallende Licht gänzlich abfängt und in dem Raume hinter ihr völliger Schatten ist, solange man also die Beugungserscheinungen vernachlässigt. Das ist bei grossen Kugeln gestattet. Bei Kugeln aber, deren Durchmesser von der Grössenordnung der Wellenlängen des Lichts ist, — und gerade solche kommen nach den eben angeführten Zahlen für die Erklärung der Cometenschweife in Betracht — können

[1]) Vgl. Bredichin, Annales de l'Observatoire de Moscou. 1886.

Sitz. Ber. math. phys. Classe d. Kgl. bayer. Akad. d. Wiss. *31* (1901) 293–338

durch die Beugung des Lichts die Verhältnisse sehr wesentlich geändert werden. Um ganz kleine Kugeln, deren Radius auch gegen die Wellenlänge sehr klein ist, schlägt die Lichtwoge herum, ohne in ihrem Verlaufe merklich gestört zu werden, und man kann sich hier durch einfache Ueberlegungen analog denen, auf welche Lord Rayleigh[1]) seine Theorie der blauen Farbe des Himmels gründet, überzeugen, dass mit fortdauernder Verringerung des Kugelradius der Druck des Lichts sogar wieder mehr und mehr unter die Schwerkraft herabsinken muss. Das Verhältnis des Lichtdrucks zur Schwerkraft wächst also nicht immerfort mit Verkleinerung des Kugelradius, sondern erreicht ein gewisses Maximum für Kugeln von der Grössenordnung der Wellenlänge. Es ist erst noch zu untersuchen, ob das maximale mögliche Verhältnis von Lichtdruck zu Schwerkraft wirklich den Wert 18.5 erreicht, den man, wie erwähnt, zur Erklärung der gestrecktesten Cometenschweife braucht.

Ohne auf die schwer diskutabeln Fragen einzugehen, inwieweit eine Anschauung, welche die Cometenschweife aus lauter getrennten Tröpfchen im leeren Raum bestehen lässt, sich unseren sonstigen Kenntnissen von den Cometen anschliesst, will ich hier versuchen, nur über diesen einen quantitativen Punkt zu entscheiden. Irgendwo in der Welt muss es ja vorkommen, dass ein kleines Teilchen im leeren Raum dem Druck des Lichts ausgesetzt ist, und daher ist dem Resultate der Rechnung in jedem Falle die physikalische Anwendbarkeit gewiss, auch wenn sich aus andern Gründen etwa die Arrhenius'sche Theorie der Cometenschweife nicht halten lassen sollte.

Bekanntlich ist der Begriff des vollkommen schwarzen Körpers mathematisch nicht exakt zu fassen und es erscheint auch physikalisch höchst zweifelhaft, ob ein Partikelchen irgend eines gewöhnlichen schwarzen Körpers von 0.1 μ bis 1 μ Durchmesser noch im geringsten Eigenschaften aufweist, nach denen man es als „schwarz“ bezeichnen könnte. Zur exakteren Be-

[1]) Philosoph. Magazine. 1871.

rechnung der Grösse des Lichtdrucks habe ich daher die kleine Kugel, die wir in den Strahlenzug eingeschaltet denken, nicht als vollkommen schwarz, sondern als vollkommen reflektierend vorausgesetzt. Es ist anzunehmen, dass für vollkommen reflektierendes Material der Lichtdruck besonders gross wird und man daher auf Grund dieser Annahme eben das Gewünschte, nämlich eine obere Grenze für den Lichtdruck, erhält.

Abgesehen von der Beziehung zur Theorie der Cometenschweife darf das Folgende vielleicht auch ein gewisses selbstständiges Interesse für sich in Anspruch nehmen. Bevor der Druck des Lichts auf eine Kugel berechnet werden konnte, musste eine exakte Theorie der Reflexion und Beugung des Lichts durch die Kugel gegeben werden. Es ist das ein Problem, dem schon Clebsch in Bd. 61 des Crelle'schen Journals eine umfangreiche Abhandlung gewidmet hat.[1] Indessen hat Clebsch Grenzbedingungen benutzt, die wir heute nicht mehr als gültig anerkennen. Er nimmt an, dass im Aether neben den transversalen Wellen auch longitudinale auftreten können, verlangt, dass an der Oberfläche der vollkommen spiegelnden Kugel alle drei Componenten der elastischen Verschiebung verschwinden, und wird dann notgedrungen zu der Folgerung geführt, dass aus einfallenden Transversalwellen durch Reflexion an der Kugel Longitudinalwellen entstehen. Clebsch hat ausserdem in die Behandlung dieses speziellen Problems die Ableitung zahlreicher allgemeiner Sätze aus der Theorie der Kugel- und Cylinderfunktionen eingeschachtelt, die damals wohl noch weniger in den festen Wissensbestand übergegangen waren, und das belastet seine Darstellung, so wertvoll jene Ableitungen an und für sich sind. Daher wäre eine Neubehandlung des Problems schon von formalem Gesichtspunkt aus wünschenswert gewesen, wenn sie nicht durch die erwähnte Aenderung der Grenzbedingungen notwendig geworden wäre.

---

[1]) Hierauf bin ich erst nach Ausführung dieser Arbeit durch eine freundliche Mitteilung von Herrn Prof. A. Sommerfeld aufmerksam gemacht worden.

Sitz. Ber. math. phys. Classe d. Kgl. bayer. Akad. d. Wiss. *31* (1901) 293–338

§ 1. Mathematische Formulierung des Beugungsproblems.

Das zunächst zu behandelnde Beugungsproblem ist dieses. Eine Kugel vom Radius a habe ihren Mittelpunkt im Nullpunkt des Coordinatensystems. Eine ebene Welle falle in der Richtung der x-Axe von ihrer positiven Seite her ein. Wir wollen zunächst annehmen, dass dieselbe linear polarisiert sei und zwar so, dass die elektrische Kraft in der x, y Ebene schwinge. Für die Componenten X, Y, Z, L, M, N der elektrischen und magnetischen Kraft gelten überall ausserhalb der Kugel die Maxwell'schen Gleichungen:

$$\frac{1}{V}\frac{dL}{dt} = \frac{dZ}{dy} - \frac{dY}{dz} \quad \frac{1}{V}\frac{dX}{dt} = \frac{dM}{dz} - \frac{dN}{dy} \quad (V = \text{Licht-geschwindigkeit})$$

nebst den entsprechenden durch cyklische Vertauschung der Coordinaten entstehenden. Die Kugel selbst soll vollkommen reflektierend oder mit anderen Worten ein vollkommener Leiter sein. Dann ist ihre Oberfläche unter allen Umständen eine Niveaufläche der elektrischen Kraft, die in die Oberfläche fallenden Kraftcomponenten müssen verschwinden, es müssen die Beziehungen gelten:

Auf der Kugeloberfläche:

$$Xy - Yx = Xz - Zx = Yz - Zy = 0.$$

Im Unendlichen muss die Componente Y der elektrischen Kraft ein Glied der Form $\cos 2\pi\left(\frac{x}{\lambda} + \frac{t}{\tau}\right)$ enthalten, welches die ebene polarisierte einfallende Welle darstellt. Deren Wellenlänge ist zugleich durch diesen Ansatz zu λ, ihre Periode zu τ und ihre Amplitude und Intensität zu 1 festgelegt. Sonst dürfen in X, Y, Z im Unendlichen keine Glieder vorkommen, welche die physikalische Bedeutung einfallender Wellenzüge haben.

Wir setzen nun:

$$\begin{aligned} X &= \text{pars real } (\xi\, e^{iqt}) & L &= \text{pars real } (\lambda\, e^{iqt}) \\ Y &= \text{pars real } (\eta\, e^{iqt}) & M &= \text{pars real } (\mu\, e^{iqt}) \\ Z &= \text{pars real } (\zeta\, e^{iqt}) & N &= \text{pars real } (\nu\, e^{iqt}) \end{aligned} \qquad 1)$$

$$q = \frac{2\pi}{\tau},$$

wobei $\xi, \eta, \zeta, \lambda, \mu, \nu$ von der Zeit unabhängige komplexe Grössen sein sollen, und erhalten an Stelle der Maxwell'schen Gleichungen die folgenden:

$$i k \lambda = \frac{\delta \zeta}{\delta y} - \frac{\delta \eta}{\delta z} \quad i k \mu = \frac{\delta \xi}{\delta z} - \frac{\delta \zeta}{\delta x} \quad i k \nu = \frac{\delta \eta}{\delta x} - \frac{\delta \xi}{\delta y} \qquad 2)$$

$$i k \xi = \frac{\delta \mu}{\delta z} - \frac{\delta \nu}{\delta y} \quad i k \eta = \frac{\delta \nu}{\delta x} - \frac{\delta \lambda}{\delta z} \quad i k \zeta = \frac{\delta \lambda}{\delta y} - \frac{\delta \mu}{\delta x} \qquad 3)$$

$$k = \frac{2\pi}{\tau V} = \frac{2\pi}{\lambda}. \qquad 4)$$

Durch Elimination der magnetischen Componenten λ, μ, ν ergiebt sich daraus:

$$k^2 \xi + \Delta^2 \xi = 0 \quad k^2 \eta + \Delta^2 \eta = 0 \quad k^2 \zeta + \Delta^2 \zeta = 0. \qquad 5)$$

$$\frac{\delta \xi}{\delta x} + \frac{\delta \eta}{\delta y} + \frac{\delta \zeta}{\delta z} = 0. \qquad 6)$$

Die Randbedingungen auf der Kugeloberfläche gehen durch Einführung von ξ, η, ζ über in:

$$\xi y - \eta x = \xi z - \zeta x = \eta z - \zeta y = 0. \qquad 7)$$

Für das Unendliche folgt, dass η ein Glied:

$$\bar{\eta} = e^{ikx} \qquad 8)$$

entsprechend der ebenen einfallenden Welle enthalten muss und dass sonst in ξ, η, ζ keine Teile vorkommen dürfen, welche physikalisch die Bedeutung aus dem Unendlichen einfallender Wellen besitzen. Unsere Aufgabe ist jetzt, drei „Wellenpotentiale" (Lösungen der Differentialgleichung $k^2 u + \Delta^2 u = 0$) ξ, η, ζ zu finden, welche durch die Bedingung (6) verknüpft

Sitz. Ber. math. phys. Classe d. Kgl. bayer. Akad. d. Wiss. *31* (1901) 293–338

sind, auf der Oberfläche der Kugel den Bedingungen (7) und im Unendlichen der Bedingung (8) genügen.

§ 2. Ansatz der Lösung in Form von Reihenentwicklungen nach Kugel- und Bessel'schen Functionen.

Man führe Polarkoordinaten r, ϑ, φ ein durch die Beziehungen:

$$x = r\cos\vartheta \qquad y = r\sin\vartheta\cos\varphi \qquad z = r\sin\vartheta\sin\varphi. \tag{9}$$

Es ist bekannt,[1]) dass die Differentialgleichung $k^2 u + \Delta^2 u = 0$ sog. „Zerfällung" nach den drei Coordinaten r, ϑ, φ gestattet, d. h., dass ein partikuläres Integral gegeben wird durch den Ansatz:

$$u = R\cdot\Theta\cdot\Phi. \tag{10}$$

Hierbei bedeutet Φ eine Funktion von φ allein und zwar in unserem Falle $\cos n\varphi$ oder $\sin n\varphi$ (n ganzzahlig), Θ ist eine Funktion von ϑ allein und zwar die n. Zugeordnete der Kugelfunktion irgend einer Ordnung m, für die wir $P_{m,n}(\cos\vartheta)$ schreiben werden. Schliesslich bedeutet R eine Funktion von r allein und zwar eine Lösung R_m der Differentialgleichung:

$$\frac{d^2 R_m}{d r^2} + \frac{2}{r}\frac{d R_m}{d r} + R_m\left(k^2 - \frac{m(m+1)}{r^2}\right) = 0. \tag{11}$$

Nutzt man die Willkürlichkeit der beiden ganzen Zahlen m und n aus und superponiert alle aus verschiedener Wahl derselben hervorgehenden partikulären Integrale der Form (9), so erhält man das allgemeinere Integral:

$$\begin{aligned} u = &\sum_m\sum_n G_{m,n} R_m(r) P_{m,n}(\cos\vartheta)\cos n\varphi \\ &+ \sum_m\sum_n H_{m,n} R_m(r) P_{m,n}(\cos\vartheta)\sin n\varphi \end{aligned} \tag{12}$$

worin die G und H willkürliche Constanten sind, und es lässt sich zeigen, dass in Rücksicht auf die weitere Willkürlich-

[1]) Vgl. Pockels, Ueber die Gleichung $\Delta u + k^2 u = 0$. Teubner 1891, pag. 109–111.

keit der Integrationskonstanten in den Lösungen R_m der Differentialgleichung (11) dieser neue Ansatz sogar die allgemeinste Lösung der Differentialgleichung $k^2 u + \Delta^2 u = 0$ giebt, welche nebst ihren ersten Derivierten ausserhalb einer Kugel um den Nullpunkt eindeutig und stetig ist.[1]) Es folgt demnach, dass jedes der drei Wellenpotentiale ξ, η, ζ, die wir zu bestimmen haben, sich durch eine Doppelsumme der Form (12) darstellen lassen muss, und es erübrigt nur, die darin auftretenden willkürlichen Constanten so zu bestimmen, dass den Bedingungen (6) (7) (8) genügt wird.

Diese Aufgabe vereinfacht sich von vornherein beträchtlich durch folgende Bemerkung: Der einzige Umstand, welcher bewirkt, dass die Lichtbewegung in unserem Problem nicht völlig symmetrisch um die x-Axe herum, unabhängig von φ, wird, ist der, dass die Polarisationsebene der einfallenden Lichtwelle in gewisser Weise vor andern durch die x-Axe gehenden Ebenen ausgezeichnet ist. Es ist danach zu vermuten, dass die Abhängigkeit der Lösung von φ immerhin keine komplizierte sein wird, dass also Glieder mit Sinus oder Cosinus höherer Vielfacher von φ nicht vorkommen werden. In der That, versucht man zunächst für ξ, η, ζ Ausdrücke der Form (12) anzusetzen, so bemerkt man alsbald, dass die grosse Mehrzahl der Constanten G und H verschwinden müssen, und wird zu dem folgenden vereinfachten Ansatz geführt:

$$\xi = \alpha \cos \varphi \qquad \eta = e^{ikx} + \beta + \gamma \cos 2\varphi \qquad \zeta = \gamma \sin 2\varphi \qquad 13)$$

wobei α, β und γ drei von φ unabhängige, also nur r und ϑ enthaltende Grössen sind, die in Form folgender Summen dargestellt werden können:

$$\begin{aligned} \alpha &= \sum_m (2m+1) A_m R_m(r) P_{m,1}(\cos \vartheta) \\ \beta &= \sum_m (2m+1) B_m R_m(r) P_{m,0}(\cos \vartheta) \qquad 14) \\ \gamma &= \sum_m (2m+1) C_m R_m(r) P_{m,2}(\cos \vartheta) \end{aligned}$$

---

[1]) Vgl. Pockels (l. c.), pag. 63 und 111.

Die Grössen A_m, B_m, C_m bedeuten hier noch zu bestimmende Constante, denen aus Bequemlichkeitsrücksichten der Faktor $(2m+1)$ beigefügt wurde.

Der Ansatz (13), (14) liefert für ξ, η, ζ an und für sich Wellenpotentiale. Es erübrigt also, die drei Bedingungen (6) (7) (8) zu erfüllen. Die Einführung der Ausdrücke (13) in die Randbedingungen (7) giebt in Rücksicht darauf, dass eine der Bedingungen (7) eine Folge der beiden anderen ist und weggelassen werden kann:

Für $r = a$:

$$e^{ikx} + \beta - \gamma = 0 \qquad \alpha \sin\vartheta - 2\gamma\cos\vartheta = 0. \qquad 15)$$

Die Bedingungen im Unendlichen verlangen — da das Glied (8) im Ausdruck (13) für η schon abgesondert ist — einfach, dass α, β und γ im Unendlichen keine einfallenden Wellen entsprechenden Teile enthalten. Was die Bedingung (6) angeht, welche für den ganzen Raum gilt, so wollen wir zunächst nur fordern, dass sie für Punkte auf der Oberfläche der Kugel a befriedigt werde. Wir werden später zeigen, dass sie dann auch für den ganzen Raum erfüllt ist. Um noch die Grössen α, β, γ und Polarkoordinaten in die auf die Oberfläche der Kugel beschränkte Bedingung (6) einzuführen, stelle man sich ξ, η, ζ für einen Augenblick bildlich als Geschwindigkeitskomponenten einer Flüssigkeit dar. Die Gleichung (6) bedeutet dann bekanntlich, dass in jedes Volumenelement ebensoviel Flüssigkeit ein- als ausströmt. Man betrachte ein Volumenelement von der Form eines Kegelstumpfs, dessen Basis ein kleines Stück dw der Oberfläche der Kugel a, dessen Decke ein Stück dw' der Oberfläche der Kugel $a + da$ sei und dessen Mantel durch lauter Abschnitte von Kugelradien gebildet werde. Da die Bedingungen (7) im Bilde besagen, dass längs der Oberfläche der Kugel a keine tangentiale Strömung stattfindet, so kann durch die Mantelfläche keine Flüssigkeitsmasse aus- oder einströmen (oder strenger: diese Flüssigkeitsmasse wird von höherer Ordnung klein). Es müssen sich also nach (6) die Strömungen durch Basis und Decke ausgleichen. Die radiale Strömungskomponente ist nun:

$$\begin{aligned} R &= \xi \cos\vartheta + \eta \sin\vartheta \cos\varphi + \zeta \sin\vartheta \sin\varphi \\ &= \cos\varphi \{a \cos\vartheta + [e^{ikx} + \beta + \gamma] \sin\vartheta\}. \end{aligned} \quad 16)$$

Durch die Basis strömt daher die Masse $dw \cdot R$ ein. Durch die Decke $dw' = dw\left(\frac{a+da}{a}\right)^2$ strömt die Masse $dw' \cdot \left(R + da \cdot \frac{\delta R}{\delta r}\right)$ aus. Demnach verlangt (6), dass:

$$R\,dw = \left(R + da \cdot \frac{\delta R}{\delta r}\right) dw'$$

oder nach einfacher Reduktion:

$$0 = \frac{2}{a} \cdot R + \frac{\delta R}{\delta r}$$

ist. Führt man hierin (16) ein, so erhält man als neue Oberflächenbedingung, welche als Ersatz der Bedingung (6) eintritt:

Für $r = a$:

$$0 = \cos\vartheta\left(r\frac{\delta a}{\delta r} + 2a\right) + \sin\vartheta\left[r\frac{\delta}{\delta r}(\beta + \gamma + e^{ikx}) + 2(\beta + \gamma + e^{ikx})\right]. \quad 17)$$

Nunmehr ist unsere Aufgabe darauf reduziert, die Grössen A_m, B_m, C_m und die noch unbestimmten Integrationskonstanten der Lösungen R_m der Differentialgleichungen (11) so zu bestimmen, dass die Summen (14) den drei Randbedingungen (15) und (17) genügen und im Unendlichen keine Teile enthalten, welche einfallenden Wellen entsprechen.

Zur Vorbereitung der Lösung stelle ich einige Sätze aus der Theorie der Kugel- und Cylinderfunktionen zusammen, welche zum Teil bekannt sind, zum Teil ohne grosse Mühe aus bekannten Sätzen folgen.

Sitz. Ber. math. phys. Classe d. Kgl. bayer. Akad. d. Wiss. *31* (1901) 293–338

§ 3. Aus der Theorie der Kugelfunctionen.

Die gewöhnliche Kugelfunction m. Ordnung wurde durch P_m oder $P_{m,0}$ bezeichnet. Die n. Zugeordnete der Kugelfunction m. Ordnung lässt sich dann definieren durch den Ausdruck:

$$\begin{aligned} P_{m,n}(\cos\vartheta) = P_{m,n}(x) &= (1-x^2)^{n/2}\frac{d^n P_m}{d x^n} \\ &= \frac{1}{2^m \cdot m!}(1-x^2)^{n/2}\frac{d^{n+m}}{d x^{n+m}}(x^2-1)^m. \end{aligned} \tag{18}$$

Zwischen den verschiedenen Zugeordneten verschiedener Ordnung bestehen die folgenden drei Rekurrenzen (vgl. F. Neumann, Beiträge zur Theorie der Kugelfunctionen, Leipzig 1878, pag. 74):

$$(2m+1)\cos\vartheta P_{m,n} = (m+1-n)P_{m+1,n} + (m+n)P_{m-1,n} \tag{19}$$

$$(2m+1)\sin\vartheta P_{m,n-1} = P_{m+1,n} - P_{m-1,n} \tag{20}$$

$$(2m+1)\sin\vartheta P_{m,n+1} = (m+1-n)(n-m)P_{m+1,n} + (m+1+n)(n+m)P_{m-1,n}. \tag{21}$$

Man findet mit Hülfe der eben angegebenen Beziehungen leicht die nachstehende später zu verwendende Relation:

$$(2m+1)\sin\vartheta\frac{d P_{m,1}}{d\vartheta} = m^2 P_{m+1,1} - (m+1)^2 P_{m-1,1}. \tag{22}$$

Ferner lassen sich die folgenden Sätze mit Hülfe von (18)—(22) aus den bekannten Integraleigenschaften der Kugelfunctionen ableiten:

$$\left.\begin{aligned} &\int_0^\pi d\vartheta \sin\vartheta\cos\vartheta P_{m,1}P_{m',1} = 0 \\ &\qquad \text{ausser wenn } m' = m+1 \text{ oder } m' = m-1. \\ &\int_0^\pi d\vartheta\sin\vartheta\cos\vartheta P_{m,1}P_{m+1,1} = 2\cdot\frac{m(m+1)(m+2)}{(2m+1)(2m+3)} \end{aligned}\right\} \tag{23}$$

$$\left.\begin{aligned} &\int_0^\pi d\vartheta \sin\vartheta\cos\vartheta P_{m,0}P_{m',0} = 0 \\ &\qquad \text{ausser wenn } m' = m+1 \text{ oder } m-1. \\ &\int_0^\pi d\vartheta\sin\vartheta\cos\vartheta P_{m,0}P_{m+1,0} = 2\cdot\frac{(m+1)}{(2m+1)(2m+3)} \end{aligned}\right\} \tag{24}$$

§ 4. Aus der Theorie der Cylinderfunctionen.

Führt man in der Differentialgleichung (11) für r die Variable $\varrho = r\,k$ ein, so schreibt sie sich:

$$\frac{d^2 R_m}{d\varrho^2} + \frac{2}{\varrho}\frac{d R_m}{d\varrho} + R_m\left[1 - \frac{m(m+1)}{\varrho^2}\right] = 0. \qquad 25)$$

Versucht man diese Differentialgleichung durch einen Ausdruck der Form $e^{-i\varrho}\cdot S$ zu befriedigen, wobei S eine nach negativen Potenzen von ϱ fortschreitende Reihe sein soll, so gelingt dies durch die folgende mit einer endlichen Anzahl von Gliedern abbrechende Reihe:

$$K_m = \frac{e^{-i\varrho}}{\varrho}\left[1 + \frac{m(m+1)}{2}\frac{1}{i\varrho} + \frac{m(m+1)}{2}\frac{[m(m+1)-1\cdot 2]}{4}\frac{1}{(i\varrho)^2}\right.$$
$$\left. + \frac{m(m+1)}{2}\cdot\frac{[m(m+1)-1\cdot 2]}{4}\cdot\frac{[m(m+1)-2\cdot 3]}{6}\cdot\frac{1}{(i\varrho)^3} + \ldots\right] \qquad 26)$$

Eine zweite Lösung der Differentialgleichung, die wir K'_m nennen wollen, erhält man natürlich, indem man in diesem Ausdruck $+i$ mit $-i$ vertauscht. Die allgemeine Lösung würde sich dann linear aus K_m und K'_m zusammensetzen.

Andrerseits kann man Lösungen von (25) in Form aufsteigender Potenzreihen darzustellen suchen und findet dadurch die folgenden beiden Funktionen:

$$\chi_m(\varrho) = \frac{(i\varrho)^m}{1\cdot 3 \ldots (2m+1)}\left\{1 + \frac{(i\varrho)^2}{2(2m+3)} + \frac{(i\varrho)^4}{2\cdot 4(2m+3)(2m+5)} + \ldots\right\}$$
$$\varphi_m(\varrho) = \frac{1\cdot 3 \ldots (2m-1)}{(i\varrho)^{m+1}}\left\{1 - \frac{(i\varrho)^2}{2(2m-1)} + \frac{(i\varrho)^4}{2\cdot 4(2m-1)(2m-3)} - \ldots\right\} \qquad 27)$$

wobei die Coeffizienten des ersten Gliedes in Rücksicht auf die spätere Anwendung gewählt sind.

Auch durch χ_m und φ_m muss sich jede andere Lösung von (25) linear darstellen lassen. Führt man dies speziell für K_m aus, so erhält man:

$$i^{2m+1} K_m(\varrho) = \chi_m(\varrho) + (-1)^{m+1}\varphi_m(\varrho)$$

oder:

$$i^{m+1}K_m(\varrho)=\frac{\varrho^m}{1\cdot 3\ldots 2m+1}\left\{1-\frac{\varrho^2}{2(2m+3)}+\frac{\varrho^4}{2\cdot 4(2m+3)(2m+5)}-\ldots\right\}$$

$$+i\frac{1\cdot 3\ldots(2m-1)}{\varrho^{m+1}}\left\{1+\frac{\varrho^2}{2(2m-1)}+\frac{\varrho^4}{2\cdot 4(2m-1)(2m-3)}-\ldots\right\} \qquad 28)$$

ein Ausdruck, der sich auch durch direkte Entwicklung von K_m nach Potenzen von ϱ gewinnen liesse. Umgekehrt kann man aber auch z. B. χ_m linear durch K_m und K'_m ausdrücken und zwar findet man:

$$2\,i\,\chi_m(\varrho)=K'_m(\varrho)-(-1)^m K_m(\varrho). \qquad 29)$$

Ferner besteht zwischen irgend zwei Lösungen R_m und R'_m von (25) die Beziehung:

$$R_m\frac{d\,R'_m}{d\,\varrho}-R'_m\frac{d\,R_m}{d\,\varrho}=\frac{C}{\varrho^2},$$

wobei C eine Constante ist. Speziell für K_m und χ_m nimmt die Constante den Wert 1 an, sodass also gilt:

$$K_m\frac{d\,\chi_m}{d\,\varrho}-\chi_m\frac{d\,K_m}{d\,\varrho}=\frac{1}{\varrho^2}. \qquad 30)$$

Die verschiedenen K_m stehen durch folgende Rekurrenzen mit einander in Verbindung:

$$(2\,m+1)\,K_m=i\,\varrho\,(K_{m+1}-K_{m-1}) \qquad 31)$$

$$\varrho\frac{d\,K_m}{d\,\varrho}=m\,K_m-i\,\varrho\,K_{m+1}=-(m+1)\,K_m-i\,\varrho\,K_{m-1}$$

$$=-i\,\varrho\left[\frac{(m+1)\,K_{m+1}+m\,K_{m-1}}{2\,m+1}\right]. \qquad 32)$$

Schliesslich gilt nach Heine, Hdb. d. Kugelfunctionen, 2. Aufl., I, pag. 82 die Entwicklung:

$$e^{i\varrho\cos\vartheta}=\sum_{m=0}^{\infty}(2\,m+1)\,\chi_m(\varrho)\,P_{m,0}(\cos\vartheta). \qquad 33)$$

§ 5. Explicite Darstellung der Componenten der electrischen Kraft.

Wir beginnen damit, die Entwicklungen (14) für α, β, γ zuvörderst so zu spezialisieren, dass dieselben im Unendlichen keine Teile enthalten, welche einfallenden Wellenzügen entsprechen. Wir haben im vorigen Paragraphen gesehen, dass R_m sich linear aus K_m und K'_m zusammensetzt, sodass man etwa schreiben kann:

$$R_m = c_1 K_m + c_2 K'_m = c_1 e^{-i\varrho} S + c_2 e^{+i\varrho} S'. \quad (c_1, c_2 \text{ Konstante.})$$

S und S' sind beide nach absteigenden Potenzen von ϱ fortschreitende Reihenentwicklungen, welche sich in sehr grosser Entfernung vom Nullpunkt, für grosses ϱ auf ihr erstes Glied $\frac{1}{\varrho}$ reduzieren. Demnach wird z. B. ein Glied der Entwicklung (14) von α in grosser Entfernung vom Nullpunkt lauten:

$$(2m+1) A_m \left(c_1 \frac{e^{-i\varrho}}{\varrho} + c_2 \frac{e^{+i\varrho}}{\varrho} \right) P_{m,1}(\cos\vartheta).$$

Um den daraus entspringenden Beitrag zur x-Componente der elektrischen Kraft zu finden, hat man nach (1) mit e^{iqt} zu multiplizieren und vom Resultat den reellen Teil zu nehmen. Setzt man:

$$(2m+1) \mathrm{A}_m c_1 = d_1 e^{i\delta_1},\ (2m+1) A_m c_2 = d_2 e^{i\delta_2},\ d_1, d_2, \delta_1, \delta_2 \text{ reell},$$

so wird dieser reelle Teil gleich:

$$\left[\frac{d_1}{\varrho} \cos(\delta_1 + qt - \varrho) + \frac{d_2}{\varrho} \cos(\delta_2 + qt + \varrho) \right] P_{m,1}(\cos\vartheta)$$

oder:

$$\left[\frac{d_1}{kr} \cos\left(\delta_1 + \frac{2\pi t}{\tau} - \frac{2\pi r}{\lambda}\right) + \frac{d_2}{kr} \cos\left(\delta_2 + \frac{2\pi t}{\tau} + \frac{2\pi r}{\lambda}\right) \right] P_{m,1}(\cos\vartheta).$$

Man sieht sofort, dass das erste Glied eine zu grösserem r fortschreitende, das zweite hingegen eine aus dem Unendlichen einlaufende Welle darstellt. Laut der Bedingung für das Un-

Sitz. Ber. math. phys. Classe d. Kgl. bayer. Akad. d. Wiss. *31* (1901) 293–338

endliche muss also das zweite Glied verschwinden, es muss d_2 und c_2 gleich null sein und R_m sich auf die einfache Form: $R_m = c_1 K_m$ reduzieren. Die Constante c_1 wollen wir noch so bestimmen, dass R_m auf der Kugeloberfläche $r = a$ gleich 1 wird. ·Ist zur Abkürzung:

$$k r = \varrho \qquad k a = \varrho_0 \tag{34}$$

so soll also sein:

$$R_m(r) = \frac{K_m(\varrho)}{K_m(\varrho_0)}. \tag{35}$$

Hiermit ist jede Willkürlichkeit in den R_m beseitigt und die Bedingungen im Unendlichen sind erfüllt.

Nachträgliche Bemerkung zu § 2. Der Ausdruck

$$\frac{\delta \xi}{\delta x} + \frac{\delta \eta}{\delta y} + \frac{\delta \zeta}{\delta z} = \sigma$$

ist ein Wellenpotential, welches im Unendlichen keine einfallenden Wellenzüge enthält — sobald ξ, η, ζ Wellenpotentiale sind und unseren Bedingungen für das Unendliche genügen. In Rücksicht auf die eben durchgeführte Betrachtung der Funktionen R_m wird also σ nach (12) und (35) in folgender Weise entwickelbar sein:

$$\sigma = \sum_m \sum_n \frac{K_m(\varrho)}{K_m(\varrho_0)} P_{m,n}(\cos \vartheta) \left[G_{m,n} \cos n\varphi + H_{m,n} \sin n\varphi\right].$$

Für $x = a$ geht diese Entwicklung über in:

$$\sigma = \sum_m \sum_n P_{m,n}(\cos \vartheta) \left[G_{m,n} \cos n\varphi + H_{m,n} \sin n\varphi\right].$$

Fordert man jetzt, dass σ für $r = a$ verschwinde, so folgt, dass alle Coeffizienten G und H null sein müssen, und damit, dass σ überall identisch verschwindet. Hiermit ist der oben benutzte Satz erwiesen, dass es genügt, die Bedingung $\sigma = 0$ auf der Oberfläche der Kugel zu erfüllen, um das Verschwinden von σ im ganzen Raum herbeizuführen.

Wir suchen weiter durch geeignete Wahl der Constanten A_m, B_m, C_m in (14) die Randbedingungen (15) und (17) zu erfüllen. Setzt man die Entwicklungen (14) in diese Randbedingungen ein, berücksichtigt (35) und ersetzt $e^{ikx} = e^{i\varrho\cos\vartheta}$ durch die Reihe (33), so erhält man die drei Gleichungen:

$$\sum_m (2m+1)\left[P_{m,0}(B_m + \chi_m(\varrho_0)) - P_{m,2}C_m\right] = 0 \qquad 36)$$

$$\sum_m (2m+1)\left[A_m P_{m,1}\sin\vartheta - 2C_m P_{m,2}\cos\vartheta\right] = 0 \qquad 37)$$

$$\sum_m (2m+1)\left[A_m S_m P_{m,1}\cos\vartheta + \sin\vartheta\{B_m S_m P_{m,0} + C_m S_m P_{m,2} + T_m P_{m,0}\}\right] = 0. \qquad 38)$$

In der letzten dieser Gleichungen habe ich noch folgende Abkürzungen eingeführt:

$$S_m = \left[r\frac{dR_m}{dr} + 2R_m\right]_{r=a} = 2 + \left[\frac{\varrho\frac{dK_m}{d\varrho}}{K_m}\right]_{\varrho=\varrho_0} \qquad 39)$$

$$T_m = \left[\varrho\frac{d\chi_m}{d\varrho} + 2\chi_m(\varrho)\right]_{\varrho=\varrho_0}. \qquad 40)$$

Man multipliziere (36) mit $\sin\vartheta$. Dann kann man nach (20) und (21) sowohl $\sin\vartheta\, P_{m,0}$, als $\sin\vartheta\, P_{m,2}$ mit Hülfe der ersten Zugeordneten $P_{m,1}$ ausdrücken, sodass die ganze Summe dann nur noch erste Zugeordnete enthält, und zwar findet man:

$$\sum_m (B_m + \chi_m(\varrho_0))(P_{m+1,1} - P_{m-1,1}) - C_m\left[-(m-1)m P_{m+1,1} + (m+1)(m+2)P_{m-1,1}\right] = 0.$$

Ordnet man hier in der Weise, dass man Glieder, welche Grössen $P_{m,1}$ mit gleichem Index m enthalten, zusammenfasst, so folgt:

$$\sum_m P_{m,1}\left[(m+2)(m+3)C_{m+1} + B_{m+1} + \chi_{m+1} - (m-1)(m-2)C_{m-1} - B_{m-1} - \chi_{m-1}\right] = 0. \qquad 36')$$

Aehnlich kann man in der Summe (37) mit Hülfe der Rekurrenzen (19) und (20) sowohl $\sin\vartheta\, P_{m,1}$ als $\cos\vartheta\, P_{m,2}$ allein durch zweite Zugeordnete $P_{m,2}$ ausdrücken und nach diesen ordnen. Man erhält dann:

$$\sum_m P_{m,2}[A_{m-1}-2(m-2)C_{m-1}-A_{m+1}-2(m+3)C_{m+1}]=0. \quad 37')$$

Schliesslich lassen sich in (38) die drei Produkte $\cos\vartheta\, P_{m,1}$, $\sin\vartheta\, P_{m,0}$ und $\sin\vartheta\, P_{m,2}$ alle nach (19)—(21) durch erste Zugeordnete $P_{m,1}$ ausdrücken. Wiederum nach letzteren ordnend erhält man:

$$\sum_m P_{m,1}\begin{bmatrix} S_{m-1}\{(m-1)A_{m-1}+B_{m-1}-(m-1)(m-2)C_{m-1}\}+T_{m-1} \\ +S_{m+1}\{(m+2)A_{m+1}-B_{m+1}+(m+2)(m+3)C_{m+1}\}-T_{m+1}\end{bmatrix}=0. \quad 38')$$

Es ist zu beachten, dass, wie aus (18) sofort ersichtlich, die drei Grössen $P_{0,1}$, $P_{0,2}$ und $P_{1,2}$ gleich null sind. Daher beginnen die Summen (36′) und (38′) mit $m=1$, die Summe (37′) mit $m=2$.

Wir haben hier nun drei Entwicklungen nach zugeordneten Kugelfunktionen, die zur Summe die Null haben sollen. Dies ist aber nach einem bekannten Satze nicht anders möglich, als indem jeder einzelne Entwicklungkoeffizient verschwindet, und damit ergeben sich die folgenden Gleichungen zur Bestimmung der A_m, B_m, C_m:

$$(m+2)(m+3)\,C_{m+1}+B_{m+1}+\chi_{m+1}$$
$$=(m-1)(m-2)\,C_{m-1}+B_{m-1}+\chi_{m-1} \quad m\geqq 1 \qquad 41)$$

$$A_{m+1}+2\,(m+3)\,C_{m+1}=A_{m-1}-2\,(m-2)\,C_{m-1} \quad m\geqq 2 \qquad 42)$$

$$S_{m+1}\,[(m+2)\,A_{m+1}-B_{m+1}+(m+2)\,(m+3)\,C_{m+1}]-T_{m+1}$$
$$+\,S_{m-1}\,[(m-1)\,A_{m-1}+B_{m-1}-(m-1)\,(m-2)\,C_{m-1}]$$
$$+\,T_{m-1}=0 \quad m\geqq 1. \qquad 43)$$

Zur Auflösung dieses unendlichen Gleichungssystems erweist es sich als nützlich, die folgenden Grössen als Unbekannte einzuführen:

$$A_m+2\,(m+2)\,C_m=p_m$$
$$B_m+(m+1)\,(m+2)\,C_m=q_m \qquad 44)$$
$$2\,(2\,m+1)\,C_m=r_m$$

Es ist hier voraus zu bemerken, dass man, da $P_{0,1}$, $P_{0,2}$ und $P_{1,2}$ verschwindet, in den Entwicklungen (14) von vornherein auch A_0, C_0 und C_1 gleich null setzen darf, woraus dann folgt:

$$p_0 = r_0 = r_1 = 0 \tag{45}$$

Durch Einführung dieser neuen Unbekannten verwandeln sich die Gleichungen (41) bis (43) in die nachstehenden:

$$q_{m+1} + \chi_{m+1} = q_{m-1} + \chi_{m-1} - r_{m-1} \quad m \geqq 1 \tag{46}$$

$$p_{m+1} = p_{m-1} - r_{m-1} \quad m \geqq 2 \tag{47}$$

$$S_{m+1}[(m+2)\,p_{m+1} - q_{m+1}] + S_{m-1}[(m-1)\,p_{m-1} + q_{m-1} - m\,r_{m-1}] + T_{m-1} - T_{m+1} = 0 \quad m \geqq 1. \tag{48}$$

Subtrahiert man hier zunächst (47) von (46), so folgt:

$$q_{m+1} + \chi_{m+1} - p_{m+1} = q_{m-1} + \chi_{m-1} - p_{m-1} \quad m \geqq 2$$

und daraus ergiebt sich, dass für

$$\left.\begin{array}{ll} \text{gerades } m: & q_m + \chi_m = p_m + g_2 \quad m \geqq 2 \\ \text{ungerades } m: & q_m + \chi_m = p_m + g_1 \quad m \geqq 1 \end{array}\right\} \tag{49}$$

sein muss, wobei g_1 und g_2 zwei vom Index m unabhängige Constante sind.

Schreiben wir (47) noch in der Form:

$$r_m = p_m - p_{m+2} \quad m \geqq 1, \tag{50}$$

so sehen wir, dass die Bestimmung der q_m und r_m auf die der p_m zurückgeführt ist. Ersetzt man jetzt in dem Gleichungssystem (48) die q_m und r_m durch die Ausdrücke (49) und (50), so erhält man nicht etwa Rekurrenzen zwischen den p_m, vielmehr fällt merkwürdiger Weise aus jeder Gleichung p_{m-1} heraus und es ergiebt sich zur unmittelbaren Bestimmung von p_{m+1}:

$$p_{m+1}[(m+1)\,S_{m+1} + m\,S_{m-1}] + S_{m+1}(\chi_{m+1} - g) - S_{m-1}(\chi_{m-1} - g) + T_{m-1} - T_{m+1} = 0 \quad m \geqq 2, \tag{51}$$

wobei g gleich g_1 oder gleich g_2 zu setzen ist, jenachdem m gerade oder ungerade ist.

Die erste für $m = 1$ entstehende Gleichung des Systems (48) konnte hier noch nicht ausgenutzt werden und man darf in dem vorstehenden Systeme erst mit $m = 2$ beginnen, weil q_0 nicht nach Art der Formel (49) auf p_0 zurückgeführt werden kann.

Sitz. Ber. math. phys. Classe d. Kgl. bayer. Akad. d. Wiss. *31* (1901) 293–338

Ueberblickt man das bisher erlangte Resultat, so lassen sich nach (51) alle p von p_3 an berechnen. Nach (45) verschwindet p_0. Ferner folgt aus $r_1 = 0$ und der ersten (für $m = 1$ entstehenden) Gleichung des Systems (50):

$$p_1 = p_3. \tag{52}$$

Von allen Grössen p bleibt daher allein noch p_2 unbekannt. Sowie man auch p_2 hätte, erhielte man aus (45) und (50) alle r und aus (49) alle q — abgesehen von q_0. So bleiben einerseits im Grunde nur die beiden Unbekannten q_0 und p_2, andererseits enthält das System (46) bis (48) aber auch noch zwei unbenutzte Gleichungen. Bei der Subtraktion der Gleichungen (47) von (46) konnte die erste (für $m = 1$ entstehende) Gleichung (46) nicht mitbenutzt werden, weil die entsprechende Gleichung (47) fehlt, und ähnlich lag es mit der ersten Gleichung (48). Die beiden restierenden Gleichungen lauten daher:

Aus (46) für $m = 1$:

$$q_2 + \chi_2 = q_0 + \chi_0 - r_0.$$

Aus (48) für $m = 1$:

$$S_2\,[3\,p_2 - q_2] + S_0\,q_0 + T_0 - T_2 = 0.$$

In Rücksicht auf (45) und (49) erhält man aus diesen beiden Gleichungen:

$$q_0 + \chi_0 = p_2 + g_2 \tag{53}$$

$$p_2\,[2\,S_2 + S_0] + S_2\,(\chi_2 - g_2) - S_0\,(\chi_0 - g_2) + T_0 - T_2 = 0. \tag{54}$$

Damit sind denn alle Gleichungen befriedigt und alle Unbekannten gefunden — bis auf eine eigentümliche durchgehende Unbestimmtheit, die in der Willkürlichkeit der beiden Constanten g_1 und g_2 liegt. Es ist schwer zu ersehen, woher die Fixierung dieser beiden Constanten noch kommen soll, nachdem sich bei ganz beliebiger Wahl derselben alle Randbedingungen im Endlichen und Unendlichen haben erfüllen lassen, und doch ist physikalisch evident, dass unser Problem nur eine Lösung besitzen kann.

Der Schlüssel zur Ueberwindung der Schwierigkeit und zur Bestimmung von g_1 und g_2 liegt in Folgendem: Man nehme einmal an, dass keine einfallende Welle vorhanden sei. Dann verschwinden in unseren Gleichungen alle Grössen χ_m und T_m, welche aus der Entwicklung des die einfallende Welle darstellenden Ausdrucks e^{ikx} entsprangen. Es bleiben daher in den Gleichungen (51) die Glieder übrig:

$$p_{m+1}\left[(m+1)\,S_{m+1}+m\,S_{m-1}\right]-g\,(S_{m+1}-S_{m-1})=0. \qquad 55)$$

Es soll hieraus p_{m+1} berechnet werden für den Fall, dass der Kugelradius a gegen die Wellenlänge sehr klein ist. Für sehr kleines ϱ wird K_m nach (28) näherungsweise dargestellt durch:

$$i^m\,K_m(\varrho)=\frac{1\cdot 3\,\ldots\,.(2m-1)}{\varrho^{m+1}} \qquad 56)$$

und daraus folgt nach (34), (35) und (39):

$$R_m(r)=\left(\frac{\varrho_0}{\varrho}\right)^{m+1}=\left(\frac{a}{r}\right)^{m+1} \quad S_m=1-m. \qquad 57)$$

Dies giebt in (55) eingesetzt:

$$p_{m+1}=\frac{g}{m\,(2m-1)}. \qquad 58)$$

Nunmehr kann man nach (50) r_m, nach den Gleichungen:

$$q_m=p_m+g$$

die aus (49) durch Weglassung der χ-Glieder entstehen, die q_m und, nachdem p_m, q_m und r_m bekannt sind, aus (44) die A_m, B_m, C_m berechnen. Es genügt, das Resultat für B_m anzuführen:

$$B_m=g\left[1-\frac{3}{(2m-3)\,(2m+1)^2}\right].$$

Damit erhält man nach (14) folgende Entwicklung von β:

$$\beta=\sum_m g\left[2m+1-\frac{3}{(2m-3)\,(2m+1)}\right]R_m(r)\,P_{m,0}(\cos\vartheta).$$

Sitz. Ber. math. phys. Classe d. Kgl. bayer. Akad. d. Wiss. 31 (1901) 293–338

Für $r = a$ und $\vartheta = 0$ wird $R_m = 1$ und $P_{m,0} = 1$ und damit wird β:

$$\beta = \sum_m g \left[2m + 1 - \frac{3}{(2m-3)(2m+1)}\right]$$

also unendlich. Mit anderen Worten: Aus den Grössen g entspringen Ausdrücke für die Componenten der elektrischen Kraft, welche mit Unstetigkeiten behaftet sind. Die Bedingung, welche zur Fixierung der Grössen g führt, ist daher die, dass die Componenten der elektrischen Kraft stetig bleiben sollen, und zwar verlangt sie, dass g_1 und g_2 null sind.

Lässt man jetzt alle Glieder g aus den Gleichungen (49) bis (54) weg, so kann man sie so zusammenfassen:

$$p_{m+1}\left[(m+1) S_{m+1} + m S_{m-1}\right] + (S_{m+1}\chi_{m+1} - T_{m+1}) - (S_{m-1}\chi_{m-1} - T_{m-1}) = 0 \quad m \geqq 1 \qquad 59)$$

$$q_m = p_m - \chi_m \quad m \geqq 1 \qquad r_m = p_m - p_{m+2} \quad m \geqq 2 \qquad 60)$$

$$p_0 = 0 \quad p_1 = p_3 \quad q_0 = p_2 - \chi_0 \quad r_0 = r_1 = 0. \qquad 61)$$

Die Gleichungen (59) gestatten noch eine beträchtliche Vereinfachung.

Nach der Definition (39) war:

$$S_m = \left[2 + \frac{\varrho \frac{d K_m}{d\varrho}}{K_m}\right]_{\varrho=\varrho_0} \quad T_m = \left[2\chi_m(\varrho) + \varrho \frac{d\chi_m}{d\varrho}\right]_{\varrho=\varrho_0}.$$

Daher ist:

$$S_m \chi_m - T_m = \left[\frac{\varrho}{K_m}\left(\chi_m \frac{d K_m}{d\varrho} - K_m \frac{d\chi_m}{d\varrho}\right)\right]_{\varrho=\varrho_0}$$

oder nach (30):

$$S_m \chi_m - T_m = -\left[\frac{1}{\varrho K_m}\right]_{\varrho=\varrho_0}. \qquad 62)$$

Ferner wird:

$$(m+1) S_{m+1} + m S_{m-1} = 2(2m+1) + \left[(m+1)\frac{\varrho \frac{d K_{m+1}}{d\varrho}}{K_{m+1}} + m \varrho \frac{\frac{d K_{m-1}}{d\varrho}}{K_{m-1}}\right]_{\varrho=\varrho_0}.$$

Nach (31) und (32) bestehen die Rekurrenzen:

$$\varrho \frac{d K_{m+1}}{d \varrho} = -(m+2) K_{m+1} - i \varrho K_m$$

$$\varrho \frac{d K_{m-1}}{d \varrho} = (m-1) K_{m-1} - i \varrho K_m$$

$$(2m+1) \varrho \frac{d K_m}{d \varrho} = -i \varrho [(m+1) K_{m+1} + m K_{m-1}]$$

$$(2m+1) \quad K_m = i \varrho [K_{m+1} - K_{m-1}]. \qquad 63)$$

Multipliziert man die erste dieser Gleichungen mit $\frac{m+1}{K_{m+1}}$, die zweite mit $\frac{m}{K_{m-1}}$, die dritte und vierte je mit $-\frac{K_m}{K_{m-1} K_{m+1}}$ und addiert, so erhält man:

$$\frac{(m+1)}{K_{m+1}} \varrho \frac{d K_{m+1}}{d \varrho} + \frac{m}{K_{m-1}} \varrho \frac{d K_{m-1}}{d \varrho}$$

$$-(2m+1) \frac{K_m}{K_{m-1} K_{m+1}} \left(K_m + \varrho \frac{d K_m}{d \varrho} \right) = -2(2m+1)$$

und damit:

$$(m+1) S_{m+1} + m S_{m-1} = (2m+1) \frac{K_m}{K_{m-1} K_{m+1}} \left(K_m + \varrho \frac{d K_m}{d \varrho} \right). \qquad 64)$$

Führt man jetzt (62) und (64) im Ausdruck von p_{m+1} ein, so folgt:

$$p_{m+1} = \frac{K_{m-1} K_{m+1}}{(2m+1) K_m} \frac{1}{K_m + \varrho \frac{d K_m}{d \varrho}} \left(\frac{1}{\varrho K_{m+1}} - \frac{1}{\varrho K_{m-1}} \right) \text{ für } \varrho = \varrho_0,$$

oder in Rücksicht auf (63):

$$p_{m+1} = -\left[\frac{i}{\varrho^2 \left(K_m + \varrho \frac{d K_m}{d \varrho} \right)} \right]_{\varrho = \varrho_0}. \qquad 65)$$

Es bleibt nur noch übrig, mit Hülfe der p_m nach (60) und (61) die q_m und r_m und dann nach (44) die A_m, B_m, C_m abzuleiten, um nach einer leichten Zwischenrechnung die voll-

Sitz. Ber. math. phys. Classe d. Kgl. bayer. Akad. d. Wiss. *31* (1901) 293–338

ständige Lösung des Beugungsproblems zu erhalten, welche, wenn man alles zusammenstellt, durch folgende Formeln gegeben wird:

Der Kugelradius sei a. Man setze:

$$\varrho = k\,r \quad \varrho_0 = k\,a \quad k = \frac{2\pi}{\lambda}.$$

Man berechne die endlichen Ausdrücke (26):

$$K_m = \frac{e^{-i\varrho}}{\varrho}\left[1 + \frac{m(m+1)}{2}\,\frac{1}{i\varrho} + \frac{m(m+1)}{2}\cdot\frac{m(m+1)-1\cdot 2}{4}\,\frac{1}{(i\varrho)^2} + \dots\right] \qquad 66)$$

und

$$\chi_m = -\frac{K_m(-\varrho_0) + (-1)^m K_m(\varrho_0)}{2i} \qquad 67)$$

und bilde damit:

$$p_{m+1} = -\left[\frac{i}{\varrho^2 \dfrac{d(\varrho K_m)}{d\varrho}}\right]_{\varrho=\varrho_0} \qquad 68)$$

sowie:

$$R_m(r) = \frac{K_m(\varrho)}{K_m(\varrho_0)}. \qquad 69)$$

Dann gelten die Entwicklungen:

$$\alpha = \sum_{m=1}^{\infty} (2m+1)\,A_m\,R_m(r)\,P_{m,1}(\cos\vartheta)$$

$$\beta = \sum_{m=0}^{\infty} (2m+1)\,B_m\,R_m(r)\,P_{m,0}(\cos\vartheta) \qquad 70)$$

$$\gamma = \sum_{m=2}^{\infty} (2m+1)\,C_m\,R_m(r)\,P_{m,2}(\cos\vartheta)$$

mit den folgenden Entwicklungskoeffizienten:

$$(2m+1)A_m = (m-1)p_m + (m+2)p_{m+2}$$

$$2(2m+1)B_m = (m+1)(m+2)p_{m+2} - m(m-1)p_m - 2(2m+1)\chi_m \qquad 71)$$

$$2(2m+1)C_m = p_m - p_{m+2}$$

und die Componenten der elektrischen Kraft für die durch die vollkommen reflektierte Kugel gestörte Lichtbewegung werden erhalten aus:

$$X = \text{pars real}\,(e^{iqt}\,a\cos\varphi)$$
$$Y = \text{pars real}\,(e^{iqt+ikx} + e^{iqt}\,[\beta + \gamma\cos 2\varphi]) \qquad 72)$$
$$Z = \text{pars real}\,(e^{iqt}\,\gamma\sin 2\varphi).$$

§ 6. Bemerkungen über die Verwendung dieser Darstellung.

Die Reihen, durch die wir hier die Verteilung der elektrischen Kraft und damit der Lichtintensität dargestellt haben, sind zwar unter allen Umständen konvergent, indessen wird, wie leicht zu sehen, die Convergenz um so langsamer, je grösser die Kugel ist, welche das Licht reflektiert. Dies Verhalten liegt insofern günstig, als man für grosse Kugeln die Lichtverteilung in rohen Zügen aus der geometrischen Optik, mit grösserer Schärfe aus der Kirchhoff'schen Beugungstheorie ableiten kann. In den Fällen, wo diese beiden Hülfsmittel versagen, nämlich für Kugeln, deren Durchmesser nicht über wenige Wellenlängen hinausgeht, wird aber gerade die numerische Rechnung nach obigen Formeln durchführbar und man könnte sich an ihrer Hand überzeugen, wie sich mit dem Kleinerwerden der Kugel z. B. der Schatten hinter ihr allmählich auflöst. Ich will mich hier begnügen nur die Grenzwerte anzuführen, in die ξ, η, ζ übergehen, wenn der Kugelradius äusserst klein auch gegen die Wellenlänge wird:

$$\xi = -k^3 a^3 \sin\vartheta\cos\varphi\left[\tfrac{1}{2}K_1(kr) + \cos\vartheta\, K_2(kr) - \frac{k^2a^2}{6}K_3(kr)(\cos^2\vartheta - \tfrac{1}{5})\right]$$

$$\eta = e^{ikx} + k^3a^3\left[\tfrac{2}{3}K_0(kr) + \tfrac{1}{2}K_1\cos\vartheta + K_2(\tfrac{1}{3} - \sin^2\vartheta\cos^2\varphi) + K_3\frac{k^2a^2}{6}\cos\vartheta\,(\sin^2\vartheta\cos^2\varphi - \tfrac{1}{5})\right] \qquad 73)$$

$$\zeta = -k^3a^3\sin^2\vartheta\sin\varphi\cos\varphi\left[K_2(kr) - \frac{k^2a^2}{6}K_3(kr)\cos\vartheta\right]$$

Sitz. Ber. math. phys. Classe d. Kgl. bayer. Akad. d. Wiss. 31 (1901) 293–338

wobei:

$$K_0 = \frac{e^{-ikr}}{kr} \quad K_1 = \frac{e^{-ikr}}{kr}\left(1 + \frac{1}{ikr}\right)$$

$$K_2 = \frac{e^{-ikr}}{kr}\left(1 + \frac{3}{ikr} + \frac{3}{(ikr)^2}\right) \tag{74}$$

$$K_3 = \frac{e^{-ikr}}{kr}\left(1 + \frac{6}{ikr} + \frac{15}{(ikr)^2} + \frac{15}{(ikr)^3}\right).$$

In grosser Distanz r von der Kugel gehen diese Formeln über in:

$$\xi = -\frac{k^2 a^3}{2} \sin\vartheta \cos\varphi \, (1 + 2\cos\vartheta) \frac{e^{-ikr}}{r}$$

$$\eta = e^{ikx} + k^2 a^3 \frac{e^{-ikr}}{r}\left[1 + \frac{\cos\vartheta}{2} - \sin^2\vartheta \cos^2\varphi\right] \tag{75}$$

$$\zeta = -k^2 a^3 \frac{e^{-ikr}}{r} \sin^2\vartheta \sin\varphi \cos\varphi .$$

Lord Rayleigh hat gefunden (l. c.), dass bei sehr kleinen Kugeln aus durchsichtigen Medien die Intensität des zerstreuten Lichts umgekehrt proportional der vierten Potenz der Wellenlänge wird und dass in Richtungen, die einen rechten Winkel mit der Richtung der einfallenden Welle machen, vollständige Polarisation des zerstreuten Lichts eintritt. Ersteres Verhalten gilt nach den Gleichungen (75) (da k umgekehrt proportional zur Wellenlänge ist) offenbar auch für das von einer kleinen vollkommen reflektierenden Kugel zerstreute Licht, hingegen erfolgt in keiner Richtung vollständige Polarisation. Man erhält für $\vartheta = 90^0$:

$$\xi = -\frac{k^2 a^3}{2} \frac{e^{-ikr}}{r} \cos\varphi \quad \eta - e^{ikx} = k^2 a^3 \frac{e^{-ikr}}{r} \sin^2\varphi$$

$$\zeta = -k^2 a^3 \frac{e^{-ikr}}{r} \sin\varphi \cos\varphi$$

und daraus für die Intensität der x-Componente:

$$\text{Mod}\, \xi^2 = \tfrac{1}{4} \frac{k^4 a^6}{r^2} \cos^2\varphi$$

und für die Intensität der senkrecht zur x-Axe stehenden Componente des zerstreuten Lichts:

$$\text{Mod}\,[(\eta - e^{ikx})^2 + \zeta^2] = \frac{k^4\, a^6}{r^2} \sin^2 \varphi\,.$$

Ist das einfallende Licht nicht, wie bisher vorausgesetzt wurde, polarisiert, sondern natürliches Licht, so muss man in diesen Ausdrücken alle möglichen Werte von φ einsetzen und das Mittel nehmen und findet dann:

für die Schwingungskomponente parallel zur x-Axe: $\frac{1}{4}\frac{k^4\, a^6}{r^2}$

„ „ „ senkrecht „ „ $\frac{k^4\, a^6}{r^2}$

d. h. „das von sehr kleinen vollkommen reflektierenden Kugeln in Richtungen senkrecht zur Normale der einfallenden Welle zerstreute Licht ist zu drei Vierteln polarisiert."

Schliesslich sei noch ein Umstand hervorgehoben. Da die Intensität des zerstreuten Lichts von der Wellenlänge abhängt, so wird, wenn das einfallende Licht weiss ist, das zerstreute eine andere Farbe zeigen. Nur bei sehr grossen Kugeln kommt die Abhängigkeit von der Wellenlänge nicht in Betracht und das reflektierte Licht bleibt farblos. Bei sehr kleinen Kugeln entsteht, wie Lord Rayleigh gezeigt hat, das tiefe Blau des Himmels. Bei Kugeln von der Grössenordnung der Wellenlänge bilden sich noch andere Farbennüancen aus, die sogar ein wenig von der Richtung des reflektierten Lichts abhängen und die sich alle aus den obigen Formeln ableiten liessen.

Nach diesen Andeutungen über die anderweitige Verwendbarkeit der Lösung des Beugungsproblems gehen wir an unsere eigentliche Aufgabe, die Berechnung des Maxwell'schen Drucks, welchen das Licht auf die Kugel ausübt. Hierzu ist zunächst erforderlich die

Sitz. Ber. math. phys. Classe d. Kgl. bayer. Akad. d. Wiss. *31* (1901) 293–338

§ 7. Ableitung der magnetischen Kraftcomponenten aus den electrischen.

Man wähle irgend einen Punkt 0 der Kugeloberfläche zum Nullpunkt eines neuen rechtwinklichen Coordinatensystemes x_1, y_1, z_1, dessen x-Axe in den durch 0 gehenden Radiusvektor falle, dessen y-Axe den durch 0 gehenden Meridian $\varphi =$ const. und dessen z-Axe den durch 0 gehenden Parallelkreis $\vartheta =$ const. tangiere. Es werde y_1 zu wachsendem ϑ, z_1 zu wachsendem φ positiv gezählt. Die Kraftcomponenten (oder genauer die komplexen Grössen, aus denen sich nach (1) die Kraftcomponenten ableiten) in Richtung der neuen Axen mögen heissen $\xi_1, \eta_1, \zeta_1, \lambda_1, \mu_1, \nu_1$. Für ξ_1, η_1, ζ_1 erhält man durch Coordinatendrehung die Ausdrücke:

$$\begin{aligned} \xi_1 &= \xi \cos\vartheta_0 + \eta \sin\vartheta_0 \cos\varphi_0 + \zeta \sin\vartheta_0 \sin\varphi_0 \\ \eta_1 &= -\xi \sin\vartheta_0 + (\eta \cos\varphi_0 + \zeta \sin\varphi_0) \cos\vartheta_0 \\ \zeta_1 &= -\eta \sin\varphi_0 + \zeta \cos\varphi_0. \end{aligned} \tag{76}$$

Die magnetischen Kraftcomponenten in Richtung der neuen Axen sind aus den Gleichungen, welche den (2) entsprechen, abzuleiten:

$$i k \lambda_1 = \frac{\delta \zeta_1}{\delta y_1} - \frac{\delta \eta_1}{\delta z_1} \quad i k \mu_1 = \frac{\delta \xi_1}{\delta z_1} - \frac{\delta \zeta_1}{\delta x_1} \quad i k \nu_1 = \frac{\delta \eta_1}{\delta x_1} - \frac{\delta \xi_1}{\delta y_1}. \tag{77}$$

Wir wollen diese Gleichungen verwenden, um speziell die magnetische Kraft im Punkte 0 zu bestimmen. Im Punkte 0 sind η_1 und ζ_1 gleich null, weil auf der Oberfläche der Kugel keine tangentialen elektrischen Kräfte existieren, und es gilt, da die y_1- und z_1-Axe die Oberfläche tangieren, auch:

$$\frac{\delta \eta_1}{\delta y_1} = \frac{\delta \eta_1}{\delta z_1} = \frac{\delta \zeta_1}{\delta y_1} = \frac{\delta \zeta_1}{\delta z_1} = 0.$$

Ferner ist geometrisch evident, dass im Punkte 0:

$$\frac{\delta}{\delta x_1} = \frac{\delta}{\delta r} \quad \frac{\delta}{\delta y_1} = \frac{1}{r} \frac{\delta}{\delta \vartheta} \quad \frac{\delta}{\delta z_1} = \frac{1}{r \sin\vartheta} \frac{\delta}{\delta \varphi}$$

ist. Damit geht (77) über in:

$$\lambda_1 = 0 \quad i\,k\,\mu_1 = \frac{1}{r\sin\vartheta}\frac{\delta\,\xi_1}{\delta\,\varphi} - \frac{\delta\,\zeta_1}{\delta\,r} \quad i\,k\,\nu_1 = \frac{\delta\,\eta_1}{\delta\,r} - \frac{1}{r}\frac{\delta\,\xi_1}{\delta\,\vartheta}.$$

Wir erfahren also zunächst, dass der magnetischen Kraft die radiale Componente fehlt, dass sie in der Kugeloberfläche liegt.

Drückt man nach (13) ξ, η, ζ durch α, β, γ aus, so folgt ferner:

$$\xi_1 = \cos\varphi\,\{\alpha\cos\vartheta + \sin\vartheta\,[e^{ikx} + \beta + \gamma]\}$$
$$\eta_1 = \cos\varphi\,\{-\alpha\sin\vartheta + \cos\vartheta\,[e^{ikx} + \beta + \gamma]\}$$
$$\zeta_1 = -\sin\varphi\,\{e^{ikx} + \beta - \gamma\}.$$

Auf der Kugeloberfläche wird in Folge der Relationen (15):

$$\eta_1 = \zeta_1 = 0 \quad \xi_1 = \alpha\cdot\frac{\cos\varphi}{\cos\vartheta} = 2\,\gamma\,\frac{\cos\varphi}{\sin\vartheta} \tag{78}$$

und damit:

$$ik\mu_1 = +\sin\varphi\left[-\frac{\alpha}{r\sin\vartheta\cos\vartheta} + \frac{\delta}{\delta\,r}(e^{ikx} + \beta - \gamma)\right]$$

$$ik\nu_1 = \quad \cos\varphi\left[-\frac{\delta\,\alpha}{\delta\,r}\sin\vartheta + \cos\vartheta\frac{\delta}{\delta\,r}(e^{ikx} + \beta + \gamma) - \frac{1}{r}\frac{\delta}{\delta\,\vartheta}\left(\frac{\alpha}{\cos\vartheta}\right)\right]$$

Nun gilt weiter an der Oberfläche der Kugel die Gleichung (17):

$$0 = \cos\vartheta\left(r\frac{\delta\,\alpha}{\delta\,r} + 2\,\alpha\right) + \sin\vartheta\left[r\frac{\delta}{\delta\,r}(e^{ikx} + \beta + \gamma) + 2(e^{ikx} + \beta + \gamma)\right]$$

welche sich in Rücksicht auf (15) auch so schreiben lässt:

$$0 = r\,\frac{\delta\,\alpha}{\delta\,r}\cos\vartheta + \sin\vartheta\,r\,\frac{\delta}{\delta\,r}(e^{ikx} + \beta + \gamma) + \frac{2\,\alpha}{\cos\vartheta}$$

und damit lassen sich die Ausdrücke von μ_1 und ν_1 in die folgenden verwandeln:

$$i\,k\,\mu_1 = \sin\varphi\left[-\frac{\delta\,\alpha}{\delta\,r}\operatorname{cotg}\vartheta - \frac{3\,\alpha}{r\sin\vartheta\cos\vartheta} - 2\,\frac{\delta\,\gamma}{\delta\,r}\right]$$

$$i\,k\,\nu_1 = \cos\varphi\left[-\frac{1}{\sin\vartheta}\frac{\delta\,\alpha}{\delta\,r} - \frac{2\,\alpha}{r\sin\vartheta} - \frac{1}{r}\frac{\delta}{\delta\vartheta}\left(\frac{\alpha}{\cos\vartheta}\right)\right]$$

Sitz. Ber. math. phys. Classe d. Kgl. bayer. Akad. d. Wiss. *31* (1901) 293–338

oder:

$$\left.\begin{aligned} -i\varrho_0\mu_1\frac{\sin\vartheta}{\sin\varphi} &= r\frac{\delta\alpha}{\delta r}\cos\vartheta + \frac{3\alpha}{\cos\vartheta} + 2\sin\vartheta\, r\frac{\delta\gamma}{\delta r} \\ &= \cos\vartheta\left(r\frac{\delta\alpha}{\delta r}+2\alpha\right) + 2\sin\vartheta\left(r\frac{\delta\gamma}{\delta r}+2\gamma\right) + \frac{\alpha}{\cos\vartheta} \\ -i\varrho_0\nu_1\frac{\sin\vartheta}{\cos\varphi} &= r\frac{\delta\alpha}{\delta r} + 2\alpha + \sin\vartheta\frac{\delta}{\delta\vartheta}\left(\frac{\alpha}{\cos\vartheta}\right), \end{aligned}\right\} \quad 79)$$

wobei die frühere Abkürzung $k\alpha = \varrho_0$ benutzt ist. Da wir es übrigens fortan nur mit Punkten auf der Kugel selbst zu thun haben, werden wir nicht mehr eigens auf diesen Umstand hinweisen müssen und dürfen uns erlauben, den Index 0 wegzulassen, sowie beliebig r und a zu vertauschen.

Hier führen wir nun unsere Entwicklungen (70) für α, β, γ ein, die wir aber zunächst ein wenig umformen. Es ist auf der Kugeloberfläche:

$$\alpha = \sum_{m=1}^{\infty}(2m+1)A_m P_{m,1}(\cos\vartheta),\quad \gamma = \sum_{m=1}^{\infty}(2m+1)C_m P_{m,2}(\cos\vartheta)$$

und nach der Bezeichnungsweise (39):

$$\begin{aligned} 2\alpha + r\frac{\delta\alpha}{\delta r} &= \sum\,(2m+1)A_m S_m P_{m,1}(\cos\vartheta) \\ 2\gamma + r\frac{\delta\gamma}{\delta r} &= \sum_{m=1}^{\infty}(2m+1)C_m S_m P_{m,2}(\cos\vartheta) \end{aligned} \quad 80)$$

Setzt man für A_m und C_m die Ausdrücke (71) durch die p_m ein, so wird:

$$\alpha = \sum_{m=1}^{\infty}[(m-1)p_m + (m+2)p_{m+2}]\,P_{m,1}(\cos\vartheta)$$

oder durch eine leichte Umstellung der Summe:

$$\alpha = \sum_{m=1}^{\infty} p_{m+1}[m\,P_{m+1,1}(\cos\vartheta) + (m+1)\,P_{m-1,1}(\cos\vartheta)].$$

Da aber nach (19):

$$(2m+1)\cos\vartheta\,P_{m,1} = m\,P_{m+1,1} + (m+1)\,P_{m-1,1}$$

ist, so erhält man das einfache Resultat:

$$\frac{\alpha}{\cos\vartheta} = \sum_{m=1}^{\infty}(2m+1)p_{m+1}P_{m,1}(\cos\vartheta). \quad 81)$$

Hieraus folgt noch mit Hülfe von (22):

$$\sin\vartheta\,\frac{\delta}{\delta\vartheta}\left(\frac{a}{\cos\vartheta}\right)=\sum_{m=1}^{\infty}p_{m+1}\,[m^2\,P_{m+1,1}-(m+1)^2\,P_{m-1,1}]$$
$$=\sum_{m=1}^{\infty}P_{m,1}\,[(m-1)^2\,p_m-(m+2)^2\,p_{m+2}]. \qquad 82)$$

Ferner notieren wir die nach (80) folgende Formel:

$$r\frac{\delta a}{\delta r}+2a=\sum P_{m,1}(\cos\vartheta)S_m[(m-1)\,p_m+(m+2)\,p_{m+2}]. \qquad 83)$$

Wir bilden weiter die in dem Ausdruck (79) für μ_1 auftretende Grösse:

$$\cos\vartheta\left(r\frac{\delta a}{\delta r}+2a\right)+2\sin\vartheta\left(r\frac{\delta\gamma}{\delta r}+2\gamma\right)$$

$$=\sum_{m=1}^{\infty}(2m+1)S_m[A_m\cos\vartheta\,P_{m,1}(\cos\vartheta)+2\,C_m\sin\vartheta\,P_{m,2}(\cos\vartheta)].$$

Drückt man sowohl $\cos\vartheta\,P_{m,1}$, als $\sin\vartheta\,P_{m,2}$ nach (19) und (21) durch $P_{m,1}$ aus, so geht diese Summe über in:

$$\sum_{m=1}^{\infty}S_m\,[A_m\,\{m\,P_{m+1,1}+(m+1)\,P_{m-1,1}\}$$
$$+2\,C_m\,\{(m+1)(m+2)\,P_{m-1,1}-m\,(m-1)\,P_{m+1,1}\}]$$

und wenn man A_m und C_m durch die p_m ausdrückt, in:

$$\sum_{m=1}^{\infty}S_m\,[m\,p_{m+2}\,P_{m+1,1}+(m+1)\,p_m\,P_{m-1,1}]$$

oder nach den $P_{m,1}$ geordnet:

$$\cos\vartheta\left(r\frac{\delta a}{\delta r}+2a\right)+2\sin\vartheta\left(r\frac{\delta\gamma}{\delta r}+2\gamma\right)$$
$$=\sum_{m=1}^{\infty}P_{m,1}\,p_{m+1}\,[(m+2)\,S_{m+1}+(m-1)\,S_{m-1}]. \qquad 84)$$

Die Einführung der Entwicklungen (81) bis (84) in (79) liefert jetzt:

$$-i\,\varrho\,\mu_1\sin\vartheta$$

$$=\sin\varphi\sum_{m=1}^{\infty}P_{m,1}\,p_{m+1}[2m+1+(m+2)S_{m+1}+(m-1)S_{m-1}]$$

Sitz. Ber. math. phys. Classe d. Kgl. bayer. Akad. d. Wiss. *31* (1901) 293–338

$$-i\varrho\nu_1\sin\vartheta = \cos\varphi\sum_{m=1}^{\infty} P_{m,1}\left[(m-1)^2 p_m - (m+2)^2 p_{m+2}\right.$$
$$\left.+ S_m\{(m-1)p_m + (m+2)p_{m+2}\}\right].$$

Hiermit sind die Grössen $\mu_1 \sin\vartheta$ und $\nu_1 \sin\vartheta$ nach Kugelfunktionen entwickelt. Für das folgende werden wir aber nötig haben, μ_1 und ν_1 selbst in Form solcher Entwicklungen zu besitzen, und das wird erreicht durch folgenden Ansatz. Sei:

$$(C_{m+1}-C_{m-1}) = p_{m+1}[2m+1+(m+2)S_{m+1}+(m-1)S_{m-1}] \quad 85)$$

$$(D_{m+1}-D_{m-1}) = (m-1)p_m(m-1+S_m)+(m+2)p_{m+2}(S_m-m-2). \quad 86)$$

Dann wird:

$$i\varrho\mu_1\sin\vartheta = \sin\varphi\sum P_{m,1}(C_{m-1}-C_{m+1})$$
$$= \sin\varphi\sum_{m=1}^{\infty} C_m(P_{m+1,1}-P_{m-1,1}).$$

Es gilt aber nach (20):

$$(2m+1)\sin\vartheta\, P_{m,0} = P_{m+1,1} - P_{m-1,1}$$

und damit:

$$i\varrho\mu_1 = \sin\varphi\sum_{m=0}^{\infty}(2m+1)C_m P_{m,0}. \quad 87)$$

Genau ebenso folgt:

$$i\varrho\nu_1 = \cos\varphi\sum_{m=0}^{\infty}(2m+1)D_m P_{m,0}. \quad 88)$$

Das sind die gewünschten Entwicklungen von μ_1 und ν_1 nach Kugelfunktionen.

Es erübrigt nur noch, die C_m und D_m aus den Rekurrenzen (85) und (86) zu bestimmen.

Zunächst lassen sich die rechten Seiten derselben vereinfachen. Die Gleichung (59) schreibt sich in Rücksicht auf (62):

$$p_{m+1}[(m+1)S_{m+1}+mS_{m-1}] - \frac{1}{\varrho K_{m+1}} + \frac{1}{\varrho K_{m-1}} = 0 \quad 89)$$

und damit folgt:

$$C_{m+1}-C_{m-1} = \frac{1}{\varrho K_{m+1}} - \frac{1}{\varrho K_{m-1}} + p_{m+1}[2m+1+S_{m+1}-S_{m-1}].$$

Ferner wird nach der Definition (39) von S_m:

$$2m+1+S_{m+1}-S_{m-1}=2m+1+\frac{\varrho\frac{dK_{m+1}}{d\varrho}}{K_{m+1}}-\frac{\varrho\frac{dK_{m-1}}{d\varrho}}{K_{m-1}}$$

und dieser Ausdruck lässt sich mit Hülfe von (32) reduzieren auf:

$$(2m+1)\cdot\frac{K_m^2}{K_{m+1}\cdot K_{m-1}}.$$

Demnach:

$$C_{m-1}-C_{m+1}=\frac{1}{\varrho K_{m-1}}-\frac{1}{\varrho K_{m+1}}-(2m+1)p_{m+1}\frac{K_m^2}{K_{m+1}K_{m-1}}.$$

Bildet man diese Gleichung für $m+1$, $m+3$, $m+5$ u. s. w. und addiert alle entstehenden Relationen, so ergiebt sich zur Bestimmung von C_m:

$$C_m=\frac{1}{\varrho K_m}-\sum_{\nu=0}^{\infty}(2m+4\nu+3)\frac{K_{m+2\nu+1}^2}{K_{m+2\nu}\cdot K_{m+2\nu+2}}p_{m+2\nu+2}. \quad 90)$$

Die Convergenz der hier auftretenden Summe lässt sich ohne Schwierigkeit erweisen.

Um die Grössen D_m zu finden, ersetze man in (85) m durch $m+1$ und addiere die entstehende Relation zu (86). Es ergiebt sich:

$$(D_{m+1}+C_{m+2})-(D_{m-1}+C_m)=(m-1)p_m(m-1+S_m)$$
$$+p_{m+2}[2(m+1)S_m+(m+3)S_{m+2}-(m+1)^2].$$

Subtrahiert man hiervon die Gleichung (89), nachdem man in ihr m durch $m+1$ ersetzt und mit 2 multipliziert hat, so bleibt:

$$(D_{m+1}+C_{m+2})-(D_{m-1}+C_m)$$
$$=(m-1)p_m(m-1+S_m)-(m+1)p_{m+2}(m+1+S_{m+2})$$
$$+\frac{2}{\varrho K_{m+1}}-\frac{2}{\varrho K_{m-1}}$$

und man sieht, dass dieser Gleichung genügt wird, wenn für jeden Wert von m gilt:

$$D_{m-1}+C_m=\frac{2}{\varrho K_{m-1}}-(m-1)p_m(m-1+S_m).$$

Sitz. Ber. math. phys. Classe d. Kgl. bayer. Akad. d. Wiss. *31* (1901) 293–338

Führt man wiederum für S_m seinen Wert (39) ein und drückt den Differentialquotienten von K_m nach (32) durch die K_m selbst aus, so erhält man leicht:

$$D_{m-1} + C_m = \frac{2}{\varrho K_{m-1}} + (m-1)\, i\, \varrho\, p_m \frac{K_{m-1}}{K_m}. \qquad 91)$$

Nachdem die C_m aus (90) gefunden sind, giebt diese Gleichung unmittelbar die D_m und damit kann man die Entwicklungen (87) und (88) für die Componenten der magnetischen Kraft wirklich bilden.

Es ist für die weitere Anwendung noch erforderlich, von den bisher verwandten komplexen Grössen zu den reellen Schwingungskomponenten überzugehen. Zu dem Zweck setze man:

$$\xi_1 = F e^{if} \cos\varphi \quad \mu_1 = G e^{ig} \sin\varphi \quad \nu_1 = H e^{ih} \cos\varphi$$
$$p_{m+1} = F_m e^{if_m} \quad C_m = i\,\varrho\, G_m e^{ig_m} \quad D_m = i\,\varrho\, H_m e^{ih_m},$$

wo sämtliche Buchstaben auf den rechten Seiten reelle Grössen sein sollen.

Dann erhält man einerseits:

$$\left.\begin{aligned} X_1 &= \text{pars real}\,(F e^{if} e^{iqt} \cos\varphi) = F\cos(f+qt)\cos\varphi \\ \text{ebenso: } M_1 &= G\cos(g+qt)\sin\varphi,\ N_1 = H\cos(h+qt)\cos\varphi \end{aligned}\right\} \quad 92)$$

und hierbei bedeutet X_1 die radiale Componente der elektrischen Kraft, M_1 und N_1 die tangentialen Componenten der magnetischen Kraft. Andrerseits folgt aus (78), (81), (87) und (88), indem man überall Reelles und Imaginäres trennt:

$$\left.\begin{aligned} F\cos f &= \sum_{m=1}^{\infty}(2m+1)P_{m,1}F_m\cos f_m, & F\sin f &= \sum_{m=1}^{\infty}(2m+1)P_{m,1}F_m\sin f_m \\ G\cos g &= \sum_{m=0}^{\infty}(2m+1)P_{m,0}G_m\cos g_m, & G\sin g &= \sum_{m=0}^{\infty}(2m+1)P_{m,0}G_m\sin g_m \\ H\cos h &= \sum_{m=0}^{\infty}(2m+1)P_{m,0}H_m\cos h_m, & H\sin h &= \sum_{m=0}^{\infty}(2m+1)P_{m,0}H_m\sin h_m \end{aligned}\right\}. \quad 93)$$

Damit ist der Uebergang zu reellen Schwingungskomponenten vollzogen und die Berechnung der an der Oberfläche der Kugel wirkenden elektrischen und magnetischen Kräfte ermöglicht.

§ 8. Der Maxwell'sche Druck.

Nach Maxwell herrscht im Aether ein Druck senkrecht zu den elektrischen Kraftlinien und ein Zug in Richtung der elektrischen Kraft, beide numerisch gleich der in der Volumeneinheit enthaltenen elektrischen Energie, also gleich

$$\sigma\,(X^2 + Y^2 + Z^2),$$

wo σ ein Proportionalitätsfaktor ist, der von der Wahl der Einheit der elektrischen Kraft abhängt. Ein ebensolcher Druck und Zug herrscht senkrecht und parallel zur magnetischen Kraft und ist wiederum numerisch gleich der in der Volumeneinheit enthaltenen magnetischen Energie $\sigma\,(L^2 + M^2 + N^2)$. Um die vollen Drucke im Aether zu erhalten, hat man das elektrische und das magnetische Drucksystem zu superponieren.

Nun steht die elektrische Kraft überall senkrecht auf unserer Kugel. Demnach wirkt auf jedes Oberflächenelement derselben zunächst ein senkrechter Zug, dessen Grösse gleich σX_1^2 ist, weil hier die elektrische Kraft nur aus ihrer Normalkomponente X_1 besteht. Die magnetische Kraft hingegen liegt in der Oberfläche unserer Kugel. Daher erfährt jedes Oberflächenelement derselben zweitens einen senkrechten Druck gleich $\sigma\,(M_1^2 + N_1^2)$. Der Gesamtdruck auf ein Oberflächenelement wird daher: $\sigma\,(M_1^2 + N_1^2 - X_1^2)$ oder nach (92) gleich:

$$\sigma\{G^2\cos^2(g+q\cdot t)\sin^2\varphi + H^2\cos^2(h+qt)\cos^2\varphi - F^2\cos^2(f+qt)\cos^2\varphi\}.$$

Der Druck ändert sich hiernach periodisch mit der halben Schwingungsperiode des Lichts. Der uns allein interessierende Durchschnittswert für Zeiten, welche viele Perioden umfassen, wird:

$$\frac{\sigma}{2}\,(G^2\sin^2\varphi + H^2\cos^2\varphi - F^2\cos^2\varphi)\,.$$

Bildet man durch Multiplikation mit $\cos\vartheta$ die in die x-Richtung fallende Componente dieses Drucks und integriert über die ganze Kugelfläche, so erhält man für den in Richtung der einfallenden Welle auf die Kugel wirkenden Druck D:

Sitz. Ber. math. phys. Classe d. Kgl. bayer. Akad. d. Wiss. *31* (1901) 293–338

$$D = \frac{\sigma}{2}\int a^2\,d\vartheta\,d\varphi \sin\vartheta\cos\vartheta\,(G^2\sin^2\varphi + H^2\cos^2\varphi - F^2\cos^2\varphi),$$

wobei, wie früher, a den Kugelradius bedeutet. Da F, G und H nicht von φ abhängen, lässt sich die Integration nach φ ausführen und das Integral auf folgende Form bringen:

$$D = \frac{\sigma\pi a^2}{2}\int_0^\pi d\vartheta\sin\vartheta\cos\vartheta\,[(G\cos g)^2 + (G\sin g)^2 + (H\cos h)^2 + (H\sin h)^2 - (F\cos f)^2 - (F\sin f)^2].$$

Hier sind nun die Summen (93) einzusetzen. Beginnen wir beispielsweise mit der Summe für $G\cos g$, so folgt:

$$\int_0^\pi d\vartheta\sin\vartheta\cos\vartheta\,(G\cos g)^2$$
$$= \sum_{m=0}^\infty\sum_{m'=0}^\infty (2m+1)(2m'+1)G_m G_{m'}\cos g_m\cos g_{m'}\int_0^\vartheta d\vartheta\sin\vartheta\cos\vartheta\,P_{m,0}P_{m',0}.$$

Die in der vorstehenden Formel rechts auftretenden Integrale sind null, ausser wenn $m' = m \pm 1$ ist. Daher reduziert sich die Doppelsumme auf die einfache:

$$2\sum_{m=0}^\infty (2m+1)(2m+3)G_m G_{m+1}\cos g_m\cos g_{m+1}\int_0^\vartheta d\vartheta\sin\vartheta\cos\vartheta\,P_{m,0}P_{m+1,0}$$

und wenn man hier für die Integrale nach (24) die Werte

$$\frac{2(m+1)}{(2m+1)(2m+3)}$$

einsetzt, so wird:

$$\int_0^\pi d\vartheta\sin\vartheta\cos\vartheta\,(G\cos g)^2 = 4\sum_{m=0}^\infty (m+1)\,G_m G_{m+1}\cos g_m\cos g_{m+1}.$$

Ganz ähnlicher Umformungen sind die übrigen Glieder des Ausdrucks von D fähig (bei den Gliedern mit F hat man die Integralformel (23) zu benutzen) und man erhält schliesslich durch Addition der sechs entstehenden Summen die Schlussformel für den Druck des Lichts:

$$D = 2\sigma\pi a^2 \Big\{ \sum_{m=0}^{\infty} (m+1)\, G_m\, G_{m+1} \cos(g_m - g_{m+1})$$

$$+ \sum_{m=0}^{\infty} (m+1)\, H_m\, H_{m+1} \cos(h_m - h_{m+1}) \qquad 94)$$

$$- \sum_{m=1}^{\infty} m(m+1)(m+2) F_m F_{m+1} \cos(f_m - f_{m+1}) \Big\}.$$

Der Proportionalitätsfaktor σ lässt sich leicht aus der durchschnittlichen Energie pro Volumeneinheit E der einfallenden Welle berechnen. Die elektrische Schwingung der einfallenden Welle war $Y = \cos(kx + qt)$, dementsprechend ihre elektrische Energie gleich $\sigma Y^2 = \sigma \cos^2(kx + qt)$. Der Mittelwert dieses Ausdrucks für Vielfache der Schwingungsperiode ist $\frac{\sigma}{2}$. Es ist bekannt, dass in ebenen Wellen die Energie der magnetischen und die Energie der elektrischen Schwingung gleich sind. Daher habe ich noch einmal $\frac{\sigma}{2}$ hinzuzufügen, um die ganze Energie pro Volumeneinheit zu erhalten, und finde damit:

$$\sigma = E. \qquad 95)$$

§ 9. Formeln zur numerischen Rechnung.

Um eine etwaige Controlle der späteren numerischen Angaben zu erleichtern, will ich die Formeln angeben, die ich zur numerischen Berechnung des Drucks benutzt habe und die aus den oben abgeleiteten auf die einfachste Weise hervorgehen, indem man von vorneherein für Trennung des Reellen und Imaginären sorgt.

a Kugelradius. $\qquad \varrho = \frac{2\pi a}{\lambda}.$

Man bilde die endlichen Ausdrücke:

Sitz. Ber. math. phys. Classe d. Kgl. bayer. Akad. d. Wiss. 31 (1901) 293–338

$$k_m \cos \chi_m = 1 - \frac{(m-1)\,m\,(m+1)\,(m+2)}{2\cdot 4}\frac{1}{\varrho^2} + \frac{(m-3)\,(m-2)\ldots\,(m+3)\,(m+4)}{2\cdot 4\cdot 6\cdot 8}\frac{1}{\varrho^4} - \ldots$$

$$k_m \sin \chi_m = \frac{1}{\varrho}\frac{m(m+1)}{2} - \frac{1}{\varrho^3}\frac{(m-2)(m-1)\ldots(m+2)\,(m+3)}{2\cdot 4\cdot 6} + \ldots$$

sowie:

$$l_m \cos \psi_m = 1 - \frac{1}{\varrho^2}\frac{m(m+1)}{2}\cdot\frac{m\,(m+1)+1\cdot 2}{4} + \frac{1}{\varrho^4}\frac{(m-2)(m-1)\ldots(m+2)(m+3)}{2\cdot 4\cdot 6}\cdot\frac{m(m+1)+3\cdot 4}{8} - \ldots$$

$$l_m \sin \psi_m = \frac{1}{\varrho}\frac{m(m+1)}{2} - \frac{1}{\varrho^3}\frac{(m-1)m(m+1)(m+2)}{2\cdot 4}\cdot\frac{m(m+1)+2\cdot 3}{6} + \ldots$$

und berechne hieraus k_m, l_m, χ_m, ψ_m für eine Reihe von Werten des Index m.

Wünscht man für den Druckwert etwa eine zweistellige Genauigkeit, so gehe man dabei bis zu einem Index $\overline{m}$, für welchen k_m und l_m grösser als 100 werden, und vernachlässige auch in den späteren Formeln alle Glieder, in welchen k_m oder l_m mit höherem Index als $\overline{m}$ vorkommen. Für $\varrho = \frac{1}{4}$, bei einem Kugeldurchmesser gleich etwa $^1/_{12}$ Wellenlänge, genügt es, $\overline{m} = 3$ zu nehmen, so dass die Berechnung des Drucks sehr leicht auszuführen ist; für $\varrho = 4$ hingegen, wenn also der Kugeldurchmesser die Wellenlänge schon um einiges übertrifft, muss man bereits bis zu $\overline{m} = 7$ oder 8 gehen und für Kugeln, deren Durchmesser mehrere Wellenlängen beträgt, wird die Rechnung praktisch undurchführbar, aber, wie sich unten zeigen wird, auch nicht mehr erforderlich.

Man berechne weiter die Hülfsgrössen:

$$\frac{1}{\varrho k m} = r_m \qquad \frac{2m+1}{\varrho^3 l_m}\,\frac{k_m^2}{k_{m-1}\,k_{m+1}} = s_m \qquad \frac{m}{\varrho^2 l_m}\,\frac{k_m}{k_{m+1}} = t_m \qquad \frac{1}{\varrho^2 l_m} = F_m$$

und die Hülfswinkel:

$$\chi'_m = \chi_m + \varrho - \frac{\pi m}{2} \quad \psi'_m = \frac{\pi}{2}(m+1) - \varrho - \psi_m$$

$$\tau_m = \psi'_m + \chi'_m - \chi'_{m+1} \quad \sigma_m = \chi'_{m-1} - \chi'_m - \tau_m .$$

Dann gilt:

$$G_m \cos g'_m + G_{m+2} \cos g'_{m+2} = r_m \cos \chi'_m + r_{m+2} \cos \chi'_{m+2} + s_{m+1} \cos \sigma_{m+1}$$

$$G_m \sin g'_m + G_{m+2} \sin g'_{m+2} = r_m \sin \chi'_m + r_{m+2} \sin \chi'_{m+2} + s_{m+1} \sin \sigma_{m+1}.$$

Man kann aus diesen beiden Gleichungssystemen die G_m und g'_m durch Rekurrenz berechnen, indem man mit einem hinreichend hohen Wert des Index $m = \overline{m}$ beginnt und $G_{\overline{m}+1}$ sowie $G_{\overline{m}+2}$ gleich null setzt. Ist das ausgeführt, so findet man H_m und h'_m nach den Formeln:

$$H_m \cos h'_m = t_m \cos \tau_m - 2\, r_m \sin \chi'_m + G_{m+1} \cos g'_{m+1}$$

$$H_m \sin h'_m = t_m \sin \tau_m - 2\, r_m \cos \chi'_m - G_{m+1} \sin g'_{m+1}.$$

Die F_m, G_m und H_m stimmen überein mit den im vorigen § so bezeichneten Grössen.

Das Verhältnis des Drucks der Strahlung D zu der Strahlungsenergie in der Volumeneinheit wird dann erhalten aus der Summe:

$$\frac{D}{E} = 2\pi a^2 \Big[\sum_{m=0}^{\infty} (m+1)\, G_m\, G_{m+1} \sin(g'_m - g'_{m+1})$$
$$+ (m+1)\, H_m\, H_{m+1} \sin(h'_{m+1} - h'_m)$$
$$+ m(m+1)(m+2)\, F_m\, F_{m+1} \sin(\psi'_m - \psi'_{m+1}) \Big].$$

§ 10. Grenzwerte des Drucks.

In den beiden Fällen, dass die Kugel entweder sehr klein oder sehr gross gegen die Wellenlänge ist, lassen sich einfache geschlossene Ausdrücke für den Druck angeben.

Wenn der Kugelradius und damit ϱ sehr klein ist, kann man sich in der Entwicklung (28) von K_m:

Sitz. Ber. math. phys. Classe d. Kgl. bayer. Akad. d. Wiss. *31* (1901) 293–338

$$i^{m+1} K_m = i \cdot \frac{1 \cdot 3 \ldots . (2m-1)}{\varrho^{m+1}} \left\{ 1 + \frac{\varrho^2}{2(2m-1)} + \frac{\varrho^4}{2 \cdot 4 (2m-1)(2m-3)} \cdots \right\}$$
$$+ \frac{\varrho^m}{1 \cdot 3 \ldots 2m+1} \left\{ 1 - \frac{\varrho^2}{2(2m+3)} + \frac{\varrho^4}{2 \cdot 4 (2m+3)(2m+5)} \cdots \right\}$$

auf die paar ersten Glieder beschränken. Wo dann in den obigen Formeln ein Faktor oder Divisor K_m oder ein Differentialquotient dieser Grössen auftritt, kann man stets nach Potenzen von ϱ entwickeln und enthält schliesslich den Druck D selbst in Form einer Potenzreihe nach ϱ, deren erstes Glied den gesuchten Grenzwert darstellt. Die Rechnungen, die hierzu führen, sind ziemlich umständlich, aber ganz elementar; man muss sich nur hüten, zu früh höhere Glieder wegzulassen, da sich zum Schluss solche niederer Ordnung herausheben. Man erhält als Grenzwert des Drucks für sehr kleine Kugeln:

$$\frac{D}{E} = \frac{14}{3} \pi a^2 \varrho^4 = \frac{224}{3} \pi^5 \frac{a^6}{\lambda^4}. \qquad 96)$$

Für sehr grosse Kugeln findet man den Druck folgendermassen. Fällt eine ebene Welle unter einem Winkel ψ auf eine vollkommen reflektierende ebene Platte auf, so erleidet letztere einen senkrechten Druck: $P = 2 E \cos^2 \psi$, wobei E wiederum die Energie pro Volumeneinheit der einfallenden Welle bedeutet. (Vgl. für die einfache Ableitung dieser Regel aus Maxwell's Druckannahmen Goldhammer, Annalen der Physik, Bd. 4, 1901, pag. 844 und Boltzmann, Wied. Annalen, 22). Betrachten wir die Oberflächenelemente unserer Kugel als eben, so folgt, dass überall auf dieselbe ein senkrechter Druck gleich $2 E \cos^2 \vartheta$ — in unserer Bezeichnung — wirkt. Die x-Componente dieses Drucks ist $2 E \cos^3 \vartheta$ und der Gesamtdruck auf die Kugel wird durch Integration über die Vorderfläche gewonnen, da die beschattete Hinterfläche natürlich keine Einwirkung erleidet. Es ergiebt sich daher:

$$D = a^2 \cdot \int_0^{2\pi} d\varphi \int_0^{\pi/2} d\vartheta \sin \vartheta \cdot 2 E \cos^3 \vartheta$$

oder:

$$\frac{D}{E} = \pi a^2. \qquad 97)$$

Es ist im ersten Augenblick auffällig, dass dieser für eine vollkommen reflektierende grosse Kugel geltende Wert genau übereinstimmt mit dem von Herrn Arrhenius benutzten für eine vollkommen schwarze Kugel gültigen. Doch lässt sich dieses Resultat leicht aufklären. An der Stelle, welche von der einfallenden Welle zuerst getroffen wird ($\vartheta = 0$), erleidet eine vollkommen reflektierende Kugel allerdings einen doppelt so starken Druck, als eine vollkommen schwarze. Indessen nimmt nach den Seiten hin bei der vollkommen reflektierenden Kugel der Druck viel rascher ab, weil bei flachen Incidenzen die Druckwirkung der reflektierten Welle die der einfallenden zum Teil wieder aufhebt.

§ 11. Ergebnis.

Nachdem der Druck für sehr grosse wie sehr kleine Kugeln aus den Formeln des vorigen Paragraphen bekannt war, erübrigte noch die Berechnung für Kugeln von der Grössenordnung der Wellenlänge. Ich will zur Bequemlichkeit das Verhältnis von D zu $\pi a^2 E$ mit V bezeichnen. Die Berechnung von V habe ich nach den Formeln § 9 für einige Werte des Kugelradius numerisch ausgeführt mit folgendem Resultat:

$\frac{2\pi a}{\lambda} = \varrho$	$1/4$	$1/2$	$\sqrt{1/2}$	1	$\sqrt{2}$	2	4·
$2a$	$0.08\ \lambda$	$0.16\ \lambda$	$0.22\ \lambda$	$0.32\ \lambda$	$0.45\ \lambda$	$0.64\ \lambda$	$1.27\ \lambda$
$V = \frac{D}{\pi a^2 E}$	0.018	0.35	1.07	2.42	2.16	1.31	1.22
$\frac{14}{3}\varrho^4$	0.018	0.29	1.17	4.67			

Zur Vergleichung wurden die Werte von V, wie sie aus der Näherungsformel (96) folgen, mitangesetzt. Man sieht: „Beschränkt man sich auf eine Genauigkeit von etwa 20%, so genügt die Formel: $D = \frac{14}{3}\pi a^2 \varrho^4 E$ zur Berechnung des Drucks von unendlich kleinen Kugeln

Sitz. Ber. math. phys. Classe d. Kgl. bayer. Akad. d. Wiss. *31* (1901) 293–338

an bis herauf zu Kugeln von ein Viertel Wellenlänge Durchmesser." Andrerseits muss nach (97) für sehr grosse Kugeln das Verhältnis von D zu $\pi a^2 E$ gleich 1 werden. Man sieht aus der Tabelle: Bei derselben Genauigkeit gilt die Formel $D = \pi a^2 E$ bis herab zu Kugeln von etwa anderthalb Wellenlängen Durchmesser.

Das Verhalten des Drucks für dazwischen liegende Werte wird durch Curve Fig. 1 veranschaulicht. Um dieselbe etwas sicherer zeichnen zu können, habe ich einige Punkte zwischen den oben berechneten bestimmt, indem ich die einfach verlaufende Grösse $\log V - 4 \log \varrho$ als Funktion von $\log \varrho$ numerisch interpolierte. Das Verhalten des Drucks lässt sich hiernach etwa so beschreiben: „Das Verhältnis V des Drucks zur „auffallenden Energiemenge" $\pi a^2 E$ steigt von dem für grosse Kugeln gültigen und auch von Herrn Arrhenius benutzten Werte 1 zunächst langsam an, wenn man den Kugelradius verkleinert. Ist der Kugeldurchmesser auf etwa $^2/_3$ Wellenlänge herabgesunken, so erfolgt ein merkwürdiges rapides Anwachsen von V, welches bei etwa $^1/_3$ Wellenlänge Durchmesser zu einem Maximum gleich 2.5 führt. Bei weiterer Verkleinerung des Kugeldurchmessers sinkt V noch plötzlicher ab, als es vorher angestiegen ist. Für $2a = {}^1/_5 \lambda$ ist es bereits wieder unter die Einheit zurückgegangen und nimmt alsbald verschwindend kleine Werte an."

Vergleichen wir nun den Druck des Lichts mit der Schwerkraft. Ist G die auf die Masseneinheit, also z. B. das Gramm, wirkende Schwerkraft der Sonne, s das spezifische Gewicht des Kugelmateriales, so hat die auf die ganze Kugel wirkende Schwerkraft den Betrag:

$$S = \tfrac{4}{3} \pi a^3 s \cdot G,$$

wobei a in Centimetern zu messen ist.

Der Druck des Lichts hat den Wert $\pi a^2 E \cdot V$ und das Verhältnis beider wird:

$$\frac{D}{S} = W = \tfrac{3}{4} \frac{E}{G} \cdot \frac{V}{a s}.$$

Nimmt man die Solarkonstante zu 2.5 (Grammkalorien pro Minute und Quadratcentimeter) an, so findet man mit Herrn Arrhenius (l. c. pag. 83) die Energiedichte der Sonnen-

Fig. 1.

Verhältnis des Drucks zur auffallenden Energie.

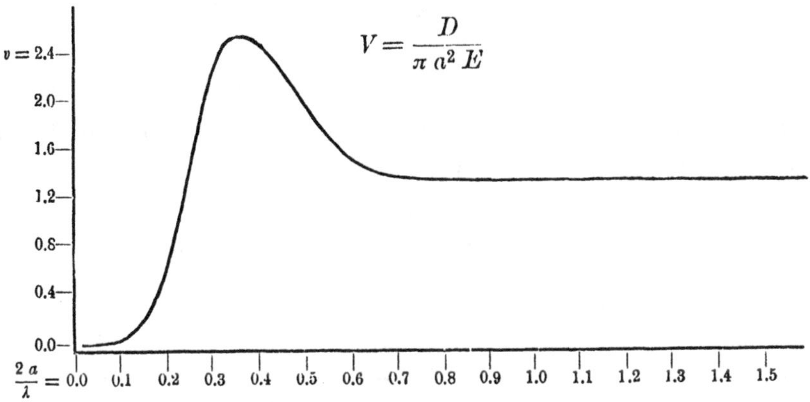

Fig. 2.

Verhältnis des Drucks zur Schwerkraft für $\lambda = 0.6\,\mu$ und $s = 1$.

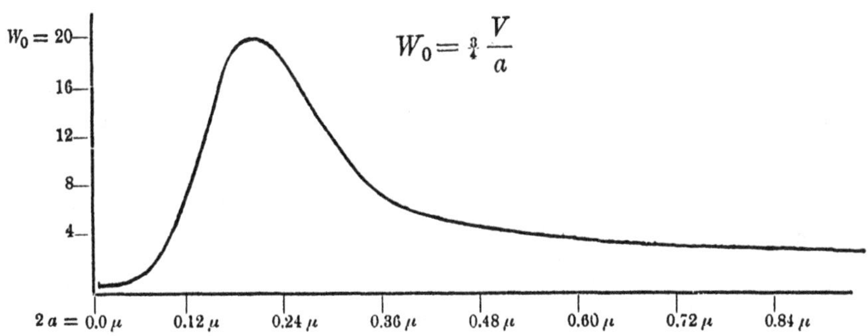

strahlung an der Oberfläche der Sonne zu $27.5 \cdot 10^{-4}\, g\, \mathrm{cm}^{-2}$ (g bedeutet das Grammgewicht). Andrerseits ist die Schwere auf der Sonnenoberfläche 27.5 mal grösser als auf der Erde, daher:

$$G = 27.5\, g$$

Sitz. Ber. math. phys. Classe d. Kgl. bayer. Akad. d. Wiss. *31* (1901) 293–338

und damit:

$$\frac{D}{S} = \tfrac{3}{4} \cdot \frac{27.5\, 10^{-4}}{27.5} \frac{V}{a\, s} = 0.75\, 10^{-4} \frac{V}{a\, s},$$

wobei immer noch a in cm zu messen ist. Zieht man es vor, a in Tausendstel mm (μ) zu messen, so gilt:

$$W = \frac{D}{S} = 0.75 \frac{V}{a\, s}.$$

Wir wollen zunächst den Normalfall betrachten, dass die ganze Sonnenstrahlung aus Wellen der Länge 0.6 μ bestehe, wie sie der hellsten Stelle des Spektrums entsprechen, und das spezifische Gewicht s gleich 1 sei, und wollen das diesem Fall entsprechende Verhältnis von Druck zu Schwere mit W_0 bezeichnen.

Dann ist:

$$W_0 = \frac{D}{S} = \tfrac{3}{4} \frac{V}{a}. \tag{98}$$

Die Werte des Verhältnisses W_0, welche aus Formel (98) hervorgehen, sind in Fig. 2 graphisch dargestellt. Für grössere Werte von a, als die Figur giebt, kann man einfach $V = 1$ und:

$$W_0 = \frac{3}{4\, a}$$

setzen. Aus dieser Formel und der Figur entnimmt man folgendes: Im Normalfall wird der Druck des Lichts gleich der Schwerkraft, sobald der Kugeldurchmesser bis auf $2.5\,\lambda = 1.5\,\mu$ herabsinkt. Bei weiterer Verkleinerung der Kugel wächst der Druck über die Schwerkraft hinaus, bis er sie bei einem Kugeldurchmesser von $0.3\,\lambda = 0.18\,\mu$ um das 18-fache übertrifft. Von diesem Maximalwert sinkt er schnell wieder ab und wird bereits für den Kugeldurchmesser $0.12\,\lambda = 0.07\,\mu$ wieder der Schwerkraft gleich, um sich sodann rasch der Null zu nähern.

„Ein Ueberwiegen des Drucks der Strahlung über die Schwerkraft findet also nur für gewisse zwischen verhältnismässig engen Grenzen liegende Kugelgrössen ($0.07 - 1.5\,\mu$) statt, innerhalb dieses Bereichs wächst aber der Druck bis auf das 18-fache der Schwerkraft an."

Bei der practischen Anwendung des vorstehenden Ergebnisses hat man vor allem zu berücksichtigen, dass die Strahlung der Sonne in Wirklichkeit nicht auf die Wellenlänge $0.6\,\mu$ konzentriert, sondern über alle möglichen Wellenlängen verteilt ist. Setzt man die Intensität der Sonnenstrahlung zwischen den Wellenlängen λ und $\lambda + d\lambda$ gleich $E\,J(\lambda)\,d\lambda$, wobei:

$$\int_0^\infty J(\lambda)\,d\lambda = 1 \qquad 99)$$

sein muss, damit der richtige Betrag der Gesamtenergie herauskommt, so hat man strenge den Druck zu berechnen nach der Formel:

$$D = \pi a^2 E \int_0^\infty J(\lambda)\,V\,d\lambda$$

und erhält für das Verhältnis von Druck zu Schwerkraft:

$$\frac{D}{S} = \frac{3}{4\,a\,s}\int_0^\infty J(\lambda)\,V\,d\lambda. \qquad 100)$$

Die genauere Kenntnis der Funktion $J(\lambda)$, der Intensität der Sonnenstrahlung für die verschiedenen Wellenlängen vor ihrer Absorption durch die irdische Atmosphäre, ist noch ein Desideratum. Um einen beiläufigen Anhalt zu gewinnen, setze ich für $J(\lambda)$ die nach Wien für den vollkommenen Radiator geltende Formel an:

$$J(\lambda) = \frac{1}{30\,\lambda_m}\left(\frac{5\,\lambda_m}{\lambda}\right)^5 e^{-5\frac{\lambda_m}{\lambda}}, \qquad 101)$$

wobei der willkürliche Faktor, mit dem $J(\lambda)$ im allgemeinen multipliziert ist, gleich so bestimmt ist, dass die Gleichung (99) erfüllt wird, und wobei für λ_m die Wellenlänge maximaler Intensität, also bei der Sonne $0.6\,\mu$ einzusetzen ist. Die Mes-

Sitz. Ber. math. phys. Classe d. Kgl. bayer. Akad. d. Wiss. *31* (1901) 293–338

sungen von Langley[1]) zeigen, dass man durch diesen Ansatz den wirklichen Verlauf der Funktion $J(\lambda)$ jedenfalls im Groben trifft. Ich habe nun den Ausdruck (101) für $J(\lambda)$ in (100) eingeführt und den Wert des Integrals durch mechanische Quadratur abgeschätzt. Dabei hat sich herausgestellt, dass durch die Verteilung der Sonnenenergie auf verschiedene Wellenlängen der Maximalwert des Verhältnisses von Druck zu Schwerkraft etwa auf die Hälfte des für den obigen Normalfall gültigen Wertes, also beiläufig auf 10 reduziert wird.

Auf der andern Seite sind aber auch Umstände in Betracht zu ziehen, welche die Druckwerte vergrössern. Es ist erstens möglich, dass die Materie der Cometenschweife ein geringeres spezifisches Gewicht hat als 1, etwa das spezifische Gewicht 0.8 der Kohlenwasserstoffe. Zweitens ist die Solarkonstante mit 2.5 für die Strahlung der Atmosphäre ausserhalb der Sonne zweifellos zu gering angesetzt und nach neueren Versuchen etwa auf 3.5—4 zu vermehren. Beides zusammen bewirkt eine Vergrösserung des Drucks auf nahezu das Doppelte, sodass man schliesslich auf einen Maximalwert von W in der Nähe von 20 zurückkommt.

Dass die Teilchen eine Constitution besitzen sollten, bei welchen grössere Drucke auftreten, als bei vollkommen reflektierenden Kugeln, ist, wenn nicht unmöglich, so doch unwahrscheinlich. Fasst man alles zusammen, so kommt man daher zu folgendem Schluss: „Die Theorie der Cometenschweife von Arrhenius erfährt insofern eine Bestätigung, als eine Zurückführung der grössten beobachteten abstossenden Kräfte auf den Druck der Sonnenstrahlung eben noch möglich erscheint. Noch grössere derartige Kräfte, welche die Schwere um mehr als das 20- oder 30-fache übertreffen, würde man aber nicht erklären können, ohne unwahrscheinlich kleine spezifische Gewichte für die Schweifteilchen anzunehmen.

[1]) Memoirs of the National Akad. of Science. Washington Vol. IV.

Schliesslich noch eine Bemerkung über einen Punkt der Sonnenphysik. Herr Arrhenius führt auch das radial gefaserte Aussehen der Sonnenkorona auf den Druck der Sonnenstrahlung zurück, indem er annimmt, dass der aus der Sonne bei Eruptionen emporgeschleuderte Staub unter seinem Einfluss nahezu radial von der Sonne wegströmt (l. c. pag. 85). Ein Zurücksinken der Teilchen in mehr parabolischen Curven, wie sie schon beobachtet worden sind, kann nach der Ansicht von Herrn Arrhenius dadurch bewirkt werden, dass zwei Teilchen sich zufällig treffen und zusammenbacken, mit dem Effekt, dass die Schwerkraft nun den Strahlendruck überwindet. Nach dem oben festgestellten rapiden Absinken des Drucks für sehr kleine Kugeln ist es einfacher, anzunehmen, dass die emporgeschleuderten Körperchen sich durch Verdunstung verkleinern und sich dadurch dem Strahlendruck mehr und mehr entziehen, sodass sie schliesslich samt ihren gasförmigen Verdunstungsprodukten unter dem Einfluss der Schwerkraft zur Sonne zurückfallen.

With E. Kron in Nachr. Kgl. Ges. d. Wiss. zu Gött., Math. Phys. Klasse 1911, *No. 3*, 197–208

Über die Helligkeitsverteilung im Schweif des Halley'schen Kometen.

Von

K. Schwarzschild und **E. Kron.**

Mit 1 Figur im Text.

Eingereicht am 4. April 1911.

Die Expedition des Potsdamer astrophysikalischen Observatoriums, welche im Frühjahr 1910 unter Leitung von Geheimrat Müller nach Teneriffa ging, nahm zur Photographie des Halley'schen Kometen nur eine kleine Stereoskopkamera mit, in die zwei Zeißaplanate von 23.3 mm Öffnung und 110 mm Brennweite oder ein Goerzanastigmat von 20 mm Öffnung und 180 mm Brennweite eingesetzt werden konnten. Die Kamera wurde auf einem von einem kräftigen Uhrwerk getriebenen parallaktischen Stativ montiert; indessen besaß das Stativ keine Feinbewegungen, um der Bewegung des Kometen genauer zu folgen; auch waren die Fokussierungseinrichtungen der Kamera unvollkommen. Was man in Deutschland vom Halley'schen Kometen in den ersten Monaten des Jahres zu sehen bekommen hatte, war so wenig verheißungsvoll, daß es keinen Zweck zu haben schien, die Expedition hierfür reicher auszurüsten, und ihr Schwerpunkt auf andre Aufgaben gelegt wurde.

Die Aufnahmen des Halley'schen Kometen, welche die Expedition erhielt, waren eine Überraschung für jeden, der die Enttäuschung über die Erscheinung des Kometen in Deutschland miterlebt hatte; mit den Aufnahmen, welche besonders auf amerikanischen Sternwarten erhalten wurden, sind sie freilich nicht vergleichbar. Jedenfalls aber ließ sich aus ihnen ein — wie uns scheint — interessantes Resultat ableiten, weil für die nötigen photometrischen Vergleichsdaten gesorgt worden war, um aus den Schwärzungen der Bilder auf der photographischen Platte auf die Helligkeitsverhältnisse im Schweif des Kometen zurückschließen zu können. Mit den beiden ganz gleichen Zeißaplanaten wurden immer gleichzeitig auf derselben Platte zwei Aufnahmen des Kometen gemacht, wobei nur das eine Objektiv einem Helligkeitsverlust von $0.^m4$ entsprechend abgeblendet war[1]). Zu den Aufnahmen mit dem Goerzanastigmaten wurden mit einem von Prof. Eberhard konstruierten Röhrenphotometer Vergleichsskalen hergestellt auf Platten derselben Emulsion, die auch gleichzeitig mit den Kometen aufnahmen entwickelt wurden.

Bei der Unschärfe der Aufnahmen hätte es keinen Zweck gehabt, die Helligkeit an jeder Stelle des Schweifes aus ihnen ableiten zu wollen. Auch ist es fraglich, ob aus einer derartigen Ausmessung auch bei den besten Aufnahmen bei der verwickelten Struktur des Schweifes eine Erkenntnis hervorginge.

Hingegen ergab sich folgende mehr summarische Fragestellung:

Die Helligkeit des Schweifes nimmt, von Unregelmäßigkeiten abgesehen, vom Kopf des Kometen aus ständig ab. Das liegt notwendig zum Teil daran, daß die Dichte im Schweif abnimmt. Die Abnahme der Dichte geschieht erstens dadurch, daß die vom Kopf ausströmende Materie sich über einen immer größeren Querschnitt verteilt, und zweitens dadurch, daß gleichzeitig die Geschwindigkeit, mit der die Schweifteilchen den Querschnitt durchsetzen, infolge der Repulsivkraft, welche die Sonne auf sie ausübt, zunimmt. Die Frage ist nun, ob diese beiden Umstände die Abnahme der Schweifhelligkeit bereits quantitativ vollständig erklären, woraus dann folgen würde, daß die Leuchtkraft pro Masseneinheit der Schweifmaterie während der Wanderung durch den Raum ungeschwächt bleibt — oder aber, ob die Abnahme der Schweifhelligkeit eine viel schnellere ist als die der Dichte, was man a priori bei einem wesentlich Gasspektrum zeigenden, also sog. selbstleuchtenden Schweif für wahrscheinlicher halten möchte.

Daß man übrigens an jeder Stelle des Schweifes einfach die Summe der Intensitäten der in dem betreffenden Visionskegel liegenden Teilchen beobachtet, daß von Verdeckung und Absorption keine Rede sein kann, ist bei der enormen Verdünnung der Schweifmaterie (s. unten) offenbar und wird im folgenden stets vorausgesetzt. Auch sei gleich darauf hingewiesen, daß die Än-

1) Infolge eines Versehens wurde diese Abblendung benutzt statt der beabsichtigten von 1^m.

With E. Kron in Nachr. Kgl. Ges. d. Wiss. zu Gött., Math. Phys. Klasse 1911, *No. 3,* 197–208

derung des Abstandes von der Sonne und damit der Beleuchtung durch die Sonne für das relativ kurze der Messung zugängliche Schweifstück wenig ausmacht.

Zur Entscheidung der obigen Frage genügt es, „Querschnittsintensitäten" des Schweifes festzustellen, den Schweif an einigen Stellen auf einer Linie senkrecht zur Schweifaxe durchzuphotometrieren und die Summe oder das Integral der längs jeder solchen Linie gefundenen Intensitäten zu bilden. Die seitliche Ausbreitung des Schweifes beeinflußt solche Querschnittsintensitäten nicht mehr, sie sind nur noch von der Strömungsgeschwindigkeit im Schweif abhängig, und über letztere hat man anderweitig genügende Angaben.

Die folgende Tabelle (1) enthält nun eine Anzahl von Querschnittsintensitäten, die aus den Aufnahmen der Teneriffa-Expedition abgeleitet wurden. Die Einheit ist eine willkürliche, von Tag zu Tag wechselnde. Der Abstand a jedes Querschnitts vom Kopf, wie er auf der Platte gemessen wurde, ist in Graden angegeben. Etwas unscharfe Aufnahmen sind für die Bestimmung der Querschnittsintensitäten eher von Vorteil, da sie die Integration über kleinere Unregelmäßigkeiten der Schweifstruktur gleich von selbst ausführen.

An die Zahlen sind bereits folgende Korrekturen angebracht: Die Intensität des Himmelsgrundes wurde aus der Schwärzung der Platte außerhalb des Kometen, für jeden Querschnitt mit abgeleitet und von der ganzen gemessenen Intensität subtrahiert. Die Extinktion in der Erdatmosphäre wurde für die Wellenlänge 430 $\mu\mu$ den Beobachtungen der Sonne entnommen, welche die Herren Müller und Kron an denselben Tagen auf der betreffenden Station angestellt haben. Nur am 14. Mai war das Wetter unsicher, die Extinktion war anormal groß und es ist leicht möglich, daß der tief stehende Kometenkopf an diesem Tag durch Dunst noch stärker geschwächt war, als aus der Tagesextinktion folgt, so daß die Intensitäten für die höher befindlichen Schweifpartien trotz der Extinktionskorrektion noch zu groß sind. Die Resultate vom 14. Mai haben daher später nur halbes Gewicht erhalten.

Es wurde weiter beachtet, daß die Distanz Δ' der verschiedenen Schweifteile von der Erde nicht die gleiche ist, und daß der Querschnitt, welcher von der Erde aus gesehen senkrecht zur scheinbaren Schweifaxe gelegt wird, in Wirklichkeit wechselnde Winkel mit der Schweifaxe bildet. Eine Änderung der Distanz Δ' bewirkt nun zwar keine Änderung der Flächenhelligkeit, aber der Schweifquerschnitt ändert sich umgekehrt proportional zu Δ'

Es sind daher die Querschnittsintensitäten mit $\varDelta'$ zu multiplizieren, um sie alle auf dieselbe Entfernung zu reduzieren. Ist ferner der Winkel der Schweifaxe mit dem Visionsradius ψ', so wird der Querschnitt gegenüber normaler Durchsicht vergrößert im Ver-

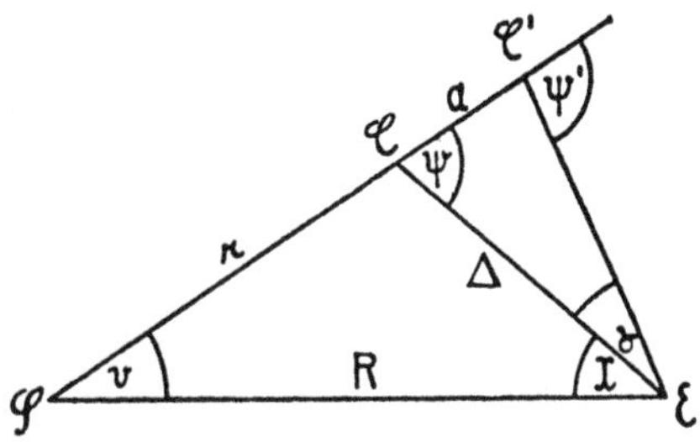

hältnis $1/\sin\psi'$, wenn man den Schweif auf eine kurze Strecke als zylindrisch ansieht. Es sind also die Querschnittsintensitäten mit $\sin\psi'$ zu multiplizieren, um sie auf Querschnitte senkrecht zur Schweifaxe zu reduzieren. Die Krümmung des Schweifes beträgt auf die hier vorkommenden Distanzen nur wenige Grade, sodaß die Schweifaxe CC' als mit der Verlängerung des Radiusvektors Sonne-Komet (SC) zusammenfallend angesehen werden darf. Aus dem Dreieck Sonne-Erde-Schweifteilchen SEC' ergibt sich aber: $\varDelta' \sin\psi' = R \sin v$. Für jede Aufnahme ist R und v und damit auch die Größe $\varDelta' \sin\psi'$ konstant, wenn man an dem Schweif entlang geht. Daraus folgt, daß die beiden eben betrachteten Einflüsse zusammen die Verhältnisse der Querschnittsintensitäten nicht ändern, sich, soweit sie hier zunächst in Betracht kommen, gegenseitig aufheben.

Schließlich wurde noch untersucht (durch direkte Aufnahme einer Fläche konstanter Helligkeit), wie die Flächenhelligkeit der Abbildung bei den verwandten Objektiven vom Abstand von der Axe abhängt, und hierfür eine kleine Korrektion angebracht.

Die Zahlen der Tabelle 1 geben hiernach die wirklichen Verhältnisse der Querschnittsintensitäten für die Zeit jeder Aufnahme.

Nunmehr ist zu untersuchen, wie weit die in Tabelle 1 enthaltenen Querschnittsintensitäten durch die wachsende Geschwindigkeit der Strömung im Schweif beeinflußt werden. Ist I die Querschnittsintensität und v die durchschnittliche Geschwindigkeit, mit der der Querschnitt durchströmt wird, so muß:

$$I \,.\, v = \text{const.} \tag{1}$$

sein, falls die Strömung stationär ist und jedes Massenteilchen des

With E. Kron in Nachr. Kgl. Ges. d. Wiss. zu Gött., Math. Phys. Klasse 1911, *No. 3,* 197–208

Tabelle 1.

Datum	Mai 1.	Mai 3.	Mai 7.	Mai 8.	Mai 9.	Mai 9.	Mai 10.	Mai 12.	Mai 14.
Objektiv	Zeiß	Zeiß	Zeiß	Zeiß	Zeiß	Goerz	Goerz	Zeiß	Zeiß
Kopfabstand a	Querschnittsintensität I								
$0^\circ.26$		1.57							
0.42	1.06								
0.52		1.11	2.33	2.78	2.52	2.89	2.65	2.23	(2.82)
0.68	1.27								
0.78		0.76	1.87	1.71	2.00	2.30	2.05	1.69	(2.56)
1.04		0.72	1.38	1.29	1.42	1.93	1.64	1.47	
1.20	0.75								
1.30		0.70	1.30	1.30	1.32	1.64	1.41	1.31	(2.23)
1.56			1.24	1.06	1.04	1.38	1.25	1.21	
1.72	0.61								
1.82		0.58							
2.08			1.01	1.07	0.81	0.97	1.03	1.02	(2.35)
2.34		0.44							
3.13			1.00	0.88	0.69	0.48	0.77	0.77	(2.05)
4.17			0.80	0.70	0.53	0.30	0.61	0.74	(2.09)
5.21			0.74	0.55	0.41	0.26	0.50	0.64	(1.60)
6.25			0.74	0.45	0.32	0.28	0.45	0.56	(1.70)
7.29			0.68						(2.01)
8.34								0.49	

Schweifes seine Leuchtkraft unverändert beibehält[1]). Was die Geschindigkeit v angeht, so kann man für das kurze hier in Betracht kommende Schweifstück die abstoßende Kraft der Sonne als konstant und in unveränderlicher Richtung wirkend annehmen, sodaß die Bewegung der Schweifteilchen relativ zum Kometenkopf einfach die freie Fallbewegung — nur in der Richtung von der Sonne weg — wird. Für dieselbe gilt dann, wenn man die Ausgangsgeschwindigkeit gleich Null setzt:

$$v = \text{const.}\,.\,\sqrt{a}$$

1) Die Gleichung gilt auch noch, wenn sich mehrere Ströme verschiedener Geschwindigkeit überlagern, falls die Geschwindigkeit mit wachsendem Abstand vom Kopfe bei ihnen allen in demselben Verhältnis ansteigt.

Herr H. D. Curtis hat in den Publications of the Astronomical Society of the Pacific 1910 pag. 125 die in Tabelle 2 folgende Zusammenstellung beobachteter Geschwindigkeiten v einzelner Schweifknoten mitgeteilt. (Die als unsicher bezeichneten Werte sind ausgelassen.)

Tabelle 2.

Datum	a	v km/sec	$\frac{v}{\sqrt{a}}$
Mai 2.—3.	0	0.9	—
Mai 13.—14.	0.0032	4.9	87
Mai 27.—28.	0.0042	12.8	198
Mai 25.—26.	0.0100	19.3	193
Juni 2.—3.	0.0147	31.8	263
Mai 28.—29.	0.0186	36.8	270
Mai 31.—Juni 1.	0.0191	33.3	241
Juni 6.	0.0237	43.8	285
Mai 26.—27.	0.0270	38.4	234
Juni 6.—7.	0.0438	70.1	331
Mai 30.—31.	0.0712	72.4	271
Juni 7.—8.	0.0905	91.2	303

Bildet man die Quotienten $\frac{v}{\sqrt{a}}$, so sieht man, daß man für Distanzen vom Kopf > 0.01 angenähert eine Konstante erhält, und daß man aus den Beobachtungen höchstens auf eine Neigung zu stärkerem Anwachsen der Geschwindigkeit mit der Entfernung vom Kopf, verglichen mit der parabolischen Formel, schließen kann.

Nimmt man die parabolische Formel an und bildet dementsprechend die Produkte $I \cdot \sqrt{a}$, so erhält man die Zahlen der Tabelle 3 (siehe folgende Seite), wobei zur Erhöhung der Übersicht der Wert des Produktes im Abstand $0^0.52$ (1 mm auf den Zeißaplanataufnahmen) vom Kopf stets gleich 1 gesetzt ist. In der letzten Spalte steht das geometrische Mittel aus den Ergebnissen der einzelnen Aufnahmen, wobei die wenigen Zahlen vom 1. Mai weggelassen wurden und die vom 14. Mai nur halbes Gewicht erhielten.

Wider unsere Erwartung folgt aus diesen Zahlen, daß die Helligkeitsabnahme des Kometenschweifes zum größten Teil durch die Dichteabnahme erklärt wird. Eine genaue Konstanz der Zahlen ist nicht zu fordern. Dazu sind die wirklichen Verhältnisse im Kometenschweif zu ver-

With E. Kron in Nachr. Kgl. Ges. d. Wiss. zu Gött., Math. Phys. Klasse 1911, *No. 3,* 197–208

Tabelle 3.

Datum	Mai 1.	Mai 3.	Mai 7.	Mai 8.	Mai 9.	Mai 9.	Mai 10.	Mai 12.	Mai 14.	Mai 3.—14.
Objektiv	Zeiß	Zeiß	Zeiß	Zeiß	Zeiß	Goerz	Goerz	Zeiß	Zeiß	Geometr. Mittel
Kopfabstand a	$I\sqrt{a}$									
0°.26		0.63								
0.42	0.84									
0.52		1.00	1.00	1.00	1.00	1.00	1.00	1.00	(1.00)	1.00
0.68	1.29									
0.78		0.83	0.98	0.75	0.97	0.97	0.95	0.93	(1.11)	0.92
1.04		0.91	0.84	0.65	0.79	0.94	0.88	0.93		0.84
1.20	1.20									
1.30		0.99	0.88	0.74	0.83	0.90	0.84	0.93	(1.25)	0.89
1.56			0.92	0.66	0.71	0.83	0.82	0.93		0.81
1.72	1.16									
1.82		0.98								
2.08			0.95	0.77	0.64	0.67	0.78	0.91	(1.66)	0.84
2.34		0.85								
3.13			1.05	0.78	0.67	0.41	0.71	0.84	(1.77)	0.77
4.17			0.97	0.71	0.60	0.30	0.65	0.93	(2.09)	0.71
5.21			1.10	0.56	0.44	0.34	0.59	0.87	(2.08)	0.66
6.25			1.00	0.63	0.51	0.29	0.60	0.91	(1.79)	0.66
7.29			1.09						(2.66)	
8.34								0.88		

wickelt und zu ungeklärt, die Strömung ist vor allem nicht stationär, wie bei Ableitung der Gleichung (1) vorausgesetzt wurde. Bei zunehmender Tätigkeit des Kometen werden vermutlich zunächst die dem Kopf benachbarten Schweifpartien heller, bei abnehmender wird das Schweifende relativ größere Helligkeit festhalten. Auch läßt sich die mögliche Größe der systematischen Messungsfehler schwer abschätzen.

Die Konstanz der Zahlen würde übrigens im Mittel noch besser werden, wenn man die Zunahme der Geschwindigkeit gemäß den Daten von Herrn Curtis stärker als beim freien Fall, ansetzen würde.

Wenn man aber auch die am stärksten abnehmende Zahlenreihe, die vom 9. Mai, herausgreift, so folgt daraus immer noch, daß die Leuchtkraft des Massenteilchens des Kometen in einer Zeit von 3 Tagen — so lange braucht das Teilchen nach Herrn Curtis' Geschwindigkeiten etwa bis zu einer Entfernung von 6° vom Kometenkopf — nur auf ein Drittel ihres Wertes abgenommen hätte. Es scheint wenig wahrscheinlich, daß ein derartiges Dauerleuchten aus der Energie der Schweifmassen selbst bestritten werden kann. Hingegen erklärt sich unser Ergebnis ohne weiteres, wenn man annimmt, daß es sich bei den Kometenschweifen um Fluoreszenz- oder Resonanzstrahlung handelt, die durch die Sonnenstrahlung hervorgebracht wird. Herr Wright (Lick Observatory Bulletin Nr. 174) hat im Schweifspektrum des Kometen 1910 a sehr wahrscheinlich bereits die Resonanzstrahlung des Natriums beobachtet. Einen analogen Ursprung würde man für das *CO*- und *CH*-Bandenspektrum des Schweifes des Halleyschen Kometen vorauszusetzen haben. Die Intensität der Fluoreszenz pro Masseneinheit wird dann übrigens längs des Schweifes nicht konstant sein, sondern mit dem Quadrat der Entfernung von der Sonne abnehmen. Dies gibt im Durchschnitt für die obigen Beobachtungstage etwa eine Abnahme von 15 % bei 6° scheinbarem Abstand vom Kopf, macht also quantitativ nicht viel aus, liegt aber im Sinne der beobachteten Zahlen.

Während im Vorstehenden nur die relativen Helligkeiten der verschiedenen Teile des Schweifes untersucht wurden, ermöglichen die Aufnahmen der Teneriffaexpedition auch noch absolute Helligkeiten abzuleiten infolge des Umstandes, daß auf den meisten Bildern bei der ungenauen Fokussierung die Sterne Scheibchen von hinreichender Größe sind, um sowohl den Scheibendurchmesser (0,1 bis 0,3 mm) als auch die Schwärzung der Scheibe messen zu können. Die folgende Tabelle (4) gibt für jedes Datum den benutzten Vergleichstern und dessen Größe nach der Göttinger Aktinometrie (aber auf Harvard-System reduziert) an. Als Einheit der Flächenhelligkeit wurde die mittlere Flächenhelligkeit der Sonne gewählt und als photographische Sterngröße der Sonne im Harvard-System nach O. Birck (Dissertation, Göttingen 1908) der Betrag $-25^{m}.8$ angenommen. Damit ergab sich die in Tabelle 4 in der 3. Spalte stehende Flächenhelligkeit des Kometenschweifs an der hellsten Stelle des Querschnitts in 0°.52 Abstand vom Kopf. Die Querschnittsintensitäten ließen sich natürlich durch Anbringen derselben Faktoren auf die Flächenhelligkeit der Sonne reduzieren.

Unter α findet man angegeben, welchen scheinbaren Abstand vom Kometenkopf ein Punkt auf dem Radiusvektor des Kometen in 0.015 Astron. Einheiten wirklichem Abstand von demselben hat. Unter I findet man die Querschnittsintensitäten aus Tabelle (1) für diesen Abstand vom Kopf interpoliert und auf die Flächenhelligkeit der Sonne bezogen.

Tabelle 4.

Datum	Vergleichstern	Größe	i	α	I	E_0
Mai 1.	10 Piscium	$4^m.38$	$0.84 . 10^{-11}$	$1^\circ.0$	$1.8 . 10^{-11}$	$1.2 . 10^{10}$
Mai 3.	"	"	0.70	1.2	1.4	1.7
Mai 7.	35 Piscium	5.93	1.14	1.5	1.8	1.9
Mai 8.	"	"	0.89	1.6	1.1	1.1
Mai 9.	"	"	1.43	1.7	1.7	1.7
Mai 12.	x Piscium	6.28	1.41	2.0	2.1	1.5
Mai 14.	"	"	0.75	2.2	2.3	1.2

Sobald man das Leuchten des Schweifs auf Reflexion oder Resonanz der Sonnenstrahlung zurückführt, gibt die absolute Flächenhelligkeit des Schweifs einen Anhaltspunkt für die Dichte der Schweifmaterie, und die absoluten Querschnittsintensitäten lassen sich in Verbindung mit den Geschwindigkeitsbestimmungen verwerten, um einen gewissen Aufschluß über die in der Zeiteinheit vom Kometen zur Schweifbildung abgegebene Masse, über die „Ergiebigkeit" des Kometen, zu erhalten.

Man stelle sich zunächst einmal vor, daß der Schweif aus lauter kugelförmigen Partikeln vom Radius ϱ bestehe, die das Sonnenlicht ohne Absorptionsverlust gleichförmig nach allen Seiten[1]) diffundieren. Die Anzahl der Partikel auf der Flächeneinheit, von der Erde aus gesehen, sei n. Der Sonnenradius sei A, der Abstand des anvisierten Raumteiles von der Sonne sei, wie oben, r. Dann ist die Flächenhelligkeit verglichen mit der Flächenhelligkeit der Sonne:

$$i = n\pi\varrho^2 \frac{A^2}{2r^2}. \qquad (2)$$

Ist q die senkrecht zum Radiusvektor und zur Gesichtslinie gezogene Koordinate, so wird die Querschnittsintensität:

1) Und zwar vorwärts sowohl, als auch rückwärts nach der Art von Teilchen, deren Durchmesser unter der Wellenlänge des Lichts liegt.

$$I = \int i dq = \pi\varrho^2 \frac{A^2}{2r^2} \int n dq. \tag{3}$$

Faßt man nun die wirklichen Teilchen ins Auge, welche den Kometenschweif bilden und welche einen Radius ϱ_0 und ein spezifisches Gewicht s haben mögen, so kann man jedem Teilchen doch immer einen „effektiven" Radius ϱ zuschreiben, sodaß die obigen Formeln gelten. Die Größe $\frac{\varrho^2}{\varrho_0^2}$ stellt dann eine Art Albedo dar. Die Gesamtmasse des Kometenschweifs, welche sich auf die Flächeneinheit von der Erde aus gesehen projiziert, wird:

$$\frac{4}{3}\pi\varrho_0^3 sn = \frac{8}{3}\varrho_0\left(\frac{\varrho_0}{\varrho}\right)^2 s \frac{ir^2}{A^2}.$$

Denkt man den Schweif längs des Visionsradius so komprimiert, daß er durchweg die in seinen mittleren Teilen herrschende maximale Dichte annimmt, und nennt die dann entstehende Dicke des Schweifs in Richtung des Visionsradius p, so erhält man für diese maximale Dichte:

$$\delta = \frac{4}{3}\frac{\pi\varrho_0^3 sn}{p} = \frac{8}{3}\frac{\varrho_0}{p}\left(\frac{\varrho_0}{\varrho}\right)^2 s \frac{ir^2}{A^2}. \tag{4}$$

Die Gesamtmasse, welche sich in der Zeiteinheit durch den Schweifquerschnitt bewegt, die Ergiebigkeit, wird:

$$E = \frac{4\pi}{3}\varrho_0^3 s \,.\, v\int n dq = \frac{8}{3}\varrho_0\left(\frac{\varrho^0}{\varrho}\right)^2 sv \frac{Ir^2}{A^2}.$$

Die Längeneinheit, die oben bei der Berechnung der Querschnittsintensität verwandt wurde, war 0.2 mm auf den Aufnahmen mit dem Zeißaplanaten entsprechend 6'.2. Die wirkliche Größe dieser Längeneinheit ist Δ sin 6'.2, wenn Δ der Abstand des Kometen ist. Hiermit sind also die obigen Größen I noch zu multiplizieren. Fügt man schließlich noch den Faktor sin ψ hinzu, der, wie erwähnt, auf normale Durchsicht reduziert, so lautet die vollständige Formel zur Berechnung der Ergiebigkeit aus den Querschnittsintensitäten I der Tabelle:

$$E = \varrho_0\left(\frac{\varrho_0}{\varrho}\right)^2 sE_0, \qquad E_0 = Iv\frac{r^2}{A^2}\Delta \sin\psi \,.\, \sin 6'.2. \tag{5}$$

Für die Strömungsgeschwindigkeit in der Entfernung 0.015 Astr. Einheiten vom Kopf ergibt sich aus Herrn Curtis' Zahlen etwa $v = 32$ km/sec. Mit diesen Daten sind die Werte E_0 der

Tabelle (4) und zwar gleich in Einheiten des *c. g. s.*-Systems gerechnet. Im Durchschnitt hat man also etwa:

$$E = 1.5 \, . \, 10^{10} \, \varrho_0 \left(\frac{\varrho_0}{\varrho}\right)^2 s.$$

Nimmt man den räumlichen Schweifquerschnitt von angenähert zylindrischer Form an, so ergibt sich aus der Seitenansicht des Schweifes in $0^0.5$ Abstand vom Kopf eine Dicke des Schweifes p — im oben erläuterten Sinne — von etwa 300 000 km (3.10^{10} cm). Damit folgt unter Verwendung des Durchschnittswertes $i = 1.0 \, . \, 10^{-11}$ aus Tabelle (4):

$$\delta = 2.10^{-17} \, \varrho_0 \left(\frac{\varrho_0}{\varrho}\right)^2 s$$

Für die Beschaffenheit der Teilchen wollen wir nun zwei Annahmen verfolgen:

1. Entsprechend der Arrhenius'schen Theorie des Strahlungsdrucks seien die Teilchen von der Größenordnung $\varrho_0 = 10^{-4}$ cm, das spezifische Gewicht sei 1. Der effektive Radius wird bei solchen Teilchen der Größenordnung nach gleich dem wirklichen und es sei hier einfach $\varrho = \varrho_0$ gesetzt. Wir sehen dabei darüber hinweg, daß derartige Teilchen wesentlich das Sonnenspektrum reflektieren müßten, statt ein Bandenspektrum zu geben, und für sie im Grunde nur der Bruchteil des Schweiflichtes, der aus reflektiertem Sonnenlicht besteht, in Rechnung gesetzt werden darf.

2. Wir sehen die einzelnen Teilchen als fluoreszierende Moleküle an, deren Radius ϱ_0 dann von der Größenordnung 10^{-8} cm ist. Es ist nicht unwahrscheinlich, daß der effektive Radius eines solchen Moleküls ebenfalls von der Größenordnung seines wirklichen Radius ist[1]), wir wollen daher auch hier $\varrho = \varrho_0$ setzen und für das Molekulargewicht s den Betrag 20 nehmen.

Man erhält in beiden Fällen:

Hypothese I.

$E = 1500$ kg/sec $= 130000$ Tonnen/Tag

$\delta = 2 \, . \, 10^{-21}$

Hypothese II.

$E = 150$ gr/sec $= 13$ Tonnen/Tag

$\delta = 4 \, . \, 10^{-24}$

1) Dies gilt wenigstens, wenn die Fluoreszenz darauf beruht, daß einige wenige Elektronen im Molekül — nur durch Ausstrahlung gedämpft — mit Lichtfrequenzen um ihre Ruhelagen schwingen.

Die freie Weglänge der Moleküle wäre bei letzterer Dichte von der Größenordnung 10^{19} cm entsprechend der Distanz der nächsten Fixsterne, Zusammenstöße würden also nicht mehr vorkommen.

Setzt man die Erde einen Tag lang einem Strom der Dichte 10^{-22} von der Geschwindigkeit 100 km/sec aus — diese Zahlen müßten nach Herrn Curtis' Geschwindigkeiten und nach Hypothese I. beim Durchgang der Erde durch den Kometenschweif etwa gegolten haben — so trifft auf jeden Quadratzentimeter der Erde eine Masse von 10^{-10} gr, welche eine Energie von 1/10000 Grammkalorie mitbringt. Die gesamte von der Erde abgefangene Masse beträgt dann 250 Tonnen. Da diese Zahlen obere Grenzwerte bilden dürften, so sieht man, wie wenig materielle Einwirkung von dem Durchgang durch den Kometenschweif zu erwarten war.

Extinktion) ergab, daß der Kern etwa um 1 Größenklasse heller war als der Stern *d* Piscium. Am 30. April, 1. und 3. Mai wurden zahlreiche Aufnahmen des Kometen mit der Stereoskop-Kamera bei verschiedenen Expositionszeiten gemacht. Die Helligkeit des Kerns nahm von Tag zu Tag zu, sie wurde am 3. Mai um etwa $0^m.1$ bis $0^m.2$ höher geschätzt als *α* Pegasi. Der Schweif war trotz des hellen Mondscheins zuletzt über die Verbindungslinie »Venus-*α* Pegasi« hinaus zu verfolgen.

Auf unserer höheren Beobachtungsstation »Alta Vista« (3300 Meter) war die Witterung so günstig, daß vom 7. bis zum 14. Mai, als wir die Beobachtungen abbrachen, ohne Ausnahme an jedem Tag Aufnahmen des Kometen teils mit der Stereoskop-Kamera, teils mit einem Görzschen Anastigmaten von 180 mm Brennweite gemacht werden konnten. Während dieser ganzen Zeit war der Komet in der reinen und durchsichtigen Höhenluft schon bald nach seinem Aufgang eine prachtvolle Erscheinung. Die Helligkeit des Kerns wurde an den letzten Tagen weit höher als die eines Sterns 1. Größe geschätzt, die Länge des Schweifes erreichte am 7. Mai nach den Schätzungen mit bloßem Auge mindestens 20°, am 10. Mai mindestens 30° und am 14. Mai bei etwas dunstiger Luft etwa 50° bis 60°. In der Nacht vom 18. zum 19. Mai, die wir in Orotava zubrachten, war der Himmel mit leichtem Gewölk bedeckt; irgend eine auffällige Erscheinung konnte nicht wahrgenommen werden.

Die Teneriffa-Expedition hat von neuem einen Beitrag zu der Lehre geliefert, daß man bei ähnlichen Erscheinungen wie der Halleysche Komet keine Mühen und Kosten scheuen sollte, um günstige Beobachtungsstationen insbesondere in größeren Höhen über dem Meere aufzusuchen und die betreffenden Expeditionen mit den allerbesten instrumentellen Hilfsmitteln auszurüsten. *G. Müller.*

Der Komet wurde in Potsdam am Steinheilschen Refraktor von 13.5 cm Öffnung und 216 cm Brennweite mit nicht ganz 80-facher Vergrößerung seit April 26 an allen Morgen, an denen es die Witterung überhaupt gestattete, von seinem Aufgange an beobachtet, bis er in der Dämmerung verschwand, also gewöhnlich von etwa $14^{3}/_{4}^{h}$ bis etwas nach 16^h M. Z. Potsdam. Da der Osthimmel fast immer von dichtem Dunst oder Wolken mehr oder weniger bedeckt war, konnte der Komet häufig nur in Wolkenlücken gesichtet werden, und gewöhnlich war die Dämmerung, wenn er sich über die Dunstschicht erhob, schon zu weit vorgeschritten, als daß noch Beobachtungen des Schweifes gemacht werden konnten. Aus gleichem Grunde gelang es nur an zwei Morgen, Mai 1 und 7, für photometrische Zwecke brauchbare photographische Aufnahmen des Kometen zu erhalten.

Im allgemeinen war nur der Kern des Kometen in rötlich gelbem Lichte mehr oder weniger verwaschen mit der Koma sichtbar, so April 26, 29, 30, Mai 1. Schweifansätze waren meist nur angedeutet und konnten einigermaßen deutlich nur an einem Morgen, Mai 5, verfolgt werden und auch da nur, wenn das ganze Bild im Gesichtsfelde bewegt wurde. Ihre Länge an diesem Morgen betrug nur einige Bogenminuten, der Kern war verwaschen und oval, der innere Teil der Koma scharf abgegrenzt. Eine deutlich sternartige Abgrenzung des Kernes gegen die Koma zeigte der Komet erst Mai 6, als der Kern durch Anschluß an nahe stehende Sterne zu etwa 6^m geschätzt wurde. Mai 7 wurde er von Herrn Prof. *Schwarzschild* und mir ebenfalls zu etwa 6^m geschätzt und konnte der Komet leicht von beiden Beobachtern als rötlicher Stern mit bloßem Auge wahrgenommen werden. Mai 11 konnte er nur zwischen Wolken beobachtet werden, er war wieder verwaschen und schien wesentlich schwächer geworden zu sein. Auch Mai 13 war er nur noch schwach zu sehen und ebenso Mai 14 und 15. Mai 16 wurde er nicht mehr gefunden.

Der Beobachter hatte den Eindruck, als ob der Komet fortschreitend an Helligkeit abnahm. Da der Komet an allen Morgen bis zu heller Dämmerung beobachtet wurde, scheint es ausgeschlossen, daß die beobachtete Helligkeitsabnahme nur eine Täuschung gewesen ist, hervorgerufen von der zunehmenden Sonnennähe. Der Komet war Anfang Mai auch auf hellem Hintergrunde hell zu sehen, während er später auch auf dunklerem Hintergrunde wesentlich schwächer erschien. Die zunehmende Extinktion allein reichte sicher nicht aus, um den Unterschied zu erklären.

Das Spektrum war bei der relativ geringen Helligkeit des Kernes mit einem Okularspektroskope gewöhnlich nur mit Mühe zu sehen und schien im wesentlichen ein kontinuierliches zu sein. Banden oder Intensitätsdifferenzen im Spektrum konnten nicht mit Sicherheit erkannt werden.

Mai 6 und 7, als der Komet am hellsten erschien, wurde mit einem Nicolschen Prisma und mit einem Savartschen Polariskop das Licht auf Polarisation untersucht. Doch war der Himmelshintergrund an beiden Morgen bereits so hell und die Konstellation derart, daß der Himmelshintergrund polarisiertes Sonnenlicht in den Apparat sandte und nicht entschieden werden konnte, wie weit die beobachteten Polarisationserscheinungen von ihm oder vom Kometen herrührten. *W. Münch.*

Während des Durchganges des Halleyschen Kometen über die Sonnenscheibe sind von Herrn Prof. *Schwarzschild* am großen Refraktor des Potsdamer Observatoriums mit dem Spektrographen Nr. 1 zwölf Aufnahmen des Sonnenspektrums gemacht worden, zwecks Entscheidung der Frage, ob das Sonnenlicht beim Durchgange durch den Kometen eine merkliche selektive Absorption erleidet. Diese zwölf Aufnahmen zerfallen in zwei Serien zu je 6 Aufnahmen, von denen die erste zu einer Zeit gemacht wurde, als sich der Komet, gemäß den von Prof. Kobold (A. N. 4413) angegebenen Durchgangszeiten in der Nähe des Mittelpunktes der Sonne befand, während sich die zweite Serie auf das Ende des Durchganges bezieht, als der Komet in unmittelbarer Nähe des Sonnenrandes stand. Da die der Ephemeride entnommenen Deklinationen des Kometen genauer waren als die Rektaszensionen, wurde der Spalt des Spektrographen in der Richtung der stündlichen Bewegung eingestellt, und dann in rascher Aufeinanderfolge eine Reihe von Aufnahmen bei konstanter Rektaszension und in etwas verschiedenen Deklinationen gemacht. Die Länge des Spaltes in Bogenmaß betrug ungefähr $3^1/_2{}'$, so daß ein ziemlich großes Gebiet der Sonne überdeckt wurde. Die Expositionsdauer betrug etwa $^1/_4{}^s$. Die 6 Aufnahmen der I. Serie wurden in der Zeit von $4^h 55^m$ bis $5^h 10^m$ (mitteleuropäische Zeit), die der zweiten in der Zeit von $5^h 28^m$ bis $5^h 30^m$ gemacht.

Um die Genauigkeit der Einstellung auf den unsichtbaren Kometen zu kontrollieren, sind aus den Elementen der Bahn (Lick Observatory Bulletin 179) zwei Kometenörter, vor und nach dem Durchgange, berechnet worden, für welche Beobachtungen vorlagen. (Es ist nicht möglich gewesen, aus den Ephemeriden von Dr. Smart, A. N. 4379, und von Fischer-Petersen, A. N. 4413, durch Interpolation brauchbare Werte zu erhalten).

Die Korrektionen der berechneten Örter ergaben sich zu:

	$\Delta\alpha$	$\Delta\delta$	Beobachtung von:
Mai 15.9960 Gr. M. Z.	$-35^s.1$	$-1'\ 3''$	Aitken, Mt. Hamilton (A. N. 4413)
Mai 20.3015 »	-55.0	$+2\ 28$	Battermann, Königsberg (A. N. 4414)

Durch Interpolation wurden für die Zeit des Durchganges die Korrektionen erhalten: $-47^s.6$ $+0'\ 59''$. Die damit berechneten Positionen des Kometen zur Zeit der Beobachtung, sowie die jeweilige Lage des Spaltmittelpunktes des Spektrographen sind in der nachstehenden Tabelle vereinigt:

	M. Z. Gr.	$\alpha_{☄}$	α_{Spalt}	$\delta_{☄}$	δ_{Spalt}
Mai 18	$15^h 55^m$	$3^h 39^m 37^s$	$3^h 40^m 0^s$	$+19° 36'.8$	$19° 38' - 19° 31'$
	16 10	40 15		37.3	
	16 28	3 41 0	3 41 15	+19 37.9	19 38 − 19 32
	16 30	41 5		38.0	

Daraus ergibt sich, daß der Spalt sehr nahe am Kopfe des Kometen gestanden hat und ihn auch wahrscheinlich mehrmals berührt hat. Die Spektralaufnahmen, auf welchen das Sonnenspektrum vom ultravioletten Teile bis zu den *D*-Linien sichtbar ist (AGFA Chromo-Platten), wurden mit Hülfe des neuen Toepferschen Meßapparates untersucht. Aus drei Fraunhoferschen Linien (λ 5890.1, 5167.5 und 4005.4) wurde eine Hartmannsche Dispersionsformel abgeleitet, mit deren Hülfe im Spektrum alle Stellen aufgesucht werden konnten, an denen vom Kometen herrührende Absorptionslinien zu erwarten waren, jedoch ohne daß es gelungen wäre, ihre Anwesenheit zu konstatieren.

Besondere Aufmerksamkeit erhielten die folgenden Stellen, an denen nach Angabe des Lowell Observatory Bulletin Nr. 47, im Spektrum des Halleyschen Kometen helle Emissionslinien auftreten:

$$\lambda = 3883,\ 4214,\ 4737,\ 5165,\ 5635$$

und auch die Wellenlängen der von Dr. Rosenberg (Astroph. Journ. vol. 30, p. 271) im Spektrum des Kometen Morehouse beobachteten Strahlungen

$$\lambda = 3804,\ 3898,\ 4035,\ 4277,\ 4563,\ 4682,\ 5100,\ 5600.$$

An keiner dieser Stellen ist eine Abweichung vom normalen Sonnenspektrum zu sehen, so daß auch die selektive Absorption des Lichtes durch den Kometenkörper nur außerordentlich schwach sein kann. *v. d. Pahlen.*

Die Bedeckung des Sterns AG Lpz II 4615[1]), über die Herr Miethe in A. N. 4417 berichtet, konnte von keinem der hiesigen Beobachter mit voller Konzentration verfolgt werden, doch ergeben die Wahrnehmungen, daß der Komet um Bruchteile einer Bogensekunde nördlich von dem Stern vorbeiging und daß ein eine größere Reihe von Sekunden dauerndes Verschwinden des Sterns nicht erfolgt ist. Aus Herrn Miethes Beobachtung, die sich aus der verhältnismäßig geringen von ihm verwandten Vergrößerung erklärt, darf also nicht auf eine starke Absorptionswirkung des Kometen geschlossen werden. *K. Schwarzschild.*

Im Anschluß hieran seien noch die folgenden bei der Redaktion eingegangenen Mitteilungen gegeben:

1. Von Professor *H. Battermann*, Direktor der Sternwarte Königsberg.

In Königsberg war in der Nacht vom 18. zum 19. Mai der Himmel großenteils von tiefliegenden Wolken bedeckt, über denselben anscheinend durchsichtige Luft; außergewöhnliche Erscheinungen nicht zu bemerken. Dämmerung und Mondschein hinderten Beobachtung schwacher cölestischer Objekte.

In den Tagen vom 19. bis 23. Mai war hier fast immer vollständig wolkenloser Himmel bei bemerkenswert durchsichtiger Luft und verhältnismäßig tiefblauer Färbung. Außergewöhnliche Erscheinungen fehlten durchaus; die Dämmerung war weniger auffällig als hier häufig der Fall ist, abgesehen natürlich von der normalen Helligkeit des Nordhimmels. Vom Bishopschen Ring um Sonne oder Mond sowie von leuchtenden Nachtwolken war keine Spur zu bemerken.

Königsberg, 1910 Juli 4.

2. Brieflich teilt Dr. *Meyermann*, Direktor des Observatoriums in Tsingtau, 1910 Juni 20, mit:

»Wie ich erst jetzt erfahre, sah Herr Dr. *Brücker*, Direktor der Schantung-Bergbau-Gesellschaft, in Hungschan, etwa 180 km westlich von Tsingtau, am Abend des 19. Mai nach Sonnenuntergang um $10\frac{1}{2}$ Uhr den Kometenschweif etwa an der Stelle, an der die Sonne untergegangen war. Dr. Brücker sah auch in den nächsten Nächten den Kometen regelmäßig wieder und verfolgte ihn in seiner Lichtabnahme und Bewegung. Das Vorliegen einer Verwechslung des Datums scheint durch andere Umstände ausgeschlossen.«

Telegramm aus Santiago de Chile, 1910 Juli 4.

Korrektion der Ephemeride Ebell (A. N. 184.317): 1910 Juli 1 $\Delta\alpha = +2^s.5$ $\Delta\delta = -0'.2$. Mittlerer Durchmesser der Nebelmasse 1', 7^m, keine Verdichtung, Schweif 2°. *F. Ristenpart.*

---

[1]) Durch Herrn *Pechüle* wurde schon darauf aufmerksam gemacht, daß in A. N. 4417 statt: AG Lpz I zu lesen sei: AG Lpz II. *Kb.*

Section 4

Structure, Kinematics and Dynamics of Stellar Systems

Structure, Kinematics and Dynamics of Stellar Systems

Commentary by R. Wielen, Heidelberg

Karl Schwarzschild was a pioneer in stellar dynamics. His most important contribution in this field was undoubtedly the description of the velocity distribution of stars as a velocity ellipsoid [4.2,3], which is still substantially valid today, and which has been named "the Schwarzschild distribution" after him. He also indicated the way in which the relaxation time of a stellar system could be calculated from the mutual gravitational perturbations of individual stars [4.11]. Other papers concerned with stellar systems dealt with the determination of the stellar density in our Milky Way system [4.5,6,10], as well as certain additional individual queries [4.1,7–9]. He also wrote popular articles about the "System of Fixed Stars", which give a particularly clear idea of the basic picture held by Schwarzschild [4.4].

The Velocity Distribution of Stars

After astronomers had realized from measurements of proper motions and radial velocities that individual stars were moving with different space velocities, the question arose of the distribution function of these "random" motions.

At first it seemed plausible to assume that, by analogy with the Maxwellian velocity distribution of atoms in a gas, the "peculiar" velocities of stars had an isotropic distribution. This hypothesis did not agree, however, with the ever more accurate and numerous observations of stellar motions: the stars' peculiar velocities are, on average, significantly higher in one direction, which we now know to be the direction of the galactic centre, than they are in other directions. It was the German astronomer Kobold, who pointed out the anisotropic distribution of stellar velocities from 1890 onwards. The Dutch astronomer Kapteyn then proposed the two-stream theory, in 1904–1905. He postulated two, mutually interpenetrating, stellar streams. Each stream (frequently known in older books as a "swarm" or a "drift") should show an *isotropic* distribution relative to its mean velocity. The mean velocities of the two stellar streams should, however, differ considerably, by about 40 km/s. The superimposition of the two isotropic distributions, displaced relative to one another, would then explain the *anisotropic* distribution of the peculiar velocities. The English astronomer Eddington (1906, 1908) at first strongly supported this theory and developed it further. However, it remained problematic how these streams differed physically from one another.

Karl Schwarzschild's great achievement was to establish the unity of the Milky Way through his "unitarian" ellipsoid hypothesis [4.2]. He postulated a unified distribution of the peculiar velocities $f(U, V, W)$ that had a Gaussian distribution in all directions, but where the velocity dispersion σ was a simple function of direction:

$$f(U,V,W) = \frac{1}{(2\pi)^{3/2}\sigma_U\sigma_V\sigma_W} \exp\left\{-\frac{(U-U_m)^2}{2\sigma_U^2} - \frac{(V-V_m)^2}{2\sigma_V^2} - \frac{(W-W_m)^2}{2\sigma_W^2}\right\}.$$

Here U, V, W are the rectangular components of the peculiar velocities, for example U in the direction of the galactic centre, V in the direction of galactic rotation, and W in the direction of the North Galactic Pole. Schwarzschild used arbitrary directions. U_m, V_m and W_m are the average stellar velocities in these three directions. The above velocity distribution f, which today we call the "Schwarzschild distribution", gives rise to a series of ellipsoidal surfaces of constant frequency in velocity space. The three velocity dispersions σ_U, σ_V and σ_W represent the "velocity ellipsoid". Schwarzschild then showed [4.2,3] that the ellipsoidal theory agreed with observations just as well as the two-stream theory. He also developed special procedures for testing his ellipsoidal theory when no space velocities U, V and W are available, only proper motions. A comparison of the ellipsoid and two-stream theories that is still worth reading on account of its vividness, is given in Eddington's textbook (1914).

Karl Schwarzschild also tried to find a dynamical explanation for this ellipsoidal distribution of stellar peculiar velocities, but was not successful [4.11]. To Schwarzschild, the Milky Way was more like an oversized, flattened star cluster. He was essentially correct in his suggestion that the Sun was "offset" from the centre of the Milky Way (but was approximately in the galactic plane), and that the stars orbited the centre in approximately circular orbits. However, he was wrong in his idea that the direction of the greatest peculiar-velocity dispersion agreed with the direction of this circular motion, because he thought that approximately as many stars were moving round the centre in one direction as in the opposite direction. He did, however, consider other possible explanations. A correct explanation for the velocity ellipsoid was first offered by the theory of galactic rotation put forward by Lindblad (1926, 1927, 1933) and Oort (1927, 1928). We now know that practically all the stars in the galactic disk rotate in the same direction around the distant galactic centre, and the observed peculiar velocities, i.e. the departures from exact circular velocity, are relatively small in comparison with the high circular velocity (about 220 km/s). The so-called epicyclic theory correctly describes the ratio between the axes of the velocity ellipsoid, σ_U/σ_V, and thus the fact that σ_U is significantly greater than σ_V (see for example, Scheffler, Elsässer 1982; Binney, Tremaine 1987).

We now know that, to a close approximation, the Gaussian Schwarzschild distribution is only valid for one generation of stars of equal ages. As the velocity dispersion increases markedly with age, the velocity distribution integrated over all values of age differs considerably from a Schwarzschild distribution (Wielen 1977; Wielen, Fuchs 1985, 1989; Fuchs, Wielen 1987). The quantitative explanation of the observed velocity distribution still remains a central problem in stellar dynamics.

Estimation of Relaxation Times in Stellar Systems

In gases, after any disturbance, the Maxwellian distribution of velocities is restored through collisions between the atoms on a time-scale known as the "relaxation time". Schwarzschild [4.11] was one of the first to investigate how the regular, Gaussian, stellar velocity-distribution could have been produced, and the length of time – the stellar relaxation time – required for it to take place. He considered the successive perturbations $\Delta\varepsilon$ to the orbital energy ε of a star, that occur during close encounters (but not direct collisions) between two stars. He estimated the relaxation time (essentially correctly) by summing the values of $(\Delta\varepsilon)^2$ over time. The value that he obtained for the relaxation time was approximately 10^{12} years. As no independent value for the age of the Milky Way (or for that of the whole universe) was available, however, he was not able to draw any further definite conclusions from this.

In deriving the relaxation time, Schwarzschild used an invalid simplification: obviously unintentionally he took it for granted that the barycentric velocity of two stars in the overall stellar system lay in the plane of the relative motion of the two stars, but in general this is not true. Despite this, the determination of the relaxation time from the changes in the orbital energy of stars through close encounters with other stars was pioneering work by Schwarzschild, and which was only overtaken, after 1940, by Chandrasekhar's work (1942).

Schwarzschild's paper on the relaxation time [4.11] was published posthumously. The actual period when the fragment was written can be set as somewhere between 1911 and 1916 (the year of his death) on the basis of the references quoted. It was probably written between April 1912 (Turner) and the end of 1914 (relevant papers by Jeans (1913, 1916) not being quoted).

We now know that in galaxies (but not in star clusters or clusters of galaxies) relaxation through mutual encounters between stars is negligible, because the age of the galaxies, including the Milky Way, is only about 15 billion years, and is therefore much shorter than the stellar relaxation time. On the other hand, the diffusion of stellar orbits caused by other inhomogeneities in the galactic gravitational field is often much greater. It gives rise to an effective relaxation time of only 10^8 to 10^9 years (Wielen 1977). The basic cause of this irregular galactic field is still uncertain. Primary candidates for being the source of the inhomogeneities are giant molecular cloud complexes, portions of spiral arms, or perhaps even massive black holes from the dark galactic corona (Wielen, Fuchs 1985, 1989; Fuchs, Wielen 1987).

Stellar Statistics

The basic concept of stellar statistics consists of obtaining the density distribution $\varrho(\vec{r})$ of stars in the Milky Way from counting the number of stars of different apparent magnitude m. As stars have differing absolute magnitudes M, it is necessary to know the luminosity function $\psi(M)$ as a distribution function of M. In addition, interstellar extinction $A(\vec{r})$ complicates the task. Mathematically, the observed number of stars

$N(m)$ turns out to be the integral over a suitable product of ϱ and ψ. In two papers [4.5,6], Schwarzschild examined how to solve this integral equation, i.e. how to determine $\varrho(\vec{r})$ from $N(m)$ for a known $\psi(M)$. The extinction A was neglected. The papers show Schwarzschild's mathematical skill and his realistic mistrust of the observational data then available. The motion of the stars [4.9,10] is also included to some extent, so that, for example, the predictions of the model distribution can be compared with observed proper motions. A clear description of Schwarzschild's contribution to the study of stellar statistics is given in the textbook by von der Pahlen and Gondolatsch (1937).

Nowadays more direct methods of determining the density distribution of stars are used. Photometric and spectroscopic data – for example of special subgroups of stars – can often be used to derive the distances of individual stars, taking interstellar extinction fully into account. It is then possible to obtain the density distribution directly by simple counts, without having to solve difficult integral equations, often with inappropriate simplifications, as was previously the case. Another method frequently used, is to set up a model, making certain assumptions about the stellar density distribution, luminosity function and velocity distribution, and also the interstellar-extinction distribution. Then the function $N(m)$ is calculated and compared with observations, making iterative corrections to the model. This procedure thus again avoids having to solve the basic equation of stellar statistics as an integral equation.

References*

Binney, J., Tremaine, S. (1987) *Galactic Dynamics* (Princeton University Press)

Chandrasekhar, S. (1942) *Principles of Stellar Dynamics* (University of Chicago Press) (Reprinted 1960)

Eddington, A.S. (1906) *Monthly Not. Roy. Astron. Soc.* **67**: 34; (1908) *Monthly Not. Roy. Astron. Soc.* **68**: 588; (1914) *Stellar Movements and the Structure of the Universe* (Macmillan, London)

Fuchs, B., Wielen, R. (1987) in *The Galaxy* edited by G. Gilmore et al. (Reidel, Dordrecht) p. 375

Jeans, J.H. (1913) *Monthly Not. Roy. Astron. Soc.* **74**: 109; (1916) *Monthly Not. Roy. Astron. Soc.* **76**: 554

Kapteyn, J.C. (1904) Address before the St. Louis Exposition Congress; (1905) Report of the British Ass. for the Advancement of Science, p. 257

Kobold, H. (1890) *Astron. Nachr.* **125**: 65; (1897) *Astron. Nachr.* **144**: 33; (1899) *Astron. Nachr.* **150**: 271; (1906) *Der Bau des Fixsternsystems* (Vieweg, Braunschweig)

Lindblad, B. (1926) *Arkiv f. Math., Astron. o. Fysik* **19A**, Nos. 21, 27 and **19B**, No. 7; (1927) *Arkiv f. Math., Astron. o. Fysik* **20A**, No. 17; *Monthly Not. Roy. Astron. Soc.* **87**: 553; (1933) in *Handbuch der Astrophysik*, Vol. V, Part 2 (Springer, Berlin) p. 937

Oort, J.H. (1927) *Bull. Astron. Inst. Netherlands* **3**: 275; **4**: 79, 91; (1928) *Bull. Astron. Inst. Netherlands* **4**: 269

van der Pahlen, E., Gondolatsch, F. (1937) *Lehrbuch der Stellarstatistik* (Barth, Leipzig)

Scheffler, H., Elsässer, H. (1982) *Bau und Physik der Galaxis* (Bibl. Inst., Mannheim)

Turner, H.H. (1912) *Monthly Not. Roy. Astron. Soc.* **72**: 387, 474

Wielen, R. (1977) *Astron. Astrophys.* **60**: 263

Wielen, R., Fuchs, B. (1985) in IAU Symposium No. 106 *The Milky Way Galaxy* edited by H. van Woerden et al. (Reidel, Dordrecht) p. 481; (1989) in *Evolutionary Phenomena in Galaxies* edited by J.E. Beckmann, B.E.J. Pagel (Cambridge University Press) p. 244

* The reference numbers in the text for Schwarzschild's papers correspond to the contents on page XIII and to the bibliography on page 493.

Astron. Nachr. *136* (1894) 81–84

ASTRONOMISCHE NACHRICHTEN.

Band 136. № 3246. 6.

Ueber den Einfluss der saecularen Aberration auf die Fixsternörter.

Von *Karl Schwarzschild.*

Der Einfluss der saecularen Aberration auf die Fixsternörter ist einer geometrischen Darstellung fähig, die vor der üblichen analytischen vielleicht den Vorzug grösserer Deutlichkeit hat.

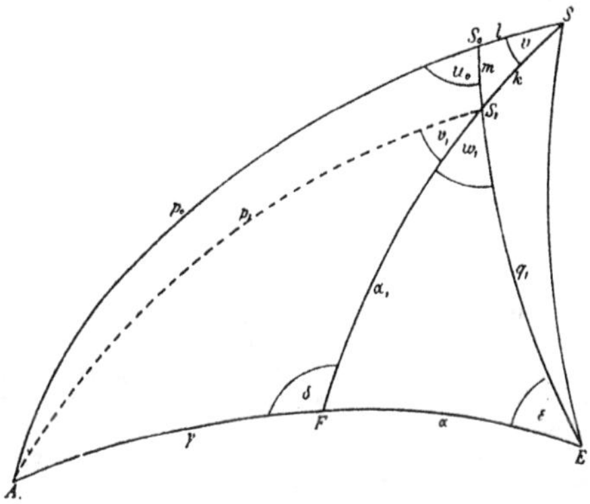

Nennt man c und a die Geschwindigkeit der Erdbewegung in der Bahn um die Sonne und die Geschwindigkeit des ganzen Sonnensystems, sind E und A die Punkte, nach denen diese Geschwindigkeiten gerichtet sind, und bezeichnet F den Apex der Gesammtbewegung der Erde und f die Gesammtgeschwindigkeit derselben, so liegt F auf dem A und E verbindenden grössten Kreis in der Art, dass:

$$\frac{a}{c} = \frac{\sin\alpha}{\sin\gamma} \tag{1}$$

wo α und γ die Bogen grössten Kreises FE und FA sind. Zugleich gilt:

$$f = a\frac{\sin(\alpha+\gamma)}{\sin\alpha} \tag{2}$$

Sei nun S der wahre Ort eines Sterns. Dann besteht die ganze Wirkung der Aberration in einer Verschiebung des Sternbildes auf dem grössten Kreis SF nach dem Punkt S_1, dessen Abstand k von S bestimmt ist durch:

$$\sin k = \frac{f}{V}\sin\alpha_1 \tag{3}$$

wenn man mit V die absolute Lichtgeschwindigkeit und mit α_1 den Abstand $S_1 F$ bezeichnet. Diese Verschiebung SS_1 zerlegen wir, indem wir den grössten Kreis ES_1 bis zum Schnittpunkt S_0 mit dem grössten Kreis AS verlängern, und bestimmen jetzt getrennt die Verschiebungscomponenten $SS_0 = l$ und $S_1 S_0 = m$.

Vernachlässigt man, wie es im Folgenden durchaus geschehen soll, Grössen dritter Ordnung, so kann man das Dreieck SS_1S_0 als gradlinig betrachten und hat nach den Bezeichnungen der Figur:

$$l = k\frac{\sin w_1}{\sin u_0}$$

oder nach (2) und (3):

$$l = \frac{a}{V}\cdot\frac{\sin(\alpha+\gamma)}{\sin\alpha}\cdot\frac{\sin\alpha_1\sin w_1}{\sin u_0}$$

Dreieck FES_1 liefert aber:

$$\sin w_1\sin\alpha_1 = \sin\varepsilon\sin\alpha$$

und Dreieck AES_0:

$$\sin p_0\sin u_0 = \sin\varepsilon\sin(\alpha+\gamma)$$

wo p_0 der Bogen AS_0 ist. Das giebt zusammen:

$$l = \frac{a}{V}\sin p_0 \tag{4}$$

Für die Componente m gilt:

$$m = l\frac{\sin v}{\sin w_1} = \frac{a}{V}\sin p_0\frac{\sin v}{\sin w_1}$$

In diesen Ausdruck sollen lauter vom Punkte S_1 aus gerechnete Bogen und Winkel, die den unteren Index 1 erhalten, eingeführt werden. Man hat in den angegebenen Genauigkeitsgrenzen:

$$m = \frac{a}{V}\sin p_1\frac{\sin v_1}{\sin w_1}\left[1+\operatorname{ctg}p_1(p_0-p_1)+\operatorname{ctg}v_1(v-v_1)\right]$$

Die Dreiecke AFS_1 und EFS_1 liefern, wenn q_1 den Bogen S_1E bedeutet:

$$\sin v_1\sin p_1 = \sin\gamma\sin\delta \qquad \sin w_1\sin q_1 = \sin\alpha\sin\delta$$

woraus unter Berücksichtigung von (1) folgt:

$$a \sin p_1 \frac{\sin v_1}{\sin w_1} = c \sin q_1$$

Setzt man ferner $AS = p$, so gilt:

$$\sin v \sin p = \sin v_1 \sin p_1$$

und durch Differentiation:

$$(v_1 - v) \operatorname{ctg} v_1 + (p_1 - p) \operatorname{ctg} p_1 = 0 .$$

Indem man dies alles in den letzten Ausdruck von m einsetzt, ergiebt sich:

$$m = \frac{c}{V} \sin q_1 \left[1 + \operatorname{ctg} p_1 (p_0 - p) \right]$$

Bedenkt man nun noch, dass

$$p_0 - p = -l = -\frac{a}{V} \sin p_0$$

ist, wofür man hier $-\frac{a}{V} \sin p_1$ einsetzen darf, so bildet sich:

$$m = \frac{c}{V} \sin q_1 \left[1 - \frac{a}{V} \cos p_1 \right] \qquad (5)$$

Die Grösse $c \sin q_1$ ist die alleinige Wirkung der jährlichen Aberration, während die Klammer die von Villarceau gefundene Abhängigkeit der Aberrationsconstante vom Sternort darstellt. Da die letztere nach den bisherigen Erfahrungen verschwindend ist, kann man behaupten, dass die gebräuchlichen Aberrationsformeln den scheinbaren Ort S_1 streng in den Ort S_0 überführen. Dies ist dann der Ort, den unsere Cataloge geben, den man für Praecession und Nutation corrigirt und dessen restirende Veränderlichkeit man als Eigenbewegung auffasst.

Um wahre Oerter zu erhalten, hätte man an die Catalogörter nun noch die Verschiebungen l nach (4) anzubringen. Wie man aus (4) sieht, hängt l nur von der relativen Lage der Punkte A und S ab und ist daher constant, so lange S keine Eigenbewegung besitzt. In diesem Fall ist auch der Catalogort S_0 ohne Eigenbewegung. Für einen bewegten Stern dagegen seien S und S' wahre Oerter zu zwei verschiedenen Zeiten und S_0 und S_0' die zugehörigen Catalogörter. Man hat dann aus der Figur, wenn $AS = p$ und $AS' = p'$ ist:

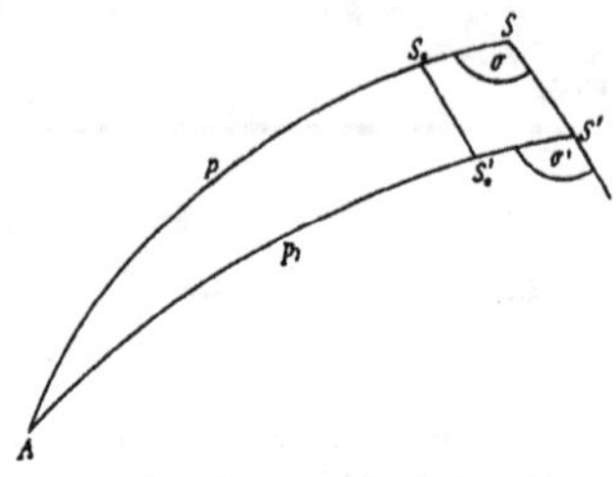

$$\sin p \sin \sigma = \sin p' \sin \sigma'$$

Da man ferner das Viereck $SS'S_0S_0'$ als eben betrachten darf und hinreichend genau

$$S_0S = \frac{a}{V} \sin p , \qquad S_0'S' = \frac{a}{V} \sin p'$$

ist, so sagt diese Gleichung nichts anderes, als dass S_0S_0' parallel SS' ist, d. h. die Richtung der Eigenbewegung bleibt unverändert. Was ihre Grösse anbetrifft, so liefert eine leichte Rechnung:

$$S_0S_0' = SS' \left[1 - \frac{a}{V} \cos p \right] \qquad (6)$$

Die Eigenbewegungen multipliciren sich also mit demselben Factor wie die Aberrationsconstante, der hier mit noch grösserem Recht als oben gleich Eins gesetzt werden darf. Demnach folgt, dass unsere Cataloge die wahren Eigenbewegungen geben.

Als einzige Wirkung der Aberration bleibt eine constante Verzerrung des Himmelsgewölbes übrig, auf deren Beseitigung man stets verzichten wird. Sollte sich später einmal eine Verschiedenheit der Aberrationsconstanten von Stern zu Stern bemerkbar machen, so hätte man nur den Villarceau'schen Factor an Aberrationsconstante und Eigenbewegung anzubringen.

Producte der Praecession und der saecularen Aberration können, wie Herr Dr. Battermann Astr. Nachr. Bd. 118 bemerkt hat, schon deshalb nicht auftreten, weil l nur von der relativen Lage von Stern und Apex, nicht aber von der Wahl eines Coordinatensystems abhängt.

München 1894 März 28. *Karl Schwarzschild*, stud. rer. nat.

Ueber die Identität der Lambert'schen und Olbers'schen Methode zur Berechnung parabolischer Bahnen.

Von *Robert Vogel*.

Es ist bekannt, dass mehrere wichtige Entdeckungen von verschiedenen Personen fast gleichzeitig gemacht wurden. Dass so etwas in Bezug auf Thatsachen vorkommen kann, ist sehr leicht verständlich, aber viel seltener trifft es ein, dass zwei Verfasser nicht gleichzeitig und dennoch unabhängig von einander zwei identische Methoden zur Lösung einer gewissen Aufgabe aufstellen. Das letztere war der Fall mit der Methode, welche von Lambert (Johann Heinrich) im Jahre 1771 und von Olbers im Jahre 1797, also ungefähr ein Vierteljahrhundert später, gegeben wurde.

Vor sechs Jahren fiel mir diese Thatsache beim Lesen des Lambert'schen Mémoire's auf und ich veröffentlichte damals einen Aufsatz in russischer Sprache, worin ich unter anderem auch auf die Identität beider Methoden hinwies;

Nachr. Kgl. Ges. d. Wiss. zu Gött., Math. Phys. Klasse 1907, *No. 5*, 614–632

Ueber die Eigenbewegungen der Fixsterne.

Von

K. Schwarzschild.

Vorgelegt in der Sitzung vom 7. Dezember 1907.

Die Hypothese von der rein zufälligen Verteilung der Spezialbewegungen der Sterne hat sich als unhaltbar erwiesen. Die Sterne bewegen sich nicht durcheinander, wie die Moleküle eines Gases, etwa nach dem Maxwell'schen Gesetz der Geschwindigkeitsverteilung. Herr Kobold hat zuerst gezeigt, daß es eine Reihe ausgezeichneter Bewegungsrichtungen giebt, denen die Sterne mit Vorliebe folgen und welche übrigens alle der Ebene der Milchstraße parallel liegen [1]). Eine zusammenfassende mathematische Formulierung der thatsächlichen Anordnung der Bewegungen hat Herr Kobold nicht versucht, da er offenbar die Erscheinungen für zu verwickelt hielt. In dieser Hinsicht hat Herr Kapteyn einen bedeutsamen Schritt vorwärts gethan, indem es ihm glückte, die wesentlichsten Züge der Verteilung der Eigenbewegungen aus dem Bilde zweier Sternschwärme, die sich durchsetzen, abzuleiten [2]). Die Bevorzugung der Milchstraßenebene kommt dabei dadurch zum Ausdruck, daß die Bewegungsrichtung des einen Schwarmes relativ zum andern parallel zur Milchstraße liegt. Die Kapteyn'sche Vorstellung ist kürzlich von Herrn Eddington [3]) an dem Material von Eigenbewegungen geprüft worden, das aus der Neubearbeitung der Beobachtungen von Groombridge hervor-

1) Vgl. H. Kobold. Der Bau des Fixsternsystems. Braunschweig 1906, pag. 78—150.

2) Report of the British Association for the Advancement of Science 1905, pag. 257.

3) Monthly Notices of the Roy. Astron. Society 1906, Bd. 67, pag. 34.

Nachr. Kgl. Ges. d. Wiss. zu Gött., Math. Phys. Klasse 1907, *No. 5*, 614–632

gegangen ist. Herr Eddington setzt ausdrücklich innerhalb jedes der beiden einander durchdringenden Schwärme das Maxwell'sche Gesetz der Geschwindigkeitsverteilung voraus. Aus der Superposition der beiden Schwärme mit Maxwell'scher Geschwindigkeitsverteilung ergiebt sich bei geeigneter Wahl ihrer Relativbewegung eine so vortreffliche Wiedergabe der beobachteten Anordnung der Geschwindigkeiten, daß man — zumal unter der Wucht von Herrn Eddington's prägnanter Darstellung — den Eindruck hat, hier sei mehr, als eine Interpolation: hier sei eine wirkliche Erkenntnis der Bewegungsvorgänge im Fixsternsystem gewonnen, mit der man sich, wenn sie überraschend erscheint, eben abfinden müsse.

Indessen ist das Abfinden mit der Vorstellung zweier sich durchsetzender „Sternengase" schwer. Man wird an sich nur ungern den Glauben an die durch das Phänomen der Milchstraße dokumentierte Einheit des Fixsternsystems aufgeben. Außerdem zeigen Herrn Eddington's Rechnungen, daß der eine Schwarm in derselben Himmelsrichtung eine Vermehrung der Sternanzahl, eine Verdichtung, aufweist, wie der andere. Auch die Größe der scheinbaren Eigenbewegungen und damit vermutlich die durchschnittliche Entfernung der Sterne beider Schwärme ergiebt sich als gleich. Die beiden „Sternengase" müßten also, während sie durcheinander strömen, gemeinsamen Verdichtungen unterliegen, und es bleibt die Frage nach plausibeln Ursachen offen, welche dies bewirken könnten.

Ich habe daher geglaubt, dasselbe Beobachtungsmaterial, auf das sich Herr Eddington stützt, noch einmal auf Grund einer unitarischen Hypothese über die Sternbewegungen bearbeiten zu sollen. Die notwendige Abänderung der alten unzureichenden Vorstellung regelloser Verteilung der Spezialbewegungen wurde in folgendem gesucht. Bezeichnet man die Komponenten der Geschwindigkeiten eines Sterns gegen einen als ruhend betrachteten Punkt mit U, V, W, so ist die Anzahl der innerhalb des Geschwindigkeitsbereiches $U+\delta U$, $V+\delta V$, $W+\delta W$ liegenden Geschwindigkeiten nach Maxwell gleich: $dU dV dW . e^{-A(U^2+V^2+W^2)}$ (A eine Konstante). Statt dieser Formel habe ich gesetzt:

$$dU dV dW . e^{-AU^2-BV^2-CW^2},$$

wo A, B, C drei verschiedene Konstante bedeuten. Es ist also die kugelförmige Maxwell'sche Dichte-Verteilung der Geschwindigkeitsvektoren durch eine ellipsoidische ersetzt. Die Fläche: $1 = AU^2+BV^2+CW^2$ selbst soll als „Geschwindigkeitsellipsoid" bezeichnet werden. Die Axen dieses Ellipsoides stellen ausgezeichnete Richtungen dar, ohne welche man eben nach unsern jetzigen Kenntnissen von der Verteilung der Eigenbewegungen nicht auskommen kann.

Es war zu vermuten, daß die Annahme einer ellipsoidischen Verteilung der Geschwindigkeitsvektoren eine verhältnismäßig gute Darstellung der beobachteten Verteilung der Eigenbewegungen liefern würde. Denn denkt man sich die Geschwindigkeitsvektoren nach der Eddington'schen Annahme zweier Maxwell'scher Verteilungen unter Berücksichtigung seiner quantitativen Ergebnisse räumlich aufgetragen, so erhält man zwei kugelförmige Haufen, die beträchtlich ineinander eingreifen und daher, indem man die Einkerbung zwischen den beiden Kugeln ausfüllt, einigermaßen durch eine ellipsoidische Verteilung ersetzt werden können.

Für die Ausführung eines Teiles der folgenden Rechnungen bin ich Fräulein J. Lehmann zu Dank verpflichtet.

Die ellipsoidische Geschwindigkeitsverteilung auf die Sphäre projiziert.

Wir wollen also einen begrenzten Raum innerhalb des Fixsternsystems ins Auge fassen und annehmen, daß die Anzahl der Sterne dieses Raumes, welche Geschwindungen zwischen U und $U+\delta U$, V und $V+\delta V$, W und $W+\delta W$ haben, gegeben sei durch:

$$(1)\quad Ke^{-AU^2-BV^2-CW^2}\,dU\,dV\,dW \quad (K, A, B, C \text{ Konstante})$$

wofern man das Koordinatensystem seinem Nullpunkt und seiner Richtung nach geeignet wählt. Was wir beobachten, ist nur die Projektion des Geschwindigkeitsvektors auf die Himmelssphäre. Es ist also zu überlegen, welches Gesetz für die Häufigkeit von Projektionen bestimmter Größe und Richtung folgt. Diese Ueberlegung sei hier nur für den Fall des Rotationsellipsoides ($B = C$) durchgeführt.

Der Winkel, welchen der Visionsradius nach der betreffenden Himmelsgegend mit der Rotationsaxe (x-Axe) des Ellipsoides macht, heiße χ. Die z-Axe kann man wegen der Symmetrie um die x-Axe in die Ebene hineingedreht denken, welche durch die Rotationsaxe und den Visionsradius geht. Nennt man nun u die Geschwindigkeitskomponente senkrecht zum Visionsradius in dieser Ebene, v diejenige senkrecht zum Visionsradius und senkrecht zu dieser Ebene, w die dem Visionsradius parallele Komponente, so gilt:

Nachr. Kgl. Ges. d. Wiss. zu Gött., Math. Phys. Klasse 1907, *No. 5*, 614–632

$$U = u \sin\chi + w \cos\chi$$
$$V = v$$
$$W = u \cos\chi - w \sin\chi.$$

Es folgt:

$$AU^2 + B(V^2+W^2) = (A\sin^2\chi + B\cos^2\chi)u^2 + Bv^2 + (A\cos^2\chi + B\sin^2\chi)w^2 + 2uw\sin\chi\cos\chi(A-B).$$

Für die durch die Komponenten u, v, w ausgedrückten Geschwindigkeiten tritt daher an Stelle von (1) der Ausdruck:

(2) $$K.du dv.dw.e^{-(A\sin^2\chi + B\cos^2\chi)u^2 - Bv^2 - (A\cos^2\chi + B\sin^2\chi)w^2 - 2\sin\chi\cos\chi(A-B)uw}.$$

Da alle in dasselbe Bereich $du.dv$ fallenden Geschwindigkeitsvektoren gleiche Eigenbewegung an der Sphäre geben und nicht nach dem Werte der dritten unbekannten Komponente w getrennt werden können, so haben wir, um eine mit der Beobachtung vergleichbare Häufigkeitsfunktion zu erhalten, über alle Werte von w von $-\infty$ bis $+\infty$ zu integrieren. Spaltet man den von w unabhängigen Faktor:

(3) $$K.e^{-(A\sin^2\chi + B\cos^2\chi)u^2 - Bv^2}$$

ab, so ist also das Integral:

(4) $$\int_{-\infty}^{+\infty} e^{-(A\cos^2\chi + B\sin^2\chi)w^2 - 2\sin\chi\cos\chi(A-B)uw}\,dw$$

auszuführen.

Der Exponent läßt sich so umschreiben:

$$(A\cos^2\chi + B\sin^2\chi)w^2 + 2\sin\chi\cos\chi(A-B)uw$$
$$= (A\cos^2\chi + B\sin^2\chi)\left[w + \frac{(A-B)\sin\chi\cos\chi.u}{A\cos^2\chi + B\sin^2\chi}\right]^2 - \frac{(A-B)^2\sin^2\chi\cos^2\chi u^2}{A\cos^2\chi + B\sin^2\chi}.$$

Beachtet man, daß, wenn p und q beliebige Konstante bedeuten:

$$\int_{-\infty}^{+\infty} e^{-p(w+q)^2}\,dw = \sqrt{\frac{\pi}{p}}$$

ist, so ergiebt sich hiernach für das ganze Integral (4) der Wert:

$$\sqrt{\frac{\pi}{A\cos^2\chi + B\sin^2\chi}}\; e^{\frac{(A-B)^2\sin^2\chi\cos^2\chi u^2}{A\cos^2\chi + B\sin^2\chi}}.$$

Fügt man hierzu den vorher abgespalteten Faktor (3), so erhält man durch einfache Umstellung den Ausdruck für die Verteilung der Geschwindigkeitsprojektionen:

$$(5)\begin{cases} ke^{-au^2-bv^2}du\,dv, \\ \text{wobei:} \\ k=\dfrac{K\sqrt{\pi}}{\sqrt{A\cos^2\chi+B\sin^2\chi}},\quad a=\dfrac{AB}{A\cos^2\chi+B\sin^2\chi},\quad b=B \end{cases}$$

ist.

Die Verteilung der Geschwindigkeitsprojektionen kann also ebenfalls als eine ellipsoidische bezeichnet werden und die Kurve $au^2+bv^2=1$ soll kurz „Geschwindigkeitsellipse" heißen. Die Beziehung der Bestimmungsstücke A, B des ursprünglichen „Geschwindigkeitsellipsoides" zu den Bestimmungsstücken a, b der „Geschwindigkeitsellipse" läßt sich einfacher in der Form schreiben:

$$(6)\qquad \frac{b}{a}-1=\left(\frac{B}{A}-1\right)\sin^2\chi.$$

Berücksichtigung der Eigenbewegung der Sonne. Häufigkeit der Eigenbewegungen bestimmter Richtung.

Bisher war als Nullpunkt ein Punkt gewählt, um den die Eigenbewegungen der Fixsterne symmetrisch verteilt sind. Unser Beobachtungspunkt, die Erde oder Sonne, bewegt sich gegen diesen Symmetriepunkt mit einer gewissen Geschwindigkeit u_0 in Richtung auf einen bestimmten Zielpunkt (Apex). Ist der Winkel zwischen dem Visionsradius und der Richtung nach dem Apex gleich σ, so erhält man die thatsächlich zu beobachtenden Eigenbewegungen an der Sphäre, indem man von dem Vektor u, v den Vektor $h=u_0\sin\sigma$ abzieht. Die Richtung nach dem Apex möge den Positionswinkel φ_0 haben in dem oben benutzten an der Sphäre gezeichneten Axensysteme, welches die Richtung von $u\,(\varphi=0)$ zur x-Axe, die von $v\,(\varphi=90^0)$ zur y-Axe hat. Die Komponenten des beobachteten Vektors sind also:

$$(7)\qquad x=u-h\cos\varphi_0,\quad y=v-h\sin\varphi_0.$$

Die Häufigkeit einer zu beobachtenden Eigenbewegung mit den Komponenten x, y wird daher:

$$(8)\qquad ke^{-a(x+h\cos\varphi_0)^2-b(y+h\sin\varphi_0)^2}dx\,dy.$$

Nachr. Kgl. Ges. d. Wiss. zu Gött., Math. Phys. Klasse 1907, *No. 5*, 614–632

Nun ist schließlich noch zu berücksichtigen, daß wir nicht die absoluten, sondern nur die angulären Beträge der Eigenbewegungen direkt beobachten. Um keine Hypothese über die Entfernungen der Sterne machen zu müssen, wollen wir uns in der üblichen Weise auf die Betrachtung der Richtung der Eigenbewegungen beschränken, also:

$$x = r\cos\varphi \quad y = r\sin\varphi \tag{9}$$

setzen, und gleich nach der Anzahl von Eigenbewegungen fragen, welche in den Winkelraum φ bis $\varphi + d\varphi$ fallen.

Der Ausdruck (8) geht durch Einführung der Polarkoordinaten (9) über in:

$$ke^{-a(r\cos\varphi + h\cos\varphi_0)^2 - b(r\sin\varphi + h\sin\varphi_0)^2}\, r\,dr\,d\varphi.$$

Hierüber hat man von $r = 0$ bis $r = \infty$ zu integrieren, um die Häufigkeit $J(\varphi)$ der Eigenbewegungen in der Richtung φ auf Grund unserer Annahmen zu erhalten. Die Integration läßt sich auf das Wahrscheinlichkeitsintegral zurückführen.

Setzt man zur Abkürzung:

$$\text{(I)}\quad \left\{\begin{aligned} F &= a\cos^2\varphi + b\sin^2\varphi, \quad F_0 = a\cos^2\varphi_0 + b\sin^2\varphi_0 \\ \xi &= h\frac{(a\cos\varphi\cos\varphi_0 + b\sin\varphi\sin\varphi_0)}{\sqrt{a\cos^2\varphi + b\sin^2\varphi}} \end{aligned}\right.$$

und führt an Stelle von r die neue Variable z ein durch:

$$z = r\sqrt{F} + \xi$$

so erhält das auszuführende Integral die Form:

$$J(\varphi) = ke^{-h^2F_0 + \xi^2}\int_\xi^\infty e^{-z^2}\,\frac{dz\,(z-\xi)}{F}$$

oder:

$$J(\varphi) = \frac{ke^{-h^2F_0}}{F}\left[\tfrac{1}{2} - \xi e^{\xi^2}\int_\xi^\infty e^{-z^2}dz\right]$$

oder auch:

$$\text{(II)}\qquad J(\varphi) = \frac{ke^{-h^2F_0}}{F}\left[\tfrac{1}{2} - \tfrac{1}{2}\sqrt{\pi}\,\xi e^{\xi^2} + \xi e^{\xi^2}\int_0^\xi e^{-z^2}dz\right].$$

Vermöge der Gleichungen (I) und (II) läßt sich also mit Hülfe einer Tafel des Wahrscheinlichkeitsintegrals die Anzahl $J(\varphi)$ der Eigenbewegungen berechnen, welche unter Annahme der ellipsoidischen Geschwindigkeitsverteilung in jeder Richtung φ zu erwarten ist.

Anschluss an eine beobachtete Verteilung. Für jede Himmelsgegend liefert die Anordnung der beobachteten Eigenbewegungen nach ihrer Richtung φ und die Abzählung der in jeden Einheitswinkelraum (nach Herrn Eddington im folgenden zu 10° gewählt) fallenden Anzahl von Eigenbewegungen die Funktion $J(\varphi)$. Dabei ist allerdings noch zu beachten, daß φ vom Schnitt der durch Visionsradius und Rotationsaxe des Ellipsoids gehenden Ebene mit der Himmelskugel an zu zählen ist und daß diese Richtung zunächst unbekannt ist. Zählt man mit Herrn Eddington Positionswinkel ϑ am Himmel von der Richtung nach Osten beginnend über Süden, Westen, Norden, so möge die Ausgangsrichtung der Zählung von φ, die frühere x-Axe, den Positionswinkel ϑ_0 haben. Es sei also allgemein:

$$\varphi = \vartheta - \vartheta_0. \tag{10}$$

Die aus der Beobachtung unmittelbar hervorgehende Häufigkeitsfunktion $J(\vartheta)$ wird daher gegeben durch:

$$J(\vartheta) = \frac{ke^{-h^2 F_0}}{F}\left[\tfrac{1}{2} - \tfrac{1}{2}\sqrt{\pi}\,\xi e^{\xi^2} + \xi e^{\xi^2}\int_0^{\xi} e^{-z^2}\,dz\right] \tag{II'}$$

wobei:

$$\left\{\begin{aligned} F &= a\cos^2(\vartheta-\vartheta_0) + b\sin^2(\vartheta-\vartheta_0), \quad F_0 = a\cos^2\varphi_0 + b\sin^2\varphi_0, \\ \xi &= h\cdot\frac{a\cos\varphi_0\cos(\vartheta-\vartheta_0) + b\sin\varphi_0\sin(\vartheta-\vartheta_0)}{\sqrt{a\cos^2(\vartheta-\vartheta_0) + b\sin^2(\vartheta-\vartheta_0)}}\end{aligned}\right. \tag{I'}$$

ist. Das Problem besteht darin, diejenigen Werte der Konstanten $a, b, \vartheta_0, \varphi_0, h, k$ zu bestimmen, welche die beobachtete Verteilung möglichst gut wiedergeben. Man bemerke übrigens, daß sich J nicht ändert, wenn man a, b, k und $1/h^2$ mit demselben Faktor multipliziert, daß wir also eine dieser Größen willkürlich wählen dürfen und nur 5 unabhängige Konstanten zur Verfügung haben. Wir wollen die Willkür gleich ausnutzen und setzen:

$$ke^{-h^2 F_0} = 1. \tag{11}$$

Es wurde folgendes Verfahren eingeschlagen. Man fasse die zu zwei gegenüberliegenden Werten von ϑ, nämlich ϑ und $\vartheta + 180^0$ gehörigen Anzahlen:

$$J = J(\vartheta), \quad J' = J(\vartheta + 180^0)$$

Nachr. Kgl. Ges. d. Wiss. zu Gött., Math. Phys. Klasse 1907, *No. 5*, 614–632

ins Auge. Bei Vermehrung von ϑ um 180° ändert sich F nicht, ξ wechselt nur sein Zeichen. Indem man daher die geraden und ungeraden Bestandteile des Ausdrucks von J trennt, erhält man:

$$J' + J = \frac{1}{F}\left[1 + 2\xi e^{\xi^2}\int_0^{\xi} e^{-z^2}\,dz\right]$$

$$J' - J = \frac{\sqrt{\pi}}{F}\,\xi e^{\xi^2}$$

Bildet man die Funktionen:

$$M(\xi) = 1 + 2\xi e^{\xi^2}\int_0^{\xi} e^{-z^2}dz \quad \text{und} \quad N(\xi) = \frac{\sqrt{\pi}.\xi e^{\xi^2}}{M(\xi)}$$

so ergiebt sich:

(12) $$\frac{J' - J}{J' + J} = N(\xi), \quad F = \frac{M(\xi)}{J' + J}.$$

Die Funktionen $M(\xi)$ und $N(\xi)$ sind in dem Täfelchen I auszugsweise tabuliert.

Tabelle I.

ξ	$M(\xi)$	$N(\xi)$	ξ	$M(\xi)$	$N(\xi)$
0.00	1.00	0.000	0.35	1.27	0.553
0.05	1.00	0.087	0.40	1.36	0.614
0.10	1.02	0.175	0.45	1.46	0.666
0.15	1.04	0.259	0.50	1.59	0.715
0.20	1.08	0.341	0.55	1.74	0.757
0.25	1.13	0.416	0.60	1.90	0.794
0.30	1.19	0.488	0.65	2.12	0.827
0.35	1.27	0.553	0.70	2.37	0.857

$M(\xi)$ ist stets positiv. $N(\xi)$ hat dasselbe Vorzeichen wie ξ.

Aus jedem Paar gegenüberliegender Werte J und J' kann man nach den Formeln (12) mit Hülfe der Tafel zunächst ξ und dann F finden. Führt man diese Rechnung für jeden Wert von ϑ durch, so müssen die erhaltenen Werte ξ und F in einer leicht zu übersehenden Weise verlaufen, falls unsre Annahmen überhaupt zur Darstellung der Beobachtungen ausreichen. Setzt man nämlich:

(13) $$F = \frac{1}{\varrho^2}, \quad \frac{\xi}{\varrho} = \sigma,$$

wonach also ϱ und σ unmittelbar aus F und ξ zu berechnen sind, so gelten für ϱ und σ die Beziehungen:

(14) $$1 = a\,\varrho^2\cos^2(\vartheta - \vartheta_0) + b\,\varrho^2\sin^2(\vartheta - \vartheta_0)$$

$$1 = ha \cos \varphi_0 \, \sigma \cos (\vartheta - \vartheta_0) + hb \sin \varphi_0 \, \sigma \sin (\vartheta - \vartheta_0). \tag{15}$$

Diese Gleichungen haben folgende geometrische Bedeutung: Trägt man zu jedem Positionswinkel ϑ den berechneten Wert ϱ als Radiusvektor auf, so müssen die entstehenden Punkte auf einer Ellipse liegen. Verfährt man ebenso mit den berechneten Werten σ, so muß eine Gerade entstehn. Die Ellipse ist unmittelbar die „Geschwindigkeitsellipse." Die Richtung ihrer großen Axe liefert den Positionswinkel ϑ_0. Die Beträge α und β ihrer Halbaxen ergeben a und b nach den Relationen:

$$a = \frac{1}{\alpha^2}, \quad b = \frac{1}{\beta^2}. \tag{16}$$

Die 3 Konstanten a, b, ϑ_0 sind also aus der Ellipse zu bestimmen. Was die Gerade angeht, so haben ihre Abschnitte p und q auf den beiden Axen der Ellipse die Werte:

$$p = \frac{1}{ha \cos \varphi_0}, \quad q = \frac{1}{hb \sin \varphi_0}, \tag{17}$$

sodaß aus den Gleichungen:

$$h \cos \varphi_0 = \frac{1}{pa}, \quad h \sin \varphi_0 = \frac{1}{qb} \tag{18}$$

auch noch die beiden übrigen Konstanten h und φ_0 folgen.

Ein Beispiel, für das ich die von Herrn Eddington als F bezeichnete Region wähle, soll das Verfahren näher erläutern.

Tabelle II. Bearbeitung des Gebietes F.

ϑ	J	J'	$N(\xi)$	ξ	$M(\xi)$	ϱ	σ	$\bar{J}$	$\bar{J'}$	$\bar{J}-J$	$\bar{J'}-J'$
5°	4	14	+ 0.555	+ 0.351	1.27	3.76	+ 10.7	5	16	+ 1	+ 2
15	5	16	+ 0.523	+ 0.328	1.23	4.13	+ 12.6	6	19	+ 1	+ 3
25	6	21	+ 0.555	+ 0.351	1.27	4.60	+ 13.1	7	22	+ 1	+ 1
35	9	27	+ 0.500	+ 0.309	1.20	5.46	+ 17.7	9	25	0	− 2
45	10	29	+ 0.486	+ 0.299	1.19	5.72	+ 19.2	11	27	+ 1	− 2
55	14	26	+ 0.300	+ 0.174	1.06	6.14	+ 35.3	12	27	− 2	+ 1
65	14	19	+ 0.152	+ 0.087	1.01	5.80	+ 66.5	13	23	− 1	+ 4
75	14	17	+ 0.097	+ 0.055	1.01	5.54	+ 100.	14	18	0	+ 1
85	13	12	− 0.040	− 0.022	1.00	5.00	− 230.	13	14	0	+ 2
95	12	11	− 0.044	− 0.025	1.00	4.79	− 190.	13	10	+ 1	− 1
105	10	11	+ 0.048	+ 0.028	1.00	4.58	− 160.	12	8	+ 2	− 3
115	11	8	− 0.158	− 0.090	1,02	4.30	− 47.7	11	6	0	− 2
125	10	7	− 0.176	− 0.100	1.02	4.08	− 40.8	10	5	0	− 2
135	10	6	− 0.250	− 0.145	1.04	3.92	− 27.0	10	5	0	− 1
145	7	6	− 0.077	− 0.044	1.00	3.60	− 82.	10	4	+ 3	− 2
155	9	5	− 0.286	− 0.166	1.06	3.63	− 21.8	11	4	+ 2	− 1
165	9	5	− 0.286	− 0.166	1.06	3.63	− 21.8	12	5	+ 3	0
175	14	4	− 0.555	− 0.351	1.27	3.76	− 10.7	14	5	0	+ 1

Nachr. Kgl. Ges. d. Wiss. zu Gött., Math. Phys. Klasse 1907, *No. 5*, 614–632

Die ersten Spalten der Tafel II geben zu jedem Positionswinkel ϑ die von Herrn Eddington abgezählte Anzahl von Eigenbewegungen dieser Region J und daneben gleich die Anzahl J' für den Positionswinkel $\vartheta + 180^0$. Es folgen die aus den Formeln (12) und (13) mit Hülfe von Tafel I sich ergebenden Werte von $N(\xi)$, ξ, $M(\xi)$, ϱ und σ. Diese Werte ϱ und σ sind als Radien zu den Positionswinkeln ϑ in Fig. 1 aufgetragen. Die Punkte ϱ schließen sich sehr scharf der durch Probieren ermittelten, in die Figur eingezeichneten Ellipse an. Die Punkte σ liegen unregelmäßiger, was daher kommt, daß sie wesentlich von den prozentual schlechter bestimmten kleineren Differenzen $J' - J$ herrühren, während die Größen ϱ wesentlich durch die größeren Summen $J + J'$ bestimmt werden. Immerhin werden sie durch die Gerade der Figur einigermaßen wiedergegeben.

Der Positionswinkel der großen Axe der Ellipse findet sich zu 57°7. Ihre Halbaxen sind $\alpha = 6.00$ und $\beta = 3.60$. Die Abschnitte der Geraden auf diesen Halbaxen sind $p = -29.9$ und $q = +14.8$. Daraus berechnen sich nach (16) und (18) die 5 Konstanten unsrer Formel:

$$a = 0.0278, \quad b = 0.0773, \quad h = 1.49, \quad \vartheta_0 = 57^\circ\!.7, \quad \varphi_0 = 324^\circ\!.0.$$

Um zu prüfen, wie die Beobachtungen bei dieser Wahl der Konstanten dargestellt werden, wurden nun umgekehrt der eingezeichneten Ellipse und Geraden die zu jedem ϑ gehörigen theoretischen Werte σ und ϱ entnommen und aus diesen durch Umkehrung der Formeln (12) und (13) der Reihe nach F, ξ, $M(\xi)$, $N(\xi)$, $J' + J$, $J' - J$ und schließlich J und J' berechnet. Die Vergleichung der so berechneten Werte $\overline{J}$ und $\overline{J'}$ mit den beobachteten zeigt, daß ein guter Anschluß erreicht ist.

Aehnlich, wie hier, waren auch in allen andern Fällen die Ellipsen sehr exakt, die Geraden weniger gut. Stets wurde aber durch die nach Gefühl eingezeichneten Kurven eine genügende Darstellung der beobachteten Häufigkeiten erreicht. Einige Male fügte ich, um den Anschluß zu verlassen, zum Schluß noch eine kleine Drehung oder Veränderung des Maßstabs der ganzen Kurve hinzu. Ich habe nicht versucht, die Darstellung aufs äußerste zu treiben. Die geometrische Methode ist direkt hierzu nicht geeignet, da sich die Abweichungen von der Ellipse und der Geraden in den Werten J and J' etwas unübersichtlich superponieren. Zu einer endgültigen Bearbeitung würde man daher doch wohl Differentialformeln benutzen müssen.

Ergebnisse der einzelnen Gebiete.

Herr Eddington hat das vom Groombridge-Greenwich-Katalog überdeckte Areal in 7 Gebiete eingeteilt. Das Gebiet A enthält alle Sterne der Poldistanz 0^0—20^0, die Gebiete B—G enthalten die Sterne von 20^0—52^0 Poldistanz aus je 4 A. R. Stunden, wobei B mit den Stunden 22^h—2^h beginnt.

Die folgende Tabelle giebt für alle 7 Regionen die beobachtete Anzahl J der Eigenbewegungen von 10^0 zu 10^0 des Positionswinkels ϑ[1]), darauf die Unterschiede $B-R$ für die Zweischwarmhypothese nach Eddington und für die hier gemachte Hypothese[2]). Für das Gebiet E, welches Herr Eddington wegen seiner geringen Sternzahl nicht weiter bearbeitet, wurden die beobachteten Zahlen der von Herrn Eddington gezeichneten kleinen Kurve entnommen und sind daher etwas unsicher. Die unter Σ gegebene Summe der absoluten Abweichungen zwischen Beobachtung und Rechnung zeigt, daß beide Hypothesen gleich guten Anschluß an die Beobachtung liefern mit Ausnahme des Gebietes C, wo die Zweischwarmhypothese mehr leistet, ohne daß die unitarische Hypothese unzulässige Abweichungen gäbe. Die Sache liegt so, daß Herr Eddington eine Unbekannte mehr hat, indem bei ihm die Verteilung der Sternanzahl zwischen beiden Schwärmen als sechste Unbekannte hinzukommt. Ist die Sternanzahl beider Schwärme sehr ungleich, so giebt das eine Unsymmetrie in der Verteilung der Eigenbewegungen, welche durch unsre Ellipsoidhypothese nicht wiedergegeben wird. Man sieht, daß die Einführung dieser sechsten Unbekannten bei den meisten Gebieten nicht mehr leistet, als unsre fünf Konstanten, und in der That findet auch Herr Eddington die Sternanzahl beider Schwärme nahezu gleich, abgesehn von dem Gebiet C. Hier herrscht eine deutliche, aber vielleicht doch nur zufällige Unsymmetrie eben von der Art, wie sie durch eine Verschiedenheit der Sternanzahl beider Schwärme wiedergegeben wird.

Man kann also sagen, daß sich die unitarische Hypothese der Verteilung der Eigenbewegungen in den einzelnen Gebieten trotz ihrer geringeren Konstantenzahl völlig befriedigend anschließt.

---

1) Für Gebiet A ist ϑ vom Stundenkreis 6^h an gerechnet.

2) Die theoretischen Werte sind ebenso, wie das Herr Eddington bei den beobachteten gethan hat, ausgeglichen, indem das Mittel aus je 3 Nachbarwerten genommen wurde.

Nachr. Kgl. Ges. d. Wiss. zu Gött., Math. Phys. Klasse 1907, *No. 5*, 614–632

Tabelle III.

Region	*A*			*B*			*C*			*D*			*E*			*F*			*G*		
ϑ	*B*	*E—B*	*S—B*	*B*	*E—B*	*S—B*	*B*	*E—B*	*S—B*	*B*	*E—B*	*S—B*	*B*	*E—B*	*S—B*	*B*	*E—B*	*S—B*	*B*	*E—B*	*S—B*
5°	49	—2	—7	67	—2	—2	11	0	+3	4	+1	+1	5	—	0	4	+2	+1	39	—4	—2
15	39	+1	—1	65	+1	—2	15	+1	+6	5	0	0	5	—	0	5	+2	+1	39	—6	—5
25	27	+4	+4	54	+5	+7	24	—1	+4	5	+1	+1	5	—	0	6	+2	+1	34	—5	—2
35	20	+1	+5	47	0	+1	34	—2	+1	7	+1	—1	6	—	—1	9	+1	0	28	—1	+2
45	18	—3	+1	37	—2	+2	42	—1	—2	8	—2	—1	6	—	—1	10	+1	+1	27	0	+2
55	15	—3	+1	31	—4	0	43	+1	—3	9	—2	0	6	—	—1	14	—2	—2	28	—2	+1
65	14	—4	—1	30	—8	—3	38	+4	—4	12	—3	0	7	—	—2	14	—2	—1	30	—2	—1
75	12	—3	0	28	—7	—5	35	0	—4	12	0	+2	6	—	0	14	—1	0	29	0	0
85	12	—2	—2	25	—5	—4	27	+1	—1	16	0	+2	5	—	—1	13	0	0	31	0	—1
95	11	0	—2	19	+2	+1	23	—1	—1	20	+2	+2	6	—	0	12	0	+1	34	—1	—2
105	10	+1	—1	20	+3	—1	16	+2	+3	30	—2	—2	6	—	+1	10	+3	+2	33	0	0
115	10	+2	0	20	+4	—1	14	+2	+3	35	—1	—3	7	—	+2	11	+1	0	32	+2	+2
125	9	+3	+2	26	—1	—6	12	+3	+3	37	—1	—3	11	—	+1	10	+1	0	34	0	+1
135	13	0	—2	25	0	—4	13	+3	0	30	+4	+3	15	—	0	10	—1	0	35	—2	—1
145	11	+2	+1	26	—3	—5	14	0	—2	27	+2	+3	18	—	+1	7	+2	+3	33	—1	+1
155	14	—1	—1	22	0	—1	14	—1	—2	22	+2	+4	20	—	+4	9	0	+2	27	+2	+3
165	12	+1	+1	20	0	0	12	0	—1	19	—1	+3	25	—	+5	9	+2	+3	28	—1	0
175	14	—1	—1	18	0	0	11	0	0	16	—2	+2	33	—	+1	14	—2	0	27	—3	—2
185	12	0	0	18	—2	—3	10	0	0	15	—3	0	38	—	—5	14	+1	+2	24	—3	—1
195	13	—1	—2	15	—1	—2	10	—1	0	15	—5	—3	32	—	—3	16	+3	+3	22	—2	—1
205	11	0	—1	16	—4	—6	9	0	0	10	—1	0	20	—	+3	21	+1	+1	20	—1	0
215	10	0	—1	10	0	—2	8	0	0	8	+1	0	14	—	+3	27	—1	—2	20	—3	—1
225	7	+2	+1	5	+4	+2	7	0	0	6	+2	+1	10	—	+4	29	—2	—3	17	+1	+2
235	7	+1	0	7	+1	—1	6	+1	0	7	0	0	8	—	+3	26	0	+1	18	—1	+1
245	9	—1	—2	6	+2	0	6	0	0	8	—1	—2	8	—	0	19	+3	+4	19	0	+1
255	9	—2	—2	6	+1	—1	4	+2	+1	8	—1	—2	6	—	+1	17	+1	+1	20	+1	+2
265	7	0	+1	7	0	—2	4	+1	0	6	0	0	6	—	0	12	+2	+2	25	—1	—1
275	7	—1	+2	6	0	0	4	+1	0	5	+1	0	6	—	0	11	—1	—1	29	0	—2
285	9	—3	+1	8	—2	—1	5	0	—1	5	+1	0	5	—	+1	11	—3	—3	32	0	—2
295	11	—4	+1	8	—1	0	5	0	—1	5	+1	0	5	—	0	8	—2	—2	34	+4	0
305	12	—2	+2	10	0	0	6	—1	—2	5	0	0	5	—	0	7	—2	—2	38	+4	+1
315	15	—1	+3	13	0	0	7	—2	—2	5	0	0	4	—	0	6	—2	—1	46	0	—4
325	21	—1	+3	18	+2	+1	7	—1	—2	5	0	0	5	—	0	6	—1	—2	48	0	—4
335	29	+1	+2	27	+3	0	6	0	0	5	0	0	6	—	—1	5	0	—1	46	+1	—2
345	39	0	0	42	+1	—3	6	0	+2	6	—1	—1	6	—	—1	5	+1	0	40	+4	+2
355	47	0	—4	60	—3	—7	8	0	+3	5	0	0	5	—	0	4	+2	+1	37	+3	+3
Σ		54	61		74	76		32	57		45	42		—	46		53	49		61	58

Die folgende Tabelle enthält die für die 7 Regionen gefundenen Konstanten.

Tabelle IV. Konstanten der Regionen.

Region	α	β	ϑ_0	p	q	a	b	h	φ_0	$\varphi_0+\vartheta_0$
A	6.75	4.22	175°.0	+ 19.8	+ 40.0	0.0219	0.0561	2.34	11°.0	186°.0
B	7.84	4.45	176.0	+ 25.6	+ 11.9	0.0163	0.0505	2.93	34.8	210.8
C	5.67	3.81	216.5	+ 13.0	+ 11.4	0.0311	0.0688	2.79	27.3	243.8
D	4.80	3.87	302.0	+ 9.2	+ 40.4	0.0433	0.0668	2.54	8.4	310.4
E	4.84	3.39	0.0	− 8.8	+ 82.0	0.0426	0.0870	2.67	357.0	357.0
F	6.00	3.60	57.7	− 29.9	+ 14.8	0.0278	0.0773	1.49	324.0	21.7
G	8.80	6.82	139.5	+ 130.0	+ 60.0	0.0129	0.0215	1.02	52.6	192.1

Fig. 2 verdeutlicht noch das Verhalten beider Theorien zur Beobachtung, indem sie für das Gebiet *F* die Werte von *J* als Radien zu den Positionswinkeln ϑ giebt.

Zusammenfassung der Gebiete.

Der Winkel $\varphi_0 + \vartheta_0$ giebt den Positionswinkel der Richtung, in welcher sich die Eigenbewegung der Sonne auf die Sphäre projiziert. Trägt man diese Richtung in dem Centrum jedes Gebietes an und verlängert sie alle zu größten Kreisen, so müßten sich alle diese Kreise in einem Punkte schneiden, welcher den Apex der Sonnenbewegung darstellt. Das ist mit einiger Annäherung der Fall. Durch graphisches Auftragen auf einer stereographischen Projection der Kugel fanden sich als Koordinaten des Apex:

$$\delta = +41^{0}, \quad A.R. = 259^{0}.$$

Die Abstände, in welchen die größten Kreise an diesem Punkt vorbeigingen, waren für die einzelnen Regionen *A* — *G* der Reihe nach 3°, 4°, 12°, 11°, 16°, 4°, 6°.

Nimmt man ferner an, daß die Rotationsaxen der Geschwindigkeitsellipsoide in allen Teilen des Raumes parallel liegen, so müssen auch die Richtungen ϑ_0 verlängert durch einen Punkt gehen. Dieser Punkt wird nahe übereinstimmen müssen mit dem Zielpunkt der Relativbewegung der beiden Kapteyn'schen Schwärme, und man wird daher auf ihn auch den Namen Vertex übertragen dürfen, den Kapteyn für letzteren Zielpunkt eingeführt hat. Die graphische Aufzeichnung ergab als Koordinaten des Vertex:

$$\delta = +3^{0}, \quad A.R. = 272^{0}.$$

Nachr. Kgl. Ges. d. Wiss. zu Gött., Math. Phys. Klasse 1907, *No. 5*, 614–632

Die Abstände der größten Kreise vom Vertex waren für die einzelnen Regionen:

$$2^0,\ 1^0,\ 10^0,\ 3^0,\ 1^0,\ 4^0,\ 14^0.$$

Fügt man die weitere Annahme hinzu, daß die Geschwindigkeitsellipsoide in allen Teilen des Raumes gleiche Größe haben, daß die mittlere Geschwindigkeit einer bestimmten Richtung überall dieselbe ist, so muß man einerseits verlangen, daß die Abplattung der gefundenen Ellipsen gemäß Gleichung (6) mit dem Abstand χ vom Vertex geht

$$\frac{b}{a}-1 = \left(\frac{B}{A}-1\right)\sin^2\chi,$$

oder wenn man die Halbaxen α, β, A, B der Geschwindigkeitsellipse resp. des Geschwindigkeitsellipsoids einführt:

$$\frac{\beta}{\alpha} = \frac{1}{\sqrt{1+\left(\frac{A^2}{B^2}-1\right)\sin^2\chi}}. \tag{19}$$

Andrerseits folgt aus dieser Annahme, daß die durch die Projektion auf die Sphäre nicht veränderte kleine Axe $\beta = B$ der Ellipsen als vereinheitlichendes Maß für die Bewegungskomponente h der Sonne zu dienen hat und daß daher die Größe $\frac{h}{\beta}$ mit der Gesamtgeschwindigkeit u_0 der Sonne durch die Gleichung:

$$\frac{h}{\beta} = u_0 \sin\sigma \tag{20}$$

zusammenhängen muß, wo σ den Winkelabstand vom Apex bedeutet.

Die folgende Tabelle giebt die Rektaszension und Deklination der Centren der einzelnen Regionen nach Herrn Eddington, die Abstände σ und χ dieser Centren von Vertex und Apex, die aus der Beobachtung folgenden Werte von $\frac{\beta}{h}$ und $\frac{\beta}{\alpha}$ und nach (19) und (20) die mit den Konstanten $\frac{A}{B} = 0.635$ und $u_0 = 0.750$ abgeleiteten theoretischen Werte.

Region	α	δ	σ	$\frac{h}{\beta}$	$0.75 \sin\sigma$	χ	$\frac{\beta}{\alpha}$ beob.	$\frac{\beta}{\alpha}$ theor.
A	—	90°	49°	0.55	0.57	87°	0.63	0.63
B	0°	50	65	0.66	0.68	86	0.57	0.63
C	60	50	87	0.73	0.75	124	0.67	0.70
D	120	50	83	0.66	0.75	122	0.80	0.69
E	180	50	54	0.79	0.61	89	0.70	0.63
F	240	54	19	0.41	0.24	54	0.60	0.71
G	300	50	29	0.15	0.36	53	0.77	0.72

Man sieht, daß $\frac{h}{\beta}$ und $\frac{\beta}{\alpha}$ im großen und ganzen das von der Theorie geforderte Verhalten zeigen, daß aber im einzelnen noch starke Abweichungen bestehn. Ich habe daher versucht, Apex und Vertex noch auf eine andere Art zu bestimmen, welche von vorneweg auf den Betrag dieser Größen Rücksicht nimmt.

Sind α und δ die $A.R.$ und Deklination des Centrums des betreffenden Gebietes, bedeuten A und D dieselben Größen für den Apex, so folgt aus dem Dreieck zwischen Pol, Apex und dem Centrum des Gebietes:

$$\sin\sigma\cos(\varphi_0+\vartheta_0) = \cos D \sin(A-\alpha)$$
$$\sin\sigma\sin(\varphi_0+\vartheta_0) = \cos D \sin\delta\cos(A-\alpha) - \sin D\cos\delta.$$

Multipliziert man hier mit u_0 und setzt:

$$x = u_0\cos D\cos A, \quad Y = u_0 \cos D \sin A, \quad z = u_0 \sin D, \tag{21}$$

so folgt:

$$\left\{\begin{aligned} u_0\sin\sigma\cos(\varphi_0+\vartheta_0) &= \frac{h}{\beta}\cos(\varphi_0+\vartheta_0) = Y\cos\alpha - X\sin\alpha \\ u_0\sin\sigma\sin(\varphi_0+\vartheta_0) &= \frac{h}{\beta}\sin(\varphi_0+\vartheta_0) = X\sin\delta\cos\alpha \\ &\qquad + Y\sin\delta\sin\alpha - Z\cos\delta. \end{aligned}\right. \tag{22}$$

Aus den beobachteten Werten von $\frac{h}{\beta}$ und $\varphi_0+\vartheta_0$ entstehen so 14 Gleichungen, aus denen nach der Methode der kleinsten Quadrate X, Y, Z berechnet wurden. Es ergab sich:

$$X = -0.044, \quad Y = -0.589, \quad Z = +0.377.$$

und damit:

Koordinaten des Apex: $A = 266^0$, $D = +33^0$, $u_0 = 0.70$.

Nachr. Kgl. Ges. d. Wiss. zu Gött., Math. Phys. Klasse 1907, *No. 5*, 614–632

Eine völlig analoge Rechnung ließ sich in Bezug auf die Größe $\sqrt{\frac{\alpha^2}{\beta^2}-1}$ und den Positionswinkel ϑ_0 für den Vertex durchführen und ergab das Resultat:

Koordinaten des Vertex: $A' = 273^0$, $D' = -6^0$, $\frac{\mathrm{B}}{\mathrm{A}} = 0.67$.

Folgende Tabelle stellt die Werte, die sich aus diesen neuen Bestimmungsstücken von Vertex und Apex ergaben, mit ihren Differenzen gegen die beobachteten zusammen:

	Apex					Vertex				
	σ	$\varphi_0+\vartheta_0$	$B-R$	$\frac{h}{\beta}$	$B-R$	χ	ϑ_0	$B-R$	$\frac{\beta}{\alpha}$	$B-R$
A	57°	184°	+ 2°	0.59	− 0.04	96°	177°	− 2°	0.63	0.00
B	68	205	+ 6	0.65	− 0.01	92	174	+ 2	0.63	− 0.06
C	94	249	− 5	0.70	+ 0.03	128	227	− 10	0.72	− 0.05
D	90	295	+ 15	0.70	− 0.04	128	303	− 1	0.72	+ 0.08
E	63	338	+ 19	0.62	+ 0.17	97	357	+ 3	0.64	+ 0.06
F	28	38	− 16	0.33	+ 0.08	66	54	+ 4	0.65	− 0.05
G	31	159	− 33	0.36	− 0.21	60	122	+ 18	0.68	+ 0.09

Abgesehen von der Region *G*, welche sich auch bei Herrn Eddington als anormal erweist, erscheint der Anschluß für den Vertex nunmehr ein befriedigender. Beim Apex zeigt außer *G* noch die Region *E* eine starke Abweichung, welche indessen auf die geringe Sternzahl und ungenaue Entnahme der Zahlen aus der Figur zu schieben sein mag. Unter den übrigen Gebieten giebt *F* die stärksten Abweichungen.

Was das Verhalten zur Zweischwarmhypothese angeht, so ist zwar kein ganz exakter Vergleich mit den analogen Zahlen Herrn Eddington's möglich, es ist aber doch deutlich, daß die Uebereinstimmung der verschiedenen Gebiete bei der Bearbeitung nach der Zweischwarmhypothese eine bessere ist, als hier.

Man muß sich daher die Frage vorlegen, ob der Anschluß der unitarischen Hypothese an die Beobachtungen, wenn auch etwas schlechter, nicht doch noch völlig ausreichend und der Zuverlässigkeit des Materials entsprechend ist. Für die Bejahung dieser Frage wird folgendes sprechen:

1) Ich habe für das Gebiet *F* mit den eben gefundenen theoretischen Werten der Konstanten die Verteilung der Eigenbewegungen zurückgerechnet und durch Zusammenfassung der Zahlen über Winkelräume von 60° folgendes erhalten:

ϑ	beob.	berechnete Anzahl
20°— 80°	67	65
80 —140	66	53
140 —200	69	69
200 —260	139	130
260 —320	55	55
320 — 20	29	42.

Die Annahme rein zufälliger Verteilung der Eigenbewegungen würde erfordern, daß die Eigenbewegungen symmetrisch zur Richtung nach dem Apex, welche hier durch den Positionswinkel 20°—200° gegeben wird, verteilt seien. Es müßten also die Gebiete 140°—200° und 200°—260° die gleiche Anzahl von Eigenbewegungen aufweisen. Man sieht, wie wenig das erfüllt ist, und wie groß der Fortschritt ist, der durch die Hypothese der ellipsoidischen Geschwindigkeitsverteilung erzielt wird, sodaß der eventuelle weitere Gewinn durch den Uebergang zur Zweischwarmhypothese dem gegenüber nicht sehr beträchtlich sein kann.

2) Unmittelbar zu übersehen ist die Bedeutung einer Verschiebung der Größe des motus parallacticus der Sterne des betreffenden Gebietes. Der Durchschnittswert von β beträgt (vgl. Eddington l. c. pag. 57) etwa 3″ im Jahrhundert. Die größte Abweichung in h zwischen Rechnung und Beobachtung ergab sich oben (für Gebiet G) zu 0.2 β. Das bedeutet also 0″.6 im Jahrhundert oder 0″.5 für die 80 Jahre Zwischenzeit, aus denen die Eigenbewegungen abgeleitet wurden. Man ist daher hier der Grenze nahe, wo restierende systematische Fehler der beiden verglichenen Kataloge trotz der sorgfältig differentiellen Reduktion wirksam geblieben sein können. Z. B. würde durch eine Aenderung des Aequinoktiums von Groombridge um 0ˢ.02 eine merkliche Verminderung der Differenzen $B-R$ für $\varphi_0+\vartheta_0$ und $\frac{h}{\beta}$ zu erzielen sein.

3) Schließlich ist daran zu denken, daß Gebiete von 60° Ausdehnung in $A.R.$ zusammengefaßt wurden und daher Eigenbewegungen beträchtlich verschiedener Richtung zu ein und demselben Positionswinkel ϑ gezählt worden sind.

Daher glaube ich, unter Einfügung der numerischen Ergebnisse das Resultat dahin zusammenfassen zu können: Die Verteilung der Eigenbewegungen des Groombridge-Greenwich-Katalogs zwingt uns, soweit sich bisher übersehen läßt, noch nicht dazu, eine Spaltung des Sternsystems in zwei verschiedene Schwärme vorzu-

Nachr. Kgl. Ges. d. Wiss. zu Gött., Math. Phys. Klasse 1907, *No. 5*, 614–632

nehmen, vielmehr wird eine befriedigende Darstellung der beobachteten Verteilung erreicht durch folgende unitarische Annahme: Die Häufigkeit einer Geschwindigkeit der Größe V bis $V+dV$ in einer bestimmten räumlichen Richtung hängt ab von dem Winkel χ, den diese Richtung mit der Richtung nach dem Vertex des Systemes macht, gemäß der Formel:

$$e^{-\frac{V^2}{V_0^2}(1-0.55\cos^2\chi)} V^2 dV \quad (V_0 \text{ eine Konstante}).$$

Der in der Milchstraße gelegene Vertex hat die Koordinaten:

$$A.R. = 273^0, \quad \delta = -6^0.$$

Die Bewegung der Sonne relativ zu einem Punkt, um den sich die Geschwindigkeiten der Sterne symmetrisch gruppieren, ist gerichtet nach dem Apex:

$$A.R. = 266^0, \quad \delta = +33^0$$

und hat die Geschwindigkeit: 0,70 V_0.

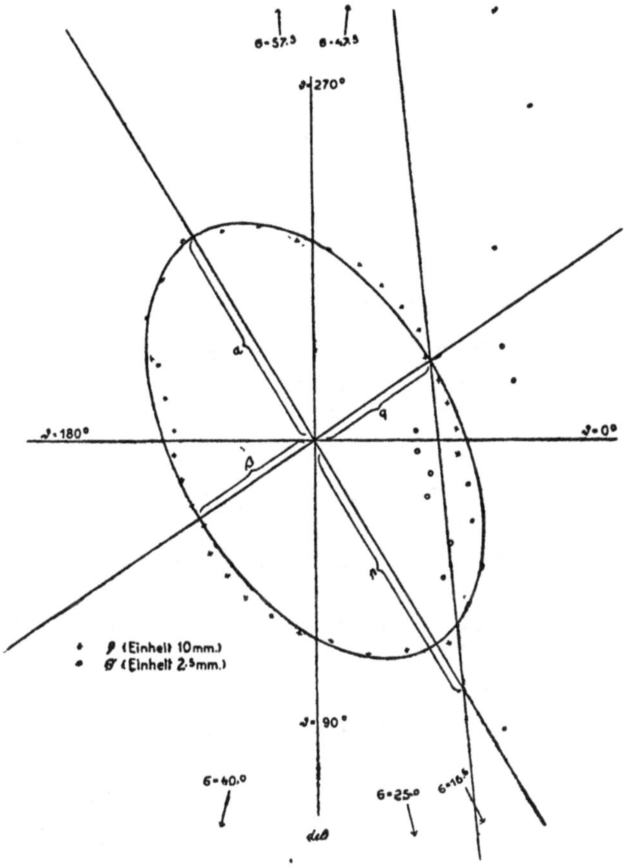

ϑ = 270°
ϑ = 180°
ϑ = 0°
ϑ = 90°
ρ (Einheit 10mm.)
σ (Einheit 2.5mm.)

Fig. 1.

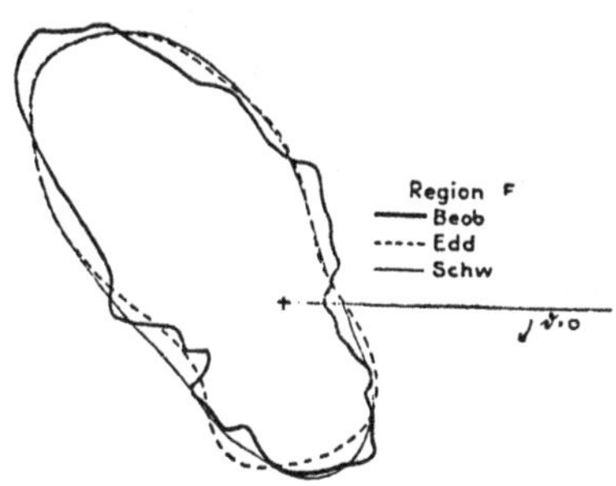

Region F
Beob
Edd
Schw

Fig. 2.

Nachr. Kgl. Ges. d. Wiss. zu Gött., Math. Phys. Klasse 1908, *No. 2*, 191–200

Ueber die Bestimmung von Vertex und Apex nach der Ellipsoidhypothese aus einer geringeren Anzahl beobachteter Eigenbewegungen.

Von

K. Schwarzschild.

Vorgelegt in der Sitzung vom 21. Februar 1908.

§ 1. In dem in diesen Nachrichten (7. Dez. 1907) erschienenen Aufsatz „Über die Eigenbewegungen der Fixsterne" wurde gezeigt, daß man die scheinbare Verteilung der Eigenbewegungen der Fixsterne an der Sphäre durch Annahme einer „ellipsoidischen" Verteilung ihrer räumlichen Geschwindigkeiten darstellen kann, über welche sich dann noch der Einfluß der Eigenbewegung der Sonne lagert. Die Aufgabe, aus einer beobachteten Verteilung die Konstanten des „Geschwindigkeitsellipsoids" und die Eigenbewegung der Sonne abzuleiten, wurde dort für den Fall gelöst, daß man in einem begrenzten Himmelsareal mehrere hundert Sterne beobachteter Eigenbewegung zur Verfügung hat. Bei weiteren Untersuchungen erhob sich das Bedürfnis nach einem Verfahren, welches dieselbe Konstantenbestimmung auch bei spärlicherem Beobachtungsmaterial erlaubt. Es ergab sich die folgende Methode, die rechnerisch von äußerster Einfachheit ist und infolge der Bequemlichkeit, mit der sie Vertex und Apex abzuleiten gestattet, der Ellipsoidhypothese in praktischer Hinsicht — abgesehen von jeder Erwägung über die etwaigen Vorzüge ihres „unitarischen" Charakters — zur Empfehlung gereichen kann. Die Bearbeitung der Eigenbewegungen der Bradley'schen und Porter'schen Sterne nach diesem Verfahren ist im Gange.

§ 2. Man zerlege die auf die Sphäre projizierten Eigenbewegungen der Sterne eines begrenzten Himmelsareals in eine nach dem Vertex gerichtete Komponente x und eine dazu senkrechte y. Es sei u_0 die Geschwindigkeit der Sonnenbewegung, σ der Abstand des betreffenden Gebietes vom Apex, φ_0 der Winkel zwischen den Richtungen nach dem Vertex und Apex. Ferner sei $h = u_0 \sin \sigma$ die auf die Himmelssphäre projizierte Komponente der Sonnenbewegung. Für die Häufigkeit einer Eigenbewegung der Komponenten x bis $x + dx$, y bis $y + dy$ ergab sich dann die Formel (l. c. Gleichung (8)):

$$(1) \qquad W = k e^{-a(x + h \cos \varphi_0)^2 - b(y + h \sin \varphi_0)^2} dx\, dy$$

wo a, b, k Konstante sind.

Es soll nun, wie früher: $x = r \cos \varphi$, $y = r \sin \varphi$ gesetzt und die Anzahl der Eigenbewegungen, welche in einen bestimmten Winkelraum φ bis φ' fallen, untersucht werden, dabei soll indessen nicht ein sehr kleiner Winkelraum $\varphi' - \varphi = d\varphi$ behandelt werden, wie bei der früheren Methode, sondern es sollen die Eigenbewegungen aus größeren Winkelräumen zusammengefaßt werden. Denn wenn die Gesamtzahl der zur Verfügung stehenden Sterne gering ist, muß man zu größern Winkelräumen greifen, um einigermaßen verläßliche Häufigkeitsangaben zu erhalten.

Welche Winkelräume nun im Speziellen zusammen zu fassen sind, wird entschieden aus dem rein formalen Gesichtspunkt, daß die Integration von W über den betreffenden Winkelraum einen möglichst einfachen geschlossenen Ausdruck geben soll. Die gewünschte Einfachheit wird erreicht, wenn man für φ und φ' zwei in Bezug auf die Ellipse:

$$a(x + h \cos \varphi_0)^2 + b(y + h \sin \varphi_0)^2 = \text{const.}$$

konjugierte Richtungen wählt.

Solche konjugierte Richtungen werden bekanntlich bestimmt durch die Bedingung:

$$(2) \qquad b \operatorname{tg} \varphi \,.\, \operatorname{tg} \varphi' + a = 0,$$

wofür man unter Einführung des Hülfswinkels σ auch schreiben kann:

$$(3) \qquad \begin{aligned} &\operatorname{tg} \varphi = \sqrt{\frac{a}{b}} \operatorname{tg} \sigma, \qquad \operatorname{tg} \varphi' = \sqrt{\frac{a}{b}} \operatorname{tg} \sigma' \\ &\operatorname{tg} \sigma \,.\, \operatorname{tg} \sigma' = -1, \qquad \sigma' = 90^\circ + \sigma. \end{aligned}$$

Man führe schiefwinklige Koordinate ξ, η ein durch die Gleichungen:

$$\xi = x\sqrt{a} \cos \sigma + y\sqrt{b} \sin \sigma, \quad \eta = -x\sqrt{a} \sin \sigma + y\sqrt{b} \cos \sigma,$$

Nachr. Kgl. Ges. d. Wiss. zu Gött., Math. Phys. Klasse 1908, *No. 2*, 191–200

Man erkennt mit Hülfe der Gleichungen (3) sofort, daß die ξ-Axe ($\eta = 0$) die Richtung φ und die η-Axe ($\xi = 0$) die Richtung φ' hat, daß also konjugierte Richtungen zu Axen gewählt sind. Ersetzt man in W die rechtwinkligen durch die schiefwinkligen Koordinaten, so erhält man als Häufigkeit einer Eigenbewegung der schiefwinkligen Komponenten ξ bis $\xi + d\xi$, η bis $\eta + d\eta$:

$$W' = \frac{k}{\sqrt{ab}} e^{-(\xi+\mu)^2 - (\eta+\nu)^2} d\xi \, d\eta,$$

wobei zur Abkürzung:

$$\mu = h(\sqrt{a}\cos\varphi_0 \cos\sigma + \sqrt{b}\sin\varphi_0 \sin\sigma)$$
$$\nu = h(-\sqrt{a}\cos\varphi_0 \sin\sigma + \sqrt{b}\sin\varphi_0 \cos\sigma)$$

gesetzt ist. Führt man noch die Hülfsgrößen m und σ_0 ein durch:

(4) $$m\cos\sigma_0 = \sqrt{a}\cos\varphi_0, \quad m\sin\sigma_0 = \sqrt{b}\sin\varphi_0,$$

so schreiben sich μ und ν einfacher:

(5) $$\mu = hm\cos(\sigma_0 - \sigma), \quad \nu = hm\sin(\sigma_0 - \sigma).$$

Um die Anzahl L aller im Winkelraum φ, φ' liegenden Eigenbewegungen zu erhalten, hat man offenbar W' von $\xi = 0$ bis $\xi = \infty$ und von $\eta = 0$ bis $\eta = \infty$ zu integrieren. Das entstehende Integral läßt sich sofort auf das Wahrscheinlichkeitsintegral

$$v(x) = \frac{2}{\sqrt{\pi}} \int_0^x e^{-x^2} dx$$

zurückführen. Es folgt nämlich durch einfache Zerlegung des Integrationsintervalls in Rücksicht darauf, daß $v(\infty) = 1$ ist:

$$L = \frac{\pi k}{4\sqrt{ab}} [1 - v(\mu)] [1 - v(\nu)].$$

Hiermit ist die Anzahl der zwischen zwei konjugierten Richtungen φ und φ' eingeschlossenen Eigenbewegungen gegeben.

Man sieht unmittelbar, daß man durch bloße Vorzeichenvertauschung von μ und ν auch die Anzahlen für die drei übrigen durch die Ganzstrahlen φ und φ' begrenzten Winkelräume erhält, und zwar findet sich folgende Zusammenstellung $\left(c = \frac{\pi k}{4\sqrt{ab}}\right)$:

	Winkelraum	Anzahl
	φ bis φ'	$L_1 = c\,[1 - v(\mu)]\,[1 - v(\nu)]$
(6)	φ' bis $\varphi + 180^0$	$L_2 = c\,[1 + v(\mu)]\,[1 - v(\nu)]$
	$\varphi + 180^0$ bis $\varphi' + 180^0$	$L_3 = c\,[1 + v(\mu)]\,[1 + v(\nu)]$
	$\varphi' + 180^0$ bis φ	$L_4 = c\,[1 - v(\mu)]\,[1 + v(\nu)]$

Es ergibt sich daraus der einfache Satz:

(7) $$L_1 \cdot L_3 = L_2 \cdot L_4$$

oder in Worten:

„Zählt man die Anzahlen von Eigenbewegungen ab, welche in die vier durch zwei konjugierte Richtungen begrenzten Winkelräume fallen, so sind die beiden Produkte aus den Anzahlen je zweier einander gegenüberliegender Winkelräume gleich."

Man sieht, daß es auf Grund dieses Satzes immer leicht möglich sein muß, konjugierte Richtungen für die Geschwindigkeitsellipse aufzufinden, und wenn erst konjugierte Richtungen bekannt sind, bietet die Ableitung der gesuchten Konstanten selbst, wie sich alsbald zeigt, keine Schwierigkeit mehr.

Aus (6) ergeben sich noch die später zu benutzenden Relationen:

(8) $$\begin{aligned} v(\mu) &= \frac{L_2 - L_1}{L_2 + L_1} = \frac{L_3 - L_4}{L_3 + L_4} = \frac{L_2 + L_3 - L_1 - L_4}{L_1 - L_2 + L_3 + L_4} \\ v(\nu) &= \frac{L_3 - L_2}{L_3 + L_2} = \frac{L_4 - L_1}{L_4 + L_1} = \frac{L_3 + L_4 - L_2 - L_1}{L_1 + L_2 + L_3 + L_4} \end{aligned}$$

Es sind ferner noch Spezialfälle hervorzuheben:

1) $\varphi = \varphi_0$, ξ-Axe in Richtung nach dem Apex.

In diesem Falle folgt:

(9) $$\sigma = \sigma_0, \quad \nu = 0, \quad \mu = hm, \quad L_1 = L_4, \quad L_2 = L_3$$

und daraus:

$$L_1 + L_2 = L_3 + L_4.$$

Dies ist die von vorne weg ersichtliche Eigenschaft der Richtung nach dem Apex, daß zu beiden Seiten dieser Richtung gleich viel Eigenbewegungen liegen.

2) $\varphi = \sigma = 0$. Die ξ- und η-Axe fallen mit den Hauptaxen der Ellipse zusammen. Die ξ-Axe, der die größere Axe der Ellipse entsprechen möge, gibt die Richtung nach dem Vertex. Es folgt natürlich $\varphi' = 90^0$ und ferner:

Nachr. Kgl. Ges. d. Wiss. zu Gött., Math. Phys. Klasse 1908, *No. 2*, 191–200

(10) $$\mu = h\sqrt{a}\cos\varphi_0, \quad \nu = h\sqrt{b}\sin\varphi_0.$$

§ 3. Auf Grund der hier entwickelten Beziehungen ergibt sich folgender bequemer Weg der Konstantenbestimmung.

Die Richtung der großen Axe der Ellipse möge den Positionswinkel ϑ_0 haben, und es sei:

(11) $$\varphi = \vartheta - \vartheta_0, \; \varphi' = \vartheta' - \vartheta_0, \; \varphi_0 = \vartheta_1 - \vartheta_0,$$

sodaß ϑ und ϑ' die Positionswinkel zweier konjugierter Richtungen sind und ϑ_1 der Positionswinkel der Richtung nach dem Apex ist.

Man lege ein rechtwinkliges drehbar gedachtes Axenkreuz durch den Nullpunkt. Die Lage des Axenkreuzes werde immer durch den Positionswinkel ϑ der einen Axe charakterisiert. Man zähle ab, wieviel Eigenbewegungen in jeden der Quadranten des Axenkreuzes fallen, indem man die Quadranten im Sinne wachsender ϑ nummeriert. Die Anzahlen seien:

$$l_1, \; l_2, \; l_3, \; l_4.$$

Diese Abzählung führe man für alle möglichen Lagen des Axenkreuzes aus (in Praxis etwa von 10^0 zu 10^0 wachsende Werte von ϑ). Die Anzahlen l erscheinen dann als Funktionen von ϑ. Man suche nun erstens diejenigen Werte von ϑ, für welche

(12) $$l_1 + l_2 = l_3 + l_4$$

ist. Es gibt zwei um 180^0 verschiedene Werte, für welche diese Bedingung erfüllt ist. Davon gibt, wie leicht zu sehn, diejenige die Richtung ϑ_1 nach dem Apex, für welche $l_3 > l_1$, $l_2 > l_4$ ist. Man suche zweitens diejenigen Werte von ϑ, für welche

(13) $$l_1 \cdot l_3 = l_2 \cdot l_4$$

ist. Jede solche Richtung ϑ hat offenbar die Eigenschaft, daß sie zu ihrer Normalen konjugiert ist, da von vorneweg die Anzahlen zwischen senkrecht aufeinander stehenden Richtungen abgezählt wurden und nun speziell diejenigen orthogonalen Richtungen herausgesucht sind, welche die Bedingung (7) des Konjugiertseins erfüllen. Da aber konjugierte Richtungen, welche zugleich normal zu einander stehen, die Hauptaxen der Ellipse sind, so hat man damit die Richtung der Hauptaxen herausgefunden.

Man wähle eine beliebige dieser Richtungen aus und bezeichne sie vorläufig mit ϑ_0, wobei wir uns vorstellen wollen, daß diese Richtung grade der größeren Axe der Ellipse entspricht, also nach dem Vertex gerichtet ist. Die Anzahlen in den vier von ϑ_0 aus gezählten Quadranten mögen die speziellen Werte: $l_1 = k_1$, $l_2 = k_2$,

$l_3 = k_3$, $l_4 = k_4$ haben, wobei also gemäß der Bestimmung von ϑ_0: $k_1 \cdot k_3 = k_2 \cdot k_4$ ist. In diesem Falle rechtwinkliger und zugleich konjugierter Durchmesser sind die Anzahlen k auch mit den oben L genannten Zahlen identisch. Die Größe der Hauptaxen findet man daher nach den Formeln (8) und (10) so:

Man rechne:

$$(14) \qquad v(\mu_0) = \frac{k_2 - k_1 + k_3 - k_4}{k_1 + k_2 + k_3 + k_4}, \quad v(\nu_0) = \frac{k_3 - k_2 + k_4 - k_1}{k_1 + k_2 + k_3 + k_4}$$

und entnehme daraus mit Hülfe einer Tabelle des Wahrscheinlichkeitsintegrales die Werte μ_0 und ν_0. Dann hat man nach (10):

$$h\sqrt{a} = \frac{\mu_0}{\cos\varphi_0}, \quad h\sqrt{b} = \frac{\nu_0}{\sin\varphi_0}$$

wobei:

$$\varphi_0 = \vartheta_1 - \vartheta_0$$

ist. Bezeichnet man, wie in dem früheren Aufsatz, die Halbaxen der Ellipse mit α und β, so ist also:

$$(15) \qquad \frac{\alpha}{h} = \frac{\cos\varphi_0}{\mu_0}, \quad \frac{\beta}{h} = \frac{\sin\varphi_0}{\nu_0}.$$

Stellt sich hier heraus, daß $\frac{\alpha}{h} > \frac{\beta}{h}$ ist, so war die Wahl von ϑ_0 richtig getroffen. Ergibt sich aber $\frac{\alpha}{h} < \frac{\beta}{h}$, so hat man ϑ_0 um 90^0 zu vermehren, um die Richtung nach dem Vertex zu erhalten, und im übrigen nur die erhaltenen Werte von α und β zu vertauschen.

Im allgemeinen genügt die Kenntnis von $\frac{\alpha}{h}$ und $\frac{\beta}{h}$; will man aber h noch selbst kennen in Übereinstimmung mit den in dem früheren Aufsatz getroffenen Festsetzungen, so beachte man, daß sich für die Gesamtzahl der Eigenbewegungen N nach (6) ergibt:

$$N = L_1 + L_2 + L_3 + L_4 = \frac{\pi k}{\sqrt{ab}} = \pi k \alpha \beta$$

Nach der früheren Festsetzung (l. c. Gl. (11)) ist:

$$k = e^{h^2(a\cos^2\varphi_0 + h\sin^2\varphi_0)} = e^{\mu_0^2 + \nu_0^2},$$

also:

$$N = \pi\alpha\beta e^{\mu_0^2 + \nu_0^2}$$

und daher:

Nachr. Kgl. Ges. d. Wiss. zu Gött., Math. Phys. Klasse 1908, *No. 2*, 191–200

$$\frac{N\mu_0\nu_0}{\pi\cos\varphi_0\sin\varphi_0}e^{-\mu_0^2-\nu_0^2}=h^2,$$

womit h gegeben ist.

Damit ist die Aufgabe der Konstantenbestimmung erledigt.

§ 4. Die vorstehende Methode versagt nur in dem einen Fall, daß die Richtung nach dem Apex mit einer der Hauptaxen der Ellipse nahe zusammenfällt. Dann werden in $\frac{\alpha}{h}$ oder in $\frac{\beta}{h}$ Zähler und Nenner gleichzeitig sehr klein und die Größe der einen, nämlich der senkrecht zur Apexrichtung liegenden, Axe der Ellipse wird unbestimmt. Man muß dann ein Paar anderer konjugierter Durchmesser zur Hülfe nehmen. Man wähle einen beliebigen Positionswinkel ϑ, am besten einen, für den $\operatorname{tg}(\vartheta-\vartheta_0)$ nahe gleich dem vorher erhaltenen Werte von $\frac{\alpha}{\beta}$ ist, und suche die konjugierte Richtung ϑ'. Dies geschieht in folgender Weise. Die Gleichung (7) läßt sich so schreiben:

$$\frac{L_1}{L_4}=\frac{L_1+L_2}{L_3+L_4}. \tag{17}$$

L_1+L_2 ist die Zahl der zwischen ϑ und $\vartheta+180^0$ liegenden Eigenbewegungen, L_3+L_4 die der zwischen $\vartheta+180^0$ und ϑ liegenden. Beide Zahlen sind unmittelbar mit ϑ gegeben. Man muß nun der Reihe nach verschiedene Richtungen ϑ' probieren, indem man für sie die Anzahl der Eigenbewegungen, welche zwischen ϑ und ϑ' liegt (L_1), und die Anzahl zwischen $\vartheta'-180^0$ und ϑ (L_4) abzählt, und durch Interpolation diejenige Richtung ϑ' bestimmt, für welche das durch (17) gegebene Verhältnis $\frac{L_1}{L_4}$ stattfindet. Sind so konjugierte Richtungen ϑ und ϑ' gefunden, so folgt noch (2) und (11):

$$\frac{\beta^2}{\alpha^2}=-\operatorname{tg}(\vartheta-\vartheta_0)\cdot\operatorname{tg}(\vartheta'-\vartheta_0), \tag{18}$$

womit das gewünschte Verhältnis $\frac{\beta}{\alpha}$ bestimmt ist.

§ 5. Beispiel. Für ein Areal, das im ganzen 91 Sterne enthielt, ergab sich folgende Verteilung J der Eigenbewegungen auf je 10^0 Positionswinkel, wobei unter ϑ immer der Anfang des Intervalls angegeben ist.

ϑ	J	ϑ	J	ϑ	J	ϑ	J
0°	0	90°	0	180°	13	270°	0
10	1	100	3	190	6	280	4
20	1	110	3	200	3	290	2
30	2	120	2	210	1	300	0
40	1	130	2	220	1	310	1
50	1	140	7	230	3	320	2
60	0	150	6	240	2	330	4
70	0	160	2	250	0	340	3
80	0	170	8	260	1	350	6

Es wurden nun die Anzahlen von Eigenbewegungen abgezählt, die in die 4 Quadranten eines rechtwinkligen Axenkreuzes fielen, dessen eine Axe durch den Positionswinkel ϑ festgelegt ist. Die erste Zeile der folgenden Tabelle gibt demgemäß die Anzahlen für die Winkelräume 0°—90°, 90°—180°, 180°—270°, 270°—0°, die zweite Zeile für 10°—100°, 100°—190°, 190°—280°, 280°—10° u. s. w. Hiermit wurde dann $l_1+l_2-l_3-l_4$ und $l_1\cdot l_3:l_2\cdot l_4$ gebildet. Es

ϑ	l_1	l_2	l_3	l_4	$l_1+l_2-l_3-l_4$	$\frac{l_1\,l_3}{l_2\,l_4}$
0°	6	33	30	22	−13	0.25
10	6	46	17	22	+13	0.10
20	8	49	15	19	+23	0.13
30	10	49	14	18	+27	0.16
40	10	48	13	20	+25	0.13
50	11	47	13	20	+25	0.15
60	17	43	12	19	+29	0.25
70	23	39	14	15	+33	0.55
80	25	37	17	12	+33	0.96
90	33	30	22	6	+35	4.0
100	46	17	22	6	+35	9.9
110	49	15	19	8	+37	7.8
120	49	14	18	10	+35	6.3
130	48	13	20	10	+31	7.5
140	47	13	20	11	+29	6.6
150	43	12	19	17	+19	4.0
160	39	14	15	23	+15	1.82
170	37	17	12	25	+17	1.05

genügt, ϑ die Werte von 0°—180° durchlaufen zu lassen, da bei Vermehrung von ϑ um 180° erstere Größe nur ihr Vorzeichen wechselt, letztere unverändert bleibt. Für $\frac{l_1\,l_3}{l_2\,l_4}$ würde die Tabelle bis zu 90° schon ausreichen, da diese Größe bei Vermehrung

Nachr. Kgl. Ges. d. Wiss. zu Gött., Math. Phys. Klasse 1908, *No. 2*, 191–200

von ϑ um 90^0 den reziproken Wert annimmt. Man findet durch Interpolation, daß $l_1 + l_2 - l_3 - l_4$ verschwindet für:

$$\vartheta = \vartheta_1 = 3^0.$$

Dies — und nicht 183^0 — ist die Richtung nach dem Apex, weil die Bedingung $l_1 (= 6) < l_3 (= 27)$ und $l_4 (= 22) < l_2 (= 40)$ erfüllt ist. Ferner findet man durch Interpolation der Logarithmen von $\frac{l_1\, l_3}{l_2\, l_4}$, daß diese Größe für $\vartheta = \vartheta_0 = 80^0$ durch 1 geht, und die Anzahlen l nehmen für dieses ϑ die speziellen Werte an:

$$\vartheta_0 = 80^0\colon\ k_1 = 25,\ k_2 = 37,\ k_3 = 17,\ k_4 = 12.$$

Damit folgt:

$$\varphi_0 = \vartheta_1 - \vartheta_0 = -77^0$$

$$v(\mu_0) = \frac{k_2 - k_1 + k_3 - k_4}{k_1 + k_2 + k_3 + k_4} = \quad 0.187,\quad \mu_0 = \quad 0.167$$

$$v(\nu_0) = \frac{k_3 + k_4 - k_2 - k_1}{k_1 + k_2 + k_3 + k_4} = -0.362,\quad \nu_0 = -0.333$$

und hiermit:

$$\frac{\alpha}{h} = \frac{\cos\varphi_0}{\mu_0} = 1.35,\quad \frac{\beta}{h} = \frac{\sin\varphi_0}{\nu_0} = 2.92.$$

Man sieht, daß sich $\beta > \alpha$ ergeben hat. Es sind daher α und β zu vertauschen und es ist ϑ_0 um 90^0 zu vermehren, sodaß das E n d r e s u l t a t wird:

$$\vartheta_0 = 170^0,\ \vartheta_1 = 3^0,\ \frac{\alpha}{h} = 2.92,\ \frac{\beta}{h} = 1.35,\ \frac{\beta}{\alpha} = 0.46.$$

Da hier ϑ_0 und $\vartheta_1 + 180^0$ nahe zusammenfallen und $\frac{\beta}{\alpha}$ schlecht bestimmt ist, wird es gut sein, auch noch das Kontrollverfahren mit Hülfe zweier beliebiger konjugierter Durchmesser anzuwenden. Ich wähle als Ausgangsrichtung $\vartheta = 20^0$, da dann $\operatorname{tg}\varphi = \operatorname{tg}(\vartheta - \vartheta_0) = 0.58$ wird, nicht allzuweit von dem Wert $\frac{\beta}{\alpha} = 0.46$ entfernt.

Es folgt dann:

$L_1 + L_2$ (Anzahl der E.-B. zwischen 20^0 und 200^0) $= 57$ $\frac{L_1 + L_2}{L_3 + L_4} = 1.67$
$L_3 + L_4$ („ „ „ „ 200^0 „ 20^0) $= 34$

Ich stelle nun einige Versuche mit verschiedenen ϑ' an:

1) $\vartheta' = 140^\circ$	L_1 (Anzahl der E.-B. zwischen 20° und 140°) = 15		
	L_4 („ „ „ „ 320° „ 20°) = 16	$\frac{L_1}{L_4} = 0.93$	
2) $\vartheta' = 150^\circ$	L_1 („ „ „ „ 20° „ 150°) = 22		
	L_4 („ „ „ „ 330° „ 20°) = 14	$\frac{L_1}{L_4} = 1.57$	
3) $\vartheta' = 160^\circ$	L_1 („ „ „ „ 20° „ 160°) = 28		
	L_4 („ „ „ „ 340° „ 20°) = 10	$\frac{L_1}{L_4} = 2.80$	

Durch Interpolation findet sich, daß $\frac{L_1}{L_4}$ den gewünschten Wert 1.67 annimmt für:

$$\vartheta' = 151^\circ.$$

Es ist daher:

$$\vartheta - \vartheta_0 = 20^\circ - 170^\circ = 210^\circ, \quad \vartheta' - \vartheta_0 = 151^\circ - 170^\circ = 341^\circ,$$

und damit nach (18):

$$\frac{\beta}{\alpha} = 0.445,$$

sodaß die definitiven Werte der Konstanten sind:

$$\frac{\alpha}{h} = 2.92, \quad \frac{\beta}{h} = 2.92 \cdot 0.445 = 1.30.$$

Naturwiss. Vorträge u. Schriften *No. 1* (Teubner Leipzig 1909) 20–38

III. Über das System der Fixsterne.[1])

I.

Wer den Himmel ein wenig zu beobachten gewohnt ist, dem prägen sich gewisse Richtungen ein, die durch die reguläre Anordnung des Planetensystems bedingt sind, – der richtet den Blick hinauf zum Himmelspol, welcher die Erdachse andeutet, beobachtet die Gestirne auf ihrem ständigen Gange von Ost nach West und sieht, wie sich Sonne, Mond und Planeten stets in der Nähe des Himmelsäquators in der Ost-Westebene bewegen. Durch diesen Reigen der Planeten hindurch zieht sich der leuchtende Kranz der Milchstraße. Als ob man Licht und Schatten auf den Blättern des Kranzes wechseln sähe, so sind helleuchtende Flecken und dunkle Stellen nebeneinander gestreut. Die schönsten Sterne des Himmels besetzen diesen Kranz oder hängen wie schimmernde Beeren an hervorragenden Zweigen. Aber was den nicht nur ästhetisch, sondern auch geometrisch empfindenden Beobachter nicht weniger in Erstaunen setzt als die Existenz und seltsame Konstitution dieses den Himmel umschließenden Bandes leuchtender Wolken, das ist die merkwürdige, aller sonstigen Regel widersprechende Orientierung des ganzen Milchstraßenrings. Die Milchstraße kümmert sich nicht um die allgemeine Ost-Westbewegung im Planetensystem, sie steigt steil auf gegen den Äquator, streift im W der Kassiopeia den Nordpol und senkt sich dicht zum Südpol hinab. Schon diese eigentümliche Lage der Milchstraße deutet an, daß wir es hier mit etwas im höchsten Sinne Überirdischem, über die Grenzen unseres in der Ost-Westrichtung orientierten Planetensystems Hinausgehendem zu tun haben. Wer für Richtungen sensibel ist, dem kann es sein, wie wenn hier ein verzaubertes Pendel, statt zur Vertikalen zurückzukehren. in schrägem Ausschlag stehen bliebe. Wie ein Geisterfinger ragt das Phänomen der Milchstraße in unseren engeren Vorstellungskreis hinein und weist wiederum hinaus auf eine höhere Ordnung, die im Fixsternsystem walten muß.

Daß das anscheinende Chaos der glitzernden Sterne sich in Ordnung lösen muß, diese Behauptung wird, so hoffe ich, durch die Tatsachen, von denen ich zu berichten habe, über den Rang einer leeren Alltagsweisheit emporgehoben werden. W e l c h e s diese Ordnung ist, auch darüber lassen sich gewisse Angaben machen, die gegenwärtig allgemein angenommen sind und sozusagen den Rahmen der zukünftigen Forschung bilden werden. Diese feststehenden Dinge möchte ich Ihnen kurz vor Augen führen.. Als meine eigentliche Aufgabe sehe ich es aber an, von den Bemühungen der Astronomen zu erzählen, den großen Rahmen mit lebendigen Farben auszufüllen. Sind es auch noch lauter skizzenhafte Striche

1) Aus „Himmel und Erde" Bd. XXI. Nach einem Vortrag im „Wissenschaftlichen Verein" zu Berlin vom 16. Dez. 1908.

und Farbenkleckse, so mag gerade das Unbestimmte, Halbsichtbare, Halbverschleierte, wenn es den Verstand beleidigt, das Gemüt zu um so ahnungsvolleren Erwartungen anregen.

Ein kurzer Überblick soll uns das äußere Gewand zeigen, in das sich unser Rätsel kleidet.

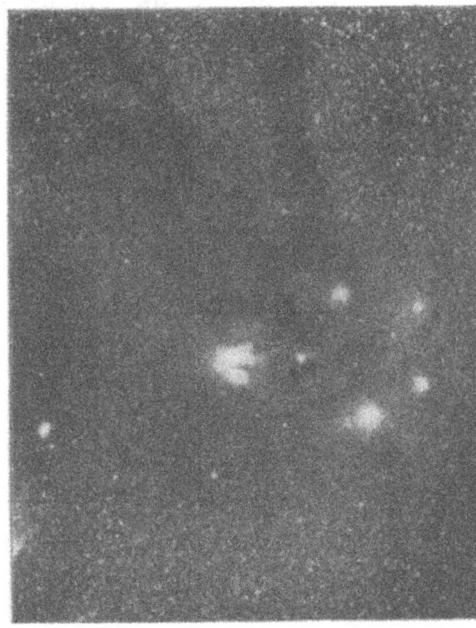

Fig. 1. Milchstraße bei ϱ Ophiuchi. (Aufnahme der Yerkes-Sternwarte.)

Betrachten wir die Milchstraße zu Beginn der Dunkelheit, so sehen wir sie gerade im Zenit durch das W der Kassiopeia gehen. Nach Westen zu, im Schwan, beginnt sie sich zu teilen. Hier liegt bei dem Sterne Deneb eine aufallende dunkle Höhle, die wie ein Loch in dem leuchtenden Bande aussieht. Die glänzendsten Sterne stehen deutlich in der Nachbarschaft der Milchstraße. Wir finden in dem östlichen Arm die Wega in der Leier, in dem westlichen den Adler mit dem leuchtenden Atair. Wer in Italien war, hat beobachten können, wie die Milchstraße in hellen Wolken, am Antares im Sternbilde des Skorpion vorbei, nach Süden zieht. Dicht beim Südpol, im Sternbilde des südlichen Kreuzes vereinigen sich die beiden Arme, die sich im Schwan getrennt hatten. In einheitlichem Zuge steigt sie nun wieder hinauf vom Südpol, wird uns, wenn wir jetzt um Mitternacht beobachten, im Osten des Sirius wieder sichtbar und durchzieht auf dem Rückzuge zur Kassiopeia die große Gruppe von Sternen erster Größe, die den Glanz unserer Winternächte bildet. An ihrem Ostrande liegen der kleine Hund mit Prokyon, die Zwillinge Kastor und Pollux, der Fuhrmann mit Kapella. Westlich liegen der Orion mit der rötlichen Beteigeuze und dem weißen Rigel, dann der weitzerstreute Sternhaufen der Hyaden und die enggedrängte Gruppe der Plejaden.

Wenn man die Milchstraße im Fernrohr beobachtet oder photographiert, so erweist sie sich aus unzählig vielen Sternen zusammengesetzt, die in der merkwürdigsten Weise bald zu Flocken gehäuft sind, bald dunkle Risse zwischen sich lassen. An einzelnen Stellen tritt ein kontinuierlicher Lichtschein auf. Dieser ist nicht, wie man zuerst vermuten möchte, durch äußerste Zusammendrängung zahlloser Sternchen bewirkt: wie Prof. Wolf[1]) kürzlich auch für diese ganz zarten Gebilde durch das Spektroskop nachweisen konnte, sind das wirkliche Gasmassen, die sich zwischen den Sternen ausbreiten. In Fig. 1 sehen Sie eine von solchem Nebel erfüllte Stelle der Milchstraße im Sternbild des Ophiuchus. Das Auffällige der Erscheinung ist, daß um diese Nebel herum ausgeprägt dunkle Stellen, Sternleeren, sind, als ob die Nebel weitergewandert wären und

1) M. Wolf, Astron. Nachrichten, Bd. 178, S. 379.

Naturwiss. Vorträge u. Schriften *No. 1* (Teubner Leipzig 1909) 20–38

im Wandern einen die Sterne verhüllenden Schleier hinter sich gelassen hätten.

Es liegt ein höchst suggestiver Reiz in all dem wechselnden Detail, den Furchen und Löchern, in der ganzen bunten Draperie der Milchstraße. Um aber in der Erkenntnis weiterzukommen, muß man das Detail verwischen, die wallenden Schleier zu einem gleichförmigen, kontinuierlichen Lichtband ausbreiten und zunächst nur das Typische der Erscheinung, den einen geschlossenen, den Himmel umsäumenden Ring, sich vor Augen halten.

Die Astronomen haben allgemein die Verteilung der Sterne in bezug auf die Milchstraße untersucht. Dabei hat sich sofort ein viel weitergreifender Einfluß der Milchstraße auf die Anordnung der Sterne gezeigt, als ihn das immerhin verhältnismäßig schmale, dem Auge sichtbare Band andeutet. Es gilt allgemein, daß die weit von der Milchstraße entfernt liegenden Gebiete des Himmels arm an Sternen sind, und daß die Sternfülle mit der Annäherung an die Milchstraße ständig zunimmt, auch da, wo sich der Sternschimmer unserem Auge noch längst nicht zu einem kontinuierlichen verdichtet. Die großen Flocken der Milchstraße sind nur die Kulmination dieses allmählichen Ansteigens. Die Milchstraße ist für die Anordnung der Sterne nicht von lokaler, sondern von universaler Bedeutung.

Wir wollen auf die beherrschende Stellung der Milchstraße dadurch Rücksicht nehmen, daß wir uns den Ring der Milchstraße von jetzt ab immer horizontal ausgebreitet denken. Die Bahnen der Planeten, die man gewöhnlich horizontal auf das Papier zeichnet, steigen dann unter einem Winkel von 60 Grad an, und die Erdachse, die man sich gewöhnlich senkrecht denkt, wird ziemlich horizontal zu liegen kommen; aber es darf uns nicht stören, wenn wir für das Planetensystem auf ungewohnte Vorstellungen kommen, denn es handelt sich hier darum, die Auffassung des Fixsternsystems möglichst zu erleichtern.

Die Tatsache, daß die Milchstraße die ganze Anordnung der Sterne beherrscht, zeigt von vornherein, daß unsere Sternenwelt etwas Einheitliches ist, daß sie zusammengehört. Es ist durch scharfe numerische Erfassung der oben qualitativ geschilderten Verhältnisse, durch Abzählen der Sterne verschiedener Helligkeit gelungen, diese allgemeine Anschauung in eine präzise Form zu fassen. In der letzten Gestalt, welche diese Forschungen durch Seeliger[1]) erhalten haben, lautet das Ergebnis: Überhaupt alle uns als einzelne Punkte sichtbaren Sterne bilden zusammen das einheitliche Milchstraßensystem. Die Gestalt dieses Systems ist die einer runden flachen Linse, oder auch eines Raumes, wie er von zwei mit den Rändern aufeinandergelegten Suppentellern eingeschlossen wird. Der horizontale Durchmesser dieser Linse, in dessen Richtung wir uns ja nach Verabredung immer die Milchstraße denken wollen, ist etwa doppelt so groß als der vertikale. Die Sterne erfüllen diesen ellipsoidischen Raum

1) H. Seeliger, Untersuchungen über die räumliche Verteilung der Sterne. Abhandl. der Münchner Akad. d. Wissenschaften. Mathem.-Physikalische Klasse. Bd. 19.

nicht in konstanter Dichte, sondern drängen sich nach der Mittelebene, der Milchstraßenebene, und nach dem Zentrum des Systems, von dem wir uns nicht allzuweit entfernt befinden, zusammen. Die Anreicherung der Sterne, die wir am Himmel nach der Milchstraße zu gewahren, ist also zum Teil eine scheinbare und dadurch hervorgerufen, daß wir in Richtung der Milchstraße durch eine längere mit Sternen besetzte Strecke hindurchsehen.

Das Wunderbarste an dieser Vorstellung ist, daß die Milchstraße das ganze Heer der sichtbaren Sterne in einen endlichen, begrenzten Bezirk einschließt. Man kann auch die Größe dieses Bezirkes einigermaßen abschätzen. Das Licht durchläuft seinen Längsdurchmesser in etwa 20000 und seinen Querdurchmesser in etwa 10000 Jahren. Das ganze System ruht abgeschlossen im leeren Raume, und nur in Entfernungen, die groß sind gegen die Dimensionen des Systems selbst, mögen sich wieder neue Sternensysteme zu neuen Milchstraßen zusammenballen.

Die Erkenntnis von der Endlichkeit und Abgeschlossenheit des ganzen Systems der sichtbaren Sterne ist fundamental und bedeutet den Abschluß einer Epoche. Aber sie ist eben doch nur der Rahmen für die weitere Forschung, von dem ich sprach. Es sind gewissermaßen nur die Zollgrenzen unseres Gebietes abgesteckt. Wir wissen nicht, ob das nur eine räumliche Einheit ist oder eine organische, und, wenn es eine organische ist, ob die Sterne sich gleichen wie die Ameisen eines Ameisenhaufens, oder ob sie etwa an Formenreichtum mit der ganzen Entwicklungslinie von der Amöbe bis zum Menschen zu vergleichen sind. Wir verlangen nach einer lebendigeren Ausfüllung des großen Schemas von der ellipsoidischen Sternschar, und ich komme nun an die eigentliche Aufgabe, Ihnen auseinanderzusetzen, welche Erkenntnisse in dieser Hinsicht die neuere Astronomie gezeitigt hat. Die räumliche Einheit des Sternsystems betrachten wir also als festgestellt, wir fragen weiter nach seiner organischen Einheit.

Bevor wir an die Arbeit gehen, muß ich Sie noch durch einen Vergleich auf die unheimliche Natur dieser Aufgabe aufmerksam machen.

Sie wissen, daß die Millionen Fixsterne Sonnen sind, ähnlich wie unsere Sonne, die den 100fachen Durchmesser der Erde hat. Wir wollen die Welt mit einem überirdischen Auge betrachten, dem eine Million Kilometer so groß erscheint wie uns ein Millimeter. Dann sind die Fixsterne lauter Kugeln von 1 mm Durchmesser, Stecknadelknöpfe. Die Distanz, in der sich diese Stecknadelknöpfe befinden, beträgt durchschnittlich 100 km. Wenn wir also von der Einheit des Sternsystems sprechen, so sprechen wir von der Zusammengehörigkeit von Stecknadelknöpfen, die sich 100 km weit voneinander im Raum befinden. Das ist eine ganz gewaltige Trennung. Die Materie ist so dünn verteilt, wie wenn man einen einzigen Liter Wasser durch die ganze Erde versprengte. So wenig wir die Existenz von Wasser ahnen würden, wenn nur ein durch die Erde versprengter Liter vorhanden wäre, so wenig wüßten wir etwas von den Sternen, falls nicht zu ihrer ungeheuren Entfernung und Seltenheit etwas ebenso Wunderbares hinzukäme: die fast absolute Leerheit der Zwischenräume. Die leuchtenden

Naturwiss. Vorträge u. Schriften *No. 1* (Teubner Leipzig 1909) 20–38

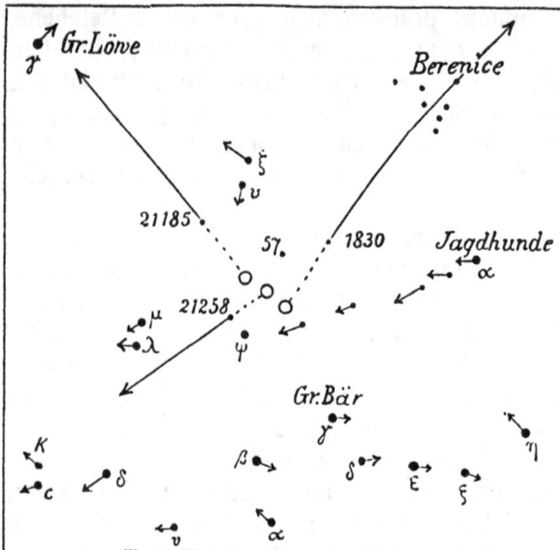

Fig. 2. Eigenbewegungen von Fixsternen.

Stecknadelknöpfe stehen in einem fast völlig staubfreien Raum. Nur dadurch wird es möglich, daß uns die Lichtstrahlen von den Sternen unverfälschte Kunde bringen, und daß überhaupt das Problem vorliegt, zwischen 100 km weit entfernten Stecknadelknöpfen nach einem organischen Zusammenhang zu suchen.

Wir wollen damit beginnen, die Bewegungen der Sterne zu studieren. Es ist bekannt, daß auch die Fixsterne nicht feststehen, sondern eine langsame Verschiebung am Himmelsgewölbe erleiden. Fig. 2 zeigt Ihnen, wie sich in 10000 Jahren die Sterne des großen Bären verschieben. Sie sehen, daß das Sternbild noch als solches erkennbar sein würde (die mit angezeichneten schwächeren Sterne sind Ausnahmen, Schnellläufer). Im allgemeinen sind die Verschiebungen so klein, daß die Sternbilder auch schon vor 10000 Jahren ungefähr ihr jetziges Aussehen haben mußten, und daß daher in ihren Verschiebungen keine Hinderung liegt, den ersten Ursprung der Namengebung der Gestirne in prähistorische Zeiten zurückzuverlegen.

Offenbaren sich irgendwelche Regelmäßigkeiten in den Eigenbewegungen der Fixsterne? Es ist zu bemerken, daß die Fig. 2 die Bewegung im allgemeinen nach links geht, und daß einige merkwürdig parallele Bewegungen auftreten, aber ich möchte Ihnen zunächst an eklatanteren Fällen eine Gesetzmäßigkeit der Bewegungen zeigen. Die Sterne jenes Sternhäufchens der Plejaden, von denen wir sprachen, bewegen sich seit 100 Jahren in einer Richtung gemeinsam weiter, so daß trotz merklicher Verschiebung der ganzen Gruppe am Himmel die Anordnung der Sterne in der Gruppe absolut unverändert geblieben ist, also ein höchst absonderlicher Fall gemeinsamer Bewegung. Aber Sie werden einwenden, daß dies ein Ausnahmefall ist, daß bei einer räumlich so dicht gedrängten Gruppe von Sternen ein organischer Zusammenhang nicht allzu überraschend ist. Nehmen wir ein zweites Beispiel. Sie erinnern sich an die Gruppe der Hyaden, die schon kaum mehr ein Sternhaufen zu nennen ist, da sie sich über viele Grade am Himmelsgewölbe erstreckt. Herr Lewis Boss[1]) hat die Bewegungen dieser Sterne studiert und folgendes Bild (Fig. 3) gefunden: die Pfeile, welche die Bewegungen in 10000 Jahren andeuten, sind nicht parallel, aber sie konvergieren nach einem

1) Lewis Boss. Astronomical Journal vol. 26. 1908.

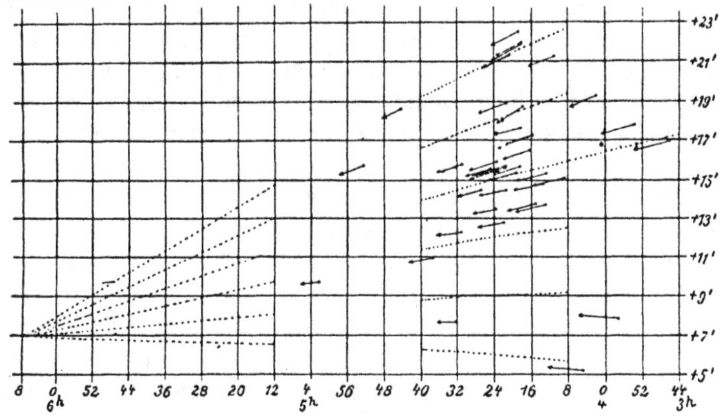

Fig. 3. Eigenbewegung der Hyaden.

Punkte. Was bedeutet dies? Es heißt, daß die Bewegungen nichtsdestoweniger im Raume parallel sind, nur daß sie nicht quer zu unserer Blickrichtung verlaufen, sondern daß die ganze Bewegung in die Himmelstiefe hinein gerichtet ist. Der Konvergenzpunkt ist der scheinbare perspektivische Fluchtpunkt dieser Bewegungen, in welchem sie zusammenzulaufen scheinen, genau wie zwei parallele Hauskanten auf einer perspektivischen Zeichnung. Auch hier könnte man noch immer den Einwand erheben, daß dies eine dicht gedrängte Gruppe von Sternen sei, die nur durch größere Nachbarschaft zu unserer Sonne scheinbar ein größeres Gebiet des Himmels einnähme. Aber die besondere Eigentümlichkeit unseres Falles verhilft zu einer direkten Bestimmung der Entfernung, die diesen Einwand widerlegt und um so bemerkenswerter ist, als dies die größte Entfernung ist, die im Weltraum bisher mit numerischer Sicherheit festgelegt worden ist. Wir wissen aus der Lage des Konvergenzpunktes in Fig. 3, unter welchem Winkel die Bewegung der Hyaden schräg gegen unsere Blickrichtung erfolgt. Die Bewegung läßt sich hiernach aus zwei Komponenten zusammensetzen, einer quer am Himmel und einer etwa doppelt so großen senkrecht in den Himmel hinein gerichteten. Erstere Komponente ist diejenige, die wir wahrnehmen, und von der wir beobachten, daß sie im Jahrhundert 11 Bogensekunden ausmacht. Wir können aber auch die andere, in den Himmel hinein gerichtete Bewegung feststellen. Die spektroskopische Beobachtung gibt uns vermöge des sog. Dopplerschen Prinzips direkten Aufschluß über die Geschwindigkeit, mit der ein Stern sich uns nähert oder von uns entfernt, und zwar ergibt sich für die Hyadensterne, daß diese sich mit einer Geschwindigkeit von 40 km in der Sekunde von uns entfernen. Ist also die Bewegung in die Tiefe gleich 40 km in der Sekunde, so muß die seitliche Bewegung 20 km in der Sekunde betragen, da beide Komponenten, wie gerade erwähnt, im Verhältnis 1 : 2 stehen, und nun haben wir folgende Beziehung: die seitliche Bewegung der Hyaden beträgt 20 km in der Sekunde, das macht auf 3100 Millionen Sekunden eines Jahrhunderts 62 Milliarden km. Die Hyaden müssen so weit ent-

Naturwiss. Vorträge u. Schriften *No. 1* (Teubner Leipzig 1909) 20–38

fernt sein, daß diese Strecke unter einem Winkel von 11 Sekunden erscheint. Das gibt eine Entfernung der Hyaden, welche das Licht in 120 Jahren zurücklegt.[1]) Damit ist die Frage der Entfernung erledigt. Rechnen wir jetzt die gegenseitigen Distanzen der Hyadensterne aus, so finden wir, daß die Hyadensterne ein wenig dichter stehen als die Sterne unserer Umgebung. Reduzieren wir sie wieder auf Stecknadelknöpfe, so wird ihre Entfernung etwa 30 km. Es ist also nachgewiesen, daß sich 40 Stecknadelknöpfe, die sich in Abständen von 30 km befinden, in einem geheimnisvollen Zusammenhang gemeinsam gleichförmig durch den Raum bewegen. In diesem gemeinsamen stillen Wandern der Sterne fühlt man, so scheint mir, aufs eindringlichste das höhere Prinzip, das sie beherrscht, so schwer es ist, dasselbe in eine präzise Vorstellung zu fassen. Man möchte sich am liebsten denken, daß die Sterne gemeinsam losgeschossen sind, der Explosion eines großen Zentralkörpers ihren Ursprung verdanken. Diese Explosion müßte aber den Sternen eine große Anfangsgeschwindigkeit erteilt haben, um sie ihrer gegenseitigen Gravitation zu entreißen, und es wäre ein merkwürdiger Zufall, wenn die Anfangsgeschwindigkeit genau ausgereicht hätte, um die Sterne bis zu ihren jetzigen relativen Ruhelagen zu führen. Viel wahrscheinlicher wäre es bei dieser Hypothese, daß das System auch jetzt noch expandierte, was nicht der Fall zu sein scheint. Man wird daher vorziehen, den Ursprung des Systems aus einem großen Urnebel anzunehmen, der sich anfänglich über die ganze Ausdehnung des jetzigen Systems erstreckte und Teile seiner Masse — jeden Teil an seinem Ort — in die jetzigen Sterne konzentrierte.

Noch ein drittes solches Sternsystem wird von jenen fünf Sternen des großen Bären gebildet, deren Bewegungen in Fig. 2 nach rechts gerichtet sind.[2])

Wenn wir versuchen, von Einzelresultaten, die sich immer nur auf Gruppen von wenigen Sternen beziehen, zu Ergebnissen für die Gesamtheit der Sterne überzugehen, so sieht das Unternehmen zunächst ziemlich hoffnungslos aus. Die Pfeile der Eigenbewegungen, wenn man das Bild irgendeiner Himmelsgegend entwirft, gehen wild durcheinander. Erst eindringenderer Betrachtung offenbaren sich gewisse Durchschnittsgesetzmäßigkeiten. Man beobachtet, daß die Sterne im allgemeinen von dem Sternbild des Herkules wegrücken nach dem entgegengesetzten Punkte des Himmels zu. So sind die Pfeile in Fig. 2 trotz aller Ausnahmen doch vorwiegend nach links gerichtet. Diese Erscheinung hat im Grunde nichts Verwunderliches. Sie beruht einfach darauf, daß die Sonne sich so gut wie jeder andere Fixstern im Raume bewegt, und daß diese Bewegung gerade nach dem Sternbilde des Herkules hin gerichtet ist. Die neueste Bestimmung[3]) liefert für den Punkt, nach dem sich die Sonne

1) Die genauen Zahlen sind nach Boos folgende: Eigenbewegung im Jahrhundert 11″3. Radialgeschwindigkeit 39.8 km/sec. Winkel der Bewegung mit dem Visionsradius 27° 1. Parallaxe 0″025, entsprechend einer Distanz von 120 Lichtjahren.

2) Vgl. H. Ludendorff, Astron. Nachr. 180 S. 265.

3) H. Weetsma, Publikationen des astronomischen Laboratoriums Groningen. Nr. 21. 1908.

hinbewegt, die Rektaszension 286 Grad und die Deklination + 31 Grad. Was die Beobachtung der scheinbaren Bewegung der Sterne am Himmel ergibt, wird durch die Ergebnisse der Beobachtungen nach dem Dopplerschen Prinzip bestätigt. Die Sterne des Herkules nähern sich uns mit einer Geschwindigkeit von 20 km in der Sekunde, ebenso schnell entfernen sich die gegenüberliegenden Sterne; es ergibt sich also zugleich, daß die Geschwindigkeit der Sonne 20 km in der Sekunde beträgt.

Wenn man nun den Einfluß der Eigenbewegung der Sonne von den beobachteten Bewegungen der Sterne abziehen will, um deren wirkliche, nicht auf einen so zufälligen Standpunkt bezogene Bewegungen zu erhalten, so ist das nicht ganz einfach; denn der Reflex der Bewegung der Sonne in der Bewegung eines Sternes hängt ganz davon ab, wie weit der Stern von uns entfernt ist; er ist groß für uns benachbarte, klein für entfernte Sterne und daher nicht scharf zu bestimmen, da man die Entfernungen der meisten in Betracht kommenden Sterne nicht kennt. Die Verhältnisse sind so verwirrend, daß bis vor wenigen Jahren die Ansicht galt, die Bewegungen der Sterne seien völlig irregulär. Wie die Moleküle in einem Gas hin- und herschwirren, so sollte auch die Bewegungen der Sterne ein Gesetz nach Art des Maxwellschen beherrschen, welches gerade ein Ausdruck dafür ist, daß absoluter Zufall regiert.

Erst in den letzten Jahren ist man zu einem positiven Ergebnis gelangt. Wenn man die Milchstraße betrachtet und an die Analogie des Planetensystems denkt, in welchem alle Körper die Sonne in einem Sinne umkreisen, so wird man vermuten, daß auch die Sterne der Milchstraße in Rotation begriffen sind um eine zur Milchstraße senkrechte Achse, um den kleinsten Durchmesser unserer Linse. Diese Vermutung ist aber irreführend, sie hat dadurch, daß sie so nahelag, die Entdeckung der wahren Gesetzmäßigkeit direkt aufgehalten. Es existiert im Milchstraßensystem keine Rotation in einem einzigen bestimmten Sinne.

Die richtige Vorstellung wird uns durch die Betrachtung der Sternzüge, die wir vorhin kennengelernt haben, nahegelegt.[1]) Wir haben die Bewegung der Hyadengruppe im Raum studiert. Sie war dabei relativ zur Sonne gerechnet. Wenn wir den Betrag der Eigenbewegung der Sonne berücksichtigen, so finden wir, daß die resultierende Bewegung, welche als wirkliche Bewegung der Hyaden anzusehen ist, nach einem Punkte der Milchstraße im Fuhrmann hin gerichtet ist. Führen wir dieselbe Rechnung für jene parallel ziehenden Sterne des großen Bären, die „Bärenfamilie", aus, so finden wir als Zielpunkt dieser Bewegung, einen gerade gegenüberliegenden Punkt der Milchstraße im Adler. Wir sehen also hier zwei Sternzüge, die sich in entgegengesetzter Richtung längs der Milchstraße bewegen. Wenn wir uns die Milchstraße wie

1) Im folgenden ist ein Versuch gemacht, in populärer Form die Hypothese des „Geschwindigkeitsellipsoids" der Sterne (vgl. Nachrichten der Kgl. Gesellschaft der Wissenschaften zu Göttingen, 1907) wiederzugeben, durch welche ich die ursprüngliche Annahme Kapteyns (British Association Report, 1906) von den zwei einander durchdringenden Sternschwärmen ersetzt habe. Beiden Auffassungen gemeinsam und völlig sichergestellt ist die Existenz der „Heerstraße", der Vorzugsrichtung der Sternbewegungen, deren Entdeckung in der zitierten Untersuchung von Kapteyn enthalten ist.

Naturwiss. Vorträge u. Schriften *No. 1* (Teubner Leipzig 1909) 20–38

früher horizontal liegend vorstellen, so wandern diese beiden Sternschwärme parallel zu einem Durchmesser ebenfalls horizontal. Denken Sie sich nun zahllose ähnliche Sternschwärme hinzu, welche alle ungefähr längs derselben Straße wandern, die einen in der einen Richtung nach dem Fuhrmann zu, etwa ebensoviele in der entgegengesetzten Richtung nach dem Adler zu, so bekommen Sie die richtige Erscheinung. Die Wege sind nicht scharf aneinander gebunden, sondern laufen zum Teil erheblich auseinander. Schon die Hyaden weichen ein wenig von der mittleren Richtung ab, welche etwas südlich vom Fuhrmann nahe auf die Beteigeuze im Orion gerichtet ist.[1]) Es gibt auch Sterne, die quer zu unserer Straße und auch solche, die aus der Ebene der Milchstraße herauswandern, aber als Haupttatsache bleibt bestehen: es existiert eine ungeheuere Heerstraße, der die Sterne mit Vorliebe folgen, in der sie sich begegnen und wieder aneinander vorbeiziehen, und diese Straße ist parallel einem Durchmesser des Milchstraßensystems.

Was man sich auch unter dieser Erscheinung denken mag, es ist damit ein ganz neuer Rhythmus in unsere Vorstellung vom Sternengebäude gekommen. Wir glauben die Stimme des Gesetzes zu hören, das die große Herde ordnet und die Sterne, ob sie nun rechts oder links gehen mögen, längs einer Straße hält.

Wenn ich einen Versuch machen soll, diese neue Tatsachc gewohnteren Bildern einzuordnen, so möchte ich folgendes andeuten: man nehme an, daß die Sternmassen der Milchstraßenlinse um die vertikale Achse derselben rotieren. Man setze aber nicht voraus, daß die Rotation in einem Sinne stattfinde, sondern es mögen ebensoviel Sterne im Sinne des Uhrzeigers, wie im entgegengesetzten, umlaufen. Wir wissen, daß die Gefahr von Zusammenstößen bei unseren Stecknadelknöpfen in 100 km Entfernung dabei keine Rolle spielen kann. Es ist hier also eine Analogie mit dem Planetensystem vorausgesetzt insofern, als alle Bewegungen einigermaßen in Kreisform und annähernd in der Ebene der Milchstraße gedacht sind. Die Analogie mit dem Planetensystem ist aufgehoben, insofern als völliges Durcheinander von rechtläufigen und rückläufigen Körpern angenommen wird. Denken wir uns mit unserer Sonne nun irgendwie in die Milchstraßenfläche, aber seitlich vom Zentrum, hinein, so werden die Sterne annähernd in zwei Richtungen an uns vorübergehen, nämlich in den beiden Richtungen, die senkrecht stehen zu der Verbindungslinie der Sonne mit dem Mittelpunkt des Systems. Die dynamische Ursache zu diesen Kreisbewegungen hat man in der Gravitationswirkung des ganzen Milchstraßensystems. Man kann abschätzen, daß diese Gravitation genügt, um den einzelnen Stern mit Geschwindigkeiten der beobachteten Größenordnung in etwa 20 Millionen Jahren im Kreise

1) Die genaueren Richtungen sind:
Zielpunkt der absoluten Bewegung der Hyaden: A. R. 95° Decl. + 45°
" " " " " Bärenfamilie: A. R. 285° Decl. — 2°
Mittlere Heerstraße der Sterne (Vertex) nach einer unpublizierten Neubearbeitung der Bradleysterne auf Grund der Ellipsoidhypothese durch Herrn K. Rudolph: A. R. 275° Decl. — 10°.

herumzuführen. Ich muß dabei noch einen Punkt hervorheben, damit Sie keinen flagranten Widerspruch zwischen der Wirklichkeit und dieser Hypothese vermuten. All die Sterne, deren Bewegungen wir kennen, und von denen ich daher vorhin allein sprechen konnte, machen nur einen ganz kleinen Teil des Milchstraßensystems aus. Und in der Tat können wir ja nach unserer Hypothese nur für unsere Nachbarschaft auf das Vorwalten einer bestimmten Bewegungsbahn schließen. Ist unsere Hypothese richtig, so muß sich die Heerstraße der Sterne drehen, wenn wir zu ferneren Gebieten des Himmels übergehen. Bis jetzt hat sich etwas Derartiges nicht nachweisen lassen, aber es eröffnet sich damit doch eine ganz unerwartete Perspektive. Bisher meinte man, daß der Mechanismus des Fixsternensystems so lange verhüllt bleiben müsse, als man nicht die Abweichung der Bewegung eines Sternes von der geraden Linie beobachten und aus der Krümmung der Bahn die Kraftwirkung anderer Sterne ableiten hönnte. Auf diesem Wege ist aber vor 100 Jahren kaum etwas Ersprießliches zu erwarten. Jetzt kann man hoffen, daß durch die bloße statistische Zusammenstellung der geradlinigen Bewegungen der Sterne sich die Zusammenhänge offenbaren werden. Während wir zur Zeit kaum über 10000 Sterne bekannter Eigenbewegung (und diese fast alle über der 9. Größenklasse) verfügen, werden wir in wenigen Jahrzehnten auf Grund der photographischen Aufnahmen Eigenbewegungen von 100000 Sternen bis zur 13. Größenklasse herab kennen gelernt haben und aus ihnen die Heerstraßen der Sterne für weit entlegene Gebiete des Milchstraßensystems ableiten können.

Wo liegt schließlich das Zentrum des Systems, wenn wir immer von dieser Hypothese Gebrauch machen? 90 Grade entfernt von Fuhrmann und Adler in der Milchstraße. Das ist entweder zwischen Schwan und Kassiopeia oder in der Gegend des südlichen Kreuzes.[1]) Nach Betrachtungen, die Herr Easton schon vor der Entdeckung der Heerstraße der Sterne über das Aussehen der Milchstraße angestellt hat, hat man von diesen beiden Möglichkeiten den Schwan zu wählen, und es sei in diesem Zusammenhang auch eine eigentümliche Wahrnehmung wiedergegeben, die Herr Courvoisier an einigen Nebeln in der Milchstraße gemacht hat[2]), daß nämlich die Sternhöhlen in der Nachbarschaft der Nebel eine Bewegung der Nebel vom Sternbilde des Schwans weg andeuten.

Sagen und glauben Sie nun nicht, daß das von Kant, Lambert, Mädler gesuchte Zentrum des Sternsystems entdeckt sei. Solange man die Wahrheit nicht weiß, sucht man sie durch irgendein Bild zu ersetzen. Was entdeckt worden ist, das ist die große Heerstraße unter den Sternen. Ich werde Ihnen nachher von einer ganz anderen Auffassung zu reden haben, bei der vom Zentrum des Systems keine Rede ist. Wie aber auch das Ganze zirkulieren mag, an einer Stelle sind wir dazu gelangt, den Kreislauf des Blutes festzustellen in dem großen Organismus, den das Fixsternsystem bildet.

---

1) Genauere Richtung: A. R. $= 347^0$ Decl. $= + 57^0$.
2) L. Courvoisier, Astron. Nachr., Bd. 170, S. 325.

Naturwiss. Vorträge u. Schriften *No. 1* (Teubner Leipzig 1909) 20–38

II.

Die Entdeckungen, die wir bisher besprachen, bedeuten einen Erfolg der alten klassischen Astronomie, welche ausschließlich die Richtungen und Bewegungen der Gestirne untersucht. Der neuere Zweig der Astronomie, die Astrophysik, trägt mit Hilfe der Spektralanalyse vielleicht in noch höherem Maße dazu bei, das System der Fixsterne als organische Einheit erkennen zu lassen. Es kommt die neuere Physiologie der älteren Anatomie des Sternsystems zu Hilfe.

Auch auf astrophysikalischem Gebiet liegen wieder gewisse allgemeine, wohlbekannte Ergebnisse vor, die ich zunächst in Kürze schildern will.

Die Spektralanalyse der Gestirne nimmt ihren Ausgangspunkt von der Spektralanalyse der Sonne als dem uns benachbartesten aller selbstleuchtenden Sterne. Wenn man das Sonnenspektrum betrachtet, so erscheinen darin auf den ersten Blick zwei Bestandteile des Sonnenkörpers ausgeprägt. Was man vor sich hat, ist im wesentlichen das bekannte, durch alle 7 Spektralfarben hindurchgehende Farbenband, wie es jeder glühende feste oder hinreichend stark komprimierte Körper aussendet. Die schwarzen Linien, welche dieses Farbenband unterbrechen, die sog. Frauenhoferschen Linien, verdanken ihren Ursprung der Sonnenatmosphäre. Jedes Gas absorbiert aus der dem Innern der Sonne entströmenden Lichtstrahlung ganz bestimmte Farbensorten, die für das Gas charakteristisch sind, die wir zudem aus irdischen Experimenten kennen. Man kann daher aus den Frauenhoferschen Linien auf die Existenz des betreffenden Gases in der Sonnenatmosphäre zurückschließen. Je schärfer man das Sonnenspektrum untersucht, um so zahlreichere Freuenhofersche Linien offenbaren sich in demselben, um so mehr von den auf der Erde bekannten Stoffen werden nachgewiesen. Ich will gleich hinzufügen, daß dasselbe auch von den Sternen gilt. So liefert uns die Spektralanalyse das große allgemeine Ergebnis: Überall im Universum ist die chemische Konstitution der Materie ein und dieselbe. Es finden sich dieselben Elemente vor, und sie geben zu denselben Lichtschwingungen Anlaß, mag das betreffende Atom auf der Erde, dem Sirius oder dem fernsten Milchstraßenstern in Schwingungen geraten. Freilich finden sich viele unbekannte Linien, die noch unbekannten Elementen oder unbekannten Zuständen von Elementen entsprechen. Aber es ist kein Zweifel, daß auch diese Elemente auf der Erde entdeckt werden, so gut die Linien des Heliums auf der Sonne erkannt und benannt wurden, bevor man das Helium auf der Erde entdeckt hatte. Auf jeden Fall ist es nicht so, daß auf anderen Sternen ganz andere chemische Elementensysteme existierten, daß sich etwa dort ein bei gewöhnlicher Temperatur flüssiges Gold aus Bechern von fester Luft kredenzen ließe, oder was man sich an tollen Widersprüchen gegen irdische Möglichkeiten sonst ausdenken mag. Überall innerhalb der Milchstraße hat sich die Urmaterie zu denselben chemischen Elementen mit denselben physikalischen Eigenschaften kristallisiert.

Wir wollen uns mit dieser allgemeinen Erkenntnis wiederum nicht begnügen, sondern ins einzelne zu dringen suchen und zunächst einmal auf der Sonne selbst nachsehen, wie dort die Elemente nach der Höhe und Tiefe geordnet sind. Wenn man von außen an die Sonne herankommt, so begegnet man vielleicht schon 20 Erddurchmesser, ehe man die eigentliche Sonnenoberfläche erreicht, gewaltigen Gasausbrüchen (s. Fig. 4), den Protuberanzen. Es wird wichtig sein, uns zu merken, daß diese Ausbrüche wesentlich aus drei Gasen, dem Helium, dem Wasserstoff und dem Kalzium bestehen. Wasserstoff und Helium als die leichtesten Gase wird man von vornherein erwarten; merkwürdig ist, daß das schwere Kalizium im reichsten Maße mit emporgeschleudert wird. Nähert man sich der Oberfläche der Sonne mehr und tritt in die sog. Chromosphäre ein, so trifft man auf die leuchtenden Dämpfe einer größeren Zahl von Elementen. Das Spektrum der Chromosphäre läßt sich am besten bei totalen Sonnenfinsternissen untersuchen. Ich habe auf einer Aufnahme von der totalen Sonnenfinsternis 1905 alle Elemente aufgesucht, die eine größere Höhe über dem Sonnenrand erreichten, und dieselben in einer Darstellung des periodischen Systems der Elemente eingetragen (vgl. Fig. 5).

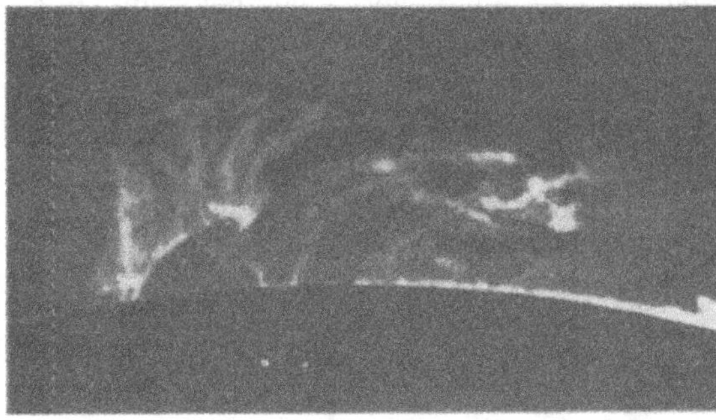

Fig. 4. Kalziumprotuberanzen. (Yerkes-Sternwarte.)

Li	Be	B	C	N	O	F										
Na	Mg	Al	Si	P	S	Cl										
K	Ca	Sc	Ti	V	Cr	Mn	Fe	Co	Ni	Cu	Zn	Ga	Ge	As	Se	Br
Rb	Sr	Y	Zr	Nb	Mo		Ru	Rh	Pd	Ag	Cd	In	Sn	Sb	Te	J
Cs	Ba	La	Ce	Di									Er			
		Yb		Ta	W		Os	Ir	Pt	Au	Hg	Tl	Pb	Bi		
	Ra		Th		Ur											

Fig. 5. Elemente der Chromophäre (Spektralgebiet 460–330 $\mu\mu$) im periodischen System. (Die Vorkommenden Elemente befinden sich im umrahmten Gébiét.)

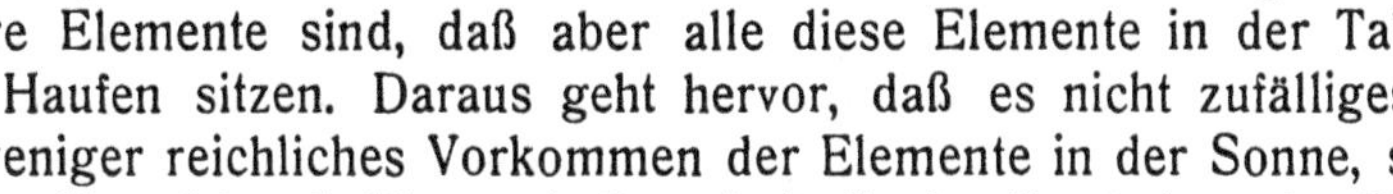

Fig. 6. Fleckenspektrum (Mitte) und normales Sonnenspektrum nach Hale.

Man sieht, daß darunter Elemente mit sehr hohem Atomgewicht, sehr schwere Elemente sind, daß aber alle diese Elemente in der Tabelle in einem Haufen sitzen. Daraus geht hervor, daß es nicht zufälliges, mehr oder weniger reichliches Vorkommen der Elemente in der Sonne, sondern gesetzmäßig wirkende Eigenschaften sind, die das Erscheinen der Elemente in den Höhen der Chromosphäre bedingen. Die meisten Frauenhoferschen Linien, wie man sie für gewöhnlich beobachtet, entspringen einer noch tieferen Zone des Sonnenballs und verraten, wie erwähnt, sehr zahlreiche Elemente, unter denen sich Eisen und Titan durch die große Zahl ihrer Linien hervortun. Die tiefste Stufe bilden die Sonnenflecken (Fig. 6).

Naturwiss. Vorträge u. Schriften *No. 1* (Teubner Leipzig 1909) 20–38

Hier werden die Linien des Spektrums noch zahlreicher; es erscheinen diejenigen Linien, die in irdischen Experimenten bei einer tieferen Temperatur der leuchtenden Gase auftreten, ganz besonders verstärkt, und es kommen hier auch Verbindungen vor, während sonst auf der Sonne die Elemente dissoziiert sind. Außer Cyan ist namentlich Titaniumoxyd nachgewiesen, dessen wunderbar regelmäßige Spektralbanden Sie in dem Spektrum eines Sonnenflecks (Fig. 7) wiederfinden.

Fig. 7. Titaniumoxydbanden im Sonnenfleck (oben) und Lichtbogen nach Hale.

Betrachtet man die Sternspektren, so stellt sich die Aufgabe praktisch zunächst so, daß man viele hundert kleine photographische Platten mit einem Spektrum je eines Sternes hat und versuchen muß, über dieses Material Herr zu werden. Das macht man, wie wenn man eine Mannschaft nach der Größe ordnet. Man nimmt zwei nicht allzu verschiedene Spektren und sucht andere Spektren, die ihrer ganzen Beschaffenheit nach zwischen diese beiden passen, also einen Übergang zwischen ihnen bilden, und fährt mit dem Einordnen der Platten fort, bis jedes Spektrum zwischen zwei anderen liegt, zwischen denen es einen Übergang bildet. Dies Einordnen in eine Serie gelingt, abgesehen von einzelnen Sternen, die aus dem Glied fallen, über Erwarten gut. Es ist höchst suggestiv, zu beachten, wie sich dabei die Elemente in ganz ähnlicher Weise einstellen, in der wir sie beim Eindringen in die Sonne vorgefunden haben.

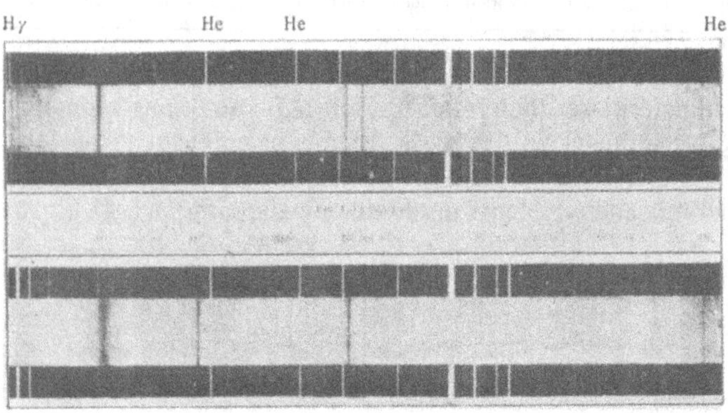

Fig. 8. Spektra von β Orionis (oben) und β Canis majoris (unten). (Frost und Adams.)

An der Spitze der Serie finden wir ein Spektrum, wie das von β Canis majoris (Fig. 8 unten). Man sieht, daß das Farbenband des glühenden Kerns nur von ganz wenigen Linien unterbrochen ist. Diese Linien entstammen (von ganz schwachen, anderen Elementen entspringenden abgesehen) dem Helium und dem Wasserstoff, den Gasen, die wir weit außen in Sonnenprotuberanzen angetroffen haben. Es folgen Spektren, in denen Heliumlinien allmählich zurücktreten (Fig. 8 oben). Beim Sirius sind sie ganz verschwunden, und es herrscht ausschließlich der Wasserstoff, der sich immer leicht durch sein rhythmisches Liniensignal kundgibt (Fig. 9 unten.) Nun tritt der Wasserstoff zurück, und das Kalzium tritt allmählich mehr und mehr hervor (Fig. 9 Mitte). Dann beginnen auch die Eisenlinien und zahlreiche Metallinien zu erscheinen; wir erhalten dasselbe Bild, wie es das Sonnenspektrum bietet (Fig. 9 oben). Gehen wir bis zum Schluß der Serie, so finden wir ähnliche Verhältnisse, wie wir sie in den Sonnenflecken antreffen. Die Linien werden kräftiger und breiter, es treten die Bandenspektra auf, und schließlich hat die Absorptionswirkung der Sternatmosphäre die Strahlung des Kerns bis auf einzelne getrennte Fetzen beseitigt.

Was ist das für eine Art von Anordnung, die wir in diesem natürlichen System der Sterne vorfinden? Es ist zweifellos ein Temperaturgang. Für die letzten Stufen der Reihe lehrt uns das die enge Analogie mit den Sonnenflecken und den Laboratoriumsexperimenten über den Einfluß der Temperatur auf die Spektra der Gase. Für den Anfang der Reihe führen uns andere Betrachtungen zu demselben Schluß. Es fiel von vornherein auf, daß die Sterne am Ende der Spektralreihe rot aussehen, und daß die Farbe des Gestirns sich um so mehr dem Weiß nähert, je näher sein Spektrum dem Anfang der Reihe steht. Das erinnert unmittelbar daran,

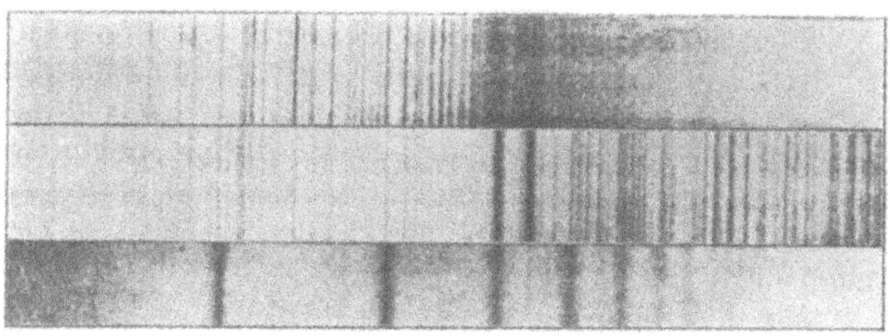

Fig. 9. Hauptspektraltypen nach Huggins.

daß ein Metall bei allmählicher Erhitzung aus der Rot- in die Gelb- und Weißglut übergeht. Also — scheint der bündige Schluß — sind die roten Sterne kühl und die weißen heiß. Leider gibt es aber noch eine zweite ebenso enge Analogie. Wenn die Sonne zum Horizont herabsinkt, geht sie von der gelben in die rote Farbe über; zunehmende Stärke der Atmosphäre, welche die Strahlen absorbiert, bewirkt hiernach ebenfalls größere Röte. Unsere Sternserie könnte auch eine Serie immer dichterer und dichterer Atmosphären sein. Dies mag zutreffen. Dadurch wird aber die Behauptung nicht hinfällig, daß man es in der Spektralserie vor allem mit einer Temperaturskala zu tun hat. Für die untersten Stufen ist das ja, wie erwähnt, durch Vergleich mit irdischen Spektren und Sonnenflecken nachgewiesen. Für die obersten Stufen leuchtet das Farbenband des Kerns so ununterbrochen, von so wenigen Frauenhoferschen Linien gestört, daß man da, wo nicht gerade solche Linien sind, gewiß voraussetzen kann, die Strahlung des eigentlichen Kerns vor sich zu haben. Hier gilt also die Analogie des glühenden Metalls und verlangt, den weißesten Sternen, den Heliumsternen, die höchste Temperatur zuzuschreiben. Dies läßt sich noch exakter verfolgen. Man hat in dem Planckschen Gesetz die Regel, nach der das Weißer- und Weißerwerden mit wachsender Temperatur erfolgt.

Naturwiss. Vorträge u. Schriften *No. 1* (Teubner Leipzig 1909) 20–38

Es treten mit wachsender Temperatur die langen roten Wellen gegenüber den kurzen violetten immer mehr zurück in einem bestimmten, durch das Gesetz gegebenen numerischen Verhältnis, aber es steigert sich dieses Verhältnis mit wachsender Temperatur nicht etwa derartig, daß der glühende Körper schließlich blau würde und zum Schluß nur ultraviolette, dem Auge unsichtbare Strahlung aussendete. Vielmehr nähert sich die Farbe des glühenden Körpers mit wachsender Temperatur einem reinen Weiß, und ich hege keinen Zweifel, daß man zukünftig auf die Frage: was ist Weiß? nicht antworten wird: Weiß ist, was Weiß ist, oder nach Helmholtz: Weiß ist die vorherrschende Farbe der Umgebung, sondern die Definition wird sein: Weiß ist die Farbe des glühenden Körpers[1]) unendlich hoher Temperatur.

Das Weiß der Hiliumsterne ist nun, soweit das die bisherigen Messungen zu beurteilen gestatten, gerade das Weiß des unendlich heißen Körpers, und die genauere Beobachtung zeigt, daß die Intensitätsverteilung in den Spektren, die Farbe, mit dem Hervortreten des Wasserstoffs und dann des Kalziums gesetzmäßig ganz ähnlich heruntergeht, wie bei einem glühenden Körper sinkender Temperatur.[1]) Man kann hiernach auch ungefähre Werte der Temperatur für jede Spektralgattung berechnen, die von 3000 Grad bei den roten Sternen beginnen und bis zu 25000 Grad für die Heliumsterne hinaufgehen. Ob sie noch weiter hinaufreichen, läßt sich schwer entscheiden, weil sich zwischen 25000 Grad und unendlich hoher Temperatur Farbe und Intensitätsverteilung im Spektrum nicht mehr merklich ändert.

Wie die Pflanzen in dem natürlichen System der Botanik, so erscheinen also hier die Sterne in einem natürlichen System nach ihrer Temperatur geordnet. Man wird jetzt die Sterne weder den Ameisen eines Ameisenhaufens, noch den tausend Spezies der ganzen Tierreihe vergleichen wollen. So bunt, wie die ganze Welt der Organismen, ist das Heer der Sterne nicht, aber auch nicht so gleichförmig, wie die Individuen einer Spezies. Wir haben vorhin die chemische Einheit des Universums hervorgehoben. Durch die Existenz der Sternserie wird die Einheit der Sterne noch enger geschlossen und wieder eine Menge phantastischer Möglichkeiten abgewiesen. An sich würde uns nichts hindern, uns einen Stern nur aus Gold oder nur aus Eisen zu denken, dessen Spektrum dann nur die Gold- oder Eisenlinien zeigen könnte. In Wirklichkeit kommt ein solcher Fall nicht vor. Das Material der Sterne ist gleichmäßig aus allen Elementen aufgebaut, wenigstens so gleichmäßig, daß bei einer bestimmten Temperatur immer dieselben Linien in den Spektren der Sterne die Herrschaft gewinnen. Es ist eigentümlich, wie sich die Elemente in der Spektralreihe folgen, wie der Wasserstoff abgelöst wird und mit dem Eisen zugleich das Titan zur Herrschaft kommt. Die hier auftretende Verknüpfung und Folge der Elemente ist noch lange nicht genügend er-

1) Genauer des sog. vollkommenen Radiators.

2) Daß die Heliumsterne die weißesten Sterne sind und dem Weiß des unendlich heißen vollkommenen Radiators sehr nahekommen, geht aus den Arbeiten der „Göttinger Aktinometrie" und spektralphotometrischen Messungen von Dr. H. Rosenberg hervor.

forscht und auf ihre Ursachen zurückgeführt. Wenn die Elemente nicht auf jedem Stern von neuem aus den Uratomen in derselben Weise gebraut werden, dann muß die große Masse, aus der das ganze System entstanden ist, so gleichmäßig durchmischt gewesen sein, daß jedes von den Millionen Sternpünktchen, in die sie sich gespalten hat, seine Portion von jedem der Elemente abbekommen hat. Hier verliert sich unser Blick in noch undurchdringliche zeitliche Entfernungen.

Aber lassen Sie uns noch einmal zu konkreten Forschungen zurückkehren. Wir haben uns noch nicht aller Mittel bedient, um die Rätsel der Milchstraße zu enthüllen. Die Armeen sind noch getrennt marschiert. Auf der einen Seite wurden die Bewegungen der Gestirne, auf der anderen ihre Spektra studiert. Mit der Vereinigung beider Kräfte muß sich ein noch nachdrücklicherer Anschlag ausführen lassen. Freilich ist bis jetzt nur die Avantgarde im Vormarsch, doch diese wollen wir noch begleiten, um einen letzten Ausblick zu gewinnen.

Spektra unserer Spektralserie, welche derselben Temperatur angehören, unterscheiden sich noch durch feinere Charaktere. Insbesondere zeigen Sterne, welche die Wasserstofflinien aufweisen, diese in sehr verschiedener Breite. Man muß daher den höheren Stufen unserer Skala seitliche Sprossen anfügen, welche bezeichnen, daß die Linien in dem Sternspektrum schmäler und schmäler werden.[1]) Derartige sekundäre Unterschiede der Spektren werden nicht überraschen, wenn man daran denkt, daß außer der Temperatur jedenfalls noch andere Verschiedenheiten zwischen den Sternen bestehen, selbst abgesehen von allen etwaigen Unterschieden ihrer chemischen Konstitution. Sowohl ihrer Masse als ihrer durchschnittlichen Dichte nach können sich die Sterne unterscheiden. Die Beobachtung von Doppelsternen zeigt, daß die Massen nicht allzu sehr verschieden sind. Als wesentlich unterscheidendes Merkmal wird daher außer der Temperatur die Dichte, der Raum, auf den die Masse des Sternes verteilt ist, in Betracht kommen. Die Frage wird also: Können wir etwas über die Dichte, die Ausdehnung, den Radius der Sterne angeben, und besteht dabei ein Zusammenhang mit den Spektren?

Hier greift die Vereinigung zwischen dem Studium der Bewegungen und der Spektren Platz.

Man kennt durch Parallaxenmessung die Entfernung von etwa 200 Sternen und durch den Reflex der Eigenbewegung des Sonnensystems im Raume erhält man noch gewisse durchschnittliche Angaben über die Entfernungen von einigen 1000 anderen Sternen. Ist die Entfernung bekannt, und dazu bekannt die scheinbare Helligkeit des Sterns, so kann man seine wirkliche Helligkeit z. B. im Verhältnis zur Sonne berechnen. Was sollten wir für die wirkliche Helligkeit der Sterne unserer Serie erwarten? Offenbar große Leuchtkraft für die heißen Sterne, geringe für die kühlen. Dabei ist freilich die Voraussetzung, daß die Oberfläche der

1) Von Miß A. C. Maury sind diese Unterschiede in Harvard Annals vol. 28. Part. I durch die Indices a, b, c bezeichnet. Die Sterne mit dem Index c sind es, deren Gigantennatur (s. unten) Hertzsprung entdeckt hat.

Naturwiss. Vorträge u. Schriften *No. 1* (Teubner Leipzig 1909) 20–38

Sterne einigermaßen gleich ist, denn die Temperatur bestimmt nur die Lichtmenge, die von der Oberflächeneinheit, dem Quadratmeter, ausgeht; und ein kühler Stern kann daher trotzdem eine große Gesamtleuchtkraft haben, wofern er eben großen Durchmesser und eine große Oberfläche besitzt.

Die Erwartung bestätigt sich nun zu einem guten Teil. Berechnet man für die Sterne, deren Entfernung zuverlässig bestimmt ist, ihre wirkliche Leuchtkraft, so sieht man dieselbe regulär mit abnehmender Temperatur geringer werden, — mit ein paar ganz eigentümlichen Ausnahmen: unter den gelben und roten Sternen sind einige von ganz ausnahmsweise großer Leuchtkraft.[1]) Da man nicht annehmen kann, daß die Oberflächeneinheit bei diesen kühlen Sternen viel Licht aussendet, so folgt, daß die Gesamtoberfläche ungeheuer groß sein muß. Man kann die Durchmesser einigermaßen abschätzen und findet für die normalen Sterne sehr konstante Zahlen, die zwischen der Hälfte und dem Doppelten des Sonnendurchmessers liegen. Für die exzeptionellen Sterne aber ergibt sich der 10—100fache Sonnendurchmesser. Indessen nicht nur unter den roten Sternen gibt es solche Giganten, sondern auch unter den weißen, an sich hell leuchtenden Sternen gibt es vereinzelte noch ganz unverhältnismäßig hellere, und zwar sind es hier, wie Herr Hertzsprung entdeckt hat[2]), gerade die Sterne mit schmalen Gaslinien, welche die Gigantennatur haben. Es ist an sich höchst merkwürdig und durch keine Theorie über die Entwicklung der Sterne vorauszusehen, daß so zerstreut zwischen den gewöhnlichen Sternen diese Giganten liegen. Wieder einmal erweist sich die Welt als vornehmstes Kunstwerk, niemals willkürlich und doch stets überraschend. Aber noch verheißungsvoller erscheint die Aussicht, die sich von diesem Endpunkt unseres Vormarsches aus eröffnet.

Wir wollen uns einmal die Giganten ausgeschlossen denken von den übrigen Sternen. Für die weißen Sterne läßt sich das bereits einigermaßen durchführen, indem man die Sterne mit schmalen Wasserstofflinien ausschließt; für die roten wird hoffentlich auch noch die Eigentümlichkeit des Spektrums entdeckt werden, welche die Giganten von den gewöhnlichen Sternen unterscheidet. Was dann übrigbleibt, sind lauter Sterne ungefähr von Sonnengröße, deren wirkliche Leuchtkraft wir aus ihrer Temperatur, aus ihrem Spektrum, entnehmen können. Beobachten wir ihre scheinbare Leuchtkraft, so haben wir dann ihre Entfernung, das A und O der Angaben, die wir brauchen, wenn wir Gesetze der Anordnung der Sterne im Raume und ihrer Bewegungen erkennen wollen. Verbindet man die Entfernungsbestimmung mittels der Spektren mit der Beobachtung der Bewegung, so müssen sich die Züge der Sterne entwirren. Überall zeigen sich schon unverstandene Gesetzmäßigkeiten, die zum Teil wieder in Beziehungen zur Milchstraße stehen. Die weißen Sterne, und unter diesen wieder besonders die weißen Giganten, stehen vorzugsweise in der Milchstraße, während die roten Sterne

1) Vgl. E. Hertzsprung. Zeitschrift für wissenschaftliche Photographie, Bd. III, S. 429, Bd. V, S. 86.

2) l. c. Bd. III, S. 435.

außerhalb der Milchstraße relativ häufiger sind. Von der Ferne gesehen, würde unsere Milchstraßenscheibe wie eine gelbliche Masse mit einem weißeren Rand aussehen. Höchst wunderbar ist, daß die Geschwindigkeit der Sterne im Raume auch mit ihrer Temperatur zusammenzuhängen scheint, indem die kalten Sterne die rascher bewegten sind.[1])

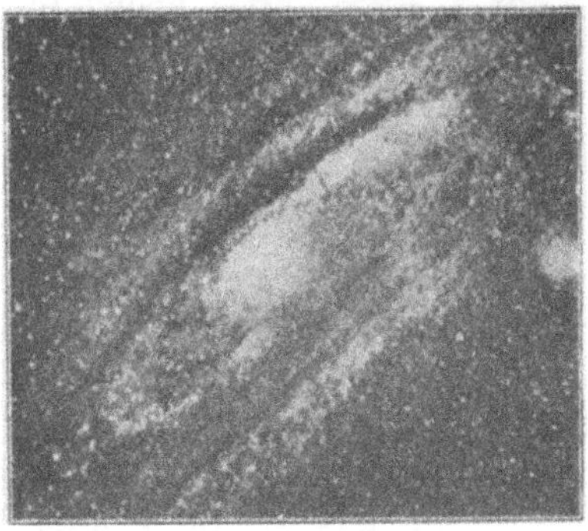

Fig. 10. Andromedanebel.

Es besteht also die Hoffnung, daß die einheitliche aber auch einförmige Linse des Milchstraßensystems sich in nicht allzu ferner Zeit auch räumlich in all die Windungen wird trennen lassen, über deren Existenz uns der bloße Anblick der Milchstraße keinen Zweifel läßt, und daß dann endlich auch der Schlüssel zu all dem beziehungsreichen Detail gefunden wird, das den Anblick der Milchstraße so anziehend macht. Wenn ich über all diese dunklen Furchen und Löcher und jene aus Sternen gewobenen Schleier Ihnen nichts Näheres sagen konnte und die Wissenschaft da die Antwort schuldig bleibt, wo die Neugierde am lebhaftesten erregt ist, so möchte ich wenigstens mit einer Hypothese schließen, die uns vielleicht ein richtiges Bild der zukünftigen Aufspaltung des Milchstraßensystems in seine einzelnen Züge vor Augen führt, die auch dem Detail der Milchstraße eine Bedeutung zuweist und uns mit einer neuen Ahnung der Kräfte erfüllt, die die Bewegung der Sterne beherrschen mögen. Es handelt sich um die Analogie der Milchstraße mit einem Spiralnebel.

Fig. 11. Spiralnebel Messier 33 Trianguli.

Fig. 10 zeigt den bekannten Andromedanebel. Schon Herschel hat denselben für eine zweite, ungeheuer weit entfernte Milchstraße erklärt. Für die Spiralnebel hält man auch heute noch an dieser Möglichkeit fest: sie gelten als e i n z i g e z u l ä s s i g e A u s n a h m e von dem Satz, daß alle

1) Campbell (Astrophys. Journal vol. 13. 1901) hat bemerkt, daß die scheinbar schwächeren Sterne die rascheren sind. Darüber scheint sich ein Einfluß des Spektraltypus zu lagern. Aus 168 von E. B. der Sonne befreiten Radialgeschwindigkeiten, die Hertzsprung aus der Literatur gesammelt hat, findet sich

Typus (Maury)	1–5.	6–9.	11–13.	14–15.	16–18.
Mittlere Radialgeschwindigkeit	9	18	20	22	36 km/sec.

Naturwiss. Vorträge u. Schriften *No. 1* (Teubner Leipzig 1909) 20–38

Fig. 12. Spiralnebel in den Jagdhunden.

uns überhaupt sichtbaren Objekte unserem Milchstraßensystem angehören. Der Andromedanebel zeigt im ganzen die unserm Milchstraßensystem zukommende Linsenform. Daß er nicht kreisrund erscheint, beruht offenbar darauf, daß wir von der Seite auf ihn blicken. Die allgemeine Linsenform scheint hier aber entstanden durch den Zug zahlreicher Spiralen um einen verdichteten Kern. Aus ähnlichen Spiralen gebildet hätten wir uns die Milchstraße vorzustellen und fänden damit eine prachtvolle Erklärung ihres äußeren Anblicks. Denken wir uns in den Andromedanebel hinein, so würden uns all diese Sterne, Knoten und Überschneidungen seiner Spiralen genau wie das Blätterwerk der Milchstraße erscheinen. Halten wir diese Analogie fest, so verhilft uns die Betrachtung einfacherer Spiralnebel zu einer Ahnung der Dynamik dieser Systeme. Der Andromedanebel und vielleicht unsere Milchstraße scheinen die letzte Entwicklungsform der Spiralnebel zu bilden. Eine viel übersichtlichere Gestalt treffen wir in den in Fig. 11 und 12 wiedergegebenen Spiralnebeln. Hier haben die einzelnen Arme den Kern noch nicht so oft umkreist wie bei dem Andromedanebel, und es tritt die merkwürdige Polarität fast aller Spiralnebel hervor. Es gehen fast immer von entgegengesetzten Stellen des Kerns zwei einander zugeordnete Spiralarme aus. Am deutlichsten ist diese Symmetrie aus dem Anfang eines Spiralnebels zu erkennen, den Fig. 13 zeigt. Man sieht, daß sich vom Zentrum aus zwei Arme in entgegengesetzter Richtung erstrecken und nur erst eine viertel Windung zurückgelegt haben. Die Bilder dieser Spiralnebel hintereinander betrachtet erscheinen wie ein in Gang kommendes Feuerrad. Der Kern entsendet Raketen nach zwei einander gegenüberliegenden Punkten, die die äußersten Spitzen der Nebelarme geben. Während der Kern langsam rotiert, geht der Ausschleuderungsvorgang weiter und bildet die nachfolgende Masse der Arme, die um so länger werden und um so zahlreichere Windungen aufweisen, je länger die Rotation und das Ausschleudern angedauert hat. Und hier komme ich auf die Entdeckung der großen Heerstraße unter den Sternen zurück, längs deren sie in entgegengesetzten Richtungen wandern: Vielleicht befinden wir uns nahe der Mitte eines solchen Spiralnebels und verspüren in den Bewegungen der Sterne die Nachwirkung der Kraft, die die Arme des Nebels auseinander treibt, und die Sterne, die an uns nach rechts und links vorüberziehen, schreiten hinaus, um den alten Spiralästen der Milchstraße ein neues Stück an der Wurzel hinzuzufügen.

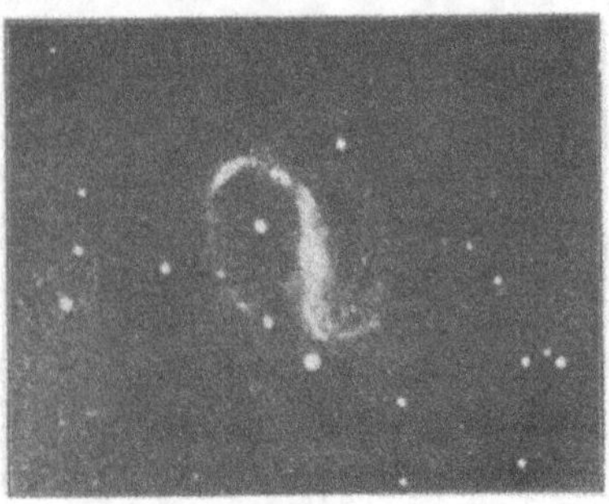

Fig. 13. Spiralnebel H. I. 55. Pegasi.

Astron. Nachr. *185* (1910) 81–88

ASTRONOMISCHE NACHRICHTEN.

Band 185. Nr. 4422. 6.

Über die Integralgleichungen der Stellarstatistik.

Von *K. Schwarzschild.*

1. Mit H. v. Seeliger[1]) nenne man $D(r)$ die Anzahl der Sterne pro Volumeneinheit in der Entfernung r von der Sonne, $\varphi(i)\,di$ denjenigen Bruchteil dieser Sterne, deren absolute Helligkeit zwischen i und $i+di$ liegt, und setze voraus, daß die Häufigkeitsfunktion $\varphi(i)$ überall im Raume dieselbe ist, daß Sterne bestimmter Helligkeit in jedem Raumvolumen prozentual gleich häufig sind. Von einer etwaigen Absorption des Lichtes im Raume werde abgesehen, so daß die scheinbare Helligkeit h eines Sterns zu seiner absoluten Helligkeit i in der Beziehung stehe:

$$h = \frac{i}{r^2}\,.$$

Die Anzahl der Sterne zwischen den Entfernungen r und $r+dr$ ist dann $4\pi\,D(r)\,r^2\,dr$. Von diesen Sternen liefern scheinbare Helligkeiten zwischen h und $h+dh$ diejenigen, deren absolute Helligkeiten zwischen $h\cdot r^2$ und $(h+dh)\,r^2$ liegen, das ist der Bruchteil: $\varphi(h r^2)\,r^2\,dh$. Damit ergibt sich für die Gesamtzahl $B(h)\,dh$ der Sterne der scheinbaren Helligkeit h bis $h+dh$ durch Integration über den ganzen Raum:

$$B(h)\,dh = 4\pi \int_{r=0}^{\infty} D(r)\,r^2\,dr\,\varphi(h r^2)\,r^2\,dh$$

oder:

$$B(h) = 4\pi \int_0^{\infty} D(r)\,\varphi(h r^2)\,r^4\,dr\,. \qquad (1)$$

Für die mittlere reziproke Entfernung (»mittlere Parallaxe«) derselben Sterne $\pi(h)$ erhält man:

$$\pi(h)\,B(h) = 4\pi \int_0^{\infty} D(r)\,\varphi(h r^2)\,r^3\,dr\,. \qquad (2)$$

Für die Funktion $\varphi(i)$ gilt noch:

$$\int_0^{\infty} \varphi(i)\,di = 1\,. \qquad (3)$$

Bei der Anwendung der Integralgleichungen (1) und (2) auf das Problem der räumlichen Verteilung der Sterne ist stets die Anzahl $B(h)$ der Sterne jeder scheinbaren Helligkeit empirisch gegeben, die Dichte $D(r)$ im Sternsystem als Funktion der Entfernung r gesucht. Je nachdem man weiter die Häufigkeitsfunktion $\varphi(i)$ oder die mittlere Parallaxe $\pi(h)$ den vorhandenen Daten für die uns näher stehenden Sterne empirisch entnimmt, hat man zwei verschiedene mathematische Aufgaben:

1) Es ist $B(h)$ und $\varphi(i)$ gegeben. Man soll durch Auflösung der Integralgleichung (1) hieraus $D(r)$ finden.

2) Es ist $B(h)$ und $\pi(h)$ gegeben. Man soll durch gleichzeitige Auflösung der Integralgleichungen (1) und (2) sowohl $D(r)$ als $\varphi(i)$ finden.

Ich möchte hier zeigen, daß sich diese Aufgaben auf die Ausführung von bestimmten Integralen zurückführen lassen.

2. Zunächst sollen die beiden Integralgleichungen auf möglichst einfache Form gebracht werden[2]). Man setze:

$$r = e^{-\rho} \qquad i = e^{-2\lambda} \qquad h = e^{-2\mu} \qquad (\lambda = \mu + \rho)$$

$$4\pi\,D(r)\,r^5 = 4\pi\,D(e^{-\rho})\,e^{-5\rho} = f(\rho)$$

$$\varphi(i) = \varphi(e^{-2\lambda}) = g(\lambda)$$

$$B(h) = B(e^{-2\mu}) = b(\mu)$$

$$\pi(h)\,B(h) = \pi(e^{-2\mu})\,B(e^{-2\mu}) = c(\mu)\,.$$

Dann gehen die beiden Integralgleichungen sofort über in:

$$b(\mu) = \int_{-\infty}^{+\infty} f(\rho)\,g(\mu+\rho)\,d\rho \qquad (\mathrm{I})$$

$$c(\mu) = \int_{-\infty}^{+\infty} f(\rho)\,g(\mu+\rho)\,e^{\rho}\,d\rho \qquad (\mathrm{II})$$

und es ist, um die Aufgaben in der neuen Bezeichnung zu wiederholen, stets $b(\mu)$ gegeben und $f(\rho)$ gesucht. Je nachdem außer $b(\mu)$ noch $g(\lambda)$ oder $c(\mu)$ gegeben ist, hat man nur aus der ersten Gleichung f oder aus beiden Gleichungen f und g zu finden.

3. Der Fouriersche Integralsatz lautet:

$$\varphi(x) = \int_0^{\infty} A(q)\cos qx\,dq + \int_0^{\infty} B(q)\sin qx\,dq$$

[1]) Betrachtungen über die räumliche Verteilung der Fixsterne I. München 1898 und II. München 1909.

[2]) Eine ganz ähnliche einfache Form, welche zugleich Absorption und endliche Begrenzung des Sternsystems mit berücksichtigt, gibt H. v. Seeliger II. Gleichungen (8).

Astron. Nachr. *185* (1910) 81–88

wobei die »Fourierkoeffizienten« $A(q)$ und $B(q)$ durch

$$A(q) = \frac{1}{\pi}\int_{-\infty}^{+\infty} \varphi(x)\cos qx\,dx$$

$$B(q) = \frac{1}{\pi}\int_{-\infty}^{+\infty} \varphi(x)\sin qx\,dx$$

gegeben sind.

Wenn man:

$$\frac{A(q)-i\,B(q)}{2} = \psi(q) \qquad \frac{A(q)+i\,B(q)}{2} = \psi(-q)$$

setzt, so nimmt er die bequemere komplexe Form an:

$$\varphi(x) = \int_{-\infty}^{+\infty} \psi(q)\,e^{iqx}\,dq \qquad \psi(q) = \frac{1}{2\pi}\int_{-\infty}^{+\infty} \varphi(x)\,e^{-iqx}\,dx.$$

$\psi(q)$ soll der »komplexe Fourierkoeffizient von $\varphi(x)$« heißen.

4. Um den Fourierschen Integralsatz zunächst auf die Gleichung (I) anzuwenden, multipliziere man sie mit $e^{-iq\mu}\,d\mu$ und integriere über μ von $-\infty$ bis $+\infty$. Man erhält:

$$\int_{-\infty}^{+\infty} b(\mu)\,e^{-iq\mu}\,d\mu = \int_{-\infty}^{+\infty} f(\varrho)\,d\varrho \int_{-\infty}^{+\infty} g(\mu+\varrho)\,e^{-iq\mu}\,d\mu = \int_{-\infty}^{+\infty} f(\varrho)\,e^{iq\varrho}\,d\varrho \cdot \int_{-\infty}^{+\infty} g(\mu+\varrho)\,e^{-iq(\mu+\varrho)}\,d\mu\,.$$

Im letzten Integral kann man ohne Änderung der Integrationsgrenzen $\mu+\varrho$ durch μ ersetzen und hat somit:

$$\int_{-\infty}^{+\infty} b(\mu)\,e^{-iq\mu}\,d\mu = \int_{-\infty}^{+\infty} f(\varrho)\,e^{iq\varrho}\,d\varrho \cdot \int_{-\infty}^{+\infty} g(\mu)\,e^{-iq\mu}\,d\mu\,. \tag{4}$$

Das sind offenbar lauter komplexe Fourierkoeffizienten. Entwickelt man nämlich alle hier auftretenden Funktionen in Fourierintegrale:

$$b(\mu) = \int_{-\infty}^{+\infty} \mathfrak{b}(q)\,e^{iq\mu}\,dq \qquad f(\varrho) = \int_{-\infty}^{+\infty} \mathfrak{f}(q)\,e^{iq\varrho}\,dq \qquad g(\mu) = \int_{-\infty}^{+\infty} \mathfrak{g}(q)\,e^{iq\mu}\,dq \tag{5}$$

so gilt für die Fourierkoeffizienten:

$$\mathfrak{b}(q) = \frac{1}{2\pi}\int_{-\infty}^{+\infty} b(\mu)\,e^{-iq\mu}\,d\mu \qquad \mathfrak{f}(q) = \frac{1}{2\pi}\int_{-\infty}^{+\infty} f(\varrho)\,e^{-iq\varrho}\,d\varrho \qquad \mathfrak{g}(q) = \frac{1}{2\pi}\int_{-\infty}^{+\infty} g(\mu)\,e^{-iq\mu}\,d\mu\,. \tag{6}$$

Damit lautet Gleichung (4):

$$\mathfrak{b}(q) = 2\pi\,\mathfrak{f}(-q)\cdot\mathfrak{g}(q)\,. \tag{III}$$

Genau ebenso läßt sich die Integralgleichung (II) behandeln. Hier hat man $c(\mu)$ in ein Fourierintegral zu entwickeln, also zu bilden:

$$c(\mu) = \int_{-\infty}^{+\infty} \mathfrak{c}(q)\,e^{iq\mu}\,dq \qquad \mathfrak{c}(q) = \frac{1}{2\pi}\int_{-\infty}^{+\infty} c(\mu)\,e^{-iq\mu}\,d\mu\,. \tag{7}$$

Statt des Integrals $\int_{-\infty}^{+\infty} f(\varrho)\,e^{iq\varrho}\,d\varrho$ tritt hier das Integral $\int_{-\infty}^{+\infty} f(\varrho)\,e^{\varrho}\,e^{iq\varrho}\,d\varrho$ auf. Schreibt man dieses in der Form:

$$\int f(\varrho)\,e^{-i\varrho(i-q)}\,d\varrho$$

so sieht man, daß es durch $\mathfrak{f}(i-q)$ gegeben wird, falls man sich die oben nur für reelle q erhaltene Funktion $\mathfrak{f}$ ins komplexe Gebiet fortgesetzt denkt. Hiernach folgt aus der zweiten Integralgleichung:

$$\mathfrak{c}(q) = 2\pi\,\mathfrak{f}(i-q)\,\mathfrak{g}(q)\,. \tag{IV}$$

Die einfachen Beziehungen (III) und (IV) zwischen den Fourierkoeffizienten der in Frage kommenden Funktionen sind es, die die Lösung der eingangs gestellten Aufgaben ermöglichen.

5. Was zunächst die erste Aufgabe angeht, bei bekannten b und g die Funktion f zu ermitteln, so hat man nach (6) die Fourierkoeffizienten $\mathfrak{b}(q)$ und $\mathfrak{g}(q)$ zu berechnen und hat dann nach (III) für die Fourierkoeffizienten von f:

$$\mathfrak{f}(q) = \frac{1}{2\pi}\,\frac{\mathfrak{b}(-q)}{\mathfrak{g}(-q)}$$

und damit für f selbst:

$$f(\varrho) = \frac{1}{2\pi}\int_{-\infty}^{+\infty} \frac{\mathfrak{b}(-q)}{\mathfrak{g}(-q)}\,e^{iq\varrho}\,dq = \frac{1}{2\pi}\int_{-\infty}^{+\infty} \frac{\mathfrak{b}(q)}{\mathfrak{g}(q)}\,e^{-iq\varrho}\,dq$$

womit die Aufgabe gelöst ist.

6. Zur Lösung der zweiten Aufgabe, bei der b und c gegeben, f und g gesucht sind, hat man zunächst die Fourierkoeffizienten $\mathfrak{b}(q)$ und $\mathfrak{c}(q)$ zu ermitteln. Dann folgt aus (III) und (IV):

$$\frac{\mathfrak{c}(q)}{\mathfrak{b}(q)} = \frac{\mathfrak{f}(i-q)}{\mathfrak{f}(-q)} .$$

Setzt man:

$$iq = p$$
$$\log \mathfrak{f}(-q) = \log \mathfrak{f}(ip) = F(p)$$
$$\log \frac{\mathfrak{c}(q)}{\mathfrak{b}(q)} = \log \frac{\mathfrak{c}(-ip)}{\mathfrak{b}(-ip)} = C(p)$$

so ist $C(p)$ bekannt und die vorstehende Gleichung geht in die einfache Differenzengleichung für $F(p)$ über:

$$F(p+1) - F(p) = C(p) .$$

Dieselbe wird nach der Methode von Laplace (Théorie analytique de la Probabilité) durch das bestimmte Integral:

$$F(p) = \frac{i}{2} \int_{-i\infty}^{+i\infty} C(p') \operatorname{cotg} \pi (p'-p)\, dp' \quad .$$

gelöst, wobei man den Integrationsweg in der komplexen p'-Ebene längs der imaginären Achse und zwar auf der Seite negativer p' zu wählen hat. Ist $F(p)$ berechnet, so erhält man ohne weiteres $\mathfrak{f}$, $\mathfrak{g}$ und damit f und g.

7. Die Anwendung der vorstehenden Formeln wird trotz ihrer formalen Einfachheit im allgemeinen eine erhebliche Rechenarbeit nach sich ziehen und lohnt sich daher im Grunde erst, wenn man über ein zuverlässigeres Beobachtungsmaterial verfügt, als es bis jetzt der Fall ist. Die Konvergenzfragen, insbesondere die bei Ableitung der Gleichung (III) vorgenommene Vertauschung der Integrationsordnung, machen keine Schwierigkeit, da man durch wiederholte Integration der ursprünglichen Integralgleichungen nach μ neue Integralgleichungen derselben Form erhält, wobei die Stetigkeit der auftretenden Funktionen und die Konvergenz der Fourierentwicklungen verstärkt ist. Eine tiefere Schwierigkeit ist die, daß man negative Werte für die Dichte oder auch für die Häufigkeitsfunktion $\varphi(i)$ erhält, wenn man von ungeeigneten Daten über die der Beobachtung zu entnehmenden Größen ausgeht. Würde sich das auch späterhin auf Grund verläßlicher Daten ergeben, so würde man daraus freilich den wichtigen Schluß auf die Unzulänglichkeit der zugrunde liegenden Annahmen, insbesondere der Unabhängigkeit der Funktion $\varphi(i)$ von der Entfernung, zu ziehen haben.

8. Es soll indessen wenigstens noch für die erste einfachere Aufgabe ein Beispiel gegeben werden, das sich den gegenwärtigen Beobachtungsdaten anschließt und zu äußerst durchsichtigen Ausdrücken führt.

Herr J. C. Kapteyn [1]) hat aus den Beobachtungsdaten für die uns benachbarten Sterne als Anzahl der Sterne zwischen den absoluten Größen M und $M + dM$ [2]) erhalten [3]):

$$\text{konst.} \cdot \frac{dM}{i} \cdot e^{-\alpha^2 (\log i - T)^2} \qquad (8)$$

($\alpha^2 = 0.385$, $T = 1.400$, $i = 1$ für die Sonne).

Berücksichtigt man, daß $M = -2.5 \log i$ ist, und führt, wie oben, $i = e^{-2\lambda}$ ein, so erhält man für die Anzahl $\varphi(i)\, di$ der Sterne zwischen den absoluten Helligkeiten i und $i + di$:

$$\varphi(i) = g(\lambda) = \text{konst.} \cdot e^{4\lambda - \alpha^2 (2\varkappa\lambda + T)^2} \qquad (\varkappa = \log e)$$

oder numerisch, wenn man die Konstante so bestimmt, daß $\int_0^\infty \varphi(i)\, di = 1$ wird [4]):

$$g(\lambda) = 0.0577\, e^{3.06\lambda - 0.290\lambda^2} .$$

Andrerseits findet Herr Kapteyn [5]) für die Anzahl der Sterne $N(m)\, dm$ zwischen den scheinbaren Größen m und $m + dm$ die nachstehenden Werte:

m	1	2	3	4	5	6	7	8	9	10	11	12	13	14	15
$\log N(m)$	1.179	1.708	2.232	2.752	3.260	3.764	4.253	4.732	5.202	5.660	6.105	6.538	6.961	7.374	7.779
Berechnet	1.151	1.696	2.230	2.752	3.254	3.765	4.255	4.733	5.202	5.658	6.103	6.538	6.952	7.375	7.776
Beob. − Rechn.	+28	+12	+2	0	+6	−1	−2	−1	0	+2	+2	0	+9	−1	+3

Dieselben werden durch die Formel:

$$\log N(m) = 0.596 + 0.5612\, m - 0.0055\, m^2 = \beta + \gamma m - \delta m^2$$

wie man aus den Differenzen Beob. − Rechn. sieht, völlig genügend dargestellt [6]). Berücksichtigt man hier, daß

$$m = -2.5 \log h$$

ist, und führt $h = e^{-2\mu}$ ein, so erhält man für die Anzahl der Sterne zwischen den scheinbaren Helligkeiten h und $h + dh$:

$$B(h) = b(\mu) = 2.5\, e^{\frac{\beta}{\varkappa} + (5\gamma + 2)\mu - 2.5\varkappa\delta\mu^2}$$

oder numerisch:

$$b(\mu) = 9.86\, e^{4.806\mu - 0.0598\mu^2} .$$

---

[1]) Astronomical Journal Nr. 566.

[2]) Strenggenommen bezeichnet dieser Ausdruck bei Herrn Kapteyn die Anzahl der Sterne zwischen den absoluten Größen $M - 1/2$ und $M + 1/2$, ebenso wie der unten folgende die Anzahl der Sterne zwischen den scheinbaren Größen $m - 1/2$ und $m + 1/2$. Doch läßt sich aus den Integralgleichungen leicht ableiten, daß für diese endlichen Intervalle dieselben Beziehungen gelten, wie für die hier benutzten unendlich kleinen Intervalle.

[3]) log bezeichnet im folgenden den Dezimal-Logarithmus.

[4]) Strenggenommen ist die Summe $\Sigma \varphi(i)$, über die einzelnen Größenklassen genommen, gleich 1 zu machen.

[5]) Publications of the Astronomical Laboratory of Groningen, Nr. 18, Tabelle II, letzte Spalte mit $\frac{4\pi}{\sin^2 1^\circ}$ multipliziert.

[6]) Diese Bemerkung haben nach brieflichen Mitteilungen bereits Herr A. Pannekoek und Herr J. C. Kapteyn selbst gemacht.

Die Beobachtungsdaten, von denen wir ausgehen, sind also zusammengefaßt folgende:

$$g(\lambda) = 0.0577\, c^{3.06\lambda - 0.290\lambda^2}$$
$$b(\mu) = 9.86\, c^{4.806\mu - 0.0598\mu^2}$$

wofür wir abgekürzt:

$$g(\lambda) = g_0\, c^{g_1\lambda - g_2\lambda^2}$$
$$b(\mu) = b_0\, c^{b_1\mu - b_2\mu^2}$$

schreiben wollen.

Als Einheit der Distanz ist durch die obigen Ansätze bereits die der Parallaxe 1″ entsprechende festgelegt, da die Sonne ($i = 1$) in dieser Entfernung die scheinbare Helligkeit $h = 1$ ($m = 0$) besitzt.

Die Funktionen g und b sind nun in Fourierintegrale zu entwickeln. Die Integrale (6) lassen sich hier sofort auf das bekannte Wahrscheinlichkeitsintegral zurückführen und ergeben für die Fourierkoeffizienten beider Funktionen:

$$\mathfrak{g}(q) = \frac{g_0}{\sqrt{4\pi g_2}}\, c^{\frac{(g_1 - iq)^2}{4g_2}}$$
$$\mathfrak{b}(q) = \frac{b_0}{\sqrt{4\pi b_2}}\, c^{\frac{(b_1 - iq)^2}{4b_2}}.$$

Es folgt aus (III):

$$\mathfrak{f}(q) = \frac{1}{2\pi}\frac{b_0}{g_0}\sqrt{\frac{g_2}{b_2}}\, c^{\frac{(b_1 + iq)^2}{4b_2} - \frac{(g_1 + iq)^2}{4g_2}}$$

oder abgekürzt:

$$\mathfrak{f}(q) = f_0\, c^{if_1 q - f_2 q^2}.$$

Damit folgt für die Funktion $f(\varrho)$ selbst durch Ausführung des Integrals (5):

$$f(\varrho) = f_0 \cdot \sqrt{\frac{\pi}{f_2}}\, c^{-\frac{(f_1 + \rho)^2}{4f_2}}.$$

Für die Dichte im Sternsystem erhält man:

$$D(\varrho) = \frac{1}{4\pi} c^{5\rho} f(\varrho) = \frac{f_0}{\sqrt{16\pi f_2}}\, c^{25f_2 - 5f_1 - \frac{(\rho + f_1 - 10f_2)^2}{4f_2}}.$$

Die numerische Ausrechnung ergibt:

$$D(\varrho) = 0.212\, c^{-\frac{(\rho + 1.72)^2}{13.27}}.$$

Wenn man hier noch ϱ durch $-\dfrac{\log r}{x}$ ersetzt, so folgt:

$$\log D(r) = 9.326 - 0.174\,(\log r - 0.75)^2.$$

Hiernach hat die Dichte ein Maximum für $\log r = 0.75$, $r = 5.6$ oder $\pi = 0''.18$. Von da an fällt sie nach innen und außen ab. Es ändert aber nichts an den Abzählungsresultaten, wenn man die Dichte innerhalb der Kugel $r = 5.6$ konstant annimmt, da die Zahl der Sterne einer Parallaxe $> 0''.18$ zu gering ist. Als wirkliches Resultat der Rechnung bleibt der durch die Formel gegebene Dichteabfall nach außen. Multipliziert man $D(r)$ mit 1000, so erhält man folgende Sternzahlen im Würfel, dessen Kantenlänge der Parallaxe 0″.1 entspricht:

r	5.6	20	100	300	1000	10000
π	0″.18	0″.05	0″.01	0″.0033	0″.0010	0″.0001
1000 D	212	187	113	64	28	3

Für die mittlere Parallaxe $\pi(\mu)$ erhält man durch einfache Ausführung des Integrals (II):

$$\pi(\mu) = c^{\pi_0 - \pi_1\mu}$$
$$\pi_0 = \frac{f_2 - f_1 + 2f_2 g_1}{1 + 4f_2 g_2} = -2.324$$
$$\pi_1 = \frac{4f_2 g_2}{1 + 4f_2 g_2} = 0.794$$

oder durch Übergang auf gewöhnliche Logarithmen und Größenklassen:

$$\log \pi_m = -1.009 - 0.159\,m.$$

Das gibt folgende Vergleichung mit den von Herrn Kapteyn[1]) aus den Beobachtungen abgeleiteten Zahlen:

m	2.7	4.1	5.1	6.0	6.9	8.6	10.5
$\log \pi_m$ (Kapteyn)	8.583	8.312	8.167	8.111	7.952	7.799	7.686
$\log \pi_m$ (berechnet)	8.561	8.338	8.179	8.037	7.894	7.624	7.321
Beob. — Rechn.	+0.022	−0.026	−0.012	+0.074	+0.058	+0.175	+0.365

Es herrscht ein deutlicher systematischer Gang in den Abweichungen, der zu beseitigen wäre, wenn man annähme, daß die Zahl der absolut schwachen Sterne im Verhältnis zur Zahl der absolut hellen größer ist, als der Kapteynsche Ausdruck (8) angibt.

Potsdam, 1910 April 9. *K. Schwarzschild.*

[1]) Astronomical Journal Nr. 566.

Leuchtende Nachtwolken. (Telegramm aus Seebad Bansin, 1910 Juni 25 Vorm.). Beobachtete leuchtende Nachtwolken gestern Freitag $10^{1}/_{2}{}^{h}$ bis $13^{1}/_{2}{}^{h}$ am Nordhimmel. *Archenhold.*

Correzione all'Annuario Astronomico pel 1910 del R. Osservatorio di Torino. Tutti i valori della declinazione apparente della stella 6 Lyncis (p. 21) devono essere diminuiti di 3″.1. La declinazione media è esatta.

G. Boccardi.

Astron. Nachr. 190 (1912) 361–376

ASTRONOMISCHE NACHRICHTEN.
Band 190. Nr. 4557. 21.

Zur Stellarstatistik. Von *K. Schwarzschild.*

§ 1. Die theoretischen Abzählungen und Mittelwertbildungen der Stellarstatistik hängen von drei Häufigkeitsfunktionen ab, nämlich von der räumlichen Verteilung der Sterndichte, der Verteilung der Leuchtkräfte und der Verteilung der Geschwindigkeiten. Die Abhängigkeiten von der Sterndichte und der Verteilung der Leuchtkräfte sind namentlich von *H. v. Seeliger* eingehend behandelt. Viel weniger beachtet sind hingegen die Aufgaben, in welche auch die Geschwindigkeitsverteilung eingeht. Charakteristisch dafür ist, daß sich *H. Gyldén* und *J. C. Kapteyn* bei der Darstellung der mittleren Parallaxe der Sterne als Funktion von scheinbarer Sterngröße und Eigenbewegung, wo es sich um eine Aufgabe letzterer Art handelt, mit rein empirischen Formeln begnügt haben.

Im folgenden werden zunächst die Abhängigkeiten der wichtigsten stellarstatistischen Größen von allen drei Häufigkeitsfunktionen zusammengestellt. Dabei wird von einer Absorption des Lichts im Weltraum abgesehen. Das wesentliche ist dann die Ausführung der allgemeinen Formeln für eine spezielle Gestalt der drei Häufigkeitsfunktionen, bei der alle Integrationen durch Exponentialformeln einfachster Art zu leisten sind. Die Einfachheit der Formeln läßt den Zusammenhang zwischen Verteilungsfunktionen und Beobachtungsdaten so leicht überblicken, daß die Formeln vielleicht auch später, wenn sie zur Darstellung eines größeren und besseren Beobachtungsmaterials nicht mehr ausreichen, zur Orientierung und zur Abschätzung des Einflusses systematischer Beobachtungsfehler nützlich bleiben werden. Besonders überraschend ist, daß sich genau Herrn *Kapteyns* empirischer Ansatz für die Parallaxe als Funktion von Sterngröße und Eigenbewegung und dazu auch die von ihm angenommene Formel für die Streuung der Parallaxen um den so gerechneten Wert ergibt, womit die theoretische Begründung der *Kapteyn*schen Formeln nachgeliefert ist. Die zum Schluß folgende zahlenmäßige Durchrechnung zeigt, daß die *Kapteyn*schen Formeln auch numerisch in sich konsequent sind. Sie führt dabei auf eine Verteilung der Geschwindigkeiten der Sterne, deren beträchtliche Abweichung vom *Maxwell*schen Gesetz bemerkenswert ist.

§ 2. Bezeichnungen, Einheiten, Grundformeln.[1])

Es sollen folgende Bezeichnungen verwandt werden:

- r Entfernung (Einheit entsprechend einer jährlichen Parallaxe von 1″),
- i scheinbare Helligkeit (Einheit die scheinbare Helligkeit eines Sterns nullter Größe),
- J absolute Helligkeit oder Leuchtkraft (Einheit die absolute Helligkeit eines Sterns, welcher bei einer jährl. Parallaxe von 1″ die nullte Größe hat. Die Sonne hat sehr nahe diese Leuchtkraft 1),
- μ Eigenbewegung (scheinbare auf die Sphäre projizierte Geschwindigkeit) nach Bogensekunden im Jahr gezählt,
- V absolute auf die Sphäre projizierte Geschwindigkeit (Einheit diejenige Geschwindigkeit, welche aus der Entfernung 1 gesehen die Eigenbewegung 1 gibt, d. i. ein Erdbahnradius im Jahr oder 4.737 km/sec).

Es werde ferner gesetzt:

$$\begin{aligned} \varrho &= -5.0 \log r \\ m &= -2.5 \log i \qquad g = -5.0 \log \mu \\ M &= -2.5 \log J \qquad G = -5.0 \log V. \end{aligned} \qquad (1)$$

Es ist dann m die scheinbare Größe in der üblichen Zählweise, M die absolute Helligkeit in Größenklassen ausgedrückt oder die absolute Größe. ϱ, g, G könnten analog als Entfernungsklasse, Eigenbewegungsklasse und Geschwindigkeitsklasse bezeichnet werden.

Bei der obigen Wahl der Einheiten gelten die Grundformeln:

$$J = i r^2 \qquad V = \mu r \qquad (2)$$

oder in die Klassenbezeichnung übertragen:

$$M = m + \varrho \qquad G = g + \varrho. \qquad (3)$$

§ 3. Die Verteilungsfunktionen.

a) Die Dichte (Anzahl der Sterne in der Volumeneinheit, im Würfel, dessen Seitenlänge der Parallaxe 1″ entspricht) als Funktion der Entfernung r heiße $D(r)$.

b) Die Verteilung der Leuchtkräfte werde bestimmt durch Angabe des Prozentsatzes der Sterne in der Volumeneinheit, deren Leuchtkraft zwischen den Grenzen J und $J + dJ$ liegt, und zwar sei dieser Prozentsatz gegeben durch den Ausdruck $\varphi(J)\,dJ$.

c) Die Verteilung der Geschwindigkeiten werde bestimmt durch Angabe des Prozentsatzes der Sterne jeder Leuchtkraft in der Volumeneinheit, deren Geschwindigkeit zwischen den Grenzen V und $V + dV$ liegt, und zwar werde dieser Prozentsatz gegeben durch den Ausdruck $\psi(V)\,dV$. Es handelt sich dabei, wie gesagt, nicht um die ganzen Geschwindigkeiten, sondern nur um ihre Projektionen auf die Sphäre. Die Geschwindigkeiten sind relativ zur Sonne zu nehmen, solange man nicht die Eigenbewegungen von der parallaktischen Bewegung befreien kann.

Jede von den drei Verteilungsfunktionen soll nur als abhängig von ihrem einen Argument angesehen werden. Es soll namentlich die Verteilung der Leuchtkräfte unabhängig von der Entfernung r sein und es soll auch außer Acht gelassen werden, daß nach *W. W. Campbell* die mittlere absolute Geschwindigkeit der Sterne sich mit ihrem Spektraltypus und damit ihrer absoluten Helligkeit ändert, daß daher in Wirklichkeit die Verteilung $\psi(V)$ auch von J abhängt.

[1]) Abgesehen von den Buchstaben im Anschluß an *E. Anding*, Kritische Untersuchungen über die Bewegung der Sonne durch den Weltraum. Zweiter Abschnitt, Seite 87.

Die Ansätze sind im folgenden immer im Durchschnitt für die ganze Himmelskugel gemacht. Es wäre aber ohne weiteres möglich, den Einfluß der Milchstraße und den Einfluß irgend einer Vorzugsrichtung der Eigenbewegungen in die Formeln aufzunehmen, indem man sie nur für ein kleines Himmelsareal und in diesem wieder nur für die Sterne einer bestimmten Richtung der Eigenbewegung gelten läßt und indem man dabei die drei Verteilungsfunktionen als vom Ort am Himmel, etwa galaktischer Länge und Breite, sowie vom Positionswinkel der Eigenbewegung abhängig denkt.

Infolge der Definition von φ und ψ als prozentualer Häufigkeiten gilt:

$$\int_0^\infty \varphi(\mathfrak{J})\,\mathrm{d}\mathfrak{J} = 1 \qquad \int_0^\infty \psi(V)\,\mathrm{d}V = 1\,. \tag{4}$$

§ 4. Die allgemeinen Formeln für die wichtigsten Anzahlen und Mittelwerte.

a) Anzahl der Sterne der scheinbaren Helligkeit i bis $i+\mathrm{d}i$. Will man die Ableitung streng machen, so hat man so vorzugehen: Die Anzahl der Sterne zwischen den Entfernungen r und $r+\mathrm{d}r$ auf der scheinbaren Himmelsfläche ω ist $\omega D(r)\, r^2\,\mathrm{d}r$.[1] Von diesen hat eine absolute Helligkeit $\mathfrak{J}$ bis $\mathfrak{J}+\mathrm{d}\mathfrak{J}$ der Bruchteil $\varphi(\mathfrak{J})\,\mathrm{d}\mathfrak{J}$. Die Zahl der Sterne von absoluter Helligkeit $\mathfrak{J}$ bis $\mathfrak{J}+\mathrm{d}\mathfrak{J}$ in der Entfernung r bis $r+\mathrm{d}r$ ist daher:

$$\omega D(r)\, r^2\, \varphi(\mathfrak{J})\,\mathrm{d}r\,\mathrm{d}\mathfrak{J} = \varDelta\,\mathrm{d}r\,\mathrm{d}\mathfrak{J} \tag{5}$$

wo $\varDelta$ zur Abkürzung eingeführt ist.

Man denke sich nun r und $\mathfrak{J}$ als Abszissen und Ordinaten in einem rechtwinkligen Koordinatensystem und die Größe $\varDelta$ als Dichte einer ungleichförmig über die Koordinatenebene verteilten Belegung. Die Kurven konstanter scheinbarer Helligkeit $i =$ konst. lauten: $\mathfrak{J} =$ konst. r^2, bilden also in diesem Koordinatensystem eine Schar von Parabeln durch den Nullpunkt. Um die Zahl aller Sterne zwischen den scheinbaren Helligkeiten i und $i+\mathrm{d}i$ zu finden, ist die Belegung in der Koordinatenebene über den Flächenstreifen zwischen den beiden Parabeln $\mathfrak{J} = i\,r^2$ und $\mathfrak{J} = (i+\mathrm{d}i)\,r^2$ zu integrieren. Die Höhe des Flächenstreifens bei der Abszisse r ist $\mathrm{d}i\,r^2$. Das Integral lautet also:

$$\int_0^\infty \varDelta\,\mathrm{d}r\,\mathrm{d}i\,r^2$$

oder ausgeschrieben, wenn man die gesuchte Anzahl der Sterne der scheinbaren Helligkeit i bis $i+\mathrm{d}i$ mit $B_i\,\mathrm{d}i$ bezeichnet:

$$B_i\,\mathrm{d}i = \omega\,\mathrm{d}i \int_0^\infty \mathrm{d}r\, r^4\, D(r)\, \varphi(i r^2)\,. \tag{6}$$

Da nach (1): $\mathrm{d}m = -2.5\log e \frac{\mathrm{d}i}{i}$ ist, so folgt noch für die Anzahl $b_m\,\mathrm{d}m$ der Sterne zwischen den scheinbaren Größen m und $m+\mathrm{d}m$:

$$b_m\,\mathrm{d}m = -B_i \frac{\mathrm{d}i}{\mathrm{d}m}\,\mathrm{d}m$$

oder

$$b_m\,\mathrm{d}m = \frac{\omega}{2.5\log e}\,\mathrm{d}m \int_0^\infty \mathrm{d}r\, r^4\, D(r)\, i\, \varphi(i r^2)\,. \tag{6a}$$

Wenn man überlegt, daß man bei der Integration über den bezeichneten Flächenstreifen denselben auch in lauter, relativ zu ihrer Breite, äußerst schmale vertikale Streifen zerlegen, oder $\mathrm{d}r$ sehr klein gegen $\mathrm{d}i$ nehmen darf, so erkennt man die Richtigkeit der folgenden kürzeren Ableitung, die später in den analogen Fällen allein durchgeführt werden soll: Die Anzahl der Sterne zwischen den Entfernungen r und $r+\mathrm{d}r$ auf der scheinbaren Himmelsfläche ω ist

$$\omega\, D(r)\, r^2\,\mathrm{d}r.$$

Von diesen liefern scheinbare Helligkeiten zwischen i und $i+\mathrm{d}i$ diejenigen, deren absolute Helligkeiten zwischen $i r^2$ und $(i+\mathrm{d}i)\, r^2$ liegen, das ist der Bruchteil: $\varphi(i r^2)\, r^2\,\mathrm{d}i$. Der Beitrag der Entfernungen r bis $r+\mathrm{d}r$ zur gesuchten Anzahl ist daher $\omega\, D(r)\, r^4\, \varphi(i r^2)\,\mathrm{d}i\,\mathrm{d}r$, und man erhält die gesuchte Anzahl, wie oben, durch Integration dieses Ausdrucks über alle Entfernungen r.

b) Mittlere Parallaxe π_m der Sterne der scheinbaren Größe m.

Die Parallaxe π ist gleich $1/r$. Die Summe aller Parallaxen der Sterne der scheinbaren Größe m findet sich offenbar, indem man unter dem Integranden in (6a) die Parallaxe als Faktor einführt. Diese Summe ist dann gleich dem gesuchten Mittelwert π_m der Parallaxe multipliziert mit der Anzahl $b_m\,\mathrm{d}m$ der Sterne. Es folgt also:

$$\pi_m\, b_m\,\mathrm{d}m = \frac{\omega}{2.5\log e}\,\mathrm{d}m \int_0^\infty \mathrm{d}r\, r^3\, D(r)\, i\, \varphi(i r^2)$$

und damit:

$$\pi_m = \frac{\int_0^\infty \mathrm{d}r\, r^3\, D(r)\, \varphi(i r^2)}{\int_0^\infty \mathrm{d}r\, r^4\, D(r)\, \varphi(i r^2)}\,. \tag{7}$$

c) Mittlere absolute Helligkeit $\mathfrak{J}_i$ der Sterne gegebener scheinbarer Helligkeit i.

Man hat hier genau, wie oben, zu verfahren, nur daß man statt des Faktors $\pi = 1/r$ den Faktor $\mathfrak{J} = i r^2$ unter dem Integranden im Zähler einführt. Das Resultat ist:

$$\mathfrak{J}_i = i \cdot \frac{\int_0^\infty \mathrm{d}r\, r^6\, D(r)\, \varphi(i r^2)}{\int_0^\infty \mathrm{d}r\, r^4\, D(r)\, \varphi(i r^2)}\,. \tag{8}$$

Zu Größenklassen übergehend, soll $M_m = -2.5\log \mathfrak{J}_i$ geschrieben werden ($m = -2.5\log i$).

[1]) Bei der numerischen Anwendung werde ich im folgenden, wo nichts anderes gesagt ist, immer die ganze Himmelsfläche nehmen und $\omega = 4\pi$ setzen. In den Formeln ist das meist nicht geschehen, um eine Verwechslung von $\pi = 3.14\ldots$ mit der Parallaxe π zu vermeiden.

d) Verteilung der absoluten Helligkeiten unter den Sternen gegebener scheinbarer Helligkeit.

Die Anzahl der Sterne der Leuchtkraft $\mathfrak{J}$ bis $\mathfrak{J} + d\mathfrak{J}$ bei der scheinbaren Helligkeit i bis $i + di$ wird in dem unter a) verwandten Koordinatensystem durch die Belegung eines Parallelogramms gegeben, das seitlich durch die Parabeln $i =$ konst. und $i + di =$ konst., nach oben und unten durch die Parallelen zur Abszissenachse, welche $\mathfrak{J}$ und $\mathfrak{J} + d\mathfrak{J}$ entsprechen, begrenzt wird. Die Höhe des Parallelogramms ist demnach $d\mathfrak{J}$, die Basis folgt aus der Differentiation von $r = \sqrt{\mathfrak{J}/i}$ bei festgehaltenem $\mathfrak{J}$ zu: $dr = \frac{1}{2}\sqrt{\mathfrak{J}/i} \cdot di/i$. Multipliziert man das Produkt aus Basis und Höhe des Parallelogramms noch mit der Dichte Δ der Belegung, so erhält man als Resultat für die gesuchte Anzahl $\varphi_{\mathfrak{J},i}\, d\mathfrak{J}\, di$ der Sterne von absoluter Helligkeit $\mathfrak{J}$ bis $\mathfrak{J} + d\mathfrak{J}$ bei einer scheinbaren Helligkeit i bis $i + di$:

$$\tfrac{1}{2}\Delta \sqrt{\mathfrak{J}/i} \cdot di/i \cdot d\mathfrak{J}$$

oder ausgeschrieben:

$$\varphi_{\mathfrak{J},i}\, d\mathfrak{J}\, di = \frac{d\mathfrak{J}\, di}{2i}\, \varphi(\mathfrak{J})\, D(\sqrt{\mathfrak{J}/i})\, (\sqrt{\mathfrak{J}/i})^3 . \qquad (9)$$

Als Anzahl der Sterne der absoluten Größe M bis $M + dM$ bei der scheinbaren Größe m bis $m + dm$ folgt hieraus:

$$\begin{aligned} &\varphi_{M,m}\, dM\, dm = \\ &= \tfrac{1}{2}(1/2.5 \log e)^2\, dM\, dm\, \mathfrak{J}\, \varphi(\mathfrak{J})\, D(\sqrt{\mathfrak{J}/i})\, (\sqrt{\mathfrak{J}/i})^3 \\ &(\mathfrak{J} = 10^{-0.4M}, \quad i = 10^{-0.4m}) . \end{aligned} \qquad (9a)$$

e) Zahl der Sterne von gegebener Eigenbewegung.

Die Überlegung ist ganz analog wie unter a). Scheinbare Eigenbewegungen zwischen den Grenzen μ und $\mu + d\mu$ liefern die Sterne, deren absolute Geschwindigkeiten bei der Entfernung r zwischen $r\mu$ und $r\mu + r d\mu$ liegen. Die Gesamtzahl dieser Sterne wird:

$$b_\mu\, d\mu = \omega\, d\mu \int_0^\infty dr\, r^3\, D(r)\, \psi(\mu r) . \qquad (10)$$

f) Mittlere Parallaxe π_μ der Sterne von gegebener Eigenbewegung μ. Ganz analog zu (7) folgt:

$$\pi_\mu = \frac{\int_0^\infty dr\, r^2\, D(r)\, \psi(\mu r)}{\int_0^\infty dr\, r^3\, D(r)\, \psi(\mu r)} . \qquad (11)$$

g) Zahl der Sterne von gegebener Eigenbewegung und scheinbarer Größe.

Unter a) wurde die Anzahl der Sterne in der Entfernung r bis $r + dr$ berechnet, welche scheinbare Helligkeiten zwischen i und $i + di$ haben. Unter diesen sind nun noch speziell diejenigen auszusondern, deren scheinbare Eigenbewegung zwischen μ und $\mu + d\mu$, deren absolute also zwischen $r\mu$ und $r\mu + r\,d\mu$ liegt. Der Prozentsatz letzterer Sterne ist: $\psi(r\mu)\, r\, d\mu$. Für die gesuchte Anzahl $B_{i,\mu}\, di\, d\mu$ der Sterne, deren scheinbare Helligkeit zwischen i und $i + di$, deren scheinbare Eigenbewegung zwischen μ und $\mu + d\mu$ liegt, findet sich daher:

$$B_{i,\mu}\, di\, d\mu = \omega\, di\, d\mu \int_0^\infty dr\, r^5\, D(r)\, \varphi(ir^2)\, \psi(\mu r) . \qquad (12)$$

Die Anzahl der Sterne der Eigenbewegung μ bis $\mu + d\mu$ zwischen den scheinbaren Größen m und $m + dm$ wird analog zu (6a):

$$b_{m,\mu}\, dm\, d\mu = \frac{\omega}{2.5 \log e}\, dm\, d\mu \int_0^\infty dr\, r^5\, D(r)\, i\, \varphi(ir^2)\, \psi(\mu r) . \qquad (13)$$

h) Mittlere Parallaxe $\pi_{m,\mu}$ der Sterne der scheinbaren Größe m und der Eigenbewegung μ.

In Analogie zu (7) findet sich hierfür nach (12):

$$\pi_{m,\mu} = \frac{\int_0^\infty dr\, r^4\, D(r)\, \varphi(ir^2)\, \psi(\mu r)}{\int_0^\infty dr\, r^5\, D(r)\, \varphi(ir^2)\, \psi(\mu r)} . \qquad (14)$$

i) Häufigkeit einer bestimmten Parallaxe unter den Sternen gegebener scheinbarer Eigenbewegung und Größe.

Die Zahl der Sterne in der Entfernung r bis $r + dr$, welche scheinbare Eigenbewegung zwischen μ und $\mu + d\mu$ und scheinbare Größe m bis $m + dm$ haben, ist nach der unter g) gegebenen Ableitung:

$$\frac{\omega}{2.5 \log e}\, dm\, d\mu\, dr\, r^5\, D(r)\, i\, \varphi(ir^2)\, \psi(\mu r).$$

Da $dr = d(1/\pi) = -d\pi\, r^2$ ist, so folgt als Häufigkeit der Sterne mit Parallaxen zwischen π und $\pi + d\pi$:

$$X_{\pi,m,\mu}\, dm\, d\mu\, d\pi = \frac{\omega}{2.5 \log e}\, dm\, d\mu\, d\pi \cdot r^7\, D(r)\, i\, \varphi(ir^2)\, \psi(\mu r).$$

Da ferner $d \log \pi = \log e \cdot d\pi/\pi$ ist, so folgt als Häufigkeit für einen Logarithmus der Parallaxe zwischen den Grenzen $\log \pi$ und $\log \pi + d \log \pi$:

$$\chi_{\pi,m,\mu} \cdot d \log \pi = (\pi/\log e)\, X_{\pi,m,\mu} \cdot d \log \pi . \qquad (15)$$

k) Mittlere Eigenbewegung der Sterne jeder scheinbaren Größe.

Der Mittelwert der Eigenbewegung für die Sterne der scheinbaren Größe m wird nach (13) gegeben durch:

$$\mu_m = \frac{\int_0^\infty b_{m,\mu} \cdot \mu\, d\mu}{\int_0^\infty b_{m,\mu} \cdot d\mu} = \frac{\int_0^\infty dr\, r^5\, D(r)\, \varphi(ir^2) \int_0^\infty \psi(\mu r)\, \mu\, d\mu}{\int_0^\infty dr\, r^5\, D(r)\, \varphi(ir^2) \int_0^\infty \psi(\mu r)\, d\mu}$$

oder:

$$\mu_m = \frac{\int_0^\infty dr\, r^3\, D(r)\, \varphi(ir^2)}{\int_0^\infty dr\, r^4\, D(r)\, \varphi(ir^2)} \cdot \frac{\int_0^\infty \psi(V)\, V\, dV}{\int_0^\infty \psi(V)\, dV} .$$

Der Quotient der beiden ersten Integrale ist die mittlere Parallaxe π_m der Sterne der Größe m. Der zweite Quotient ist die mittlere Absolutgeschwindigkeit der Sterne. Bezeichnet man letzere mit $\overline{V}$, so gilt also:

$$\mu_m = \pi_m \cdot \overline{V} \tag{16}$$

wie natürlich auch leicht einzusehen ist, ohne Integrale anzuschreiben. Wenn man die Abhängigkeit der Geschwindigkeitsverteilung ψ von der Leuchtkraft berücksichtigen wollte, so müßte man allerdings auf das Integral zurückgehn, und es würde dann auch die Proportionalität zwischen mittlerer Eigenbewegung und Parallaxe für verschiedene Größenklassen aufhören.

l) Gravitationspotential des Sternsystems. Das Gravitationspotential im Zentrum des kugelförmig gedachten Sternsystems wird, wenn man allen Sternen die Sonnenmasse 1 gibt:

$$P = 4\pi k^2 \int_0^\infty D(r)\, r\, dr\,. \tag{17}$$

Die beim Fall aus dem Unendlichen nach dem Zentrum entstehende Maximalgeschwindigkeit W hängt mit P durch die Gleichung zusammen:

$$\tfrac{1}{2} W^2 = P\,. \tag{17a}$$

Da die Erdgeschwindigkeit in unseren Einheiten gleich 2π, der Abstand Erde-Sonne gleich $\sin 1''$ wird, so wird in unseren Einheiten:

$$k^2 = 4\pi^2 \sin 1''.$$

m) Helligkeit des Himmelsgrundes.

Die mittlere Leuchtkraft $\overline{\mathfrak{J}}$ der Sterne der Volumeneinheit hat den Wert:

$$\overline{\mathfrak{J}} = \int_0^\infty \mathfrak{J}\,\varphi(\mathfrak{J})\, d\mathfrak{J}\,. \tag{18}$$

Bei der Dichte $D(r)$ ist die Leuchtkraft der Volumeneinheit $\overline{\mathfrak{J}}\,D(r)$. Die gesamte scheinbare Helligkeit aller auf der scheinbaren Himmelsfläche ω liegenden Sterne wird daher:

$$H = \omega \int_0^\infty \frac{\overline{\mathfrak{J}}\,D(r)}{r^2} \cdot r^2\, dr = \omega \cdot \overline{\mathfrak{J}} \int_0^\infty D(r)\, dr\,. \tag{19}$$

Nimmt man für ω einen Quadratgrad ($\log \omega = 6.4838 - 10$) und bildet

$$S = -2.5 \log H \tag{19a}$$

so gibt S die Größe eines Sterns an, der so viel Licht gibt, wie ein Quadratgrad des Himmelsgewölbes.

§ 5. Spezieller Ansatz für die Verteilungsfunktionen.

Der Ansatz für die Verteilungsfunktionen, welcher durchgeführt werden soll, ist der folgende:

Dichte:	$D(r) = 10^{a_0 - a_1\varrho - a_2\varrho^2}$	$(\varrho = -5.0 \log r)$	(20)
Verteilung der Leuchtkräfte:	$\varphi(\mathfrak{J}) = 10^{b_0 - b_1 M - b_2 M^2}$	$(M = -2.5 \log \mathfrak{J})$	(21)
Verteilung der Geschwindigkeiten:	$\psi(V) = 10^{c_0 - c_1 G - c_2 G^2}$	$(G = -5.0 \log V)$.	(22)

Die a, b, c sind zunächst willkürliche Konstanten. Nur verlangen die Forderungen $\int_0^\infty \varphi(\mathfrak{J})\, d\mathfrak{J} = 1$ und $\int_0^\infty \psi(V)\, dV = 1$, wie aus der unten folgenden Gl. (29) leicht zu entnehmen ist, die Beziehungen:

$$b_0 = \log(2.5\sqrt{\log e/\pi}) + \tfrac{1}{2}\log b_2 - (b_1 + 0.4)^2/4b_2 \qquad \log(2.5\sqrt{\log e/\pi}) = -0.0317 \tag{23}$$

$$c_0 = \log(5.0\sqrt{\log e/\pi}) + \tfrac{1}{2}\log c_2 - (c_1 + 0.2)^2/4c_2 \qquad \log(5.0\sqrt{\log e/\pi}) = +0.2693\,. \tag{24}$$

Der Ansatz $\varphi(\mathfrak{J})$ wurde von Herrn *Kapteyn* empirisch gewonnen, der Ansatz $D(r)$ in A. N. 4422 daraus mit Hilfe der beobachteten Sternanzahlen deduziert. Der Ansatz $\psi(V)$ ist zunächst rein formal der Bequemlichkeit der Rechnung wegen eingeführt.

Es seien gleich folgende Bemerkungen hinzugefügt. Man kann schreiben:

$$\varphi(\mathfrak{J})\, d\mathfrak{J} = \varphi(\mathfrak{J}) \frac{d\mathfrak{J}}{dM}\, dM = -\frac{dM}{2.5 \log e}\, 10^{-0.4M + b_0 - b_1 M - b_2 M^2} = dM \cdot \text{konst.} \cdot 10^{-b_2[M + (b_1 + 0.4)/2b_2]^2}\,.$$

Aus letzterer Schreibweise erkennt man, daß, wenn man M als unabhängige Variable nimmt, das Verteilungsgesetz zur Abszisse:

$$\underline{M} = -(b_1 + 0.4)/2b_2 \tag{25}$$

symmetrisch wird. $\underline{M}$ ist daher sowohl der Mittelwert der absoluten Größe M der Sterne, als auch ihr Mittewert, über dem und unter dem gleich viele Werte liegen. Rechnet man aus $\underline{M} = -2.5 \log \underline{\mathfrak{J}}$ den Wert $\underline{\mathfrak{J}}$ aus, so gibt $\underline{\mathfrak{J}}$ das geometrische Mittel von $\mathfrak{J}$. Verschieden davon ist natürlich das arithmetische Mittel $\overline{\mathfrak{J}}$, das durch das Integral $\int_0^\infty \mathfrak{J}\,\varphi(\mathfrak{J})\, d\mathfrak{J}$ gegeben wird. Die Ausführung letzteren Integrals nach dem unten folgenden Schema liefert, wenn man auch $\overline{\mathfrak{J}}$ in Größenklassen $\overline{M}$ ausdrückt:

$$-2.5 \log \overline{\mathfrak{J}} = \overline{M} = -(b_1 + 0.6)/2b_2\,. \tag{25a}$$

Ganz analog findet man für die Geschwindigkeitsklassen $\underline{G}$ und $\overline{G}$, welche dem geometrischen und arithmetischen Mittel $\underline{V}$ resp. $\overline{V}$ der Geschwindigkeit entsprechen:

$$\underline{G} = -5 \log \underline{V} = -(c_1 + 0.2)/2c_2 \qquad \overline{G} = -5 \log \overline{V} = -(c_1 + 0.3)/2c_2\,. \tag{26}$$

§ 6. Berechnung der Anzahl der Sterne jeder Größe aus diesem Ansatz als Beispiel.

Da die Ausführung der allgemeinen Formeln nach diesem Ansatz durchweg nach demselben Schema verläuft, wird es genügen, nur die Berechnung der Anzahl $b_m\,dm$ der Sterne der scheinbaren Größe m bis $m + dm$ als Beispiel genauer anzugeben. Die allgemeine Formel (6a) lautet:

$$b_m = \frac{\omega}{2.5 \log e} \int_0^\infty dr\, r^4\, D(r)\, i\, \varphi(i r^2)\,.$$

Führt man hier den obigen Ansatz für D und φ und die Klassenzählung für die Variablen ein, setzt also:

$$r = 10^{-0.2\rho}, \qquad i = 10^{-0.4m}, \qquad M = m + \varrho$$

so hat man:

$$dr\, r^4 = -\frac{0.2}{\log e} d\varrho \cdot 10^{-\rho}$$

und damit:

$$b_m = \frac{0.2\,\omega}{2.5\,(\log e)^2} \int_{-\infty}^{+\infty} d\varrho \cdot 10^{-\rho + a_0 - a_1\rho - a_2\rho^2 - 0.4m + b_0 - b_1(m+\rho) - b_2(m+\rho)^2}.$$

Ordnet man den Exponenten nach Potenzen von ϱ, so lautet er:

$$a_0 + b_0 - (b_1 + 0.4)\,m - b_2 m^2 - \varrho\,(1 + a_1 + b_1 + 2 b_2 m) - \varrho^2 (a_2 + b_2) \tag{27}$$

wofür zur Abkürzung geschrieben werden soll:

$$A_0 - A_1 \varrho - A_2 \varrho^2. \tag{28}$$

Das auszuführende Integral lautet damit:

$$b_m = \frac{0.2\,\omega}{2.5\,(\log e)^2} \int_{-\infty}^{+\infty} d\varrho \cdot 10^{A_0 - A_1\rho - A_2\rho^2}.$$

Führt man hier die Substitution ein: $\sqrt{\frac{A_2}{\log e}}\left(\varrho + \frac{A_1}{2A_2}\right) = z$, so wird das Integral bis auf einen konstanten Faktor gleich dem Wahrscheinlichkeitsintegral $\int_{-\infty}^{+\infty} e^{-z^2} dz$, dessen Wert bekanntlich $\sqrt{\pi}$ ist, und man erhält:

$$\int_{-\infty}^{+\infty} d\varrho \cdot 10^{A_0 - A_1\rho - A_2\rho^2} = \sqrt{\frac{\pi \log e}{A_2}}\, 10^{A_0 + {}^1/_4 (A_1{}^2/A_2)} \tag{29}$$

und damit:

$$b_m = \frac{0.2\,\omega \sqrt{\pi}}{2.5\,(\log e)^{3/2}} \cdot \frac{10^{A_0 + {}^1/_4 (A_1{}^2/A_2)}}{\sqrt{A_2}}$$

oder auch:

$$\log b_m = 0.7942 - {}^1/_2 \log A_2 + A_0 + {}^1/_4 (A_1{}^2/A_2)\,.$$

Führt man hier die Werte der Größen A nach (27), (28) ein, so wird:

$$\log b_m = 0.7942 - {}^1/_2 \log (a_2 + b_2) + a_0 + b_0 - (b_1 + 0.4)\,m - b_2 m^2 + {}^1/_4 (1 + a_1 + b_1 + 2 b_2 m)^2/(a_2 + b_2)$$

oder nach Potenzen von m geordnet:

$$\begin{aligned} \log b_m = {} & 0.7942 - {}^1/_2 \log (a_2 + b_2) + a_0 + b_0 + {}^1/_4 (1 + a_1 + b_1)^2/(a_2 + b_2) \\ & + [b_2 (a_1 + 0.6) - a_2 (b_1 + 0.4)]\, m/(a_2 + b_2) - a_2 b_2 m^2/(a_2 + b_2)\,. \end{aligned}$$

Das Resultat ist also, daß $\log b_m$ eine quadratische Funktion von m wird, deren Koeffizienten in einfacher Weise mit den Koeffizienten der Verteilungsfunktionen zusammenhängen.

§ 7. Ausführung der allgemeinen Formeln für den speziellen Ansatz der Verteilungsfunktionen.

Analoge Rechnungen wurden nun auch für die andern allgemeinen Formeln ausgeführt. Die Resultate sind im folgenden zusammengestellt. Die Reihenfolge der Formeln wurde der Übersichtlichkeit wegen geändert, indessen ist bei jedem Resultat die Nummer der allgemeinen Formel, die zugrunde liegt, angegeben.

(6a)	Anzahl $b_m\,dm$ der Sterne der scheinbaren Größe m bis $m + dm$:	$\log b_m = \alpha_0 - \alpha_1 m - \alpha_2 m^2$
(10)	Anzahl $b_\mu\,d\mu$ der Sterne, deren Eigenbewegung in den Grenzen μ und $\mu + d\mu$ liegt ($\mu = -5.0 \log g$):	$\log b_\mu = \beta_0 - \beta_1 g - \beta_2 g^2$
(13)	Anzahl $b_{m,\mu}\,dm\,d\mu$ der Sterne der scheinbaren Größe m bis $m + dm$ und der Eigenbewegung μ bis $\mu + d\mu$:	$\log b_{m,\mu} = \gamma_0 - \gamma_1 m - \gamma_2 g - \gamma_3 m^2 - \gamma_4 m g - \gamma_5 g^2$
(7)	Mittlere Parallaxe der Sterne der scheinbaren Größe m:	$\log \pi_m = P_0 - P_1 m$
(11)	Mittlere Parallaxe der Sterne der Eigenbewegung μ:	$\log \pi_\mu = Q_0 - Q_1 g$

(14) Mittlere Parallaxe der Sterne der scheinbaren Größe m und der Eigenbewegung μ: $\log \pi_{m,\mu} = R_0 - R_1 m - R_2 g$

(8) Mittlere absolute Helligkeit (in Größenklassen ausgedrückt) der Sterne der scheinbaren Größe m: $M_m = N_0 + N_1 m$

(9a) Anzahl $\varphi_{M,m}\,dM$ der Sterne von der absoluten Größe M bis $M+dM$ bei der scheinbaren Größe m: $\log \varphi_{M,m} = K_0 - K_1 (M - M_m - K_2)^2$

(15) Anzahl $\chi_{\pi,m,\mu}\,d\log\pi$ der Sterne der scheinbaren Größe m und der Eigenbewegung μ, für die der Logarithmus der Parallaxe zwischen $\log\pi$ und $\log\pi + d\log\pi$ liegt: $\log \chi_{\pi,m,\mu} = L_0 - L_1 (\log\pi - \log\pi_{m,\mu} - L_2)^2$

Den Zusammenhang zwischen den hier auftretenden Koeffizienten α, β, γ, P, Q, R, N, K, L und den Koeffizienten a, b, c der Verteilungsfunktionen gibt die folgende

Koeffiziententafel.

$\alpha_0 = 0.7942 - \tfrac{1}{2}\log(a_2+b_2) + a_0 + b_0 + \frac{(1+a_1+b_1)^2}{4(a_2+b_2)}$	$\alpha_1 = \frac{a_2(b_1+0.4) - b_2(a_1+0.6)}{a_2+b_2}$	$\alpha_2 = \frac{a_2 b_2}{a_2+b_2}$
$\beta_0 = 0.8299 - \tfrac{1}{2}\log(a_2+c_2) + a_0 + c_0 + \frac{(0.8+a_1+c_1)^2}{4(a_2+c_2)}$	$\beta_1 = \frac{a_2 c_1 - c_2(a_1+0.8)}{a_2+c_2}$	$\beta_2 = \frac{a_2 c_2}{a_2+c_2}$
$\gamma_0 = 0.7942 - \tfrac{1}{2}\log(a_2+b_2+c_2) + a_0 + b_0 + c_0 + \frac{(1.2+a_1+b_1+c_1)^2}{4(a_2+b_2+c_2)}$	$\gamma_1 = \frac{(a_2+c_2)(b_1+0.4) - b_2(a_1+c_1+0.8)}{a_2+b_2+c_2}$	$\gamma_3 = \frac{b_2(a_2+c_2)}{a_2+b_2+c_2}$
	$\gamma_2 = \frac{(a_2+b_2)c_1 - c_2(a_1+b_1+1.2)}{a_2+b_2+c_2}$	$\gamma_4 = \frac{-2b_2 c_2}{a_2+b_2+c_2}$
		$\gamma_5 = \frac{c_2(a_2+b_2)}{a_2+b_2+c_2}$
$K_1 = a_2+b_2$ $K_2 = \frac{0.1}{a_2+b_2}$	$P_0 = -0.1\,\frac{a_1+b_1+0.9}{a_2+b_2}$	$P_1 = \frac{0.2\,b_2}{a_2+b_2}$
$L_1 = 25(a_2+b_2+c_2)$ $L_2 = -\frac{0.01}{a_2+b_2+c_2}$	$Q_0 = -0.1\,\frac{a_1+c_1+0.7}{a_2+c_2}$	$Q_1 = \frac{0.2\,c_2}{a_2+c_2}$
$\log\frac{W^2}{2} = -2.8881 + a_0 + \frac{(0.4+a_1)^2}{4a_2} - \tfrac{1}{2}\log a_2$	$R_0 = -0.1\,\frac{a_1+b_1+c_1+1.1}{a_2+b_2+c_2}$	$R_1 = \frac{0.2\,b_2}{a_2+b_2+c_2}$
$S = 9.46 - \frac{b_1+0.6}{2b_2} - 2.5\left(a_0 + \frac{(a_1+0.2)^2}{4a_2} - \tfrac{1}{2}\log a_2\right)$		$R_2 = \frac{0.2\,c_2}{a_2+b_2+c_2}$
	$N_0 = -0.5\,\frac{a_1+b_1+1.2}{a_2+b_2}$	$N_1 = \frac{a_2}{a_2+b_2}$

Die gleichgültigen Konstanten K_0, L_0 sind weggelassen. Hingegen ist der Wert des Gravitationspotentials, resp. der zugehörigen Maximalgeschwindigkeit W nach (17) und die Helligkeit S des Himmelsgrundes nach (19a) mit aufgenommen.

Die hier gefundenen Ausdrücke für π_m, $\pi_{m,\mu}$ und $\chi_{\pi,m,\mu}$ sind es, die mit den *Kapteyn*schen empirischen Ansätzen der Form nach übereinstimmen, wie gleich näher gezeigt werden wird. In diesen Formeln liegt daher die theoretische Begründung der *Kapteyn*schen Ansätze.

§ 8. Anschluß an die Beobachtungsdaten und numerische Auswertung.

Die beobachtete Sternanzahl b_m läßt sich nach A. N. Nr. 4422 darstellen durch die Formel:

$$\log b_m = 0.596 + 0.5612\,m - 0.0055\,m^2 .$$

Für die mittlere Parallaxe π_m der Sterne der Größe m, sowie für die mittlere Parallaxe $\pi_{m,\mu}$ der Sterne der Größe m und der Eigenbewegung μ findet Herr *Kapteyn* (Publications of the Astronomical Laboratory at Groningen, No. 8 Seite 24) aus den Beobachtungen die Ausdrücke:

$$\pi_m = 0''.01605\,(0.75)^{m-5.5}$$

$$\pi_{m,\mu} = (0.905)^{m-5.5}\,\sqrt[1.405]{0.0387\,\mu}\ .$$

Beachtet man, daß $g = -5\log\mu$ ist, so schreiben sich diese Ausdrücke

$$\log\pi_m = -1.108 - 0.125\,m$$

$$\log\pi_{m,\mu} = -0.766 - 0.0434\,m - 0.1423\,g$$

in offensichtlicher formaler Übereinstimmung mit den Resultaten des vorigen Paragraphen.

Die 7 fettgedruckten Koeffizienten, α_0, α_1, α_2, P_0, P_1, R_0, R_2 in der obigen Bezeichnungsweise, sollen nun der Beobachtung entnommen werden. In Verbindung mit den Gleichungen (23) und (24) bestimmen sie dann die 9 Koeffizienten der drei Verteilungsfunktionen und damit alle übrigen Größen. Es wird genügen, das Resultat der kurzen völlig elementaren Rechnung anzugeben. Es ergab sich:

$$\begin{aligned}
\log D(r) &= 0.488 - 0.097\,\varrho - 0.0088\,\varrho^2 & (\varrho &= -5\log r)\\
\log \varphi(\mathcal{J}) &= -2.879 + 0.737\,M - 0.0147\,M^2 & (M &= -2.5\log \mathcal{J})\\
\log \psi(V) &= -0.922 - 0.165\,G - 0.0581\,G^2 & (G &= -5\log V)\\
\log b_m &= 0.596 + 0.561\,m - 0.0055\,m^2 & &\\
\log b_\mu &= 5.198 + 0.758\,g - 0.0077\,g^2 & (g &= -5\log \mu)\\
\log b_{m,\mu} &= -0.365 + 0.468\,m + 0.350\,g - 0.0121\,m^2 + 0.0208\,mg - 0.0168\,g^2 & &\\
\log \pi_m &= -1.108 - 0.125\,m & &\\
\log \pi_\mu &= -1.44 - 0.174\,g & &\\
\log \pi_{m,\mu} &= -0.767 - 0.036\,m - 0.142\,g & &\\
M_m &= -11.9 + 0.374\,m & &\\
\log \varphi_{M,m} &= K_0 - 0.0235\,(M - M_m - 4.25)^2 & &\\
\log \chi_{\pi,m,\mu} &= L_0 - 2.04\,(\log\pi - \log\pi_{m,\mu} + 0.123)^2 & &\\
\log W &= 3.00 \qquad S = 4^{\mathrm{m}}.0\,. & &
\end{aligned}$$

§ 9. Bemerkungen zu den numerischen Resultaten.

Während die Sternanzahlen und mittleren Parallaxen für jede Größenklasse b_m und π_m ganz der Beobachtung, resp. den *Kapteyn*schen Ergebnissen entnommen sind, gilt das für den Zusammenhang der mittleren Parallaxe mit Größe und Eigenbewegung $\pi_{m,\mu}$ nur für das konstante und das von der Eigenbewegungsklasse g abhängige Glied. Für das von m abhängige Glied findet Herr *Kapteyn* empirisch den Faktor 0.043, während sich hier deduktiv in völlig genügender Übereinstimmung 0.036 ergeben hat. Der Ausdruck für die Verteilung der Logarithmen der Parallaxen läßt sich auch so schreiben:

$$\chi_{\pi,m,\mu} = \text{konst.}\; e^{-{}^1/_2(\log\pi - \log\pi_{m,\mu} + 0.123)^2/0.326^2}\,.$$

Er besagt, daß das geometrische Mittel der Parallaxe $\underline{\pi}_{m,\mu}$ bei gegebener Sterngröße und Eigenbewegung aus dem arithmetischen Mittel $\pi_{m,\mu}$ hervorgeht nach der Beziehung:

$$\log \underline{\pi}_{m,\mu} = \log \pi_{m,\mu} - 0.123\,, \qquad \underline{\pi}_{m,\mu} = 0.75\,\pi_{m,\mu}\,.$$

Er besagt ferner, daß die Logarithmen von $\pi/\underline{\pi}$ nach dem *Gauß*schen Fehlergesetz verteilt sind mit einem mittleren Fehler des Dezimal-Logarithmus von 0.326, einem wahrscheinlichen von 0.22. Herr *Kapteyn* hat, wie erwähnt, empirisch denselben Ansatz gemacht und für den wahrscheinlichen Fehler (l. c. S. 23, dort mit ϱ bezeichnet) 0.19 ± 0.02 gefunden. Das zugehörige Verhältnis zwischen geometrischem und arithmetischem Mittel der Parallaxe $\underline{\pi}/\pi_{m,\mu}$ ist 0.81. Auch hier ist die numerische Übereinstimmung sehr gut.

Man sieht daher, daß die Sternanzahlen b_m, die Säkularparallaxen, auf denen wesentlich Herrn *Kapteyns* Formel für π_m beruht, und die einzelnen direkt gemessenen Parallaxen, aus denen die Formel für $\pi_{m,\mu}$ mitsamt dem wahrscheinlichen Fehler des Logarithmus gewonnen ist, durch unsern Ansatz der Verteilungsfunktionen in ihrer Gesamtheit numerisch genügend dargestellt werden.

Die Formel für $\varphi_{M,m}$ läßt sich in der Gestalt schreiben:

$$\varphi_{M,m} = \text{konst.}\; e^{-{}^1/_2(M - M_m - 4.25)^2/3.03^2}\,.$$

Sie besagt, daß die absoluten Größen der Sterne gegebener scheinbarer Größe nach dem *Gauß*schen Fehlergesetz verteilt sind, entsprechend einem mittleren Fehler von $3^{\mathrm{m}}.03$. Das stimmt überein mit dem Resultat, welches Herr *Hertzsprung* (A. N. 4422 S. 92) aus den Radialgeschwindigkeiten in Verbindung mit den Eigenbewegungen abgeleitet hat. Die mittlere absolute Größe $\underline{M}_m$ (entsprechend dem geometrischen Mittel der Leuchtkräfte) wird hiernach ferner:

$$\underline{M}_m = M_m + 4.25 = -7^{\mathrm{m}}.7 + 0.374\,m\,.$$

Nach dem Mittelwert der absoluten Größe beurteilt sind also die mit bloßem Auge sichtbaren Sterne um 5 bis 6 Größenklassen lichtstärker als die Sonne. Mit abnehmender scheinbarer Helligkeit der Sterne nimmt ihre mittlere absolute Helligkeit nur sehr langsam ab. Erst für $m = 20.5$ liefert die Formel $\underline{M}_m = 0$. Nach dieser Extrapolation würde also erst bei den Sternen 20. Größe die mittlere absolute Größe auf die der Sonne herabsinken.

Wir betrachten nun noch besonders die numerischen Resultate für die drei Verteilungsfunktionen.

Den Verlauf der Dichte nach der erhaltenen Formel gibt folgende kleine Tabelle:

r	10	20	40	80	160	320	640	1280	2560	5120
π	0".1	0".05	0".025	0".0125	0".0062	0".0031	0".0016	0".0008	0".0004	0".0002
$D(r)$	5.7	5.6	5.0	4.1	3.1	2.1	1.3	0.7	0.4	0.2 .

Die Verteilung der Leuchtkräfte schreibt sich, wenn man gleich die absolute Größe als unabhängige Variable einführt:

$$\varphi(\mathcal{J})\,d\mathcal{J} = \varphi(\mathcal{J})\,(d\mathcal{J}/dM)\,dM = dM\cdot\text{konst.}\cdot e^{-{}^1/_2(M - 11.5)^2/3.83^2}\,.$$

Die absolute Größe M ist demnach um den Mittelwert $11^{\mathrm{m}}.5$ mit einem mittleren Fehler von $3^{\mathrm{m}}.8$ nach dem *Gauß*schen Fehlergesetz gestreut.

Astron. Nachr. *190* (1912) 361–376

Natürlich können diese Werte für die Verteilung der Dichte und der Leuchtkräfte in keiner Weise beanspruchen richtiger zu sein, als die Resultate von *Seeliger* [1]), *Kapteyn* [2]), *Hertzsprung* [3]). Sie wurden abgeleitet, weil sie so mühelos zu gewinnen waren. Besonders zu beachten ist, daß die Dichte jenseits $r = 40000$ und die Anzahl φ für dunkle Sterne mit $M > 5$ in jedem Falle reine Extrapolationen sind, da so entfernte oder so schwache Sterne in den bisherigen Abzählungen numerisch keine Rolle spielen. Eine Andeutung dafür, daß die Dichte in großen Entfernungen und die Anzahl der lichtschwachen Sterne durch die obigen Ansätze überschätzt ist, liegt in dem unwahrscheinlich hohen Wert des Gravitationspotentials, der aus ihnen folgt. Die Maximalgeschwindigkeit wird nämlich $W = 1000 = 4700$ km/sec. Die Helligkeit des Himmelsgrundes G ergibt sich indessen in hinlänglicher Übereinstimmung mit den Beobachtungen gleich der eines Sternes $4^m\!.0$ pro Quadratgrad. [4])

Schließlich ist noch die gefundene Verteilung der Sterngeschwindigkeiten zu betrachten. Das geometrische Mittel $\underline{V}$ der (auf die Sphäre projizierten) Sterngeschwindigkeit wird nach (26) gleich 4.25 entsprechend 20.1 km/sec, das arithmetische Mittel $\bar{V}$ wird 6.30 entsprechend 29.8 km/sec. Die mittlere jährliche Eigenbewegung der Sterne jeder scheinbaren Größe wird demnach gemäß Formel (16) das 6.3fache ihrer mittleren Parallaxe.

Von Interesse ist die Vergleichung der gefundenen Geschwindigkeitsverteilung mit dem *Maxwell*schen Gesetz. Für die Projektionen der Pekuliarbewegungen auf die Sphäre würde letzteres lauten:

$$\psi_1(v) = (v/k^2)\, e^{-v^2/2k^2}$$

wo k die wahrscheinlichste Geschwindigkeit ist. Die oben gefundene Verteilung schließt aber die parallaktische Bewegung der Sterne mit ein. Um den Vergleich zutreffend zu machen, hat man daher auch der *Maxwell*schen Verteilung die Eigenbewegung der Sonne zu überlagern, und zwar im Durchschnitt für alle Richtungen und Himmelsgegenden. Ich habe das durch einen rohen Überschlag ausgeführt. Für die Sonnengeschwindigkeit wurde der Wert 4.0 (= 19 km/sec) angesetzt. Für k wurde der Wert 3 gewählt. Die folgende Figur bringt die oben gefundene stellare Verteilung $\psi(v)$ und die für Sonnenbewegung korrigierte *Maxwell*sche Verteilung für die in km/sec ausgedrückten Geschwindigkeiten zur Darstellung. Die Wahl $k = 3$ wurde gerade so getroffen, daß die Maxima beider Kurven annähernd gleich hoch wurden. Man erkennt, daß die stellaren Geschwindigkeiten viel stärker gestreut sind, als das *Maxwell*sche Gesetz angibt. Es gibt weniger mittlere und mehr große Geschwindigkeiten. Das beobachtete Vorkommen verhältnismäßig so vieler großer Radialgeschwindigkeiten ist damit in Übereinstimmung.

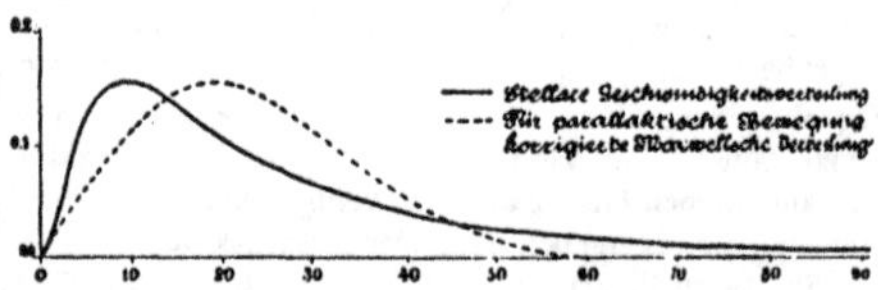

Potsdam, 1911 Dez. 18.

K. Schwarzschild.

[1]) *H. v. Seeliger.* Über die räumliche Verteilung der Sterne im schematischen Sternsystem. Sitzungsberichte der Bayr. Akademie der Wissenschaften. Math.-Phys. Klasse. 1911.

[2]) *J. C. Kapteyn.* Publications of the Astronomical Laboratory at Groningen. No. 11.

[3]) *E. Hertzsprung.* Astronomische Nachrichten Nr. 4422.

[4]) *C. G. Abbot.* Astronomical Journal No. 627.

Une nouvelle variable 3.1912 Coronae borealis.

Ce premier février, Mme. *L. Ceraski* a trouvé une nouvelle variable dont les coordonnées approchées sont les suivantes:

$$\alpha = 16^h\,13^m\,34^s \quad \delta = +30^\circ\ 4' \quad (1855.0)$$
$$16\ \ 15\ \ 21 \qquad +29\ \ 57 \quad (1900.0).$$

Nous avons en tout quinze photographies de cette région du ciel (1907–1911), à en juger l'éclat de cette étoile change de $9^1/_2$ à $10^1/_2$ gr. environ. M. *Blažko* regrette de ne pouvoir établir avec certitude le type de la variable.

Moscou, 1912 février 8. *W. Ceraski.*

Literarische Anzeige.

E. Anding. Sechsstellige Tafeln der *Bessel*schen Funktionen imaginären Argumentes. Leipzig. Wilhelm Engelmann. 1911. 72 + IV S. Preis 7 M., geb. 8.80 M.

Während es für die reellen *Bessel*schen Funktionen schon seit langer Zeit hinreichend bequeme Tafeln zur Entnahme der Funktionswerte gibt, fehlte es an entsprechenden Hilfsmitteln für die analogen Funktionen imaginären Argumentes fast gänzlich. Die vorliegenden Tafeln füllen diese für Arbeiten im Gebiete der angewandten Mathematik schon oft empfindlich fühlbar gewordene Lücke aus und werden, da sie das ganze Gebiet der Werte des Argumentes von o bis ∞ umfassen, sich sehr nützlich erweisen.

Komet 1905 VI. Die definitive Bahnbestimmung ist von Herrn *A. A. Michailoff*, Sternwarte Moskau, übernommen worden.

Geschlossen 1912 Febr. 22. Herausgeber: H. Kobold. Druck von C. Schmidt. Expedition: Kiel, Moltkestr. 80.

ASTRONOMISCHE NACHRICHTEN.

Band 191. Nr. 4561. 1.

Ein Theorem zur Verteilung der Sterngeschwindigkeiten. Von *K. Schwarzschild.*

Die Geschwindigkeit eines Sterns relativ zur Sonne habe, nach drei festen Achsenrichtungen zerlegt, die Komponenten x, y, z. Man fasse alle Sterne eines bestimmten Raumvolumens ins Auge und zähle ab, wie viele von ihnen eine Geschwindigkeit mit Komponenten zwischen den Grenzen x bis $x+dx$, y bis $y+dy$, z bis $z+dz$ haben. Diese Anzahl, dividiert durch die Gesamtzahl der betrachteten Sterne, werde $W(x,y,z)\,dx\,dy\,dz$ genannt. Die Funktion W der drei Variabeln x, y, z charakterisiert die Geschwindigkeitsverteilung der Sterne in dem betreffenden Raumvolumen. Um sie zu veranschaulichen, kann man sich W als Dichte einer Massenverteilung denken in der Art, daß $W(x,y,z)$ die Dichte in dem Punkt der Koordinaten x, y, z bedeutet. Die so entstehende — bildliche — Massenverteilung soll auch kurz als »Geschwindigkeitskörper« bezeichnet werden.

Eine sehr einfache Annahme, die man über die wirkliche Geschwindigkeitsverteilung der Sterne machen kann, ist die, daß die Geschwindigkeitsverteilung W in allen Raumteilen, aus denen die beobachteten Eigenbewegungen stammen, die gleiche sei. Diese einfache Annahme stellt auch jedenfalls eine Annäherung an die wirklichen Verhältnisse dar. Denn indem sie festgehalten wurde, haben sich sowohl nach der Zweischwarmhypothese, wie nach der Ellipsoidhypothese die beobachteten Anzahlen der Eigenbewegungen jeder Richtung gut darstellen lassen. Strenge gilt sie freilich nicht. Sie wird z. B. schon gestört durch die Existenz von auf ein kleines Himmelsareal konzentrierten Sternströmen, wie es die Hyaden sind. Jedenfalls ist es aber gut, ihre Konsequenzen zu verfolgen, um desto sicherer entscheiden zu können, wo sie unrichtig wird.

Eine besonders einfache Konsequenz der obigen Annahme haben die Herren *Hough* und *Halm* hervorgehoben [1]). Es folgt aus ihr, daß entgegengesetzte Himmelsareale spiegelbildlich gleiche Verteilung der scheinbaren Eigenbewegungen zeigen müssen (die Spiegelung erfolgt an einer Ebene senkrecht zum Visionsradius). Für entgegengesetzte Himmelsareale wird nämlich der Geschwindigkeitskörper aus entgegengesetzten Richtungen betrachtet, man betrachtet dieselbe Projektion des Geschwindigkeitskörpers einmal von der Vorder- und einmal von der Rückseite. Diese Konsequenz wurde auch von den Herren *Hough* und *Halm* an den Eigenbewegungen der *Bradley*-Sterne geprüft (l. c.), und es fanden sich Abweichungen, die aber verhältnismäßig sehr klein sind.

Hier soll nun eine zweite Konsequenz einfacher Art abgeleitet und einer vorläufigen Prüfung unterzogen werden. Es sei nochmals hervorgehoben, daß dabei über die spezielle Beschaffenheit des Geschwindigkeitskörpers keine Annahme gemacht und eben nur vorausgesetzt wird, daß er in allen Raumteilen derselbe sei.

Man bestimme in der bei diesen Untersuchungen üblichen Art den Prozentsatz der Sterne in einem kleinen Areal der scheinbaren Himmelsfläche, deren Eigenbewegung zwischen zwei bestimmten Positionswinkeln ϑ und $\vartheta+d\vartheta$ liegt. Theoretisch kommt diese Aufgabe offenbar auf folgendes hinaus: Man denke sich den Nullpunkt des Systems der Geschwindigkeitskoordinaten x, y, z auf irgend eine Stelle des Visionsradius gesetzt und lege durch den Visionsradius zwei Ebenen entsprechend den Positionswinkeln ϑ und $\vartheta+d\vartheta$. Diese beiden Ebenen schneiden einen Doppelkeil aus dem Geschwindigkeitskörper aus. Von dem Doppelkeil entspricht die eine Hälfte dem Positionswinkel ϑ, die andre dem Positionswinkel $\vartheta+180°$. Das Integral von W über das Innere des dem Positionswinkel ϑ entsprechenden Keils liefert die gesuchte Anzahl.

Es sei speziell der Visionsradius r in der xy-Ebene verlaufend und zwar in einem Winkel ϱ mit der x-Achse. Dann ist die Gleichung des Visionsradius:

$$z = 0 \qquad x = r\cos\varrho \qquad y = r\sin\varrho\,. \tag{1}$$

Wir wollen ferner speziell die der xy-Ebene (bis auf den infinitesimalen Winkel $d\vartheta$) parallelen Eigenbewegungen untersuchen. Der Positionswinkel ϑ sei daher der Positionswinkel der xy-Ebene. Die erste der beiden zu betrachtenden schneidenden Ebenen ist also in diesem Falle die xy-Ebene selbst, d. i. die Ebene der Gleichung $z = 0$. Die zweite Ebene soll durch den eben angegebenen Visionsradius gehen und den Winkel $d\vartheta$ mit der xy-Ebene bilden. Daraus folgt leicht als Gleichung dieser zweiten Ebene:

$$z = d\vartheta(-x\sin\varrho + y\cos\varrho)\,. \tag{2}$$

Die gesuchte Anzahl wird nun:

$$N d\vartheta = \int W(x,y,z)\,dx\,dy\,dz \tag{3}$$

wobei über den Raum zwischen der xy-Ebene und der durch (2) bestimmten Ebene zu integrieren ist.

Da z sich nur von $z = 0$ bis zu einem zu $d\vartheta$ proportionalen infinitesimalen Betrage ändert, so ist die Integration nach z ohne weiteres ausführbar. Sie gibt:

$$N\cdot d\vartheta = d\vartheta\int W(x,y,0)(-x\sin\varrho + y\cos\varrho)\,dx\,dy\,. \tag{4}$$

Es ist noch zu überlegen, über welche Grenzen nach x und y zu integrieren ist. Integriert man über alle Werte $y > x\,\mathrm{tg}\varrho$, für welche also z ($d\vartheta$ als positiv angenommen) positiv ist, so erhält man alle Eigenbewegungen, die in Richtung wachsender ϱ verlaufen. Die Anzahl derselben soll mit $\overset{+}{N}\cdot d\vartheta$ bezeichnet werden. Integriert man umgekehrt über diejenige Hälfte der xy-Ebene, für welche $z < 0$ ist, ($y < x\,\mathrm{tg}\varrho$), so erhält man die Anzahl der entgegengesetzt laufenden Eigenbewegungen, welche mit $\underset{-}{N}\cdot d\vartheta$ bezeichnet

[1]) Monthly Notices Vol. 70 S. 578.

werden möge. Man beachte aber, daß man letztere Anzahl infolge des negativen Wertes von $y\cos\varrho - x\sin\varrho$ mit negativem Vorzeichen erhält, wenn man das Integral so ausführt, wie es in (4) dasteht.

Wir wollen nun das obige Integral über alle Werte von x, y, über die ganze xy-Ebene, ausführen. Dann erhält man nach dem gesagten den Wert von $\underset{+}{N}\cdot d\vartheta - \underset{-}{N}\cdot d\vartheta$, und zwar findet man:

$$\underset{+}{N} - \underset{-}{N} = \cos\varrho \int W(x,y)\, y\, dx\, dy - \sin\varrho \int W(x,y)\, x\, dx\, dy\,.$$

Die beiden hier auftretenden über die ganze xy-Ebene genommenen Integrale sind von der Verteilung der Sterngeschwindigkeiten abhängige Konstante, die wir mit B und A bezeichnen wollen. Das Resultat lautet also:

$$\underset{+}{N} - \underset{-}{N} = B\cos\varrho - A\sin\varrho\,. \tag{5}$$

Das ist der Satz, der abgeleitet werden sollte. Da die xy-Ebene in keiner Weise ausgezeichnet ist, gilt er für jede andere Ebene genau so, nur nehmen die Konstanten A, B dann andre Werte an. In Worten läßt er sich so aussprechen: Man wähle einen beliebigen größten Kreis am Himmel und rechne längs desselben von irgend einem Ausgangspunkte an den Winkel ϱ. Man zähle für kleine Areale längs dieses Kreises, die verschiedenen Werten von ϱ entsprechen, die Sterne ab, deren Eigenbewegungen dem Kreise parallel sind (bis auf einen kleinen Winkel $d\vartheta$), und zwar getrennt für die im einen Sinne und die im entgegengesetzten Sinne laufenden Sterne. Die Differenz beider Anzahlen, natürlich auf gleiche Gesamtsternzahl jedes Areals bezogen, ist eine einfache Sinusfunktion des Winkels ϱ.

Zur Prüfung des Satzes habe ich Herrn *Eddingtons* Abzählungen [1]) aus dem P. G. Catalogue von *Boss* benutzt. Es wurden drei größte Kreise betrachtet: 1) der Kreis 0^h, 12^h durch Nord- und Südpol, 2) der Kreis 6^h, 18^h durch Nord- und Südpol, 3) der Äquator. Die für jede Stelle dieser Kreise benutzten Areale sind unten in Herrn *Eddingtons* Bezeichnung angegeben [2]). Der Winkelbereich der abgezählten Eigenbewegungen betrug $\pm 20°$ zu beiden Seiten der Richtung des betreffenden größten Kreises. Die Anzahlen $\underset{+}{N}$ und $\underset{-}{N}$ sind in Prozenten der Gesamtsternzahl ausgedrückt, welche für jedes Areal in der letzten Zeile angegeben ist.

1) Meridian 0^h, 12^h.

α	—	0^h	0^h	0^h	0^h	—	12^h	12^h	12^h	12^h
δ	$+90°$	$+50°$	$+17°$	$-17°$	$-50°$	$-90°$	$-50°$	$-17°$	$+17°$	$+50°$
ϱ	$0°$	$40°$	$73°$	$107°$	$140°$	$180°$	$220°$	$253°$	$287°$	$320°$
Areal	I	II	(VIII+XVII)	(XII′+XIII′)	V′	I′	II	(VIII′+XVII′)	(XII+XIII)	V
$\underset{+}{N}$	7.2	8.3	12.2	10.4	7.5	2.5	1.8	3.9	1.5	3.5
$\underset{-}{N}$	3.7	2.0	2.0	4.2	3.9	6.3	10.3	10.2	12.4	7.3
$\underset{+}{N} - \underset{-}{N}$	+3.5	+6.3	+10.2	+6.2	+3.6	−3.8	−8.5	−6.3	−10.9	−3.8
Berechnet	+2.6	+7.6	+ 9.2	+7.6	+3.6	−2.6	−7.6	−9.2	− 7.6	−3.6
Differenz	+0.9	−1.3	+ 1.0	−1.4	0.0	+1.2	−0.9	+2.9	− 3.3	−0.2
Sternanzahl	217	192	271	316	138	154	162	282	263	142

2) Meridian 6^h, 18^h.

α	—	6^h	6^h	6^h	6^h	—	18^h	18^h	18^h	18^h
δ	$+90°$	$+50°$	$+17°$	$-17°$	$-50°$	$-90°$	$-50°$	$-17°$	$+17°$	$+50°$
ϱ	$0°$	$40°$	$73°$	$107°$	$140°$	$180°$	$220°$	$253°$	$287°$	$320°$
Areal	I	(III+IV)	X	XV′	(VI′+VII′)	I′	(III′+IV′)	X′	XV	(VI+VII)
$\underset{+}{N}$	37.1	32.6	32.5	14.3	8.8	5.7	2.4	0.9	6.9	18.2
$\underset{-}{N}$	9.9	3.8	1.7	15.1	18.4	38.0	34.5	29.3	25.2	13.6
$\underset{+}{N} - \underset{-}{N}$	+27.2	+28.8	30.8	− 0.8	−9.6	−32.3	−32.1	−28.4	−18.3	+4.6
Berechnet	+25.3	+32.6	+26.9	+12.3	−6.4	−25.3	−32.6	−26.9	−12.3	+6.4
Differenz	+ 1.9	− 3.8	+ 3.9	−13.1	−3.2	− 7.0	+ 0.5	− 1.5	− 6.0	−1.8
Sternanzahl	217	338	174	154	373	154	260	182	131	369

[1]) Monthly Notices Vol. 71 S. 10.
[2]) Die Areale mit südlicher Deklination sind durch einen Akzent bezeichnet.

3) Äquator ($\delta = 0°$).

$\alpha = \varrho$	18°	54°	90°	126°	162°	198°	234°	270°	306°	342°
Areal	(VIII+XIII')	(IX+XIV')	(X+XV')	(XI+XVI')	(XII+XVII')	(VIII'+XIII)	(IX'+XIV)	(X'+XV)	(XI'+XVI)	(XII'+XVII)
$\underset{+}{N}$	25.1	21.1	7.8	3.7	3.7	4.2	6.1	10.5	19.9	26.5
$\underset{-}{N}$	5.3	3.8	6.4	18.3	30.3	28.1	13.2	6.0	2.2	4.8
$\underset{+}{N} - \underset{-}{N}$	+19.8	+17.3	+1.4	−14.6	−26.6	−23.9	− 7.1	+4.5	+17.7	+21.7
Berechnet	+22.6	+12.9	−1.9	−15.9	−23.8	−22.6	−12.9	+1.9	+15.9	+23.8
Differenz	− 2.8	+ 4.4	+3.3	+ 1.3	− 2.8	− 1.3	+ 5.8	+2.6	+ 1.8	− 2.1
Sternanzahl	270	297	328	341	250	275	275	313	335	317

Vermittels graphischer Auftragung der Werte $\underset{+}{N} - \underset{-}{N}$ und Entnahme äquidistanter Werte (von 30° zu 30°) aus den entstehenden Kurven wurden für die drei größten Kreise folgende Darstellungen gefunden:

$$1)\ 8.7 \sin\varrho + 2.6 \cos\varrho$$

$$2)\ 20.4 \sin\varrho + 25.3 \cos\varrho \qquad 3)\ -1.9 \sin\varrho + 24.3 \cos\varrho\,.$$

Während die Übereinstimmung zwischen der Beobachtung und der Sinusformel für den Äquator und den Meridian 0^h, 12^h so gut ist, wie nur zu erwarten, zeigt der die Milchstraße unter einem kleinen Winkel (30°) schneidende Meridian 6^h, 18^h Abweichungen von stärkerem Betrage und ein Überwiegen des negativen Zeichens. Es liegt hierin wiederum eine Andeutung, wie solche auf anderen Wegen von den Herren *Hough*, *Halm* und *Eddington* gefunden worden sind, daß die Geschwindigkeitsverteilung nicht überall im Raume dieselbe ist. Doch steht man dabei so dicht an der Grenze dessen, was das bisherige Material an Eigenbewegungen überhaupt geben kann, daß eine Deutung der Abweichungen schwer und unsicher wird [1]).

Potsdam, 1912 Jan. 14. *K. Schwarzschild.*

[1]) Bei der Beurteilung der obigen Abweichungen ist übrigens auch zu berücksichtigen, daß die Areale z. T. recht groß sind, während die Theorie unendlich kleine Areale voraussetzt.

Marsbeobachtungen 1911.

Während der Monate September bis Dezember 1911 haben wir mit dem 246 mm Refraktor der Urania-Sternwarte zu Kopenhagen eine Reihe von Marsbeobachtungen angestellt. Leider sind die Wetterverhältnisse während der ganzen Opposition ziemlich ungünstig gewesen, und noch dazu verteilten sich die günstigen Abende so über die Periode, daß es uns ganz unmöglich war, die Planetenoberfläche gleichmäßig zu beobachten. Die Gegend um den Meridian 100° haben wir überhaupt nicht gesehen. Wenn wir trotzdem im Stande sind einige sichere Beobachtungen vorzulegen, müssen wir es der angewandten Beobachtungsmethode verdanken. Die Methode war die folgende: Die beiden Beobachter zeichneten erst jeder, ganz unabhängig voneinander, eine Skizze von dem augenblicklichen Aussehen der Planetenscheibe. Erst wenn beide sich mit ihren Zeichnungen zufrieden erklärt hatten, wurden diese untereinander verglichen, und sie zeigten meistens gute Übereinstimmung. War eine Nichtübereinstimmung vorhanden, so wurden die Beobachtungen wiederholt. Bei ruhiger Luft haben wir oft mit Vorteil ein cosinrotes Glas anwenden können. Dieses Glas vergrößert beträchtlich den Kontrast zwischen den dunklen und den hellen Partien der Planetenoberfläche. Eine 200–300 fache Vergrößerung ist stets zur Anwendung gekommen. Alle unsere Skizzen haben wir in die auf Tafel 1 reproduzierte Generalkarte zusammengearbeitet, welche also alle von uns beobachteten Details enthält. Von dem Aussehen des Planeten im Jahre 1911 bemerken wir, daß alle Einzelheiten viel deutlicher hervortraten als im Jahre 1909. Die Küstenlinien der großen Kontinente waren alle gut definiert. Die Inseln im »Südmeere« dagegen waren alle außerordentlich verschwommen. Besonders war dies bei Hellas auffallend, die 1909 gut und scharf definiert war. In den Meeren haben wir viele hellere und dunklere Partien bemerkt; sie sind alle auf unserer Karte angegeben. Eine kleine »Insel« befindet sich zwischen Deucalionis regio und dem Kontinente, eine andere an der Mündung des Cyclops.

Anfangs September waren mehrere Landschaften hellweiß gefärbt. Sie sind alle auf unserer Karte von einer punktierten Linie umgeben. Besonders war dies mit Libya der Fall, die wir auch später sowohl am Rande als in der Mitte der Scheibe hellweiß sahen. Von den Kanälen haben wir 29 gesehen. Sirenius und Erinnys ausgenommen, ist keiner der Kanäle auf der Grenze der Sichtbarkeit gewesen. Verdopplungen haben wir nicht bemerkt. Die in die Karte eingezeichneten »Lacus« waren alle sehr auffällig und dunkel. Nodus Gordii war rötlich gefärbt. Die beiden Polarflecke sind stets sichtbar gewesen. — Eine geplante mikrometrische Durchmusterung der ganzen Marsscheibe haben wir leider der gar zu geringen Anzahl von günstigen Nächten wegen aufgeben müssen.

Urania-Sternwarte, Kopenhagen, 1912 Jan. 14. *C. Luplau Janssen, E. Buch Andersen.*

Über die räumliche Bewegung der Praesepe. Von *K. Schwarzschild.*

Im letzten Frühjahr wurden einige Sterne der Praesepe mit dem großen Refraktor und Spektrograph III mit kurzer Camera aufgenommen, um etwas über die Radialgeschwindigkeit dieses Sternhaufens zu erfahren. Es stellte sich heraus, daß genaue Resultate aus den erhaltenen Spektrogrammen nicht abzuleiten waren. Die Sterne zeigten, von dem einen Stern 39 Cancri von spätem Spektraltypus abgesehen, ähnlich, wie die meisten Hyaden, sehr verwaschene Linien, welche manchmal der Verdoppelung verdächtig waren. Auch wurde mit weitem Spalt (0.07 oder 0.10 mm) gearbeitet und die Spektra wurden sehr schmal gehalten, um mit der Expositionszeit nicht über 90^m hinaus gehen zu müssen. Das Gesamtergebnis scheint indessen sicher genug, um eine Mitteilung zu rechtfertigen.

Die Daten über die aufgenommenen Sterne sind in folgender Tabelle zusammengestellt.

Bezeichnung des Sterns	Nr. Hall [1])	BD	Größe visuell [2])	Größe photogr. [3])	Spektrum [4])	Rad.-G. km/sec
38 Cancri	62	+20°2149	$6^m.79$	$6^m.72$	A5	+33
39 »	78	+20 2158	6.54	7.42	K	+33
40 »	81	+20 2159	6.78	6.42	A0	+38
	88	+19 2069	7.04	6.83	A5	+21
41 »	96	+20 2171	6.56	6.23	A2	+33
42 »	102	+20 2172	7.07	6.83	A5	+59

[1]) Nach *A. Hall*, Washington Observations 1867, Appendix 4. — [2]) Nach der PD. — [3]) Nach Publikationen der v. Kuffnerschen Sternwarte Wien-Ottakring, Bd. V. Durch Subtraktion von $0^m.25$ auf Göttinger Aktinometrie B und damit auf internationales System reduziert. — [4]) Nach Potsdamer Objektivprisma-Aufnahmen.

Die Aufnahmen stammen größtenteils von Herrn Dr. *Münch*, die Ausmessungen größtenteils von Herrn Dr. *Jantzen*. Die einzelnen erhaltenen Radialgeschwindigkeiten sind folgende:

Hall Nr.	1913	km/sec	Hall Nr.	1913	km/sec
62	April 18	+33	96	März 22	+17
78	März 9	+32		» 25	+73
	» 12	+31		April 4	+48
	April 15	+35		» 11	+16
81	Febr. 28	+32		» 12	+41
	März 13	+33		» 14	+27
	April 14	(+58)		» 15	+24
88	April 22	+21		» 16	+18
96	Febr. 28	+32	102	April 21	+59

Der wahrscheinliche Fehler der einzelnen Platte dürfte für 39 Cancri 2–3 km/sec, für die übrigen Sterne 5–7 km/sec betragen. Bei Stern 88 wurde nur die Wasserstofflinie $H\gamma$ gemessen. Die letzte Platte des Sterns Hall Nr. 81 ist unterexponiert und hat nur halbes Gewicht, sodaß eine Veränderlichkeit seiner Radialgeschwindigkeit noch nicht sicher behauptet werden kann. Auch bei Stern Hall Nr. 96 ist die Veränderlichkeit der Radialgeschwindigkeit nur als sehr wahrscheinlich zu bezeichnen, da nur bei der einen Platte vom 25. März die Abweichung vom Mittel den mittleren Fehler stark überschreitet.

Eine starke positive Radialgeschwindigkeit der Gruppe ist nach diesen Ergebnissen offenbar. Die Mittelwerte für die einzelnen Sterne sind bereits in der obigen Tabelle angegeben. Das Mittel aus den 6 Sternen beträgt +36 km/sec.

Was die Eigenbewegung der Praesepe an der Sphäre angeht, so enthält der Preliminary General Catalogue von *Boss* zwischen $8^h 28^m$ und $8^h 38^m$ in α, sowie $+19° 0'$ und $+21° 0'$ in δ im ganzen 10 Sterne, die alle nahe gleiche Eigenbewegung (zwischen $-0^s.0012$ und $-0^s.0029$ in α, zwischen $-0''.012$ und $-0''.022$ in δ) haben und die zweifellos physisch zur Praesepe gehören. Von den obigen sechs Sternen fehlt darunter nur Hall Nr. 88, dessen physische Zugehörigkeit indessen aus der Vermessung der Praesepe von *Schur* *) zu entnehmen ist. Im Mittel ergibt sich aus den 10 *Boss*schen Sternen als Eigenbewegung der Praesepe $-0^s.0022$ in α und $-0''.018$ in δ oder $0''.036$ im Positionswinkel 240°.

Prof. *Hertzsprung* und mir fiel bei gemeinsamer Betrachtung vorstehender Zahlen auf, daß hier eine gewisse Ähnlichkeit mit der Bewegung der Hyaden vorliegt. Man nehme hypothetisch an, daß Hyaden und Praesepe gleiche und gleichgerichtete Bewegung im Raume hätten. Rechnet man dann nach *Boss* mit dem Zielpunkt $6^h 9^m$, $+7° 0'$ und der Geschwindigkeit 44.0 km/sec der Hyaden, so folgt für den Ort der Praesepe ($8^h 34^m$, $+20° 4'$) die Radialgeschwindigkeit +34.8 km/sec und die Eigenbewegung $5.7\,\pi$, wobei π die Parallaxe der Praesepe in Bogensekunden bedeutet. Der Positionswinkel der Eigenbewegung wird 255°. Man erhält die beobachtete Eigenbewegung der Praesepe von $0''.036$, wenn ihre Parallaxe zu $0''.0063$ angesetzt wird. Das ist ein a priori plausibler Wert der Parallaxe, da die Hyadensterne bei der Parallaxe $0''.025$ um etwa zwei Sterngrößen heller sind, als die Praesepesterne von gleichem Spektraltypus. Die Radialgeschwindigkeit stimmt genauer als zu erwarten. Der Unterschied von 15° im Positionswinkel der Eigenbewegung entspricht einer Verschiebung von $0''.9$ im Jahrhundert, und zwar müßte die Eigenbewegung in δ nur $-0''.009$ statt $-0''.018$ betragen.

Es ergibt sich aus dieser Vergleichung rückwärts, daß Hyaden und Praesepe jedenfalls in grober Annäherung gleiche Größe und Richtung ihrer räumlichen Bewegung zukommt. Ob diese Übereinstimmung so eng ist, daß man an physische Zusammengehörigkeit beider im Spektralcharakter ja verwandten Sterngruppen glauben müßte, oder ob im Gegenteil eine Verschiedenheit beider Bewegungen bestimmt vorhanden ist, das zu entscheiden ist mehr Material an Radialgeschwindigkeiten, besonders aber eine schärfere Kenntnis der Eigenbewegung der Praesepe erforderlich.

Astrophysikalisches Observatorium, Potsdam, 1913 Juni 28. *K. Schwarzschild.*

*) Astronomische Mitteilungen der Kgl. Sternwarte Göttingen, Teil IV.

Astron. Nachr. *198* (1914) 217–24

ASTRONOMISCHE NACHRICHTEN.

Band 198. Nr. 4740. 12.

Über das Verhältnis der mittleren Eigenbewegung zur mittleren Parallaxe der Sterne.

Von *K. Schwarzschild.*

§ 1. Bezeichnet man die mittlere Parallaxe der Sterne der scheinbaren Größe m mit π_m, die mittlere Eigenbewegung dieser Sterne mit μ_m, so ist bekannt, daß unter sehr allgemeinen Voraussetzungen Proportionalität zwischen μ_m und π_m, Konstanz des Verhältnisses μ_m/π_m für alle Größenklassen zu erwarten ist. Die Überlegung ist folgende:

Ist r die Entfernung eines Sterns, so findet man seine (auf die Sphäre projizierte, relativ zur Sonne gerechnete) Geschwindigkeit V aus der Eigenbewegung μ nach der Relation:

$$V = \mu r \ .$$

Man betrachte die Sterne der scheinbaren Größe m. Unter diesen werden die leuchtkräftigen in großer Entfernung stehen, die schwächeren in geringerer. Wir bilden die mittlere Eigenbewegung $\bar{\mu}_1$ aller in der Entfernung r_1 stehenden Sterne. Es gilt:

$$\bar{\mu}_1 = (1/r_1)\cdot \bar{V}_1$$

wenn $\bar{V}_1$ die mittlere Geschwindigkeit dieser Sterne ist. Für eine andere Entfernung gilt entsprechend:

$$\bar{\mu}_2 = (1/r_2)\cdot \bar{V}_2 \ .$$

Wenn nun die mittlere Geschwindigkeit einer beliebig aus dem Raum gegriffenen größeren Zahl von Sternen unabhängig vom Ort und von der Leuchtkraft der Sterne ist, so ist:

$$\bar{V}_1 = \bar{V}_2 = \bar{V}$$

gleich der mittleren Geschwindigkeit der Sterne überhaupt. Führt man statt der Entfernung noch die Parallaxe $p = 1/r$ ein, so gilt also:

$$\bar{\mu}_1 = p_1\cdot \bar{V} \qquad \bar{\mu}_2 = p_2\cdot \bar{V}$$

und ähnlich für alle möglichen Entfernungen. Mittelt man nun diese Gleichungen über alle möglichen Entfernungen, so folgt:

$$\bar{\mu} = \bar{p}\cdot \bar{V} \ .$$

Links steht die mittlere Eigenbewegung aller Sterne der scheinbaren Größe m, rechts die mittlere Parallaxe $\bar{p}$ derselben Sterne multipliziert mit der mittleren Geschwindigkeit der Sterne überhaupt. Es folgt also Proportionalität zwischen mittlerer Eigenbewegung und Parallaxe der Sterne jeder scheinbaren Größe, wobei der Proportionalitätsfaktor die mittlere Geschwindigkeit der Sterne überhaupt ist.

§ 2. Wenn man nun den Proportionalitätssatz an den in der Literatur vorhandenen Angaben über die mittleren Parallaxen und Eigenbewegungen prüft, so findet man ihn keineswegs erfüllt. Die erste Zeile folgender Tabelle gibt die von Herrn *J. C. Kapteyn* (Astron. Journ. Nr. 566) bestimmten Parallaxen, die zweite die mittleren Eigenbewegungen nach den Angaben von Herrn *v. Auwers* in der Einleitung der AG-Zone +15° bis +20° Seite (142) und nach der Fortsetzung der *Auwers*schen Tabelle zu schwächeren Sternen, die Herr *Comstock* in Astron. Journ. Nr. 655 Seite 55 mitgeteilt hat.

	$2^m.7$	$4^m.1$	$5^m.1$	$6^m.0$	$6^m.9$	$8^m.6$	$10^m.5$
π	0.″0383	0.″0205	0.″0147	0.″0129	0.″0089	0.″0063	0.″0048
μ	0.″104	0.″079	0.″063	0.″055	0.″054	0.″045	0.″030
μ/π	2.7	3.9	4.3	4.3	6.1	7.1	6.3

Man sieht, wie das Verhältnis μ/π beim Übergang zu schwächeren Sternen ansteigt. Es ist ferner zu erwähnen, daß Herr *Charlier* aus einer Diskussion des *Boss*schen Preliminary General Catalogue (Studies in Stellar Statistics I Seite 45) für die Eigenbewegung der Sterne der scheinbaren Größe m findet:

$$\log \mu_m = \text{const.} - 0.2\,\lambda_1\, m$$

wobei $\lambda_1 = 0.35$ ist, während die oben angeführten *Kapteyn*schen Parallaxen sich durch eine entsprechende Formel mit $\lambda_1 = 0.6$ darstellen lassen. Also auch hier ein langsameres Absinken der Eigenbewegung als der Parallaxe.

Eine Erklärung der Abweichung von der Proportionalität kann man in der Ableitung der Zahlen aus den Beobachtungsdaten selbst suchen. Es werden nämlich bei den Mittelbildungen stets große Eigenbewegungen, die man als sporadische Fälle ansieht, weggelassen, und wenn die Prinzipien hierfür bei Bearbeitung der Eigenbewegungen ganz andere als bei der Ableitung der mittleren Parallaxen sind — wie das bei den oben angeführten Zahlen in der Tat der Fall war — so kann dadurch eine wesentliche Entstellung des Verhältnisses μ/π eintreten.

Man kann aber auch die Abweichung von der Proportionalität wenigstens teilweise als reell auffassen und muß dann annehmen, daß die Voraussetzungen des Proportionalitätssatzes im Sternsystem nicht erfüllt sind, daß eine Abhängigkeit der Geschwindigkeit vom Ort oder von der Leuchtkraft der Sterne besteht. Für eine Abhängigkeit vom Ort hat man bisher noch keine Andeutung; hingegen darf man seit Herrn *Campbells* Nachweis eines Zusammenhanges zwischen der Geschwindigkeit und dem Spektraltypus der Sterne vermuten, daß eine Beziehung zwischen der Geschwindigkeit und der Leuchtkraft bestehe und zwar in dem Sinne, daß Sterne größerer Leuchtkraft im Durchschnitt kleinere Geschwindigkeit haben. Der Sinn ist gerade derjenige, wie er zur Erklärung des beobachteten Verhaltens von μ/π erfordert wird. Denn geht man zu Sternen geringerer scheinbarer Helligkeit, höherer Größenklasse m über, so ist auch deren durchschnittliche absolute Helligkeit kleiner, ihre mittlere Geschwindigkeit also größer zu vermuten. Der Quotient: mittlere Eigenbewegung durch mittlere säkulare Parallaxe

Astron. Nachr. *198* (1914) 217–224

der Sterne wird daher mit der scheinbaren Sterngröße wachsen müssen, wie das die obigen Zahlen zeigen. Eine Bekräftigung für eine solche Vermutung findet man in den Untersuchungen von Herrn *v. Seeliger* in A. N. 4617 und 4640. Ferner sei darauf hingewiesen, daß Herr *Hertzsprung* aus ganz anderem Material, aus der Statistik von Doppelsternen, in A. N. 4543 eine Wahrscheinlichkeit für eine Abnahme der Geschwindigkeit mit der Leuchtkraft gefunden hat, die sich mit unserem Problem noch näher deckt, da Herr *Hertzsprung* auch mit den totalen Eigenbewegungen inklusive Sonnenbewegung rechnet. Herrn *Campbells* Resultat bezicht sich auf die von Sonnenbewegung befreiten Geschwindigkeiten, bei Einschluß der Sonnenbewegung wird der Gang natürlich sehr vermindert.

Ich möchte hier nun den Einfluß einer Abhängigkeit der Geschwindigkeit der Sterne von ihrer Leuchtkraft im engsten Anschluß an die Untersuchungen zur Stellarstatistik in A. N. 4557 quantitativ behandeln, um zu entscheiden, wie weit man auf diese Weise die Unproportionalität von μ und π erklären kann.

§ 3. Die sehr einfache Ergänzung der Ansätze in A. N. 4557, durch welche eine Abhängigkeit der Geschwindigkeit der Sterne von ihrer Leuchtkraft berücksichtigt werden kann, läßt sich folgendermaßen angeben. Der Prozentsatz der Sterne irgend eines Raumelements, deren Leuchtkraft zwischen $\mathfrak{J}$ und $\mathfrak{J}+d\mathfrak{J}$, deren (auf die Sphäre projizierte) Geschwindigkeit zwischen V und $V+dV$ liegt, war dort angesetzt zu:

$$\varphi(\mathfrak{J})\,\psi(V)\,d\mathfrak{J}\,dV$$

wobei:

$$\varphi(\mathfrak{J}) = 10^{b_0 - b_1 M - b_2 M^2} \qquad M = -2.5 \log \mathfrak{J}$$
$$\psi(V) = 10^{c_0 - c_1 G - c_2 G^2} \qquad G = -5.0 \log V$$

war. Wir wollen jetzt für diesen Prozentsatz den Ausdruck wählen:

$$\Phi(\mathfrak{J}, V)\,d\mathfrak{J}\,dV$$

wobei:

$$\Phi(\mathfrak{J}, V) = 10^{b_0 - b_1 M - b_2 M^2 - c_1 G - c_2 G^2 - c_3 MG}$$

sein soll. Damit $\Phi(\mathfrak{J}, V)$ eine prozentuale Zahl darstellt, muß gelten:

$$\int_0^\infty\int_0^\infty \Phi(\mathfrak{J}, V)\,d\mathfrak{J}\,dV = 1\,.$$

Das Wesentliche der Erweiterung ist die Aufnahme des Gliedes $c_3 MG$ in den Exponenten von Φ. Dieses Glied bewirkt, daß Φ nicht mehr in ein Produkt zweier Funktionen nur von $\mathfrak{J}$ resp. V zerlegt werden kann, und gibt eine Abhängigkeit der mittleren Geschwindigkeit der Sterne von ihrer Leuchtkraft. Bildet man nämlich die mittlere Geschwindigkeit $\bar{V}$ der Sterne der Leuchtkraft $\mathfrak{J}$:

$$\bar{V} = \frac{\int_0^\infty \Phi(\mathfrak{J}, V)\,V\,dV}{\int_0^\infty \Phi(\mathfrak{J}, V)\,dV}$$

so erhält man durch Ausführung der Integrale gemäß dem l. c. § 6 gegebenen Beispiel:

$$\log \bar{V} = \varkappa M + \lambda$$

wobei gilt:

$$\varkappa = c_3/(10\,c_2) \qquad \lambda = (c_1 + 0.3)/(10\,c_2)\,.$$

Der Dezimallogarithmus der mittleren Sterngeschwindigkeit vermehrt sich also um $\varkappa$, wenn die absolute Sterngröße um eine Größenklasse zunimmt.

Folgende beiden Bemerkungen verdeutlichen noch den Sinn unseres neuen Ansatzes:

a) Die Verteilung der Leuchtkräfte unter der Gesamtheit der Sterne der Volumeneinheit findet man aus dem neuen Ansatz durch das Integral:

$$\varphi(\mathfrak{J}) = \int_0^\infty \Phi(\mathfrak{J}, V)\,dV$$

Es ist dabei, wie früher, $\varphi(\mathfrak{J})\,d\mathfrak{J}$ die prozentuale Anzahl der Sterne der Leuchtkraft $\mathfrak{J}$ bis $\mathfrak{J}+d\mathfrak{J}$ in der Volumeneinheit, ohne Trennung der Sterne nach ihrer Geschwindigkeit.

Die Ausführung des Integrals gibt:

$$\log \varphi(\mathfrak{J}) = b'_0 - b'_1 M - b'_2 M^2$$

mit:

$$b'_0 = b_0 + (c_1 + 0.2)^2/(4\,c_2)$$
$$b'_1 = b_1 - 5\varkappa(c_1 + 0.2)$$
$$b'_2 = b_2 - 25\,c_2\,\varkappa^2\,.$$

Die Verteilung der Leuchtkräfte hat also dieselbe Gestalt wie früher, nur daß die Koeffizienten b' an Stelle von b treten.

b) Faßt man speziell die Sterne einer bestimmten Leuchtkraft $\mathfrak{J}$ ins Auge und fragt nach der Anzahl, für die der Logarithmus der Geschwindigkeit zwischen $\log V$ und $\log V + d\log V$ liegt, so erhält man (e Basis der natürl. Log.):

$$d\log V \cdot V\Phi(\mathfrak{J}, V)/\log e$$

und es folgt, da $G = -5 \log V$ ist:

$$\log[V\Phi(\mathfrak{J}, V)] = b_0 - b_1 M - b_2 M^2 - (c_1 + 0.2)\,G - c_2 G^2 - c_3 MG$$

und durch eine einfache Umsetzung:

$$\log[V\Phi(\mathfrak{J}, V)] = \log \varphi(\mathfrak{J}) - 25\,c_2(\log V - \log \bar{V} + 0.01/c_2)^2\,.$$

Diese Gleichung besagt, daß die Logarithmen der Geschwindigkeiten um ihr Mittel nach dem *Gauß*schen Fehlergesetz gestreut sind mit dem mittleren Fehler $\sqrt{[\log e/(50\,c_2)]}$.

Es galt nunmehr, die Ableitungen in A. N. 4557 unter Berücksichtigung der Abhängigkeit der Geschwindigkeit von der Leuchtkraft zu wiederholen. Es bleiben dabei unverändert alle Größen, bei denen keine Einteilung nach der Eigenbewegung erfolgt, also die Anzahl, die mittlere Parallaxe, die mittlere absolute Größe und die Streuung der absoluten Größe für die Sterne jeder scheinbaren Helligkeit, man muß nur in $\varphi(\mathfrak{J})$ die Koeffizienten $b_0'\,b_1'\,b_2'$ statt der früheren ungestrichenen verwenden. Kleine Modifikationen treten bei den von den Eigenbewegungen abhängigen Größen ein. Es folgt, wenn wir uns auf das Wichtigste beschränken, für die mittlere Eigenbewegung der Sterne jeder Größe:

$$\mu_m = \frac{\int_0^\infty \mathrm{d}r\, r^3 D(r) \int_0^\infty V \mathrm{d}V\, \Phi(ir^2, V)}{\int_0^\infty \mathrm{d}r\, r^4 D(r) \int_0^\infty \mathrm{d}V\, \Phi(ir^2, V)} \; {}^{*}),$$

für die mittlere Parallaxe der Sterne der Größe m und der Eigenbewegung μ:

$$\pi_{m,\mu} = \frac{\int_0^\infty \mathrm{d}r\, r^4 D(r)\, \Phi(ir^2, \mu r)}{\int_0^\infty \mathrm{d}r\, r^5 D(r)\, \Phi(ir^2, \mu r)}$$

für die Streuung der Parallaxen um diesen Mittelwert oder schärfer für die Zahl der Sterne der Größe m und Eigenbewegung μ, für die der Logarithmus der Parallaxe zwischen $\log \pi$ und $\log \pi + \mathrm{d} \log \pi$ liegt:

$$\chi_{\pi,m,\mu}\, \mathrm{d}\log \pi = \mathrm{d}\log \pi\, (\omega/[2.5\, (\log e)^2])\, r^6 D(r)\, i\, \Phi(ir^2, \mu r).$$

Die spezielle Ausführung für die logarithmischen Ansätze der Verteilungsfunktionen liefert folgendes, wenn ich alle Formeln unter Wiederholung der bereits angegebenen und der aus A. N. 4557 zu übernehmenden, zusammenstelle:

Anzahl $b_m \mathrm{d}m$ der Sterne der scheinbaren Größe m bis $m + \mathrm{d}m$:

$$\log b_m = \alpha_0 - \alpha_1 m - \alpha_2 m^2 .$$

Mittlere Parallaxe der Sterne der scheinbaren Größe m:

$$\log \pi_m = P_0 - P_1 m .$$

Mittlere Eigenbewegung der Sterne der scheinbaren Größe m:

$$\log \mu_m = S_0 - S_1 m .$$

Mittlere Parallaxe der Sterne der scheinbaren Größe m und der Eigenbewegung μ:

$$\log \pi_{m,\mu} = R_0 - R_1 m - R_2 g \qquad (g = -5 \log \mu)$$

Streuung um diese mittlere Parallaxe:

$$\log \chi_{\pi,m,\mu} = L_0 - L_1 (\log \pi - \log \pi_{m,\mu} - L_2)^2 .$$

Mittlere Geschwindigkeit der Sterne der absoluten Größe M:

$$\log \bar{V} = \varkappa M + \lambda .$$

Dabei ist der Zusammenhang der hier auftretenden Koeffizienten mit den Koeffizienten a, b, c der Verteilungsfunktionen $D(r)$ und $\Phi(\mathfrak{J}, V)$ der nachstehende:

$$\alpha_0 = 0.7942 - {}^1/_2 \log (a_2 + b_2') + a_0 + b'_0 + \frac{(1 + a_1 + b_1')^2}{4(a_2 + b_2')}$$

$$\alpha_1 = \frac{a_2(b_1' + 0.4) - b_2'(a_1 + 0.6)}{a_2 + b_2'} \qquad \alpha_2 = \frac{a_2 b_2'}{a_2 + b_2'}$$

$$P_0 = -0.1 \frac{a_1 + b_1' + 0.9}{a_2 + b_2'} \qquad P_1 = \frac{0.2 b_2'}{a_2 + b_2'}$$

$$S_0 = \lambda + P_0(1 + 5\varkappa) + \frac{\varkappa(1 + 5\varkappa)}{20(a_2 + b_2)} \qquad S_1 = (1 + 5\varkappa)P_1 - \varkappa$$

$$R_0 = -0.1 \frac{1.1 + a_1 + b_1 + c_1}{a_2 + b_2 + c_2 + c_3} \qquad R_1 = \frac{0.1(2 b_2 + c_3)}{a_2 + b_2 + c_2 + c_3}$$

$$R_2 = \frac{0.1(2 c_2 + c_3)}{a_2 + b_2 + c_2 + c_3}$$

$$L_1 = 25(a_2 + b_2 + c_2 + c_3) \qquad L_2 = -\frac{0.01}{a_2 + b_2 + c_2 + c_3}$$

$$\varkappa = \frac{c_3}{10 c_2} \qquad \lambda = \frac{c_1 + 0.3}{10 c_2}$$

$$b_0' = b_0 + \frac{(c_1 + 0.2)^2}{4 c_2} \qquad b_2' = b_2 - 25 c_2 \varkappa^2$$

$$b'_1 = b_1 - (c_1 + 0.2)\, 5\varkappa .$$

Es seien noch folgende Bemerkungen an die Formeln angeschlossen:

a) Die Formel für die Streuung der Parallaxen bei gegebener Größe und Eigenbewegung besagt, daß die Logarithmen der Parallaxe nach dem *Gauß*schen Fehlergesetz verteilt sind mit dem mittleren Fehler: $\sqrt{[\log e/(2 L_1)]}$.

b) Das Verhältnis von Eigenbewegung zu mittlerer Parallaxe der Sterne jeder Größe folgt der Formel:

$$\log (\mu_m/\pi_m) = S_0 - P_0 - (S_1 - P_1)\, m$$

und es ist: $\quad S_1 - P_1 = \varkappa(5 P_1 - 1)$.

Kennt man also die Abnahme der mittleren Parallaxe mit der scheinbaren Größe, den Koeffizienten P_1, und die Abhängigkeit der Absolutgeschwindigkeit $\bar{V}$ von der absoluten Größe, also den Koeffizienten $\varkappa$, so kann man daraus sofort berechnen, wie sich das Verhältnis von Eigenbewegung und Parallaxe mit der scheinbaren Größe der Sterne ändert.

c) Verschwindet der Koeffizient c_3, der die Abhängigkeit der Geschwindigkeit von der absoluten Helligkeit bedingt, so verschwindet $\varkappa$, alle Formeln kehren in die des früheren Aufsatzes zurück, das Verhältnis μ_m/π_m wird, wie dort, dasselbe für alle Größenklassen.

§ 4. Die vorstehenden Formeln sollen nunmehr verwandt werden zu der beabsichtigten quantitativen Untersuchung über den Einfluß einer Abhängigkeit der Geschwindigkeit von der Leuchtkraft der Sterne. Diese Abhängigkeit drückt sich durch den Koeffizienten c_3 oder bequemer durch $\varkappa$ aus. Wir wollen so vorgehen, daß wir $\varkappa$ als einen Parameter behandeln, den wir zur Verfügung halten, und alle andern Größen als Funktionen von $\varkappa$ darzustellen suchen. Wir stehen dann mit geringer Modifikation vor derselben Aufgabe, die in A. N. 4557 § 8 behandelt wurde: Sieben Koeffizienten sind den empirischen Daten über die Anzahl und Parallaxe der Sterne zu entnehmen, daraus sind die Verteilungsfunktionen d. h. die Koeffizienten a, b resp. b', c, sowie alle weiteren interessierenden Größen gemäß den Formeln des vorigen Paragraphen abzuleiten. Als empirische Daten sollen wieder dieselben benutzt werden, wie in A. N. 4557, nämlich die auf Herrn *Kapteyn*s Angaben gegründeten Formeln für die Anzahl und Parallaxe der Sterne der Größe m:

$$\log b_m = 0.596 + 0.5612\, m - 0.0055\, m^2$$

$$\log \pi_m = -1.108 - 0.125\, m$$

sowie der erste und letzte Koeffizient aus der Formel für die mittlere Parallaxe der Sterne jeder Größe und Eigenbewegung:

$$\log \pi_{m,\mu} = -0.766 - 0.0434\, m - 0.1423\, g$$

unter Beiseitelassung des schlecht bestimmten mittleren. Es werden also die Koeffizienten $\alpha_0, \alpha_1, \alpha_2, P_0, P_1, R_0, R_2$ der Beobachtung entnommen. Die ganz elemen-

*) Wie in A. N. 4557 bezeichnet $D(r)$ die Dichte in der Entfernung r und es ist gesetzt: $\log D(r) = a_0 - a_1 \rho - a_2 \rho^2$, $\rho = -5 \log r$.

tare Auflösung der gemäß den Formeln des vorigen Paragraphen entstehenden Gleichungen nach den Unbekannten a_0, a_1, a_2, b_0 usw. und die Durchrechnung der übrigen Formeln, wobei $\varkappa$ immer als gegeben angesehen wird, liefert folgendes: Unverändert ergeben sich die früheren Verteilungen der Dichte und der Leuchtkräfte der Sterne (die früher b genannten Koeffizienten der Verteilung der Leuchtkräfte sind numerisch gleich den jetzt b' genannten Größen), wie das natürlich ist, da hier keine Unterteilung nach der Geschwindigkeit erfolgt, auf die allein man die Änderung des allgemeinen Ansatzes schieben kann. Für die Verteilung der Geschwindigkeiten aber (die der allgemeinen Form nach immer eine *Gauß*sche Verteilung der Geschwindigkeitslogarithmen bleibt) und die davon abhängigen Größen ergeben sich nachstehende den Parameter $\varkappa$ enthaltende Ausdrücke. Die mittlere Geschwindigkeit der Sterne der absoluten Größe M wird:

$$\log \bar{V} = \varkappa M + 0.80 + 4.3\varkappa - 10.7\varkappa^2 .$$

Die mittlere Streuung des Logarithmus der Geschwindigkeit für Sterne ein und derselben Leuchtkraft wird:

$$\pm \sqrt{[\log c/(50\, c_2)]} = \pm 0.39 \sqrt{[(1+5\varkappa)(1-12.3\varkappa)]} .$$

Die *Kapteyn*sche Parallaxenformel lautet mit ihrem nunmehr deduktiv bestimmten mittleren Koeffizienten:

$$\log \pi_{m,\mu} = -0.766 - (0.036 + 0.267\varkappa)m - 0.1423 g$$

und der mittlere Fehler des Logarithmus einer hiernach berechneten Parallaxe wird:

$$\pm \sqrt{[\log c/(2\, L_1)]} = \pm 0.33 \sqrt{(1-12.3\varkappa)} .$$

Das Verhältnis von Eigenbewegung und Parallaxe wird:

$$\log(\mu_m/\pi_m) = 0.80 + 0.375\varkappa(m+2.3) .$$

Wir sind damit zu einem auffälligen Ergebnis gelangt. Wir sehen nämlich, daß für $\varkappa = {}^1/_{12.3}$ die Streuung der Geschwindigkeiten und zugleich der mittlere Fehler der aus scheinbarer Größe und Eigenbewegung berechneten Parallaxen zu Null würde. Man kann verstehen, wie das zusammenhängt. Wenn die Geschwindigkeit so stark von der Leuchtkraft abhängt, wie der Wert $\varkappa = {}^1/_{12.3}$ verlangt, so gibt bei der großen Verschiedenheit der Leuchtkräfte der Sterne dies eine so starke Verschiedenheit der Geschwindigkeiten, daß für eine weitere Streuung der Geschwindigkeiten unter den Sternen derselben Leuchtkraft nach den Beobachtungsdaten kein Raum mehr bleibt. Die Geschwindigkeit jedes Sterns müßte direkt durch seine Leuchtkraft gegeben sein, und dann würde man bei Kenntnis der scheinbaren Helligkeit und Eigenbewegung eines Sterns drei Gleichungen für drei Unbekannte, für die Entfernung, die absolute Helligkeit und Geschwindigkeit haben, könnte also die Parallaxe direkt berechnen.

In Wirklichkeit ist aber eine Streuung der Geschwindigkeiten unter den Sternen gleicher Leuchtkraft und ein mittlerer Fehler der Parallaxe $\pi_{m,\mu}$ vorhanden. Herr *Kapteyn* findet in Groningen Publ. Nr. 8 Gleichung (47) als Minimum der wirklichen Streuung des $\log \pi$ bei reichlichem Ansatz für die Beobachtungsfehler ± 0.24. Daraus berechnet sich die obere Grenze für $\varkappa$:

$$\varkappa < 0.038 .$$

Mit diesem Grenzwert von $\varkappa$ folgt:

$$\log \bar{V} = 0.038\, M + 0.95 .$$

Das ist immerhin eine Verdoppelung der Geschwindigkeit auf eine Abnahme der Leuchtkraft um 8 Größenklassen. Mit dem Grenzwert von $\varkappa$ folgt ferner:

$$\log(\mu_m/\pi_m) = 0.89 + 0.014\, m .$$

Das ist ein äußerst langsames Anwachsen des Verhältnisses μ/π, welches beträchtlich zurückbleibt hinter dem Anwachsen der oben mitgeteilten Zahlen.

Es ergibt sich somit als Resultat der quantitativen Rechnung: 1. Die vermutete Abhängigkeit der Geschwindigkeit von der Leuchtkraft kann nicht über eine Verdoppelung der Geschwindigkeit bei Abnahme der Leuchtkraft um 8 Größenklassen hinausgehen (es sei dabei daran erinnert, daß es sich um die Projektion der Geschwindigkeiten relativ zur Sonne auf die Sphäre handelt). 2. Diese Abhängigkeit bewirkt im Logarithmus des Verhältnisses von mittlerer Eigenbewegung zu mittlerer Parallaxe nur ein Ansteigen um 0.014 pro Größenklasse. Das viel stärkere Anwachsen der oben angeführten Verhältniszahlen kann also nicht reell sein und muß darauf zurückgeführt werden, daß bei der Bestimmung der mittleren Eigenbewegungen, vor allem der hellen Sterne, ein stärkerer Ausschluß anscheinend sporadischer großer Bewegungen erfolgt ist als bei Herrn *Kapteyns* Bestimmung der mittleren Parallaxe.

Eine direkte Ableitung des Zusammenhangs zwischen Geschwindigkeit und Leuchtkraft aus dem Verhältnis μ/π würde wohl nur möglich sein, wenn man über gutes Material bis zu sehr schwachen Sternen herab verfügte. Dann müßte man aber vielleicht auch die Möglichkeit einer Änderung der Geschwindigkeiten mit der Entfernung berücksichtigen. Es sei noch bemerkt, daß dies im Rahmen der obigen Ansätze sehr leicht möglich wäre, indem man ganz analog zu dem Gliede $c_3\, G M$ noch ein Glied $c_4\, G \varrho$ in der Verteilung der Geschwindigkeiten hinzufügte.

Potsdam, 1914 März 6. *K. Schwarzschild.*

Visual observations of Phoebe, the ninth Satellite of Saturn. By *E. E. Barnard.*

The following observations of Phoebe have been obtained with the large telescope. The unfortunate cloudy condition of the skies this winter prevented any other observations of the satellite.

When best seen in these observations, Phoebe was as bright as the 14th magnitude. It was difficult from bad seeing, however, in the measures of December 16. The estimated brightness on December 21 and 23 was 14th magnitude. It was estimated at this last observation (23rd) that the satellite was of exactly the same brightness as a 14th magnitude star $10''$–$15''$ south of it. On this date clouds prevented a second set of measures of $\Delta\alpha$.

On December 21, to identify the satellite by motion, two sets of $\Delta\alpha$ were made with a 12th magnitude star with

Sitz. Ber. Kgl. Preuss. Akad. d. Wiss. Berlin 1914, 489–510

Über die Häufigkeit und Leuchtkraft der Sterne von verschiedenem Spektraltypus.

Von K. Schwarzschild.

§ 1. Für unsre Kenntnis vom Bau und von der Entwicklung des Sternsystems ist es wertvoll, zu wissen, nicht nur wieviel Sterne in der genügend groß gewählten Einheit des Raumes enthalten sind, sondern auch, wie die Sterne in der Raumeinheit gemischt sind, wie sie sich prozentual auf die verschiedenen Spektraltypen verteilen. Die scheinbare Häufigkeit der verschiedenen Spektraltypen am Himmel, etwa unter den dem bloßen Auge sichtbaren Sternen, gibt nur ein äußerst verzerrtes Bild ihrer wirklichen relativen Häufigkeit in der Raumeinheit. Denn Sterne von großer absoluter Helligkeit, wie die Heliumsterne, sind dem Auge bis in weite Fernen hinaus sichtbar, man sammelt sie gewissermaßen aus einem größeren Volumen auf und sieht sie in einer größeren Zahl als lichtschwächere, an sich gleich häufige Sterne; die lichtstarken Sterne spielen am Himmelsgewölbe eine viel größere Rolle als ihrer Zahl in der Raumeinheit entspricht. Man erkennt so unmittelbar, daß ein Rückschluß aus der scheinbaren Verteilung der Spektraltypen am Himmel auf ihre Verteilung in der Raumeinheit nur möglich ist, wenn man die Leuchtkraft der Sterne jedes Spektraltypus kennt, oder genauer genommen, da auch die Sterne desselben Spektraltypus noch sehr an Leuchtkraft verschieden sind, wenn man für jeden Spektraltypus das Gesetz der Verteilung der Leuchtkräfte kennt. Die Bestimmung der räumlichen Dichte und der Verteilung der Leuchtkräfte für die Sterne jedes Spektraltypus sind daher zwei gemeinsam zu behandelnde Probleme.

Man ist in der Lage, diese Probleme etwas schärfer anzugreifen, seitdem Hr. W. W. Campbell die Radialgeschwindigkeiten von etwa 1600 Sternen aller Spektraltypen, die größtenteils auf dem Lickobservatorium bestimmt worden sind, mitgeteilt hat. Aus Hrn. Campbells Zahlen kann man entnehmen, wie die Geschwindigkeiten der Sterne jedes Spektraltypus verteilt sind bezüglich der dem Visionsradius parallelen Komponente. Man mache die Annahme, daß alle Komponenten der Sterngeschwindigkeiten gleich verteilt sind. Dann gibt die aus

Sitz. Ber. Kgl. Preuss. Akad. d. Wiss. Berlin 1914, 489–510

Hrn. Campbells Zahlen gewonnene Verteilung der Radialgeschwindigkeiten auch die Verteilung einer Komponente der Bewegung der Sterne senkrecht zum Visionsradius an. Mit dieser Verteilung der wirklichen Bewegungen vergleiche man die Verteilung einer Komponente der scheinbaren Bewegung, der Eigenbewegung, der Sterne des betreffenden Spektraltypus. Befänden sich diese Sterne alle in derselben Entfernung, so wäre das Resultat der Vergleichung ein sehr einfaches. Beide Verteilungen müßten identisch sein, wenn man nur die Eigenbewegungen mit einem gewissen Faktor multiplizierte, der dann auch gleich ein Maß der Entfernung der Sterne ist. Ähnlich einfach wird die Vergleichung auch noch, wenn die Sterne zwar in verschiedenen Entfernungen stehen, aber ein und dieselbe Leuchtkraft besitzen. Man kann sich dann nämlich alle Sterne in diejenige Entfernung gebracht denken, in welcher sie von einer bestimmten scheinbaren Helligkeit, z. B. von nullter Sterngröße, sind. Dabei verändern sich — bei festgehaltener absoluter Geschwindigkeit — die scheinbaren Eigenbewegungen, und zwar multipliziert sich die Eigenbewegung eines Sterns der scheinbaren Größe m mit $10^{0.2\,m}$, wenn man den Stern durch Entfernungsänderung auf nullte Größe bringt. Die mit diesem Faktor multiplizierte Eigenbewegung jedes Sterns soll seine (auf die nullte Größe) reduzierte Eigenbewegung heißen. In unserem Falle müßten nun die reduzierten Eigenbewegungen, mit einem geeigneten Faktor multipliziert, wieder dieselbe Verteilung zeigen wie die Radialgeschwindigkeiten, und dieser Faktor würde die Entfernung messen, in welcher alle unsre Sterne von der scheinbaren Größe Null wären.

Da nun in Wirklichkeit die Sterne weder in einer Entfernung stehen noch alle gleiche Leuchtkraft haben, so müssen die reduzierten Eigenbewegungen offenbar stärker gestreut sein als die Radialgeschwindigkeiten. Was die verschiedene Entfernung ausmacht, kann man berechnen, wenn man die Dichte der Sterne als unabhängig von der Entfernung oder ihrem Verlaufe nach sonst bekannt annimmt. Die Eigenbewegungen sollen aus dem Preliminary General Catalogue von L. Boss entnommen werden, der etwa bis $6^{m}.3$ geht. Bis zu den Entfernungen, in denen diese Sterne stehen, dürfte man die Dichte noch ohne bedenklichen Fehler als konstant ansehen, doch soll, um etwas genauer zu sein, der aus den allgemeinen stellarstatistischen Untersuchungen folgende Dichteverlauf benutzt werden.

Was nach Berücksichtigung der verschiedenen Entfernungen noch an Streuung der reduzierten Eigenbewegungen gegenüber der Streuung der Radialgeschwindigkeiten übrigbleibt, das ist auf die Verschiedenheit der Leuchtkräfte zurückzuführen. Die Vergleichung im einzelnen — mathematisch die Lösung einer Integralgleichung — muß die genauere Verteilung der Leuchtkräfte ergeben.

Ist die Verteilung der Leuchtkräfte bekanntgeworden, so ist der Rückschluß von der scheinbaren Häufigkeit auf die Häufigkeit in der Raumeinheit für die Sterne von jedem Spektraltypus leicht.

Dies der allgemeine Gedankengang der Untersuchung. Ihre Grundlage ist die erwähnte Annahme, daß alle Komponenten der Sterngeschwindigkeiten gleich verteilt sind. Diese Annahme ist nun zweifellos im allgemeinen nicht richtig, einesteils wegen der Bewegung der Sonne im Weltraum, andernteils wegen der Existenz einer Vorzugsrichtung in den von Sonnenbewegung befreiten Sternbewegungen. Man muß sich solche speziellen Komponenten der Bewegungen aussuchen oder schaffen, auf welche die Annahme zutrifft. Die einfachste Art, dies zu erreichen, besteht darin, daß man die Radialgeschwindigkeiten zunächst alle von Sonnenbewegung befreit und daß man die so entstehenden absoluten Geschwindigkeiten vergleicht mit derjenigen Komponente der Eigenbewegung der Sterne, welche senkrecht steht zur Richtung nach dem Apex der Sonnenbewegung, welche also ebenfalls frei ist von Sonnenbewegung. Wir wollen diese Komponente der Eigenbewegung der Sterne kurz als »Querbewegung« bezeichnen. Für die Geschwindigkeitskomponente der Sterne, die sich als Querbewegung an den Himmel projiziert, und für die absoluten Radialgeschwindigkeiten würde die Annahme gleicher Verteilung streng gültig sein, wenn nicht die Vorzugsrichtung in den absoluten Sternbewegungen bestände. Indessen liegen die quantitativen Verhältnisse so, daß die gleiche Verteilung in diesem Falle nicht sehr gestört wird. Denkt man sich die Verteilung der absoluten Sterngeschwindigkeiten gegeben durch ein verlängertes Revolutionsellipsoid, dessen große Achse nach dem Vertex gerichtet ist, so wird die Radialgeschwindigkeit über den größeren Teil des Himmels in der Nähe der kleinen Achse liegen. Die Querbewegung würde, wenn Vertex und Apex genau zusammenfielen, immer strenge mit der kleinen Achse zusammenfallen; da tatsächlich Apex und Vertex etwa 40° auseinanderliegen, so wird die Querbewegung, ähnlich wie die Radialgeschwindigkeit, nur in der Nähe der kleinen Achse des Geschwindigkeitsellipsoids liegen. Querbewegung und Radialgeschwindigkeit werden also trotz der Existenz der Vorzugsrichtung ziemlich ähnlich verteilten Geschwindigkeitskomponenten entsprechen.

Natürlich bedürfte dieser Punkt noch einer genaueren quantitativen Durchrechnung. Aber ich betrachte die ganze folgende Untersuchung nur als eine erste Orientierung, bei der ich mich auch nicht um eine peinliche Genauigkeit bei den Abzählungen und sonstigen Reduktionen bemüht habe. Ich beabsichtige auf eine zweite Durchführung mit Benutzung der nach dem Sonnenapex gerichteten Komponente der Eigenbewegung bald zurückzukommen.

Sitz. Ber. Kgl. Preuss. Akad. d. Wiss. Berlin 1914, 489–510

§ 2. Das Beobachtungsmaterial.

Die Radialgeschwindigkeiten wurden den Katalogen von Campbell (*B*-Sterne Lick Observatory Bulletin Nr. 195, *A*-Sterne Nr. 211, übrige Sterne Nr. 229) entnommen. Dieselben wurden in absolute Radialgeschwindigkeiten durch Abzug des Einflusses der Sonnenbewegung verwandelt. Ferner wurde auch die von Hrn. Campbell gefundene, von ihm mit *K* bezeichnete systematische Korrektion für jeden Spektraltypus berücksichtigt. Für die *B*- und *A*-Sterne hat Hr. Campbell die absoluten Radialgeschwindigkeiten an den genannten Stellen unter der Bezeichnung V_2 selbst mitgeteilt. Die Elemente der Sonnenbewegung und die Konstante *K*, die er zugrunde legte, waren:

	Sonnenapex α	δ	Sonnengeschwindigkeit	*K*
B-Sterne	270°0	+30°0	20.2 km/sec	+4.1 km/sec
A-Sterne	270°0	+30°0	19.5 »	0 »

Für die übrigen Sterne wurde die Reduktion neu ausgeführt mit folgenden Konstanten:

F-Sterne	269°7	+30°8	20.0 km/sec	0 km/sec
G-Sterne	269°7	+30°8	20.0 »	0 »
K-Sterne	269°7	+30°8	20.0 »	+2 »
M-Sterne	269°7	+30°8	20.0 »	+4 »

Die Radialgeschwindigkeiten wurden dann für jeden Spektraltypus (unter Zusammenfassung von B_0 bis B_9, A_0 bis A_9 usw.) nach ihrem absoluten Betrage abgezählt[1], und zwar wurde bestimmt, wieviel Radialgeschwindigkeiten über 5, 10, 20 km/sec usw. vorkommen. Diese Werte zunächst ziffernmäßig, dann in Prozenten der Gesamtzahl ausgedrückt, gibt Tabelle 1.

Tabelle 1.

Anzahl und Prozentsatz der Radialgeschwindigkeiten über *V*.

V km/sec	*G*	Anzahl *B*	*A*	*F*	*G*	*K*	*M*	Prozentsatz *B*	*A*	*F*	*G*	*K*	*M*
0.0	+∞	226	211	198	146	439	80	100.0	100.0	100.0	100.0	100.0	100.0
5.0	−0°12	115	146	148.5	111	350.5	63	51.7	69.2	75.0	76.0	79.9	78.7
10.0	−1.63	44.5	94.5	108	85.5	258	48	19.7	44.8	54.5	58.6	58.8	60.0
20.0	−3.13	11	36	56	40	143	25	4.6	17.1	28.3	27.4	32.6	31.2
40.0	−4.64	0	0	8	14	36	6	0.0	0.0	4.0	9.6	8.2	7.5
80.0	−6.14			0	3	3	0			0.0	2.1	0.7	0.0
160.0	−7.65				1	0					0.7	0.0	

Die zweite Spalte wird später erklärt werden.

[1] Es sind die nicht im P. G. C. von L. Boss vorkommenden Sterne weggelassen, da dies durchweg schwächere, aus einem exzeptionellen Grunde beobachtete Sterne sind.

Die Querbewegungen wurden dem Preliminary General Catalogue von L. Boss entnommen, und zwar stand mir hierfür nicht nur eine von Hrn. J. C. KAPTEYN gütigst überlassene Umrechnung der Bossschen Eigenbewegungen nach Apizialkomponenten zur Verfügung, in der die Spektraltypen nach den besten Angaben der Harvard Annals eingetragen waren, sondern ich konnte auch von einer Reduktion der Eigenbewegungen auf die nullte Größe mit Hilfe des erwähnten Faktors $10^{0.2m}$ Gebrauch machen, die Frl. J. LEHMANN für einen andern Zweck ausgeführt hatte.

Das Resultat für die auf die Größe 0.0 reduzierten Querbewegungen gibt folgende Tabelle:

Tabelle 2.

Verteilung der auf $0^m.0$ reduzierten Querbewegungen.

Grenzen	*B*	*A*	*F*	*G*	*K*	*M*
0".00 bis 0".05	287	258	91	105	322	50
0.06 » 0.10	140	212				
0.11 » 0.20	155	338	53	77	209	46
0.21 » 0.40	95	411	102	99	320	49
0.41 » 0.80	14	305	154	96	286	49
0.81 » 1.60	2	107	144	64	166	21
1.61 » 3.20	0	21	111	51	50	0
3.21 » 6.40	0	3	35	22	16	0
6.41 » 12.80	0	0	12	15	10	0
12.81 » 25.60	0	0	3	7	3	0
25.61 » 51.20	0	0	1	1	3	0
51.21 » 102.40	0	0	0	0	1	0
Gesamtsternzahl	693	1655	706	537	1386	215

Diese Zahlen bedürfen z. T. einer erheblichen Korrektion, weil die Eigenbewegungen mit Beobachtungsfehlern behaftet sind. Ich habe aus den Angaben von L. Boss roh entnommen, daß der wahrscheinliche Fehler einer auf $0^m.0$ reduzierten Eigenbewegung 0".05 beträgt und habe die Abzählungen nach der Formel von Hrn. EDDINGTON (Monthly Not. of the Roy. Astr. Soc. Bd. 73 S. 360) korrigiert. Für die *B*-Sterne sind die Querbewegungen so wenig größer als die Beobachtungsfehler, daß die Reduktion für die Bewegungen bis 0".1 nicht mehr nach Hrn. EDDINGTONS Formel ausgeführt werden konnte und in andrer Weise abgeschätzt wurde, wobei das Resultat immer sehr unsicher bleibt. Die späteren Ableitungen sind also speziell für die *B*-Sterne nur mit Vorbehalt aufzunehmen. Die korrigierte Verteilung der Querbewegungen wurde, wie bei den Radialgeschwindigkeiten, in der Weise zum Ausdruck gebracht, daß der Prozentsatz der den Betrag μ' über-

Sitz. Ber. Kgl. Preuss. Akad. d. Wiss. Berlin 1914, 489–510

steigenden reduzierten Querbewegungen berechnet wurde (Tab. 3). Die zweite Spalte der Tabelle findet unten ihre Erklärung.

Tabelle 3.

Prozentsatz der reduzierten Querbewegungen über μ'.

μ'	g'	B	A	F	G	K	M
0″00	∞	100	100.0	100.0	100.0	100.0	100.0
0.05	+ 6.50	50	83.8				
0.10	+ 5.00	32	71.3	86.7	80.4	76.6	76.7
0.20	+ 3.50	14	50.5	79.7	65.0	61.8	54.7
0.40	+ 1.99	2	25.3	65.1	47.3	38.4	32.5
0.80	+ 0.49	0.3	7.9	43.4	29.9	17.9	9.4
1.60	− 1.02	0.0	1.5	22.9	17.9	6.0	0.0
3.20	− 2.52		0.2	7.2	8.4	2.4	
6.40	− 4.03		0.0	2.2	4.3	1.2	
12.80	− 5.53			0.5	1.5	0.5	
25.60	− 7.04			0.1	0.2	0.3	
51.20	− 8.54			0.0	0.0	0.1	
102.40	−10.05					0.0	

§ 3. Einheiten und Klassenzählung.

Es ist für die weitere Behandlung von Vorteil, eine Zählweise der vorkommenden Größen zu wählen, welche sich mehrfach in dieser oder ähnlicher Art in neueren stellarstatistischen Arbeiten findet. Es seien hier im genauen Anschluß an meine Arbeit in Nr. 4557 der Astronomischen Nachrichten folgende Bezeichnungen und Einheiten verwandt:

r Entfernung (Einheit entsprechend einer jährlichen Parallaxe von 1″),

i scheinbare Helligkeit (Einheit die scheinbare Helligkeit eines Sterns nullter Größe),

J absolute Helligkeit oder Leuchtkraft (Einheit die absolute Helligkeit eines Sterns, welcher bei einer jährlichen Parallaxe von 1″ die nullte Größe hat. Die Sonne hat sehr nahe diese Leuchtkraft 1),

μ Eigenbewegung in Bogensekunden im Jahr — in gegenwärtiger Arbeit ist speziell mit μ die Querbewegung gemeint,

V absolute Geschwindigkeitskomponente (Einheit diejenige Geschwindigkeit, welche aus der Entfernung 1 gesehen die Eigenbewegung 1 gibt, d. i. ein Erdbahnradius im Jahr oder 4.737 km/sec).

Es werde ferner gesetzt (unter log ist durchweg der Dezimallogarithmus verstanden):

$$\rho = -5.0 \log r$$

$$m = -2.5 \log i \qquad g = -5.0 \log \mu$$

$$M = -2.5 \log J \qquad G = -5.0 \log V. \qquad (1)$$

Es ist dann m die scheinbare Größe in der üblichen Zählweise, M die absolute Helligkeit in Größenklassen ausgedrückt. ρ, g, G können analog als Entfernungsklasse, Eigenbewegungsklasse und Geschwindigkeitsklasse bezeichnet werden. Ich werde den Ziffern für m und M gelegentlich den üblichen Index m (magnitudo), den Ziffern für ρ, g, G den Index c (Klasse) geben.

Es gilt:

$$i = \frac{J}{r^2} \qquad \mu = \frac{V}{r}$$

oder in die Klassenbezeichnung übertragen:

$$M = m + \rho \qquad G = g + \rho. \qquad (2)$$

Die Entfernung ρ', in welcher ein Stern die scheinbare Größe $m = 0$ hat, ist gegeben durch: $\rho' = M$. Die Eigenbewegung in dieser Entfernung, also die oben eingeführte reduzierte Eigenbewegung wird:

$$g' = G - \rho' = G - M = g - m.$$

Die Formel für die Reduktion der Eigenbewegung auf die nullte Größe lautet also in der Klassenbezeichnung:

$$g' = g - m, \qquad (3)$$

was natürlich mit dem oben angeführten Reduktionsfaktor $\mu'/\mu = 10^{0.2m}$ identisch ist. Die Geschwindigkeit und reduzierte Eigenbewegung sind bereits in Tabelle 1 und 3 in Klassen G und g' verwandelt angegeben.

§ 4. Analytische Darstellung der Verteilung der Radialgeschwindigkeiten.

Für jeden Spektraltypus wurde der Zentralwert $\underline{V}$ der Radialgeschwindigkeit, das ist der Wert, unter und über dem 50 Prozent der Radialgeschwindigkeiten liegen, bestimmt. Es ergab sich graphisch:

Tabelle 4.
Zentralwerte der Radialgeschwindigkeit.

Typus	B	A	F	G	K	M
$\underline{V}$ in km/sec	5.1	8.5	11.4	12.3	12.7	12.0
$\underline{G}$	-0^c17	-1^c28	-1^c91	-2^c07	-2.14	-2^c02

Die Zahlen sind gleich in Geschwindigkeitsklassen $\underline{G}$ umgerechnet.

Sitz. Ber. Kgl. Preuss. Akad. d. Wiss. Berlin 1914, 489–510

Die Prozentzahlen aus Tabelle 1, welche die Verteilung der Radialgeschwindigkeiten angeben, wurden nun für jeden Spektraltypus als Funktion von $V/\underline{V}$ oder, was auf dasselbe hinauskommt, von $G-\underline{G}$ aufgetragen. Wie Fig. 1 zeigt, liegen die den verschiedenen Typen

Fig. 1.

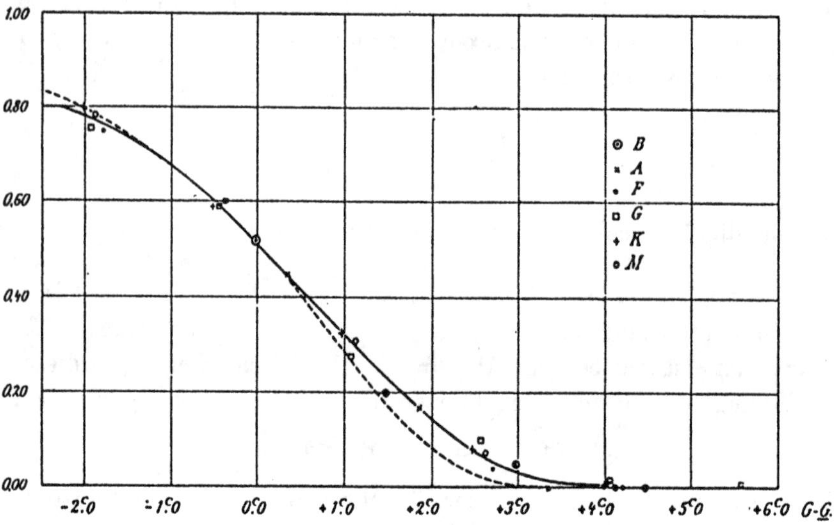

entsprechenden Punkte außerordentlich nahe auf ein und derselben Kurve. Soweit es also die jetzigen Beobachtungen zu beurteilen gestatten, folgt die Verteilung der Radialgeschwindigkeiten für alle Spektraltypen sehr nahe demselben Gesetz, wofern man sie nur nach Vielfachen des — von Typus zu Typus verschiedenen — Zentralwerts mißt. Die Radialgeschwindigkeiten unterhalb des Zentralwerts schmiegen sich nahe einer Gaussschen Verteilung an, die den gleichen Zentralwert hat — die gestrichelte Kurve der Figur 1 stellt eine solche Gausssche Verteilung dar. Dagegen sind bei allen Typen mehr große Geschwindigkeiten vorhanden, als der Gaussschen Verteilung entspricht. Die Verteilung der großen Geschwindigkeiten kann man andererseits sehr gut wiedergeben durch die Annahme, daß die Logarithmen der Geschwindigkeiten oder, was auf dasselbe hinauskommt, die Geschwindigkeitsklassen nach dem Gaussschen Fehlergesetz verteilt sind, indem man also für den Bruchteil der Sterne mit Geschwindigkeiten zwischen den Klassen G und $G+dG$ einen Ausdruck ansetzt:

$$\frac{\gamma}{\sqrt{\pi}} e^{-\gamma^2(G-G_0)^2}\, dG \quad (\gamma \text{ und } G_0 \text{ Konstante}).$$

Ich will eine GAUSSsche Verteilung der Logarithmen oder Klassen kurz als »logarithmische Verteilung« bezeichnen. Solche logarithmische Verteilungen sind für die weitere Rechnung sehr bequem. Deswegen habe ich mich bemüht, die gesamte Verteilung auch für Geschwindigkeiten unterhalb des Zentralwerts durch Kombination logarithmischer Verteilungen darzustellen und bin durch Versuche zu folgender Formel gelangt. Es sei $\Psi(G)\,dG$ die zwischen den Klassen G und $G+dG$ liegende relative Sternanzahl. Dann ist:

$$\Psi(G) = 0.216\, e^{-0.213(G-\underline{G}+0.45)^2} + 0.044\, e^{-0.213(G-\underline{G}-4.0)^2}. \quad (4)$$

Dies sind zwei logarithmische Verteilungen mit gleicher Streuung (gleichem γ), aber verschiedenem Zentralwert ($\underline{G}-0.45$ bzw. $\underline{G}+4.0$) superponiert. Als relative Anzahl $\Sigma(G)$ der über der Geschwindigkeit V oder der entsprechenden Klasse G liegenden Sterne ergibt sich daraus:

$$\Sigma(G) = \int_{-\infty}^{G} \Psi(G)\,dG = 0.83\, W[0.46(G-\underline{G}+0.45)] + 0.17\, W[0.46(G-\underline{G}-4.0)], \quad (5)$$

wobei W das Wahrscheinlichkeitsintegral

$$W(x) = \frac{1}{\sqrt{\pi}} \int_{-\infty}^{x} e^{-x^2}\,dx \quad (6)$$

bedeutet. Es ist dafür gesorgt, daß $\Sigma(G)$ für $G = +\infty$ gleich 1 wird, wie es die Definition relativer Häufigkeiten verlangt. Die Funktion $\Sigma(G)$ soll die Zahlen aus der zweiten Hälfte der Tabelle 1 oder auch die aus allen Spektraltypen im Durchschnitt folgende, in Fig. 1 gezeichnete Kurve, wiedergeben.

Vor Ausführung der Vergleichung sei noch eine Verlängerung der beobachteten Kurven zu kleinen Werten von V ausgeführt unter einer Voraussetzung, die nur falsch wäre, wenn die Sterne deutlich in zwei Gruppen sich uns nähernder oder von uns entfernender Sterne zerfielen. Es sei nämlich angenommen, daß sich die kleinen Geschwindigkeiten auf gleiche Geschwindigkeitsintervalle gleichförmig verteilen. Für die Funktionen $\Psi(G)$ und $\Sigma(G)$ besagt diese Voraussetzung folgendes. Es ist nach (1):

$$\Psi(G)\,dG = 5 \log e\, \Psi(G) \frac{dV}{V}.$$

Es gibt also $5 \log e\, \Psi(G)/V$ die Verteilung der Geschwindigkeiten auf gleiche Intervalle dV. Dieser Ausdruck soll für kleine V (das sind große G) konstant sein. Mithin für große G:

$$\Psi(G) = \frac{V}{5 \log e} \cdot \mathrm{const} = \frac{10^{-0.2\,G}}{5 \log e} \cdot \mathrm{const}$$

und daraus:

$$\Sigma(G) = \int_{-\infty}^{G} \Psi(G)\, dG = \text{const} \cdot 10^{-0.2\,G}.$$

Die Konstante wurde natürlich so gewählt, daß Anschluß bei den größten beobachteten Werten von G hergestellt wurde.

Die folgende Tabelle gibt die aus Fig. 1 entnommenen und auf diese Weise noch extrapolierten beobachteten Werte verglichen mit der Rechnung nach Formel (5).

Tabelle 5.

Beobachtete Verteilung der Radialgeschwindigkeiten verglichen mit Formel (5).

$V/\underline{V}$	$G-\underline{G}$	B	R	$B-R$	$V/\underline{V}$	$G-\underline{G}$	B	R	$B-R$
0.031	+7.52	98.4	99.8	−1.4	0.71	+0.75	64.0	65.2	−1.2
0.044	+6.77	97.7	99.4	−1.7	1.00	0.00	50.7	51.7	−1.0
0.062	+6.02	96.7	98.4	−1.7	1.41	−0.75	36.0	35.0	+1.0
0.088	+5.26	95.4	96.5	−1.1	2.00	−1.51	22.3	20.4	+1.9
0.125	+4.51	93.4	93.6	−0.2	2.82	−2.25	10.5	9.9	+0.6
0.176	+3.76	90.7	90.3	+0.4	4.00	−3.01	3.2	4.0	−0.8
0.25	+3.01	86.6	86.4	+0.2	5.65	−3.76	1.0	1.3	−0.3
0.35	+2.25	81.5	81.9	−0.4	8.00	−4.51	0.5	0.3	+0.2
0.50	+1.51	73.5	75.2	−1.7					

Fig. 2.

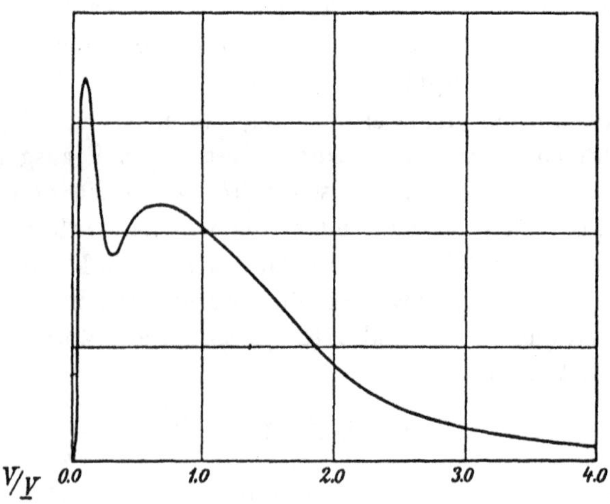

Die aus (4) oder (5) resultierende Verteilung V von auf gleiche Intervalle dV, also die Funktion $\frac{1}{V}\Psi(G)$, ist in Fig. 2 aufgezeichnet. Für kleine V pendelt unsre interpolatorische Darstellung noch beträchtlich, statt, wie zu fordern ist, sich einer Parallelen zur Abszissenachse anzunähern. Indessen schien die Summenfunktion $\Sigma(G)$ nach den Resten $B-R$ in Tab. 5 sich den beobachteten Werten genügend anzuschließen, um vorläufig bei dieser Darstellung stehenbleiben zu können.

§ 5. Theorie[1]. Die Anzahl der Sterne in der Kugelschale zwischen den Entfernungsklassen ρ und $\rho+d\rho$ sei:

$$\Delta(\rho)\,d\rho\,.$$

Unter diesen Sternen seien von der absoluten Größe M bis $M+dM$ der Bruchteil

$$\Phi(M)\,dM$$

und unter diesen wieder von der Geschwindigkeitsklasse G bis $G+dG$ der Bruchteil:

$$\Psi(G)\,dG\,.$$

Da Φ und Ψ relative Anzahlen ausdrücken sollen, ist:

$$\int_{-\infty}^{+\infty}\Phi(M)\,dM = 1\,. \qquad \int_{-\infty}^{+\infty}\Psi(G)\,dG = 1\,. \tag{7}$$

Die Beziehung zwischen $\Delta(\rho)$ und der Anzahl $D(r)$ der Sterne in der Volumeneinheit in der Entfernung r ist offenbar:

$$\Delta(\rho)\,d\rho = -4\pi\,D(r)\,r^2\,dr\,,$$

und da nach (1):

$$d\rho = -5\log e\,\frac{dr}{r}$$

ist, so folgt:

$$\frac{4\pi}{5\log e}\,r^3 D(r) = \Delta(\rho)\,. \tag{8}$$

Die Anzahl der Sterne in der Kugelschale ρ bis $\rho+d\rho$, der Größe M bis $M+dM$ und der Geschwindigkeitsklasse G bis $G+dG$ wird:

$$\Delta(\rho)\,\Phi(M)\,\Psi(G)\,d\rho\,dM\,dG\,.$$

Faßt man nun die Sterne einer bestimmten scheinbaren Größe m bis $m+dm$ und einer bestimmten Eigenbewegung g bis $g+dg$ ins Auge, so ist, infolge der Beziehungen (2): $M = m+\rho$, $G = g+\rho$, die Gesamtzahl dieser Sterne zwischen den Entfernungen ρ bis $\rho+d\rho$ gleich:

$$\Delta(\rho)\,\Phi(m+\rho)\,\Psi(g+\rho)\,d\rho\,dm\,dg\,.$$

---

[1] Man vergleiche Astr. Nachr. Nr. 4557. Die hier gewählte Schreibweise der Ausdrücke nähert sich mehr der von Hrn. C. V. L. CHARLIER in seinen Studies on Stellar Statistics (Lund 1912) benutzten.

Die Gesamtzahl der Sterne aller Entfernungen der scheinbaren Größe m bis $m+dm$ und der Eigenbewegung g bis $g+dg$ wird daher:

$$b_{m,g}\,dm\,dg = dm\,dg\int_{-\infty}^{+\infty}\Delta(\rho)\,\Phi(m+\rho)\,\Psi(g+\rho)\,d\rho\,.$$

Diese Anzahl ist identisch mit der Zahl der Sterne der Größe m, die die reduzierte Eigenbewegung $g' = g - m$ haben. Führen wir der Deutlichkeit wegen einen neuen Buchstaben b' für diese Anzahl ein, so ist:

$$b'_{m,g'}\,dm\,dg' = dm\,dg'\int_{-\infty}^{+\infty}\Delta(\rho)\,\Phi(m+\rho)\,\Psi(g'+m+\rho)\,d\rho\,. \qquad (9)$$

Das ist die gesuchte allgemeine Formel für die Verteilung der reduzierten Eigenbewegungen unter den Sternen der scheinbaren Größe m. Wir haben b' und Ψ aus der Beobachtung bestimmt, $\Delta(\rho)$ entnehmen wir anderweitigen Daten. Es bleibt dann eine Integralgleichung für Φ zu lösen.

Eine Bemerkung, die sich an Gleichung (9) knüpft, ist vorauszunehmen. H. von Seeliger hat gezeigt, daß man die Dichte in den hier in Frage kommenden Teilen des Sternsystems sehr nahe einer (negativen) Potenz von r proportional setzen kann. Daraus folgt, daß $\Delta(\rho)$ von der Form $ce^{\lambda\rho}$ wird, wo c und λ Konstanten sind. Bei dieser Form von $\Delta(\rho)$ kann man aber $b'_{m,g'}$ schreiben

$$b'_{m,g'} = \text{const.}\ e^{-\lambda m}\int_{-\infty}^{+\infty}e^{\lambda(m+\rho)}\,\Phi(m+\rho)\,\Psi(g'+m+\rho)\,d\rho$$

oder indem man $m+\rho$ durch eine neue Variable ersetzt, die man wieder ρ nennen kann:

$$b'_{m,g'} = \text{const.}\ e^{-\lambda m}\int_{-\infty}^{+\infty}e^{\lambda\rho}\,\Phi(\rho)\,\Psi(g'+\rho)\,d\rho\,.$$

Das Integral ist eine Funktion von g', welche m nicht enthält, d. h. die Verteilung der reduzierten Eigenbewegungen ist bis auf einen von der scheinbaren Sterngröße abhängigen Proportionalitätsfaktor für alle scheinbaren Sterngrößen dieselbe, die prozentuale Verteilung wird identisch für alle scheinbaren Sterngrößen. Wir dürfen daher unsere Auszählung der Eigenbewegungen, die alle Sterne bis etwa zur Größe 6.3 zusammen betrifft, so behandeln, als ob sie nur Sterne einer einzigen bestimmten scheinbaren Größe beträfe. Da die ganze Dichteänderung in dem für uns in Betracht kommenden Gebiet nur gering ist, gilt derselbe Satz noch angenähert, auch wenn $D(r)$ nicht genau von der Seeligerschen Form genommen wird, und zwar namentlich dann, wenn man die gefundene

Verteilung als für die mittlere Größe der abgezählten Sterne (etwa $5^{\mathrm{m}}.7$) für den Bossschen Katalog) gültig annimmt. Demgemäß wird unten verfahren werden.

Wir wollen einige spezielle Ausführungen der allgemeinen Formeln vornehmen für die besonders bequemen logarithmischen Verteilungsfunktionen:

$$\begin{aligned} \Delta(\rho) &= a e^{-\alpha^2(\varrho-\varrho_0)^2} \\ \Phi(M) &= \frac{\beta}{\sqrt{\pi}} e^{-\beta^2(M-M_0)^2} \\ \Psi(G) &= \frac{\gamma}{\sqrt{\pi}} e^{-\gamma^2(G-G_0)^2}. \end{aligned} \tag{10}$$

ρ_0, M_0, G_0 sind dabei die Zentralwerte, zu denen die Verteilung symmetrisch ist; $\frac{1}{\sqrt{2\alpha^2}}$, $\frac{1}{\sqrt{2\beta^2}}$, $\frac{1}{\sqrt{2\gamma^2}}$ geben die dem mittleren Fehler entsprechende »mittlere Streuung« um den Zentralwert an. Die Bedingungen (7) sind erfüllt. Für $D(r)$ findet man aus (8) durch einfache Umsetzung:

$$D(r) = A\, e^{-\alpha^2(\varrho-\varrho_1)^2}, \tag{11}$$

wobei gilt:

$$\rho_1 = \rho_0 + \frac{0.3}{\log e}\frac{1}{\alpha^2} \qquad a = \frac{4\pi}{5\log e} A\, 10^{-0.3(\varrho_0+\varrho_1)}. \tag{12}$$

Die Einführung der logarithmischen Ansätze in (9) gibt:

$$b'_{m,g'} = \frac{a\beta\gamma}{\pi}\int_{-\infty}^{+\infty} d\rho\, e^{-P},$$

wobei P den Wert hat:

$$P = \alpha^2(\rho-\rho_0)^2 + \beta^2(m+\rho-M_0)^2 + \gamma^2(g'+m+\rho-G_0)^2.$$

Daraus wird durch einfache Umsetzung:

$$P = \varkappa^2(\rho-\rho_2)^2 + \sigma^2(g'-g'_0)^2 + \tau^2(m-m_0)^2,$$

wobei die Abkürzungen eingeführt sind:

$$\varkappa^2 = \alpha^2+\beta^2+\gamma^2 \qquad \sigma^2 = \frac{(\alpha^2+\beta^2)\gamma^2}{\varkappa^2} \qquad \tau^2 = \frac{\alpha^2\beta^2}{\alpha^2+\beta^2} \tag{13}$$

$$\begin{aligned} \rho_2 &= \frac{\alpha^2\rho_0 + \beta^2(M_0-m) + \gamma^2(G_0-g'-m)}{\varkappa^2} \\ g'_0 &= G_0 - M_0 - \frac{\alpha^2}{\alpha^2+\beta^2}(m+\rho_0-M_0) \\ m_0 &= M_0 - \rho_0. \end{aligned} \tag{14}$$

Sitz. Ber. Kgl. Preuss. Akad. d. Wiss. Berlin 1914, 489–510

Die Ausführung der Integration nach ρ gibt:

$$b'_{m,g'} = \frac{a}{\sqrt{\pi}} \frac{\beta\gamma}{\varkappa} e^{-\sigma^2(g'-g'_0)^2 - \tau^2(m-m_0)^2}. \tag{15}$$

Die Anzahl aller Sterne der Größe m bis $m+dm$, welche $b_m dm$ heißen möge, erhält man, indem man hier über alle Werte von g' integriert:

$$b_m = \frac{a}{\alpha} \tau e^{-\tau^2(m-m_0)^2}. \tag{16}$$

Wenn man logarithmiert und an Stelle von a und ρ_0 nach (12) A und ρ_1 einführt, erhält man die bei kleinen Werten von α für die numerische Rechnung geeignetere Formel:

$$\begin{aligned}\log b_m = \log\left(\frac{4\pi}{5\log e} A\right) + \frac{1}{2}\log\frac{\beta^2}{\alpha^2+\beta^2} + \frac{0.09}{\log e}\frac{1}{\alpha^2+\beta^2} \\ -0.6\,\rho_1 + \frac{0.6\,\beta^2}{\alpha^2+\beta^2}(m+\rho_1-M_0) - \log e\,\frac{\alpha^2\beta^2}{\alpha^2+\beta^2}(m+\rho_1-M_0)^2.\end{aligned} \tag{17}$$

Für die relative Häufigkeit der Eigenbewegungen unter den Sternen der scheinbaren Größe m erhält man aus (15) und (16):

$$\frac{b'_{m,g'}}{b_m} = \frac{\sigma}{\sqrt{\pi}} e^{-\sigma^2(g'-g'_0)^2}. \tag{18}$$

Es sind also die reduzierten Eigenbewegungsklassen nach dem Gaussschen Gesetz verteilt mit der mittleren Streuung $\frac{1}{\sqrt{2\sigma^2}}$. Die Formel (13) für σ^2 kann man so schreiben:

$$\frac{1}{2\sigma^2} = \frac{1}{2\gamma^2} + \frac{1}{2(\alpha^2+\beta^2)}.$$

Im Falle $\alpha = 0$, konstanter Dichte, ist hiernach das Quadrat der mittleren Streuung der Eigenbewegung gleich der Summe dieser Quadrate für die absoluten Geschwindigkeiten und die Leuchtkräfte.

Es seien noch die auf die mittlere Parallaxe der Sterne jeder scheinbaren Größe m bezüglichen Formeln hinzugefügt.

Die Anzahl der Sterne der scheinbaren Größe m bis $m+dm$ in der Entfernung ρ bis $\rho+d\rho$ ist allgemein:

$$\Delta(\rho)\,\Phi(m+\rho)\,d\rho\,dm$$

oder nach den logarithmischen Ansätzen:

$$\frac{a\beta}{\sqrt{\pi}} e^{-Q}\,d\rho\,dm, \tag{19}$$

wobei Q den Wert hat:

$$Q = \alpha^2(\rho - \rho_0)^2 + \beta^2(m + \rho - M_0)^2$$

und umgesetzt werden kann in die Form:

$$Q = (\alpha^2 + \beta^2)(\rho - \rho_3)^2 + \tau^2(m - m_0)^2, \qquad (20)$$

wobei

$$\rho_3 = \frac{\alpha^2 \rho_0 + \beta^2(M_0 - m)}{\alpha^2 + \beta^2}$$

ist.

Man sieht aus (19) und (20), daß die Sterne der scheinbaren Größe m bei den logarithmischen Ansätzen in bezug auf Entfernungsklassen nach dem GAUSSschen Gesetz verteilt sind. Der Zentralwert ist ρ_3, die mittlere Streuung $\frac{1}{\sqrt{2(\alpha^2+\beta^2)}}$. Der Entfernungsklasse ρ_3 entspricht eine wahrscheinliche Parallaxe der Sterne der scheinbaren Größe m:

$$\log \underline{\pi}_m = 0.2\rho_3 = 0.2\frac{\alpha^2 \rho_0 + \beta^2 M_0}{\alpha^2 + \beta^2} - 0.2\frac{\beta^2}{\alpha^2+\beta^2} m \qquad (21)$$

und der mittlere Fehler des Logarithmus dieser Parallaxe wird:

$$\frac{0.2}{\sqrt{2(\alpha^2+\beta^2)}} \qquad (22)$$

Die mittlere Parallaxe der Sterne der scheinbaren Größe m ergibt sich aus der Formel:

$$\bar{\pi}_m = \frac{\int_{-\infty}^{+\infty} \frac{1}{r}\Delta(\rho)\Phi(m+\rho)\,d\rho}{\int_{-\infty}^{+\infty} \Delta(\rho)\Phi(m+\rho)\,d\rho},$$

deren Ausführung für die logarithmischen Verteilungsfunktionen den einfachen Zusammenhang zwischen mittlerer und wahrscheinlicher Parallaxe gibt:

$$\log \bar{\pi}_m = \log \underline{\pi}_m + \frac{0.01}{\log e}\frac{1}{\alpha^2+\beta^2} \qquad (23)$$

§ 6. Bestimmung der Verteilung der Leuchtkräfte.

Es sollen nun die vorstehenden Rechnungen mit den logarithmischen Verteilungsfunktionen zur Lösung unsres Problems verwertet werden. Für die drei in Betracht kommenden Funktionen $\Delta(\rho)$, $\Psi(G)$, $\Phi(M)$ machen wir folgende Ansätze:

Sitz. Ber. Kgl. Preuss. Akad. d. Wiss. Berlin 1914, 489–510

1. Für $\Delta(\rho)$: Aus Untersuchungen über die Verteilung aller Sterne zusammen — ohne Trennung nach dem Spektraltypus — ergibt sich als ein genäherter Ausdruck für die Anzahl der Sterne in der Volumeneinheit (Astr. Nachr. Nr. 4557): $\log D(r) = 0.488 - 0.097\rho - 0.0088\rho^2$, der sich leicht umsetzen läßt in:

$$D(r) = 5.70\, e^{-0.0203\,(\rho+5.5)^2}.$$

Dieser Ausdruck hat ein Maximum von 5.70 für $\rho = -5.5$ ($r = 12.6$, $\pi = 0''.080$). Dieses Maximum ist als die in unsrer Umgebung herrschende Dichte aufzufassen. Das Absinken von $D(r)$ nach kleinerem r zu, welches die Formel gibt, ist eine bedeutungslose und unschädliche Extrapolation.

Ich will nun annehmen, daß sich das Mischungsverhältnis der Spektraltypen in den in Betracht kommenden Entfernungen nicht ändert, daß die Dichte der Sterne jedes Spektraltypus der Dichte aller Sterne zusammen proportional ist, daß mithin für jeden einzelnen Spektraltypus unter Einführung einer besonderen Konstanten A gilt:

$$D(r) = A e^{-0.0203\,(\rho+5.5)^2}. \tag{24}$$

Dabei ist A wieder als die Dichte in unserer Umgebung aufzufassen. Damit sind die Konstanten α und ρ_1 in (11) festgelegt zu:

$$\alpha^2 = 0.0203 \qquad \rho_1 = -5.5\,, \tag{24'}$$

und es folgt aus (12):

$$\rho_0 = -39.6 \qquad \log a = \log A + 14.30$$
$$\Delta(\rho) = A \cdot 10^{14.30} \cdot e^{-0.0203\,(\rho+39.6)^2}.$$

2. Für $\Psi(G)$ genügte zur Darstellung der beobachteten Verteilung der Radialgeschwindigkeiten nicht die einfache logarithmische Verteilung, vielmehr war die Superposition zweier solcher nötig. Es ergab sich die Formel (4), die wir so schreiben können:

$$\Psi(G) = p_1 \frac{\gamma}{\sqrt{\pi}} e^{-\gamma^2 (G-G_1)^2} + p_2 \frac{\gamma}{\sqrt{\pi}} e^{-\gamma^2 (G-G_2)^2} \qquad p_1 + p_2 = 1 \tag{25}$$

mit den numerischen Werten:

$$\gamma = 0.46 \qquad p_1 = 0.83 \qquad p_2 = 0.17$$
$$G_1 = \underline{G} - 0.45 \qquad G_2 = \underline{G} + 4.0\,. \tag{25'}$$

3. Was die Verteilung der Leuchtkräfte $\Phi(M)$ angeht, so wollen wir den Versuch machen, die einfache logarithmische Verteilung:

$$\Phi(M) = \frac{\beta}{\sqrt{\pi}} e^{-\beta^2 (M - M_0)^2} \tag{26}$$

beizubehalten. Die Auflösung unsrer Integralgleichung reduziert sich dann auf eine Bestimmung solcher Werte der Konstanten β und M_0, welche die beobachtete Verteilung der Eigenbewegungen möglichst gut wiedergeben.

Mit diesen Ansätzen der Verteilungsfunktionen bleiben die Formeln (16), (17), (21) bis (23) für die Anzahl und die mittlere Parallaxe der Sterne jeder Größe ungeändert. Die Formel (18) für die Verteilung der Eigenbewegungen geht über in die Summe zweier logarithmischer Verteilungen:

$$\frac{b'_{m,g'}}{b_m} = p_1 \frac{\sigma}{\sqrt{\pi}} e^{-\sigma^2 (g' - g'_1)^2} + p_2 \frac{\sigma}{\sqrt{\pi}} e^{-\sigma^2 (g' - g'_2)^2}, \tag{27}$$

wobei ist:

$$\sigma^2 = \frac{(\alpha^2 + \beta^2)\gamma^2}{\alpha^2 + \beta^2 + \gamma^2}, \quad g'_1 = G_1 - M_0 - \frac{\alpha^2}{\alpha^2 + \beta^2}(m + \rho_0 - M_0), \quad g'_2 = g'_1 + G_2 - G_1. \tag{28}$$

Die prozentuale Summe aller reduzierten Eigenbewegungen über g' wird:

$$\frac{1}{b_{m_0}} \int_{-\infty}^{g'} b'_{m,g'} \, dg' = p_1 W[\sigma(g' - g'_1)] + p_2 W[\sigma(g' - g'_2)]. \tag{29}$$

Dieser Ausdruck ist mit den in Tabelle 3 gegebenen beobachteten Anzahlen zu vergleichen, und zwar ist, wie gesagt, durch geeignete Wahl von β und M_0 möglichster Anschluß herzustellen. Statt β und M_0 kann man auch die unmittelbar davon abhängenden Konstanten σ und g'_1 zuerst zu bestimmen suchen. Das läßt sich zeichnerisch leicht ausführen. Der Ausdruck (29) lautet unter Benutzung der numerischen Werte aus (25'):

$$0.83\, W[\sigma(g' - g'_1)] + 0.17\, W[\sigma(g' - g'_1 + 4.45)]. \tag{30}$$

Mit Hilfe einer Tabelle des Wahrscheinlichkeitsintegrals habe ich diesen Ausdruck für eine Anzahl Werte von σ als Funktion von $g' - g'_1$ aufgetragen und dann durch Auswahl aus den entstehenden Kurven und gleichzeitige Parallelverschiebung längs der Abszissenachse (Veränderung von g'_1) möglichsten Anschluß an die beobachteten Zahlen herbeizuführen gesucht.

Die gefundenen Werte von σ und g'_1 und den Vergleich zwischen Rechnung und Beobachtung gibt folgende Tabelle.

Sitz. Ber. Kgl. Preuss. Akad. d. Wiss. Berlin 1914, 489–510

Tabelle 6.

Darstellung der beobachteten Verteilung der Querbewegungen durch Formel 30.

Typus	*B*			*A*			*F*			*G*			*K*			*M*		
σ	0.32			0.35			0.30			0.225			0.30			0.36		
g'_1	+5.75			+3.00			+0.48			+1.55			+2.22			+2.70		
g'	*B*	*R*	*B−R*	*B*	*R*	*B−R*	*B*	*R*	*B−R*	*B*	*R*	*B−R*	*B*	*R*	*B−R*	*B*	*R*	*B−R*
6ᵐ50	50	53.5	−3.5	83.8	85.0	−1.2												
5.00	32	30.7	+1.3	71.3	71.5	−0.2	86.7	89.7	−3.0	80.4	78.5	+1.9	76.6	77.0	−0.4	76.7	75.4	+1.3
3.50	14	12.8	+1.2	50.5	49.9	+0.6	79.7	79.5	+0.2	65.0	64.7	+0.3	61.8	59.8	+2.0	54.7	55.1	−0.4
1.99	2	3.7	−1.7	25.3	25.9	−0.6	65.1	63.6	+1.5	47.3	47.8	−0.5	38.4	38.6	−0.2	32.5	30.1	+2.4
0.49	0.3	0.7	−0.4	7.9	8.7	−0.8	43.4	42.4	+1.0	29.9	31.2	−1.3	17.9	19.2	−1.3	9.4	11.0	−1.6
− 1.02	0.0	0.1	−0.1	1.5	1.9	−0.4	22.9	21.9	+1.0	17.9	17.3	+0.6	6.0	7.1	−1.1	0.0	2.4	−2.4
− 2.52				0.2	0.3	−0.1	7.2	8.4	−1.2	8.4	8.0	+0.4	2.4	1.8	+0.6	0.0	0.3	−0.3
− 4.03				0.0	0.0	0.0	2.2	2.3	−0.1	4.3	3.1	+1.2	1.2	0.3	+0.9			
− 5.53							0.5	0.5	0.0	1.5	1.0	+0.5	0.5	0.0	+0.5			
− 7.04							0.1	0.0	+0.1	0.2	0.2	0.0	0.3		+0.3			
− 8.54							0.0		0.0	0.0	0.0	0.0	0.1		0.1			
−10.05													0.0		0.0			

Die Darstellung ist befriedigend genug, um die Gausssche Verteilung der Größenklassen für das unten noch näher abzugrenzende, überhaupt in Betracht kommende Größenbereich als eine brauchbare Annäherung an die Verteilung der Sterne jedes Spektraltypus nach ihrer absoluten Größe ansehen zu können.

Nachdem σ und g'_1 bekannt sind, kann man mit den in Tab. 4 wie in (24'), (25') angeführten Werten aller in Betracht kommenden Konstanten und der mittleren Größe der Sterne des Bossschen Katalogs $m = 5.7$ aus den Gleichungen (28) ohne weiteres M_0 und β berechnen. M_0 ist, wie wiederholt sei, die mittlere absolute Größe, $\frac{1}{\sqrt{2\beta^2}}$ die mittlere Streuung um diese absolute Größe. Die Werte für jeden Spektraltypus gibt Tab. 7.

Tabelle 7.

	Mittlere Größe M_0	β	Mittl. Streuung $1/\sqrt{2\beta^2}$	Log. der mittl. Parallaxe $\log \pi_m$	$\log \pi_m/\underline{\pi}_m$	Mittl. Fehler von $\log \underline{\pi}_m$
B	−3.22	0.421	1ᵐ68	−1.276 − 0.179 m	0.116	±0.32
A	−2.52	0.520	1.36	−0.947 − 0.186 m	0.079	0.26
F	+1.81	0.370	1.91	−0.569 − 0.174 m	0.147	0.36
G	+9.15	0.217	3.29	−0.817 − 0.139 m	0.347	0.55
K	−0.46	0.370	1.91	−0.963 − 0.174 m	0.147	0.36
M	−3.31	0.558	1.26	−1.035 − 0.188 m	0.067	0.24

§ 7. Mittlere Parallaxe der Sterne von jedem Spektraltypus.

Durch die Bestimmung von M_0 und β ist man in den Stand gesetzt, auch die mittlere und wahrscheinliche Parallaxe der Sterne von bestimmter scheinbarer Größe m nebst dem mittleren Fehler ihres Logarithmus für jeden Spektraltypus zu berechnen. Man hat nur die gefundenen Zahlenwerte in Formel (21)—(23) einzusetzen. Die Ergebnisse sind in Tab. 7 aufgenommen. Es ist bemerkenswert, wie sicher die Parallaxe für Sterne von bekanntem Spektraltypus — vom G-Typus abgesehen — aus der scheinbaren Helligkeit abgeleitet werden kann.

Eine noch wesentlich sicherere Parallaxe würde natürlich bei Mitberücksichtigung der Eigenbewegung gefunden werden können. Indessen wird man hierfür am besten die Komponente der Eigenbewegung nach dem Apex heranziehen, welche die parallaktische Verschiebung enthält, und nicht die vorstehend benutzte Querbewegung. Ich möchte deshalb auf diese Frage bei Behandlung der andern Komponente der Eigenbewegung zurückkommen.

§ 8. Anzahl der Sterne von jedem Spektraltypus.

Durch die Bestimmung der mittleren Größe und mittleren Streuung der Sterne jedes Spektraltypus ist die relative Häufigkeit der Sterne verschiedener Leuchtkraft in der Volumeneinheit festgelegt. Um unsere Aufgabe zu Ende zu führen, bedarf es nun noch der Bestimmung der Gesamtzahl der Sterne von jedem Spektraltypus in der Volumeneinheit, der Konstante A. Mit den im vorstehenden bestimmten sonstigen Konstanten rechnet man aus Formel (17) folgende Ausdrücke für die Anzahl der Sterne der scheinbaren Größe m am Himmel:

Typus						
B	$\log b_m = \log A$	$+3.82$	$+0.574\,m$	$-0.00790\,m^2$		
A	» = »	$+3.02$	$+0.607$	-0.00818		
F	» = »	$+1.12$	$+0.634$	-0.00766		
G	» = »	-0.30	$+0.594$	-0.00609	(31)	
K	» = »	$+2.52$	$+0.599$	-0.00766		
M	» = »	$+3.39$	$+0.599$	-0.00826 .		

Wir bestimmen A, indem wir mit der beobachteten Anzahl der Sterne 5.0ter Größe vergleichen. Für alle Sterne zusammen hat man sehr nahe (vgl. A. N. Nr. 4422)

$$\log b_m = 0.596 + 0.5612\,m - 0.0055\,m^2 .$$

Sitz. Ber. Kgl. Preuss. Akad. d. Wiss. Berlin 1914, 489–510

Daraus folgt für $m = 5$:

$$\log b_5 = 3.214. \qquad b_5 = 1640.$$

Anderseits finden sich aus den Angaben von Hrn. E. C. Pickering in Harv. Ann. Vol. 64, S. 144 als Prozentsätze des Vorkommens der einzelnen Spektraltypen unter den mit bloßem Auge sichtbaren Sternen (bis $6^m.25$) folgende Zahlen:

B	*A*	*F*	*G*	*K*	*M*
0.117	0.308	0.118	0.099	0.281	0.075.

Multipliziert man diese Zahlen mit 1640, so erhält man die Anzahl b_5 für jeden einzelnen Spektraltypus, wie in Tabelle 8 angegeben. Der Vergleich mit den Formeln (31) für $m = 5$ ergibt sofort A. Die Zahlen A bedeuten definitionsgemäß die Anzahl Sterne im Würfel, dessen Seite einer Parallaxe von 1" entspricht. Um bequemere Zahlen zu haben, ist noch der Wert $\frac{4\pi}{3}\,1000\,A$ angegeben, welcher die Zahl der Sterne mit Parallaxen über $0''.1$ bezeichnet.

Tabelle 8.

	B	*A*	*F*	*G*	*K*	*M*
b_5	192	500	194	162	461	123
log *A*	−4.21	−3.14	−1.81	−0.31	−2.67	−4.09
Zahl der Sterne mit Parallaxe > $0''.10$	0.26	3.0	65	2000	9.0	0.34

Es sei gleich davor verwarnt, diese Gesamtzahlen der Sterne jedes Spektraltypus für mehr als Rechnungsgrößen zu nehmen.

§ 8. Grenzen der Gültigkeit der Verteilungsformeln. Ergebnis.

Die aufgestellten Gaussschen Verteilungen für die absoluten Größen der Sterne jedes Spektraltypus können nämlich nur über ein begrenztes Größenbereich beanspruchen, eine Annäherung an die wirkliche Zahl der Sterne zu geben. Nach großen Werten von M, dunklen Sternen, hin läßt sich die Grenze folgendermaßen abschätzen. Der Bossche Katalog, aus dem wir die Eigenbewegungen entnehmen, reicht rund bis zur scheinbaren Größe $m = 6$. Denken wir uns Sterne von Sonnenhelligkeit, $M = 0$. Diese werden dann bei der Abzählung der Eigenbewegungen vorkommen bis zur Entfernungsklasse $\rho = -6$. In größeren Entfernungen werden diese Sterne zu schwach werden, um im Bossschen Katalog aufzutreten. Das Volumen der Kugel $\rho = -6$ ist das 4fache des Volumens der Kugel $\rho = -5$, welche der Parallaxe 0."1 entspricht. Nun sind die Abzählungen der Eigenbewegungen, wie sie Tabelle 2 enthält, jedenfalls auf ± 4 Sterne zufällig. Man darf daher für jeden Spektraltypus auf die Kugel der Parallaxe 0."1 sicher einen Stern der Leuchtkraft $M = 0$ zufügen oder wegnehmen, ohne die beobachtete Verteilung der Eigenbewegungen merklich zu ändern. Für Sterne, die um 1, 2 usw. Größen dunkler sind als die Sonne ($M = +1$, $+2$ usw.) wird die Zahl der Sterne für die Kugel der Parallaxe 0."1 nach einer entsprechenden Überlegung unbestimmt um 4, 16 usw. Sterne, so daß die aus den Verteilungsfunktionen hervorgehenden Zahlen für dunklere Sterne bald unbrauchbar werden. Für Sterne von größerer Leuchtkraft als die Sonne nimmt die Unbestimmtheit hingegen um einen Faktor 4 für jede Größenklasse ab und wird dadurch mehr und mehr einflußlos. Für die sehr hellen Sterne versagen die Verteilungsfunktionen aber aus einem andern Grunde. Eigenbewegungen unter 0."005 für Sterne fünfter Größe ($g = 11.5$), die, auf die nullte Größe reduziert, der Klasse $g' = 6.5$ entsprechen, können ihrem Betrage nach nicht mehr verbürgt werden, es können also solche Eigenbewegungen immer als beliebig klein angesetzt und beliebig entfernten Sternen zugeschrieben werden. Je entfernter man die Sterne wählt, um so weniger kommen auf die Volumeneinheit. Da die absolute Geschwindigkeit der Sterne im Mittel von der Klasse $\underline{G} = -1$ ist, so wird diese Unbestimmtheit etwa bei Sternen der absoluten Helligkeit $M_0 = \underline{G} - g' = -7.5$ beträchtlich.

Nach diesen Überlegungen ist also z. B. für den G-Typus das Maximum der Häufigkeit, das nach der gefundenen GAUSSschen Verteilung der absoluten Größen bei $M_0 = +9.15$, also bei ganz dunklen Sternen eintreten müßte, eine bereits ganz bedeutungslose Extrapolation, brauchbar ist nur ein relativ kleines Stück des Anstiegs der GAUSSschen Kurve, das bei den Sternen über Sonnenhelligkeit liegt.

Innerhalb eines demgemäß abgegrenzten Bereichs der Brauchbarkeit wird die Bedeutung der gefundenen Verteilungsfunktionen wohl am deutlichsten, wenn die Anzahl der Sterne über einer vorgegebenen Helligkeit M mit Parallaxen über 0."1 angegeben wird, in Formeln also die Zahl:

$$N_M = \frac{4\pi}{3} 10^3 A \frac{\beta}{\sqrt{\pi}} \int_{-\infty}^{M} e^{-\beta^2 (M - M_0)^2} d\beta . \quad \begin{pmatrix} M_0 \text{ und } \beta \text{ aus Tab. 7.} \\ A \text{ aus Tab. 8.} \end{pmatrix}$$

Das ist in nachstehender Tabelle geschehen, die also das Resultat der Untersuchung wiedergibt:

Sitz. Ber. Kgl. Preuss. Akad. d. Wiss. Berlin 1914, 489–510

Tabelle 9.

Zahl der Sterne heller als die absolute Größe *M*, mit einer Parallaxe über 0."1.

Typus \ *M*	0.0	−2.0	−4.0	−6.0
B	0.3	0.2	0.09	0.013
A	2.9	1.9	0.42	0.016
F	11.2	1.5	0.08	0.001
G	5.1	0.6	0.05	0.003
K	5.5	2.0	0.31	0.017
M	0.3	0.3	0.10	0.006
Summe	25.3	6.5	1.05	0.056

Die Zahlen verlaufen etwa so, wie nach den früheren mehr qualitativen Untersuchungen, insbesondere von E. Hertzsprung[1], zu erwarten ist: Sondert man in der Volumeneinheit alle Sterne aus, welche die Sonne an Leuchtkraft um 4 Größenklassen oder mehr übertreffen, so sind unter diesen Sternen die Spektraltypen in ähnlichem Verhältnis vertreten wie am Himmel unter den mit bloßem Auge sichtbaren Sternen. Dabei besteht die bekannte auffällige Erscheinung, daß die einer mittleren Entwicklungsstufe entsprechenden Typen *F* und *G* seltener sind als die extremeren Typen *A* und *K*. Beschränkt man sich auf die äußerst hellen Sterne, die die Sonne um 6 Größenklassen und mehr übertreffen, so sieht man diese hellen Sterne noch schärfer in zwei Gruppen zerlegt, die der weißen *B*- und *A*-Sterne und die der gelben *K*-Sterne. Faßt man hingegen alle Sterne bis herab zur Sonnenhelligkeit zusammen, so ist deren Verteilung in der Volumeneinheit gänzlich verschieden vom Anblick des Himmels. Die Lücke bei den mittleren Entwicklungsstadien ist verschwunden, es überwiegt der *F*-Typus, die extremen Typen *B* und *M* erscheinen als seltene Ausnahmen.

[1] E. Hertzsprung, Zur Strahlung der Sterne. Archiv für wissenschaftliche Photographie. Bd. III und V.

Ausgegeben am 7. Mai.

Stationäre Geschwindigkeitsverteilung im Sternsystem.

Fragment von K. Schwarzschild (†).

Das Planetensystem wird durch die Gravitation für außerordentlich lange Zeiten stabil in seinem jetzigen Zustand erhalten. Es ist möglich, daß auch im Fixsternsystem der Gravitation dieselbe Rolle zufällt, daß sich durch ihre Wirkung die verhältnismäßig reguläre linsenförmige Gestalt des Sternsystems dauernd erhält. Wie im Planetensystem müssen dann auch im Fixsternsystem die Anfangsgeschwindigkeiten der einzelnen Körper in besonderer Weise geordnet sein. Im Hinblick auf diese Möglichkeit soll nun folgende Aufgabe behandelt werden: Es sei ein Sternhaufen gegeben von ellipsoidischer Begrenzung und gleichförmiger Dichte (gleicher Anzahl und Masse der Sterne in der hinreichend groß genommenen Volumeinheit). Wie müssen die Anfangsgeschwindigkeiten der Sterne in einem solchen Haufen beschaffen sein, damit diese räumliche Verteilung unter der gegenseitigen Attraktion der Sterne dauernd erhalten bleibt?

Zur Begründung und Präzisierung der Aufgabe sei folgendes bemerkt: Der Sternhaufen soll seinen Dimensionen und seiner Massendichte nach natürlich etwa unserem Milchstraßensystem entsprechen. Ohne Gravitation würde die Anordnung des Milchstraßensystems bei den tatsächlichen Geschwindigkeiten der Sterne verhältnismäßig sehr vergänglich sein. Wenn z. B. die Sonne mit ihrer Geschwindigkeit von 20 km/sec geradeaus liefe, würde sie innerhalb 10^8 Jahren eine Strecke durchmessen, die einer Parallaxe von $0''.0005$ entspricht und damit etwa dem Halbmesser des Sternsystems nach SEELIGER gleichkommt. Das Alter von Erde und Sonne ist zweifellos höher als 10^8 Jahre anzusetzen. Daß andererseits die Gravitation des Sternsystems erheblich sein muß, lehrt eine einfache Rechnung wie diese: Man nehme eine Kugel, deren Radius der Parallaxe $0''.003$ entspricht. Die Massendichte innerhalb dieser Kugel sei so bemessen, daß je eine Sonnenmasse auf eine Kugel kommt, deren Radius der Parallaxe $1''$ entspricht. Eine Kreisbewegung an der Oberfläche der großen Kugel unter der Gravitation derselben erfolgt mit 20 km/sec Geschwindigkeit und innerhalb 10^8 Jahren.

Bei den Gravitationswirkungen selbst ist zu unterscheiden zwischen den *regulären Kräften*, die der Gesamtattraktion des Systems ent-

In Festschrift für Hugo von Seeliger (Springer Berlin 1924) 94–105

springen, wenn man die Sternmassen durch eine kontinuierliche Dichteverteilung ersetzt, und den *irregulären Kräften*, die von den sich zufällig nähernden Sternen der Umgebung ausgeübt werden. Letztere Kräfte wirken wie die Zusammenstöße der Moleküle in der Gatheorie, wenn auch bei der Seltenheit erheblicher Annäherungen zwischen zwei Fixsternen außerordentlich viel langsamer. (Direkte Zusammenstöße zwischen zwei Fixsternen sind so selten, daß sie nicht berücksichtigt zu werden brauchen.) Nach der unten folgenden Abschätzung würden die irregulären Kräfte erst nach Zeiten von der Größenordnung 10^{12} Jahren einen merklichen Energieaustausch und erst nach noch längeren Zeiten eine gleichmäßige Verteilung der Energie auf alle Freiheitsgrade zuwege bringen. Da nach dem oben Gesagten die regulären Kräfte viel erheblicher sind, so darf man in erster Annäherung die irregulären Kräfte neben den regulären vernachlässigen und jedem Stern eine bestimmte, nur wenig gestörte — damit ist nicht gemeint geschlossene — Bahn durch das Sternsystem zuschreiben.

Herr Eddington [1]) hat ausgeführt, daß ein unmittelbarer Beweis für das Überwiegen der regulären über die irregulären Kräfte durch die Existenz von Sternströmen in der Art des Hyadenstroms gegeben ist. Wie man aus der Übertragung bekannter Betrachtungen über die Auflösung von Kometen [2]) erkennt, vermögen die regulären Kräfte einen solchen Sternstrom nicht zu zerstören, solange die Dichte des Sternstroms einige Male größer ist als die durchschnittliche Dichte der Sternverteilung, welche er durchsetzt. Die irregulären Kräfte hingegen wirken zerstörend auf ihn. Wenn also solche Sternströme noch jetzt existieren, so beweist dies, daß sich die Wirkung der irregulären Kräfte bisher noch nicht zu einem erheblichen Betrag hat summieren können.

Mit Hilfe unserer Abschätzung der irregulären Kräfte kann man über diesen Schluß Herrn Eddingtons noch hinausgehen, wenn man außerdem annimmt, daß die Sternströme ebenso alt sind wie das ganze Sternsystem, und nach dem Spektraltypus der Sterne des Hyadenstromes liegt kein Grund vor, ihn für besonders jung zu halten. Es folgt dann, daß das Alter des ganzen Sternsystems die Größenordnung von 10^{12} Jahren nicht überschreitet.

Es soll jetzt die eben benutzte Abschätzung der irregulären Kräfte ausgeführt werden. Dieselbe ist namentlich in prinzipieller Hinsicht lückenhaft, dürfte aber doch zu einer richtigen Größenordnung führen.

Man betrachte die Begegnung zweier Sterne der Massen m_1 und m_2. Die Geschwindigkeit ihres Schwerpunktes sei s; die ursprüngliche Relativgeschwindigkeit von m_1 gegen m_2 sei v. Der Winkel zwischen

[1]) Brit. Assoc. 1911. Stellar Distribution and Movements.

[2]) Tisserand: Méc. celeste IV. Chap. XVI.

s und v sei φ. Dann gilt für die Geschwindigkeiten v_1 und v_2 beider Massen vor der Begegnung:

$$v_1^2 = s^2 + 2sv\frac{m_2}{m_1+m_2}\cos\varphi + v^2\left(\frac{m_2}{m_1+m_2}\right)^2$$

$$v_2^2 = s^2 - 2sv\frac{m_1}{m_1+m_2}\cos\varphi + v^2\left(\frac{m_1}{m_1+m_2}\right)^2$$

Die beiden Sterne beschreiben unter ihrer gegenseitigen Anziehung eine Hyperbel umeinander. Dabei bleibt die Schwerpunktsgeschwindigkeit s unverändert. Auch die Relativgeschwindigkeit v kehrt zu ihrem Ausgangswert zurück, nur ist sie aus der Richtung der einen Asymptote in die der anderen getreten. Ist $2w$ der (den Brennpunkt enthaltende) Winkel zwischen den beiden Asymptoten, so wird der Winkel φ' zwischen Relativ- und Schwerpunktsgeschwindigkeit nach der Begegnung: $\varphi' = \varphi + 180° - 2w$. Die Geschwindigkeiten v_1' und v_2' der Sterne nach der Begegnung werden damit:

$$v_1' = s^2 + 2sv\frac{m_2}{m_1+m_2}\cos\varphi' + v^2\left(\frac{m_2}{m_1+m_2}\right)^2$$

$$v_2' = s^2 - 2sv\frac{m_1}{m_1+m_2}\cos\varphi' + v^2\left(\frac{m_1}{m_1+m_2}\right)^2$$

Der Energieaustausch bei der Begegnung ist daher:

$$\Delta\varepsilon = m_1\frac{v_1^2 - v_1'^2}{2} = m_2\frac{v_2'^2 - v_2^2}{2} = sv\frac{m_1 m_2}{m_1+m_2}(\cos\varphi - \cos\varphi')$$

$$= 2sv\frac{m_1 m_2}{m_1+m_2}\sin\frac{\varphi'+\varphi}{2}\cdot\sin\frac{\varphi'-\varphi}{2} = 2sv\frac{m_1 m_2}{m_1+m_2}\sin\frac{\varphi+\varphi'}{2}\cdot\cos w$$

Der Asymptotenwinkel w läßt sich nun noch ausdrücken mittels der Relativgeschwindigkeit v und des Abstandes D der Asymptote vom Brennpunkt, wobei zu beachten ist, daß dieser Abstand zugleich die kleinste Entfernung ist, die die Sterne ohne Ablenkung durch die gegenseitige Gravitation erreicht hätten.

Wir wollen Sonnenmasse und Erdbahnhalbmesser als Einheit wählen und die Zeiteinheit gleich 58 Tagen setzen, so daß die GAUSSsche Konstante gleich 1 und die Einheit der Geschwindigkeit gleich der Erdgeschwindigkeit (30 km/sec) wird.

Bezeichnet man mit a und e die Halbachse und Exzentrizität der Hyperbel, so gilt nunmehr:

$$v^2 = \frac{m_1+m_2}{a}, \qquad \cos w = \frac{1}{e}, \qquad D = ae\sin w = a\,\mathrm{tg}\,w = \frac{m_1+m_2}{v^2}\mathrm{tg}\,w$$

Daraus folgt:

$$\mathrm{tg}\,w = \frac{Dv^2}{m_1+m_2}, \qquad \cos w = \frac{m_1+m_2}{\sqrt{(m_1+m_2)^2 + D^2 v^4}}$$

In Festschrift für Hugo von Seeliger (Springer Berlin 1924) 94–105

Setzt man dies in den obigen Ausdruck für den Energieaustausch ein, so nimmt derselbe die für die weitere Rechnung geeignete Gestalt an:

$$\Delta\varepsilon = 2\,m_1\,m_2 \sin\left(\frac{\varphi+\varphi'}{2}\right)\frac{s\,v}{\sqrt{(m_1+m_2)^2+D^2v^4}}$$

Damit ist die einzelne Begegnung, so weit als hier erforderlich, erledigt, und es ist die Summenwirkung einer großen Reihe von Begegnungen zu bilden. Da für den einzelnen Stern von Begegnung zu Begegnung das Vorzeichen von $\Delta\varepsilon$ wechselt und — wenigstens für Sterne mittlerer Geschwindigkeit — auf beide Vorzeichen gleichmäßig verteilt ist, so hat man die Quadratsumme $\sum(\Delta\varepsilon)^2$ zu bilden und aus dieser die Wurzel zu ziehen.

Bei der Summation der Begegnungen ist ferner noch zu überlegen, bis zu welcher Größe D_0 der Minimaldistanz D man die Begegnungen als solche rechnen will. Es ist dabei zu bedenken, daß von der Anziehung jedes Sternes ein Teil in Abrechnung zu bringen ist, der durch die regulären Kräfte des kontinuierlichen Systems, durch welches wir das wirkliche ersetzen, wiedergegeben wird. Insofern ist also der obige Ausdruck für den Energieaustausch zu groß. Auf der anderen Seite werden auch die Begegnungen jenseits der Distanz D_0 noch kleine Beiträge zum Energieaustausch liefern. Man hat daher für D_0 einen Wert zu nehmen, bei dem der eine Fehler den anderen gerade aufhebt. Der richtige Wert von D_0 wird ein kleines Vielfaches von der Größenordnung der mittleren Sterndistanz sein. Es soll im folgenden $D_0 = 10^6$ (entsprechend 0″,2 Parallaxe) gesetzt werden. Es wird sich zeigen, daß es auf eine Zehnerpotenz in der Wahl von D_0 für unseren Zweck nicht ankommt.

Wir wollen nun die Wirkung der Begegnungen für die Zeiteinheit summieren. Zur Vereinfachung sollen alle Sternmassen gleich 1 gesetzt werden. Die mittlere Geschwindigkeit der Sterne sei V. Dann ist die mittlere Schwerpunktsgeschwindigkeit zweier Sterne $V/\sqrt{2}$, die mittlere Relativgeschwindigkeit $\sqrt{2}\cdot V$. Ist N die Anzahl der Sterne pro Volumeinheit, so wird bei dieser mittleren Relativgeschwindigkeit die Zahl der Begegnungen pro Zeiteinheit, bei welchen die Distanz D zwischen den Grenzen D und $D + dD$ liegt, gleich:

$$2\pi\sqrt{2}\,V N D\,dD$$

Die Häufigkeit jeder Relativgeschwindigkeit in den Grenzen v bis $v + dv$ werde nach dem MAXWELLschen Gesetz angesetzt zu:

$$\frac{3}{2}\sqrt{\frac{3}{\pi}}\cdot\frac{1}{V^3}e^{-\frac{3}{4}\frac{v^2}{V^2}}\cdot v^2\,dv$$

Dabei sind die Konstanten so gewählt, daß das mittlere Geschwindigkeitsquadrat gleich $2\,V^2$ wird, und daß die Integration über alle Werte von v die Summe 1 ergibt.

Als Mittelwert von $\sin^2\frac{\varphi+\varphi'}{2}$ werde der Wert $\frac{1}{2}$, als Mittelwert von s^2 der Wert $\frac{1}{2}V^2$ eingeführt. Mit alledem erhält man als Summation für die Zeiteinheit:

$$\sum(\Delta\varepsilon)^2 = 3\sqrt{6\pi}\,N\iint\frac{e^{-\frac{3}{4}\frac{v^2}{V^2}}\cdot v^4\,dv\,D\,dD}{4+v^4D^2}$$

wobei das Integral über v von 0 bis ∞ und über D von 0 bis D_0 zu erstrecken ist. Die Integration nach D läßt sich ohne weiteres ausführen und ergibt:

$$\sum(\Delta\varepsilon)^2 = \tfrac{3}{2}\sqrt{6\pi}\,N\int_0^\infty e^{-\frac{3}{4}\frac{v^2}{V^2}}\cdot\log\left(1+\frac{v^4D_0^2}{4}\right)dv$$

Setzt man $z^2 = \frac{3}{4}v^2/V^2$ und $Y = \frac{2}{3}\cdot V^2\cdot D_0$, so schreibt sich dies:

$$\sum(\Delta\varepsilon)^2 = 3\sqrt{2\pi}\,NV\int_0^\infty e^{-z^2}\log(1+Y^2z^4)\,dz$$

Die Ausführung letzteren Integrals wird einfach in Rücksicht auf den Umstand, daß Y eine große Zahl von der Ordnung 10^6 ist. Schreibt man nämlich:

$$\int_0^\infty e^{-z^2}\log(1+Y^2z^4)\,dz = 2\log Y\int_0^\infty e^{-z^2}dz + \int_0^\infty e^{-z^2}\log\left(\frac{1}{Y^2}+z^4\right)dz$$

so nähert sich mit wachsendem Y das zweite Integral dem Wert:

$$C = 4\int_0^\infty e^{-z^2}\log z\,dz$$

wofür ich durch eine rohe mechanische Quadratur finde: $C = -3{,}5$. Es gilt somit nahe:

$$\int_0^\infty e^{-z^2}\log(1+Y^2z^4)\,dz = \sqrt{\pi}\log Y - 3{,}5$$

und:

$$\sum(\Delta\varepsilon)^2 = 3\sqrt{2\pi}\,NV(\sqrt{\pi}\log Y - 3{,}5)$$

Man sieht, daß Y und damit die Distanz D_0 nur logarithmisch eingeht, so daß es, wie erwähnt, auf ihren genauen Wert nicht ankommt. Die Anzahl N der Sterne in der Volumeinheit werde bestimmt durch den Radius R der Kugel, auf welche durchschnittlich ein Stern kommt. Dann ist: $\frac{1}{N} = \frac{4\pi}{3}R^3$ und die vorstehende Formel geht über in:

$$\sum(\Delta\varepsilon)^2 = \frac{9}{4}\sqrt{\frac{2}{\pi}}\frac{V}{R^3}(\sqrt{\pi}\log Y - 3{,}5)$$

Das ist das Ergebnis für die Summation über die Zeiteinheit (58 Tage).

Die Summation über t Tage ergibt, wenn man die numerischen Koeffizienten ausrechnet und Dezimallogarithmen einführt:

$$\sum (\Delta\varepsilon)^2 = 46 \frac{V}{R^3} [\log(V^2 D_0) - 1{,}03] \cdot t$$

Es sind schließlich noch numerische Werte für V, D_0 und R einzuführen. Die mittlere Sterngeschwindigkeit V wird gleich der Erdgeschwindigkeit 1 gesetzt, D_0 sei gleich 10^6 (entsprechend 0″.2 Parallaxe) und R gleich $2 \cdot 10^5$ (entsprechend 1″ Parallaxe). Dann folgt:

$$\sum (\Delta\varepsilon)^2 = 2{,}8 \cdot 10^{-14}\, t$$

Nach 10^{12} Jahren wird $\sum(\Delta\varepsilon)^2 = 0{,}028$, die Wurzel daraus $= 0{,}17$ und mithin vergleichbar mit der durchschnittlichen Energie $\frac{1}{2} m V^2$ des einzelnen Sternes, welche nach den obigen Annahmen $= 0{,}5$ ist. Das ist das oben benutzte Resultat.

Wir kommen nun zu unserem eigentlichen Problem des Aufsuchens einer stationären Geschwindigkeitsverteilung in einem homogenen ellipsoidischen Sternhaufen.

Die Begrenzung des Sternhaufens habe in rechtwinkligen Koordinaten x, y, z die Gleichung:

$$\frac{x^2}{a^2} + \frac{y^2}{b^2} + \frac{z^2}{c^2} = 1 \tag{1}$$

so daß a, b, c die Achsen des begrenzenden Ellipsoids sind. Indem die Dichte innerhalb dieses Ellipsoids als konstant vorausgesetzt wird, folgt bekanntlich für das Gravitationspotential:

$$V = -\left(\frac{A^2}{2} x^2 + \frac{B^2}{2} y^2 + \frac{C^2}{2} z^2\right) + D \tag{2}$$

wo A, B, C, D Konstanten sind. Die Bewegungsgleichungen eines Punktes, der sich im Innern des Ellipsoids bewegt, sind daher:

$$\frac{d^2x}{dt^2} = \frac{\partial V}{\partial x} = -A^2 x, \quad \frac{d^2y}{dt^2} = \frac{\partial V}{\partial y} = -B^2 y, \quad \frac{d^2z}{dt^2} = \frac{\partial V}{\partial z} = -C^2 z \tag{3}$$

Die Integration derselben ergibt:

$$x = \alpha \sin\vartheta_1, \qquad y = \beta \sin\vartheta_2, \qquad z = \gamma \sin\vartheta_3, \tag{4}$$

$$\vartheta_1 = At + k_1, \qquad \vartheta_2 = Bt + k_2, \qquad \vartheta_3 = Ct + k_3, \tag{5}$$

wobei α, β, γ, k_1, k_2, k_3 Integrationskonstanten sind. Wir wählen α, β, γ positiv. Jede Koordinate führt also eine Sinusschwingung aus, deren Periode gleich $\frac{2\pi}{A}$ resp. $\frac{2\pi}{B}$ und $\frac{2\pi}{C}$ ist, deren Amplitude α resp. β, γ ist.

Es werde nun ausdrücklich vorausgesetzt, daß diese drei Perioden inkommensurabel untereinander seien. Man denke sich die drei Variablen ϑ für einen Augenblick als rechtwinklige Koordinaten aufgetragen und ziehe von ihnen stets solche Vielfache von 2π ab, daß sie immer zwischen 0 und 2π bleiben. Der Punkt der Koordinaten ϑ_1, ϑ_2, ϑ_3 wird dann innerhalb eines Würfels der Seitenlänge 2π bleiben und wird nach hinreichend langer Zeit diesen Würfel in gleichmäßiger Dichte ausfüllen. Der Stern, dessen Koordinaten x, y, z sind, wird demgemäß das Parallelepiped $|x|<|\alpha|$, $|y|<|\beta|$, $|z|<|\gamma|$ ausfüllen, dieses aber mit ungleichförmiger Dichte. Man hat:

$$dx\,dy\,dz = \alpha\beta\gamma\cos\vartheta_1\cos\vartheta_2\cos\vartheta_3\,d\vartheta_1\,d\vartheta_2\,d\vartheta_3 \tag{6}$$

oder

$$d\vartheta_1\,d\vartheta_2\,d\vartheta_3 = \Delta\,dx\,dy\,dz$$

wobei

$$\Delta = \frac{1}{\sqrt{(\alpha^2 - x^2)(\beta^2 - y^2)(\gamma^2 - z^2)}} \tag{7}$$

ist. Der hier auftretende Faktor Δ von $dx\,dy\,dz$ gibt die — im Mittel über eine hinreichend lange Zeit genommene — Dichte an, mit welcher der Stern das von ihm durchlaufene Parallelepiped erfüllen würde. Da bei stationärer Dichteverteilung die Dichte in jedem Augenblick mit dem Mittelwert der Dichte über beliebig lange Zeiten identisch ist, so ist das zugleich der Beitrag, welchen der betrachtete Stern zur Massendichte in jedem Punkte des Systems liefert.

Wir betrachten jetzt die Gesamtheit aller Sterne des Systems. Die Anzahl derjenigen Sterne, deren Amplituden zwischen α und $\alpha + d\alpha$ resp. β und $\beta + d\beta$, γ und $\gamma + d\gamma$ liegen, sei:

$$\varphi(\alpha, \beta, \gamma)\,d\alpha\,d\beta\,d\gamma \tag{8}$$

Dann ist die gesamte Dichte an dem Punkte der Koordinaten x, y, z

$$\varrho(x, y, z) = \int \frac{\varphi(\alpha, \beta, \gamma)\,d\alpha\,d\beta\,d\gamma}{\sqrt{(\alpha^2 - x^2)(\beta^2 - y^2)(\gamma^2 - z^2)}} \tag{9}$$

Das Integral ist zu erstrecken über alle Amplituden, die größer sind als die betreffende Koordinate, weil Sterne kleinerer Amplitude die betreffende Koordinate überhaupt nicht erreichen, mithin auch nicht zur Dichte beitragen würden. Die Integrationen sind also von den Grenzen $\alpha^2 = x^2$, $\beta^2 = y^2$, $\gamma^2 = z^2$ bis ins Unendliche zu erstrecken.

Damit nun unser Sternsystem wirklich stationär seine Dichteverteilung beibehält, muß diese Dichte ϱ mit der zu Anfang angenommenen übereinstimmen. Es muß das vorstehende Integral für Punkte im Innern des Ellipsoids (1) gleich 1, außerhalb gleich 0 werden. Die Gleichung (9) ist demnach eine Integralgleichung, aus der die Funktion $\varphi(\alpha, \beta, \gamma)$ gemäß der vorgegebenen Funktion ϱ zu bestimmen ist.

In Festschrift für Hugo von Seeliger (Springer Berlin 1924) 94–105

Betrachtet man zunächst den eindimensionalen Fall

$$\varrho(x) = \int_{\alpha=x}^{\infty} \frac{\varphi(\alpha)\,d\alpha}{\sqrt{\alpha^2 - x^2}}$$

so ist das eine bereits von ABEL behandelte Integralgleichung, deren Lösung folgendermaßen geschieht: Man multipliziere mit $\frac{x\,dx}{\sqrt{x^2-\xi^2}}$ (ξ ein Parameter) und integriere von $x = \xi$ bis $x = \infty$. Das gibt

$$\int_{x=\xi}^{\infty} \frac{\varrho(x)\,x\,dx}{\sqrt{x^2-\xi^2}} = \int_{x=\xi}^{\infty}\int_{\alpha=x}^{\infty} \frac{\varphi(\alpha)\,x\,dx\,d\alpha}{\sqrt{(\alpha^2-x^2)(x^2-\xi^2)}}$$

Vertauscht man rechts die Integrationsordnung, so hat man bei festgehaltenem α zunächst über x von $x = \xi$ bis $x = \alpha$ zu integrieren, und dann über α von $\alpha = \xi$ bis $\alpha = \infty$. Da aber

$$\int_{x=\xi}^{x=\alpha} \frac{x\,dx}{\sqrt{(\alpha^2-x^2)(x^2-\xi^2)}} = \frac{\pi}{2}$$

ist, so folgt:

$$\int_{x=\xi}^{\infty} \frac{\varrho(x)\,x\,dx}{\sqrt{x^2-\xi^2}} = \frac{\pi}{2}\int_{\alpha=\xi}^{\infty} \varphi(\alpha)\,d\alpha$$

oder:

$$\varphi(\xi) = -\frac{2}{\pi}\frac{d}{d\xi}\int_{x=\xi}^{\infty} \frac{\varrho(x)\,x\,dx}{\sqrt{x^2-\xi^2}}$$

womit die Lösung gegeben ist.

Die dreidimensionale Integralgleichung läßt sich ganz analog behandeln und hat die Lösung:

$$\varphi(\xi,\eta,\zeta) = -\frac{8}{\pi^3}\frac{d^3}{d\xi\,d\eta\,d\zeta}\iiint_{|x|\geqq\xi,\,|y|\geqq\eta,\,|z|\geqq\zeta} \varrho(x,y,z)\,\frac{x\,dx\cdot y\,dy\cdot z\,dz}{\sqrt{(x^2-\xi^2)(y^2-\eta^2)(z^2-\zeta^2)}}\,. \tag{10}$$

In unserem speziellen Falle hat man $\varrho = 1$ zu setzen und bis zu den durch das Ellipsoid

$$\frac{x^2}{a^2} + \frac{y^2}{b^2} + \frac{z^2}{c^2} = 1$$

bestimmten Grenzen zu integrieren.

Diese Integration ist leicht. Setzt man nämlich:

$$x^2 - \xi^2 = X^2\,, \qquad y^2 - \eta^2 = Y^2\,, \qquad z^2 - \zeta^2 = Z^2\,,$$

so geht das auszuführende Integral über in:

$$\iiint dX\,dY\,dZ$$

zu nehmen zwischen den Grenzen $X = Y = Z = 0$ und:

$$\frac{X^2}{a^2} + \frac{Y^2}{b^2} + \frac{Z^2}{c^2} = 1 - \frac{\xi^2}{a^2} - \frac{\eta^2}{b^2} - \frac{\zeta^2}{c^2}$$

Das ist aber nichts anderes als das Volumen eines Ellipsoids, dessen Achsen gleich resp. a, b, c multipliziert mit $\sqrt{1 - \frac{\xi^2}{a^2} - \frac{\eta^2}{b^2} - \frac{\zeta^2}{c^2}}$ sind. Das Volumen dieses Ellipsoids hat den Wert:

$$\frac{4}{3}\pi\,a\,b\,c\sqrt{1 - \frac{\xi^2}{a^2} - \frac{\eta^2}{b^2} - \frac{\zeta^2}{c^2}}^{\,3}$$

Setzt man diesen Wert für das Integral in (10) ein, so erhält man:

$$\varphi(\xi\eta\zeta) = \frac{32}{\pi^2 a\,b\,c} \cdot \frac{\xi\eta\zeta}{\sqrt{1 - \frac{\xi^2}{a^2} - \frac{\eta^2}{b^2} - \frac{\zeta^2}{c^2}}^{\,3}}$$

oder:

$$\varphi(\alpha, \beta, \gamma) = \frac{32}{\pi^2 a\,b\,c} \cdot \frac{\alpha\beta\gamma}{\sqrt{1 - \frac{\alpha^2}{a^2} - \frac{\beta^2}{b^2} - \frac{\gamma^2}{c^2}}^{\,3}} \qquad (11)$$

Damit ist diejenige Verteilung der Anfangsbedingungen gefunden, bei welcher das homogene ellipsoidische Sternsystem stationär ist.

Es interessiert besonders, wie dabei die Geschwindigkeitsverteilung in jedem Punkte des Systems ist. Für die Geschwindigkeitskomponenten hat man:

$$\frac{dx}{dt} = u = A\sqrt{\alpha^2 - x^2}\,, \quad \frac{dy}{dt} = v = B\sqrt{\beta^2 - y^2}\,, \quad \frac{dz}{dt} = w = C\sqrt{\gamma^2 - z^2}\,,$$

und damit, wenn man x, y, z festhält, u, v, w als unabhängige Variable statt α, β, γ einführt:

$$\frac{\varphi(\alpha, \beta, \gamma)\,d\alpha\,d\beta\,d\gamma}{\sqrt{(\alpha^2 - x^2)(\beta^2 - y^2)(\gamma^2 - z^2)}}$$

$$= \frac{32}{\pi^2 a\,b\,c} \cdot \frac{du\,dv\,dw}{\sqrt{1 - \frac{x^2}{a^2} - \frac{y^2}{b^2} - \frac{z^2}{c^2} - \frac{u^2}{a^2 A^2} - \frac{v^2}{b^2 B^2} - \frac{w^2}{c^2 C^2}}}$$

Gleich häufig sind hiernach Geschwindigkeiten, für welche der Ausdruck:

$$W^2 = \frac{u^2}{a^2 A^2} + \frac{v^2}{b^2 B^2} + \frac{w^2}{c^2 C^2}$$

In Festschrift für Hugo von Seeliger (Springer Berlin 1924) 94–105

denselben Wert hat. Die Geschwindigkeitsverteilung ist demnach eine ellipsoidische. Die Achsen des Geschwindigkeitsellipsoids sind in jedem Raumpunkte den Achsen des Sternhaufens parallel. Die Häufigkeit nimmt mit wachsendem W zu und wird unendlich für $W^2 = 1 - \frac{x^2}{a^2} - \frac{y^2}{b^2} - \frac{z^2}{c^2}$. Letzterer Fall entspricht den Sternen, die bis an die Grenze des Systems gelangen. Über diesen Maximalwert von W hinausgehende Geschwindigkeiten kommen nicht vor. Der Maximalwert von W und damit die obere Grenze der vorkommenden Geschwindigkeiten wird um so kleiner, je mehr man sich der Grenze des Systems nähert. Auf der Grenze selbst werden alle Geschwindigkeiten Null.

Die Achsen des Geschwindigkeitsellipsoids findet man aus dem Ausdruck[1])

$$a^2 A^2 = 2\pi \int_0^\infty \frac{1}{\left(1+\frac{s}{a^2}\right)} \cdot \frac{ds}{\sqrt{\left(1+\frac{s}{a^2}\right)\left(1+\frac{s}{b^2}\right)\left(1+\frac{s}{c^2}\right)}}$$

nebst den entsprechenden. Aus diesen Ausdrücken erkennt man leicht, daß das Geschwindigkeitsellipsoid im selben Sinne, aber weniger stark abgeplattet ist wie der Sternhaufen selbst.

Wenn man zum Fall des Rotationsellipsoids ($a = b$) übergehen will, so kann man voraussetzen, daß dabei immer noch ein kleiner Unterschied zwischen den Perioden $\frac{2\pi}{A}$ und $\frac{2\pi}{B}$ bleibt, welcher dieselben inkommensurabel macht. Unter dieser Voraussetzung geschieht der Übergang, indem man in der Tat einfach in den obigen Formeln $a = b$ setzt. Will man hingegen von vornweg strenge die Voraussetzung $a = b$, verbunden mit genauer Gleichheit der Perioden $\frac{2\pi}{A}$ und $\frac{2\pi}{B}$ machen, so ändert das Problem vollständig seinen Charakter. Es beschreibt dann jeder Stern in der Projektion auf die x, y-Ebene eine feste Ellipse und erfüllt im Laufe der Zeit nicht ein Parallelepiped, sondern den Mantel eines elliptischen Zylinders. Da in diesem Falle eine viel geringere Durchmischung eintritt, so wird die stationäre Geschwindigkeitsverteilung nicht mehr eindeutig bestimmt. Vielmehr zeigt sich, daß man die Abhängigkeit der Verteilungsfunktion $\varphi(\alpha, \beta, \gamma)$ von einer der beiden Variablen α, β bis auf gewisse Stetigkeitsforderungen beliebig ansetzen kann, worauf dann die von der anderen bestimmt ist. Man kann in diesem Falle eine stationäre Geschwindigkeitsverteilung angeben, bei der die Sterne — auf die x, y-Ebene projiziert — genau oder angenähert kreisförmige Bahnen beschreiben. Ebensowohl gibt es aber auch Geschwindigkeitsverteilungen, bei welchen die angenähert radialen Bewegungen vorwiegen.

[1]) Tisserand: Méc. céleste II, S. 45.

In Wirklichkeit wird nun die Forderung $a = b$ gewiß nicht strenge erfüllt sein. Wenn die Rotationssymmetrie aber sehr nahe erfüllt ist, so werden sich die Perioden $\frac{2\pi}{A}$ und $\frac{2\pi}{B}$ sehr wenig unterscheiden, und die oben vorausgesetzte Durchmischung der Sterne wird so lange Zeit in Anspruch nehmen, daß es eine Übertreibung wäre, das Prinzip des stationären Zustandes auch noch auf diese Zeit anzuwenden. Es ist also im Falle guter Rotationssymmetrie des Fixsternsystems sehr wohl möglich, daß die strenge Voraussetzung $a = b$ besser mit der Wirklichkeit stimmende Folgerungen gibt als der Grenzübergang aus dem dreiachsigen Ellipsoid.

Ganz abgesehen von dem hypothetischen Charakter unserer Forderung stationärer Geschwindigkeitsverteilung wäre eine Übertragung der obigen Ableitung auf das wirkliche Sternsystem schon deshalb bedenklich, weil die Dichte in demselben nicht gleichförmig ist. Es würde interessant sein, eine in dieser Beziehung sich der Wirklichkeit näher anschließende Verteilung zu bilden. Indessen glaube ich, daß die obige Ableitung zur Fixierung der Vorstellungen nützlich und in folgender Beziehung lehrreich ist.

Die bekannte Erscheinung einer der Milchstraße parallelen Vorzugsrichtung in den Eigenbewegungen der Fixsterne habe ich mir durch die Annahme zu erklären gesucht, daß die Sonne exzentrisch im Sternsystem steht und daß die Sterne angenähert kreisförmige Bahnen parallel der Milchstraßenebene beschreiben, wobei direkte und retrograde Bewegungen als gleichhäufig vorausgesetzt werden. Nach dieser Anschauung wäre die Richtung nach dem Zentrum des Systems senkrecht zur Richtung des bei $7^h + 10°$ liegenden Vertex der Sternbewegungen zu suchen.

Herr HERTZSPRUNG hat mir mündlich die Meinung ausgesprochen, daß die Sternbewegungen parabolischen Charakter haben und im wesentlichen radial gerichtet sein könnten, so daß wir das Zentrum in der Richtung der Vertexlinie selbst zu suchen hätten. Letztere Anschauung ist kürzlich von Herrn H. H. TURNER publiziert und mit großem Schwung ausgeführt worden.

Die obige Rechnung zeigt nun, daß die stationäre Geschwindigkeitsverteilung in einem homogenen Ellipoisd die Richtung nach dem Zentrum nirgends verrät, sondern nur die Richtung der Hauptachsen des Ellipsoids wiedergibt. Wenn etwas Ähnliches für das wirkliche Sternsystem gilt, so muß die Richtung nach dem Zentrum des Systems weder mit der Vertexlinie noch mit einer Senkrechten dazu zusammenfallen.

Die geringe mittlere Geschwindigkeit der Heliumsterne führt Herr EDDINGTON auf den Umstand zurück, daß sich diese im Durchschnitt in großen Entfernungen näher der äußeren Grenze des Sternsystems befinden. Herr KAPTEYN hat die Schwierigkeiten dieser Anschauung

In Festschrift für Hugo von Seeliger (Springer Berlin 1924) 94–105

hervorgehoben. Herr TURNER verweist die Heliumsterne umgekehrt mehr in das Zentrum des Systems. Halten wir uns an die oben gefundene Geschwindigkeitsverteilung, so ergibt sich folgendes: Kleine Geschwindigkeiten herrschen vor in den äußeren Teilen des Systems. Indessen kleine Energien (kleine Werte der Summen $\frac{\alpha^2}{a^2} + \frac{\beta^2}{b^2} + \frac{\gamma^2}{c^2}$, von kinetischer und potentieller Energie) finden sich nur im Innern des Systems. Wenn man also annimmt, daß die Heliumsterne von Anfang an kleine Gesamtenergie gehabt haben, so kommt man auf die TURNERsche Anschauung, gegen welche Widersprüche nicht vorliegen.

Bibliography*

1890

– Zur Bahnbestimmung nach Bruns, *Astron. Nachr.* **124**: 211–216 .... 1.1
– Methode zur Bahnbestimmung der Doppelsterne, *Astron. Nachr.* **124**: 215–218 .... 1.2

1894

– Ueber den Einfluss der saecularen Aberration auf die Fixsternörter, *Astron. Nachr.* **136**: 81–84 .... 4.1

1895

– Definitive Bahnbestimmung des Cometen 1842 II, *Astron. Nachr.* **137**: 177–190 .... 1.3

1896

– Ueber die Stabilität der Bewegung eines durch Jupiter gefangenen Cometen, *Astron. Nachr.* **141**: 1–8 .... 1.4

1898

– Ueber eine Classe periodischer Lösungen des Dreikörperproblems, *Astron. Nachr.* **147**: 17–24 .... 1.5
– Ueber weitere Classen periodischer Lösungen des Dreikörperproblems, *Astron. Nachr.* **147**: 289–298 .... 1.6

1901

– Der Druck des Lichts auf kleine Kugeln und die Arrhenius'sche Theorie der Cometenschweife, *Sitz. Ber. math. phys. Classe d. Kgl. bayer. Akad. d. Wiss.* **31**: 293–338 .... 3.1

* The numbers on the right-hand side of each entry refer to the section in the *Collected Works* where the paper can be found (e.g. 1.5 refers to the 5th paper reprinted in Sect. 1).

1903

- Über die periodischen Bahnen vom Hecubatypus, *Astron. Nachr.* **160**: 385–400 .... 1.7
- Über Himmelsmechanik, *Ges. dt. Naturforscher und Ärzte, Verhandl. 1903, allg. Teil*, 1–14 .... 1.8
  Also in: *Die Umschau* **7**: 861–866; *Phys. Zeitschr.* **4**: 765–773; *Naturw. Rundschau* **18**: 637–639, 649–651; *Jahresber. d. dt. Mathematiker-Vereinigg.* Vol. **XIII**, No. 3/4 (1904) 145–156

1905

- (With W. Villiger) Über die Helligkeitsverteilung des ultravioletten Lichtes auf der Sonnenscheibe, *Phys. Zeitschr.* **6**: 737–744 .... 2.3
- Zur Sonnenfinsternis vom 29. August in Nordafrika, *Umschau* **9**: 901–907 .... 2.2

1906

- (With W. Villiger) On the Distribution of Brightness of the Ultra-Violet Light on the Sun's Disk, *Astrophys. J.* **23**: 284–305 .... 2.4
- (With W. Villiger) Note on the Ultra-Violet Radiation of Sun-Spots and Faculae, *Astrophys. J.* **23**: 345–346 .... 2.5
- Ueber das Gleichgewicht der Sonnenatmosphäre, *Nachr. Kgl. Ges. d. Wiss. zu Gött., Math. Phys. Klasse 1906*, No. 1, 41–53 .... 2.6

1907

- Ueber die Eigenbewegungen der Fixsterne, *Nachr. Kgl. Ges. d. Wiss. zu Gött., Math. Phys. Klasse 1907*, No. 5, 614–632 .... 4.2
- Ueber die totale Sonnenfinsternis vom 30. August 1905, *Abhandl. Kgl. Ges. d. Wiss. zu Gött.* NF **5**(2): 3–73 = *Astron. Mitt. Kgl. Sternw. zu Gött.* **13** (1906) 3–73 .... 2.1

1908

- Ueber die Bestimmung von Vertex und Apex nach der Ellipsoidhypothese aus einer geringen Anzahl beobachteter Eigenbewegungen, *Nachr. Kgl. Ges. d. Wiss. zu Gött., Math. Phys. Klasse 1908*, No. 2, 191–200 .... 4.3
- Book Review: R. Emden, *Gaskugeln* (Teubner, Leibniz 1907), *Vierteljahrsschrift d. Astron. Ges.* **43**: 26–55 .... 2.12

1909

- *Über das System der Fixsterne. Naturwiss. Vorträge u. Schriften No. 1; zu H. von Seeligers 60. Geburtstag*, edited by Berliner Urania (Teubner, Leipzig), III. Über das System der Fixsterne, pp. 20–38 .... 4.4
  Second (unaltered) edition 1916; slightly different version also in *Himmel und Erde* **21**: 433–451; (*New Ideas in Astronomy*), St. Petersburg 1913 (Russian translation)

1910

- Über die Integralgleichungen der Stellarstatistik, *Astron. Nachr.* **185**: 81–88 .... 4.5
- Remark on the occultation of the star AG Lpz II 4615 reported by Miethe, in: Beobachtungen des Halleyschen Kometen von Seiten des Astrophysikalischen Observatoriums bei Potsdam, *Astron. Nachr.* **185**: 135–140, (140) 3.3

1911

- Vierte Versammlung der Internationalen Vereinigung für kooperative Sonnenforschung 1910, *Astron. Nachr.* **186**: 37–40 .... 2.13
- (With E. Kron) Über die Helligkeitsverteilung im Schweif des Halley'schen Kometen, *Nachr. Kgl. Ges. d. Wiss. zu Gött., Math. Phys. Klasse 1911*, No. 3, 197–208 .... 3.2
  English translation: *Astrophys. J.* **34**: 342–352

1912

- Zur Stellarstatistik, *Astron. Nachr.* **190**: 361–376 .... 4.6
- Ein Theorem zur Verteilung der Sterngeschwindigkeiten, *Astron. Nachr.* **191**: 1–6 .... 4.7

1913

- Tagung der Solar Union zu Bonn. 31. Juli – 5. August 1913, *Naturwiss.* **1**: 865–866 .... 2.14
  Different version also in *Astron. Nachr.* **195**: 411–414
- Über die räumliche Bewegung der Praesepe, *Astron. Nachr.* **196**: 9–10 . 4.8
- Bemerkung zur Berechnung des Strahlungsgleichgewichtes der Atmosphäre, *Meteorolog. Zeitschr.* **30**: 454–456 .... 2.7
- (With G. Eberhard) Über Umkehrungen der Calciumlinien H und K in Sternspektren, *Sitz. Ber. Kgl. Preuss. Akad. d. Wiss. Berlin 1913*, 308–310 .... 2.9
  English translation: *Astrophys. J.* **38**: 292–295

1914

1924

Contents for Volume 2

5. Astronomical Positioning

6. Photographic Photometry

6.3 Ueber Abweichungen vom Reciprocitätsgesetz für Bromsilbergelatine / On Departures from the Reciprocity Law for Silver-Bromide Gelatine

6.4 Ueber die Wirkung intermittirender Belichtung auf Bromsilbergelatine / On the Effects of Intermittent Exposures on Silver-Bromide Gelatine

6.5 Bemerkungen zur Sensitometrie / Remarks on Sensitometry

6.6 Über die photographische Vergleichung der Helligkeit verschiedenfarbiger Sterne / On the Photographic Comparison of the Magnitudes of Stars of Different Colours

6.7 Ueber sensitometrische Regeln und ihre astronomische Anwendung / On Sensitometry Laws and Their Astronomical Application

6.8 Professor G. Jägers Theorie des photographischen Prozesses / Professor G. Jäger's Theory of the Photographic Process

6.9 Plan zur Durchführung einer photographisch-photometrischen Durchmusterung des nördlichen Himmels / Plan for Carrying Out a Photographic-Photometric Survey of the Northern Sky

6.10 Über eine Schraffierkassette zur Aktinometrie der Sterne / On a Schraffierkassette for Stellar Actinometry (with Br. Meyermann)

6.11 Über eine Interpolationsaufgabe der Aktinometrie / On an Interpolation Problem in Actinometry

6.12 Aufnahmen des Sternhaufens *h* Persei mit Spiegeln von sehr großem Öffnungsverhältnis / Exposures of the Cluster *h* Persei Using Mirrors with Very Large Aperture Ratios (with W. Villiger)

6.13 Über eine neue Schraffierkassette / On a New Schraffierkassette (with Br. Meyermann)

6.14 Über die Farbentönung der Sterne / On the Colour Tints of the Stars

6.15 Remarque sur la détermination des grandeurs photographiques absolues / Note on the Determination of Absolute Photographic Magnitudes

6.16 Über die Bestimmung absoluter photographischer Helligkeiten / On the Determination of Absolute Photographic Magnitudes

6.17 Aktinometrie der Sterne der B.D. bis zur Grösse 7.5 in der Zone 0° bis +20° Deklination. Teil A / Actinometry of B.D. Stars down to Magnitude 7.5 in the Zone between Declinations 0° and +20°, Part A (with Br. Meyermann, A. Kohlschütter and O. Birck)

6.18 Aktinometrie der Sterne der B.D. bis zur Grösse 7.5 in der Zone 0° bis +20° Deklination. Teil B / Actinometry of B.D. Stars down to Magnitude 7.5 in the Zone between Declinations 0° and +20°, Part B (with Br. Meyermann, A. Kohlschütter, O. Birck and W. Dziewulski)

6.19 Buchbesprechung / Book Review: J.A. Parkhurst, *Yerkes Actinometry, Zone +73° to +90°*

Contents for Volume 3

8. Optics

9. Physical Papers

10. Miscellaneous

10.13 Buchbesprechung / Book Review: K.W. Zenger, *Die Meteorologie der Sonne und das Wetter im Jahre 1887, zugleich Wetterprognose für das Jahr 1897 / Meteorology of the Sun and the Weather in 1887, together with a weather forecast for 1897*

10.14 Buchbesprechung / Book Review: Kurd Laßwitz, *Auf zwei Planeten / On Two Planets*

10.15 Buchbesprechung / Book Review: W. Förster, *Himmelskunde und Weissagung / Sky Lore and Prophesy*

10.16 Buchbesprechung / Book Review: H. Poincaré, *Leçons sur les hypothèses cosmogoniques / Lessons on Cosmogonal Theories* and T.J.J. See, *Researches on the Evolution of the Stellar Systems.* Vol. II

10.17 Ueber das in der Festschrift zur Feier des hundertfünfzigjährigen Bestehens der K. Gesellschaft der Wissenschaften mit dem Gaußischen Tagebuche reproducierte Porträt des „26-jährigen Gauß" / On the Portrait of the "26-year-old Gauss" reproduced with Gauss' Diary in the Commemorative Volume for the 150th Anniversary of the Royal Society for Science (with F. Klein)

10.18 Über die Errichtung eines Gaußturmes bei Göttingen / Appeal for the Erection of the Gauss Tower near Göttingen (with Gieseke and F. Klein)

10.19 Antrittsrede des Hrn. SCHWARZSCHILD; Erwiderung des Sekretars Hrn. PLANCK / Inaugural Address by Herr Schwarzschild; Reply by the Secretary Herr Planck

Bibliography

GPSR Compliance
The European Union's (EU) General Product Safety Regulation (GPSR) is a set of rules that requires consumer products to be safe and our obligations to ensure this.

If you have any concerns about our products, you can contact us on

ProductSafety@springernature.com

In case Publisher is established outside the EU, the EU authorized representative is:

Springer Nature Customer Service Center GmbH
Europaplatz 3
69115 Heidelberg, Germany

www.ingramcontent.com/pod-product-compliance
Lightning Source LLC
LaVergne TN
LVHW012024041026
849558LV00003B/146
9783540524557